Spectroscopy and Modelling of Biomolecular Building Blocks

Spectroscopy and Modelling of Biomolecular Building Blocks

JEAN-PIERRE SCHERMANN

Laboratoire de Physique des Lasers
Institut Galilée Université Paris 13
Villetaneuse, France

ELSEVIER

Amsterdam • Boston • Heidelberg • London • New York • Oxford
Paris • San Diego • San Francisco • Singapore • Sydney • Tokyo

Elsevier
Radarweg 29, PO Box 211, 1000 AE Amsterdam, The Netherlands

First edition 2008

British Library Cataloguing in Publication Data
A catalogue record for this book is available from the British Library

Library of Congress Cataloging-in-Publication Data
A catalog record for this book is available from the Library of Congress

ISBN: 978-0-444-52708-0

For information on all Elsevier publications
visit our website at books.elsevier.com

Transferred to Digital Printing 2011

Contents

Preface

The purpose of this book is to provide an overview of the intertwining of different disciplines such as optical and electron spectroscopy, quantum chemistry and mass-spectrometry when they are involved in the understanding of elementary building blocks of biomolecules and their assemblies. During recent years, a rapidly increasing amount of knowledge has been accumulated about the precise structures and properties of biomolecular constituents and their non-covalent assemblies under idealized conditions met when they are isolated in the gas-phase. In the age of structural genomics, proteomics, metabolomics and more and more systemic approaches, there are still many motivations for studies to be conducted at the level of individual molecules, a methodology that may seem at first sight somewhat too reductionist.

The understanding of living systems relies in fact on studies conducted on very different scales. The inner scale is concerned by the building blocks that have been selected and assembled by Nature in order to create biologically active compounds. Biomolecules such as DNA, RNA, proteins or carbohydrates are composed of many copies of small blocks such as amino acids, nucleobases or sugars. These small blocks are strongly linked by covalent bonds but also engage into weak bonds with their neighbours and the surrounding medium, leading to the extraordinary recognition capability of biomolecular systems. The deciphering of the replication machinery and the establishment of a relationship between the three-dimensional architecture of biomolecules and their functions require investigations. Those investigations of intrinsic properties and interactions of elementary constituents are to be performed on a global as well as a very local scale. For these elementary constituents, experiments conducted under isolated conditions and quantum chemistry studies allow the unambiguous separation between their intrinsic properties and those induced by the presence of the environment.

A large fraction of the gas-phase studies of systems of biological interest are the results of the parallel development of theoretical concepts and sophisticated experimental techniques first tested on very simple model systems. Non-covalent bonds play a crucial role in biological structures and hence in biological functions. For example, proteins are long polypeptide chains made of amino acids. The 20 naturally occurring amino acids are each characterized by a different side chain which is more or less hydrophilic or hydrophobic, basic or acidic and thus can engage into different bonds with its neighbours. Even simple organic molecules such as phenol and indole, the respective side chains of tyrosine and tryptophan, can establish different hydrogen bonds with surrounding water molecules. It took many years for molecular physicists to obtain a satisfying picture of this very elementary problem through advances in predictive quantum chemical calculations and incisive experimental techniques involving high-resolution mass-spectrometry, high-performance optical sources, supersonic beams and very low temperature cluster sources. Once these methods became mature, they prompted the study of larger isolated molecules or complexes actually involved in the biophysics and biochemistry of living bodies, under the same experimental conditions that had been achieved previously for simple molecules and complexes. In parallel, the impressive progress in computer speed and memory capacity occurring at a very fast pace allowed quantum calculations to become available for molecular systems possessing a large enough number of atoms to become of biological interest. Today, not only nucleic acid bases or amino acid components are studied in beams or ion cells with their structures predicted by means of high-level quantum chemistry calculations, but oligonucleotides, oligopeptides and oligosaccharides typically containing from two to five building blocks are also within reach of such sophisticated approaches.

An important step has been the demonstration that biopolymers of a rather limited size already display structural properties that are similar to those encountered in three-dimensional architectures of much larger biomolecules. For example, the local folding or secondary structure of proteins is maintained by short-distance interactions between amino acid chemical groups. Several fundamental secondary structures such as α-helix, β-sheets, and γ-turns have been identified long ago in crystals and in liquids. However, the role of the solvent in cells, usually the ubiquitous water, in the establishment of those structures could only be unambiguously settled by removing it partially or totally. This was first accomplished by allowing peptide ions to drift under the influence of an electric field in a high-pressure gas and measuring their arrival times which strongly depend upon their folded structures. As in a liquid, the molecular systems in such an experiment are not in a single configuration but rather in an equilibrium ensemble of closely resembling folds. The size of the systems and the number of states which must then be taken into account are prohibitively large for interpretation of the measurement by means of quantum calculations. An alternative modelling method based on molecular mechanics has been developed in order to replace the costly quantum ab initio calculations in describing large biomolecular systems. In this method, physical forces are used to describe interactions between atoms. Molecules are then described as a collection of masses connected by springs. A potential energy function of atomic positions is built and its parameters are derived by fitting data from very different origins. Crystal structure geometries are used but also structures deduced from microwave or infrared spectroscopy. Even quantum calculations are in use when no experimental data are available. Today, the competition between secondary structures such as β, γ turns and 3_{10} helices can be addressed at an unprecedented degree of precision by combining new methods for vaporizing small biomolecules into the gas-phase, cooling these molecules to a limited number of conformations, and determining vibrational frequencies by using broadly-tunable infrared sources. Such structural competition already appears in di- or tri-peptides and can be disentangled with the help of the density functional theory (DFT) method. The use of the electron density function to describe the electrons allows an accurate treatment with a more affordable computational requirement than pure ab initio calculation and is now routinely applied to the study of larger and larger systems. Although there are still gaps between small molecules of biological interest and intermediate size oligomers and biomolecules such as proteins, windows are being opened in the walls which used to separate structural biology from quantum chemistry and gas-phase spectroscopy.

Following the deciphering of the Human genome, the search of new personalized treatments, which constitute the pharmacogenetic programme, aims to design drugs adapted to each human being. The discovery of a new drug is complex and requires sophisticated modelling of biomolecules as well as a deep understanding of biochemical mechanisms at the molecular level. It may seem that the size of molecules of biological interest that can be accurately scrutinized in the gas-phase by means of spectroscopy and quantum chemistry is too small compared to that of DNA with 3×10^9 base pairs or even that of small compounds such as siRNAs which only contain around 20 base pairs. In fact, it turns out that the molecular weight of most orally administered drugs is rather small (Lipinski rules). It thus appears that gas-phase studies of small molecules can contribute in the near future to the improvement of selection procedures of potential drugs. For example, very often, developments come for the design of inhibitors of expressed enzymes. The active sites of enzymes can be highly hydrophobic and only contain a very small number of water molecules which have been conserved through evolution. The design of selective inhibitors has been possible thanks to the determination of crystal structures of enzymes in complexes with inhibitors. Moreover, there is a need for rapid screening of potential drugs, and the combination of mass-spectrometry and spectroscopy may offer a possibility to obtain accurate information about non-covalent complex formation.

X-ray crystallography and NMR spectroscopy have tremendously improved their resolution and increased the size of the biomolecules they can tackle. They provide biologists with a

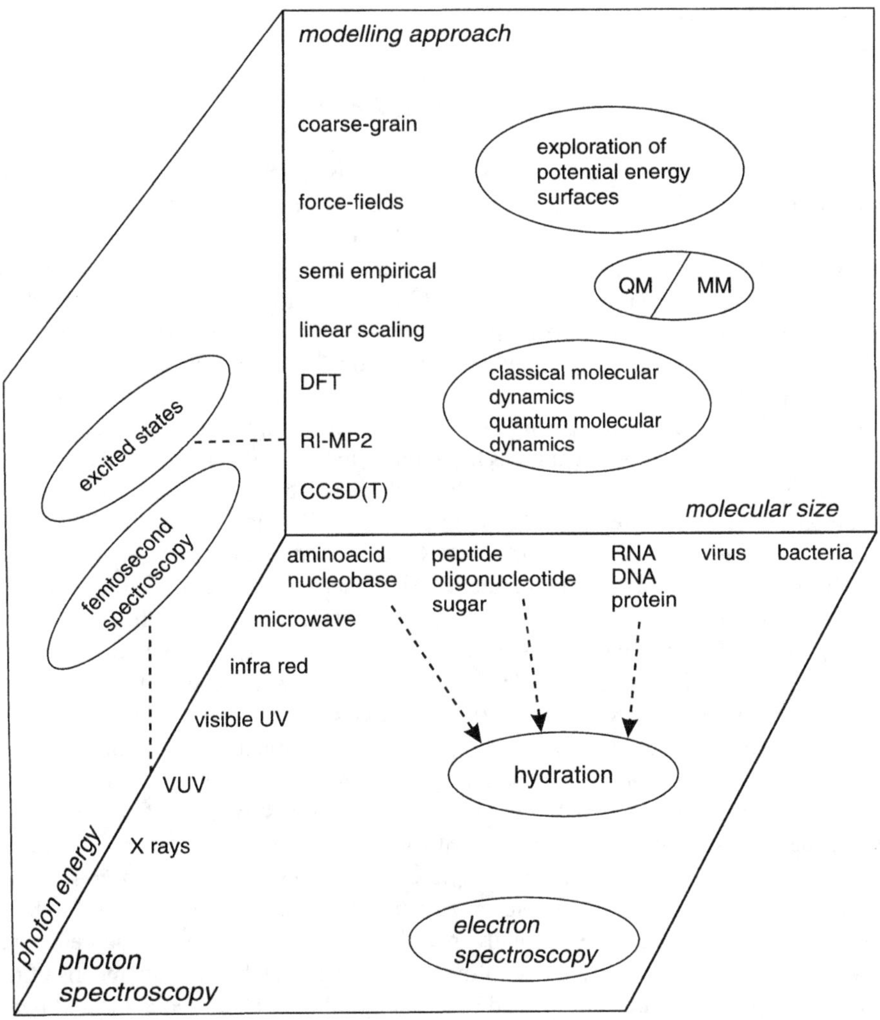

wealth of information concerning the organization of supramolecular structures which govern cell survival. However, some important biological problems fall short of full understanding by means of those methods and the theoretical tools used for modelling the experimental data they provide. For example, processes where electronic excitations intervene are not taken into account in force-fields. Photodynamical processes as well as radiation-induced damages cannot be interpreted without the help of quantum chemistry. Those processes occur on a very short timescale, typically in the femtosecond to picosecond range. The solvent then not only plays a structural role but also opens new pathways for energy or charge transfer. Gas-phase investigations offer a unique possibility for disentangling those pathways. For example, ionizing radiations produce electrons which are slowed down to thermal energies in cells and can eventually attach to nucleobases, leading to a disruption in the translation of the genetic code. A cascade of events thus occurs between the penetration of high-energy rays and observation of radiation-induced diseases. Photo-induced chemical transformation or electron scattering studies of isolated DNA components then provide direct information about processes that are the ones most likely leading to mutation, in the absence of any other environmental influence.

BOOK OVERVIEW

This book is an attempt to present the merging of three disciplines that first evolved separately and are thus usually topics of different textbooks [1–7]. Quantum chemistry and modelling on one side and mass-spectrometry and related methods on the other already constitute independently crucial and widely used tools in many aspects of biological and pharmaceutical research and development. Their common feature is that they deal with systems characterized under conditions rather close to ideal. The main difference between those disciplines arises from their long-time belonging to very distant "stellar systems" that have been explored either by means of experimental setups or computer programs. Fortunately, the number of "interplanetary voyagers" is constantly increasing and the seemingly "light-year" separations rapidly vanish. Different spectroscopic techniques have joined the concert and the score now includes a wide number of instruments. Unfortunately, each piece cannot be entirely played in this book and only excerpts are presented. From time to time, biological problems will be touched on but only brief surveys will be provided. Only a restricted number of references are given but maybe after a flight at rather high altitude, the reader may chose to land and visit some places he had never visited before and explore them further more in depth.

The beginning of each of the first three chapters provides a basic introduction to quantum chemistry and molecular modelling (Chapter 1), to frequency-resolved, time-resolved photon and electron spectroscopy (Chapter 2) and to mass-spectrometry (Chapter 3). Experts may then be only interested in recent examples of developments given at the end of each chapter. Spectroscopists might, for example be interested by recent mass-spectrometry techniques or specialists of mass-spectrometry by time-resolved femtosecond spectroscopy. Modelling is the ubiquitous ingredient for interpretation of experiments and recent developments of methods dealing, for example with excited states or very large systems are presented. The first three chapters are presented separately. They are illustrated through examples of biomolecular system experimentally investigated in the gas-phase or theoretically.

A synthetic view of the different disciplines is presented in Chapter 4 by means of recent case-studies concerning nucleobases and oligonucleotides (4.1), amino acids and peptides

(4.2), sugars (4.3), neuromolecules (4.4), non-covalent complexes (4.5), chiral systems (4.6), electron interactions (4.7) and large biomolecular systems (4.8). Finally, Chapter 5 aims to establish links between gas-phase studies and condensed-phase research. A glimpse into possible future applications concludes this book.

ACKNOWLEDGEMENTS

This book would have never been written without the constant scientific and friendly support of Seong Keun Kim who, in fact, should have been co-author. The kind help of Gilles Grégoire was several times crucial when I was faced with problems and needed his expertise. Several parts are the result of the many years of close and warm cooperation with Charles Desfrançois. The role of Charles in this book was also decisive as first organizer of the series of European Conferences on Biomolecules in the Gas Phase. The vast knowledge of the next broad-minded organizers of European and Gordon Conferences on Biomolecules in the Gas Phase: Mike Bowers, Martin Jarrold, Rainer Weinkauf, Mattanjah de Vries, John Simons, Pavel Hobza and Eugen Illenberger has been extremely precious since it allowed me to discover so many facets of this field.

Very pleasant stays in the laboratories of Kit Bowen, Tamotsu Kondow, Bernard Brutschy and Fumitaka Mafune offered me great opportunities to enhance my knowledge while taking advantage of their great hospitality.

The manuscript has widely benefited from the pedagogical skills of Thérèse Huet, Michel Mons, Xavier Assfeld, Jean Demaison, Pavel Rosmus, Daniel Borgis, Marie-Pierre Gaigeot, Gilles Ohanessian and Philippe Dugourd during the Summer School organized in Aussois by Isabelle Kleiners.

The idea of writing this book was born during the everyday laboratory life along discussions with my present or past collaborators Yves Bouteiller, Hassan Abdoul-Carime, Sophie Carles, Frédéric Lecomte and Bruno Lucas. Many topics have been developed thanks to collaborations with Ludwik Adamowicz, Hans-Dieter Barth, Alberto Modelli, Jean Liquier, François Piuzzi, Valérie Gabelica, Fernando Pfluger, Don Weaver, Joel Lemaire, Claude Dedonder-Lardeux, Jacqueline Fayeton and Christophe Jouvet.

Encouragements of Danièle, Anne-Marie, Nicolas and Anne, have been warmly present all along the course of the writing.

REFERENCES

1. Kaltashov IA, Eyles SJ: *Mass spectrometry in biophysics*: Wiley, Hoboken, NJ, 2005.
2. Lifshitz C, Laskin J (eds.): *Principles of mass spectrometry applied to biomolecules*: Wiley-Interscience Series on Mass Spectrometry, Wiley, Hoboken, NJ, 2006.
3. Gross JH: *Mass spectrometry*: Springer-Verlag, Berlin/Heidelberg, 2004.
4. Schlick T: *Molecular modeling and simulation*: Springer-Verlag, New York, 2002.
5. Leach A: *Molecular modelling: principles and applications*: Prentice Hall, Pearson, Harlow, UK, 2001.
6. Siudzak G: *Mass spectrometry for biotechnology*: Elsevier, Amsterdam, 1996.
7. Albani JR: *Structure and dynamics of macromolecules: absorption and fluorescence studies*: Elsevier, Amsterdam, 2004.

– 1 –

Modelling

GENERAL FEATURES

This chapter presents different approaches of the modelling of biomolecular systems, from pure quantum calculations concerning rather small systems up to pure classical descriptions of large biological entities such as viruses. Modelling of isolated biomolecules will be considered in the next chapters in order, for example to interpret experimental spectroscopic data, search for rules of fragmentation processes in mass-spectrometry or predict structures of non-covalent complexes. Reviews and books can be found in references [1–8]. The emphasis upon determination of structures that will be noticed all over this book is related to the observed close link between structure and biological activity. This is the basis of quantitative structure–activity relationships (QSAR) that mathematically relate numerical structural properties to bioactivity and are used in drug design. Modelling thus play a great role in the interpretation and prediction of structural features [2,3]. One must nevertheless notice that active biomolecules do not necessarily possess well-defined three-dimensional structures. Many proteins, for example possess disordered regions that are often still functional [9].

According to the desired accuracy, molecular size and power of available computers, different approaches are available. In classical mechanics representations, molecules are simply considered as ensembles of atoms held together by assemblies of springs representing chemical bonds. The quantum nature of those bonds and their rearrangements corresponding to biochemical reactions are best described by solving, as exactly as possible, the Schrödinger equation.

This chapter will first consider the different forces responsible for the structure and the extraordinary ability of biomolecules to recognize each other [10]. The terminology employed in the different methods used for exploration and optimization of molecular geometrical structures, ab initio, force-fields and semi-empirical approaches, will then be presented and illustrated through applications. In this chapter, we will not provide demonstrations that can be found in the following references [1, 11] but we will present the different terms currently used in the literature.

1.1 INTERACTIONS RESPONSIBLE FOR BIOMOLECULAR STRUCTURES AND BIOSPECIFIC RECOGNITION

In molecules, atoms share electrons in order to form strong covalent bonds with their neighbours (bonded interactions). They also interact more weakly with other atoms (non-bonded

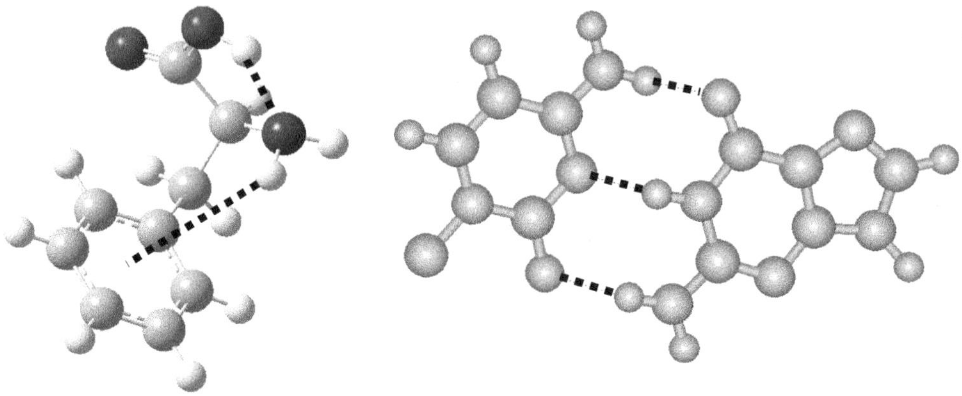

Figure 1.1.1 Left: Intramolecular bonds in the lowest energy gas-phase conformer of the phenylalanine amino acid. Balls represent atoms and sticks represent bonded interactions. Dashed lines correspond to intramolecular hydrogen bonds established between different functional groups of the same molecular system. Right: Intermolecular bonds in the guanine–cytosine (GC) pair. The dashed lines correspond to hydrogen bonds established between two different molecules.

interactions) belonging to the same molecule (intramolecular bonds) or other nearby molecules (intermolecular bonds). Structures of biomolecules result from the establishment of both bonded and non-bonded interactions. The role of non-bonded interactions is crucial since they are responsible for recognition properties between biomolecules. For example, intermolecular bonds enable the formation of A—T or G—C pairs in DNA (see Chapter 4.1) as shown in Figure 1.1.1 or the folding of peptides and proteins (see Chapter 4.2). They allow transcription factors to bind to DNA, recognition of specific peptide sequences [12] or brain receptors to bind neurotransmitters and their agonists [13].

Recognition between biomolecules relies on the formation of very specific interactions. Nevertheless, a wide variety of interactions exist and all of them are not necessarily specific. For example, human DNA contains approximately 3 billion nucleobase pairs and its regulatory proteins or designed drugs [14] must find sequences of typically 10–30 given base pairs in order to act as promoters or repressors of the transcription of genes. The exploration of DNA thus relies both on an extremely rapid and non-specific scan in order to find the target sequence followed by the very specific binding [15]. In this chapter, we will examine the different intermolecular forces responsible for recognition, starting from electrostatic interactions, followed by the highly specific hydrogen bonds and the weak van der Waals interactions. Hydrophobic interactions that play a crucial role in solution will be considered in Section 5.3.

1.1.1 Interactions between distributions of charges

The interaction potential energy $U(r)$ between molecules at distance r is due to the electric forces exerted between different distributions of charges contained by those molecules. Forces $F(r)$ are related to $U(r)$ by $F(r) = -\,\mathrm{d}U(r)/\mathrm{d}r$. Charges can be found in ions or on ionized groups. For example, in proteins, protons will preferentially tend to localize on side chains of some amino acids such as arginine (Arg) or lysine (Lys) while side chains of glutamic acid (Glu) or aspartic acid (Asp) tend to be negatively charged [16]. The interaction energy between two charges $z_1 = Z_1 \cdot e$ and $z_2 = Z_2 \cdot e$ at a distance r (e is the charge of an electron) in a medium of dielectric constant ε is given by

$$E_{\text{charges}} = \frac{Z_1 Z_2 e^2}{4\pi\varepsilon_0 \varepsilon r} \tag{1.1.1}$$

In vacuum, ε is equal to 1 and close to 78 in water. In the interior of a protein, ε is generally supposed to be in between 2 and 4. Ionic interactions are thus stronger in the gas-phase and intervene at longer distances since they are not shielded by the presence of water molecules. The electrostatic energy of n atoms with partial charges q and positions r is given by

$$U_{\text{elec}}(r_1, \ldots r_i, r_j \ldots) = \frac{1}{2}\sum_i \sum_j \frac{q_i q_j}{4\pi\,\varepsilon_0\,|r_i - r_j|} \tag{1.1.2}$$

and can be extremely difficult to evaluate. A "trick" called Ewald sum [17, 18] uses a mesh (here ignored) and consists in dividing this sum in two rapidly converging summations.

$$\frac{1}{r_{i,j}} = \left(\frac{1}{r_{i,j}} - \frac{erf(\alpha r_{ij})}{r_{ij}}\right) + \frac{erf(\alpha r_{ij})}{r_{ij}} \tag{1.1.3}$$

The erf function is the usual error function $erf(r) = (2/\sqrt{\pi})\int_0^r e^{-x^2}\,dx$. The first summation is calculated in the real space over all pairs of charged atoms within a distance cut-off. The second summation, in a Fourier reciprocal space, replaces point charges with Gaussian distributions of charges with dispersion α. By carefully trimming the separation between the two summations, a convergence can be obtained with a considerable reduction of computer time.

Some atoms (e.g. O, N, S) tend to attract electrons leading to asymmetrical distributions of charges corresponding to permanent dipole or higher (quadrupole, octopole, ...) moments. The dipole moment of a distribution of charges q_i at positions r_i is $P = \Sigma_i q_i r_i$. Usual dipole moment units are either the Debye (D) or the Coulomb · meter (C · m). Two unit charges e separated by 1 Å create a dipole of 4.8 D. The isolated water molecule possesses a dipole moment of $1.85\,\text{D} = 6.2 \times 10^{-30}\,\text{C} \cdot \text{m}$. The electric field **E** created by a charge induces in a molecule a displacement of the electronic cloud and nuclei in opposite directions leading to an induced-dipole $\mathbf{P}_{\text{ind}} = \alpha\mathbf{E}$ where α is the molecular polarizability. At a given time, the centre of gravity of negative charges does not necessarily coincide with that of positive charges and an atom or a molecule thus possesses an instantaneous dipole moment.

The potential created at a point P situated at a distance r from a dipole μ is $V(r) = \mu \cos\theta / 4\pi\mu\varepsilon_0 r^2$ where θ is the angle between μ and the direction OP. The interaction between two dipoles μ_1 and μ_2 is maximum when the dipole moments are aligned, such as peptide bonds in α-helices [11, 19] (see Section 3.3.3) or head-to-tail as in the formic acid or formamide dimers [20]. Interactions between charges and permanent dipoles play a great role in structures. In peptides and proteins, for example amino acids are linked by peptide bonds possessing large (~4 D) dipole moments. Those dipole moments interact between themselves and with the charged terminals [19] (see Section 4.2.4.2).

Charges of opposite signs, such as charged side chains of amino acids, strongly attract each other and form so-called salt-bridges (like $Na^+$and Cl^- in sodium chloride) [21–23]. For example, in an isolated Arg amino acid, the guanidinium and carboxylate groups remain neutral (the amino acid is not in its zwitterionic form, see Section 5.5.4). On the contrary, the arginine dimer adopts a favourable head-to-tail configuration because the amino acids become zwitterionic and their end-groups form salt-bridges.

Charges can be delocalized as, for example in π-clouds of aromatic groups [24–26]. Cation-π interactions (Figures 1.1.2 and 1.1.3) are important and intervene in many

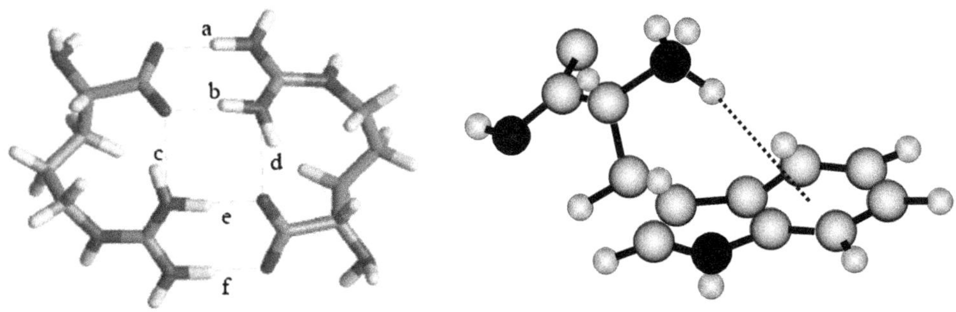

Figure 1.1.2 Left: Array of salt-bridge interactions (dotted lines) between guanidinium and carboxylate groups in an arginine dimer (reproduced with permission from reference [22] ©2001 American Chemical Society). Right: Cation-π interaction (dotted line) between the protonated amino group and the aromatic indole side chain in the lowest energy configuration of the protonated tryptophan amino acid.

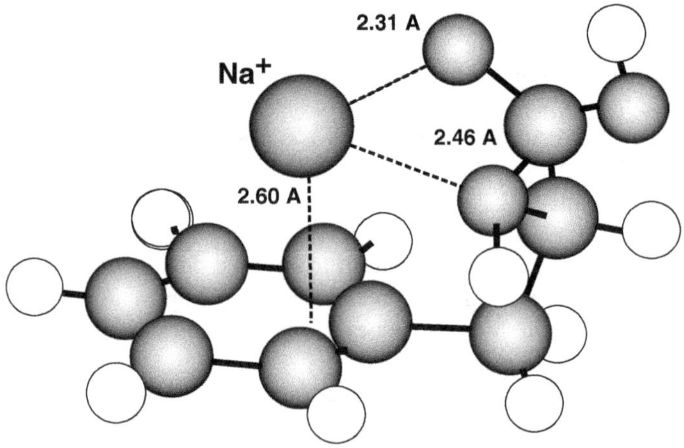

Figure 1.1.3 Cation-π interactions in the Na–phenylalanine complex (distances are in Å) (adapted with permission from reference [27] ©2000 Elsevier).

bio-recognition processes [27–29]. Interactions are thus found quite often between strongly basic groups carrying protons, such as amines or guanidinium ions (side chain of Arg), and aromatic groups either in proteins [30, 31] or protein–DNA complexes [32].

1.1.2 Hydrogen bonds

A hydrogen atom covalently linked to an electronegative atom D (donor), such as O or N, acquires a partial positive charge and can then also bind to another nearby electronegative atom A (acceptor). A —D—H···A— bond, called hydrogen bond, is formed and the main contribution is electrostatic [26]. Only hydrogen atoms do not have inner-shell electrons and strong electron density changes (polarization effects) take place whenever a polar group A—C approaches another polar group D—H. The strength of the interaction strongly depends upon the relative orientation of the dipole bonds D—H and A—C and the preferential arrangement is linear. Directionality of hydrogen bonding plays a crucial role in biomolecular structures [33] such as in DNA base pairs (see Chapter 4.1) or peptide helices (see Chapter 4.2). The equilibrium distance between D and A also results from other contributions such as the repulsions between D and A and C and the polarization of D in the field of A or vice versa. The energy of a A—H···B bond varies in a rather wide domain, from very weak (1–2 kJ/mol) to strong (40 kJ/mol) [34, 35]. In a weak hydrogen bond, the H atom is attached to either one of the O or N atoms [36] or possibly a C atom [37]. Other weak hydrogen bonds can be established with aromatic groups (e.g. indole) acting as donors or acceptors [38, 39]. When the electronegative atoms D and A become closer, the barrier for transfer of the H atom between D and A is lowered. It can eventually become low enough so that the zero-point energy (ZPE) level (see Section 2.1) lies above the barrier and the H atom can then freely move between D and A.

Typical hydrogen bonds are the —N—H···O=C– bonds encountered in secondary structures of peptides and proteins (see Chapter 4.2) and the O···N distance is comprised in between 2.85 and 3 Å [40, 41]. In a weak hydrogen bond such as —C—H···O—, the distance between C and O is less than 3.8 Å [42, 43]. Atoms belonging to charged groups establish strong hydrogen bonds (20–140 kJ/mol). For example, in an enzymatic site [34, 44], oxygen atoms in the COO^- group of an asparagine amino acid are strongly linked to the H—N group of a nearby histidine amino acid. [45]. Protons establish strong bonds and bind to molecules [46–48] (see Chapter 4.2 and Section 5.2.1.4). Hydrogen bonds can be established between connected neighbours or atoms topologically close but belonging to distant parts of a folded chain [49] (Figure 1.1.4).

1.1.3 van der Waals interactions

1.1.3.1 Dispersive interactions

Neutral molecules without any permanent dipole or multipole moment do not interact through usual electrostatic forces. However, they undergo dispersion or van der Waals interactions [50, 51] that have been explained as due to a quantum effect by London. The fluctuations of the electronic cloud of each molecule lead to instantaneous dipoles. Each instantaneous dipole induces a dipole in neighbouring atoms leading to an induced attraction. This attraction also exists if one of the molecules possesses a dipole moment. Water, for example can

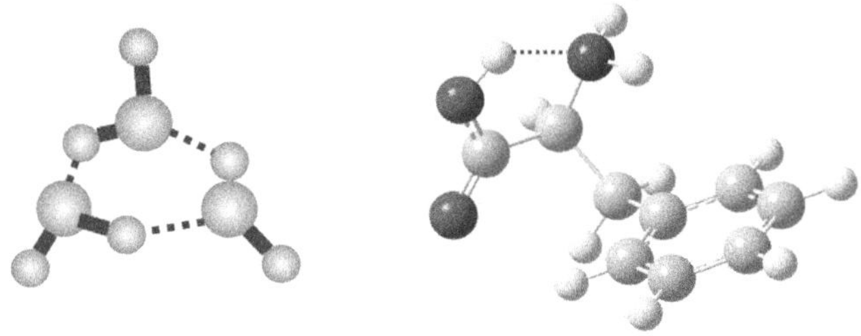

Figure 1.1.4 Left. Inter-molecular hydrogen bonding in the neutral water trimer. Right. Intra-molecular hydrogen bonding in phenylalanine amino acid (reproduced with permission from ref. [41] ©2002 Wiley).

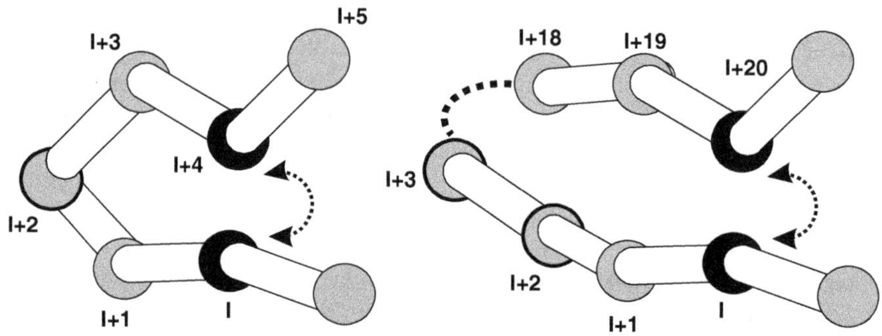

Figure 1.1.5 Spatial neighbouring atoms can establish hydrogen bonds. Those atoms can belong to closely connected regions (i and $i + 4$ on the left figure) or to widely separated regions of a biomolecule (i and $i + 20$ on the right figure).

bind to aromatic groups [52–54]. These dispersion interactions can compete with hydrogen bonding [55] (Figure 1.1.5). A striking example is given by the DNA structure (see Chapter 4.1). If the usual hydrogen bonding pattern is prevented by replacing H atoms by methyl groups, nucleobase pairs adopt a stacking configuration [56–58]. Dispersion interactions require rather high level of theory and computational costs but recent improvements facilitate their accurate calculations [59] (see Section 1.2.1).

1.1.3.2 Repulsive interactions

Atoms are neutral but when their distances become too small, electron orbitals start to overlap and electrostatic repulsion between electrons forbids them to occupy the region of space separating nuclei. If the atomic nuclei could coincide, electrons would have to share the same set of orbitals but the Pauli principle would then prohibit them to have the same set of quantum numbers, in particular electrons with the same spin (leading to "exchange forces"). The reduction of electron density decreases the electrostatic shielding between nuclei which also repel each other. This leads to short-range repulsive forces which receive different names such as overlap, Pauli exclusion, Coulomb repulsion or exchange forces.

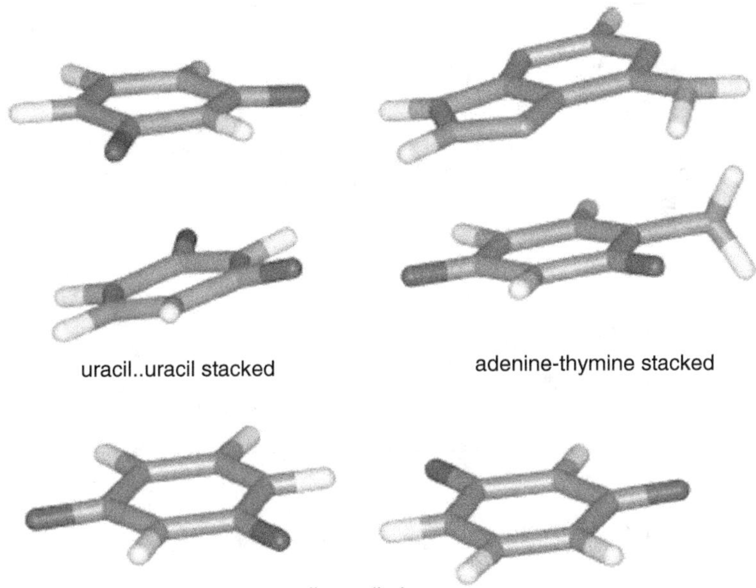

Figure 1.1.6 Competition between hydrogen bonding and dispersion interactions in nucleobase pairs. Structures of stacked and H-bonded uracil dimers and adenine-thymine stacked base pairs (see 4.1) (reproduced with permission from ref. [55] ©2005 American Chemical Society).

Electrostatic repulsion also takes place in multiply charged molecular systems. In a peptide, for example two protonated residues should not be too close from one another. However, even dipeptides can be doubly charged. In a doubly charged tryptophan–lysine peptide, one proton is on the N-terminus while the other is on the opposite side at the end of the long $-(CH)_4-$ side chain. Isolated di-anions can also be observed [60]. In the gas-phase, due to the low dielectric constant, Coulomb repulsion strongly influences the folding and unfolding of ions [61–63].

1.1.3.3 van der Waals interactions

Addition of attractive and repulsive forces leads to the creation of potential wells. A familiar expression of the interaction energy between two identical atoms is the Lennard–Jones (LJ) 6–12 potential V_{LJ}. The equilibrium internuclear distance R_{min} corresponds to the minimum of the energy ε_{LJ} when the interaction energy of the atoms at infinity is taken equal to 0 (Figures 1.1.7 and 1.1.8). The expression of the LJ potential is $V_{LJ} = \varepsilon_{LJ}[(R_{min}/r)^{12} - 2(R_{min}/r)^6]$.

The LJ potential is also null for $r = \sigma = R_{min}2^{-1/6}$. σ is the distance of closest approach of the two identical atoms.

In the case of two different atoms A and B, with respective LJ parameters σ_A, σ_B, ε_A, ε_B, the LJ potential of the A···B system is deduced from the combination rules: $\varepsilon_{AB} = \sqrt{\varepsilon_A \varepsilon_B}$ and $\sigma_{AB} = (\sigma_A + \sigma_B)/2$. Simple modelling taking into account directionality of hydrogen bonds can be performed by using LJ expressions. One then considers four atoms, the donor D, the acceptor A, the hydrogen atom H and the fourth atom C linked to A (Figure 1.1.7),

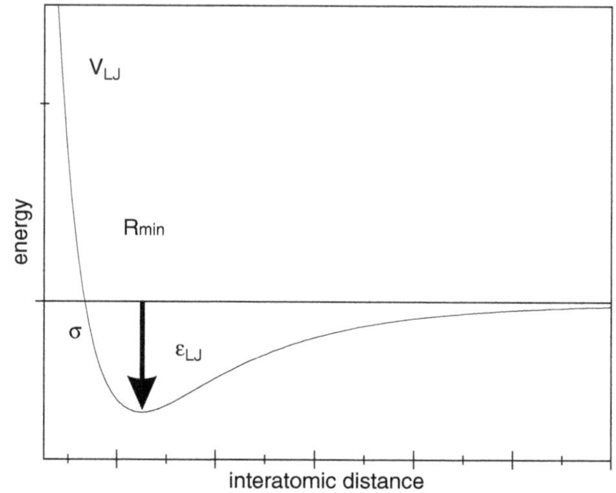

Figure 1.1.7 Lennard–Jones potential.

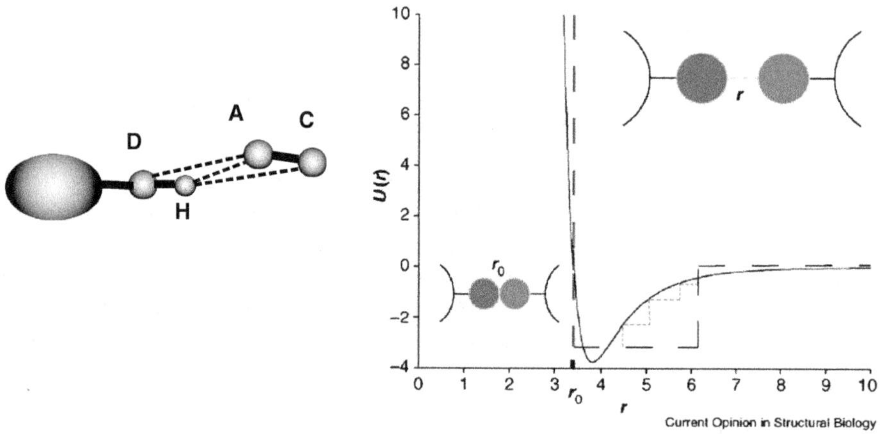

Figure 1.1.8 Left: Simple modelling of hydrogen bond D—H···A—C by means of Lennard–Jones potentials. Right: Approximation of the Lennard–Jones potential (solid line) by discrete square-well potentials (dashed lines). The *r* potential is approximated by a reflective wall at a distance corresponding to the sum of the hard-core radii of the interacting atoms (reproduced with permission from ref. [64] ©2005 Elsevier).

and uses the same combination rules for the LJ potentials for describing respectively weak bindings between D and A, H and A and H and C.

1.2 QUANTUM MECHANICS MODELLING

Quantum mechanics (QM) modelling includes very elaborate and powerful methods that allow the calculation of geometries and energies of systems containing any atomic element present in biomolecules. A comprehensive presentation is given in reference [65]. QM modelling can also provide vibrational and electronic spectra, ionization energies, proton affinities and electron affinities, dipole moments, etc., without any prior knowledge (ab initio). A molecular system is an assembly of N interacting positively charged nuclei described by a set of coordinates $\mathbf{R}$ and n negatively charged electrons described by a set of coordinates $\mathbf{r}$. The binding between those charges with both repulsive and attracting interactions arises from quantum effects and the rigorous mean for predicting the properties of this assembly is to describe it by its total wave function $\Psi_{tot}(\mathbf{r};\mathbf{R})$ and solve the Schrödinger equation $H\Psi_{tot}(\mathbf{r};\mathbf{R}) = E_{tot}\Psi_{tot}(\mathbf{r};\mathbf{R})$ by performing an ab initio calculation. In general, the Schrödinger equation cannot be solved exactly. By using quantum mechanical software packages (Gaussian, Molpro, Turbomole, …), one can obtain a reasonably good approximation to the solution of the Schrödinger equation by selecting a method and an atomic basis set among those that are available. In this chapter, we present an overlook of the most usual methods and the readers can find all necessary details in the given references.

1.2.1 Born–Oppenheimer separation: potential energy surfaces

The masses of nuclei and electrons are very different and the rapid motion of the light electrons thus instantaneously adapts itself to the motion of the heavy nuclei which is much slower. This leads to the Born–Oppenheimer approximation which states that the total wave function can be written as a simple product: $\Psi_{tot} = \Psi(\text{electrons})\,\chi\,(\text{nuclei})$. More precisely,

$$\Psi_{tot}(\mathbf{r};\mathbf{R}) = \Psi_{el}(\mathbf{r};\mathbf{R})\chi_N(\mathbf{R}) \tag{1.2.1}$$

is the product of the wave function $\chi_N(\mathbf{R})$ of the nuclei and the wave function $\Psi_{el}(\mathbf{r};\mathbf{R})$ of the electrons into which the nuclei coordinates $\mathbf{R}$ are parameters which define the molecular geometry.

The Hamiltonian H (operator associated to energy) is the sum of five terms

$$H = T_N + T_e + V_{Ne} + V_{NN} + V_{ee} \tag{1.2.2}$$

The two first terms respectively correspond to the kinetic energy of nuclei and electrons. The last terms respectively correspond to the attractive Coulomb interactions between nuclei and electrons and repulsive interactions between nuclei or electrons among themselves. The total Hamiltonian can be separated in a pure nuclear Hamiltonian $H_N = T_N$ and an electronic Hamiltonian H_{el}

$$H_{el} = T_e + V_{Ne} + V_{ee} = T_e - \sum_{A=1}^{N}\sum_{i=1}^{n}\frac{Z_A}{r_{iA}} + \sum_{i=1}^{n}\sum_{j>i}^{n}\frac{1}{r_{ij}} \tag{1.2.3}$$

The electronic wave function $\Psi_{el}(\mathbf{r};\mathbf{R})$ then satisfies the following equation:

$$H_{el}\Psi_{el}(\mathbf{r};\mathbf{R}) = E_{el}(\mathbf{R})\Psi_{el}(\mathbf{r};\mathbf{R}) \tag{1.2.4}$$

The electronic energy $E_{el}(\mathbf{R})$ is then equal to

$$E_{el}(\mathbf{R}) = \frac{\int \Psi_{el}(\mathbf{r};\mathbf{R})^* H_{el} \Psi_{el}(\mathbf{r};\mathbf{R})}{\int \Psi_{el}(\mathbf{r};\mathbf{R})^* \Psi_{el}(\mathbf{r};\mathbf{R}) d\tau} \tag{1.2.5}$$

summed over the whole space and is thus obtained by numerically calculating integrals.

The total electronic energy is then defined as

$$E_{el}^{tot}(\mathbf{R}) = E_{el}(\mathbf{R}) + \sum_{A=1}^{N} \sum_{B>A}^{N} \frac{Z_A Z_B}{R_{AB}} \tag{1.2.6}$$

In this expression, the nuclear coordinates $\mathbf{R}$ do not explicitly appear in $E_{el}(\mathbf{R})$ and the last term (interaction between nuclei) is constant when the nuclear geometry is fixed. Once the electronic wave function $\Psi_{el}(\mathbf{r};\mathbf{R})$ is known, the nuclear Schrödinger equation satisfied by the nuclear wave function $\chi_N(\mathbf{R})$ can be solved

$$(H_N + E_{el}^{tot}(\mathbf{R}))\Psi_{el}(\mathbf{r};\mathbf{R})\chi_N(\mathbf{R}) = E_{tot}\,\Psi_{el}(\mathbf{r};\mathbf{R})\chi_N(\mathbf{R}) \tag{1.2.7}$$

Quantum chemistry calculations provide the representation of $E_{el}^{tot}(\mathbf{R})$ as a function of the nuclear coordinates $\mathbf{R}$, also called potential energy surface (PES). An example of two-dimensional PES is displayed in Figure 1.2.1.

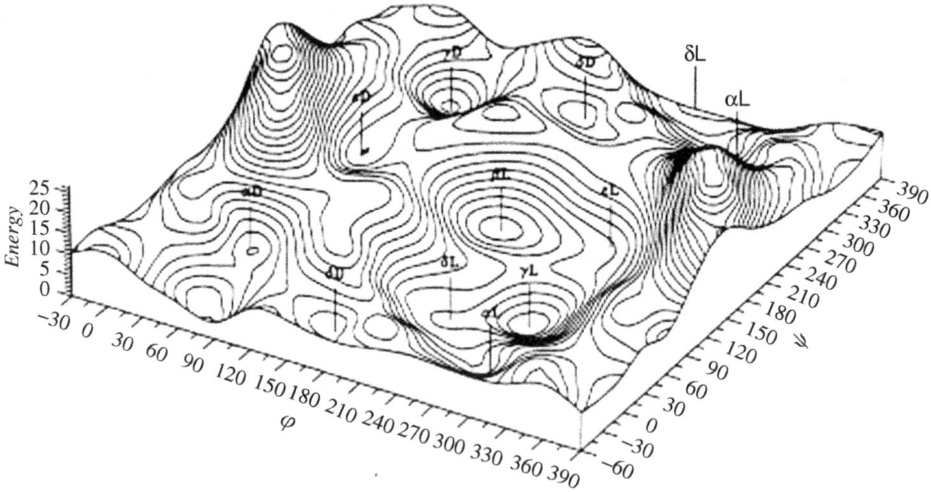

Figure 1.2.1 Two-dimension potential energy surface of the HCO—L—Ala—NH_2 peptide computed at the 3-21G RHF level. The nuclear coordinates are the two angles φ and ψ describing the peptide linkage in a Ramachandran plot (see Chapter 4.2)(reproduced with permission from reference [65] ©1999 Elsevier).

1.2.2 Energy of an electronic state

The Hamiltonian H_{el} of the n electrons can be written as

$$H_{\mathrm{el}} = \sum_{i=1}^{n} (T_{ei} + V_{Ni}) + \sum_{i=1}^{n} \sum_{j>i}^{n} \frac{1}{r_{ij}} \tag{1.2.8}$$

In this expression, the first term is a sum of n Hamiltonians, called core-hamiltonians, h_i^c, for each electron i moving in the field of the nuclei. The second term corresponds to the repulsion between the electrons. Electrons cannot be distinguished and the electronic wave function $\Psi_{\mathrm{el}}(\mathbf{r};\mathbf{R})$ can be written as a Slater determinant $|\phi_1(1)\phi_2(2)\ldots\phi_n(n)|$ built from a set of spin orbitals $\phi_i(x_i)$, each $\phi_i(x_i)$ describing a single electron in the molecular system. A Slater determinant is used to take into account the Pauli exclusion principle that forbids two electrons with the same set of quantum numbers to occupy the same region of space. One can then calculate the expression of the electronic energy which contains three terms. The first is the sum of the energies h_i^c and the repulsion term provides two terms. The Coulomb repulsion energy between electrons i and j respectively described by $\phi_i(x_i)$ and $\phi_j(x_j)$ is given by a two-electron term

$$J_{ij} = \iint \phi_i(x_i)\phi_j(x_i)\frac{1}{r_{ij}}\phi_i(x_i)\phi_j(x_i)d\tau_i d\tau_j \tag{1.2.9}$$

Electrons with the same spin tend to avoid each other and thus experience a lower Coulomb repulsion. This leads to a third two-electron term

$$K_{ij} = \iint \phi_i(x_i)\phi_i(x_i)\frac{1}{r_{ij}}\phi_j(x_i)\phi_j(x_i)d\tau_i d\tau_j \tag{1.2.10}$$

called exchange interaction. The electronic energy is then given as a sum

$$E_{\mathrm{el}} = \sum_{i=1}^{n} h_i^c + \sum_{j\neq i} J_{ij} - K_{ij} \tag{1.2.11}$$

The molecular orbitals (MOs) $\phi_i(x_i)$ are generally constructed as linear combinations of atomic orbitals (AOs) $\chi_v : \phi_i(x_i) = \Sigma_{\nu=1}^{\nu=M} c_{\nu i}\chi_\nu$. The variation theorem states that the energy calculated from an approximate wave function will always be greater than the energy calculated from the exact wave function. The aim is thus to obtain energy values as negative as possible. When the above expression of the electronic energy is calculated in the base of AOs χ_v with imposed condition of orthonormality of MOs, one obtains a set of equations called Hartree–Fock (HF) equations. Those equations are solved by an iterative variational procedure called self consistent field (SCF) scheme. The result of a HF calculation is a set of M one-electron MOs.

The essential idea of the HF or molecular orbital method is that, for a closed-shell system, electrons are assigned two at a time to a set of MOs. In a neutral, closed-shell and ground state molecular system, the n electrons are distributed into $n/2$ shells, each complete shell containing two electrons with opposite spins α and β. The expression of the restricted HF (RHF) electronic energy then becomes

$$E_{\text{el}} = 2\sum_{i=1}^{n/2} h_i^c + \sum_{i=1}^{n/2}\sum_{j=1}^{n/2}(2J_{ij} - K_{ij}) \tag{1.2.12}$$

Starting from the lowest energy orbitals, the electrons are fed in the M orbitals with two electrons per occupied orbital. The remaining orbitals are called virtual orbitals. In molecular systems containing odd numbers of electrons or in electronic states where the number of electrons with spin α is different from those with spin β, an RHF calculation introduces a phenomenon called spin contamination. One then can use single-occupied shells in a spin-unrestricted HF calculation (UHF).

In SCF-HF calculation, each electron moves independently in the average potential of all other electrons. This approximation thus ignores the correlation between motions of neighbouring electrons that tend to avoid each other. Correlation is usually divided into two parts. Static correlation represents the contribution needed to describe bond dissociation processes correctly and dynamical correlation represents the contribution to the energy brought by the instantaneous interaction between electrons.

The difference between the exact energy E_{exact} and the HF energy E_{HF} is the correlation energy ε_c. An ab initio calculation thus faces two main problems: the calculation of this correlation energy ε_c and the choice of AOs for the description of MOs. In the following part of this chapter, we briefly describe the most usual methods used for the calculation of ε_c and some widely used basis sets.

1.2.3 Electronic correlation

At the beginning of an ab initio calculation, one must choose a compromise between the expected accuracy and the computer time investment. Several methods are currently used. The HF approximation assumes that the exact wave function can be expressed by means of a single Slater determinant $|\phi_1\phi_2 \ldots \phi_n|$ built from the occupied orbitals ϕ_i. With n electrons and M orbitals, one has $n/2$ occupied MOs in the complete shell RHF case and $M-n/2$ virtual orbitals. One can try to improve the HF wave function describing the molecular system by adding other Slater determinants obtained by replacing occupied orbitals by virtual orbitals ϕ_V. One then has a single excitation (CIS) or single and double (CISD), triple, etc., excitations. For example, one can replace two occupied orbitals by two virtual orbitals, obtaining a new Slater determinant $\Psi_2 = |\phi_1\phi_2 \ldots \phi_n| \rightarrow |\phi_1\phi_2\phi_{Vi}\phi_{Vi+1} \cdots \phi_n|$. The now used wave function is a linear combination of the built determinants. There are different possible choices. The lowest energy orbitals corresponding to the inner core can remain intact in a "frozen core" approximation. The use of the whole set of M possible orbitals corresponds to the full configuration interaction (FCI). The number of ways to attribute n electrons in M orbitals becomes rapidly enormous and the FCI is only a limit (Figure 1.2.2).

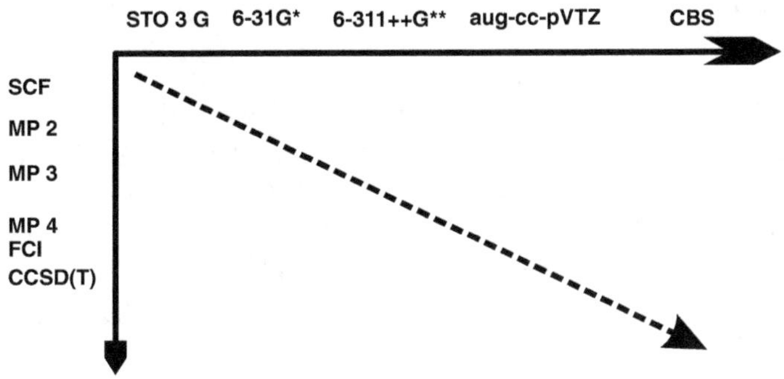

Figure 1.2.2 Evolution towards the "exact" solution of the Schrödinger equation as a function of the one-particle space (basis set) vs. the quality of the n-particle space (computational method) (adapted from reference [65] ©1999 Elsevier).

1.2.3.1 Many-body perturbation theory

In order to take into account electron correlation, Moller and Plesset have introduced a perturbation treatment. The unperturbed energies E_0 and wave functions Ψ_{0i} are obtained from a HF calculation into which the molecular wave function is represented as a product of orbitals generated from a set of (usually) Gaussian AOs. In SCF-HF method, each electron is treated in the time-average field of all the other electrons of the molecular cloud, neglecting the fact that electrons repel each other according to their distances at any given point and that electron motions are thus correlated. The second order perturbation term (MP2) method includes interactions between configurations that are doubly excited compared to the SCF orbitals: an electron from an occupied orbital i is moved to an unoccupied (or "virtual") orbital a simultaneously with another electron in orbital j excited into a virtual orbital b. A better approximation to the exact energy is then obtained by adding small corrections terms starting from the MP2, then the third order term (MP3), etc. The second order and higher order energy corrections are obtained by calculating integrals similar to the above expressions of the Coulomb J_{ij} and exchange K_{ij} integrals but now with occupied orbitals ϕ_i and virtual orbitals ϕ_v. The expression of the MP2 correction is

$$E_{\mathrm{MP2}} = \frac{1}{2}\sum_{i,j,a,b} \frac{\langle ia|jb\rangle\left[\langle ia|jb\rangle - \langle ib|ja\rangle\right]}{\varepsilon_i + \varepsilon_j + \varepsilon_a - \varepsilon_b} \tag{1.2.13}$$

The correction is negative and improves the energy value but the method is not variational and can sometimes overcorrect the HF energy. The calculation is very lengthy (the computer timescales as N^5) since it requires the calculation of four-centre two electrons integrals such as

$$\langle pq|rs\rangle = \int \phi_p(r_1)\phi_q(r_1)\frac{1}{|r_1 - r_2|}\phi_r(r_2)\phi_s(r_2)d^3r_1d^3r_2 \tag{1.2.14}$$

1.2.3.2 RI-MP2

In order to reduce computer time, an approximation called "resolution of the identity" (RI) has been introduced [66, 67]. In two-electron integrals $\langle pq|rs \rangle$, each orbital product is replaced by a linear combination of functions from a new basis (α_Λ): $\phi_p(r)\phi_q(r) \approx \Sigma_\Lambda C_\Lambda^{pq}\alpha_\Lambda$. This new basis set is optimized by minimizing the difference between the "true" E_{MP2} and the approximate E_{RI-MP2} energy values for a set of test atoms and molecules [67]. The four-centre integrals then reduce to three-centre integrals and the computer time now scales as N^4 [68].

1.2.3.3 Coupled-cluster theory

In the SCF-HF independent particle approximation, motions of electrons are not correlated. We have seen that the HF method provides a set of M molecular one-electron orbitals among which are chosen among the occupied orbitals ϕ_i which compose the Slater determinant $|\phi_1\phi_2 \dots \phi_n|$. Instead of correlating the motions of a specific pair of electrons, one can correlate the motions of any two electrons within a selected pair of occupied orbitals i and j by introducing a two-particle cluster function

$$f_{ij}(x_m, x_n) = \sum t_{ij}^{ab} \phi_{Va}(x_m)\phi_{Vb}(x_n) \tag{1.2.15}$$

The ϕ_{Va} and ϕ_{Vb} are virtual orbitals and the t_{ij}^{ab} are called the cluster coefficients. An improved Slater determinant is obtained by replacing each product of independent particle wave functions

$$\begin{aligned} &\Phi_0 = \phi_i(x_1)\phi_j(x_2)\phi_k(x_3)\phi(x_4) \text{ by} \\ &\Phi_0 + \sum t_{ij}^{ab} \phi_{Va}(x_1)\phi_{Vb}(x_2)\phi_k(x_3)\phi(x_4) \end{aligned} \tag{1.2.16}$$

Since the Slater determinant produces a linear combination of such orbital products by exchanging electrons, one now finds, for example, products such as

$$\begin{aligned} &[\phi_i(x_1)\phi_j(x_2) + f_{ij}(x_1, x_2)]\phi_k(x_3)\phi_l(x_4) \text{ and} \\ &[\phi_i(x_3)\phi_j(x_4) + f_{ij}(x_3, x_4)]\phi_k(x_1)\phi_l(x_2) \end{aligned} \tag{1.2.17}$$

Those products only differ in their distribution of electronic coordinates and the cluster function thus correlates the motion of every pair of electrons found in orbitals ϕ_i and ϕ_j. With this method, it is possible to generate all the Slater determinants containing a single (S), two (D) etc., excited orbitals [69, 70]. If the treatment of triple excitations is only perturbative, the acronym of the method is CCSD(T). Today, the CCSD(T) method is one of the most accurate methods [59] but it can still only be used for the study of rather small biomolecular systems [71–74].

Benchmark calculations provide guides [75] concerning the necessity of costly high-level calculations and the acceptable use of reasonably cheap calculations applicable to much larger systems. For example, density functional theory (DFT) considered in the next chapter usually provides satisfying result for the interpretation of infrared spectra (see Section 2.1.3), except in cases such as presence of methyl groups where dispersion becomes important. On the contrary, accurate and reliable calculations of binding energies require much higher levels of theory [75]. An example is given in a benchmark calculation comparing RI-MP2 and CCSD(T) predictions of binding energies of several formamide dimers [20] can be found in reference [76]. This model system calculation provides an important result about the evaluation of C—H···O binding energy. This interaction intervenes in many molecular recognition patterns [12, 77, 78] (see Section 4.2.3) and is usually underestimated.

1.2.4 Density functional theory

The electronic energy E_{el} of the fundamental electronic state of a molecular system is completely determined by the knowledge of the electronic density $\rho(\mathbf{r})$ [79]. In a system containing n electrons, one has $n = \int \rho(\mathbf{r})\mathrm{d}\mathbf{r}$. The energy is a functional of $\rho(\mathbf{r})$ (a functional allows a function to be mapped to a number). In the DFT introduced by Kohn, the density can be written as the sum of the square moduli of a set of one-electron orbitals: $\rho(\mathbf{r}) = \Sigma_i|\psi_i(\mathbf{r})|^2$. As in ab initio methods described above, those one-electron orbitals are also expressed as linear combinations of atomic-centred basis functions. The electronic energy E_{el} is the sum of several terms. It comprises the kinetic energy of the electrons, the Coulomb energy of the electrons in the field of fixed nuclei, the Coulomb energy of the electrons in their own field and an exchange-correlation term $E_{XC}[\rho(\mathbf{r})]$. There exist a variety of functionals for expressing this exchange-correlation term [80, 81].

In a homogeneous system such as an electron gas, the functional is known with high accuracy. If an atom or a molecule is also assumed as homogeneous, the purely local density approximation (LDA) is used. In the LDA, the contribution of each infinitesimal volume $\mathrm{d}\mathbf{r}$ to the exchange-correlation energy is taken as the value it would have if the whole space was filled with a homogeneous electron gas with the same electron density as that in $\mathrm{d}\mathbf{r}$. A better accuracy is obtained if the inhomogeneous aspect of the density is taken into account through introduction of its gradients in the generalized gradient approximation (GGA). One of the most widely used functional is the B3LYP functional that is a combination of the functional of the density introduced by Becke [82] depending upon the gradients of $\rho(\mathbf{r})$ (it is called "non-local" or "gradient-corrected") and the correlation functional introduced by Lee, Parr and Yang providing the exact exchange energy [54, 83]. In this "hybrid" functional, the exchange term is of HF type. A comparative study of a variety of DFT theories is provided in reference [54].

DFT calculations are usually less computer time-consuming than ab initio quantum calculations [84] and can provide, with caution [81], acceptable results. Tables 5.2.1 and 5.2.2 compare results obtained with the B3LYP functional and optimized basis sets to results obtained from CCSD(T) calculations and experimental results. Another comparison between DFT, MP2 and semi-empirical methods applied to structures of glycine and alanine based peptides can be found in reference [85]. Rather than directly guessing functionals, it is possible to derive correct orbital-dependent expressions and then obtain "ab initio DFT" [86]. In any case, the wave function of n electrons including $3n$ variables is replaced by the density with only 3 variables and DFT scales as n^3 (n = DFT/B3LYP) while the HF and MP2 calculations respectively scale as n^4 and n^5.

1.2.4.1 Dispersion and DFT

The attractive C_6/r^6 dispersion interaction is usually not taken into account in functionals. Whenever this dispersion term becomes crucial (e.g. in stacked configurations, Figure 1.1.6), it is generally assumed that MP2 calculations are preferable. A comparison between

DFT/B3LYP and MP2 calculations concerning the tyrosine–glycine peptide is presented in reference [87]. Dispersion can be taken into account in DFT calculations [54, 59]. A first possibility is the introduction of an empirical expression of a dispersion term

$$E_{\text{dispersion}} = -\sum_{i,j} f_{\text{damp}} C_{6i,j} r_{i,j}^{-6} \tag{1.2.18}$$

The damping term is fitted by comparison to a wide database of interaction energies of small complex systems containing nucleobase pairs and amino acid pairs [75]. On average, the contribution of dispersion missing in the DFT calculations varies from 15% in hydrogen-bonded complexes up to 100% in stacked complexes. A second possibility is offered by the use of extended functionals such as the X3LYP functional [54] linearly mixing the Becke88 and the PW1 exchange functionals.

1.2.5 Linear scaling for large systems

Experimental gas-phase studies of large molecular systems of biological interest do not provide already sufficiently well-resolved data and their interpretation thus do not require the use of sophisticated predicting methods (see IR of proteins in Section 2.1.3.2.6). Nevertheless, some approaches aim to defeat the barrier imposed by the number N of atoms. In coarse-grain models of peptides for example amino acids are only represented by points located on their C_α carbons, ignoring further subtleties [88–90].

At first sight, it would seem that quantum calculations should be a private domain reserved to small ($\approx$100 atoms) biomolecular systems. This is not entirely true since although some valence MOs can be widely spread over the whole considered molecular systems, it is also reasonable to assume that formation of chemical bonds is mostly sensitive to the local environment of the involved atoms. It exists some methods leading to computer times scaling only as N and no longer as N^3 as DFT for example [91–94]. Once admitting that the properties of a given region are not too influenced by the presence of a remote region of another molecule, one can, for example divide the system of interest into subsystems that overlap. Each subsystem consists of a core surrounded by one or two buffer regions. Only core–core or core–inner buffer interactions are accurately calculated [91] ("divide and conquer" method). Low computational costs can then be achieved for large systems. Still accurate QM calculations scaling linearly with the number of basis functions are possible. Examples concerning very fast calculations (3 min instead of 15 h) of $(\text{glycine})_{16}$ or progesterone conducted by linear scaling local coupled-cluster (CC) theory can be found in reference [95].

1.2.6 Basis sets

In this chapter, we consider the construction of MO $\phi_i(x_i)$ as linear combinations of AO $\chi_\nu : \phi_i(x_i) = \Sigma_{\nu=1}^{\nu=M} c_{\nu i}\chi_\nu$. Since the number M of AOs is finite, this representation is only an approximation. The two main conditions imposed to basis sets of AOs are that they should be physically meaningful and the cost of computation of the integrals should be minimized. A first possible choice is the use of the Slater type orbitals (STO) with analytic expressions very close to those of real AOs $\chi_\nu = R_{nl}(r)Y_{lm}(\theta,\varphi)$. The angular part $Y_{lm}(\theta,\varphi)$ is a spherical harmonic and determines whether the orbital is of s ($\ell = 0$), p ($\ell = 1$), ... type. The radial part $R_{nl}(r)$ is proportional to $r^{n-1}e^{-\xi r}$ where r is the electron-atomic centre distance. Integrals only involving one or two atomic centres can be computed with STOs but integrals involving three or four different centres cannot. For simplifying integral calculations, one uses the fact that the product of two Gaussian functions is another Gaussian function. STOs are then replaced by linear combinations of Gaussian type orbitals (GTO) $g(\xi_i)$ with radial dependences proportional to $e^{-\xi_i r^2}$. A linear combination $\chi_v = \Sigma_{i=1}^{L} c_{i\mu} g(\xi_{i\mu})$ contains L Gaussians $g(\xi_i)$, each represented by its coefficient $c_{i\mu}$ and its exponent $\xi_{i\mu}$. For example, a STO-6G orbital corresponds to $L = 6$. If the coefficients $c_{i\mu}$ and $\xi_{i\mu}$ are allowed to vary during the calculation the Gaussians are called primitive or uncontracted. Most often, those coefficients are kept constant in contracted orbitals.

In a minimum basis set, only enough AOs are used to contain the electrons of the core and valence orbitals of the neutral atoms. For the first row, only five AOs (1s, 2s, 2p) are used and for the second row, only nine AOs (1s, 2s, 2p). A first improvement consists in an increase of basis functions for valence electrons of each atom, in a so-called split-valence basis set. For a hydrogen atom, the 1s orbital is replaced by an inner 1s AO with a larger ζ exponent and an outer 1s′ more diffuse AO with a smaller ζ exponent. Ten AOs represents atoms of the second row in a double-zeta (VDZ) basis set. The core electrons are only represented by a single ζ function. Up to six zeta basis sets can be used [96].

For atoms of the second row (e.g. C, N or O), core electrons are in a 1s shell. In order to accurately compute their interaction with the valence electrons, they must be precisely described in the region near the nuclei where their density is maximum. A single Gaussian is a very inaccurate description and a contraction of six Gaussians is more appropriate. A 6-31G base means six contracted Gaussians for the inner electrons, three contracted Gaussians and an isolated Gaussian orbital for the valence electrons. Larger sets such as 6-311-G use three sizes of contracted functions for each orbital type.

The split-valence basis sets only allow changes in orbital sizes. The orbital shapes can be "polarized" by adding orbitals, denoted by *, with larger angular momenta. For example, one adds a p orbital for each H atom in a "polarized" 6-31G* base. A 6-31G* corresponds to d functions for atoms of the second row. Another improvement can be obtained by adding diffuse functions, denoted by +, with small exponents. Those diffuse functions allow orbitals to occupy larger regions of space. This is important, for example in descriptions of hydrogen bonds, lone-pairs, excited states and calculations of polarizabilities, proton affinities [97] or anion structures where excess electrons can be in extremely diffuse orbitals [98].

Different strategies can be applied for obtaining basis sets. In the popular Gaussian basis set, the core region description is optimized for atomic properties and the valence electron description for a set of molecular properties. In the Dunning correlation-consistent (cc) family of basis sets [99], a contraction scheme is used after orbitals obtained by Huzinaga in an optimization of primitive Gaussian functions. Those orbitals are denoted (aug)-cc-p(C)-VXZ(X = D,T,Q,5,6). aug means the addition of diffuse functions and C the addition of functions to obtain a high-accuracy core-valence correlation description.

In principle, it is possible to improve molecular wave functions by adding to SCF orbitals combinations of orbitals. Those combinations are obtained by replacement of occupied SCF orbitals up to the maximum possible number of unoccupied orbitals. This would then lead to a Full Configuration Interaction (FCI). For a system containing n electrons, the wave function should then contain all possible excitations through order n. With a complete basis set

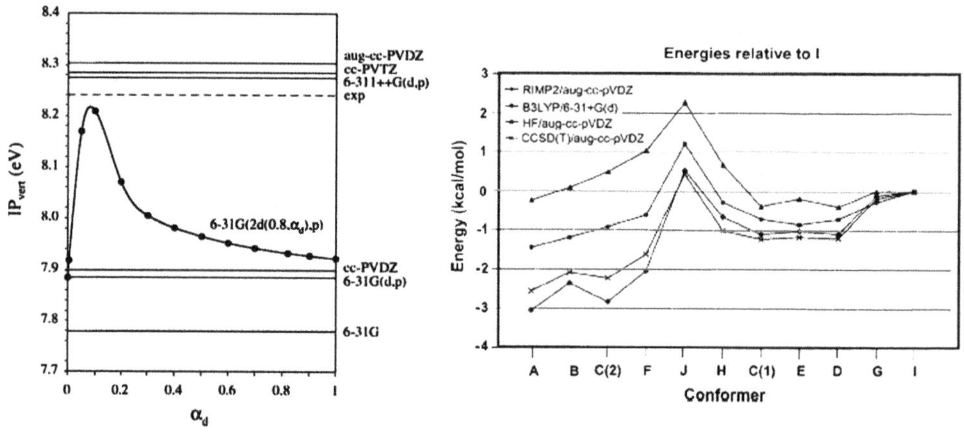

Figure 1.2.3 Left: Influence of basis sets upon the vertical ionization potential (VIP) of the guanine nucleobase (in eV). The VIP is calculated at the MP2/6-31G(2d(0.8, α_d),p) level and plotted as a function of the α_d coefficient of the d-polarization functions on C, N and O atoms. The horizontal dotted line corresponds to the experimental value. VIPs obtained with extended basis sets, respectively 6-311++G(d,p), aug-cc-PVDZ and cc-PVTZ, are represented by horizontal lines (reproduced with permission from reference [100] ©2006 American Chemical Society). Right: Influence of level of theory upon energies of conformers of tryptamine. Conformer I is chosen as zero of energy (reproduced with permission from reference [101] ©2005 American Institute of Physics).

(CBS), the limit would then be the exact solution of the Schrödinger equation. In practice, the one-electron AOs are truncated as well as the n-electron basis. Empirical parameters are used to compensate as well as possible these truncations. Those parameters are optimized to minimize the truncation error for a given set of molecules. When using (aug)-cc-p(C)-VXZ orbitals, the CBS limit is reached when X tends to infinity. Practically, one extrapolates the energies E(X) obtained by using (aug)-cc-p(C)-VXZ (X = 2 or 3) orbitals with two possible formula. For HF calculations, one obtains a good estimate of the CBS energy limit E(CBS) by means of $E(X) = E(CBS) + \exp(-bX)$ and for energies obtained from correlation methods, one can use $E(X) = E(CBS) + cX^{-3}$. Examples of CBS extrapolations are given for nucleic acid base pairs in references [102–104]. Examples of the influence of basis sets and levels of theory are displayed in Figure 1.2.3.

1.3 MOLECULAR MECHANICS: FORCE-FIELDS

In a classical mechanics description of a biomolecule, the existence of electrons that are responsible for the chemical binding between atoms is ignored. The molecule is considered as a mechanical system held by internal forces and is described by models called force-fields. In a first step, the molecule potential energy is determined by means of analytical expressions containing adjustable parameters and a minimization procedure provides the atomic positions and thus the molecular equilibrium geometry. Then, atoms receive initial velocities according to the chosen molecular temperature and a dynamical simulation deduced by integration of the Newton equation $\vec{F} = m\vec{a}$ is conducted. Information such as formation and breaking of intermolecular bonds, interactions between a biomolecule and its environment or concerted motions can then be obtained. For simplification and rapidity of calculations, the potential energy of the biomolecule arises from a rather small number of specific interactions such as stretching of bonds, twisting around single bonds, van der Waals interactions and Coulomb interactions (Figure 1.3.1). The total potential energy can be written as a sum [4]

$$E_{\mathrm{T}} = E_{\mathrm{bond}} + E_{\mathrm{angle}} + E_{\mathrm{bond-angle}} + E_{\mathrm{tors}} + E_{\mathrm{vdW}} + E_{\mathrm{Coul}} \tag{1.3.1}$$

Instead of using the Cartesian coordinates of each atom, internal variables corresponding to bond lengths, bond angles and dihedral angles are used and the contribution of each interaction is modelled by a potential function of these variables. One of the great advantages of internal coordinates is that they can be frozen in a search of the different minima of the potential energy. For example, bond lengths can be kept as constants while the different values of angles are varied. The first three terms of E_{T} correspond to atoms directly bonded or bonded to a common atom and are called bonded interactions. The van der Waals and electrostatic interactions are called non-bonded and intervene between atoms which can belong to different regions of the molecule (see Figure 1.1.5). Force-fields are built by selecting analytical expressions of the potential energy terms which must only contain a restricted number of parameters. Those parameters are then evaluated by an iterative fitting procedure of already known geometries and interaction energies of a selected group of reasonably small and representative molecules with similar chemical properties, for example a set of small peptides, nucleic acids or sugars. After parameterization, a force-field is characterized by its transferability if it can be applied to satisfactorily predict properties of large biomolecules composed of the same chemical groups and which were not included in the parameterization set. There is thus a compromise between a better transferability which may impose the choice of a larger number of parameters and the necessity to keep simple potential functions for limiting computer time and allowing the treatment of larger and larger biomolecules in presence of a solvent.

In the first generation of force-fields, CHARMM [105, 106], AMBER [107], GROMOS [108] and OPLS [109], the emphasis was set upon simplicity of potential energy functions. For example, in the most widely used force-fields, the intramolecular terms describing bonds, angle bonded and torsion interactions are, respectively,

$$E_{\mathrm{bond}} = \sum_{\mathrm{bonds}} K_b (r_{ij} - r_{ij}^0)^2, \quad E_{\mathrm{angle}} = \sum_{\mathrm{angles}} K_\theta (\theta_{ijk} - \theta_{ijk}^0)^2 \tag{1.3.2}$$

and

$$E_{\mathrm{torsion}} = \sum_{\mathrm{torsions}} K_\varphi (1 + \cos(\varphi_{ijkl} - \delta)) \tag{1.3.3}$$

Intermolecular (non-bonded) terms are

$$E_{\mathrm{vdW}} = \sum \varepsilon_{ij} \left[\left(\frac{R_{\min ij}}{r_{ij}} \right)^{12} - 2 \left(\frac{R_{\min ij}}{r_{ij}} \right)^{6} \right] \tag{1.3.4}$$

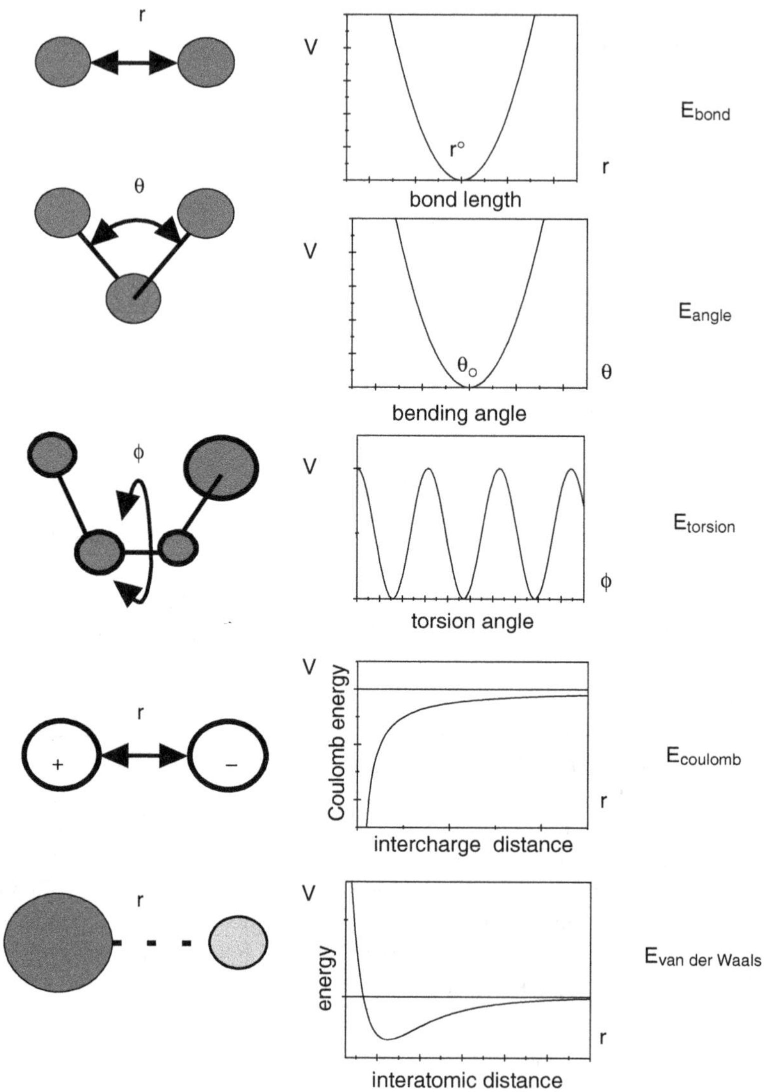

Figure 1.3.1 Empirical potential energy functions.

and

$$E_{\text{Coulomb}} = \sum \frac{q_i q_j}{\varepsilon r_{ij}} \tag{1.3.5}$$

r_{ij}^0 is the equilibrium value of the bond distance r_{ij} between atom i with charge q and atom j with charge q'.

θ^0_{ijk} is the equilibrium value of the angle θ^0_{ijk} between atoms i, j and k. φ_{ijkl} is the dihedral angle between atoms i, j, k and l. δ is a phase shift. The expression of E_{vdW} is that of a LJ potential (see Section 1.1.3.3), $R_{\min ij}$ is the equilibrium distance between atoms i and j and ε_{ij} is the well depth. In E_{coulomb}, q_i and q_j are the partial charges of atoms i and j while ε is the dielectric constant. This last value is an adjustable parameter varying from the bulk water value equal to 78 down to 1 in vacuum or in highly hydrophobic regions of proteins. In contrast with van der Waals interactions, Coulomb interactions decay very slowly with distance and a cut-off distance with a typical value of 10 Å is generally added in order to neglect interactions between atoms too far away from each other.

The other terms of the E_T expression can differ among force-fields. Following the choice of the analytical expressions, different bond lengths, angles, partial charges, etc., must be determined. Different types of atoms are introduced, for example sp^2 hybridized carbon or amide nitrogen atoms. The crucial step, called parameterization of a force-field, usually aims to predict as well as possible condensed-phase data: NMR and crystal structures or vibrational spectra in liquids. In absence of experimental data, quantum calculations, also called "gas-phase" calculations, are used for parameterization and sometimes conflicts between gas-phase and condensed-phase data can appear. For example, the model of peptide backbone is *N*-methylacetamide and the bond lengths or angles are not the same whether they are taken from gas-phase determinations, ab initio calculations or crystal structures [110]. The same situation occurs for distances and binding energies of DNA base pairs since packing effects are strongly influent in crystals but are absent in gas-phase [111]. Nevertheless, modelling of gas-phase experiments conducted with force-fields designed for condensed-phase can provide satisfactory results if a proper parameterization is conducted [112]. Comparisons between predictions of force-fields are most often conducted for a given family of biomolecules, for example DNA [113] or carbohydrates [114–117].

In order to increase transferability, force-fields have been designed with analytical potential energy expressions more complete than the harmonic approximation. For example, in the Merck molecular force-field (MMFF), the intramolecular bond interaction becomes

$$E_{\text{bond}} = \sum_{\text{bonds}} K_b (r_{ij} - r^0_{ij})^2 \cdot \left(1 + cs(r_{ij} - r^0_{ij}) + \frac{7}{12}(cs^2(r_{ij} - r^0_{ij})^2)\right) \qquad (1.3.5)$$

where cs is called the cubic-stretch constant. Quantum ab initio calculations can be used for refinement of force-fields. For example, usual force-fields predict planar amino groups in hydrogen-bonded pairs of nucleobases. On the contrary, ab initio calculations and improved force-fields show that two minima exist in the potential energy of the amino group of cytosine where the hydrogen atoms are displaced outside of the ring plane while the nitrogen atom is slightly less displaced in the opposite direction [118].

1.3.1 Polarizable force-fields

Usually, force-fields describe Coulomb (electrostatic) interactions in terms of fixed charges centred on atoms. Electron clouds which are ignored in force-field descriptions are nevertheless polarizable [119]. This means that molecular charge distributions are modified by electric fields themselves due to the presence of other charges. Permanent atomic charges (metal ions, for example) create an electric field E^0 and thus displacements of positive and negative charges leading to induced-dipoles which in turn create an electric field E^P. The overall result is that an induced point-dipole $\mathbf{P}_{ind}$ proportional to the polarizability α can appear at each atomic centre, bond and lone-pair: $\mathbf{P}_{ind} = \alpha(E^0+E^P)$. A transfer of a bond-charge increment q_{ij} from a site i to a site j is allowed. The total charge on a site becomes the sum of contributions from all bond-charge increments. This procedure into which site i looses q_{ij} while site j gains q_{ij} ensures that the total charge of the biomolecule remains constant. Polarizable force-fields can strongly improve the description of hydrogen bonds and interactions between cations and aromatic groups [120, 121]. In principle, such force-fields possess better transferability and can be used to predict both gas-phase and condensed-phase structures of biomolecules [122].

1.3.2 Reactive force-fields

Reactions taking place in large biomolecular systems such as enzymes [123, 124] cannot be modelled with usual force-fields and require the use of so-called "reactive" force-fields [125, 126]. In order to take into account dissociation of reactants and formation of new bonds in products, the minimum energy path is assumed to conserve total bond orders. For example, the total bond order of carbon, BO_C, should not exceed 4 and that of H, BO_H, should not exceed 1. Terms are added to those already present in the above eq. (1.3.1) and the fundamental assumption is that the bond order BO_{ij} between pairs of atoms can be obtained from the distance r_{ij} between atoms i and j. In a carbon–carbon bond, for example the σ bond term ($p_{bo,1}$ and $p_{bo,2}$) is unity below 1.5 Å and negligible above 2.5 Å. The first π bond term ($p_{bo,3}$ and $p_{bo,4}$) is unity below 1.2 Å and negligible above 1.75 Å. The second π bond term ($p_{bo,5}$ and $p_{bo,6}$) is unity below 1.0 Å and negligible above 1.4 Å. For carbon–hydrogen and hydrogen–hydrogen bonds, only σ bond terms are considered. The bond orders are then evaluated from

$$BO_{ij} = \exp\left[p_{bo,1}\left(\frac{r_{ij}}{r_0}\right)^{p_{bo,2}}\right] + \exp\left[p_{bo,3}\left(\frac{r_{ij}}{r_0}\right)^{p_{bo,4}}\right] + \exp\left[p_{bo,5}\left(\frac{r_{ij}}{r_0}\right)^{p_{bo,6}}\right] \quad (1.3.6)$$

After parametrization, such a force-field can be used for systems containing more than 1,000 atoms. The computer time can be divided by a factor of 10^5 as compared to a DFT calculation for a 500-atom system. The use of variable charges and partial bond orders improves the transferability of interatomic potentials from one molecule to another.

1.4 SEMI-EMPIRICAL METHODS

Due to the large computer time and memory requirements for the computation of integrals involving two electrons and before the introduction of linear scaling methods (Section 1.2.4) calculations using ab initio and DFT methods have been usually only applied to molecular systems with restricted number of atoms (Table 1.4.1). For a reasonably rapid evaluation of structures or for studying large systems out of reach from pure quantum methods, semi-empirical methods have been devised and possess the following common

Table 1.4.1

Approximate comparison between methods. In the third column, the scaling dependence of the computational cost with the number of basis functions and molecular size N is given in parenthesis. An exact full CI would scale as $N!$

Method	Size	Computational cost	Accuracy
Semi-empirical	300	1 ($\approx N^3$)	10
HF ab initio	100	20 ($\approx N^4$)	5
DFT	100	20 ($\approx N^3$)	3
Post-HF MP2	50	50 ($\approx N^5$)	2
RI-MP2	100	30 ($\approx N^4$)	2
CCSD	10	200 ($\approx N^6$)	1
Linear scaling	>1,000	<1 ($\approx N^1$)	

features. Those methods use basis sets that only contain STOs for valence shells. All integrals involving two electrons belonging to three or four centres are supposed to be null. Moreover, some two-centre integrals are neglected. Core terms are estimated by means of empirical relationships. The corresponding integrals are supposed proportional to the overlap of the involved AOs. One-centre integrals are often estimated from data extracted from electronic spectra of the concerned atoms or ions. The most familiar semi-empirical methods usually implemented in available softwares are the AM1 (Austin Model 1) and PM3 (semi-empirical parametric method number 3). Those methods have been improved and can be used for studying systems as large as chlorophyll [127–130].

Comparisons between the different methods, ab initio, DFT or semi-empirical approaches have been performed [85, 131]. It is usually worthwhile to compare theoretical predictions not only between themselves but also to experimental data [132].

For interpretation of experimental data, several steps are required. An extensive exploration of the potential energy landscape (see Section 1.5) can only be conducted at a low-level of theory. Searches using different force-fields (see Section 1.3) or semi-empirical methods do not necessarily locate the same energy minima. The order of those minima, as function of energy, as well as predicted hydrogen bonds, is usually erroneous. However, they provide initial configurations that can be further improved by means of DFT or ab initio methods. General rules are difficult to derive but it is often accepted that for comparison between spectroscopic data and predictions for systems where hydrogen bonding has an important role, simple DFT calculations using the B3LYP method and with basis sets such as 6-31+G(d,p) (with proper scaling, see Section 2.1.3) provide a qualitatively satisfying agreement. An improved quantitative interpretation requires a more accurate prediction of relative energies of the different isomers, that is generally obtained from MP2 calculations with a typical 6-311+G(d,p) basis set. The optimized structures obtained from DFT are often recalculated at a higher level of theory (e.g. MP2) without re-optimization ("single-point" procedure). Such a calculation is then described as for example MP2/6-311++G**//B3LYP/6-31+G* single-point. This "cheap" procedure can be unfortunately misleading as demonstrated by further re-optimization at the higher level of theory. Since ZPE corrections (see Section 2.1.3.1) require optimization before calculation of vibrational frequencies, they can then only be evaluated at the DFT B3LYP/6-31+G*level [133]. When hydrogen bonding is absent and dispersion forces play a crucial role (in stacking of nucleobases, for example), RI-MP2 calculations with extended basis sets are then required [134].

1.5 EXPLORATION OF POTENTIAL ENERGY LANDSCAPES

A PES resembles a mountain range with peaks and valleys connected by passes. Exploration of the PES for the determination of its important features: mimimum energy configurations or local minima (valleys), transition states or saddle points (passes) and energy barriers (peaks) is a difficult problem [135–138]. When the size of biomolecules increases, the number of possible configurations becomes extremely large. Exploration of its potential energy landscape by a large biomolecule such as a protein should in principle take an astronomical amount of time when starting from an unfolded configuration in order to find its folded biologically active structure. However, it has been pointed out as the "Leventhal paradox" that in reality, proteins fold correctly within fractions of seconds and even considerably much shorter times (see Section 4.2.6). This is explained by the existence in the PES of a funnel [135, 139, 140] consisting of favourable kinetic pathways converging rapidly towards the native structure and avoiding kinetic traps.

The different available modelling methods are explained in depth in references [1, 4, 5, 141] and this chapter will only present a brief look to some of the most widely used methods for exploration of the PES of small molecules. The connection between theory and experiment will be considered in following chapters. Usually, in an attempt to simplify as much as possible the interpretation of spectroscopic measurements, experimentalists try to restrict as much as possible the number of populated configurations by lowering the temperature. Starting from the lowest energy configuration domain, it will be shown in Section 2.1.2.4 that is also possible to experimentally determine energy barriers and to explore configurations outside those already populated. It will also be shown in Chapter 4 how different experimental methods can sample different regions of the configuration space and why accurate modelling is required to interpret those differences. An up-to-date example of conformational search for the interpretation of high-resolution experimental data can be found in reference [103].

1.5.1 Systematic energy sampling and energy minimization

If the studied molecular system possesses a rather reasonable number of internal degrees of freedom, it seems natural to search all its possible conformations by means of a systematic procedure, which samples the internal energy over the whole range of each degree of freedom [142, 143]. For example, each rotation or torsion angle [144, 145] can be set equal to $p2\pi/n$ with $0 \leq p \leq n$ with typical steps in between 10° and 30°. The major problem with systematic searching is that the number of conformations required to adequately sample the entire conformational space of a molecule can become extremely large. For example, 10° steps can be chosen for an amino acid with one degree of freedom for the side chain and two for the backbone (Figure 1.5.1). For a peptide containing N such amino acids, about 10^{3N} conformers are generated.

Following systematic energy sampling, usually with a low-level (force-field or semi-empirical) method, an energy minimization is performed with the highest possible level of theory for the largest possible number of found conformers. Very often, surprises come at

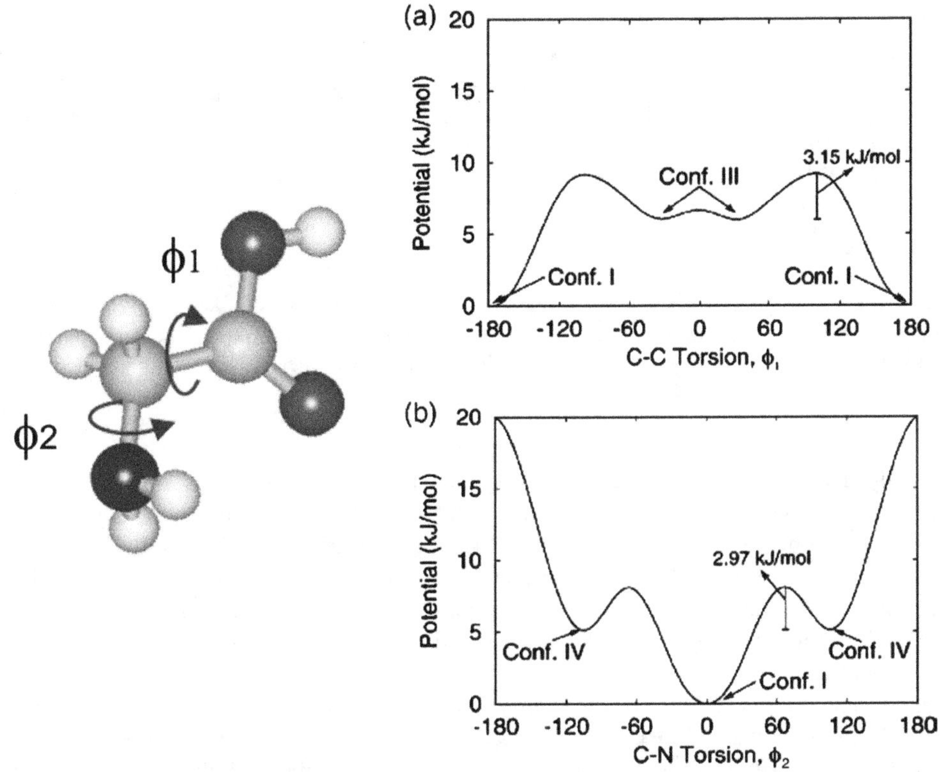

Figure 1.5.1 Glycine one-dimensional potential energy curves plotted as function of the angles ϕ_1 and ϕ_2 which are the respective coordinates of the twisting motions around the C—C (a) and C—N bonds (b) (reproduced with permission from reference [146] ©2004 Royal Society of Chemistry).

the rendezvous. There is usually no direct relationship between the order of structures derived from force-fields or semi-empirical methods on one side and optimized DFT or ab initio on the other side. Minima can appear or disappear easily. Even the order of minima as a function of energy can be different between elaborated methods, for example between structures optimized by means of DFT or MP2.

1.5.2 Representation of potential energy landscapes: Disconnectivity diagrams

The total and intimate description of the PES cannot be obtained when the size of biomolecules increases and a coarse-grained picture then provides useful insights [88–90]. For biomolecules with sizes still accessible to reasonable full-atom descriptions, the disconnectivity graph picture [147–150] classifies the different minima of the PES into basins of attraction. Ignoring tunnelling effects, a molecule can only move from one conformation to another one if its internal energy exceeds the barrier between the corresponding local minima of the PES. In a disconnectivity graph (Figure 1.5.2), the potential energy is represented

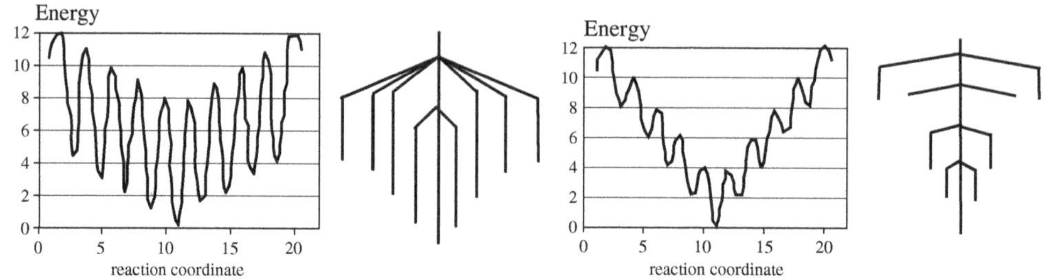

Figure 1.5.2 Schematic PES and disconnectivity diagrams respectively corresponding to a PES with large ("weeping willow" type, left) and small ("palm tree" type, right) barriers compared to the energy differences between successive local minima (adapted with permission from reference [138] ©2005 American Institute of Physics).

vertically. Two minima are in the same energy set at the threshold energy E_i if they can be inter-converted without exceeding E_i. A node on the horizontal axis at energy E_i represents such as set and nodes are joined by upward lines if the corresponding basins merge together at the higher energy E_{i+1}. Nodes are further joined by upward lines when they merge into larger basins at higher energies. Each local minimum appears on the diagram as a single node or at the lower terminus of a vertical line at a height corresponding to its energy. The vertices at which upward lines meet correspond to the energy of connecting saddle points.

According to the relative values of energy differences between local minima and barriers, disconnectivity graphs can be classified into three groups corresponding to different kinetic properties [138]. An example of disconnectivity diagram corresponding to the experimentally studied molecule *N*-acetyl tryptophan methyl amide (NATMA) (see Section 2.1.2.4) is presented in Figure 1.5.3. There are respectively 164 or 178 conformational minima and 714 or 836 transition states with the Amber or the OPLS-AA force-fields. In agreement with the experiment [152], both force-fields predict three dominant conformers. However, their respective predicted energy ordering is different and also differs from the experimental finding. Populations of conformers not only depend upon the potential energy landscape but also from kinetic cooling conditions in a supersonic expansion (Figure 3.1.2).

1.5.3 Monte Carlo exploration

In Monte Carlo (also called Metropolis) simulations, relatively large motions are randomly imposed to the molecular system which jumps abruptly from one conformation to another (Figure 1.5.4). This is in contrast with molecular dynamics (MD) (see Section 1.5.5) where systems evolve smoothly through time. The simulation does not consider barriers and the only relevant parameter is the relative energy of the initial and final conformations [153]. One considers a molecular system at a given temperature T. The probability of occupation of conformation X with energy $E(X)$ in the canonical ensemble is proportional to the Boltzmann factor $\rho_X(T) \propto \exp(-E(X)/k_B T)$. For simplicity, the PES is here represented by a one-dimension curve with two mimima X_1 and X_2 with respective energies $E(X_1)$ and $E(X_2)$. The probability of moving from conformation X to conformation Y is $\pi_{X_1 \rightarrow Y_2}$. The probability that an equilibrium is reached if micro reversibility remains satisfied is $\rho_X \pi_{X \rightarrow Y_2} = \rho_{Y_2} \pi_{Y \rightarrow Y_1}$

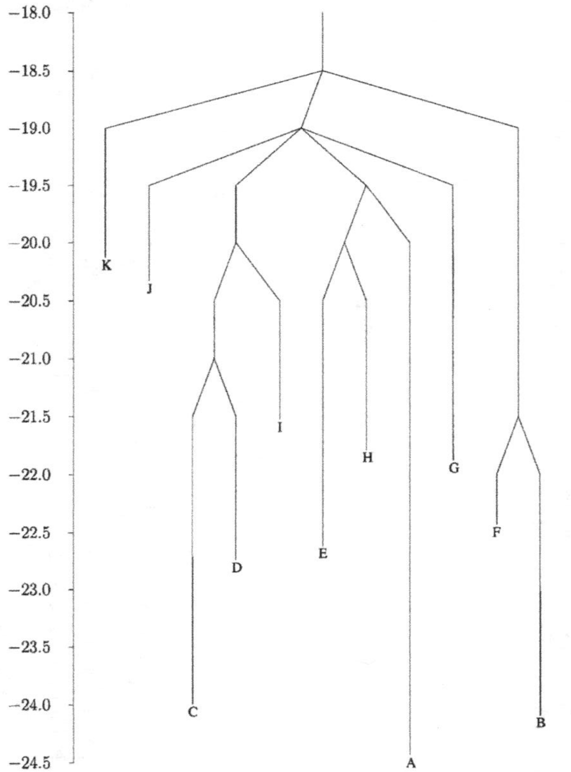

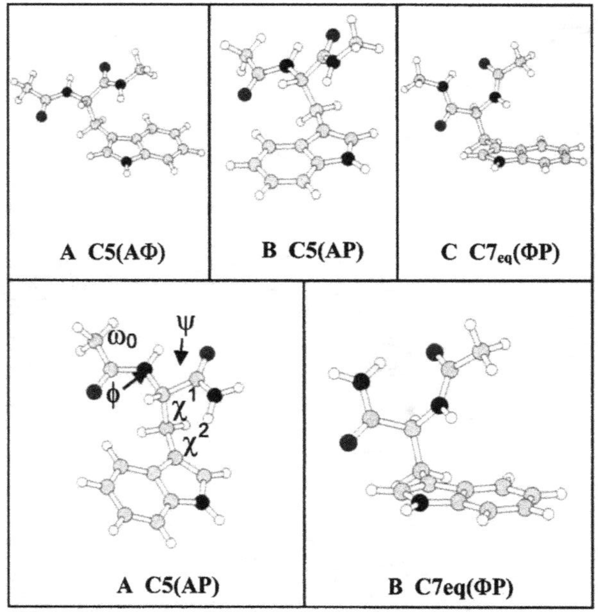

Figure 1.5.3 Left: Disconnectivity graph of *N*-acetyl tryptophan methyl amide (NATMA) using the Amber force-field. The energy scale is in kcal/mol relative to the global minimum. There are 164 conformational minima and 714 transitions states. Right: Assigned structures of the experimentally observed conformers of NATMA (a–c) and *N*-acetyl tryptophan amide (NATA) (d–e) (reproduced with permission from references [151, 152] ©2004 American Institute of Physics).

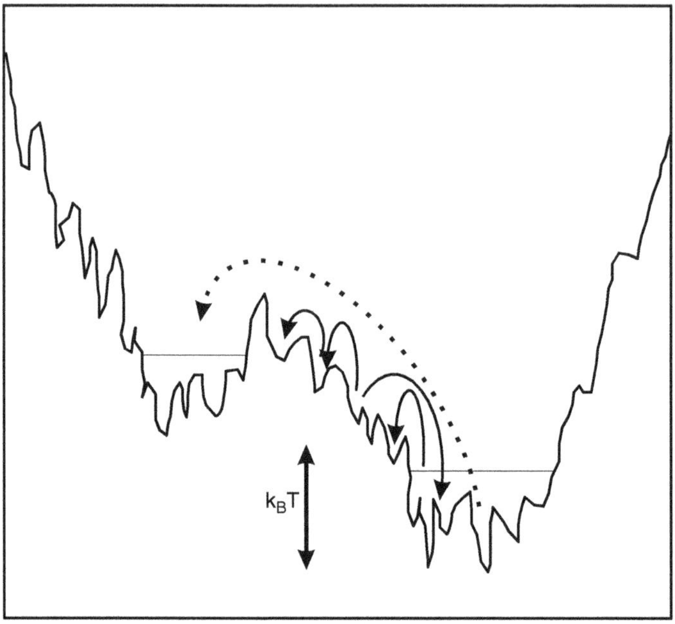

Figure 1.5.4 In a Monte Carlo simulation, a random walk in configuration space is generated according to the probability distribution $\rho_x(T) \propto \exp(-E(x)/k_BT)$. Along this walk, new configurations are generated by displacing the initial configuration by randomly chosen small steps. Uphill moves can be accepted and barriers of the order of k_BT or smaller do not hinder the random walk then allowing the system to move to configurations of high probability (adapted with permission from reference [136] ©2002 Annual Review of Physical Chemistry).

(principle of detailed balance). In order to determine the two conformations X_1 and X_2, the Metropolis algorithm starts from a random conformation X_i and generates a new trial conformation X_{i+1} by applying a random perturbation satisfying the principle of detailed balance. If $E(X_{i+1}) < E(X_i)$, X_{i+1} is immediately accepted with the probability

$$p = \exp\{-[E(X_{i+1}) - E(X_i)]/k_BT\} \tag{1.5.1}$$

If $E(X_{i+1}) > E(X_i)$, p is first compared to a random number *rand* comprised between 0 and 1 and state X_{i+1} is only kept if $p <$ *rand* and otherwise rejected. The acceptance probability p_{acc} at step i is thus

$$p_{acc} = \min\{1, \exp\{-[E(X_{i+1}) - E(X_i)]/k_BT\}\} \tag{1.5.2}$$

and this procedure is iteratively reproduced. Conformations with lower energies are always accepted but those with higher energy still have a chance to be accepted.

1.5.3.1 Parallel tempering

If large energy barriers exist in the energy landscape, the search procedure can be trapped in local wells. In order to avoid this problem, several methods have been proposed [153, 154]. Among them, a method called parallel tempering [155] uses a set of replicas (copies) of the system in its initial conformation [156–160]. Temperature is discretized and takes a set of values $T_1 < T_2 < \cdots < T_n$. Each replica is simulated as described above at a fixed temperature belonging to this set. A random walk in potential energy space is then induced by a random exploration of the temperature space. Every few steps, pairs of replicas are exchanged according to the following rule: exchange of replicas i and $j = i + 1$ is accepted with the acceptance probability

$$p_{acc} = \min\left\{1, \exp\left[-(E(X_{i+1}) - E(X_i))\left(\frac{1}{k_B T_i} - \frac{1}{k_B T_{i+1}}\right)\right]\right\} \tag{1.5.3}$$

In practice, T_1 must be chosen sufficiently low to allow exploration of the global minimum and T_n must be sufficiently high to avoid any trapping in a local minimum.

1.5.3.2 Quantum Monte Carlo methods

In classical statistical mechanics, the energy E or mean value $<H>$ of the Hamiltonian H is given by

$$E = <H> = \frac{\int P(\vec{R}) H(\vec{R}) d\vec{R}}{\int P(\vec{R}) d\vec{R}} \tag{1.5.4}$$

where the distribution function $P(\vec{R})$ of coordinates $\vec{R}$ is the Boltzmann distribution at temperature T. In a quantum description, $P(\vec{R})$ becomes the square of a trial wave function $\Psi_T(\vec{R})$

$$E = <H> = \frac{\int \Psi_T^2(\vec{R}) H(\vec{R}) d\vec{R}}{\int \Psi_T^2(\vec{R}) d\vec{R}} \tag{1.5.5}$$

Starting from a set of atomic coordinates $\vec{R}$, a new set $(\vec{R}')$ is randomly chosen following a Monte Carlo procedure and provides a new wave function. The new set is accepted if $E[\Psi(\vec{R}')] > E[\Psi(\vec{R})]$, otherwise $\vec{R}$ is kept. Once a wave function is established, expectation values of the required quantities can be evaluated. Monte Carlo methods present the important property of scaling in between N^2 or N^3 where N is the number of atoms.

Quantum Monte Carlo methods are used for the calculation of the equilibrium thermodynamics of molecules at a finite temperature T. In contrast with classical methods, they no longer ignore ZPE effects [146, 161]. In order to obtain efficient quantum mechanical simulations with shorter computer time and for elimination of fluctuations introduced by high-frequency vibrations, torsional degrees of freedom which have much lower frequencies than other intramolecular motions are frequently treated apart. The torsional configuration space is sampled by discretizing the torsion angles and a Monte Carlo procedure is applied with a Boltzmann distribution to generate the required multidimensional probability distribution function. A classical force-field has been used to investigate the enkephalin peptide which possesses 33 torsions [162]. In the case of a much smaller molecule, glycine, with only 3 torsions (Figure 1.5.5), a PES is deduced from a high-level (MP2/6-311++G**) calculation and the different conformer populations can then be obtained as a function of temperature (Figure 1.5.6).

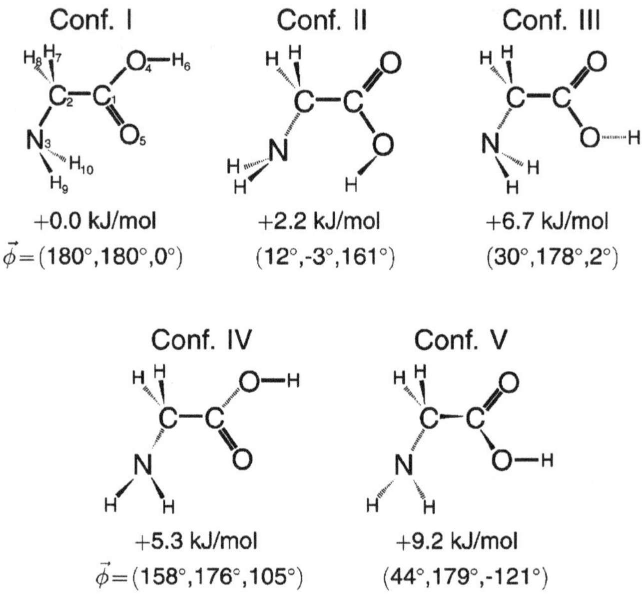

Figure 1.5.5 Most stable conformers of glycine and respective values of their torsional angles (energies are calculated at the MP2/6-311++G** level). The torsional angles (ϕ_1, ϕ_2, ϕ_3) in the figure correspond to the twisting motions about the C—C bond (ϕ_1), the C—O bond (ϕ_2) and the C—N bond (ϕ_3). (reproduced with permission from reference [146] ©2004 Royal Society of Chemistry).

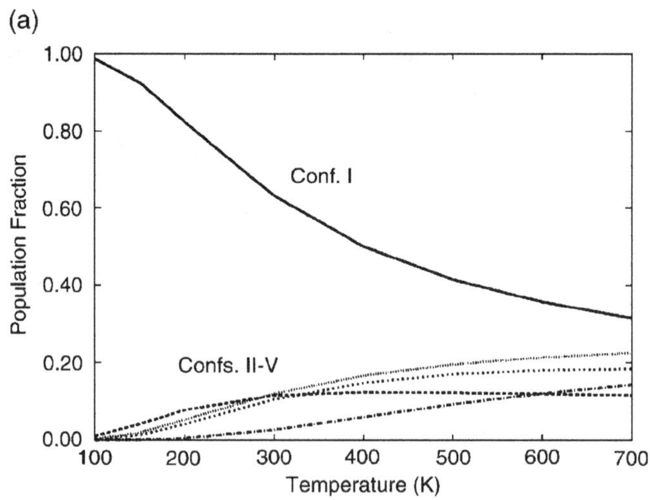

Figure 1.5.6 Respective populations of the most stable conformers of glycine calculated by the torsional path integral Monte Carlo (TPIMC) technique as a function of temperature (reproduced with permission from reference [146] ©2004 Royal Society of Chemistry).

1.5.4 Genetic algorithm

A genetic algorithm search is based on the mock eugenic hypothesis of "survival of the fittest". An initial generation of m different configurations is built by randomly selecting positions of atoms and each configuration is then optimized. Within this population, the lowest energy configuration has energy E_{min}. Each configuration with energy E_{conf} is then characterized by a "fitness" (or "beauty") function comprised between 0 and 1 and equal to

$$f = \exp\left[\frac{E_{\text{conf}} - E_{\text{min}}}{\sum_{\text{conf}=1}^{n} (E_{\text{conf}}/m)}\right] \tag{1.5.6}$$

A second generation is then built by crossbreeding configurations, each configuration acquiring a probability of becoming a parent according to its fitness. In order to produce a child configuration, some atomic positions of two parent configurations can be randomly exchanged or a weighted mean position value can be taken. Another possibility, called mutation, is a random displacement of atomic positions of children. Reasonable configurations of clusters [163] or peptides with more than 20 residues can be obtained with such a procedure [164]. Genetic algorithms are also used for computer-aided fits of rotationally resolved spectra [165] (see Section 2.1.4.1).

1.5.5 Molecular dynamics

Molecular dynamics simulations provide information such as atomic positions and velocities at the microscopic level. At a given time, the studied molecular system can adopt different conformations and macroscopic quantities are obtained by averaging over the whole ensemble of possible microscopic values. For example, a molecular spectrum in solution is obtained by averaging over possible conformations populated at the considered temperature. The assumption of ergodicity allows the replacement of an ensemble averaging by a time averaging and it is hoped that if the system evolves during a sufficient amount of time, it will sample all its possible states. In order to get a rough evaluation of how long a simulation should be run, one can roughly estimate that a typical amount of time $\tau = 10^{-12} e^{\Delta G/kT}$ (in seconds) is required to overpass a barrier ΔG and thus to explore a neighbouring region of phase space. In case of a small barrier equal to only 4 kJ/mol, a picosecond is sufficient while more than a millisecond is required for a larger barrier of 40 kJ/mol.

Continuous classical MD simulations consider the classical motion of atoms by solving their Newtonian equations of motion $\mathbf{F} = \text{m}\mathbf{a}$ where $\mathbf{F}$ is the force exerted on an atom, m its mass and $\mathbf{a}$ its acceleration. Forces are derived from the knowledge of the classical potential energy landscape. Integration of these equations of motion yields trajectories describing positions, velocities and accelerations of atoms as they vary with time. In ab initio MD simulations such as Car-Parrinello molecular dynamics (CPMD), atom motions are still computed classically but forces exerted on atoms are deduced by solving the Schrödinger equation generally by means of DFT calculations [166, 167]. In contrast with classical MD, electrons are explicitly taken into account and this allows the treatment of chemical reactions implying bond breaking and bond formation. Currently, CPMD can be run during simulation times of 10 ps for molecular systems while classical MD using empirical force-fields can be run during durations sufficiently long to follow folding of proteins towards their native state [168–171].

In order to handle very large biomolecular systems over long simulation times, discrete molecular dynamics (DMD) simulations [172, 173] use a different approach. Pairs of particles interact through spherically symmetric potentials that consist of a single or several square wells. A LJ potential, for example can be reasonably mimicked by means of a few well-chosen square wells [173] (Figure 1.1.8). A compromise must then be found between the smallest possible number of square wells to speed up simulations and the necessity to keep a realistic description of interactions. The strongly repulsive r^{-12} potential is approximated by a reflective wall at an internuclear distance corresponding to the sum of the hard-sphere radii of the interacting atoms. Instead of solving the equation of motion, $\mathbf{F} = m\mathbf{a}$, DMD simulations only require the satisfaction of conservation of energy and momentum at the borders between square wells [173, 174] and the equations become simply ballistic as in a hard-sphere collision problem. This use of discrete interaction potentials provides algorithms 10^8–10^9 faster than traditional MD simulations.

1.5.5.1 Classical molecular dynamics

In a MD simulation, the studied molecular system starts at time $t = 0$ in an optimized geometry derived from a force-field, a semi-empirical or a quantum calculation. At a very low initial temperature T, atoms are given an initial distribution of velocities corresponding to a Boltzmann distribution. The probability that an atom i of mass m_i has a velocity v_{ix} in the x direction is given by

$$p(v_{ix}) = \sqrt{\frac{m_i}{2\pi k_B T}} \exp\left(-\frac{m_i v_{ix}^2}{2 k_B T}\right) \tag{1.5.7}$$

The equations of motion are integrated (see below) and, during a heating period of time, new velocities are periodically assigned at a slightly higher temperature. When the desired temperature is reached, the simulation proceeds during an equilibration phase until molecular properties such as structure and energy are stabilized. The simulation is then run at a temperature large enough to surmount energy barriers and to hopefully explore the full potential energy landscape. A time series of conformers, called a trajectory or a path, is obtained. The search procedure then uses a simulated annealing or quenching procedure that consists in a well-controlled temperature cooling and provides global minima [143, 175]. The use of deposition of molecules on helium clusters is the equivalent experimental procedure (see Section 3.1.3). At high temperatures, the studied system occupies high-energy regions and can overpass high-energy barriers. When the temperature drops, the lower energy configurations become more probable in agreement with the Boltzmann distribution. In principle, an infinite number of temperature steps is required to ensure that the global minimum is really reached at absolute zero. In practice, it is hoped that if several independent simulated annealings are run and provide the same result, the true global minimum is most probably reached.

1.5.5.1.1 Integration of equations of motion

The equation of motion of atom i is

$$F_i = ma_i = m_i \frac{d^2 r_i}{dt^2} = -\frac{dV(r_i)}{dr_i} \quad (1.5.8)$$

where V is the potential energy of the molecular system. The position $r_i(t + \delta t)$ of atom i at time $t + \delta t$ can be obtained from the position, velocity and acceleration at time t by means of the Taylor expansion

$$r_i(t+\delta t) = r_i(t) + v_i(t)\delta(t) + \frac{1}{2}a_i(t)(\delta t)^2 + \cdots \quad (1.5.9)$$

Similarly, one has

$$r(t-\delta t)_i = r_i(t) - v_i(t)\delta(t) + \frac{1}{2}a_i(t)(\delta t)^2 + \cdots \quad (1.5.9)$$

The sum of these two equations leads to the Verlet algorithm

$$r_i(t+\delta t) = 2r_i(t) - r_i(t-\delta t) + a_i(t)(\delta t)^2 \quad (1.5.10)$$

In the Verlet algorithm, the new atomic positions at time $t + \delta t$ are derived from positions and accelerations at time t without any explicit velocities. This algorithm is often replaced by the *leap-frog* algorithm where

$$r_i(t + \delta t) = r_i(t) + v_i(t + 1/2\,\delta t)$$

and

$$v_i(t+\delta t) = v_i\left(t - \frac{1}{2}\delta t\right) + a_i(t)\delta(t) \quad (1.5.11)$$

Velocities are first calculated at time $t + (1/2\,\delta t)$ and used to calculate positions at time $t + \delta t$. Velocities leap over positions and then positions leap over velocities. Velocities at time t are given by

$$v_i(t) = \frac{1}{2}\left\{v_i\left(t - \frac{1}{2}\delta t\right) + v_i\left(t + \frac{1}{2}\delta t\right)\right\} \quad (1.5.12)$$

1.5.5.2 Classical molecular dynamics with constraints

An important difficulty of MD simulations is due to the presence of very different time-scales. For modelling a high-frequency vibration such as an O—H stretch (3,600 cm^{-1}, 1.1×10^{14} Hz), a correct sampling corresponding to time steps δt of 10^{-15} s is required while interesting motions of peptide backbones can take place on a timescale of fraction of nanoseconds or even microseconds for proteins. 10^5 to 10^9 time steps can then be required and it is often interesting to ignore the very fast hydrogen atom motions (C—H, N—H or O—H stretches). Procedures like the SHAKE algorithm that constrain the corresponding bond lengths or angles and leave torsions free allow larger time intervals δt and thus accelerate calculations. Such a procedure uses the Lagrange multiplier method. For example, the distance $r_{k_1k_2}$ between two atoms k_1 and k_2 can be constrained to be equal to the bond length $d_{k_1k_2}$. This can be written as

$$\sigma_k[(r_i)] = r_{k_1k_2}^2 - d_{k_1k_2}^2 = 0 \tag{1.5.13}$$

Thus, one has a set of N_c constraints $\sigma_k(r_i)$ for $k = 1,2, \ldots, N_c$. Multiplying each $\sigma_k[(r_i)]$ by a Lagrange factor Λ_k (to be determined) and adding these N_c null terms to $V[(r_i)]$ does not change $V(r_i)$ but introduces constraints. The equation of motion becomes

$$m_i \frac{d^2 r_i(t)}{dt^2} = -\frac{\partial}{\partial r_i}\left[V[(r_i)] + \sum_{k=1}^{k=N_c} \Lambda_k \sigma_k[(r_i)]\right] \tag{1.5.14}$$

As in the unconstrained case, the first term of the right-side of this equation of motion is the unconstrained force F_i acting on atom i but the second term represents a new force F_i^c acting on atom i to respect the imposed constraint [176]. When using such a method (e.g. SHAKE), an algorithm restores the selected internal coordinates (usually involving hydrogen atoms) to their constrained bond or angle value after a MD or minimization step [177]. For example, the ϕ/ψ angles of a dipeptide can be constrained to specific torsion angles and the molecule's energy can be computed to create a potential energy map [178].

1.5.5.3 Docking of ligands to biomolecules

Molecular recognition [179] between biomolecules has first been described as a "lock-and-key" mechanism involving interactions between rigid bodies with structural complementarity. Binding of ligands to their receptors requires some degree of flexibility and each protagonist adapts itself to the other in an "induced-fit" mechanism. For example, docking of a ligand to a protein induces a conformational rearrangement of the protein backbone generally in a limited region called the flexible loop and comprising 6 to 20 amino acids. Flexible loops are situated at the periphery of globular proteins and can modify their conformations in order to lock or favour the formation of a complex. Those flexible and hinge regions required for adaptative docking are generally identified by only considering the backbone or C_α atoms. It is possible to initiate the docking procedure from rigid structures deduced from MD simulations or NMR structures (Figure 1.5.7). The docking process is repeated from various combinations of starting structures that can span various degrees of flexibility from small side chain rearrangements up to large-scale global backbone motions. An introduction to docking can be found in reference [1] and the following references concern docking of ligands to DNA [180], protein to DNA [15] and protein to protein [181].

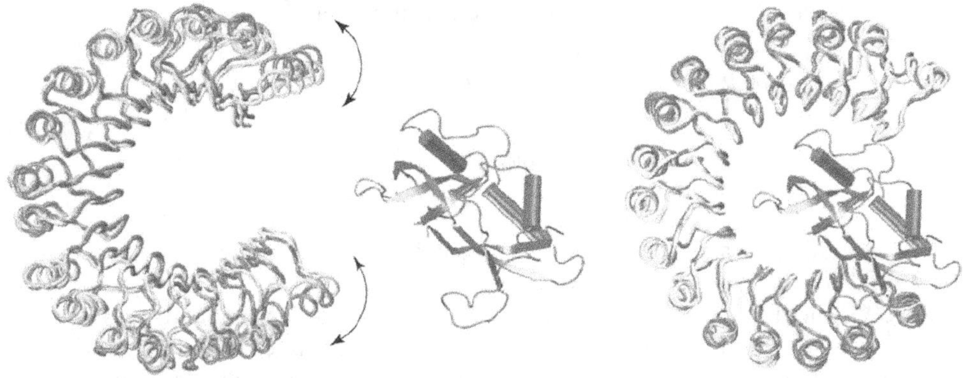

Figure 1.5.7 Simulation of the docking of RNase A to the porcine RNase A inhibitor (reproduced with permission from reference [181] ©2006 Elsevier).

1.5.5.4 Quantum molecular dynamics: Car-Parrinello method

Quantum molecular dynamics can be used to study biological processes where bond formation and breaking of bonds take place at finite temperature. Instead of using classical force-fields, the equations of motion are numerically solved by using forces derived from electronic structure calculations, most usually DFT. In a Born–Oppenheimer framework, the nuclei would be considered in a fixed geometry at a given time and the shape of electronic cloud adapted to this geometry would be calculated. This would provide the forces applied to the nuclei leading to their next geometry. This step-by-step procedure is in fact not used and electronic calculations are performed "on the fly" as the simulation proceeds [166, 182–185]. This method is called CPMD [183, 186–191]. The timescale of the slow nuclei is separated from that of the rapid electrons which are now treated as dynamical variables. As previously, nuclei are described by their positions $\{R_N\}$ and their temperature $T_N \propto \Sigma_i M_i \dot{R}_i^2$. Electrons are described by wave functions $\{\psi_i\}$ and must adiabatically follow the nuclear motions. For that purpose, electrons receive a fictitious mass μ_i (no longer equal to one but typically to a few hundreds) and a fictitious temperature$T_e \propto \Sigma_i \mu_i \langle \dot{\psi}_i | \dot{\psi}_i \rangle$. Valence electrons are the only ones taken into account and are represented by one-electron orbitals derived from pseudo-potentials.

The Car-Parrinello method makes use of a Lagrangian description in order to introduce the constraint of orthogonality of the wave functions by means of Lagrange multipliers Λ_k. The Car-Parrinello Lagrangian L_{CP} is the difference between the kinetic energy of nuclei and electrons minus the potential energy and is given by

$$L_{\text{CP}} = \sum_i \frac{1}{2} M_i \dot{R}_i^2 + \sum_i \frac{1}{2} \mu_i \left\langle \dot{\psi}_i \middle| \dot{\psi}_i \right\rangle - E_{\text{KS}}[\{\psi_i\},\{R_i\}] + \sum_{i,j} \Lambda_{ij} \left(\langle \psi_i | \psi_i \rangle - \delta_{ij} \right) \quad (1.5.15)$$

E_{KS} is the Kohn–Sham energy (see DFT) and the last term ensures the orthogonality of the wave functions. This Lagrangian is used to generate trajectories for nuclei and electronic degrees of freedom via the coupled set of equations of motion

$$\begin{aligned} M_i \overset{\infty}{R_i}(t) &= -\frac{\partial}{\partial R_i} E_{\mathrm{KS}}[\{\psi_i\},\{R_i\}] + \sum_{i,j} \Lambda_{ij} \frac{\partial}{\partial R_i} \langle \psi_i | \psi_j \rangle \\ \mu_i \overset{\infty}{\psi_i}(t) &= -\frac{\partial}{\partial \psi_i} E_{\mathrm{KS}}[\{\psi_i\},\{R_i\}] + \sum_{i,j} \Lambda_{ij} \psi_j \end{aligned} \tag{1.5.16}$$

The use of the Car-Parrinello method for biomolecular systems will be illustrated in Chapter 4.2 for the analysis of a resonant infrared multiphoton dissociation experiment, run at 300°K, in the case of the protonated dialanine Ala–Ala H^+peptide. It will be shown that, in contrast with the interpretation of very low temperature experiments where well-defined rigid structures preserved below barriers are considered, at a biologically relevant temperature quantum molecular dynamics shed new light on the time-evolution of the biomolecular systems between structures and even chemical processes such as proton transfers (see Section 4.2.3) [184].

Quantum molecular dynamics calculations can also be conducted for much larger systems [191–194] and for comparison between infrared spectra of molecules in the gas-phase or in solution with explicit solvent molecules [167, 195, 196] (see Chapter 5).

1.6 MIXED APPROACHES QM/MM

Structures of biomolecules containing a very large number of atoms such as proteins are usually modelled by means of molecular mechanics (MM) using force-fields or semi-empirical methods such as AM1 (see Section 1.4). The advantage of MM simulations in structure calculations is their applicability to very large systems and computer time efficiency. Their drawback is the requirement of parameterization, their often poor precision and their inability to deal with reactive processes, except for reactive force-fields. In a large biomolecule such as an enzyme only a small region, called the active site, participates into the reaction process [123, 197]. Local chemical phenomena such as breaking and building of bonds can be described by explicitly treating electron clouds with QM while physical intermolecular interactions taking place between widely separated regions of a molecule are correctly treated by classical MM [198, 199]. Methods combining the speed and possibility of handling very large number of atoms of structural classical calculations and the accuracy of quantum computations have thus been devised and are called QM/MM approaches [200, 201]. The two regions respectively described by QM and MM representations interact [202]. A first large difficulty arises in the definition of the border [203] between those two regions and a second one appears for connecting those regions. Moreover, including the presence of solvent is most often a necessity.

When a covalent bond X—Y between two atoms X and Y and involving the sharing of two electrons is conceptually broken at the frontier between the MM and the QM parts (Figure 1.5.6), several possibilities are opened. It is not possible to simply truncate the molecular orbital describing this bond or leave an unpaired electron on each atom. A first possibility, called "link atom" method [204], consists in linking to each atom a hydrogen atom (this H atom can also be replaced by a methyl group) or two H atoms to terminate the dangling bonds (saturation of valences) of X and Y. In this approach with artificial

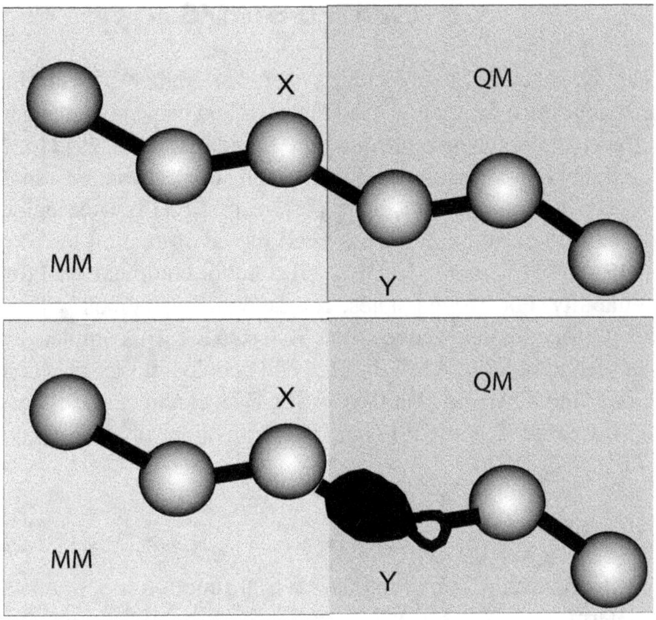

Figure 1.6.1 Top: Separation of a large biomolecular system into a classical part described classically (MM) and a part described on a quantum level (QM). X—Y is the broken bond. Bottom: Addition of a hybrid orbital to the X atom (by courtesy of X. Assfeld).

X—H—Y or ···X—H—H—Y bonds, link atoms must be calculated for each new system and the method also does not allow the treatment of reactivity. Another possibility [205] consists in adding a hybrid orbital (e.g. sp^2) on the atom belonging to the QM part (X on Figure 1.6.1).

Different levels of theory can be used for the different parts of a large system like in an onion; the inner part might be treated at the CCSD(T) level, the intermediate part at the HF or MP2 level and the external part at the MM level. The partitioning of the system into such "high" and "low" levels of theory is implemented in softwares [206, 207]. ONIOM (Our owN n-layered Integrated molecular Orbital+molecular mechanics Method) is widely used [208–210]. A "model" system representing the critical or reactive part is "cut-out" from the large system. The ONIOM energy is then given as E_{ONIOM} = energy of the large system at a low-level of computation plus energy of the model system at a high level of computation – energy of the model system at a low-level of computation.

For example, rigorously treating the effect of amino acid mutations in α-helical peptides containing 17 amino acids is still not possible. The ONIOM calculation of such a large system involves two (high and medium) levels. The high level concerns the cores of the different α-helices and does not take into account the differences between the substituted amino acids (considered as simple glycines). This high level uses DFT while the medium level only uses the semi-empirical approach AM1. The exact nature of the substituted amino acids is then correctly taken into account by considering their side chains [210]. Other examples [211, 212] of such mixed treatments are considered further in Chapter 4.7.

1.7 EXCITED STATES

Calculations of excited states of biomolecular systems are rather difficult. In the following, we will only briefly introduce the mainly used methods and provide elementary explanations about acronyms. Recent comprehensive reviews are found in references [213, 214]. The situation is somewhat parallel to that encountered for ground state calculations. In the HF approximation, we have seen that a single electron picture of MOs is used neglecting electron correlation. This electron correlation is recovered in wave function based methods in the Moller–Plesset (MPn) and CC methods. DFT offers better computational time efficiency at the expanse of reliability. For excited states, corresponding methods will be configuration interaction singles (CIS), complete active space SCF (CASSCF), complete active space perturbation theory of second order (CASPT2) [215] and coupled-cluster of second order (CC2). Geometries are often first determined at the CASSCF level and energies are then refined in a protocol named CASPT2//CASSCF [216]. The equivalent of DFT for ground states is time-dependent DFT (TD-DFT).

1.7.1 Ab initio methods

The configuration interaction (CI) method uses a trial function Ψ_{CI} that is a linear combination of Slater determinants Ψ_i^α and $\Psi_{ij}^{\alpha\beta}$. The coefficients of the linear combination are determined by applying the variational principle that leads to the lowest energy solution of the time-independent Schrödinger equation. The determinants are built by exciting electrons from orbitals i and j occupied in the HF ground state determinant into non-occupied orbitals α and β. In principle, one would like to include all possible determinants to obtain a complete configuration interaction. This is only a dream in the case of biomolecules since for the water molecule, such a calculation would already required 3×10^7 determinants with a simple 6-31G* basis set. In order to obtain realistic computational times, less ambitious methods are used. The CIS wave function is obtained through the addition of a rather small number of new Slater determinants. In the HF ground state determinant $\Phi_0(r) = |\phi_1(r) \ldots \phi_i(r) \ldots \phi_n(r)|$, an occupied orbital $\phi_i(r)$ is replaced by a virtual orbital $\phi_i^\alpha(r)$ providing a new singly excited Slater determinant $\Phi_i^\alpha(r)$ and the CIS wave function is a linear combination $\Psi_{\text{CIS}} = \Sigma_{i,\alpha} c_i^\alpha \Phi_i^\alpha(r)$. The excited state energies obtained by means of CIS are overestimated and the ordering of states is often not respected due to the lack of correlation [214]. Examples of calculated electronic spectra of amino acids containing UV chromophores can be found in references [217, 218].

The multiconfiguration self consistent field (MCSCF) methods are configuration interaction methods where determinant coefficients and wave functions are still optimized by the variational principles but the number of used determinants is carefully reduced. In a CASSCF wave function, orbitals are divided in two classes. Inactive orbitals are doubly occupied in each Slater determinant while active orbitals can hold zero, one or two electrons. There are thus N_{total} = total electrons distributed into N_{active} active electrons and $M_{\text{active}} = 2\ m_{\text{inactive}}$ paired inactive electrons. All possible Slater determinants are generated by considering the N_{active} active electrons leading to the complete active space (CAS). For example, a CASSCF ($N_{\text{total}} = 4$, $M_{\text{active}} = 6$) calculation on a closed-shell system includes the two highest occupied molecular orbitals (HOMOs) providing the four electrons. In order to complete the active space with six active orbitals, one adds to the two HOMOs the next four lowest virtual orbitals. Active spaces containing up to 14 orbitals are often implemented in available softwares. An (historical) example of CASSCF calculation of excited states of tyrosine with six active electrons distributed among six orbitals (6,6) and a 6-31G basis set can be found in reference [219]. Accurate calculations used for the optimization of local minima and saddle points of excited 9H-adenine include an active space with 12 electrons distributed over 10 valence orbitals [220] (Figure 1.7.1).

The CASPT2 method is a multiconfiguration method that improves the CASSCF wave function Ψ_{CAS} up to second order of perturbation. The coefficients of the zero-order Ψ_{CAS} are not modified. In order to include dynamical effects, the new wave function Ψ_{CASPT2} becomes a linear combination of Ψ_{CAS} and determinants obtained from Ψ_{CAS} by single and double excitations. The respective order of levels is somewhat modified when going from CASSCF to CASPT2 due to inclusion of dynamical correlation in the latter method (Table 1.7.1).

The CC2 method goes one step further with the inclusion of highly excited states. It improves the correlation treatment and is one of the most accurate methods. It is somewhat equivalent to the MP2 method for electronic excited states [221].

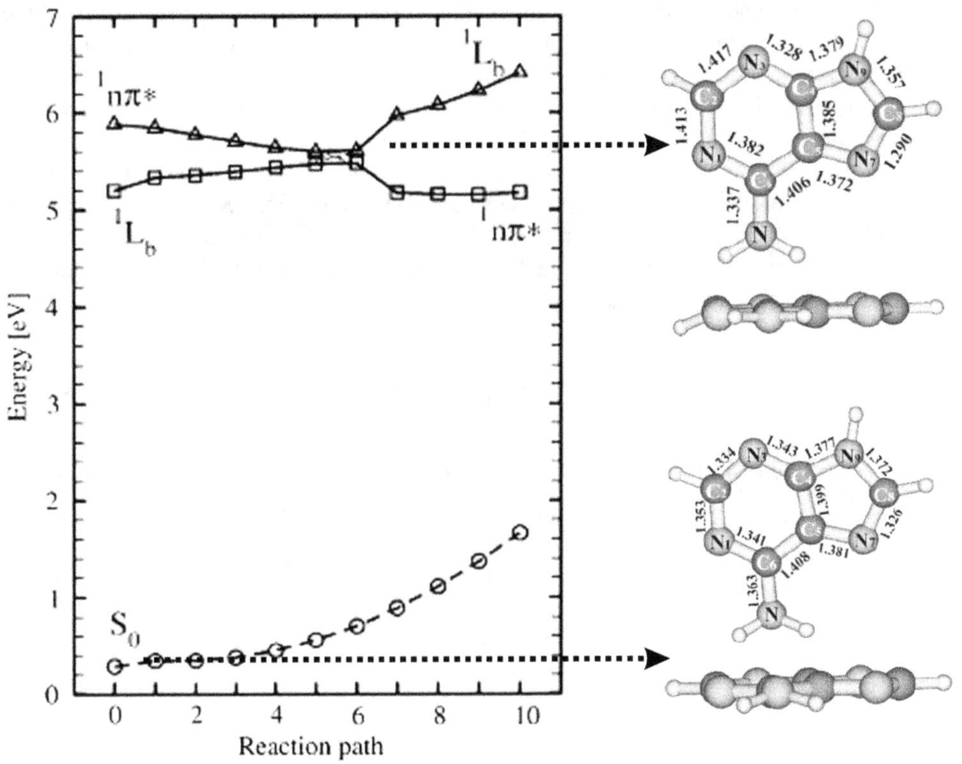

Figure 1.7.1 Left: CASSCF potential energy curve of 9H-adenine. The coordinate corresponds to the linearly interpolated internal path. This path is a straight line connecting the excited initial minimum of the ^1L_b state to the conical intersection. Right: Equilibrium geometries of the ground state calculated at the MP2/6-31G(d,p) level (bottom) and the $^1n\pi^*$ state at the conical intersection calculated at the CASSCF level (note the pyramidization of the amino group) (reproduced with permission from reference [220] ©2005 American Chemical Society).

Table 1.7.1

Vertical excitation energies of 9H-adenine calculated at the geometry of the ground state optimized at the MP2/6-31G(d,p) level [220]. Note the huge difference between the ^1L_b and ^1L_a energy levels at the CASSCF level as compared to the CASPT2 situation where those two levels are close to the $^1n\pi^*$ state

Excited state	E CASSCF (eV)	E CASPT2 (eV)
$S_1\ {}^1\pi\pi^* \ldots {}^1L_b$	5.226	4.852
$S_2\ {}^1\pi\pi^* \ldots {}^1L_a$	6.859	4.902
$S_3\ {}^1n\pi^*$	6.193	5.503
$S_4\ {}^1n\pi^*$	6.674	5.685
$S_5\ {}^1n\pi^*$	7.232	6.098

1.7.2 Time-dependent density functional theory (TD-DFT)

A widely used approach for calculation concerning the ground state is DFT that replaces the wave function $\psi_i(r)$ by the electron density $\rho(r) = \Sigma_i|\psi_i(r)|^2$. Electron-exchange and correlation (xc) are gathered in the functional (see Section 1.2.4). The extension of DFT to excited states is called TD-DFT [222, 223]. The word time is somewhat surprising in the quantum chemistry context. To get an idea where it comes from, let us consider a biomolecular system in its ground state described by its wave function $\psi(r)$. We have seen that quantum chemistry aims to solve as best as possible the time-independent Schrödinger equation $H\psi_0(r) = E_0\psi_0(r)$. A small perturbation $V(t) = E\cos\omega t$ due to an external oscillatory electric field is added and one then tries to get the linear response to this perturbation by solving the time-dependent Schrödinger equation

$$(H(r)+V(r,t))\psi(r,t) = i\hbar\frac{\partial\psi(r,t)}{\partial t} \tag{1.7.1}$$

The solution $\psi_0(r)$ is improved by adding a small time-dependent wave function and time dependence

$$\psi_0(r)\rightarrow\psi(r,t) = [\psi_0(r)+\delta\psi(r,t)]e^{-iE_0 t/\hbar} \tag{1.7.2}$$

the time-dependent equation becomes

$$V(r,t)\psi(r) = \left(\frac{i\hbar\partial}{\partial t} - H(r)+E_0\right)\partial\psi(r,t) \tag{1.7.3}$$

In order to remove the time dependence, one takes the Fourier transform and gets the time-independent equation

$$V(r,\omega)\psi(r) = (\hbar\omega - H(r)+E_0)\partial\psi(r,\omega) \tag{1.7.4}$$

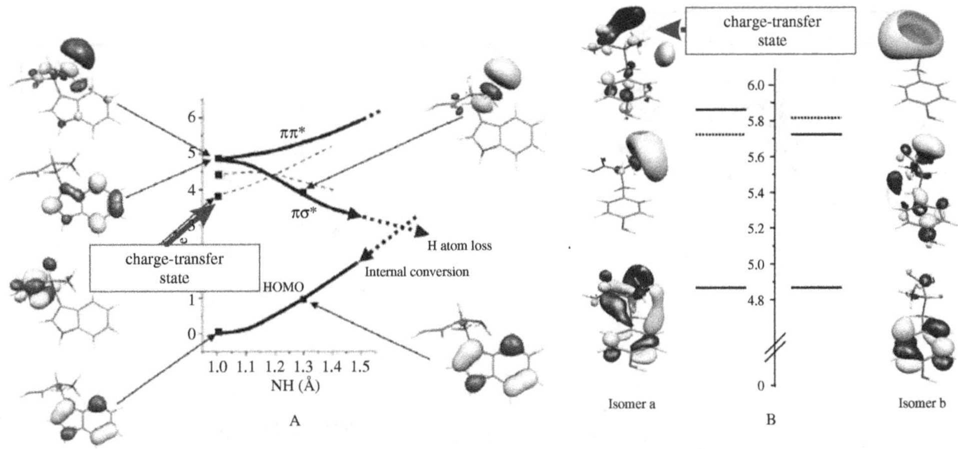

Figure 1.7.2 Left: Potential energy curves of protonated tyrosine along the dissociative N—H coordinate calculated at the TD-DFT level (reproduced with permission from reference [229] ©2004 Royal Society of Chemistry). Right: Vertical excited state energies of the two isomers of protonated tyrosine calculated at the CC2 level with an aug-cc-PVDZ basis set (straight line, $\pi\pi^*$ state; dotted line, $\pi\sigma^*$ state). Note the strong upward energy shift of the charge-transfer state corresponding to a delocalized electronic density on the carboxylic acid group (by courtesy of G. Grégoire and C. Dedonder).

In TD-DFT [224], the wave function becomes a functional of the electron density

$$\psi(r,t) = \psi[\rho(t)](t)e^{-i\omega t} \tag{1.7.5}$$

The linear response theory [213, 225] is applied to study the response of the electron density to the perturbation introduced by the external oscillatory electric field. The molecular system is initially assumed in its ground state before the external oscillatory electric field is applied and it is admitted that the electron density varies only slowly in time (adiabatic local density approximation ALDA). As in above, Fourier transformation of the time-dependent solution then provides excited state energies and oscillator strengths.

TD-DFT is unrivaled for large biomolecular systems [226] with up to 300 second row atoms. It can provide electronic state energies with a typical accuracy of 0.5 eV. TD-DFT can be also used in solution studies [227]. However, the ordering of excited states can become erroneous in the presence of charge-transfer or Rydberg states [225]. A new density functional called M06-HF has been introduced in order to solve this problem [228].

The situation for excited states is thus somewhat similar to that occurring for the ground state. A first rapid exploration of structures is more rapidly conducted with TD-DFT and must then be refined by means of CC2 in order to get more reliable excited state energies and remove possible artefacts. An example of such artefact is shown in Figure 1.7.2.

1.7.3 Excited potential energy surfaces

Following vertical photoexcitation in the Franck–Condon region, molecular systems evolve on hypersurfaces. In regions where couplings of hypersurfaces are strong, the Born–Oppenheimer

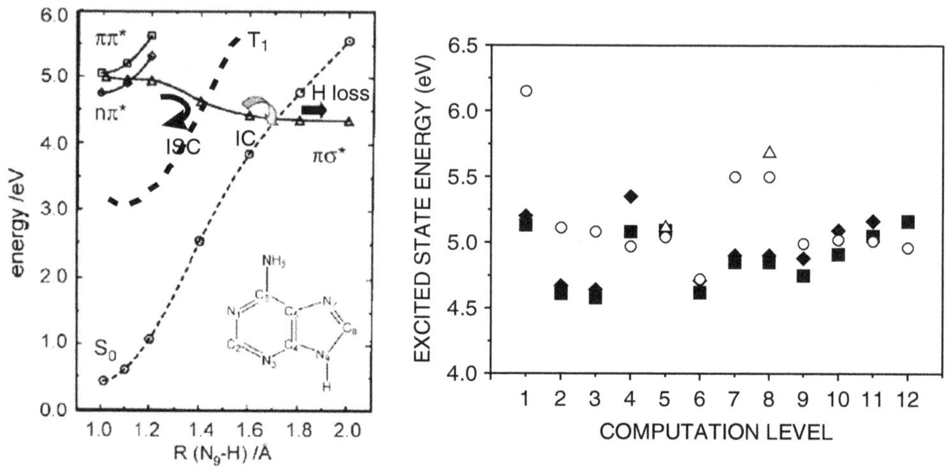

Figure 1.7.3 Left: Two-dimensional potential energy curves of 9H-adenine showing intersystem (ISC) and internal conversion (IC) crossings as well as the dissociative pathway along the $\pi\sigma^*$ state leading to hydrogen atom loss (reproduced with permission from reference [231] ©2002 Royal Society of Chemistry). Right: Energies of $\pi\pi^*$ (black squares and diamonds), $n\pi^*$ (open circles) and $\pi\sigma^*$ (triangles) excited states of adenine calculated at different levels of theory. Column (1) CASPT2/6-31G*, (2) MP2/6-31G, (3) CIS/6-31G, (4) TD-DFT/B3LYP, (5, left figure) TD-DFT/B3LYP, (6) CIS/4-31G, (7) CASPT2/6-31G(d,p), (8) MP2/6-31G(d,p), (9) CASPT2/6-311G* (10) DFT/MRCRI-TZVP (11) CASPT2/6-31G*, (12) CASPT2/6-31G(d,p) (adapted with permission from reference [232]).

approximation is no longer valid and transfers between electronic states can occur. If a transfer takes place between states of the same spin multiplicity, the system can undergo an internal conversion (IC) or a chemical reaction and the branching ratio is driven by the topology of the crossing and the kinetic energy [230]. A transfer driven by the spin-orbit interaction from one state to another one belonging to a different spin multiplicity (e.g. from a singlet to a triplet) is named intersystem crossing (ISC). Figure 1.7.3 presents an overview of those different processes.

1.7.3.1 Crossing of potential energy surfaces

1.7.3.1.1 Two-state model for curve crossing

Crossing of PESs in a multidimensional space (Figure 1.7.4) is a complex phenomenon. Most discussions on this topic employ a heuristic model that can be applied to the crossing of two PESs. For a molecular system whose spin terms in the electronic Hamiltonian are neglected so that we can have the electronic wave function in real form, let us assume that a set of wave functions ψ_1 and ψ_2 constitutes a complete orthonormal set. An eigenfunction of the type $\Psi = c_1\psi_1 + c_2\psi_2$ for the electronic Hamiltonian gives rise to an eigenvalue problem that is reduced to solving the following secular equation:

$$\begin{bmatrix} H_{11}-E, & H_{12} \\ H_{21}, & H_{22}-E \end{bmatrix} \begin{bmatrix} c_1 \\ c_2 \end{bmatrix} = 0 \qquad (1.7.6)$$

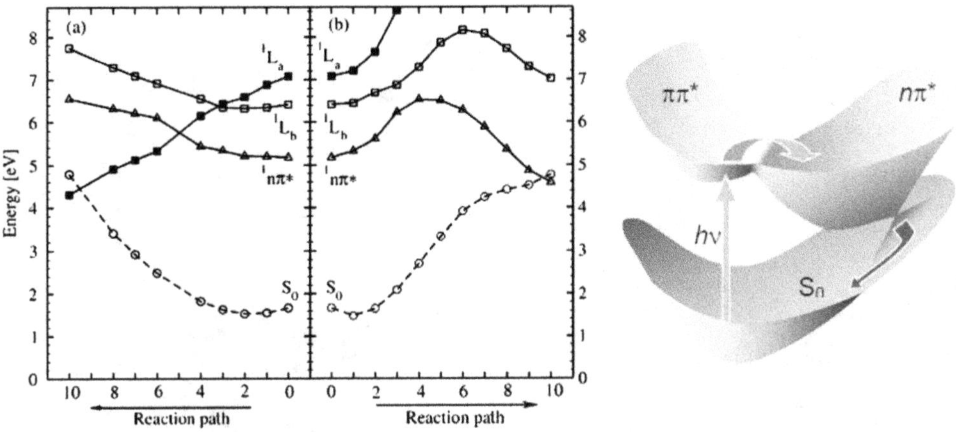

Figure 1.7.4 Left: Two-dimensional potential energy curves displaying crossings (conical intersections) between the lowest excited states of 9H-adenine, a $n\pi^*$ state and the S_0 ground state. The coordinate corresponds to the linearly interpolated internal path (see Figure 1.7.1). Right: Three-dimensional representation of a conical intersection between an excited $n\pi^*$ state and the ground state of 9H-adenine (reproduced with permission from reference [220] ©2005 and [240] ©12002 American Chemical Society).

which solutions are

$$E_{\pm} = \frac{1}{2}[H_{11} + H_{22} \pm \{(H_{11} - H_{22})^2 + 4H_{12}^2\}^{1/2}] \tag{1.7.7}$$

The energy gap between the eigenstates is $\Delta E = E_+ - E_- = (\Delta H^2 + 4H_{12}^2)$ where

$$\Delta H = H_{11} - H_{22} \tag{1.7.8}$$

If R_c is the nuclear distance at which the diagonal matrix elements are equal ($H_{11} = H_{22}$), or the "diabatic" surfaces cross each other, then $\Delta E(R_c) = 2|H_{12}(R_c)|$. In the Born–Oppenheimer regime, the nuclear motion is sufficiently slow to make valid separation of the electronic and nuclear motion. The eigenstates of the electronic Hamiltonian hence represent the energies of the system for a slow-varying nuclear configuration. Any processes for this system involving slow motions of the nuclei will follow one of these eigenstates depending on the quantum state of the electrons. Such processes are commonly called "adiabatic" processes and the electronic eigenstates are usually called the "adiabatic" states. The term "diabatic" processes refer to cases where the nuclear motions are not slow enough to remain separable from the electronic motion, so that the system no longer follows the eigenstates of the electronic Hamiltonian.

1.7.3.1.2 Non-crossing rule and conical intersections

In order for the two eigenstates to cross each other, it is necessary to satisfy two independent conditions

$$H_{11} = H_{22} \quad \text{and} \quad H_{12} = H_{21} = 0 \tag{1.7.9}$$

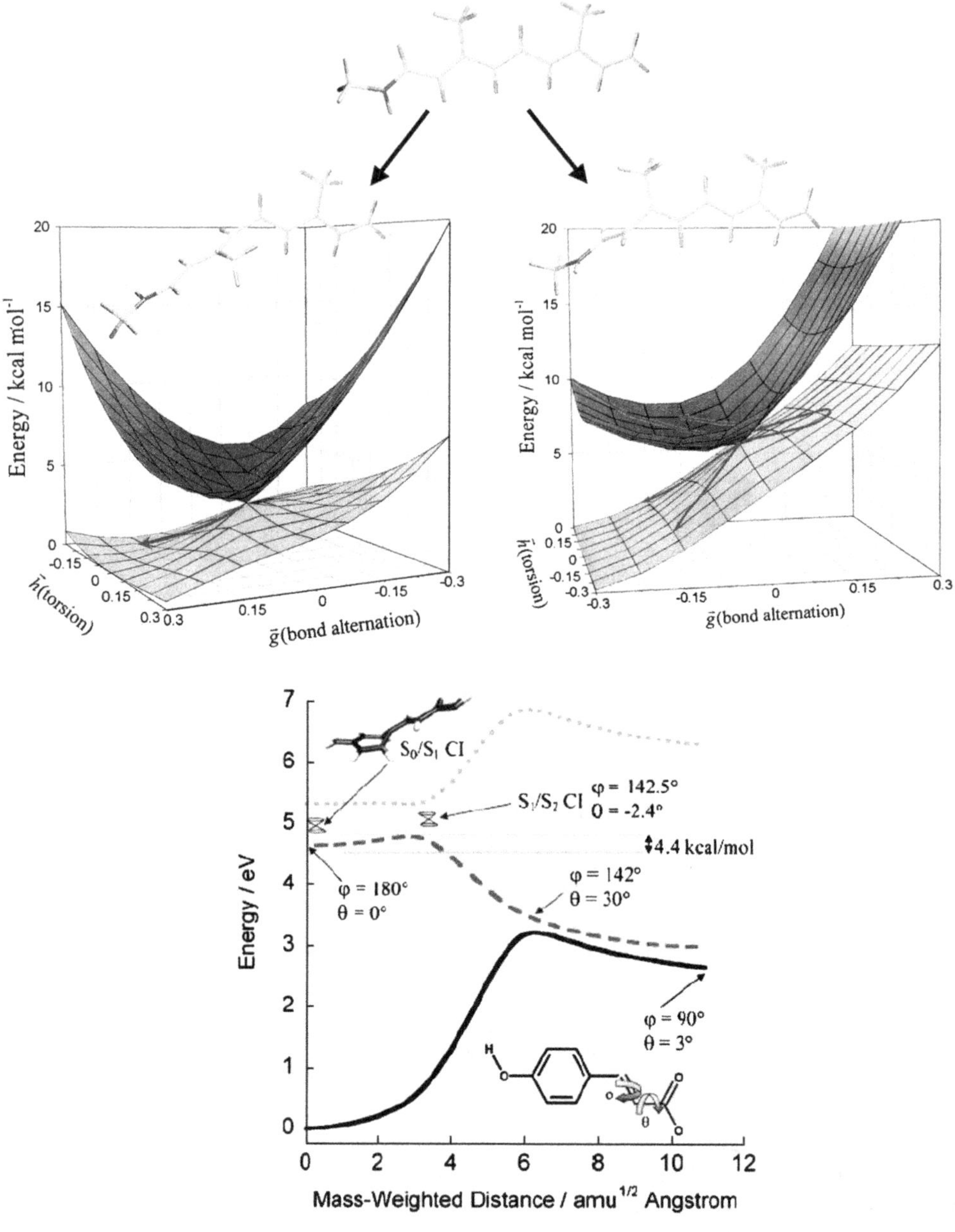

Figure 1.7.5 Top: The retinal protonated Schiff base (RPSB) found in bacterial rhodopsin and its analogue used in the model calculation. Middle: Ground and first excited electronic states, computed with ab initio electronic structure methods of the RPSB analogue as a function of displacement along the two coordinates that are most effective in promoting internal conversion from S_1 to S_0 (g and h). All other coordinates are fixed at their values at the $C_{11}{=}C_{12}$ (Left) or $C_{13}{=}C_{14}$ (Right) conical intersections. The $C_{13}{=}C_{14}$ conical intersection has a sloped topography while the $C_{11}{=}C_{12}$ conical intersection is strongly peaked (reproduced with permission from reference [241] ©2002 National Academy of Sciences). Bottom: Potential energy curves for photoisomerization of the neutral form of the photoactive yellow protein (PYP) (reproduced with permission from reference [226] ©2003 American Chemical Society).

This requires at least two independent nuclear variables. In a diatomic molecule, there would only be one such variable, the internuclear distance in a diatomic molecule. For polyatomic molecules, there are more parameters than the number of conditions (maximum of three when H_{12} is a complex number if the electronic spin becomes important and should be included in the electronic Hamiltonian). If there are $(3N-6)$ independent nuclear coordinates and if spin is neglected, eq. (1.7.9) can be satisfied in a space of $(3N-7)$ dimension when states of different symmetry are involved. If one coordinate is kept constant in a triatomic molecule, then the intersection of surfaces with different symmetries will occur along a line, and with the same symmetry at a point. Surfaces in the crossing region then form a cone. This occurs because the matrix elements H_{ij} in eq. (1.7.9) are linear functions of the displacements (x and y, say) in the neighbourhood of the crossing region. The term "conical intersection" is used to describe this situation [226, 233–238]. A comprehensive study of conical intersection can be found in reference [239].

Conical intersections are encountered in large photoactive proteins such as the photoactive yellow protein (PYP) containing 125 residues [226] (see Chapter 4.2) or rhodopsins. The retinal protonated Schiff base (RPSB) is a well-studied biochromophore possessing double bonds capable of undergoing selective *cis–trans* isomerization following photoexcitation (Figure 1.7.5). A CASSCF calculation, including the 10 conjugated π electrons, two methyl groups on the backbone and an additional methyl group replacing the link to the protein in the active space, has been conducted with the 6-31G* basis set on the model system mimicking the chromophore of the rhodopsin family [241]. Conical intersections are encountered along the minimum energy paths for torsion around the $C_{13}{=}C_{14}$ and $C_{11}{=}C_{12}$ bonds.

REFERENCES

1. Leach A: *Molecular modelling: principles and applications*: Prentice Hall, Pearson, Harlow, UK, 2001.
2. Aarons L, Karlsson MO, Mentré F, Rombout F, Steiner JL, van Peer A: Role of modelling and simulation in Phase I drug development. *European Journal of Pharmaceutical Sciences* 2001, 13:115–122.
3. Dill KA, Bromberg S: *Molecular driving forces: statistical thermodynamics in chemistry and biology*: Garland Science, Taylor and Francis, New York and London, 2002.
4. Schlick T: *Molecular modeling and simulation*: Springer-Verlag, Amsterdam, 2002.
5. Wales DJ: Energy landscapes and properties of biomolecules. *Physical Biology* 2005, 2:S86–S93.
6. Wales DJ: *Energy landscapes with applications to clusters, biomolecules and glasses*: Cambridge University Press, Cambridge, UK, 2003.
7. Janin J, Levitt M: Theory and simulation. Accuracy and reliability in modelling proteins and complexes. *Current Opinion in Structural Biology* 2006, 16:139–141.
8. Colvin ME, Cramer CJ, Dykstra CE, Jensen JH, Krimm S, Rivail JL, Thakkar AJ, Yanez M: Molecular quantum mechanics to biodynamics: essential connections. *Journal of Molecular Structure: THEOCHEM* 2006, 764:1–8.
9. Dyson HJ, Wright PE: Intrinsically unstructured proteins and their functions. *Nature Reviews Molecular Cell Biology* 2005, 6:197–211.
10. Wilchek M, Bayer EA, Livnah O: Essentials of biorecognition: the (strep)avidin-biotin system as a model for protein-protein and protein-ligand interaction. *Immunology Letters* 2006, 103:27–32.
11. Daune M: *Molecular biophysics. Structures in motion*: Oxford University Press, Oxford, 1999.

12. Suvire FD, Rodriguez AM, Mak ML, Papp JG, Enriz RD: Binding mechanism of RGD and its mimetics to receptor GPIIb/IIIa. A theoretical study. *Journal of Molecular Structure: THEOCHEM* 2001, 540:257–270.
13. Baker SR, Boot J, Brunays M, Dobson D, Green R, Hayhurst L, Keenan M, Wallace L: High affinity ligands for the 7 nicotinic receptor that show no reactivity with the 5-HT3 receptor to bind to their receptors. *Bioorganic Medicinal Chemical Letters* 2005, 15:4727–4730.
14. White S, Szewczyk JW, Turner JM, Baird EE, Dervan PB: Recognition of the four Watson-Crick pairs in the DNA minor groove by synthetic ligands. *Nature* 1998, 291:468–471.
15. Kalodimos CG, Biris N, Bonvin AMJ, Levandowski MM, Guennegues M, Boelens R, Kaptein R: Structure and flexibility adaptation on non specific and specific protein-DNA complexes. *Science* 2004, 305:386–389.
16. Kim J, Mao J, Gunner MR: Are acidic and basic groups in buried proteins predicted to be ionized. *Journal of Molecular Biology* 2005, 348:1283–1298.
17. Sagui C, Pedersen LG, Darden TA: Towards an accurate representation of electrostatics in classical force-fields: efficient implementation of multipolar interactions in biomolecular simulations. *Journal of Chemical Physics* 2003, 120:73–87.
18. Koehl P: Electrostatics calculations: latest methodological advances. *Current Opinion in Structural Biology* 2006, 16:142–151.
19. Kobko N, Paraskevas L, del Rio E, Dannenberg JJ: Cooperativity in amide hydrogen bonding chains: implications for protein-folding models. *Journal of the American Chemical Society* 2001, 123:4348–4349.
20. Lucas B, Gregoire G, Lecomte F, Reimann B, Schermann JP, Desfrancois C: Infrared spectroscopy of mass-selected neutral molecular systems without chromophore: the formamide monomer and dimer. *Molecular Physics* 2005, 103:1497–1503.
21. Barril X, Aleman C, Orozco M, Luque FJ: Salt bridge interactions: stability of the ionic and neutral complexes in the gas-phase, in solution, and in proteins. *Proteins-Structure Function and Bioinformatics* 1998, 32:67–79.
22. Julian RR, Hodyss R, Beauchamp JL: Salt bridge stabilization of charged zwitterionic arginine aggregates in the gas phase. *Journal of the American Chemical Society* 2001, 123:3577–3583.
23. Wyttenbach T, Paizs B, Barran P, Breci L, Liu B, Suhai S, Wysocki VH, Bowers MT: The effect of the initial water of hydration on the energetics, structures, and H/D exchange mechanism of a family of pentapeptides: an experimental and theoretical study. *Journal of the American Chemical Society* 2003, 125:13768–13775.
24. Kang H, Dedonder-Lardeux C, Jouvet C, Gregoire G, Desfrancois C, Schermann JP, Barat M, Fayeton JA: Control of bond-cleaving reactions of free protonated tryptophan ion by femtosecond laser pulses. *Journal of Physical Chemistry A* 2005, 109:2417–2420.
25. Gallivan JP, Dougherty DA: Cation-π interactions in structural biology. *Proceedings of the National Academy of Sciences of the United States of America* 1999, 96:9459–9464.
26. Sobczyk L, Grabowski SJ, Krygowski TM: Interrelation between H-bond and π-electron delocalization. *Chemical Reviews* 2005, 105:3513–3560.
27. Ryzhov V, Dunbar RC, Cerda B, Wesdemiotis C: Cation-π effects in the complexation of Na^+ and K^+ with Phe, Tyr, and Trp in the gas phase. *Journal of the American Society for Mass Spectrometry* 2000, 11:1037–1046.
28. Biot C, Buisine E, Kwasigroch JM, Wintjens R, Rooman M: Probing the energetic and structural role of amino acid/nucleobase cation-π interactions in protein-ligand complexes. *Journal of Biological Chemistry* 2002, 277:40816–40822.
29. Anderson MA, Ogbay B, Arimoto R, Sha W, Kisselev OG, Cistola DP, Marshall GR: Relative strength of action-pi vs. salt-bridge interactions; The G α (350–360) peptide/rhodopsin system. *Journal of the American Chemical Society* 2006, 128:7531–7541.
30. Tatko CD, Waters ML: The geometry and efficacy of cation-π interactions in a diagonal position of a designed ß-hairpin. *Protein Science* 2003, 12:2443–2452.

31. Crowley PB, Golovin A: Cation-π interactions in protein-protein interfaces. *Proteins-Structure Function and Bioinformatics* 2005, 59:231–239.
32. Gromiha MM, Santhosh C, Ahmad S: Structural analysis of cation-π interactions in DNA binding proteins. *International Journal of Biological Macromolecules* 2004, 34:203–211.
33. Steiner T: Hydrogen bonds from water molecules to aromatic acceptors in very-high resolution protein crystal structures. *Biophysical Chemistry* 2002, 95:195–201.
34. Cleland WW, Kreevoy MM: Low-barrier hydrogen bonds and enzymatic catalysis. *Science* 1994, 264:1887–1890.
35. Vishveshwara S, Madhusudhan MS, Maizel JV: Short-strong hydrogen bonds and a low barrier transition state for the proton transfer reaction in RNase A catalysis: a quantum chemical study. *Biophysical Chemistry* 2001, 89:105–177.
36. Kushwaha PS, Mishra PC: Relationship of hydrogen bonding energy with electrostatic and polarization energies and molecular electrostatic potentials for amino acids: an evaluation of the lock and key model. *International Journal of Quantum Chemistry* 2000, 76:700–713.
37. Manikandan K, Ramakumar S: The occurrence of C–H$\cdots$O hydrogen bonds in α-helices and helix termini in globular proteins. *Proteins-Structure Function and Bioinformatics* 2004, 56:768–781.
38. Munoz MA, Ferrero R, Carmona C, Balon M: Hydrogen bonding interactions between indole and benzenoid-π-bases. *Spectrochimica Acta Part A-Molecular and Biomolecular Spectroscopy* 2004, 60:193–200.
39. Georgiev S, Neusser HJ: Mass analyzed threshold ionization of hydrogen bonded clusters of biological molecules: the 3-methylindole C_6H_6 complex. *Journal of Electron Spectroscopy and Related Phenomena* 2005, 142:207–213.
40. Keutsch FN, Cruzan JD, Saykally RJ: The water trimer. *Chemical Reviews* 2003, 103:2533–2577.
41. Lee KT, Sung J, Lee KJ, Park YD, Kim SK: Conformation-dependent ionization energies of L-phenylalanine. *Angewandte Chemie-International Edition* 2002, 41:4114–4117.
42. Schiott B, Iversen B, Kent G, Madsen H, Larsen FK, Bruice TC: On the electronic nature of low-barrier hydrogen bonds in enzymatic reactions. *Proceedings of the National Academy of Sciences of the United States of America* 1998, 95:12799–12802.
43. Sessions RB, Gibbs N, Dempsey CE: Hydrogen bonding in helical polypeptides from molecular dynamics simulations and amide hydrogen exchange analysis: alamethicin and melittin in methanol. *Biophysical Journal* 1998, 74:138–152.
44. Kim SK, Oh KS, Lee JY: Catalytic role of enzymes: short H-bond-induced partial proton shuttles and charge redistributions. *Proceedings of the National Academy of Sciences of the United States of America* 2000, 12:6373–6378.
45. Meot-Mer M: The ionic hydrogen bond. *Chemical Reviews* 2005, 105:213–284.
46. Ewing RG, Eiceman GA, Stone JA: Proton-bound cluster ions in ion mobility spectrometry. *International Journal of Mass Spectrometry* 1999, 193:57–68.
47. Moore DT, Oomens J, van der Meer L, von Helden G, Meijer G, Valle J, Marshall AG, Eyler JR: Probing the vibrations of shared, OH^+ O^- bound protons in the gas phase. *ChemPhysChem* 2004, 5:740–743.
48. Armentano D, De Munno G, Di Donna L, Sindona G, Giorgi G, Salvini L: Self-assembling of cytosine nucleoside into triply-bound dimers in acid media. A comprehensive evaluation of proton-bound pyrimidine nucleosides by electrospray tandem mass spectrometry, X-rays diffractometry, and theoretical calculations. *Journal of the American Society for Mass Spectrometry* 2004, 15:268–279.
49. Dill KA: Dominant forces in protein folding. *Biochemistry* 1990, 29:7133–7154.
50. Brutschy B, Hobza P: van der Waals molecules. Introduction. *Chemical Reviews* 2000, 100:3 861–3862.
51. Braun JE, Mehnert T, Neusser HJ: Binding energy of van der Waals- and hydrogen-bonded clusters by threshold ionization techniques. *International Journal of Mass Spectrometry* 2000, 203:1–18.
52. Brutschy B: The structure of microsolvated benzene derivatives and the role of aromatic constituents. *Chemical Reviews* 2000, 100:3891–3920.

53. Fujii A, Patwari N, Ebata T, Mikami N: Vibrational spectroscopic evidence of unconventional hydrogen bonds. *International Journal of Mass Spectrometry* 2002, 200:289–312.
54. Zimmerli U, Parrinello M, Koumoutsakos P: Dispersion corrections to density functionals for water aromatic interactions. *Journal of Chemical Physics* 2004, 120:2693–2699; Xu X, Goddard WA: Bonding properties of the water dimer: a comparative study of density functional theories. *Journal of Physical Chemistry A* 2004, 108:2305–2313.
55. Dabkowska I, Gonzalez HV, Jurecka P, Hobza P: Stabilization energies of the hydrogen-bonded and stacked structures of nucleic acid base pairs in the crystal geometries of CG, AT, and AC DNA steps and in the NMR geometry of the 5′-d(GCGAAGC)-3′ hairpin: complete basis set calculations at the MP2 and CCSD(T) levels. *Journal of Physical Chemistry A* 2005, 109: 1131–1136.
56. Langlet J, Claverie P, Caron F, Boeuve JC: Interactions between nucleic acid bases in hydrogen bonded and stacked configurations: the role of molecular charge distribution. *International Journal of Quantum Chemistry* 1981, 20:299–338.
57. Elsner M, Hobza P, Frauenheim T, Suhai S, Kaxiras E: Hydrogen bonding and stacking interactions of nucleic acid base pairs: a density-functional-theory treatment. *Journal of Chemical Physics* 2001, 114:5149–5155.
58. Kabelac M, Plutzer C, Kleinermanns K, Hobza P: Isomer selective IR experiments and correlated *ab initio* quantum chemical calculations support planar H-bonded structure of the 7-methyl adenine-adenine and stacked structure of the 9-methyl adenine-adenine base pairs. *Physical Chemistry Chemical Physics* 2004, 6:2781–2785.
59. Jurecka P, Cerny J, Hobza P, Salahub DR: Density functional theory augmented with an empirical dispersion term. Interactions energies of 80 non-covalent complexes compared with *ab initio* quantum mechanics calculations. *Journal of Computational Chemistry* 2006, 28:555–569.
60. Sommerfeld T: A fresh look at aromatic dianions. *Journal of the American Chemical Society* 2002, 124:1119–1124.
61. Jarrold MF: Unfolding, refolding, and hydrations of proteins in the gas phase. *Accounts of Chemical Research* 1999, 32:360–367.
62. Jarrold MF: Peptides and proteins in the vapor phase. *Annual Review of Physical Chemistry* 2000, 51:179–207.
63. Taraszka JA, Counterman AE, Clemmer DE: Large anhydrous polyalanine ions: substitution of Na^+ for H^+ destabilizes folded states. *International Journal of Mass Spectrometry* 2001, 204: 87–100.
64. Dokholian NV: Studies of folding and misfolding using simplified models. *Current Opinion in Structural Biology* 2005, 16:1–7.
65. Csazar AG, Perczel A: *Ab initio* characterization of building units in peptides and proteins. *Progress in Biophysics and Molecular Biology* 1999, 71:243–309.
66. Bernholdt DE: Parallel RI-MP2 energies and gradients implementations and applications. *Journal of Chemical Physics* 1998, 109:1593–1600.
67. Weigend F, Häser M, Patzelt H, Ahlrichs R: RI-MP2: optimized auxiliary basis sets and demonstration of efficiency. *Chemical Physics Letters* 1998, 294:143–152.
68. Jurecka P, Nachtigall P, Hobza P: RI-MP2 calculations with extended basis sets-a promising toll for study of H-bonded and stacked DNA base pairs. *Physical Chemistry Chemical Physics* 2001, 3: 4578–4582.
69. Bartlett RJ, Stanton JF: Applications of post-Hartree-Fock methods: a tutorial. *Review of Computational Chemistry* 1994, 5:65–169.
70. Abrams ML, Sherril CD: Important configurations in configuration interaction and coupled-cluster wave functions. *Chemical Physics Letters* 2005, 412:121–124.
71. Haranczyk M, Gutowski M: Valence and dipole-bound anions of the most stable tautomers of guanine. *Journal of the American Chemical Society* 2005, 127:699–706.

72. Berdys-Kochanska J, Kruszewski J, Skurski P: Negatively charged xanthine. I. Anions formed by canonical isomers. *Journal of Physical Chemistry A* 2005, 109:11407–11412.
73. Piacenza M, Grimme S: Systematic quantum mechanical study of DNA-base tautomers. *Journal of Computational Chemistry* 2004, 25:83–98.
74. Fan Y, Ho J, Bettens RPA: Approximating coupled cluster level vibrational frequencies with composite methods. *Journal of Physical Chemistry A* 2006, 110:2796–2800.
75. Jurecka P, Sponer J, Cerny J, Hobza P: Benchmark database of accurate (MP2 and CCSD(T) complete basis set limit) interaction energies of small model complexes, DNA base pairs, and amino acid pairs. *Physical Chemistry Chemical Physics* 2006, 8:1985–1993.
76. Frey JA, Leutwyler S: An *ab initio* benchmark study of hydrogen bonded formamide dimers. *Journal of Physical Chemistry A* 2006, 110:12512–12518.
77. Takagi J: Structural basis for ligand recognition by RGD (Arg-Gly-Asp)-dependent integrins. *Biochemical Society Transactions* 2004, 32:403–406.
78. Weiss MS: More hydrogen bonds for the (structural) biologist. *Trends Biochemical Sciences* 2001, 26:521–523.
79. Geerlings P, De Proft F, Langenaeker W: Conceptual density functional theory. *Chemical Reviews* 2003, 103:1793–1873.
80. Sinnokrot MO: Density functional theory predictions of anharmonicity and spectroscopic constants of diatomic molecules. *Journal of Chemical Physics* 2001, 115:2439–2448.
81. Cerny J, Hobza P: The X3LYP extended density functional accurately describes H-bonding but fails completely for stacking. *Physical Chemistry Chemical Physics* 2005, 7:1624–1626.
82. Becke AD: Density-functional thermochemistry. IV. A new dynamical correlation functional and implications for exact-exchange mixing. *Journal of Chemical Physics* 1996, 104:1040–1046.
83. Parr RG, Yang W: Density-functional thermochemistry. IV. A new dynamical correlation functional and implications for exact-exchange mixing. *Annual Review of Physical Chemistry* 1995, 46:701–728.
84. Koné M, Illien B, Graton J, Laurence C: B3LYP and MP2 calculations of the enthalpies of hydrogen-bonded complexes of methanol with neutral bases and anions: comparison with experimental data. *Journal of Physical Chemistry A* 2005, 109:11907–11913.
85. Elstner M, Jalkanen KJ, Knapp-Mohammady M, Frauenheim T, Suhai S: Energetics and structure of glycine and alanine based model peptides: approximate SCC-DFTB, AM1 and PM3 methods in comparison with DFT, HF and MP2 calculations. *Chemical Physics* 2001, 363:203–219.
86. Bartlett RJ, Schweigert IV, Lotrich VF: *Ab initio* DFT: getting the right answer for the right reason. *Journal of Molecular Structure: THEOCHEM* 2006, 764:33–40.
87. van Mourik T, Karamertzanis PG, Price SL: Molecular conformations and relative stabilities can be as demanding of the electronic structure method as intermolecular calculations. *Journal of Physical Chemistry A* 2006, 110:8–12.
88. Doruker P, Jernigan RL, Navizet I, Hernandez R: Important fluctuation dynamics of large protein structures of large protein structures are preserved upon coarse-grained renormalization. *International Journal of Quantum Chemistry* 2002, 90:822–837.
89. Navizet I, Cailliez F, Lavery R: Probing protein mechanics: Residue-level properties and their use in defining domains. *Biophysical Journal* 2004, 87:1426–1435.
90. Tozzini V: Coarse-grained models for proteins. *Current Opinion in Structural Biology* 2005, 15: 144–150.
91. van der Vaart A, Gogonea V, Dixon SL, Merz KH: Linear scaling molecular orbital calculations of biological systems using the semiempirical divide and conquer method. *Journal of Computational Chemistry* 2000, 21:1494–1504.
92. Werner HJ, Manby FR, Knowles PJ: Fast linear scaling second-order Møller-Plesset perturbation theory (MP2) using local and density fitting approximations. *Journal of Chemical Physics* 2003, 118:8149–8160.

93. Chen X, Zhang Y, Zhang ZH: An efficient approach for *ab initio* energy calculations of biopolymers. *Journal of Chemical Physics* 2005, 122:184105.
94. Skylaris CK, Haynes PD, Mostofi AA, Payne MC: Introducing ONETEP: linear-scaling density functional simulations on parallel computers. *Journal of Chemical Physics* 2005, 122: 084119.
95. Schütz M: A new, fast, semi-direct implementation of linear scaling local coupled cluster theory. *Physical Chemistry Chemical Physics* 2002, 4:3941–3947.
96. Weigend F, Ahlrichs R: Balanced basis sets of split valence, triple zeta and quadruple zeta valence quality for H to Rn: design and assessment of accuracy. *Physical Chemistry Chemical Physics* 2005, 7:3297–3305.
97. Range K, Riccardi D, Cui Q, Elstner M, York DM: Benchmark calculations of proton affinities and gas-phase basicities of molecules important in the study of biological phosphoryl transfer. *Physical Chemistry Chemical Physics* 2005, 7:3070–3079.
98. Gutowski M, Jordan KD, Skurski P: Electronic structure of dipole-bound anions. *Journal of Physical Chemistry* 1998, 102:2624–2633.
99. Wood DE, Dunning TH: Gaussian basis sets for use in correlated molecular calculations. V. Core-valence basis sets for boron through neon. *Journal of Chemical Physics* 1995, 103:4752–4785.
100. Cauët E, Dehareng D, Liévin J: *Ab initio* study of the ionization of the DNA bases: ionization potentials and excited states of the cations. *Journal of Physical Chemistry A* 2006, 110: 9200–9211.
101. Clarkson JR, Dian BC, Moriggi L, DeFusco A, McCarthy V, Jordan KD, Zwier TS: Direct measurement of the energy thresholds to conformational isomerization in Tryptamine: experiment and theory. *Journal of Chemical Physics* 2005, 122.
102. Hobza P, Sponer J: Toward true DNA base-stacking energies: MP2, CCSD(T), and complete basis set calculations. *Journal of the American Chemical Society* 2002, 124:11802–11808.
103. Reha D, Valdes H, Vondrasek J, Hobza P, Abu-Riziq A, Crews B, de Vries MS: Structure and IR spectrum of phenylalanyl-glycyl-glycine tripetide in the gas-phase: IR/UV experiments, *ab initio* quantum chemical calculations, and molecular dynamic simulations. *Chemistry – A European Journal* 2005, 11:6803–6817.
104. Svozil D, Frigato T, Havlas Z, Jungwirth P: *Ab initio* electronic structure of thymine anions. *Physical Chemistry Chemical Physics* 2005, 7:840–845.
105. MacKerell AD, Banavali N, Foloppe N: Development and current status of the CHARMM force field for nucleic acids. *Biopolymers* 2000, 56:257–265.
106. Macias AT, McKerrel AD: CH/π interactions involving aromatic amino acids. Refinement of the CHARMM tryptophan force-field. *Journal of Computational Chemistry* 2005, 26:1452–1463.
107. Wang J, Cieplack P, Kollman KA: How well does a restrained electrostatic potential (RESP) model perform in calculating conformational energies of organic and biological molecules. *Journal of Computational Chemistry* 2000, 21:1049–1074.
108. Lins RD, Hüneberger PH: A new GROMOS force field for hexopyranose-based carbohydrates. *Journal of Computational Chemistry* 2005, 26:1400–1412.
109. Damm W, Frontera A, Tirado-Rives J, Jorgensen WL: OPLS all-atom force field for carbohydrates. *Journal of Computational Chemistry* 1997, 18:1965–1970.
110. McKerrell A: All-atom empirical potential for molecular modelling and dynamics studies of proteins. *Journal of Physical Chemistry B* 1998, 102:3586–3616.
111. Fonseca Guerra C, Bickelhaupt FM, Suijders JG, Baerends EJ: Hydrogen bonding in DNA base pairs: reconciliation of theory and experiment. *Journal of the American Chemical Society* 2000, 122:4117–4128.
112. Jin L, Barran PE, Deakin JA, Lyon M, Uhrin D: Conformation of glycoaminoglycans by ion mobility mass spectrometry and molecular modelling. *Physical Chemistry Chemical Physics* 2005, 7:3464–3471.

113. Hobza P, Kabelac M, Sponer J, Mejzlik P, Vondrasek J: Performance of empirical potentials (AMBER, CFF95, CVFF, CHARMM, OPLS, POLTEV), semiempirical quantum chemical methods (AM1, MNDO/M, PM3), and *ab initio* Hartree-Fock method for interaction of DNA bases: comparison with nonempirical beyond Hartree-Fock results. *Journal of Computational Chemistry* 1997, 18:1136–1150.
114. Petrova P, Koca J, Imberty A: Potential energy hypersurfaces of nucleotide sugars: *ab initio* calculations, force-field parametrization, and exploration of the flexibility. *Journal of the American Chemical Society* 1999, 121:5535–5547.
115. Perez S, Imberty A, Engelsen SB, Gruza J, Mazeau K, Jimenez-Barbero J, Poveda A, Espinosa JF, van Eyck BP, Johnson G, *et al.*: A comparison and chemometric analysis of several molecular mechanics force fields and parameter sets applied to carbohydrates. *Carbohydrate Research* 1998, 314:141–155.
116. Gruza J, Koca J, Perez S, Imberty A: Comparison of force-fields parametrizations as applied to conformational analysis of ribofuranosides. *Journal of Molecular Structure*: *THEOCHEM* 1998, 424:269–280.
117. Imberty A: Oligosaccharide structures: theory versus experiment. *Current Opinion in Structural Biology* 1997, 7:617–623.
118. Poltev VI, Gonzalez E, Deriabina AS, Lozano L, Martinez A, Robinson T, Gorb L, Leszczynski J: Molecular mechanics description of cytosine energy and geometry using preliminary *ab initio* results. *Journal of Molecular Structure: THEOCHEM* 2005, 729:59–64.
119. Kaminski GA, Stern HA, Berne BJ, Friesner RA, Cao YX, Murphy RR, Zhou R, Halgren TA: Development of a polarizable force-field for proteins via *ab initio* quantum chemistry: first generation model and gas phase tests. *Journal of Computational Chemistry* 2002, 23: 1515–1531.
120. Gresh N, Sponer J: Complexes of pentahydrated $Zn^{+}$with guanine, adenine and the guanine-cytosine and adenine-thymine base pairs. Structures and energies characterized by polarizable molecular mechanics and *ab initio* calculations. *Journal of Physical Chemistry B* 1999, 103: 11415–11427.
121. Rogalewicz F, Ohanessian G, Gresh N: Interaction of neutral and zwitterionic glycine with Zn^{2+} in gas phase: *Ab initio* and SIBFA molecular mechanics calculations. *Journal of Computational Chemistry* 2000, 21:963–973.
122. Halgren TA, Damm W: Polarizable force-fields. *Current Opinion in Structural Biology* 2001, 11:236–242.
123. Gao J, Truhlar DG: Quantum mechanical methods for enzyme kinetics. *Annual Review of Physical Chemistry* 2002, 53:467–505.
124. Garcia-Viloca M, Gao J, Karplus M, Truhlar DG: How enzymes work: analysis by modern theory and computer simulations. *Science* 2004, 303:186–195.
125. van Duin ACT, Dasgupta S, Lorant F, Goddard WA: ReaxFF: a reactive force field for hydrocarbons. *Journal of Physical Chemistry A* 2001, 105:9396–9409.
126. Strachan A, Kober EM, van Duin ACT, Oxgaard J, Goddard WA: Thermal decomposition of RDX from reactive molecular dynamics. *Journal of Chemical Physics* 2005, 122:054502.
127. Bernal-Uruchurtu MI, Ruiz-Lopez MF: Basic ideas for the correction of semiempirical methods describing H-bonded systems. *Chemical Physics Letters* 2000, 330:118–124.
128. Linnanto J, Korppi-Tommola J: Semiempirical PM5 molecular orbital study on chlorophylls and bacteriochlorophylls: Comparison of semiempirical, *ab initio*, and density functional results. *Journal of Computational Chemistry* 2004, 25:123–138.
129. Van Mourik T, Fruchtl HA: The potential energy landscape of noradrenaline: An electronic structure study. *Molecular Physics* 2005, 103:1641–1654.
130. Bredow T, Jug K: Theory and range of modern semiempirical molecular orbital methods. *Theoretica Chimica Acta* 2005, 113:1–14.

131. Sherer EC, York DM, Cramer CJ: Fast approximate methods for calculating nucleic acid base pair interaction energies. *Journal of Computational Chemistry* 2002, 24:57–67.
132. Chass GA, Mirasol RS, Setiadi DH, Tang TH, Chin W, Mons M, Dimicoli I, Dognon JP, Viskolcz B, Lovas S: Characterization of the conformational probability of N-acetyl-phenylalanyl-NH2 by RHF, DFT, and MP2 computation and AIM analyses, confirmed by jet-cooled infrared data. *Journal of Physical Chemistry A* 2005, 109:5289–5302.
133. Carcabal P, Jockusch RA, Hunig I, Snoek LC, Kroemer RT, Davis BG, Gamblin DP, Compagnon I, Oomens J, Simons JP: Hydrogen bonding and cooperativity in isolated and hydrated sugars: mannose, galactose, glucose, and lactose. *Journal of the American Chemical Society* 2005, 127:11414–11425.
134. Bakker JM, Compagnon I, Meijer G, von Helden G, Kabelac M, Hobza P, de Vries MS: The mid-IR absorption spectrum of gas-phase clusters of the nucleobases guanine and cytosine. *Physical Chemistry Chemical Physics* 2004, 6:2810–2815.
135. Berry RS, Elmaci N, Rose JP, Vekhter B: Linking topography of its potential surface with the dynamics of folding of a protein model. *Proceedings of the National Academy of Sciences of the United States of America* 1997, 94:9520–9524.
136. Bolhuis PG, Chandler D, Dellago C: Transition path sampling: throwing ropes over rough mountain passes, in the dark. *Annual Review of Physical Chemistry* 2002, 53:291–318.
137. Lu J, Zhang C, Berry RS: Kinetics of model energy landscapes: an approach to complex systems. *Physical Chemistry Chemical Physics* 7:3443–3456.
138. Despa F, Wales DJ, Berry RS: Archetypal energy landscapes: dynamical diagnosis. *Journal of Chemical Physics* 2005, 122, 024103.
139. Saven JG: Designing protein energy landscapes. *Chemical Reviews* 2001, 101:3113–3130.
140. Clark PL: Protein folding in the cell: reshaping the folding funnel. *Trends in Biochemical Sciences* 2004, 29:527–534.
141. Wales DJ: Exploring the energy landscape. *International Journal of Modern Physics B* 2005, 19: 2877–2885.
142. Michielin O, Zoele V, Gierash TM, Eckstein J, Napper A, Verdine G, Karplus M: Conformational analysis of a stereochemically complete set of *cis*-enediol peptide analogues. *Journal of the American Chemical Society* 2002, 124:11131–11141.
143. Bisetty K, Catalan JG, Kruger HG, Perez JJ: Conformational analysis of small peptides of the type Ac-X-NHMe, where X = Gly, Ala, Aib and Cage. *Journal of Molecular Structure: THEOCHEM* 2005, 731:127–137.
144. Kolossvary I, Guida WC: Torsional flexing: conformational searching of cyclic molecules in biased internal coordinate space. *Journal of Computational Chemistry* 1993, 14:691–698.
145. Chen J, Im W, Brooks CL: Application of torsion angle molecular dynamics for efficient sampling of protein conformations. *Journal of Computational Chemistry* 2005, 26:1565–1578.
146. Miller TF, Clary DC: Quantum free energies of the conformer of glycine on an *ab initio* potential energy surface. *Physical Chemistry Chemical Physics* 2004, 6:2563–2571.
147. Krivov SV, Karplus M: Free energy disconnectivity graphs: application to peptide models. *Journal of Chemical Physics* 2002, 117:10894–10903.
148. Mortenson PN, Evans DA, Wales DJ: Energy landscapes of model polyalanines. *Journal of Chemical Physics* 2002, 117:1363–1376.
149. Mortenson PN, Wales DJ: Energy landscapes, global optimization and dynamics of the polyalanine Ac(ala)(8)NHMe. *Journal of Chemical Physics* 2001, 114:6443–6454.
150. Evans DA, Wales DJ: The free energy landscape and dynamics of met-enkephalin. *Journal of Chemical Physics* 2003, 119:9947–9955.
151. Evans DA, Wales DJ, Dian BC, Zwier TS: The dynamics of conformational isomerization in flexible biomolecules. II. Simulating isomerizations in a supersonic free jet with master equation dynamics. *Journal of Chemical Physics* 2004, 120:148–157.

152. Dian BC, Longarte A, Winter PR, Zwier TS: The dynamics of conformational isomerization in flexible biomolecules. I. Hole-filling spectroscopy of N-acetyl tryptophan methyl amide and N-acetyl tryptophan amide. *Journal of Chemical Physics* 2004, 120:133–147.
153. Li ZQ, Scheraga HA: Monte Carlo-minimization approach to the multiple-minima problem in protein folding. *Proceedings of the National Academy of Sciences of the United States of America* 1987, 84:6611–6615.
154. Pillardy A, Czaplewski C, Liwo A, Lee J, Ripoll DR, Kazmierkiewicz R, Oldziej S, Wedemeyer WJ, Gibson KD, Arnautova YA: Recent improvements in prediction of protein structure by global optimization of a potential energy function. *Proceedings of the National Academy of Sciences of the United States of America* 2001, 98:2329–2333.
155. Earland DJ, Deem MW: Parallel tempering: theory, applications and new perspectives. *Physical Chemistry Chemical Physics* 2005, 7:3910–3916.
156. Hansmann UHE: Parallel tempering algorithm for conformational studies of biological molecules. *Chemical Physics Letters* 1997, 281:140–150.
157. Mitsutake A, Okamoto Y: Replica-exchange simulated tempering method for simulations of frustrated systems. *Chemical Physics Letters* 2000, 332:131–138.
158. Cecchini M, Rao F, Seeber M, Calflisch A: Replica exchange molecular dynamics simulations of amyloid peptide aggregation. *Journal of Chemical Physics* 2004, 121:10748–10756.
159. Schug A, Wenzel W: All-atom folding of the Trp-cage protein with an adaptative parallel tempering method. *Europhysics Letters* 2004, 67:307–313.
160. Poulain P, Antoine R, Broyer M, Dugourd P: Monte Carlo simulations of flexible molecules in a static electric field: electric dipole and conformation. *Chemical Physics Letters* 2005, 401:1–6.
161. Clary DC: Torsional diffusion Monte Carlo: a method for quantum simulations of proteins. *Journal of Chemical Physics* 2001, 114:9725–9732.
162. Miller TF, Clary DC: Torsional path integral Monte Carlo method for calculating the absolute quantum free energy of large molecules. *Journal of Chemical Physics* 2003, 119:68–76.
163. Carles S, Lecomte F, Schermann JP, Desfrançois C: Gas-phase experimental and theoretical studies of adenine, imidazole, pyrrole and water non-covalent complexes. *Journal of Physical Chemistry A* 2000, 104:10662–10669.
164. Sun S: A genetic algorithm that seeks native states of peptides and proteins. *Journal of Biophysics* 1995, 69:340–355.
165. Chervenkov S, Wang PQ, Braun JE, Neusser HJ: Fragmentation and conformation study of ephedrine by low- and high-resolution mass selective UV spectroscopy. *Journal of Chemical Physics* 2004, 121:7169–7174.
166. Carloni P: Density functional theory-based molecular dynamics of biological systems. *Quantitative Structure-Activity Relationships* 2002, 21:166–172.
167. Gaigeot MP, Sprik M: *Ab initio* molecular dynamics computation of the infrared spectrum of aqueous uracil. *Journal of Physical Chemistry B* 2003, 107:10344–10358.
168. Qiu LL, Pabit SA, Roitberg AE, Hagen SJ: Smaller and faster: The 20-residue Trp-cage protein folds in 4 μs. *Journal of the American Chemical Society* 2002, 124:12952–12953.
169. Ding F, Buldyrev SV, Dokholian NV: Folding Trp-cage to NMR resolution native structure using a coarse-grained protein model. *Biophysical Journal* 2005, 88:147–155.
170. Patriksson A, Adams C, Kjeldsen F, Raber J, van der Spoel D, Zubarev RA: Prediction of N-C-α bond cleavage frequencies in electron capture dissociation of Trp-cage dications by force-field molecular dynamics simulations. *International Journal of Mass Spectrometry* 2006, 248:124–135.
171. Bunagan MR, Yang X, Saven JG, Gal F: Ultrafast folding of a computationally designed Trp-cage mutant Trp2-cage. *Journal of Physical Chemistry B* 2006, 110:3750–3763.
172. Dokholyan V, Buldyrev SV, Stanley HE, Shakhnovich EI: Discrete molecular dynamics studies of the folding of protein-like model. *Folding and Design* 1998, 3:577–587.

173. Dokholyan V: Studies of folding and misfolding using simplified models. *Current Opinion in Structural Biology* 2006, 16:79–85.
174. Zhou Y, Karplus M, Wichert JM, Hall CK: Equilibrium themodynamics of homopolymers and clusters molecular dynamics and Monte Carlo simulation of systems with square-well interactions. *Journal of Chemical Physics* 1997, 107:10691–10708.
175. Kalivas JH: *Adaptation of simulated annealing to chemical optimization problems*: Elsevier, Amsterdam, 1995.
176. Kerr ID, Sankararamakrishnan R, Smart OS, Sansom MSP: Parallel helix bundles and ion channels: molecular modelling via simulated annealing and restrained molecular dynamics. *Biophysical Journal* 1994, 67:1501–1515.
177. Chen W, Brooks CL: Application of torsion angle molecular dynamics for efficient sampling of protein conformations. *Journal of Computational Chemistry* 2005, 26:1565–1578.
178. Arnautova YA, Jagielska A, Scheraga HA: A new force field (ECEPP-05) for peptides, proteins, and organic molecules. *Journal of Physical Chemistry B* 2006, 110:5025–5044.
179. Brooijmans N, Kuntz ID: Molecular recognition and docking algorithms. *Annual Review of Biophysical Biomolecular Structure* 2003, 32:335–373.
180. Tuttle T, Kraka E, Cremer D: Docking, triggering, and biological activity of Dynemicin A in DNA: a computational study. *Journal of the American Chemical Society* 2004, 127:9469–9484.
181. Bonvin AMJ: Flexible protein-protein docking. *Current Opinion in Structural Biology* 2006, 16: 194–200.
182. Sulpizi M, Laio A, VandeVondele J, Cattaneo A, Rothlisberger U, Carloni P: Reaction mechanism of caspases: insights from QM/MM Car-Parrinello simulations. *Proteins-Structure Function and Genetics* 2003, 52:212–224.
183. Moret ME, Tapavicza E, Guidoni L, Rohrig UF, Sulpizi M, Tavernelli I, Rothlisberger U: Quantum mechanical/molecular mechanical (OM/MM) Car-Parrinello simulations in excited states. *Chimia* 2005, 59:493–498.
184. Marinica DC, Gregoire G, Desfrancois C, Schermann JP, Borgis D, Gaigeot MP: *Ab initio* molecular dynamics of protonated dialanine and comparison to infrared multiphoton dissociation experiments. *Journal of Physical Chemistry A* 2006, 110:8802–8810.
185. Grégoire G, Gaigeot MP, Marinica DC, Lemaire J, Schermann JP, Desfrancois C: Gas-phase protonated peptides studied with IRMPD spectroscopy experiments and quantum molecular dynamics simulations conducted at 300K. *Physical Chemistry Chemical Physics* 2007, 9:3082–3097.
186. Car R, Parinello M: Unified approach for molecular dynamics and functional theory. *Physical Review Letters* 1985, 55:2471–2474.
187. Tse JS: *Ab initio* molecular dynamics with density functional theory. *Annual Review of Physical Chemistry* 2002, 53:249–290.
188. Zahn D: Car-Parinello molecular dynamics simulations of base-catalyzed amide hydrolysis in aqueous solution. *Chemical Physics Letters* 2004, 383:134–137.
189. Gaigeot MP, Vuilleumier R, Sprik M, Borgis D: Infrared spectroscopy of *N*-methylacetamide revisited by *ab initio* molecular dynamics simulations. *Journal of Chemical Theory and Computation* 2005, 1:772–789.
190. Spiegel K, Rothlisberger U, Carloni P: Duocarmycins binding to DNA investigated by molecular simulation. *Journal of Physical Chemistry B* 2006, 110:3647–3660.
191. Piana S, Bucher D, Carloni P, Rothlisberger U: Reaction mechanism of HIV-1 protease by hybrid Car-Parrinello/classical MD simulations. *Journal of Physical Chemistry B* 2004, 108:11139–11149.
192. Frigyes D, Alber F, Pongor S, Carloni P: Arginine-phosphate salt bridges in protein-DNA complexes: a Car-Parrinello study. *Journal of Molecular Structure: THEOCHEM* 2001, 574:39–45.
193. Gervasio FL, Laio FL, Iannuzzi M, Parinello M: Influence of DNA structure on the reactivity of the guanine radical cation. *Chemistry A European Journal* 2004, 10:4848–4852.

194. Raynaud C, Maron L, Jolibois F, Daudey JP, Esteves PM, Ramirez-Solis A: *Ab initio* molecular dynamics: plane waves vs. local basis. The role of energy cutoff on the convergence of molecular properties. *Chemical Physics Letters* 2005, 414:161–165.
195. Gaigeot MP, Ghomi M: Geometrical and vibrational properties of nucleic acid constituents interacting with explicit water molecules as analyzed by density functional theory calculation. 1. Uracil + n_wH_2O ($n_w = 1, \ldots, 7$). *Journal of Physical Chemistry B* 2001, 105:5007–5017.
196. Kirchner B, Hutter JJ: Solvent effects on electronic properties from Wannier functions in a dimethyl sulfoxide/water mixture. *Journal of Chemical Physics* 2004, 121:5133–5142.
197. Rosta E, Klähn A, Warshel A: Towards accurate *ab initio* QM/MM calculations of free-energy profiles of enzymatic reactions. *Journal of Physical Chemistry B* 2006, 110:2934–2941.
198. Mo Y, Gao J: Polarization and charge-transfer effects in aqueous solution via *ab initio* QM/MM simulations. *Journal of Physical Chemistry B Letters* 2006, 110:2976–2980.
199. Bergès J, Richards G, Rauk A, Houée-Levin C: QM/MM study of electron addition on protein disulfide bonds. *Chemical Physics Letters* 2006, 421:63–67.
200. Robertazzi A, Platts JA: Gas-phase DNA oligonucleotide structures. A QM/MM and atoms in molecules study. *Journal of Physical Chemistry A* 2006, 110:3992–4000.
201. Riccardi D, Schaefer P, Yang Y, Yu H, Ghosh N, Prat-Resina X, König P, Li G, Xu D, Guo H: Development of effective quantum mechanical/molecular mechanical (QM/MM) methods for complex biological processes. *Journal of Physical Chemistry B* 2006, 110:6458–6459.
202. Assfeld X, Rivail JL: Quantum chemical computations on parts of large molecules: the *ab initio* local self-consistent field method. *Chemical Physics Letters* 1996, 263:100–106.
203. Reuter N, Dejaegere A, Maigret B, Karplus M: Frontier bonds in QM/MM methods: a comparison of different methods. *Journal of Physical Chemistry A* 2000, 104:1720–1735.
204. Lin H, Truhlar DG: Charge redistribution for link atoms in combined quantum mechanics and molecular mechanics. *Journal of Physical Chemistry A* 2005, 109:3991–4104.
205. Gao J, Amara P, Field MJ: A generalized hybrid orbital (GHO) method for the treatment of boundary atoms in combined QM/MM calculations. *Journal of Physical Chemistry A* 1998, 102:4714–4721.
206. Dapprich S, Komaromi I, Byun KS, Morokuma K, Frisch M: A new ONIOM implementation in Gaussian98. Part I. The calculation of energies, gradients, vibrational frequencies and electric field derivatives. *Journal of Molecular Structure: THEOCHEM* 1999, 461:1–21.
207. Frisch M, Trucks JGW, Schlegel HB, Scuseria GE, Robb MA, Cheeseman JR, Montgomery JA, Vreven T, Kudin KN, Burant JC: Gaussian package (edn revision B03). In I. Gaussian (Ed), Pittsburg, PA, 2003.
208. Zhang RB, Zhang XD, Qu ZW, Ai XC, Zhang XK, Zhang QY: Studies of the solvent effects on the internal reorganization energy for electron transfer of uracil and its anion with ONIOM. *Journal of Molecular Structure: THEOCHEM* 2003, 624:169–176.
209. Zhang RB, Ai XC, Zhang XK, Zhang QY: Solvent effects on the excited state properties of 2-aminopurine. A theoretical study by the ONIOM and supramolecular method. *Journal of Molecular Structure: THEOCHEM* 2004, 680:21–27.
210. Wieczorek R, Dannenberg JJ: The energetic and structural effects of single aminoacid substitutions upon capped α helical peptides containing 17 aminoacid residues. An ONIOM DFT/AM1 study. *Journal of the American Chemical Society* 2005, 127:17216–17223.
211. Berdys J, Anusiewicz I, Skurski P, Simons J: Damage to model DNA fragments from very low-energy (<1 eV) electrons. *Journal of the American Chemical Society* 2004, 126: 6441–6447.
212. Berdys J, Anusiewicz I, Skurski P, Simons J: Theoretical study of damage to DNA by 0.2–1.5 eV electrons attached to cytosine. *Journal of Physical Chemistry A* 2004, 108:2999–3005.
213. Dreuw A, Head-Gordon M: Single-reference *ab initio* methods for the calculation of excited states of large molecules. *Chemical Reviews* 2005, 105:4009–4037.

214. Serrano-Andres L, Merchan M: Quantum chemistry of the excited state: 2005 overview. *Journal of Molecular Structure: THEOCHEM* 2006, 729:99–108.
215. Pei K, Ma Y, Zheng X: A combined DFT, MP2 and CASSCF study on the properties of tropolone-$(H_2O)_n$ ($n = 0$–2) complexes. *Journal of Molecular Structure: THEOCHEM* 2006, 774:43–47.
216. Sinicropi A, Andruniow T, Ferre N, Basosi R, Olivucci M: Properties of the emitting state of the green fluorescent protein resolved at the CASPT2//Cascf//CHARMM level. *Journal of the American Chemical Society* 2005, 127:11534–11535.
217. Smolyar A, Wong CF: Theoretical studies of the spectroscopic properties of tryptamine, tryptophan and tyrosine. *Journal of Molecular Structure: THEOCHEM* 1999, 488:51–67.
218. Kushwaha PS, Mishra PC: Electronic spectra, excited-state geometries and molecular electrostatic potentials of aromatic aminoacids. *Journal of Photochemistry and Photobiology A: Chemistry* 2000, 137:79–86.
219. Samuels AC, Jensen JO, Hameka HF: Theoretical studies of the fluorescence and phosphorescence of tyrosine. *Journal of Molecular Structure: THEOCHEM* 1998, 454:25–30.
220. Perun S, Sobolewski AL, Domcke W: *Ab initio* studies on the radiationless decay mechanisms of the lowest excited singlet states of 9H-adenine. *Journal of the American Chemical Society* 2005, 127:6257–6265.
221. Gregoire G, Jouvet C, Dedonder C, Sobolewski AL: On the role of dissociative $\pi\sigma^*$ states in the photochemistry of protonated tryptamine and tryptophan: an *ab initio* study. *Chemical Physics* 2006, 324:398–404.
222. Casida ME, Casida KC, Salahub DR: Excited-state potential energy curves from time-dependent density-functional theory: a cross-section of formaldehyde's 1A1 manifold. *International Journal of Quantum Chemistry* 1998, 70:933–941.
223. Tozer DJ, Amos RD, Handy NC, Roos BO, Serrano-Andres L: Does density functional theory contribute to the understanding of excited states of unsaturated organic compounds? *Molecular Physics* 1999, 97:859–868.
224. Zheng WJ, Xu SJ, Radisic D, Stokes S, Li X, Bowen KH: On the interaction of electrons with betaine zwitterions. *Journal of Chemical Physics* 2005, 122.
225. Hu C, Sugino O, Myamoto Y: Modified linear response for time-dependent density functional theory. Application to Rydberg and charge-transfer excitations. *Physical Review A* 2006, 74: 032508.
226. Ko C, Levine B, Toniolo A, Olsen S, Werner HJ, Martinez TJ: *Ab initio* excited-state dynamics of the photoactive yellow protein chromophore. *Journal of the American Chemical Society* 2003, 125:12710–12711.
227. Jacquemin D, Perpète EA, Scalmani G, Frisch MJ, Ciofini I, Adamo C: Absorption and emission spectra in gas-phase and solution using TD-DFT: formaldehyde and benzene as case studies. *Chemical Physics Letters* 2006, 421:273–276.
228. Zhao Y, Truhlar DG: Density functional for spectroscopy: no long-range self-interaction error, good performance for Rydberg and charge-transfer states, and better performance on average than B3LYP for ground states. *Journal of Physical Chemistry A* 2006, 110:13126–13130.
229. Kang H, Dedonder-Lardeux C, Jouvet C, Martrenchard S, Gregoire G, Desfrancois C, Schermann JP, Barat M, Fayeton JA: Photo-induced dissociation of protonated tryptophan $TrpH^+$: a direct dissociation channel in the excited states controls the hydrogen atom loss. *Physical Chemistry Chemical Physics* 2004, 6:2628–2632.
230. Grégoire G, Jouvet C, Dedonder C, Sobolewski A: *Ab initio* study of the excited state deactivation pathways of protonated tryptophan and tyrosine. To be published.
231. Sobolewski AL, Domcke W, Dedonder-Lardeux C, Jouvet C: Excited-state hydrogen detachment and hydrogen transfer driven by repulsive $\pi\sigma^*$ states: a new paradigm for nonradiative decay in aromatic biomolecules. *Physical Chemistry Chemical Physics* 2002, 4:1093–1100.

232. Canuel C: Photophysics of isolated nucleic bases and their hydrates: femtosecond dynamics study by detection of photoelectrons, photoions and fragments Ph.D. dissertation. University Paris11, Orsay, 2006.
233. Yarkony DR: On the construction of diabatic bases using molecular properties. Rigorous results in the vicinity of a conical intersection. *Journal of Physical Chemistry A* 1998, 102: 8073–8077.
234. Worth GA, Cederbaum LS: Mediation of ultrafast electron transfer in biological systems by conical intersections. *Chemical Physics Letters* 2001, 338:219–223.
235. Kang H, Jung B, Kim SK: Mechanism for ultrafast internal conversion of adenine. *Journal of Chemical Physics* 2003, 118:6717–6719.
236. Toniolo A, Olsen S, Manohar L, Martinez TJ: Conical intersection dynamics in solution: the chromophore of green fluorescent protein. *Faraday Discussions* 2004, 127:149–163.
237. Marian CM: A new pathway for the rapid decay of electronically excited adenine. *Journal of Chemical Physics* 2005, 122.
238. Cerny J, Spirko V, Mons M, Hobza P, Nachtigallova D: Theoretical study of the ground and excited states of 7-methyl guanine and 9-methyl guanine: comparison with experiment. *Physical Chemistry Chemical Physics* 2006, 8:3059–3065.
239. Domcke W, Sobolewski AL: Ab-initio studies on the photophysics of the guanine–cytosine base pair. *Physical Chemistry Chemical Physics* 2004, 6:2763–2771.
240. Kang H, Lee KT, Jung B, Ko YJ, Kim SK: Intrinsic lifetimes of the excited state of DNA and RNA bases. *Journal of the American Chemical Society* 2002, 124:12958–12959.
241. Ben-Nun M, Mohnar F, Schulten K, Martinez TJ: The role of intersection topography in bond selectivity of *cis-trans* photoisomerization. *Proceedings of the National Academy of Sciences of the United States of America* 2002, 99:1769–1773.

-2-

Spectroscopy

GENERAL FEATURES

In this chapter, we consider resonant interactions of biomolecular systems at first with photons and next with electrons. The different spectroscopic methods using either tunable photon or electron sources are treated independently, as generally done, in this chapter. It is nevertheless worthwhile to note that they present common features as will be shown in Chapter 4.7. Except for the terahertz region which only begins to emerge, a very sizeable fraction of the lowest energy part of the electromagnetic spectrum, ranging from the microwave region up to the vacuum ultraviolet (VUV) region is already widely used for studying biomolecular systems in the gas-phase. In the case of electrons (see Section 2.3), this range corresponds to the so-called sub-thermal and thermal energy range (0–10 eV). The electromagnetic domain will be first considered in the three spectral regions corresponding respectively to quantization of the rotation energy (molecules are considered as rigid rotors), the vibrational energy (molecules are considered as sums of anharmonic oscillators) and the electronic energy. Transitions between rotational and vibrational states respectively occur in the microwave (Section 2.1.2) and infrared (IR) (Section 2.1.3) spectral regions while transitions between electronic states take place in the visible/UV and VUV (Section 2.1.4) regions (Figure 2.1.1). Different experimental methods are used to explore those regions and will be considered in the next paragraphs. Some of these methods such as IR/UV, IR/VUV or rotationally resolved spectroscopy simultaneously involve different spectral ranges.

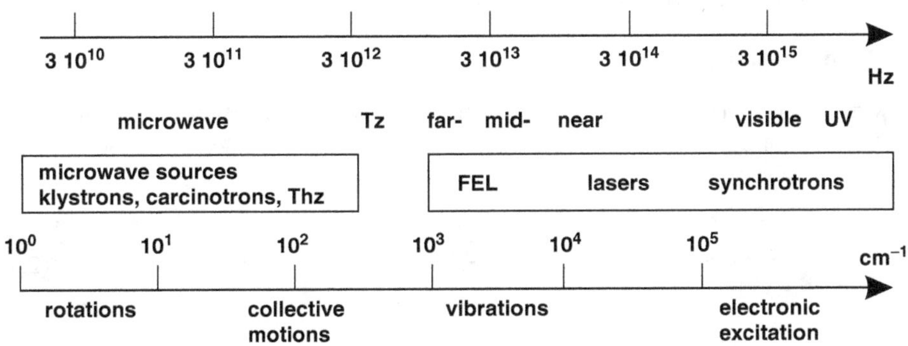

Figure 2.1.1 Electromagnetic spectrum, tunable sources, transition frequencies and energies.

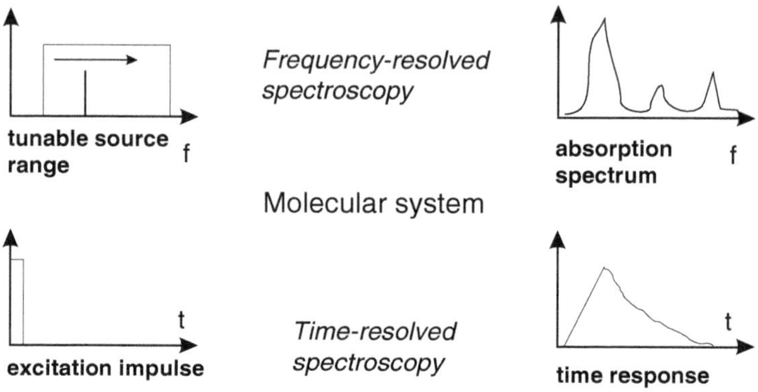

Figure 2.1.2 A molecular system can be studied by means of spectroscopy either by recording its absorption or emission spectrum as a function of the frequency of a tunable source or by monitoring its time-response to an impulse of very brief duration with respect to the nuclear motion time-constants.

Spectroscopic studies can be conducted in the frequency domain (Section 2.1) or in the time domain (Section 2.2). In both cases, it is also possible to combine electron and photon spectroscopy, for example in frequency-resolved IR-Rydberg electron transfer (RET) spectroscopy (Section 2.3.3.2) or time-resolved femtosecond photoelectron spectroscopy (Section 2.3.3.4). However, molecular systems are not linear in the mathematical sense and information obtained in the time domain is fortunately not the simple Fourier transform of information obtained in the frequency domain. Frequency-resolved spectroscopic methods provide a wealth of very accurate information on energetics whereas time-resolved spectroscopy focuses on dynamical properties, in particular chemical events (Figure 2.1.2).

2.1 FREQUENCY-RESOLVED SPECTROSCOPY

Before considering the different frequency-resolved spectroscopic methods, we first examine problems encountered in spectroscopic studies of molecules of biological interest.

2.1.1 Experimental considerations

2.1.1.1 Sensitivity

Following absorption of a photon with frequency ν_{ab}, a molecular system at absolute temperature T is brought from an initial state a into an excited state b. The intensity of the corresponding signal is given by the following expression:

$$I \propto \nu_{ab} \frac{1}{TQ(T)} g_a \exp\left(\frac{-h\nu_{ab}}{kT}\right)\left[1-\left(\exp-\frac{h\nu_{ab}}{kT}\right)\right]|\mu_{ab}|^2 \tag{2.1.1}$$

$Q(T) = Q_{rot}Q_{vib}$ is the partition function, product of the rotational and vibrational partition functions. g_a the degeneracy of the initial state and μ_{ab} the electric dipole moment. Electronic

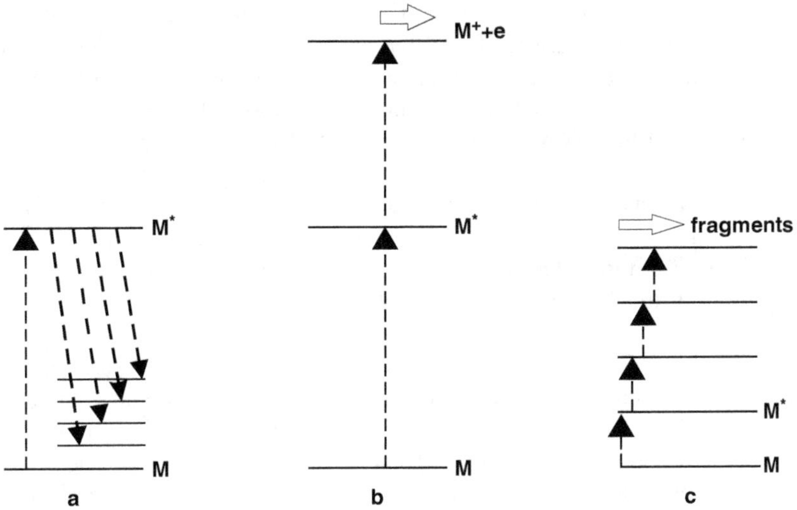

Figure 2.1.3 Examples of detection schemes used for observation of resonant interaction of molecular systems with photons. In scheme a (laser-induced fluorescence, LIF), the absorption spectrum of the molecular system M in the visible/UV range can be monitored by collecting photons reemitted from excited state M^* as a function of the excitation wavelength. Dispersing the fluorescence light with a monochromator provides the vibrational spectrum. In scheme b (REMPI), a second photon is absorbed from the intermediate excited state M^* and an ion M^+ (possibly M and an electron) is detected. This scheme provides both conformer selection and mass-spectrometry detection. In scheme c (resonant IRMPD), an ion sequentially absorbs photons with energy corresponding to resonant interaction between vibrational ground states $v = 0$ and $v = 1$. When a sufficient amount of energy is absorbed, ionic fragments are detected. This scheme does not require the existence of any visible/UV chromophore as in schemes a and b.

transitions, pure rotation transitions (permanent dipole moments) or electronic transitions in the visible or UV are more intense than vibration–rotation transitions (induced-dipole moments).

In gas-phase experiments concerning molecules of biological interest, densities of absorbing species are always extremely small as compared with those encountered in liquids or solids. Except in a few cases [1, 2], direct monitoring of absorption is usually impossible. Several detection schemes have thus been devised and some of them are presented in Figure 2.1.3. Those schemes rely on the fact that one-by-one detection of charged particles is easy. On the contrary, neutral particles must have a sufficient density for detecting their absorption or must be accelerated to be counted individually [3, 4].

Assuming a Beer–Lambert law $I(L) = I_0 \exp[-\alpha L]$ over an absorption path L, the sensitivity can be expressed by means of the minimum observable absorption coefficient α_{min} (in cm^{-1}). Detection techniques are not equally sensitive along the full spectral range and typical values of α_{min} are 10^{-7}–10^{-10} cm^{-1} in the microwave domain, 10^{-4}–10^{-5} cm^{-1} in the IR domain and 10^{-10}–10^{-11} cm^{-1} in the visible–UV domain.

2.1.1.2 Resolution and conformer selectivity

Most often, several conformers of molecules of biological interest are present even at very low experimental temperature. This then leads to the co-existence of close-lying or even overlapping spectral features. A good enough experimental resolution is thus highly desirable to allow the distinction between the different conformers that may be observable. We will see in Section 4.1.1 that even a sufficient spectral resolution may not sometimes be even enough to correctly interpret experimental data. Several effects contribute to the broadening of spectral lines. Among them, one must consider the natural broadening due to finite lifetimes of involved states, the line broadening introduced by the Doppler effect and that due to collisions. Moreover, tunable radiation sources are not perfectly monochromatic. For example, in the IR region, a pulsed optical parametric oscillator tunable over the 1,450–1,800 cm^{-1} range has a typical line-width of 1 cm^{-1} while a free electron laser (FEL) broadly tunable over the 100–2,500 cm^{-1} range has a typical line-width of 10–20 cm^{-1}.

It is somewhat difficult to compare different spectroscopic methods since each of them possesses their own capabilities and drawbacks. In the search for determination of gas-phase structures, several factors must be considered. When the experimental technique allows for mass-selection, distinction between investigated species is of course considerably easier. One must nevertheless remember that in the case of weakly bound complexes such as hydrated $M\cdots(H_2O)_N$ complexes, mass spectrometric detection of products after spectroscopic excitation can lead to ambiguities when the investigated species are likely to fragment after electronic or vibrational excitation. When several conformers are present, some spectroscopic methods such as R2PI (2.1.4) or RET (2.3.2.2) allow a separate investigation of each conformer. The problem is then unfortunately the lack of universality of those methods. The existence of a chromophore or of a large dipole moment is then required in the investigated species. Some methods are thus more "universal" than others, that is they can be applied to a much larger fraction of possible molecular systems. For example, infrared-multiphoton dissociation (IRMPD) (Section 2.1.3.2.5) can be applied to the investigation of nearly any mass-selected species but does not provide built-in isomer selectivity and cannot be used for neutrals.

2.1.1.2.1 Synergy between spectroscopy and quantum calculations

Frequency-resolved spectroscopy aims to provide structures of biomolecular building blocks through comparison between recorded spectra of isolated species and predicted spectra of the different conformers. In an ideal situation, the used experimental method provides a separate spectrum for each populated lowest energy conformer. Following an extensive exploration of the potential energy surface (PES), a set of simulated spectra is obtained together with an evaluation as reliable as possible of the relative energies of the corresponding conformers. In this way, an excellent match between simulated and observed spectra leads in principle to the totally unambiguous attribution of each experimental spectrum to a given conformer structure. One might then express some doubts about the usefulness of experiments if predictions from modelling are perfect. Unfortunately, in practice, several problems can be encountered. Experimentally, the investigated conformers are not necessarily the lowest energy ones. As it will further shown (4.1 and 5.5), it may sometimes turn out that a "reasonable" match is observed between simulated spectra of low-lying conformers and observed spectra. Although grėat cares have been taken in the interpretation, some misinterpretations can still occur and are only discovered when a different experimental method brings a more accurate vision. On the theoretical side, calculations of spectra and even more predictions of relative energies of the different conformers are not trivial. It may occur that at a given level of theory, predicted spectra of two different conformers are nearly identical or too close to provide an unambiguous attribution (see Section 2.1.3.1). A more elaborate treatment is then required. Moreover, comparisons between predicted and observed spectra can suffer from human bias since the definition of "acceptable", "good" or even "excellent" match may depend on the personal opinion of trained eyes. Reinterpretations of experimental spectra are not uncommon. One possibility to remove the human factor may then be the use of a trained neural network. A large possible number of sets of characteristic frequencies can be used as inputs and the corresponding validated structures are used as outputs for training. In the case of peptides, for example accurately measured N—H, C—H, O—H, C=O stretches and N—H bends, are used as inputs and carefully identified structures as outputs. After training, the neural network proposes a global or local (see Section 4.2.2.1) structure when a set of experimental frequencies of a new peptide is presented.

2.1.2 Microwave spectroscopy

Microwave spectroscopy has since a long time been a powerful method for precise determination of gas-phase structures [5–9] of molecules easily set into the gas-phase. With the advent of laser ablation sources (Section 3.1.2) coupled to Fourier-transform microwave spectrometers, this method is rapidly extending to a large number of thermally fragile systems with high melting points [10–16]. After describing the principles of rotational spectroscopy and examples, we will briefly describe present rotational coherence spectroscopy (RCS) and terahertz spectroscopy.

The rotational energy E_{rot} of a molecule, in absence of centrifugal distortion, can be expressed in terms of its principal moments of inertia I_{a}, I_{b} and I_{c} and the components P_{a}, P_{b} and P_{c} of the total angular momentum P:

$$E_{\text{rot}} = \frac{P_{\text{a}}^2}{2I_{\text{a}}} + \frac{P_{\text{b}}^2}{2I_{\text{b}}} + \frac{P_{\text{c}}^2}{2I_{\text{c}}}. \tag{2.1.2}$$

The moment of inertia I about any axis through the centre of gravity is defined by $I = \Sigma_i m_i r_i^2$ where m_i and r_i are the mass and distance of atom i from the considered axis. By convention, $I_{\text{c}} \geq I_{\text{b}} \geq I_{\text{a}}$. If the molecule is linear ($I_{\text{c}} = I_{\text{b}} = I$, $I_{\text{a}} = 0$), its rotational energy is simply $E_{\text{rot}} = hB_R J(J+1)$ where J is the rotational quantum number and $B_R = h/8\pi I$. The selection rule is $J \rightarrow J+1$ and absorption frequencies are $\nu_{\text{rot}} = 2B_R(J+1)$. The pure rotational spectrum of a linear molecule consists of equally spaced lines. An absorption frequency measurement provides an experimental value of I. This value can then be compared with the different possible predicted frequency values. Those values are directly related to molecular geometries deduced from structure calculations of each possible conformer of the studied species. For high values of J, a centrifugal distortion term is introduced:

$$\nu_{\text{rot}} = 2B_R(J+1) - D_J(J+1)^2. \tag{2.1.3}$$

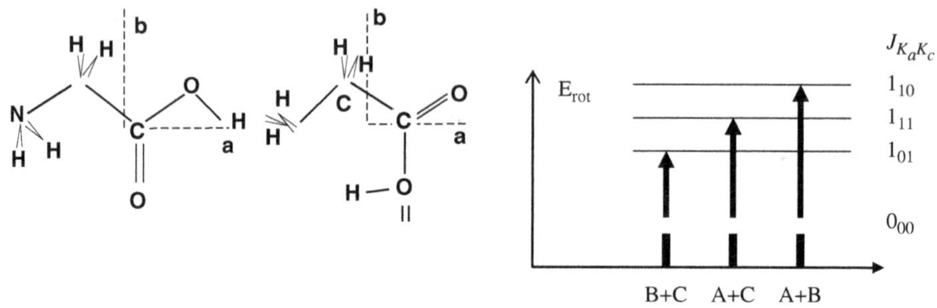

Figure 2.1.4 Left: The two lowest energy conformers of glycine and their a and b axis. Right: Schematic rotational spectrum of an asymmetric top molecule characterized by quantum numbers J, K_a and K_c.

Generally, a molecule of biological interest does not possess any symmetry axis (Figure 2.1.4) and the following three rotational constants A, B and C are considered:

$$A = \frac{h}{8\pi I_a}, \; B = \frac{h}{8\pi I_b}, \; C = \frac{h}{8\pi I_c}. \tag{2.1.4}$$

One must also consider the projections of the angular momentum along the three axis corresponding to quantum numbers K.

The gas-phase rotational spectroscopy of molecules of biological interest takes advantage of the extremely large accuracy of frequency measurements. It can be performed in the 20–200 GHz range [6, 17, 18–20]. Spectrometers using Fourier-transform techniques and supersonic expansion cooling of the studied species allow discrimination between conformers [11, 16, 21, 22]. Fourier-transform methods (Figure 2.1.5) consist in excitation of dipole moments by means of a very brief microwave impulse which is followed by the recording of the electromagnetic field emitted by those dipoles. This time-dependent so-called "free precession" signal as in nuclear magnetic resonance (NMR) is then Fourier-transformed in the frequency domain by means of a computer program (Figure 2.1.5). In order to extract molecular parameters from the obtained spectra (Figure 2.1.6), the interpretation requires a fit of a very large number of spectral line positions and the interpretation is conducted by using expressions of the rotational energy such as, for example the semi-rigid Watson Hamiltonian [13]:

$$\begin{aligned} H_R = {} & AP_a^2 + BP_b^2 + CP_c^2 - \Delta_J P^4 - \Delta_{JK} P^2 P_a^2 - \Delta_K P_a^4 - 2\delta_J P^2 (P_b^2 - P_c^2) \\ & - \delta_K [P_a^2 (P_b^2 - P_c^2) + (P_b^2 - P_c^2) P_a^2]. \end{aligned} \tag{2.1.5}$$

In this expression, Δ_J, Δ_{JK}, Δ_K, δ_J, δ_K are the quartic centrifugal distorsion constants. In order to obtain an unequivocal best fit of the experimental data, a whole set of isotopic studies can be performed by substitution of several ^{12}C or ^{14}N atoms into ^{13}C and ^{15}N atoms. A demonstration of the accuracy of rotational spectroscopy for structure determination is given in Table 2.1.1 for the proline amino acid.

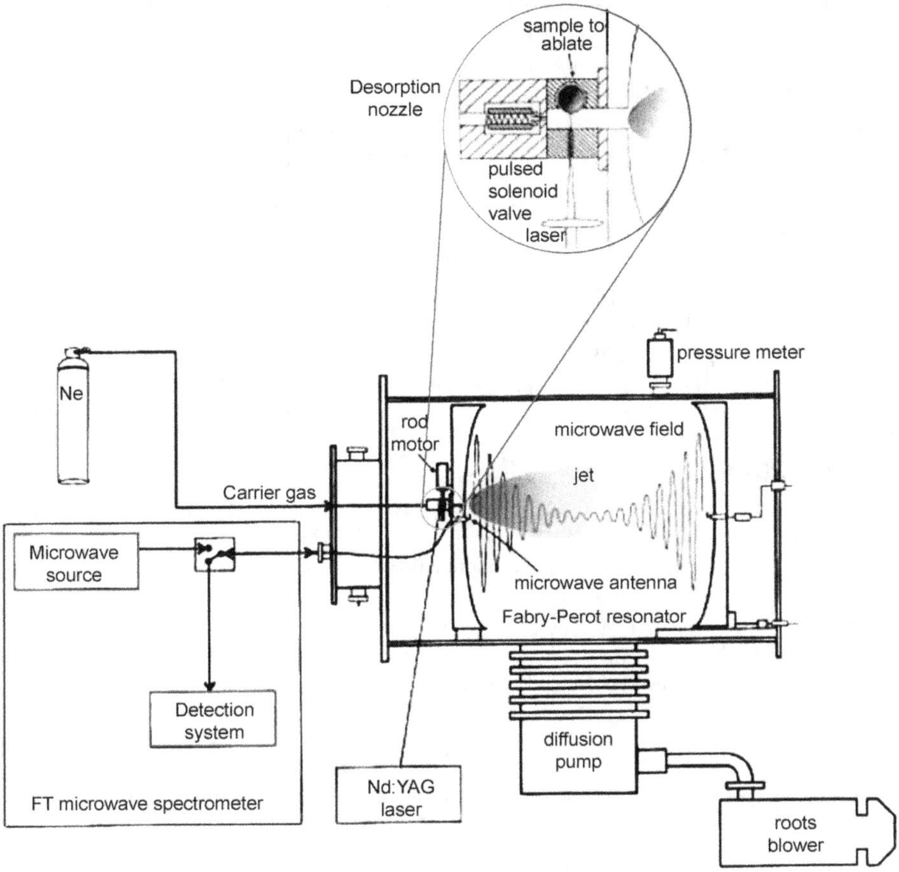

Figure 2.1.5 Schematic of a Fourier-transform microwave spectrometer [13]. The studied species are introduced by laser desorption in a supersonic expansion into a Fabry–Pérot spectrometer. A brief microwave pulse in the cavity induces a polarization of the vaporized molecules. Following that pulse, the transient emitted signal from those molecules contains the frequencies corresponding to rotational transitions that are obtained from Fourier transform of this transient signal (reproduced with permission from reference [13] ©2002 Wiley).

2.1.2.1 Rotational coherence spectroscopy

Structural information can be also obtained from rotational spectroscopy by means of RCS. RCS by itself does not belong to microwave methods but it provides results that are somewhat comparable. In RCS [23–25], a polarized picosecond light-pulse impinges on a supersonic molecular beam possessing a visible or UV chromophore. This pulse is tuned to an electronic transition of the studies species and interacts with molecules whose transition dipole moments are aligned along its polarization vector. A superposition of rotational states is created leading to a temporary alignment. This alignment rapidly dephases but due to the molecular rotation, re-phasing of the alignment takes place periodically. By measuring the time intervals between re-phasing, one obtains the rotational constants. The advantage of RCS over microwave spectroscopy is that it can be used for molecules lacking permanent dipole moment and possessing much larger masses.

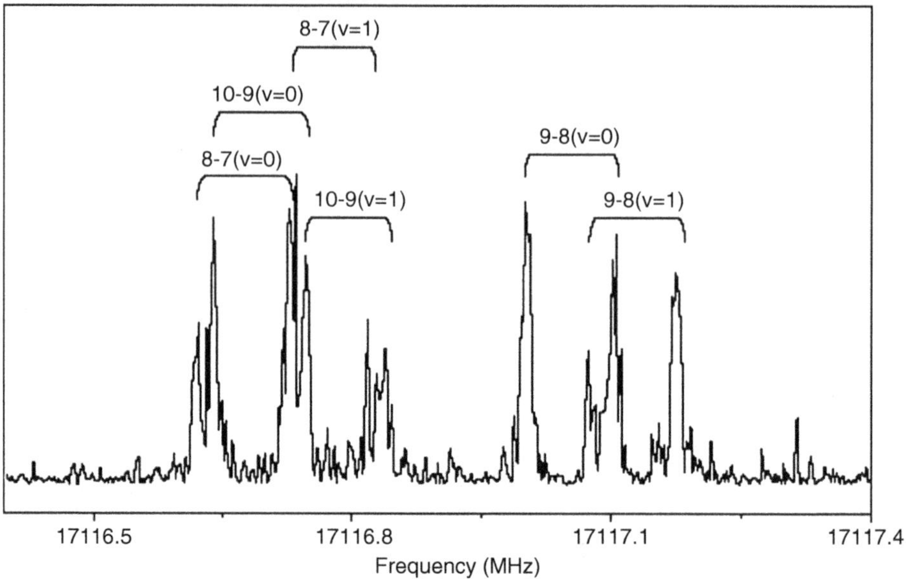

Figure 2.1.6 Fully resolved microwave spectrum of a spectral line of a conformer of *N*-phenylformamide. Note that the rotational transitions are here only within 1 MHz (reproduced with permission from reference [16] ©2006 Elsevier).

Table 2.1.1

Rotational constants and molecular parameters (bond lengths and angles) of the proline aminoacid obtained by means of a microwave Fourier-transform spectrometer

Constants	A (MHz)	B (MHz)	C (MHz)	Δ_J (kHz)	Δ_{JK} (kHz)	Δ_K (kHz)	δ_J (kHz)	δ_K (kHz)
	3673.90038	1688.42056	1407.37716	0.6341	−2.402	5.118	0.1210	0.581
Parameters	$r(C_5-N)$	$r(C-C)$	$r(C_6-O)$	$r(C_6=O)$	$<O=C_6-O$	$<C_2-C_6-O$	$N-C_2-C_6$	C_5-N-C_2
	1.451 (6) Å	1.544 (16) Å	1.340 (10) Å	1.210 (10) Å	124.9°	116.2°	111.0°	106.3°

Note: The aminoacid is produced in the gas-phase by means of laser ablation in a supersonic expansion (from reference [16]).

2.1.2.2 Terahertz and far-infrared spectroscopy

Terahertz and far-IR spectroscopy explore low-frequency motions. They correspond to the spectral region from 0.3 to 9 THz (10–300 cm^{-1}) where sources and detectors are still under development [26]. Gas-phase experiments have up to now been restricted to small molecules [27]. Systems of biological interest have mostly been studied in condensed-phase [28, 29] but extension of those investigations appear in the gas-phase.

2.1.3 Infrared spectroscopy

This chapter is a brief introduction to the main characteristics of IR spectroscopy. Full coverage of this field can be found in textbooks [30]. We first recall elementary knowledge concerning harmonic and anharmonic vibrations of a diatomic molecule. We then consider coupling of rotation and vibration as well as polyatomic molecules. Intramolecular vibrational relaxation (IVR) is described and calculations of IR spectra are considered. Experimental methods used for recording IR spectra of biomolecular systems in the gas-phase are then presented.

2.1.3.1 Vibrational spectra

The vibrational motion of a diatomic molecule A–B can be approximated to the harmonic motion of two particles of reduced mass μ connected by a spring of force constant k. The energies of the vibrational levels are given by: $E_n = (n + 1/2)h\nu$. The selection rule is $\Delta n = \pm 1$. The frequency ν is equal to $1/2\pi\sqrt{k/\mu}$. When H atoms are involved (N—H, O—H or C—H stretches), large isotopic effects are observed when hydrogen is replaced by deuterium. For $n = 0$, the energy of the lowest energy state is called the zero-point energy (ZPE). The origin of the ZPE arises from the uncertainty principle. In the lowest energy state of a quantum harmonic oscillator, the product of the uncertainty of the momentum Δp by the uncertainty of the position of nuclei Δr reaches its minimum possible value $\hbar/2$.

The potential energy curve is only close to a harmonic potential if sufficiently small oscillations are considered in the vicinity of the minimum. The potential is in fact anharmonic and is better described by a Morse potential $V(r) = D_e[1 - \exp(-\alpha(r - r_0))]^2$ where D_e is the dissociation energy and r_e the equilibrium distance (bond length) of the molecule (Figure 2.1.7). The vibrational energy can be expanded in a series of terms:

$$E_n = \left(n + \frac{1}{2}\right)h\nu - x\left(n + \frac{1}{2}\right)^2 h\nu + y\left(n + \frac{1}{2}\right)^3 h\nu \ldots \tag{2.1.6}$$

In a polyatomic molecule containing N atoms, $3N$ degrees of freedom describe the whole molecule (3 per atom). After removal of the centre of mass motion and rotation of the whole molecule, $3N-6$ vibrational degrees of freedom must be considered. In a given normal mode of vibration Q_i, all nuclei undergo harmonic motion and move in phase with different amplitudes. All nuclei have the same frequency of oscillation. The potential energy is then given by: $V(Q) = V_e + 1/2\Sigma_i\lambda_i Q_i^2$. V_e is the energy minimum usually taken as the origin ($V_e = 0$) and the λ_i are the force constants. In a first approximation, the vibrational motion of a polyatomic molecule can then be considered as the sum of $3N-6$ harmonic motions and the vibrational energy is then $E_{vib} = \Sigma_{i=1}^{3N-6}(n_i + 1/2)h\nu_i$. In the lowest energy state, $n_1 = 0$, $n_2 = 0\ldots$ and the vibrational energy is equal to the zero-point energy $\text{ZPE} = \Sigma_{i=1}^{3N-6} 1/2\, h\nu_i$. In general, normal modes of vibrations involve the movement of all atoms of a molecule but, in some cases, the movement is more or less localized on a single bond of the molecule (Figure 2.1.8).

As in a diatomic molecule, the potential energy of a polyatomic molecule is not a simple sum of harmonic energies but anharmonic terms must be considered:

$$V(Q) = V_e + \frac{1}{2}\sum_i \lambda_i Q_i^2 + \frac{1}{6}\sum_{rst} Q_r Q_s Q_t + \cdots \tag{2.1.7}$$

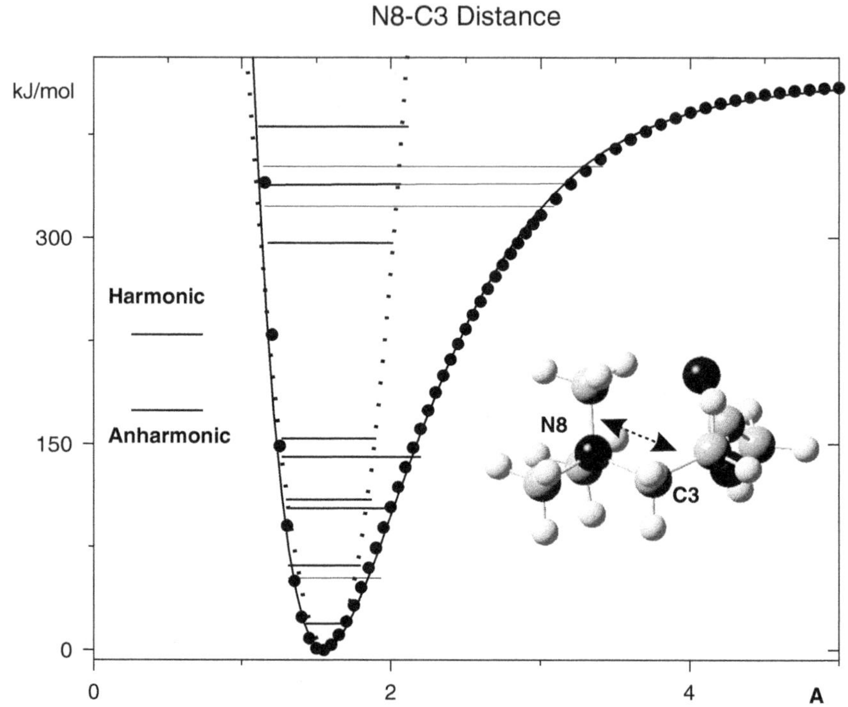

Figure 2.1.7 Morse potential (solid line) fitted to the one-dimensional potential energy curve representing the C3—N8 bond dissociation of the acetylcholine neurotransmitter (see Chapter 4.3) (distances in Å). The points are calculated at the MP2/6-31+G* level. The dotted line corresponds to the harmonic approximation. Some harmonic and anharmonic vibrational levels corresponding to the C3—N8 stretch are represented.

The vibrational energy becomes:

$$E_{\text{vib}} = \sum_{i=1}^{3N-6} \left(n_i + \frac{1}{2} \right) h\nu_i + \sum_{i<j}^{3N-6} x_{ij} \left(n_i + \frac{1}{2} \right) \left(n_j + \frac{1}{2} \right) + \cdots \tag{2.1.8}$$

The existence of the anharmonicity terms x_{ij} introduces coupling between normal modes [31]. The selection rules are no longer $\Delta n_i = 1$ but overtones ($\Delta n_i = 2$) and intercombination bands with frequencies $n_i\nu_i + n_j\nu_j$ appear. This has several consequences. From the lowest vibrational energy state $n_1 = n_2 = \cdots = n_i = 0$, the frequency of an absorbed photon is equal to one of the fundamental vibrational frequencies ν_i or one of its harmonics $n_i\nu_i$. When the internal energy of a molecule increases due to vibrational excitation, either by absorption of photons (see Section 2.1.3) or following collisions (see Section 3.2.2), the density of vibrational states (number of vibrational energy levels per frequency unit) increases extremely rapidly. The vibrational levels can no longer be considered as discrete but rather as forming a quasi-continuum. Nearly any photon energy can then be absorbed. An important consequence is that when energy is temporarily fed into a single normal mode, it is very

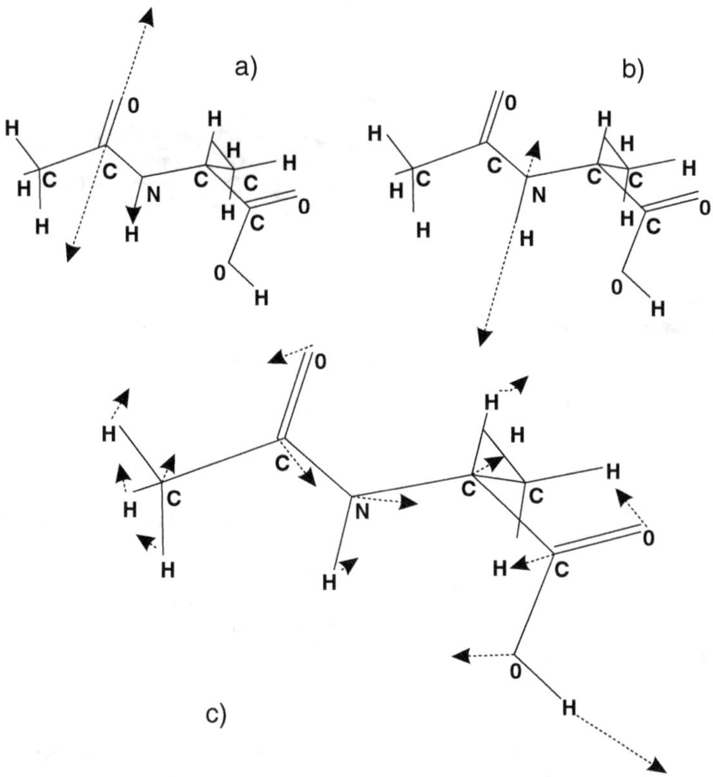

Figure 2.1.8 Vibrational normal modes of *N*-acetylalanine: (a) "local" amide I C=O stretch; (b) "local" N—H stretch; (c) "non local" amide mode.

rapidly redistributed over the different vibrational modes. This phenomenon is called intramolecular vibrational redistribution (IVR) [32–37]. It can be described by a tier model [37] (Figure 2.1.9). The energy fed into an initial rovibrational state very rapidly flows into sparse states called "doorway states", then more slowly into a dense bath of states. The IVR process is used in the so-called "spectator" or "messenger" method [38] for monitoring vibrational absorption in the gas-phase (see below).

2.1.3.1.1 Calculation of infrared spectra

Building blocks of biomolecules and their hydrated complexes present several conformers that can be identified during exploration of the PES (see Section 1.5). Since each conformer possesses a characteristic vibrational spectrum, the comparison between an observed IR spectrum and the different predicted conformer spectra can hopefully points out the presence or absence of those conformers. In fact, attribution of experimental spectra to computed conformers is not a trivial task. The number of conformers that must a priori be taken into account in those comparisons is generally deduced from energetic considerations. For example, the relative dependence of line intensities on temperature generally ascertain the determination of the lowest electronic energy state (usually denoted by X) [42]. If the expected

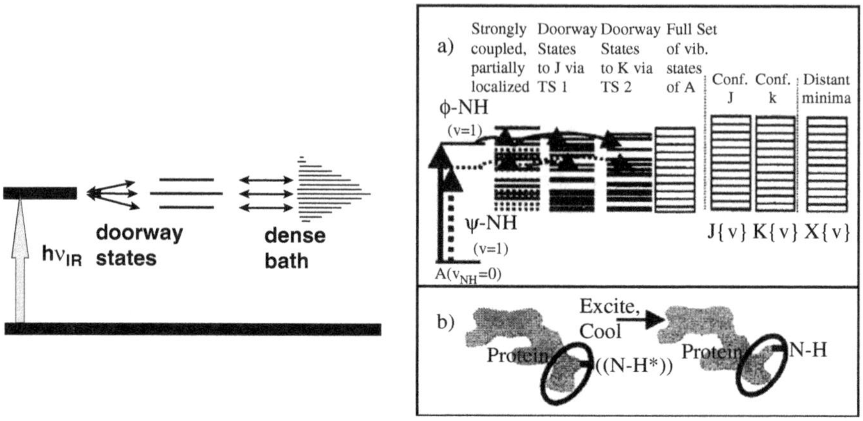

Figure 2.1.9 Left: Tier model of intramolecular vibrational redistribution (IVR). Following resonant infrared excitation of a discrete level, a rapid redistribution of internal energy occurs first through doorways states then into a dense bath of states close to a quasi-continuum in the vicinity of the energy of the initial state [39, 40]. Right: Schematic energy level diagram containing tiers of vibrational energy levels relevant to conformational isomerization of conformers of *N*-acetyl tryptophan methyl amide following excitation of N-H stretch vibrations (see Section 2.1.4.4). Arrows indicate transitions from strongly coupled, partially localized levels and doorways leading to isomerization (reproduced with permission from reference [41] ©2004 American Institute of Physics).

Boltzmann population of a conformer is too small, it is often discarded. However, as already stated above, it sometimes turns out that different conformers have similar predicted spectra, within the calculation accuracy, and misinterpretations of spectra can occur [42, 43].

The most widely used means for prediction of IR spectra of molecules and complexes are density functional theory (DFT) and ab initio calculations (see Section 1.2). These calculations are computationally expansive and are thus limited to systems containing in between 20 and 50 atoms when conducted within the harmonic approximation. Following a geometry optimization at a given level of theory (see Section 1.2), for example DFT B3LYP/6-31G* [44], B3LYP/6-31+G* [45] or MP2/6-311++G* [46, 47], the harmonic frequencies are calculated and then scaled according to recommended factors comprised in between 0.9 and 1.0. Those recommended scaling factors are available on the NIST website webbook.nist.gov/chemistry and in reference [234]. Their role is to minimize errors introduced by the neglect of anharmonicity. These scaling factors are least-square adjusted over the whole IR range. If one is only interested in comparison between frequency shifts of a given vibrational mode (e.g. N—H or O—H stretches), predictions can be improved by using a suited "local" scaling factor. A scaling factor adjusted, for example to predict experimental free OH stretch frequencies within 5 or 10 cm^{-1} then cannot be used for H-bonded OH stretches. It is interesting to note that a satisfying level of calculations for simulating IR spectra is generally lower than for prediction of relative energies of the different conformers of a given species. For example, the relative total energies of monosaccharide conformers [48] (see Chapter 4.4) are calculated at the MP2/6-311++G**//B3LYP+G* level while vibrational harmonic frequencies are satisfactorily reproduced at the B3LYP/6-31+G* level. Examples of comparisons between relative conformer energies deduced from different levels of calculations are also given in Chapters 4 and 5. It is also interesting to note that scaling factors are required for harmonic calculations of vibrational frequencies but not for electronic properties.

For small molecular systems (typically up to di- or tri-peptides), it is also possible to take into account anharmonic couplings between normal modes and then obtain vibrational frequencies without any scaling factor [49]. Improvements can sometimes be observed for some frequencies while the situation worsen for others [50]. In eq. (2.1.7), the number of coupling elements x_{ij} between normal modes can rapidly become extremely large, out of reach of computation possibilities if a high level of theory is required. However, a fast procedure consists in calculating all those coupling terms at a low-level and then only using a restricted number of them corresponding to sizable couplings between the modes of interest [49]. A striking example of the improvement brought by this procedure is provided by the interpretation of experimental results obtained on the neurotransmitter noradrenaline (see Chapter 4.3). This molecule possesses two close-lying lowest energy conformers (AG1a and GG1a) which calculated scaled free OH-stretch frequencies are identical (respectively 3,669 and 3,668 cm^{-1}) within the harmonic approximation. This level of calculation thus does not allow attribution of observed spectral features to a specific conformer. When a fast anharmonic procedure is applied, the OH-stretch frequencies of the two conformers become respectively 3,672 and 3,656 cm^{-1} and the attribution of the experimental value of 3,669 cm^{-1} to the AG1a conformer becomes possible without ambiguity [49].

These spectra calculations are derived from the knowledge of lowest energy conformers of the considered molecular systems and provide infinitely narrow predicted lines. In order to take into account line broadening due to the existence of rotational contours and to experimental conditions (e.g. excitation laser frequency stability or collisional broadening), theoretical spectra are usually convoluted by ad hoc line shapes. This situation is satisfying when the experimental temperature is such that one can reasonably admit that the studied molecular system is still in frozen structures corresponding to conformers separated by large enough barriers. This assumption is no longer valid in gas-phase experiments conducted at room temperature or in liquids. IR absorption spectra can then be derived from a molecular dynamic calculation conducted at the experimental temperature and providing the instantaneous dipole moment $M(t)$. The IR absorption spectrum (Figure 2.1.10) is given by the Fourier transform of the autocorrelation function [51, 52]:

$$S(\omega) \propto \omega^2 \int_{-\infty}^{\infty} dt \langle M(t)M(0) \rangle e^{i\omega t}. \tag{2.1.9}$$

In the case of biomolecules containing too large number of atoms and thus still out of reach of elaborate methods considered above, force-fields (see Section 1.3) and semi-empirical methods (see Section 1.4) can be used for computing vibrational spectra but their accuracy is relatively poor. Moreover, these methods have until now been devised for interpretation of the only available condensed-phase data and thus include non-explicit corrections for crystal or solvent effects. They might thus not be always very well adapted for interpretation of gas-phase experiments. An improvement of semi-empirical methods has been devised by taking the PES obtained from the PM3 semi-empirical (see Section 1.4) and deriving harmonic frequencies ω_{PM3} of normal modes of several conformers of glycine, alanine and proline. Scaling factors λ_i for each mode Q_i have then been obtained by comparing those

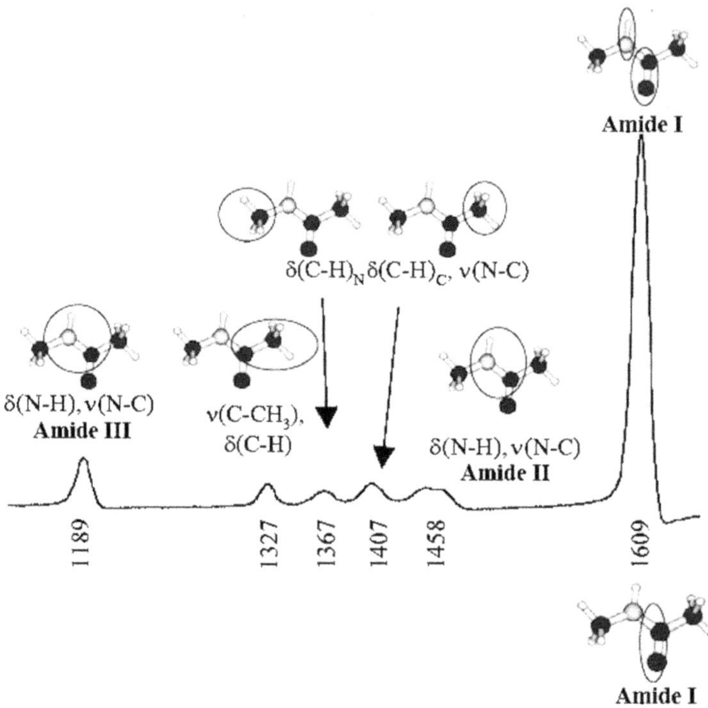

Figure 2.1.10 Infrared spectrum of isolated *trans N*-methylacetamide (NMA) calculated by means of a Car–Parinello quantum dynamical simulation conducted at 300°K (see Section 1.5.5.3) (reproduced by courtesy of MP Gaigeot).

frequencies to harmonic ab initio frequencies $\omega_{\text{ab initio}}$: $\lambda_i = \omega_{\text{ab initio},i}/\omega_{\text{PM3},i}$. From this modified PM3 potential, anharmonic frequencies can then be calculated [53, 54]. This method possesses properties of transferability and has been tested in the demanding low-frequency spectral region (500–1,600 cm^{-1}) where the role of anharmonicity is crucial in the case of guanine–cytosine complex [55] (Figure 2.1.11).

2.1.3.1.2 Hydrogen bonding and infrared spectra

Hydrogen bonding corresponds to the formation of weak D—H···A bonds (see Section 1.1.2). Most often, the attention is then focused on the three A, H and D atoms but other nearby atoms can also play an important role [56]. The donor D—H covalent bonds stretch is correlated with the strength of the H-bonds. The spectral bands corresponding to those D—H stretches most usually shift towards lower frequency (red-shifts), increase in intensity and broaden. In a few cases, blue shifts corresponding to D–H bond length contraction and increase of D—H vibrational frequency are observed [57–60]. The frequency ν (D—H···A) and strength of the D—H···A hydrogen bond (see Section 1.1.2) are correlated. Thus, after calibration within a family of related molecular systems such as peptides or proteins, the experimental measurement of the spectral red-shift of a transition corresponding to a given D—H chemical group provides direct information on the engagement of this group into

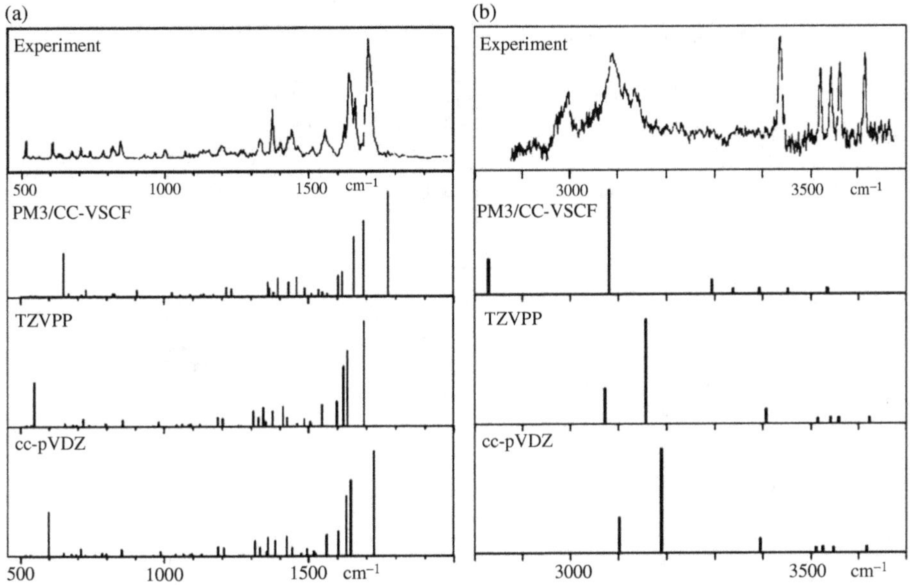

Figure 2.1.11 Comparison between the experimental mid-infrared spectrum, respectively from 500 to 1800 cm^{-1} (a) and 2800 to 3700 cm^{-1} (b), of the G···C pair, and predictions of improved PM3 calculations (PM3/CC-VCF) and ab initio calculations (reproduced with permission from reference [55] ©2005 American Chemical Society).

more or less strong hydrogen bonds. For example, the two normal modes corresponding to free N—H stretches of the amino group NH_2 of gas-phase peptides (amide A) are local modes (see Figure 2.1.9) and coupled because they share the same hydrogen atoms [61]. The splitting between symmetric (lowest frequency) and anti-symmetric (highest frequency) modes is equal to twice the coupling constant A (see Figure 2.1.13). In presence of H-bonding of one of the two N—H bonds, the splitting becomes equal to $\sqrt{\delta\nu^2 + 4A^2}$. In the case of weak H-bonding, $\delta\nu \ll A$ and the normal modes are a mixing of the local N—H stretches. They are close to the symmetric/anti-symmetric combination found in the free case in absence of H-bonding. In the case of strong H-bonding $\delta\nu \gg A$, the two modes are local, one of them being strongly red-shifted and the frequency of the other converging towards that of the free stretch local mode. A systematic study of predicted peptide configurations (see Chapter 4.2) and corresponding experimentally measured spectral shifts provides a scale of hydrogen bonding strengths and related spectral shifts (Figure 2.1.12). If an N—H group interacts with a C═O group of the same residue (C_5 contact), the H-bonding configuration is very unfavourable and the observed spectral shift is small (~30 cm^{-1}). The spectral shift increases when the configuration becomes more and more favourable for H-bonding (N—H···π, C_{10} and C_7). The most favourable configuration corresponding to a quasi-linear alignment of the N—H···O═C bonds occurs in α-helices (see Chapter 4.2).

It is interesting to compare the structural information deduced from gas-phase experiments on small size systems as above and obtained either on much larger systems in the gas-phase (see Section 2.1.3.2.6) or from experiments conducted in liquid-phase [62]

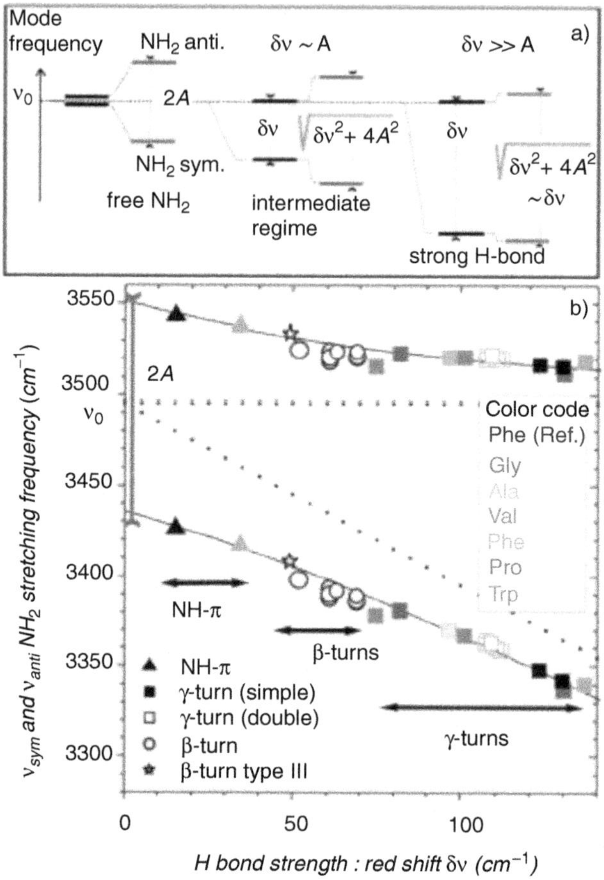

Figure 2.1.12 (a): Effect of H-bonding upon the splitting of vibrational levels of a NH_2 group. In absence of H-bonding, the two modes are separated by 2 Å. The frequency of the symmetric mode is red-shifted from $\delta\nu$ when strong H-bonding is present. (b): Evolution of the normal modes (symmetric and anti-symmetric NH stretches) frequencies of neutral di- and tri-peptides as a function of the hydrogen-bond strength deduced from experimentally observed splitting between these two modes. The different red-shift regions (NH-π, β-turns, γ-turns) correspond to the formation of different local conformations and thus different H-bonding patterns (reproduced with permission from reference [61] ©2006 Royal Society of Chemistry).

In the latter cases, the broadening of lines becomes such that only percentages of secondary structures such as α-helices or β-sheets can be obtained.

Vibrations are more efficiently coupled through H-bonds than through covalent bonds. In H-bonded chains where a hydrogen bond is formed at the neighbouring hydrogen bond site, those H-bonds mutually polarize each others leading to cooperativity. The cooperative nature of the hydrogen bond means that acting as an acceptor strengthens the hydrogen bonding ability of a molecule that also acts as a donor. Arrangements of the hydrogen bond donor (D)–acceptor (A) orientation have then an important role in the cooperativity. Cooperative effects are very important in water as demonstrated by IR studies [63] (see Section 5.6). The

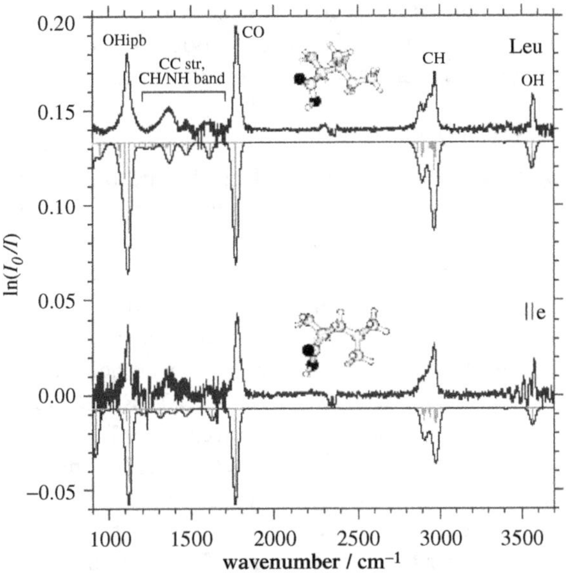

Figure 2.1.13 Fourier-transform infrared spectra of leucine (Leu) and iso-leucine (Ile) measured at 250°C. The simulated spectra underneath the experimental spectra are obtained from B3LYP/6-31G(d,p) calculations with respective scaling factors of 0.964 and 0.9546 (reproduced with permission from reference [2] ©2005 Elsevier).

O···O bond distance is equal to 2.952 Å in the neutral water dimer and decreases to 2.80 Å in the trimer, 2.79 Å in the tetramer down to 2.76 Å in the pentamer. Calculated vibrational frequencies of coupled N—H, C═O stretches (amide I), and C—N stretch/CNH bend (amide II) have been compared for covalently bound glycine molecules in β-strands and in extended chains of H-bonding formamide molecules. The three types of vibrations couple more strongly and are more red- (amide I and N—H) or blue-shifted (amide II) in the formamide chains than in polyglycine [64]. Another example of cooperativity of H-bonds will be considered in Chapter 4.4. An interesting model for describing amide I bands in peptides and proteins has been proposed and combines nearest-neighbour maps with long-range interactions. A peptide such as Leu-enkephalin, for example is then partitioned into different model floating oscillators [65].

2.1.3.2 Experimental methods used in vibrational spectroscopy of gas-phase biomolecular systems

Due to the very low available densities of gas-phase molecules of biological interest, IR spectroscopy of those systems requires highly sensitive experimental methods. A large variety of continuous or pulsed IR photon sources are now available with very different photon yields, ranges of tunability and spectral line-widths. The main problems of the different methods that are considered in the next paragraphs are (a) their range of applicability: do they require molecular systems possessing large vapour pressures, the presence of a UV chromophore or existence of a large dipole moment, etc. (b) their operating temperature that must be sufficiently low to only populate a restricted number of isomers and the spectral resolution that must allow to distinguish between the different possible isomers of a given species by means of comparison between predicted and observed IR spectra.

2.1.3.2.1 Fourier-transform spectroscopy

Fourier-transform spectroscopy allows the record of IR spectra over a wide spectral range without the need of broadly tunable sources and is routinely used as an analytical tool. It is also used for determination of biomolecular structures in condensed phases [66, 67] with the possible help of databanks of Fourier-transform spectra of proteins [68]. In contrast, gas-phase studies using Fourier-transform spectroscopy are scarce due to the difficulty for obtaining sufficiently high vapour pressures leading to a sizable absorption. Natural amino acids (Phe, Pro, Val, Leu and Ile) have been heated up to 570°K in a stainless steel cell with heated NaCl windows and the FT-IR spectra have been recorded between 1,000 and 3,800 cm^{-1} by means of a conventional FT-IR spectrometer [2]. Bare molecules and hydrogen-bonded clusters of methyl-lactate and methyl-glycolate IR spectra have been recorded in the OH and CH stretch region (2.5–3.5 μm) [1]. In this last case, the production of a large density of clusters along the IR beam path requires giant gas pulses. Those pulses last longer than 0.1 s at a repetition rate up to 0.1 Hz. The expanded gas is issued from a 15 m reservoir and flows through a 120 mm length and 0.5 mm width slit nozzle.

2.1.3.2.2 Messenger method

Since densities of biomolecular species generally do not allow for the direct observation of absorption of IR radiation, several methods, called "action methods", consist in recording the consequences of IR absorption on molecular properties of the IR-excited species. This can be achieved, for example by monitoring the IR-induced fragmentation as a function of the IR radiation frequency [69]. In order to record the IR spectrum of a molecular system AB–C, it is possible to weakly attach a messenger m (e.g. an atom, an electron or a molecule) and then study the weakly bound AB–C$\cdots m$ system. When this system absorbs photons with energy $h\nu_{B-C}$ corresponding to excitation of an intramolecular B–C vibration, the rapid redistribution of energy among all vibrational degrees of freedom (IVR, see above) is followed by the breaking of the weakest bond. If the C$\cdots m$ binding energy is less than the IR photon energy $h\nu_{B-C}$, IR absorption is monitored by observing the decrease of the AB–C$\cdots m$ signal. The presence of the messenger m should, in principle, be as less perturbative as possible. In the case of atoms, argon is generally chosen since it has a sufficient polarisability to allow attachment and its presence only induces rather small spectral shifts. The binding energy of water molecules is generally rather large. Nevertheless, the IR photo-induced loss of a water molecule has been observed in the case of hydrated glycine anions (Figure 2.1.14) following the following reaction [70]:

$$[\mathrm{Gly}\cdot(\mathrm{H_2O})_{n+1}]^- + h\nu_{\mathrm{IR}} \rightarrow [\mathrm{Gly}\cdot(\mathrm{H_2O})_{n+1}]^{*-} \rightarrow [\mathrm{Gly}\cdot(\mathrm{H_2O})_n]^- + \mathrm{H_2O}. \quad (2.1.10)$$

The excess energy brought by absorption of an IR photon can also be transferred into electronic energy. If the studied system possesses a large enough dipole and/or quadrupole

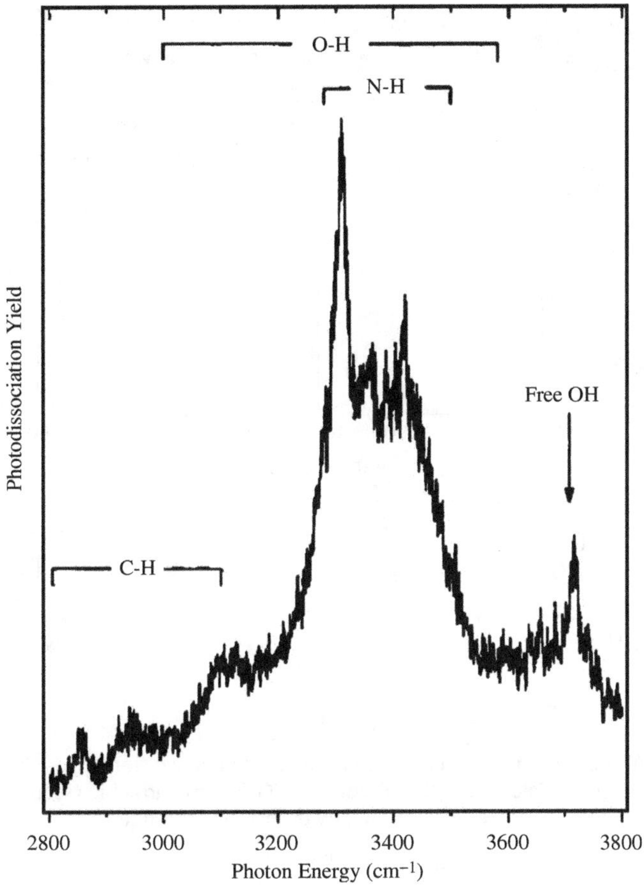

Figure 2.1.14 Vibrational spectrum of the [glycine $\cdot$ $(H_2O)_6]^-$ complex anion monitored through photodissociation of a water molecule (reproduced with permission from reference [70] ©2004 American Institute of Physics).

moment, a very weakly bound electron can be attached (see Section 2.3.2.2). It is possible to vibrationally excite the neutral parent (before electron attachment) with IR photons possessing energies larger than the electron binding energy. The vibration-to-electronic energy transfer then leads to detachment of the excess electron (acting as the "messenger") of the multipole-bound anions. The result of resonant photon absorption of the neutral parents is then a depletion of the observed anion signals [44, 50] (Figure 2.1.15).

2.1.3.2.3 Helium cluster spectroscopy

Temperatures as low as 0.37°K are reached when molecular systems are deposited on helium nanodroplets (see Section 3.3) [49]. Resonant absorption of IR photons of molecular systems imbedded in helium droplets is followed by vibrational relaxation to the surrounding medium and several hundred helium atoms evaporate. IR absorption is then

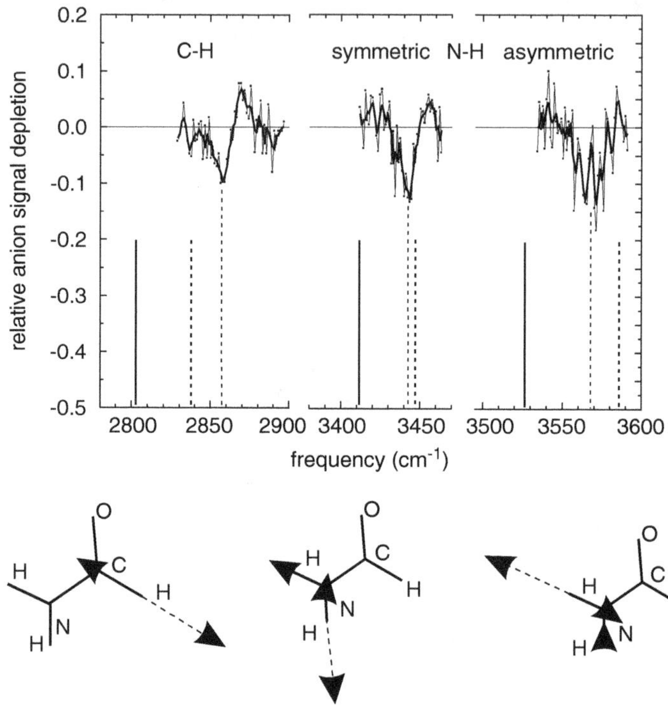

Figure 2.1.15 Infrared spectrum of jet-cooled formamide obtained from dipole-bound anion depletion (see Section 2.3.2.2). Calculated (B3LYP/6-31++G(2d,2p) harmonic (scaling factor 0.96) and anharmonic frequencies are respectively located by dashed and full lines.

monitored through the corresponding depletion of the helium flux reaching a bolometer [46]. This method is powerful since it allows for isomer selection in systems that do not necessarily include chromophores. This isomer selectivity relies on the presence of permanent dipole moments that exist in most molecules of biological interest since very low temperatures then allow their rather easy alignment in a DC field. The used IR laser beam can be linearly polarized either in the same direction as the aligned dipole moments or perpendicularly. Studied molecular systems such as nucleobases [47, 71], imidazole [72] or tryptamine [73] also possess vibrational transition moments. The angles between transition moments for a particular vibrational mode and permanent electric dipole moment are called vibrational transition moment angles (VTMA) and are characteristic of the excited vibrational modes for each isomer. VTMAs are measured by orienting the studied molecular systems in the DC field and then rotating the laser polarization direction [47, 71]. Since VTMAs can be predicted from harmonic frequency calculations, they provide useful additional information for the assignment of the various vibrational modes when several isomers are simultaneously present (Figure 2.1.16). It is worth to note that populations of the different complex configurations are imposed by either the quenching of the Boltzmann populations before deposition on the helium nanodroplets or by the growth mechanisms of the complexes on the very cold cluster surfaces [47].

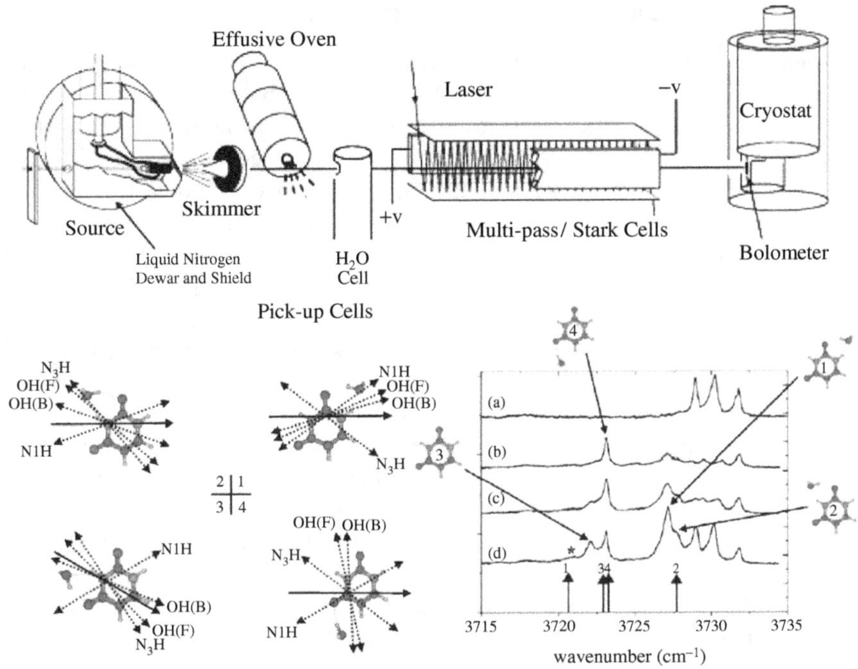

Figure 2.1.16 Top: Schematic of a helium cluster spectroscopy experiment. Bottom left: Relative directions of the permanent dipole moments (solid single headed arrows) and vibrational transition dipole moments (dashed double headed arrows) of different vibrational modes of the uracil–water complex. Right: IR spectra of neutral uracil–water complexes deposited on helium droplets. Water clusters (a); uracil–water complexes without applied DC field (c); with a DC field perpendicular (b) or parallel (d) to the DC field orienting the complexes. The vertical arrows represent the predicted frequencies and intensities (reproduced with permission from reference [72] ©2006 American Chemical Society and from reference [71] ©2005 Royal Society of Chemistry).

2.1.3.2.4 Cavity ring-down spectroscopy

In cavity ring-down spectroscopy, a pulsed-laser beam is injected into an optical resonator formed by a pair of high reflectivity ($\approx$99.9%) mirrors and bounces back and forth. Only a small fraction is transmitted through the exit mirror at each bounce and is monitored by a cooled IR detector, allowing the determination of the cavity-field decay or "ring-down lifetime". The use of a high-quality resonator optical provides a multi-path scheme equivalent to an absorption-cell of several hundred metres or more and allows very sensitive detection. In presence of molecules or complexes present inside the resonator with IR absorption lines that coincide with the introduced laser-beam frequency, the absorption brings additional losses and the ring-down lifetime is shortened. The experiment can be performed by using either pulsed-lasers [74, 75] or CW lasers ("photon trap") [76]. In the latter case, the cavity length is scanned by means of a piezoelectric ceramic and laser light can only enter the cavity under precise phase conditions. The cavity ring-down method is very sensitive and, for example the IR spectra of several uracil–water complexes produced in a 10 cm heated pulsed-jet slit have been recorded in the 2,800–3,400 cm^{-1} region [77, 78].

2.1.3.2.5 Infrared-multiphoton dissociation spectroscopy

Collision-induced dissociation (CID) and IRMPD respectively use collisions with a neutral gas or photon absorption for activation of polyatomic ions above fragmentation thresholds (see Section 3.2.2). In both cases, IVR is responsible for the breaking of the weakest bonds in molecular systems whichever intramolecular vibrational frequencies are initially excited. From the obtained similar fragmentation patterns, the sequences of polyatomic ions can be deduced [79]. The first experiments using IRMPD have been performed with fixed-wavelength IR low power CW lasers [80]. With the advent of widely tunable and powerful-pulsed IR laser, FEL [69, 81–84] and optical parametric oscillators (OPO) [85], IRMPD has been extended to resonant IRMPD. This method is becoming a nearly universal tool for the determination of polyatomic ion structures. We describe here only the main features of this method and examples of applications are given in Chapter 4.2.

In an IRMPD experiment using a tuneable source, mass-selected polyatomic ions are submitted to pulsed IR radiation. Those ions possess a set of vibrational modes with frequencies ν_i, each mode being characterized by a quantum number n_i. Although ions may be at a temperature of several hundred degrees Kelvin, only the fundamental level $n_i=0$ of each mode has a significant population. Ions are confined in a small spatial region overlapping with a tuneable IR beam. This allows long interaction times up to seconds and thus the possibility of a large number of sequential IR photon absorptions. Absorption takes place resonantly whenever the frequency of the IR beam illuminating the ions coincides with the frequency of a transition starting from the fundamental level (pump transition $n_i=0 \rightarrow n_i=1$). Absorption corresponding to overtones ($\Delta n_i > 1$) is much weaker and can be generally neglected. Anharmonicity forbids coherent multiphoton excitation $\nu_i=0 \rightarrow \nu_i=1 \rightarrow \nu_i=2 \rightarrow \cdots$ [69]. Following the initial absorption step, ions have acquired an internal energy $E_i=h\nu_i$. In between two IR excitation pulses, IVR [32], due to the anharmonic coupling between the different vibrational modes, takes place. IVR redistributes the energy E_i over other vibrational degrees of freedom in less than typically 1 ns. Ions come back to their initial fundamental level $n_i=0$ while other modes with $\nu_j \neq \nu_i$ become excited. If the experiment is performed in absence of collision (e.g. in an ICR cell) or in an amount of time shorter than the mean collision time (e.g. in a Paul trap) [69, 86], the acquired IR photon energy remains as internal energy. The absorption resonant at the pump frequency $n_i=0 \rightarrow n_i=1$ can thus take place and the ion internal energy again increases from a second IR photon energy $E_i=h\nu_i$. This process is repeated until the acquired internal energy exceeds the fragmentation energy threshold E_f. Modelling of IRMPD can be found in reference [69]. Monitoring the fragmentation yield then provides a signature of the ion absorption spectrum. However, it must be noted that the spectral dependence of the ionic fragmentation yield is not strictly equivalent to the ion IR absorption spectrum that one would record by using single IR photon absorption with the messenger method (see Section 2.1.3.2.2). Spectral lines can be shifted and broadened due to absorption taking place from combination bands. Some experimental line intensities can strongly differ from those predicted from a calculation of the IR absorption spectrum and this can be attributed to two different reasons. The IVR process taking place between two photon absorptions is more or less efficient whether a "local" mode or a mode involving motions spread over the molecular backbone is initially excited. It can also turn out that following the first photon absorption, a chemical transformation, such as a tautomerization, can occur. The spectral lines corresponding to absorption of the transformed chemical groups are then missing in the spectral dependence of the ionic fragmentation pattern [87] (Figures 2.1.17 and 2.1.18).

Resonant IRMPD experiments have been performed in FT-ICR cells and ion traps (see Section 3.2.1) with ions issued either from MALDI or electrospray sources (see Section 3.1) [86]. It is interesting to note that CID (see Section 3.2.2) and IRMPD fragmentation mass patterns are very similar due to the similar ergodic redistribution of internal energy during the ion excitation processes. CID is considerably much easier to implement but does not provide the same quantitative structural information. On the contrary, the confrontation between experimentally observed resonant IRMPD fragmentation spectra and calculated IR absorption spectra provides structural information. Resonant IRMPD spectra can be obtained for systems as large as the 28-amino acid β-amyloid peptide and its complexes with drugs. In the case of the isolated β-amyloid peptide, resonant multiphoton absorption must be used while a single photon IR absorption spectrum can be obtained for the β-amyloid peptide–drug complex using the weakly bound drug as a messenger.

2.1.3.2.6 Extension of infrared spectroscopy towards large size gas-phase biomolecular systems

While the use of IR spectroscopy for the determination of biomolecular structures in solution is well-established [88], the search for the maximum size for which valuable structural

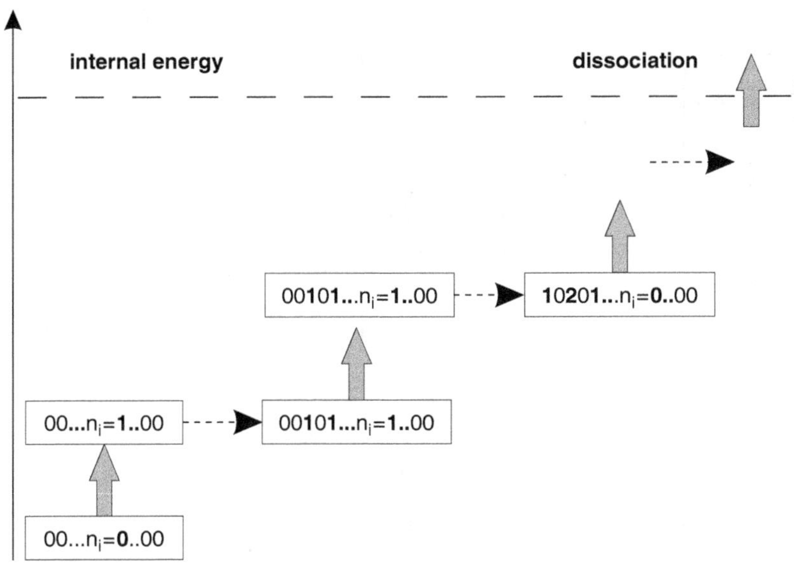

Figure 2.1.17 Principle of resonant infrared multiphoton dissociation spectroscopy. Resonant absorption of an infrared photon with frequency ν_i excites a molecular system from vibrational state $00..n_i = 0..00$ towards vibrational state $00..n_i = 1..00$ and the internal energy of the system increases from $h\nu_i$. Due to internal vibrational relaxation, this energy is redistributed among other modes and the system comes back to state $0010..1....n_i = 0..00$ from which it can again absorb a photon with frequency ν_i. Following these cycles of resonant absorption followed by energy redistribution, the system acquires a sufficient internal energy to fragment. The record of the fragment yield as a function of the IR frequency provides the IRMPD absorption spectrum.

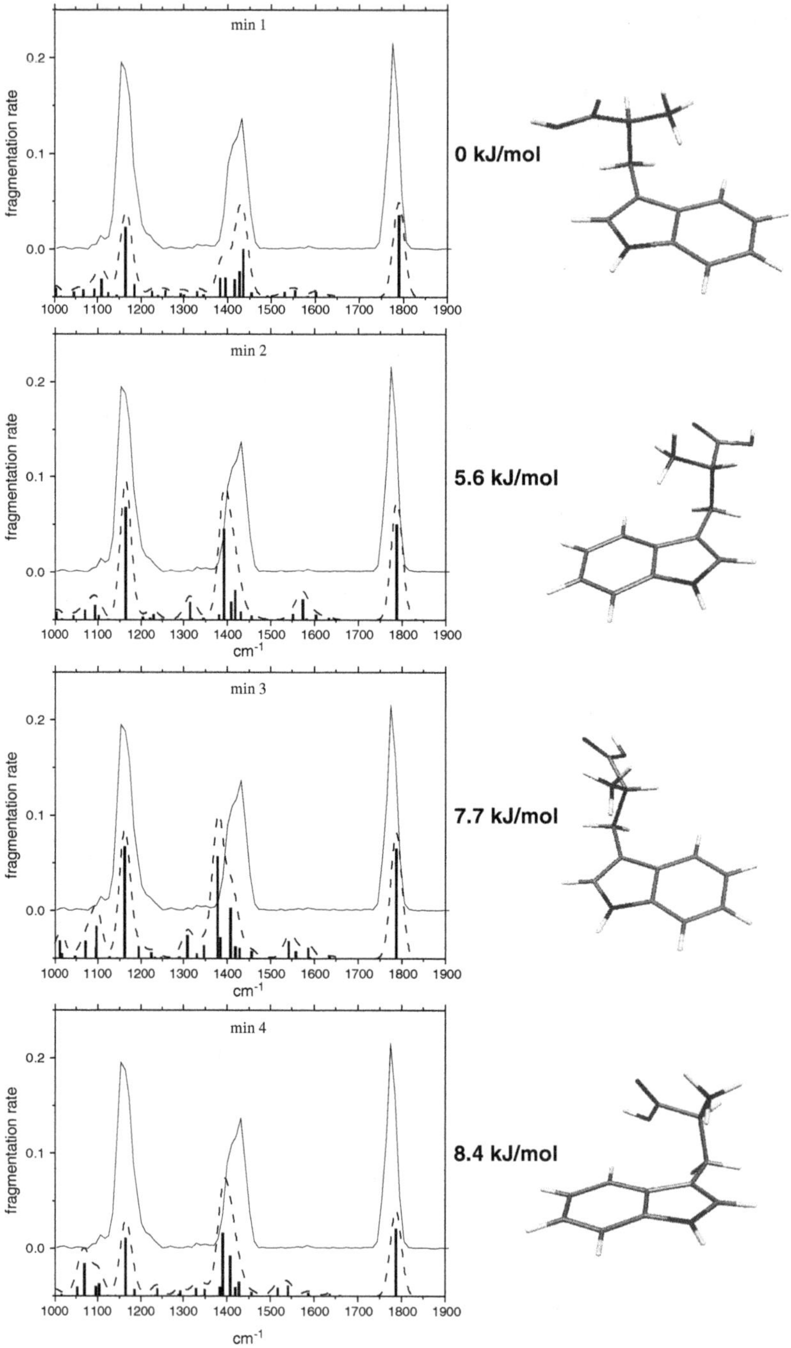

Figure 2.1.18 IRMPD spectrum of the protonated tryptophan aminoacid. Straight lines represent experimental spectra. Simulated spectra (DFT BP 86, basis set for all atoms SV(P)) (column bars) convoluted with a $30\,\mathrm{cm}^{-1}$ wide Gaussian function (dashed line) for the lowest conformers of WH^+ (left part) along with their corresponding structures (right part). Energetics are calculated at the MP2 level using aug-cc-pVDZ basis set for N and O and SV(P) for C and H (by courtesy of G. Grégoire).

information can be extracted from IR spectroscopy of gas-phase species is an active research field. Improvements in both experimental methods and theoretical interpretations rapidly push forward this size limit [89, 90]. At first sight, one would expect that the number of conformers increases tremendously with molecular size and that IR spectra should rapidly become impossible to disentangle. In fact, it turns out that even in an ensemble of 15-residue peptides such as gramicidins [90], each peptide adopts a single conformation in the gas-phase in absence of any solvent. The only difference between gramicidin peptides is a single amino acid that strongly modifies the structure. It is then possible to clearly identify a helix, a β-sheet or a random coil structure (Figures 2.1.19 and 2.1.20).

With the combination of a high-resolution ICR cell (see Section 3.2.1) and a free electron laser for IR-induced fragmentation, the mid-IR spectrum of the gas-phase cytochrome c protein has been recorded for different charge states ranging from 12^+ to 16^+(Figure 2.1.21), in absence of any solvent effect [91]. Folding properties of cytochrome in gas-phase, as a function of the electrospraying solution pH modifying the charge state, has also been followed by fluorescence [92] (see Section 2.1.4.1). Two broad spectral features, rather similar to those observed in condensed-phase, are attributed to the amide I and II modes. The amide I band near 1,660 cm^{-1} falls in the range where α-helices are encountered in proteins in solution phase [93]. This is in agreement with ion-mobility studies (see Section 3.2.3) of cytochrome c [94, 95] (see Chapter 4.8). A third feature appears around 1,480 cm^{-1} for charge states equal or larger than 13^+. This study shows that IR spectra of perfectly mass-identified species

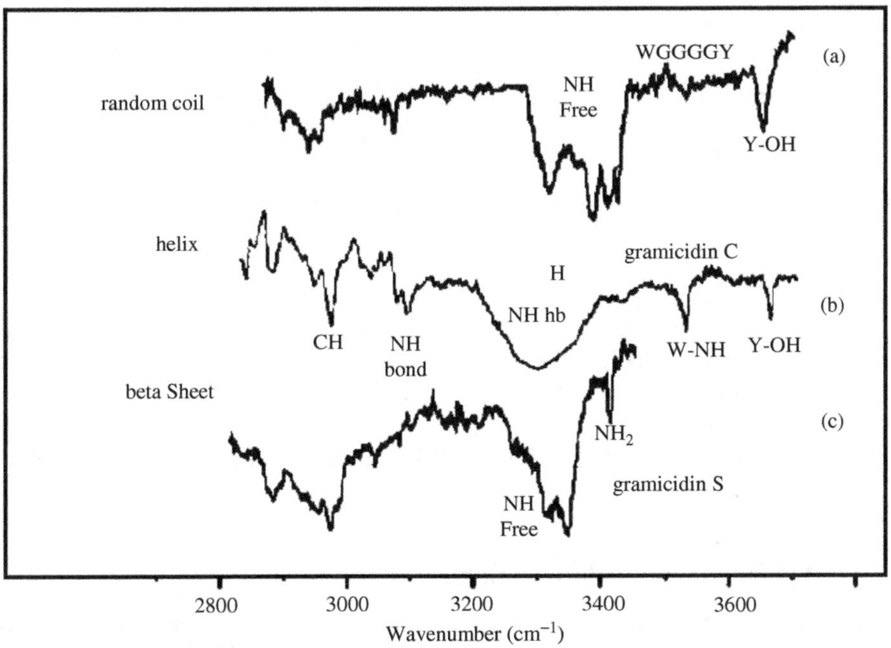

Figure 2.1.19 Infrared–UV double resonance spectrum of a hexapeptide with random coil structure (WGGGGY) and two 15-residue peptides, respectively gramicidin C with a helix structure and gramicidin S with a β-sheet structure (reproduced with permission from reference [90] ©2006 Wiley).

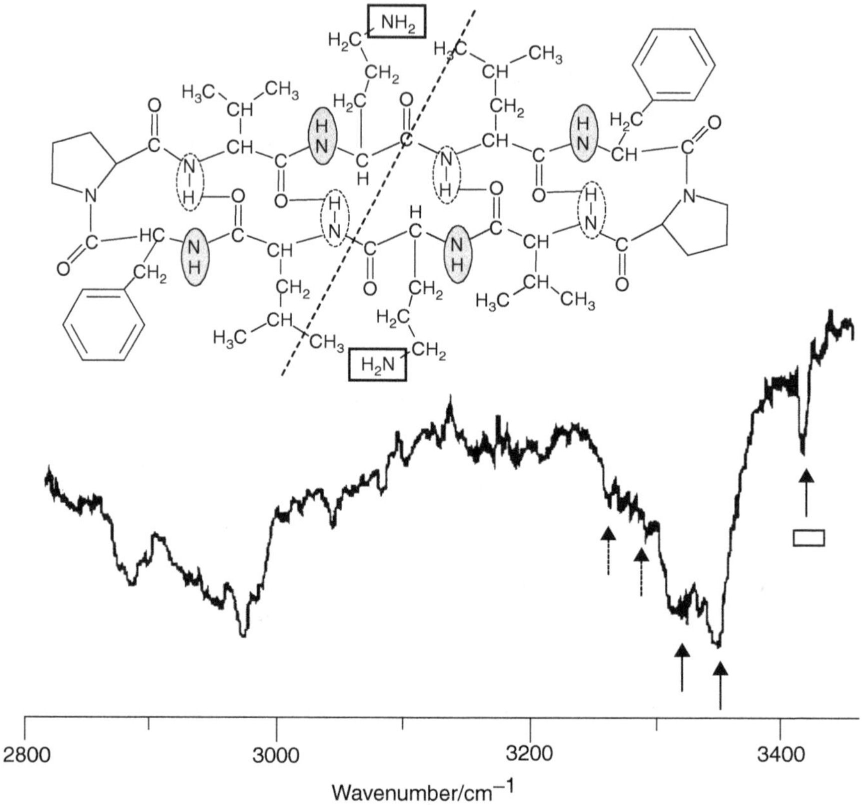

Figure 2.1.20 IR/R2PI spectrum of the laser-desorbed and jet-cooled 15-residue peptide gramicidin S [90]. The solid grey arrows indicate the spectral region of free or weakly H-bonded N—H groups, dotted black arrows correspond to H-bonded N—H groups and the rectangle marks the region of free NH_2 groups (reproduced from reference [90] ©2006 Wiley).

can be obtained and opens the route to studies of such large biomolecular systems at very low temperatures where hopefully better resolved spectra might be obtained [96, 97].

The IR spectra of microbial cells contain the contribution of all present biomolecules. The major contributions from the cell envelopes come from polysaccharides and lipids. In the cytoplasm, especially in rapidly dividing cells, spectral features belonging to DNA and RNA constitute major components [98] (Figure 2.1.22).

2.1.4 Visible and ultraviolet spectroscopy

In this chapter, we now consider spectroscopic studies conducted on molecular systems possessing a chromophore absorbing in the visible or near UV region. Nucleobases (see Chapter 4.1), three amino acids (tryptophan, tyrosine and phenylalanine, see Chapter 4.2) and some neurotransmitters (see Chapter 4.3) belong to this group. When no chromophore is naturally present, as, for example in sugars, the covalent addition of a chromophore such as benzyl or

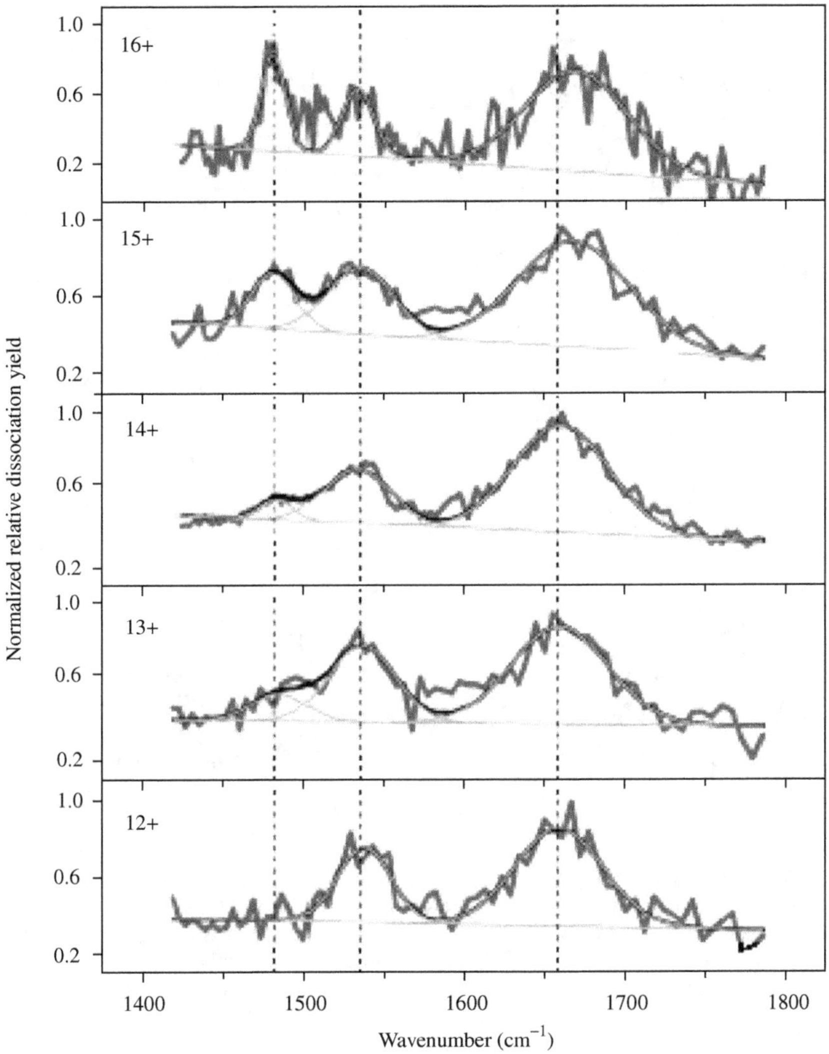

Figure 2.1.21 Infrared photodissociation spectra of different charge states of bovine cytochrome c recorded in an ICR cell with a free electron laser. The bands around 1,660 and 1,535 cm^{-1} are respectively the amide I and amide II bands (reproduced with permission from reference [91] ©2005 Royal Society of Chemistry).

phenyl group is possible and allows the use of the different methods developed around resonant absorption and emission of visible/UV photons [99, 100]. Reviews can be found in references [101, 102]. We will here first consider resonant absorption from the ground state S_0 to first excited states followed by either laser-induced fluorescence (LIF) back to S_0 or ionization through resonant two-photon absorption (R2PI). Non-radiative processes leading to transfer to a triplet state, back to S_0 or chemical reaction and followed by means of time-resolved pump–probe methods will be further examined in Section 2.2 (Figure 2.1.23).

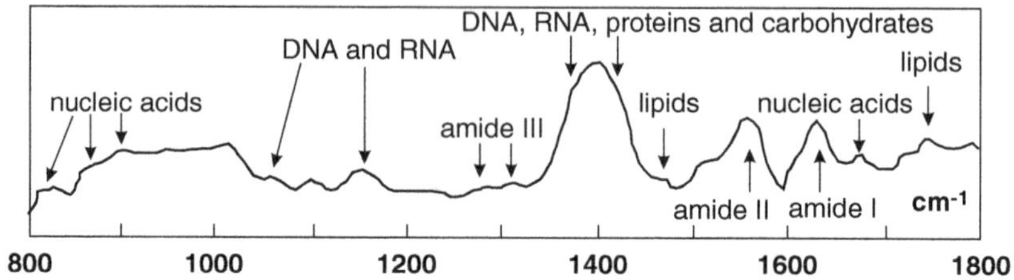

Figure 2.1.22 Infrared spectral features of a cytoplasmic extract of *Escherichia coli* cell.

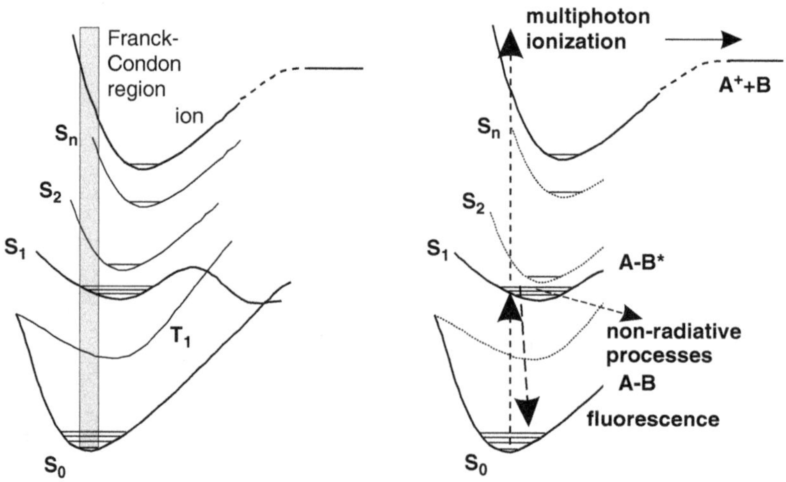

Figure 2.1.23 Left: Schematic potential energy diagram of electronic states. The photo-excited S_1 state can be coupled to the triplet state T_1 (intersystem conversion) and to the ground state S_0 (internal conversion). A reaction can also take place Right: Simplified potential energy diagram showing resonant absorption from the ground state S_0 to the first excited state S_1 followed either by fluorescence back to S_0 (LIF), non-radiative processes or ionization due to absorption of a second photon (R2PI).

2.1.4.1 Frequency-resolved visible/ultraviolet spectroscopy

2.1.4.1.1 Laser-induced fluorescence

The fate of laser-excited isolated species can be observed by monitoring their fluorescence as a function of the excitation wavelength. When those species are cooled either in a supersonic expansion [101–104] or through collisions with a buffer gas in the case of confined ions [105], more or less sharp and well-resolved spectral lines can be observed. We must note that fluorescence is widely used in living cell studies. The green fluorescence protein (GFP) and related species will be examined in Section 4.2.7. Fluorescence spectroscopy of single biomolecules in thin films or immobilized on surfaces is a powerful method allowing the monitoring of their interactions with their surrounding [106]. Multiphoton excitation of fluorescence is used in condensed-phase to eliminate the strong resonant single-photon absorption along the exciting laser path. The use of several laser crossing beams, each bringing a fraction of the required photons then allows the localization of the investigate species [107].

The fluorescence spectrum emitted from a selectively populated rovibronic level (v'_k, J'_k) consists of all allowed transitions to lower levels (v''_m, J''_m). The wave number differences of these fluorescence lines immediately yield the term differences of those terminating levels (v''_m, J''_m). With a mean lifetime $\tau_k = 1\Sigma_m A_{km}$, the excited molecules undergo spontaneous transition to lower levels $E_m(v''_m, J''_m)$. At a population density $N_k(v'_k, J'_k)$, the radiation power of a fluorescence line with frequency $\nu_{km} = (E_k - E_m)/h$ is given by:

$$P_{km} \propto \frac{N_k A_{km}(E_k - E_m)}{h}. \tag{2.1.11}$$

The spontaneous transition probability A_{km} is proportional to the square of the matrix element:

$$A_{km} \propto \left| \int \psi_k^* r \psi_m d\tau_{\text{nucl}} \, d\tau_{\text{el}} \right|^2, \tag{2.1.12}$$

where **r** is the vector of the excited electron and the integration extends over all nuclear and electronic coordinates. Within the Born–Oppenheimer approximation (see Section 1.2.1), the total wave function can be separated into a product $\psi = \psi_{\text{el}}\psi_{\text{vib}}\psi_{\text{rot}}$ of electronic, vibrational and rotational factors. If the electronic transition moment does not critically depend on the internuclear separation R, the total transition probability is then proportional to the product of three factors:

$$A_{km} \propto \left| \int \psi_{\text{el}}^* r \psi_{\text{el}} \, d\tau_{\text{el}} \right|^2 \left| \int \psi'_{\text{vib}} \psi''_{\text{evibl}} \, d\tau_{\text{vib}} \right|^2 \left| \int \psi'^*_{\text{rot}} \psi''_{\text{rot}} \, g_i d\tau_{\text{rot}} \right|^2, \tag{2.1.13}$$

where the first integral represents the electronic matrix element which depends on the coupling of the two electronic states. The second integral is the Franck–Condon factor which depends on the overlap of the vibrational wave function ψ_{vib} in the upper and lower state. The third integral depends on the orientation of the molecule relative to the electric vector of the incident wave. If a single upper level (v'_k, J'_k) has been selectively excited, each vibrational band $v'_k \to v''_m$ consists of at most three lines: P, Q and R ($J' - J'' = 0, \pm 1$).

This oversimplified analysis must of course be refined for molecules containing a rather large of atoms such as neurotransmitters but a full analysis of rovibrational spectra can still be conducted [108–110]. Molecular parameters such as rotational constants of the different conformers are then determined with help of a computer-aided fitting procedure that can include, for example the use of a genetic algorithm (see Section 1.5.4) [110, 111].

Experimentally, a fraction as large as possible of the whole fluorescence yield can be collected and recorded as function of the excitation laser wavelength. The intense sharp lines are attributed to the origin 0–0 transitions. The distance from these origins are vibrational progressions each corresponding to excitation of vibrational levels of different conformers. It is also possible to fix the excitation wavelength on each origin transition and then disperse the fluorescence arising from each conformer [104, 112, 113]. This method dubbed single vibronic level fluorescence (SVLF) provides the vibrational bands of the ground state that are then identified through comparison with calculations of the vibrational spectrum [114, 115]. The example of the chiral *cis* 1-amino-indan-2-ol molecule, a key component of the *indinavir* drug that acts as an inhibitor of the HIV protease, is presented in Figure 2.1.24.

LIF of molecules can also be used for large systems. For example, the LIF method has been applied to electrosprayed cytochrome c ions [92]. When ions are produced from a solution at neutral pH, no fluorescence is observed while dispersed fluorescence of tryptophan can be monitored in a spray obtained from a low pH solution. Moreover, this fluorescence increases with the distance from the spay needle. In solution and at the exit of spray needle, cytochrome c has a folded and compact structure (see Figure 3.3.1) and the tryptophan fluorescence is quenched by its environment. At low pH, Coulomb repulsion between protons of the highly charged cytochrome c is favoured by the decrease of the dielectric constant

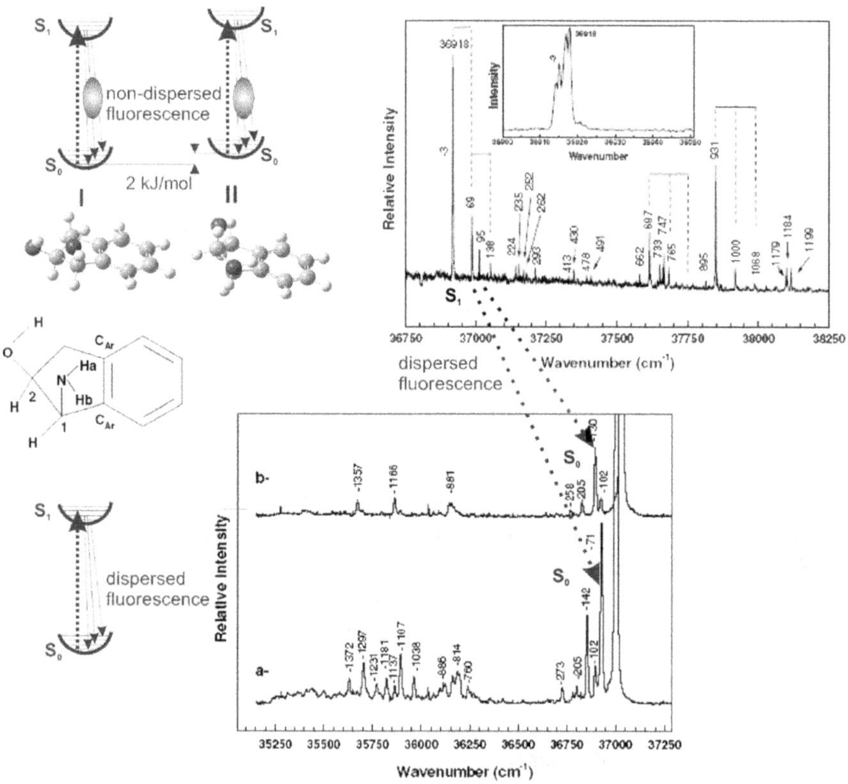

Figure 2.1.24 Top left: Lowest energy conformers of *cis* 1-aminò-indan-2-ol. The non-dispersed fluorescence spectrum of *cis* 1-amino-indan-2-ol is represented as a function of the excitation laser energy. The inset presents an expanded view of the origin 0–0 R (red-shifted) and B (blue-shifted) transitions separated from $3\,\mathrm{cm}^{-1}$ that indicate the existence of the two conformers I and II. Bottom: Dispersed emission SVLF spectra of *cis* 1-amino-indan-2-ol following excitation of (a) $0-0+69\,\mathrm{cm}^{-1}$ of or (b) the $0-0+95\,\mathrm{cm}^{-1}$ bands of the excited S_1 state. Those bands respectively correspond to the bands of the ground S_0 state observed at $-71\,\mathrm{cm}^{-1}$(a) and $-130\,\mathrm{cm}^{-1}$ (b) from the intense $0-0$ bands. Bands R and B are respectively assigned to isomers I and II (reproduced with permission from reference [115] ©2006 Elsevier).

(see Section 1.3). The protein unfolds along the course of the spray relieving tryptophan which can then fluoresce.

2.1.4.1.2 Rotationally resolved visible and ultraviolet fluorescence spectroscopy

Rotational contours of fluorescence spectra corresponding to the third term of expression 2.1.13 can be resolved with the use of high spectral purity sources such as CW or seeded pulsed dye lasers [108–110, 116, 117] (in contrast with the generally used pulsed lasers). The method presents similarities with microwave spectroscopy (see Section 2.1.2) but the presence of a visible/UV chromophore is required. The main advantages are conformer selection and possible determination of both the ground and excited state geometries. Figure 2.1.25 illustrates the case of the exploration of the conformational space of tryptamine. Among the 27 possible conformers, six of them are identified for the bare molecule on the basis of the respective rotational constants eq. (2.4) of several isotopomers calculated at the MP2/6-311G(d,p) level. In contrast, a single conformer of the monohydrated complex is observed. There are only small differences between rotational constants of the conformers. For example, the A constants eq. (2.4) vary from 1594.16 to 1767.74 MHz for the ground state and from 1592.77 to 1783.8 MHz for the excited state. An unambiguous assignment of conformers then requires the accuracy of nearly fully resolved vibration–rotation spectra [110].

2.1.4.1.3 IR–LIF depletion fluorescence spectroscopy

Once the presence of different conformers has been identified in a LIF spectrum, it is possible to tune the UV excitation laser on the different UV absorption lines corresponding to suspected different isomers. An IR laser beam then illuminates the scrutinized species and its frequency is scanned. Any time this IR laser frequency resonantly coincides with a vibrational frequency of one of the populated isomers, the corresponding IR absorption brings a sizable fraction of the ground state population of the considered isomer up to the vibrationally excited state (Figure 2.1.26). This depletion of the ground state population then appears as a dip in the electronic state fluorescence.

2.1.4.1.4 Time-resolved fluorescence

Once a molecular system has been photo-excited, radiative and non-radiative processes enter into competition. The lifetime τ of an excited state thus depends on the radiative lifetime τ_R (spontaneous emission) and the non-radiative lifetime τ_{NR}. If $\tau_{NR} \ll \tau_R$, fluorescence becomes unobservable. The competition is strongly dependent on the relative positions and couplings of the photo-excited state and its nearby excited states that correspond to potential energy curve crossings (Figures 2.1.23 and 2.1.27). After tuning a UV excitation laser on absorption lines of the different isomers, it is possible to monitor the fluorescence decay and deduce excited state lifetimes leading to information concerning couplings between excited states.

2.1.4.1.5 Fluorescence resonant energy transfer (FRET)

Chromophores can be chemically attached to biomolecules and a resonant radiative transfer can be observed from a photo-excited donor chromophore A^* and an acceptor chromophore B. This energy transfer is strongly dependent on the distance and one can thus

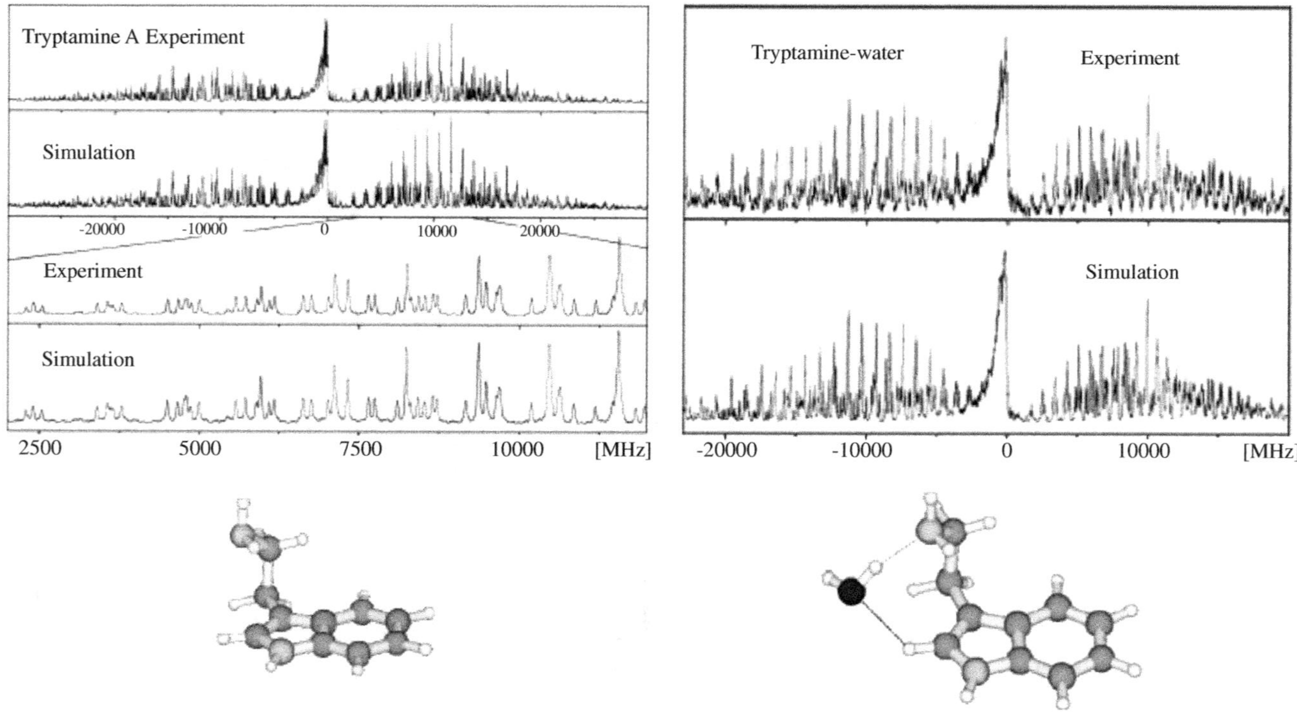

Figure 2.1.25 Left: Rotationally resolved fluorescence spectrum of the lowest energy conformer of tryptamine. The upper and lower traces respectively present the experimental and simulated spectra. An enlarged portion is shown below. Right: Rotationally resolved fluorescence spectrum of the lowest energy conformer of the monohydrated tryptamine complex (reproduced with permission from reference [110] ©2005 American Chemical Society).

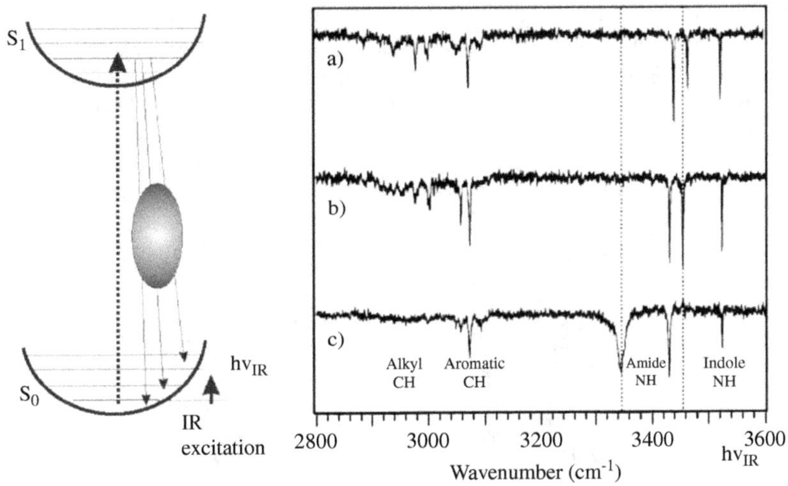

Figure 2.1.26 Fluorescence-dip spectra of three conformers of *N*-acetyl tryptophan methyl amide (NATMA) (see Section 2.1.4.4) (reproduced with permission from reference [102] ©2005 American Chemical Society).

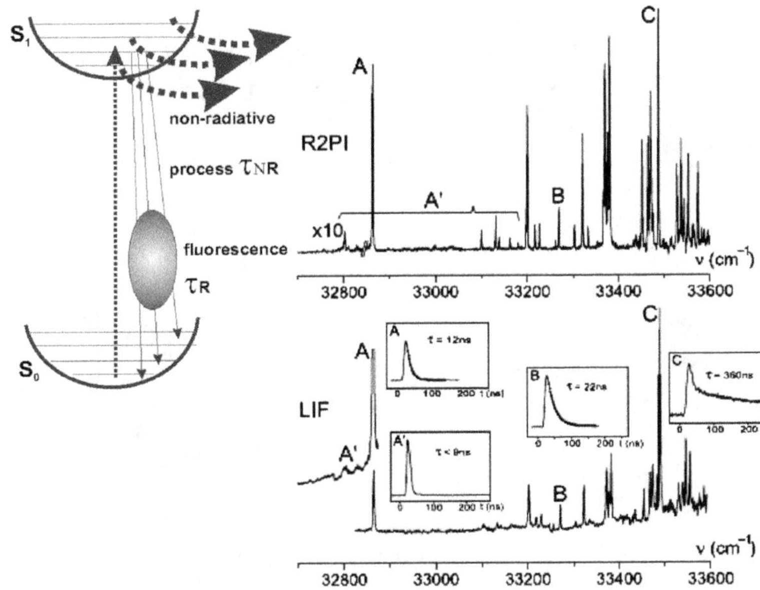

Figure 2.1.27 Left: For a given isomer, different vibrational levels can have different lifetimes according to the crossings between the photo-excited state and other states leading to non-radiative processes. The crossings can also vary according to the isomer. Right: R2PI (see Section 2.1.4.2) and LIF spectra of laser-desorbed jet-cooled guanine. Peaks A, B and C correspond to origin bands of $\pi\pi^*$ transitions of three conformers and peak A′ corresponds to the $n\pi^*$ transition of A. The insets represent fluorescence decay curves and derived lifetimes of A, A′, B and C conformers (reproduced with permission from reference [102] ©2001 Elsevier).

follow structural changes in real time through observation of the FRET process. FRET is widely used in solution studies [118–120] and has also been observed with biomolecular ions obtained either from electrospray (see Section 3.1.6) and confined in quadrupole trap (Figure 2.1.28) or from MALDI [121–123] (see Section 3.1.5) and mass-selected in an ICR cell [124].

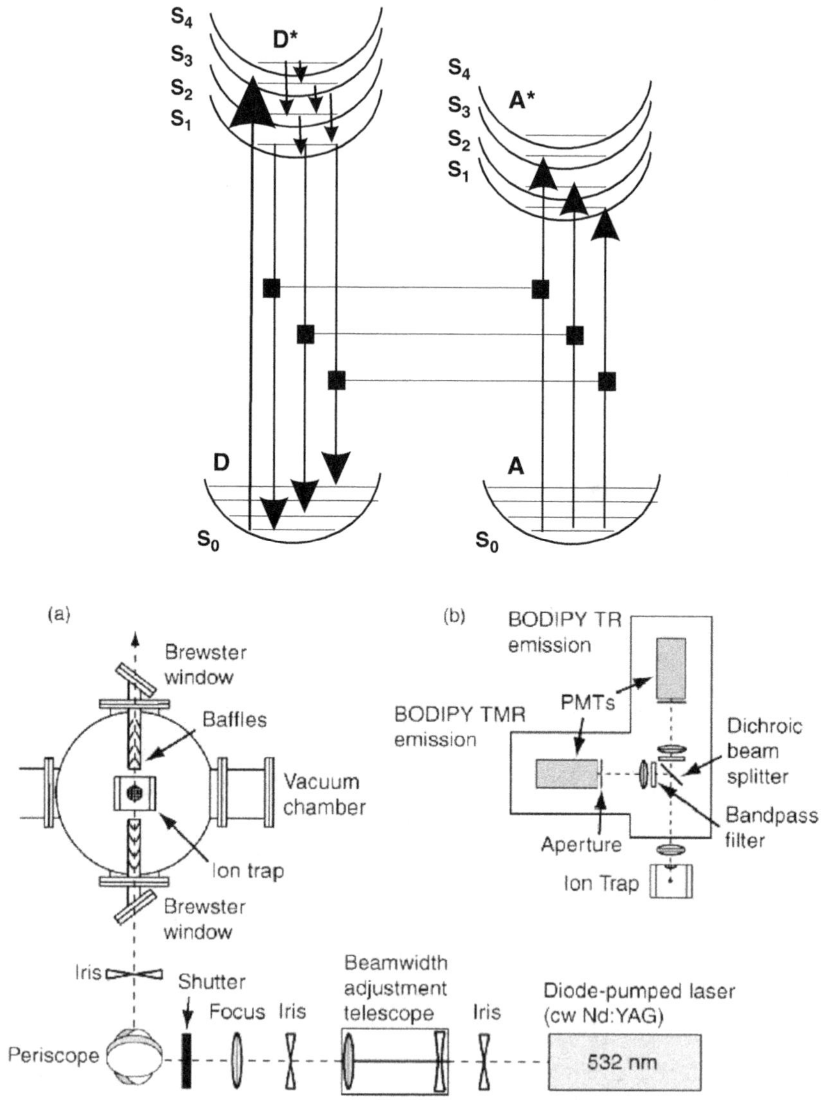

Figure 2.1.28 Left: Following laser excitation, the donor fluoresces. Fluorescence bands of the donor match absorption bands of the acceptor. Right: Schematic of the set up used for dynamical FRET studies of polypeptide ions derivatized with BODIPY dyes and stored in a trap (reproduced with permission from reference [122] ©2006 Elsevier).

The donor is photo-excited to one of its electronic state S_n that non-radiatively decays within picoseconds to the first excited state S_1. This state then radiates with typical fluorescence lifetimes of nanoseconds (Figure 2.1.28). The (A, D) pair is chosen such as the fluorescence spectrum $f_D(\lambda)$ of D and the absorption spectrum $a_A(\lambda)$ of A overlap (Figure 2.1.29). The overlap is determined by the integral is $J = \int_0^\infty f_D(\lambda) a_a(\lambda) \lambda^4 \, d\lambda$. These fluorescence pairs can be dyes such as tetramethylrhodamine and Texas red [123] or fluorescent proteins (see Section 4.2.7) expressed in biological cells. The efficiency E of the FRET process is measured by monitoring the fluorescence quantum yield of the donor I_{da} in presence of the acceptor and I_d in absence of the acceptor: $E = 1 - I_{da}/I_d$.

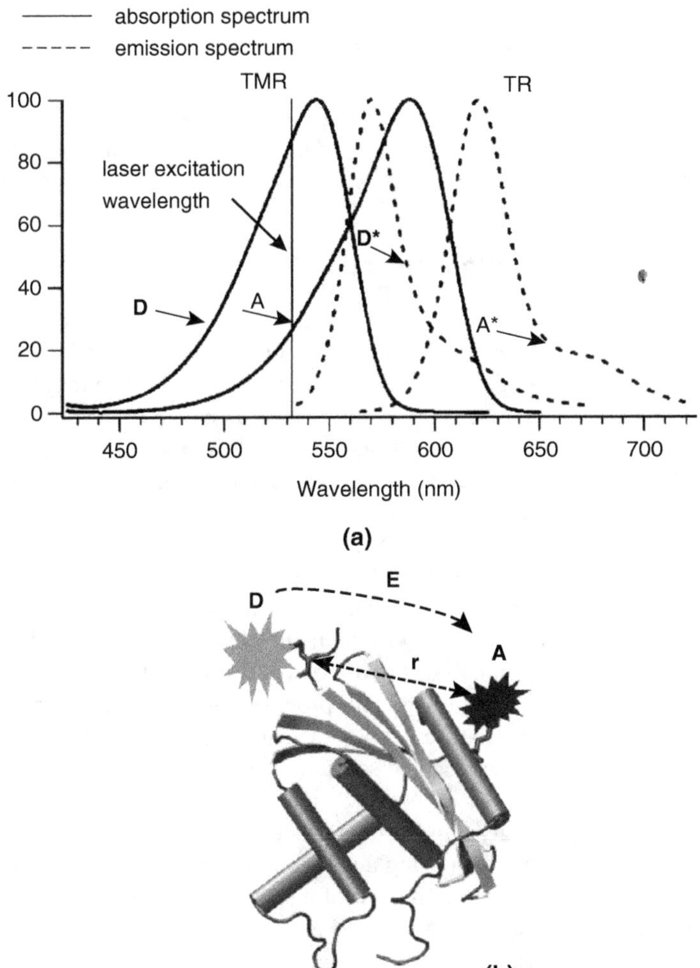

Figure 2.1.29 (a) Absorption and emission spectra of fluorophores used in FRET studies of trapped gas-phase oligonucleotide complexes (see Chapter 4.1) (reprinted with permission from reference [126] ©2003 Elsevier). (b) Ribonuclease (RNase H) enzyme with a FRET dye pair attached at residues 3 and 135 (reprinted with permission from reference [118] ©2005 National Academy of Sciences).

If the distance between the donor and acceptor probes is r and if a fast orientational averaging of the donor and acceptor radiative dipoles (the dipole–dipole interaction is taken to be that of isotropically oriented dipoles) occurs, the FRET efficiency is given by the Förster expression [118]:

$$E = \frac{1}{1+(r/R_0)^6}, \tag{2.1.14}$$

where R_0 is the Förster distance numerically given by the following expression in vacuum:

$$R_0 = 9.78 10^3 \sqrt[6]{I_d J} \text{Å}. \tag{2.1.15}$$

A typical value of the Förster distance is that of the (tryptophan-*N*-BODIPY [(4,4-difluoro-5,7-dimethyl-4-bora-3a,4a-diaza-*s*-indacene-3-yl)methyl iodoacetamide]) pair equal to 26.5 Å [125]. The simple Förster model can be refined by using TD-DFT calculations (see Section 1.7).

2.1.4.2 Resonant multiphoton ionization

Once a molecular system possessing a chromophore is excited into one of its first electronic states, we have seen that different pathways can be open (Figure 2.1.23). According to the branching ratios between those pathways, excited state lifetimes span a range from nanoseconds down to hundreds of femtoseconds. From one of its excited states with a sufficiently long lifetime, it is possible to bring a molecular system above its ionization limit with a simple widely tunable nanosecond pulsed laser (in contrast with femtosecond lasers that are usually only tunable in a small range). The resonant two-photon ionization (R2PI or REMPI) spectroscopy has thus been widely used in a large number of studies [113, 127–130] and reviews can be found in references [131, 132].

Often, the energy difference between the excited state and the ionization limit is low enough to allow the use of a single laser frequency for both photoexcitation and photoionization (one-color R2PI). The use of two tunable lasers (two-color R2PI) can allow precise measurements of ionized fragment apparition energies and thus provide binding energies in some favourable cases [133] (Figure 2.1.30, see also Chapter 4.6). The great power of the R2PI method arises from its ability to allow both spectroscopic selection between isomers and mass-analysis of ionized species.

R2PI resolution can be strongly improved by first exciting a vibrationally excited state ν' of the intermediate S_1 state and then using a resonant transition towards high n and high ℓ states converging towards the ionization limit of a specific vibrational state ν'' of the ionic core [131, 134]. The binding Coulomb interaction of the outer electron with the ionic core varies as Ryd/n^2 and can be easily counterbalanced by applying an external field. In the zero kinetic energy photoelectron (ZEKE) spectroscopy, long-lived neutrals in high n, ℓ ($150 < n < 300$) states are field-ionized by applying a pulsed (typically few microseconds and 1 kV/cm) electric field or a spatially delayed CW electric field. Only electrons with nearly zero energy are then selectively detected. In mass-analysed threshold ionization (MATI), threshold ions are detected by means of a mass-analyser [135] (Figure 2.1.31).

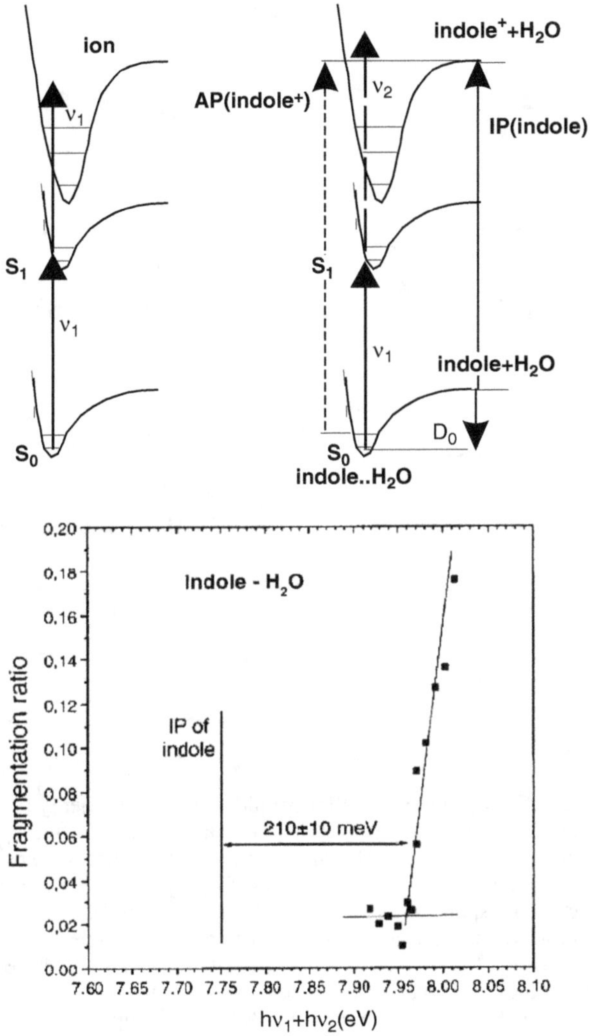

Figure 2.1.30 Left: Single-color ($\nu_1 + \nu_1$) and two-color ($\nu_1 + \nu_2$) versions of the resonant multi-photon ionization (R2PI) method. Right: Spectroscopic determination of the binding energy D_0 of a water molecule to indole (see also Figure 4.6.1). As shown in the left figure, D_0 is given by the difference between the ionization of bare indole and the apparition potential of fragmentation of the indole–water complex (reproduced with permission from reference [133] ©1999 American Chemical Society).

2.1.4.3 Depletion spectroscopy

We have already seen in Section 2.1.4.1 that the complexity of fluorescence spectra due to the presence of several isomers and conformers of investigated species can be disentangled by means of IR laser-induced depletion of the population of the ground state of each isomer. This method belongs to the ensemble of spectroscopic "hole-burning" or "double resonance"

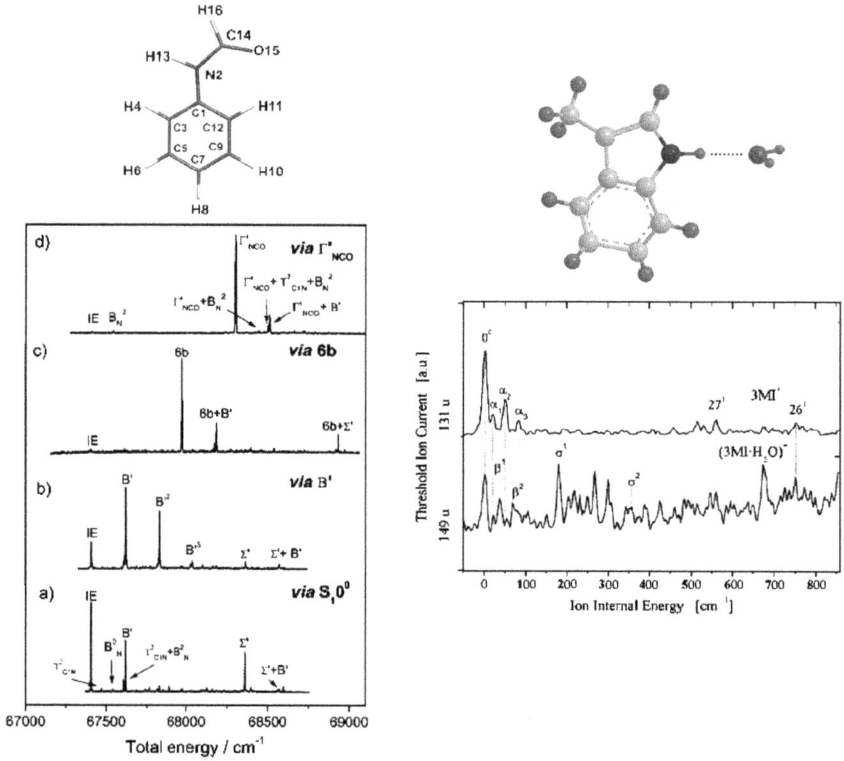

Figure 2.1.31 Left: ZEKE spectrum of *trans*-formanilide. The lowest energy feature at 67 408 cm^{-1} corresponds to the adiabatic ionization energy. In the spectrum recorded via the S_1 origin intermediate state (a), two major features correspond to 211 and 952 cm^{-1} ion internal energies. Excitation via different vibrational levels of the S_1 state extends the Franck–Condon region of the cation potential energy surface. The other spectra correspond to the in-plane side-arm bend (b), an aromatic ring mode (c) and a NCO wag (d). Right: MATI spectra of 3-methylindole and of the 3-methylindole–water complex recorded via their respective S_1 origin intermediate states. Bands denoted with β and σ are bending and stretching modes of the 3-methylindole ··· water hydrogen bond (reproduced with permission from reference [135] ©2005 Elsevier and reference [136] ©2005 Royal Society of Chemistry).

techniques reviewed in references [131, 132]. A first optical excitation ("pump") causes the depopulation of vibrational states while a second excitation ("probe") selectively probes each particular conformer or tautomer. Monitoring the probe signal tuned on a given isomer or conformer electronic transition when the pump frequency is scanned leads to observation of dips in the recorded spectra. Several possible schemes are shown in Figure 2.1.32 and examples are shown in Figure 2.1.33. The most widely used scheme is IR/R2PI (scheme a of Figure 3.1.32) [100, 137–145].

The investigated IR spectral region extends from the O—H, N—H and C—H stretches in between 3,000 and 3,800 cm^{-1} down to 500 cm^{-1} [89, 146]. Most IR/R2PI experiments are conducted with nanosecond lasers and thus very short-lived excited states sometimes cannot be observed through ionization [43]. Coupling IR and femtosecond excitation and ionization can circumvent this problem [147].

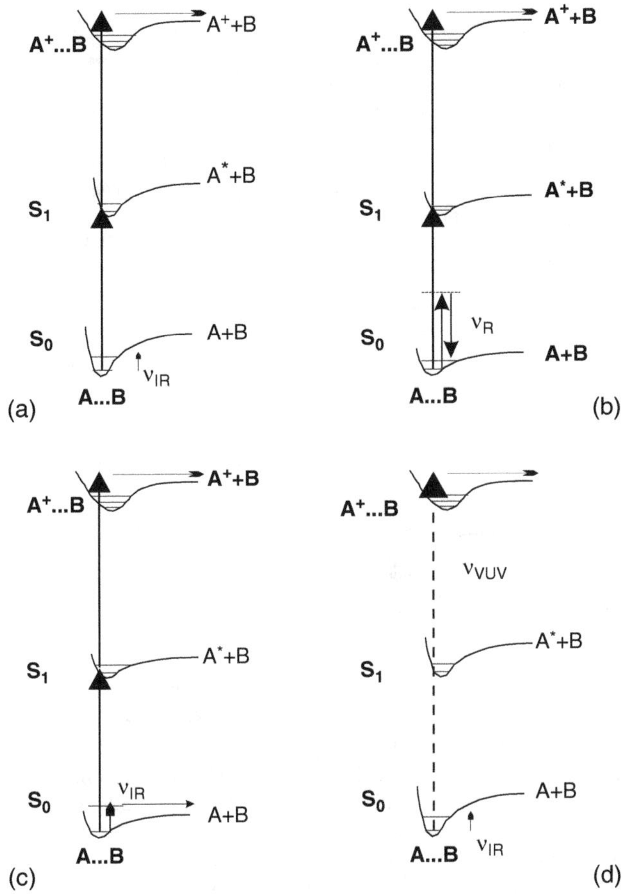

Figure 2.1.32 Different double resonance schemes involving infrared excitation. (a) IR vibrational depletion and R2PI detection. (b) Raman vibrational depletion and R2PI detection. (c) IR-induced predissociation of a van der Waals complex and R2PI detection. (d) IR vibrational depletion and VUV detection (adapted from reference [131] ©2000 American Chemical Society).

2.1.4.4 Experimental exploration of potential energy landscapes

Supersonic expansions (see Section 3.1.1) are precious for cooling down molecular systems that are initially at temperatures lying in between 300 and 500°C. If the potential energy barriers between conformers are large (the PES has a "weeping willow" disconnectivity diagram (see Section 1.5) with barriers typically greater than 10–15 kJ/mol [132], the initial Boltzmann population distributions are more or less preserved during the cooling process. On the contrary, for conformers separated by small barriers ("palm tree" type), typically less than 6 kJ/mol, collisional cooling takes place and rather only a small number of conformers with energies close to the lowest energy configuration are then populated. Nearly all spectroscopic experiments performed with supersonic expansions thus only consider the lowest lying conformers. By changing expansion conditions, cooling can be modified and somewhat

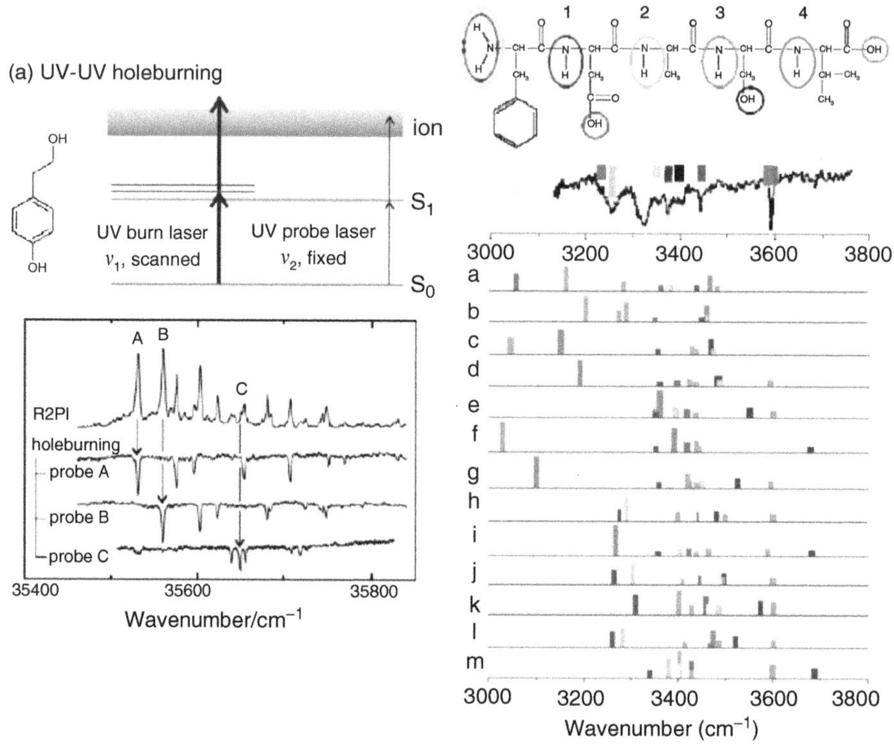

Figure 2.1.33 Double resonance spectra. Left: UV/UV hole burning scheme, R2PI and hole-burning spectra of tyrosol (reproduced with permission from reference [132] ©2001 Royal Society of Chemistry). Right: IR/UV spectrum of the neutral Phe-Asp-Ala-Ser-Val peptide (upper trace). Local modes are represented as circles. In the lower trace, spectra of 13 optimized structures calculated at the B3LYP/6-31G** level are presented. Two structures h and j exhibit an $\alpha-$turn that correspond to hydrogen-bonding of the phenylalanine C═O (i) to the N—H (i + 4) of valine (reproduced with permission from reference [148] ©2006 Elsevier).

different conformer populations can be obtained but usually without any experimental access to energy barriers.

An experimental protocol called stimulated emission pumping population transfer spectroscopy (SEP-PTS) [149–151] and shown in Figure. 2.1.34 allows the exploration of the PES in the vicinity of the lowest energy configurations populated by the supersonic expansion. The first step is the initial cooling in the high-pressure region of the supersonic expansion which populates a small number of low-lying conformers of the studied system. Selective excitation of a chosen conformer (B in Figure 2.1.34) by a pump laser tuned to its $S_1(v=0) \leftarrow S_0(v=0)$ transition is followed by stimulated emission pumping to $S_0(v=n) \leftarrow S_1(v=0)$ which populates the nth vibrational state of conformer B. If this state is situated under the energy barrier for isomerization to conformer C, re-cooling that takes place downstream (see Section 3.1.1) keeps the molecular system in configuration B. If the vibrational state $S_0(v=n)$ lies above the barrier, isomerization occurs and the re-cooling brings the molecular system in both conformers B and C. A third laser beam in the collision-free region can then probe the respective populations of the vibrational ground states of conformers B and C. The probe laser

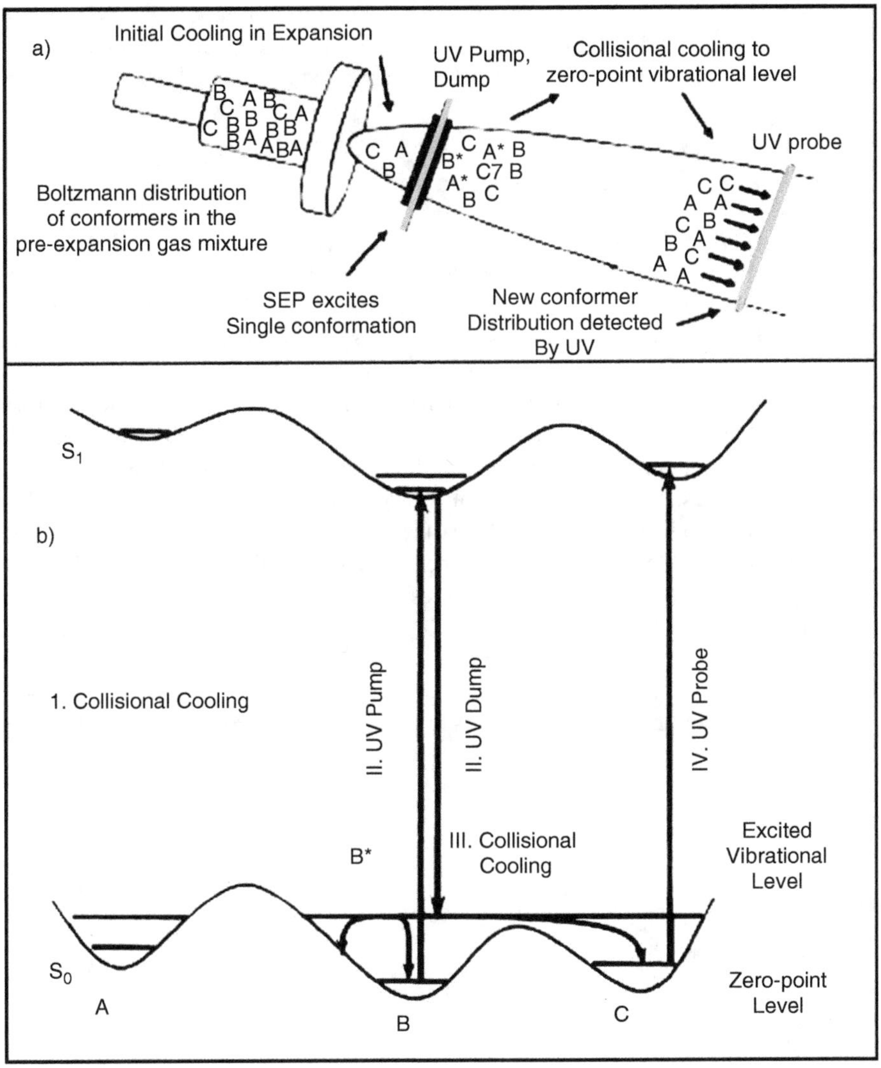

Figure 2.1.34 Selected stimulated emission pumping of a molecular system into an excited vibrational state of conformer B is performed with the UV pump and UV dump lasers in the early part of the supersonic expansion. Collision cooling takes place further in the expansion and isomerization to the C conformer is monitored downstream by means of the UV probe laser. Top (a) Scheme of the experimental set-up. Bottom (b) Pump, dump and probe lasers (reproduced by permission from reference [102] ©2006 American Chemical Society).

can be first set on the $S_1(v = 0) \leftarrow S_0(v = 0)$ line of conformer A and the different vibrational states of conformer B are progressively populated until reaching the threshold value of the energy barrier $E_{th}(B \rightarrow C)$. Since vibrational energy is quantized, the exact value cannot be exactly deduced and only a lower bound (no isomerization) and an upper bound (threshold for observation of isomerization) can be obtained. Reversing the respective roles of B and C, a similar bracketing of the threshold value of the energy barrier $E_{th}(C \rightarrow B)$ is obtained and the difference $E_{th}(B \rightarrow C) - E_{th}(C \rightarrow B)$ provides the energy difference $E_{min}(C) - E_{min}(B)$ between the minima B and C. This general method has been used, for example in the case of tryptamine which contains a very flexible ethylamine side chain connected to an indole chromophore (see Section 4.4.3).

2.1.5 VUV and IR/VUV spectroscopy

Ionization potentials (IPs) of biomolecular systems typically lie in the 8–10 eV (800–1,000 kJ/mole) energy range corresponding to the Vacuum Ultraviolet VUV spectral range. Two-photon (R2PI) ionization can be used for the determination of IPs in the case of molecular systems possessing chromophores. A two-color R2PI experiment conducted on phenylalanine has shown the dependence of IPs on conformer geometries [152] (Figure 2.1.35).

For systems lacking chromophore, direct ionization using Vacuum Ultraviolet VUV radiation is nevertheless possible but then does not benefit from the conformer selection brought by R2PI. In the VUV domain, intense photon sources are synchrotrons [153] and harmonic generation of pulsed laser radiation either in mercury or in rare gases [154]. The photoionization threshold of neutral histidine has been measured and interpreted by means of calculations of ab initio structures of the seven low-lying neutral and cationic conformers [153]. Two of these conformers have nearly the same structures as neutrals or cations. The experimentally observed adiabatic ionization energy of 8.2 ± 0.1 eV is attributed to those two conformers which calculated adiabatic ionization energies at the B3LYP/6-311G level are equal to 8.16 eV.

One of the problems raised by the absence of visible/UV chromophores in many biomolecular systems can be solved by using VUV radiation absorption as a universal mean

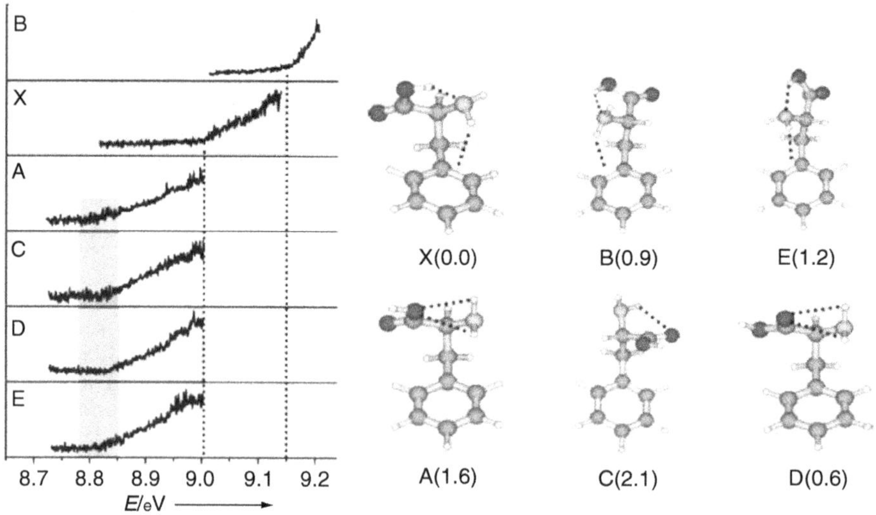

Figure 2.1.35 Left: Photoionization efficiency curves measured in a R2PI experiment of different conformers of phenylalanine displayed in the right figure. Right: Geometries of conformers of phenylalanine calculated at the MP2/6-31+G(d) level with ZPE corrections at the B3LYP 6-31+G(d) level (reproduced with permission from reference [152] ©2002 Wiley).

for ionization instead of resonant two-photon (REMPI). This has been demonstrated in the case of mass-selected neutral clusters in combination with IR spectroscopy [154].

2.2 TIME-RESOLVED SPECTROSCOPY

Time-resolved spectroscopies in the UV or IR domain are powerful tools for investigating dynamics of biomolecules in condensed phases [155–158] and can provide, for example information about polar solvent relaxation after electronic rearrangement in a solute chromophore [159, 160]. Femtosecond gas-phase studies conducted on isolated molecules and small hydrated clusters provide detailed insights and have been mostly applied to nucleobases (see Chapter 4.1 and Section 5.4) and amino acids possessing UV chromophores (see Chapter 4.2 and Section 5.5). A comprehensive review can be found in reference [161].

Following initial excitation of a molecular system from the ground state to an excited state with a first laser (pump), its time evolution can be experimentally monitored by means of time-delayed probe lasers. In studies of neutral species, ions in femtosecond time-resolved multiphoton ionization (fs-TRMPI), electrons in femtosecond time-resolved photoelectron spectroscopy (fs-TRPES) or both ions and electrons (see Section 2.3.3.4) are collected as a function of the time-delay between pump and probe lasers. Photo-excitation of neutrals can also be responsible for unimolecular dissociation leading, for example to

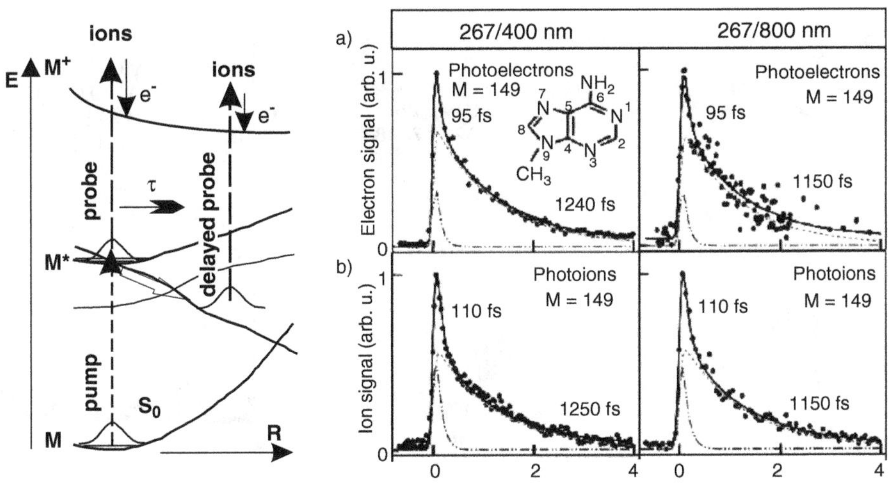

Figure 2.2.1 Left: Scheme of femtosecond time-resolved spectroscopy. A pump laser creates a molecular wave-packet that further evolves on an excited state. A probe laser photoionizes the system and photo-produced ions, electrons or both signals are then recorded as a function of time-delay τ between pump and probe. Right: Time-resolved photoion and photoelectron transients (dots) in 9-methylated adenine observed with two pump–probe schemes (267/400 and 267/800 nm). (a) Signal decays recorded by means of a photoelectron imaging technique integrated over all electron energies. (b) Signal decays recorded by detection of parent 9-Me–adenine ions. For interpretation, the dashed-dotted and dotted curves display the curve fittings obtained by convolution of an experimental Gaussian cross-correlation with two exponential decay curves (see Figure 2.2.2) (reproduced with permission from reference [168] ©2006 Royal Society of Chemistry).

hydrogen atom production. Those atoms can be detected by (1 + 1′) two-photon ionization [162] or by photo-fragment translational spectroscopy [163, 164]. In the case of ions, it is possible to mass-analyse ionic photo-fragments as a function of the time-delay between pump and probe [165–167] (Figure 2.2.1).

Femtosecond spectroscopy experiments can sometimes present difficulties of interpretation when the only available data are transient signals. Often, bi- or tri-exponential decays are observed and it is usually admitted that they are signatures of the presence of two or three excited states with different lifetimes. The presence of two lifetimes can then be, for example interpreted as due to the existence of two conformers. A molecular system is initially excited in state A^* with lifetime t_1. This state A^* is coupled to another state B^*, with lifetime t_2, that in turn can be coupled to another state C with lifetime t_3. It seems reasonable to deduce those lifetimes by means of the following fitting function (Figure 2.2.2).

$$Y(t) = y0 + A_1 \exp\left(\frac{-x}{t_1}\right) + A_2 \exp\left(\frac{-x}{t_2}\right) + A_3 \exp\left(\frac{-x}{t_3}\right). \tag{2.2.1}$$

The validity of this simple and natural approach can be questioned [169]. Lifetimes strongly depend on couplings between states A^*, B^* and C^*. These couplings are themselves strongly dependent on conformer geometries and environment. In the case of a flexible molecule such as tryptamine studied at room temperature [165], there are probably no definite conformers

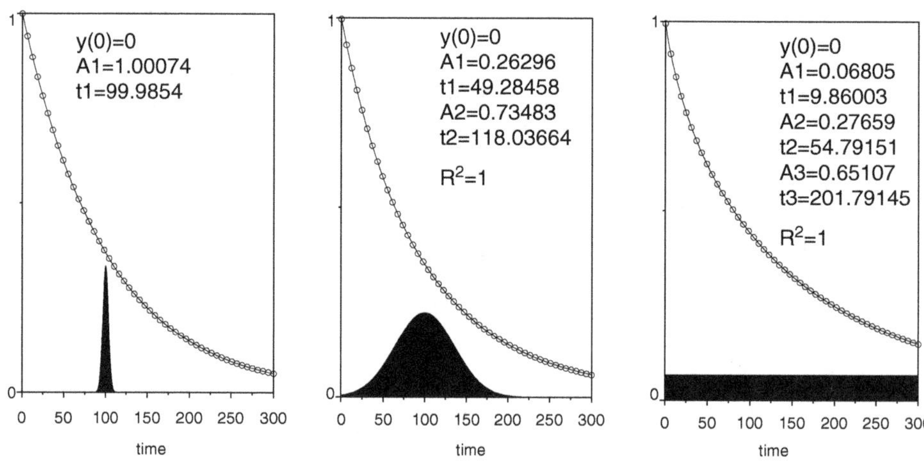

Figure 2.2.2 A hypothetical molecular system is supposedly composed of isomers with different distributions shown in black. The two Gaussian distributions (left and middle) are centered at 100 a.u. with respective time width of 5 and 50 a.u. The rectangular uniform distribution (right) extends from 0 to 300 a.u. The three curves correspond to transients that are sums of all exponential decays of individual isomers integrated from 0 to 300 a.u. These simulated "experimental" curves are respectively fitted by a single, double or triple exponential decay plotted with open circles. The fitting parameters are reported in each graph. The fits seem almost prefect with correlation coefficients $R^2 = 1$. At first sight, one could consider that the "observed" transients should be unambiguously assigned to the presence of one, two or three sets of conformers. This is obviously wrong in the two last cases (by courtesy of G. Grégoire, C. Dedonder and C. Jouvet).

but rather a continuous distribution of geometries (see in Chapter 4.2 simulations of protonated dipeptides by means of quantum dynamics). In a polar liquid such as water, the environment of an excited molecular system varies on an extremely short timescale [160] and again an excited species is in presence of a quasi-continuous distribution of solvent geometries. A quasi-continuous distribution of lifetimes can thus be expected. One should then expect transients quite different from simple bi- or tri-exponential decays. In order to evaluate the influence of a large number of very different lifetimes, simulations have been conducted with the assumption of a molecular system possessing a large number of isomers each with a different lifetime. Three simple lifetime distributions are displayed in Figure 2.2.1. Three simple lifetime distributions have been evaluated: two Gaussian distributions centred at 100 a.u. with time-windows of 5 and 50 a.u. and a square distribution from 0.1 to 300 a.u. The resulting observed transient is the sum of all exponential decays of individual isomer integrated from 0 to 300. The very surprising result of these simulations is that these decays representing extremely inhomogeneous distributions can be very well fitted by double or triple exponential decay functions [169]. Clearly, an experimental situation where only two or three excited state lifetimes exist is correctly interpreted by means of fits of transient signals by multi-exponential decays. Unfortunately, the reciprocal does not hold – a decay curve even very well fitted by a bi-exponential decay does not necessarily correspond to a situation where only two excited state lifetimes have to be considered. In order to distinguish between a "true" bi-exponential decay and a "numerical fitting artifact", transients alone are not sufficient and an additional experimental parameter such as rotation of the laser polarization or time-resolved mass-analysis must be added.

2.3 ELECTRON SPECTROSCOPY

2.3.1 Electron–molecule processes

Electron attachment to a neutral molecular system is the result of a delicate balance between attractive electrostatic, polarization interactions and electron–electron repulsion interactions. The result of the capture of an excess electron whose energy is less than the IP of a molecule AB, called parent, is a daughter transient anion AB^{-*}. In absence of any third-body, a bound anion state AB^{-*} is imbedded in the continuum $AB + e$. It can then either auto-detach $\langle 1 \rangle$, releasing an electron and leaving a more or less vibrationally excited molecule AB^* or it can break into fragments [170–172] (dissociative electron attachment (DEA), see Chapter 4.7) $\langle 2 \rangle$. In presence of a third-body, the transient AB^{-*} anion can provide some of its internal energy and become stabilized with respect to electron auto-detachment, leading to a long-lived *valence* anion (non-DEA) $\langle 3 \rangle$. This occurs, for example when a transient anion is produced by capture of an electron by a molecule imbedded inside a molecular cluster, the excess energy then flows through intermolecular bonds. Another stabilization mechanism can also occur in a collision, a fraction of the anion excess energy being transferred into kinetic or internal energy of the colliding partner:

$$
\begin{aligned}
AB + e(\varepsilon_i) \rightarrow AB^{-*} &\rightarrow AB^* + e(\varepsilon_f) && \langle 1 \rangle \\
&\rightarrow A^- + B && \langle 2 \rangle \\
&\xrightarrow{\text{third body}} AB^- && \langle 3 \rangle
\end{aligned}
\qquad (2.3.1)
$$

Two main mechanisms are responsible for the temporary trapping of an excess electron in a valence anion. In a Feschbach resonance, the incoming electron interacts at some distance with the molecular target AB and excites it either into a vibrationally excited state [173] ("vibrational Feshbach resonance") or into an electronically excited state [174] ("core-excited Feshbach resonance"). A two particle one-hole state or *core-excited* resonance is composed of two-electrons in electronically excited orbitals and a positive ionic core. The incoming electron is temporarily trapped and the anion AB^{-*} explores all its available phase-space until the initial molecular geometry of AB is recovered and the reverse process of attachment becomes allowed. The auto-detachment lifetime is then given by the following expression [175]:

$$\tau^{-1} = \frac{\sigma(\varepsilon_i)\varepsilon_i^{1/2}}{[\varepsilon_i + \mathrm{EA} + a_z]^{N-1}} \frac{m}{\pi^2 h^3} (\varepsilon_i + a'\varepsilon_z)^{N+1/2} 2^{2N} \frac{N!(N-1)!}{(2N+1)!} \tag{2.3.2}$$

$N=3n-6$ is the number of vibrational degrees of freedom and n is the number of atoms, $\sigma(\varepsilon_i)$ the electron capture cross section, a_z and a'_z are correction factors, m the electron mass. At very low incident energies of the attached electron, this auto-detachment lifetime increases rapidly with the molecular anion size and can become comparable with the observation time-window in a mass-spectrometer, even in absence of any external stabilization process. It also decreases rapidly when the incident energy ε_i of the electron increases.

An excess electron can also temporarily occupy a usually unfilled molecular orbital of the parent molecule AB, leading to a so-called single-particle or "*shape*" resonance. The shape of the potential energy as a function of the incoming electron–molecule distance r results from the sum of two terms. The polarization potential $-\alpha/2r^4$ corresponds to the attraction between the incoming electron charge and the molecular induced-dipole. The repulsive centrifugal potential $\ell(\ell+1)/r^2$ depends on the angular momentum ℓ of the relative electron–molecule motion. Shape resonances [176] correspond to temporary electron trapping in this potential which is first repulsive and then attractive and thus presents an energy barrier. For some energies, ε_i, corresponding to bound states inside the inner potential well, the incoming electron tunnels through the barrier, becomes trapped and again tunnels out. The lifetimes are then considerably shorter than in the case of Feshbach resonances and, due to the Heisenberg uncertainty relation, broad lines are observed in electron scattering experiments [177, 178]. In contrast with core-excited resonances, shape resonances are too short-lived to cause molecular dissociation.

Biomolecular building blocks possess chemical groups with different partial charges leading to the existence of permanent dipoles μ (see Section 3.2.5) and/or quadrupole moments Q. If an electron approaches a polar system, it first experiences an attraction resulting from the sum of the polarization potential (see above) and the electron–dipole (varying as $-\mu/r^2$) and/or electron–quadrupole interaction (varying as $-Q/r^3$). At even shorter distances, the excess electron then feels the repulsion from the other electrons of the molecular cloud (Pauli repulsion). If the quantum well resulting from the sum of those attractive and repulsive terms is deep enough to support a bound state, the excess electrons can be trapped in a very diffuse orbital. This orbital is not a localized valence orbital of the parent but is totally located outside the molecular cloud of the accepting molecule (Figure 2.3.1). In absence of

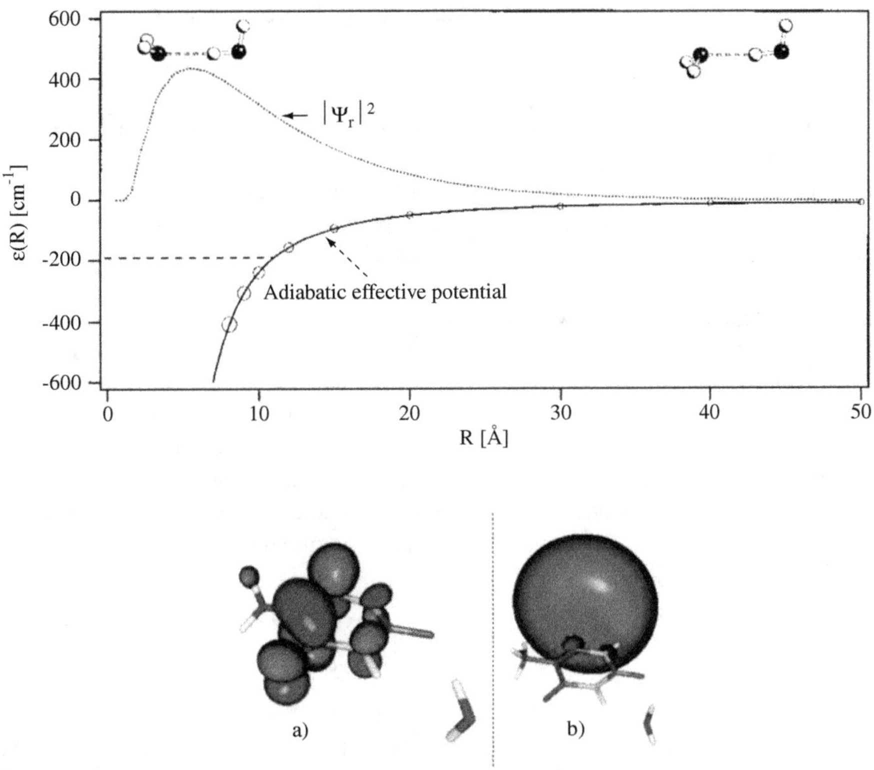

Figure 2.3.1 Top: Water dimer-electron potential responsible for the formation of a dipole-bond anion and the diffuse excess-electron orbital. The attractive and repulsive potential energies create a quantum well with a bound state (reproduced with permission from reference [190] ©1999 American Institute of Physics). Bottom: Highest occupied molecular orbital (HOMO) of the valence (a) and multipole-bound (b) thymine–water anion complex calculated at the MP2/aug-cc-pVDZ level (reproduced with permission from reference [191] ©2006 American Chemical Society).

sizable quadrupole moment, the dipole moment must exceed a value of 2.5 D [179] in order to allow the formation of so-called *dipole-bound* anions [180–182]. In absence of any dipole moment, the quadrupole moment must exceed 40 a.u. [183]. Although the excess electron is in a diffuse orbital outside the valence electron cloud, it nevertheless polarizes valence electrons and a sizable fraction of its binding energy arises from this dispersion effect [180].

For elementary energy conservation reasons, the two-body system excess electron–molecule system is, as above in the case of valence anions, unstable with respect to auto-detachment and dipole-bound anions are thus not directly observable in free electron-molecule scattering experiments. When such anions are produced under high-pressure conditions [184–187] or by electron transfer from highly excited atoms (RET, see below), a stabilization mechanism against auto-detachment is provided either by three-body collisions or through removal of vibrational energy by the atomic ionic core [188, 189].

2.3.2 Electron affinities and vertical detachment energies

2.3.2.1 Definitions

The energy difference between a neutral parent species and its anion is called the electron affinity [192, 193]. In fact, different electron affinities must be taken into account if the neutral and anion structures are considered. For simplicity, a single diatomic picture is generally used (Figure 2.3.2). The abscissa coordinate corresponds to the most strongly modified internal molecular degree of freedom in the electron attachment process (e.g., a dissociative internuclear distance).

Due to the large difference between nuclei and electron masses, the nuclear configuration of a neutral parent is conserved during the electron process that takes place during a time duration much shorter than any molecular vibration. This corresponds to the vertical electron affinity EA_V. This electron affinity is nearly always negative (the anion is unstable). The anion then relaxes and if the process is non-dissociative, one can define the adiabatic electron affinity (EA_{adia}). Since the nuclear configurations of neutrals and anions are different, the corresponding sets of vibrational frequencies ω_i are also different and one must thus take into account the zero-point vibrational energies (ZPE) equal to $ZPE = 1/2\sum_{i=1}^{i=3N-6} h\omega_i$ (see Section 2.1.3). The EA_{adia} is then defined as:

$$EA_{adia} = \left[E_{optimized}(\text{neutral}) + ZPE_{neutral}\right] - \left[E_{optimized}(\text{anion}) + ZPE_{anion}\right]. \quad (2.3.3)$$

Whether EA_{adia} is negative or positive, the anion is unstable with respect to auto-detachment or stable. In the case of dipole-bound anions, the vertical and adiabatic electron affinities are virtually identical and are also called electron binding energy. When a stable anion is photoexcited, the electron detachment process is characterized by the vertical detachment energy (VDE) (see Figure 2.3.2 and Section 2.3.3.3).

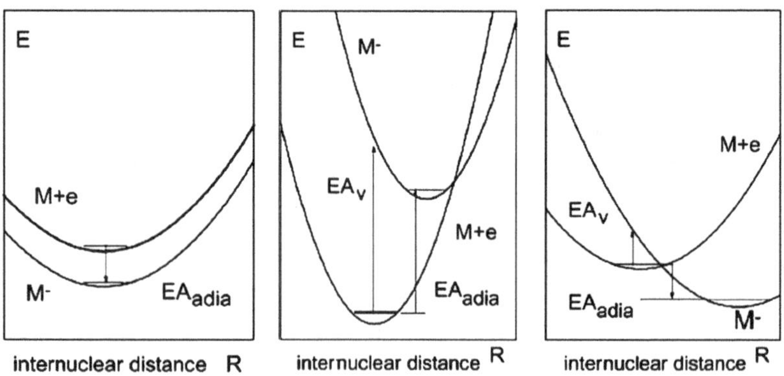

Figure 2.3.2 Schematic potential energy diagram of anions (M^-) and their neutral parents (M) plus electron at infinity. From left to right are represented dipole-bound anions, molecular systems with negative and positive valence electron affinities.

2.3.2.2 Calculation of electron affinities

Calculations of electron affinities are generally difficult. Reviews concerning these calculations can be found in reference [193]. High levels of theory such as MP2 and CCSD(T) (see Section 1.2) are strongly preferable but are still restricted to rather small biomolecular anions such as that of uracil [194, 195]. For reduction of computation time, the applicability of DFT, with several possible functionals such as BLYP, B3LYP, BP86 and B3PW91, has thus been tested and seems to be acceptable, at least for large enough systems [193]. Accuracies of 0.2 eV can be expected. This turns out to be sometimes insufficient, for example in the case of isolated nucleobases (see Section 4.1.1.3). Electron affinities of nucleobases are close to zero and even their sign is questionable, experimentally [184, 185, 196, 197] as well as theoretically [198–200]. This problem arises from the existence of dipole-bound anion states competing with the valence anion states. The prediction of electron affinities is extremely dependent on the method and basis sets [191, 201, 202]. As shown in Table 2.3.1, DFT can be employed but only with optimized basis sets [203].

The problem of prediction of electron affinities is even more difficult when dealing with molecular systems with negative electron affinities, in particular nucleobases [200, 205]. DFT and MP2 predictions of vertical electron affinities measured in electron transmission spectroscopy do not always agree with the experimental values [206] and semi-empirical approaches [207] or composite quantum methods such as G2 or G3 [208] can sometimes be preferable.

2.3.2.3 Influence of solvation upon electron affinities

Transient anions M^- created from electron attachment to molecular species M possessing negative valence electron affinities EA(M) cannot be directly observed except as very short-lived resonances in electron scattering resonances [209]. When those species are surrounded by N solvent molecules S in inhomogeneous $M \cdots (S)_N$ or in homogenous $(M)_N$ clusters, their valence electron affinity EA(N) can become positive above a size threshold N_{th}. This arises from the larger binding energy $E_{sol}^{M^-}(N)$ of a neutral solvent molecule S to an anion $M^-(S)_{N-1}$ (charge-induced dipole interaction) than its binding

Table 2.3.1

Electron affinity of the water dimer (in eV) with 6-31+G**, 6-311++G(3df,2p) and a basis set built by adding to the standard 6-31G set two *sp* and two *d* shells of diffuse functions on the oxygen centres and one *s* and two *p* on the hydrogen ones at different level of correlation

Correlation scheme	6-31+G**	6-311++G(3df,2p)	Optimized basis set
HF	−3.12	−0.88	−0.239
MP2	−2.63	−0.62	−0.473
MP4	−2.42	−0.63	−0.44
CCSD(T,E4T)	−2.62	−0.65	−0.483
BPW91	−2.04	−0.37	0.009
B3LYP	−2.06	−0.34	0.0212

Note: The general trend is that very diffuse orbitals must be introduced and optimized to obtain a value close to the experimental one, here 30 ± 4 meV [203, 204].

energy $E_{sol}^{M}(N)$ to a neutral $M(S)_{N-1}$. The solvated EA(N) and isolated EA(M) electron affinities are related by:

$$EA(N) = EA(M) + E_{sol}^{M^-}(N) - E_{sol}^{M}(N). \tag{2.3.4}$$

Following experimental observation of the size threshold N_{th} for valence anion formation and calculation of the binding energies $E_{sol}^{M}(N)$ and $E_{sol}^{M^-}(N)$, one can obtain a bracketing of the negative values of electron affinities [210]:

$$E_{sol}^{M}(N_{th}) - E_{sol}^{M^-}(N_{th}) \leq EA(M) \leq E_{sol}^{M}(N_{th}-1) - E_{sol}^{M^-}(N_{th}-1). \tag{2.3.5}$$

Due to the flow of the internal energy of the nascent anion M^- towards its solvent surrounding through intermolecular bonds, the electron affinity deduced from this method is close to the EA_{adia} of M.

2.3.3 Experimental methods

2.3.3.1 Free-electron scattering

In order to study free electron-biomolecular system scattering, an energy-selected electron beam is directed through a gas-cell or a beam containing the investigated system. When looking for extremely short-lived anions (resonances) that cannot be directly observed because their lifetime lie in the 10^{-14}–10^{-12} s range, the scattered electrons are rejected at a retarding electrode following the collision region and the transmitted electron current comprising the unscattered electrons is monitored as a function of the incident electron energy. Dips in recorded electron currents correspond to the temporary formation of temporary anion states. In order to accentuate these dips, the electron energy is modulated and the derivative of the transmitted current is recorded by means of a synchronous detector (Figure 2.3.3). The energies of the mid-point between the minimum and the maximum in the derivative signal are assigned to the electron scattering resonances [207, 211]. In the case of DEA or non-DEA, long-lived anions are directly observable and can then be mass-analysed [170, 173, 212].

2.3.3.2 Rydberg electron transfer spectroscopy

We have seen that molecules with sufficient dipole moments can give birth to dipole-bound anions that possess excess electrons in very diffuse orbitals. Those anions can thus be easily prepared in charge-exchange collisions between polar molecular systems and highly excited atomic (Rydberg atoms) species [189] already possessing electrons in diffuse orbitals. In a Rydberg atom, the atomic level energies depend on the principal quantum number n and are given by the Rydberg formula $E(n) = -\text{Ryd}/n^2$. The classical frequency of the outer electron is equal to Hartree/$\hbar n^3$. The charge-transfer from a highly excited atom to a molecular species with a dipole-bound electron affinity EA_{db} is resonant at $n = n_{max}$ when this classical frequency in the excited atomic orbital is equal to that of the excess electron in the dipole-bound anion $EA_{db}/\hbar$.

RET spectroscopy consists in monitoring the charge-transfer rate between a beam of laser-excited Rydberg atoms and a beam of polar molecular systems. In contrast with creation of

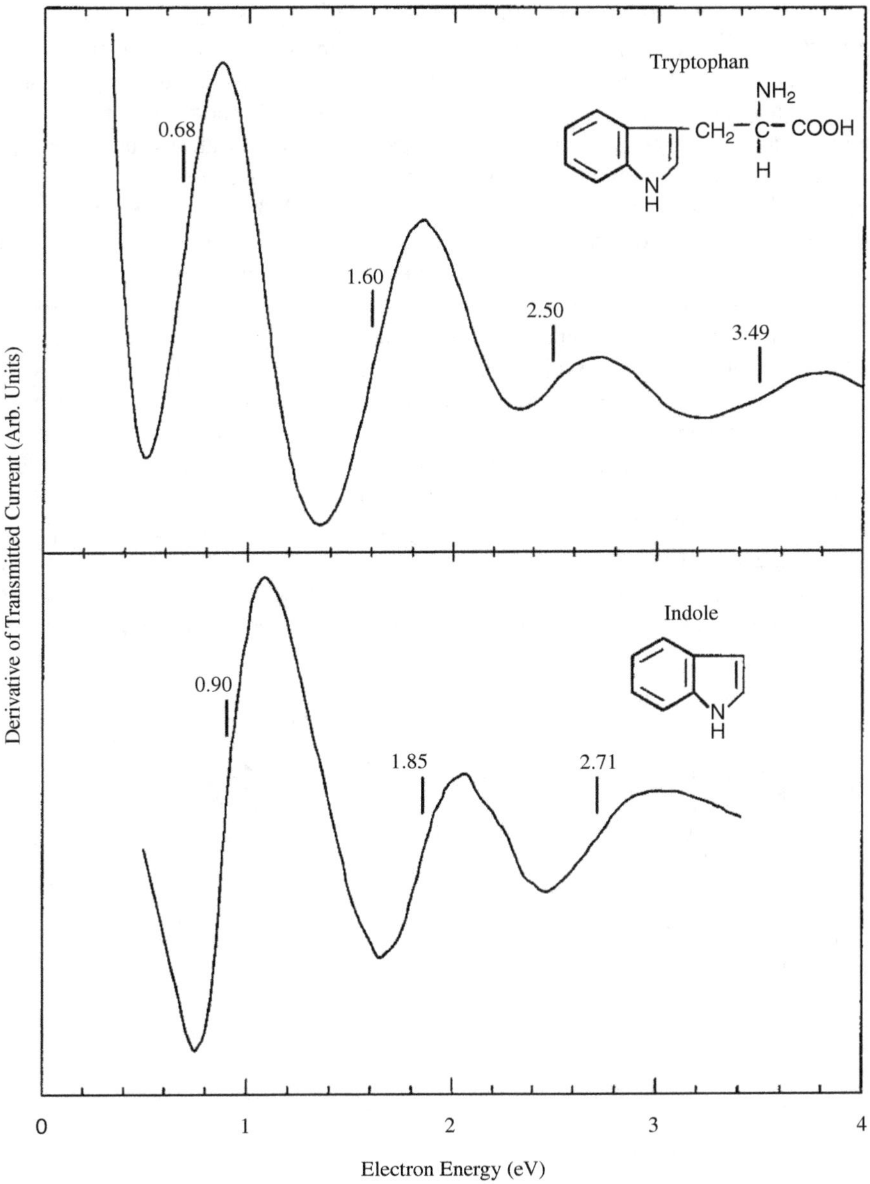

Figure 2.3.3 Derivative of transmitted electron current as a function of electron energy in collision between free electrons and tryptophan (top) or indole (bottom). The two spectra are similar and the anion states associated with the side chain of tryptophan are stabilized by ~0.2 eV with respect to those of bare indole (reproduced with permission from reference [209] ©2001 American Institute of Physics).

a valence anion, the dependence of a dipole-bound anion creation rate as a function of the principal quantum number n of the laser-excited Rydberg atoms is strongly peaked due to the resonant character of the charge-exchange process. The anion creation rate corresponding to the process: $A^{**}(n) + M \rightarrow A^+ + M^-$ is maximum for $n = n_{max}$ (the presence of the atomic ionic core A^+ provides the required stabilization process against auto-detachment). If the above picture was strictly exact, the electron binding energy EA of the electron accepting species M should then be related to n_{max} by the relationship $EA = 27.2eV/n_{max}^3$ In practice, a detailed interpretation of the experimental data through the help of a multiple curve-crossing model of charge-exchange collisions between excited atoms and polar species provides the electron binding energies which are related to the experimentally determined n_{max} values by the following semi-empirical relationship [213]: $EA = 28eV/n_{max}^{2.8}$.

The principle of the RET method is thus the following. A beam of laser-excited atoms $A^{**}(n)$ (in practice, $6<n<35$) is crossed at right angle with a supersonic beam of polar molecules or complexes and the dipole-bound anions created in the charge-transfer process are detected. The experimental n-dependencies of the charge-exchange process between the laser-excited Rydberg atoms and the studied molecular species are recorded and present one, sometimes two peaks [214], corresponding to electron transfer to one or possibly two conformers with different electron binding energies. For each possible conformer, the electron binding energy is computed and compared with the experimental values deduced from the above formula.

The RET method is illustrated with the example of the adenine–imidazole (side chain of histidine) complex [189] in Figure 2.3.4. This study was prompted by the large interest in design of artificial ligands capable of selectively binding to predetermined DNA sequences in the human genome and acting as artificial gene repressors [215–217]. Polyamide hairpins complexes containing imidazole, pyrrole and other molecules can establish very specific binding

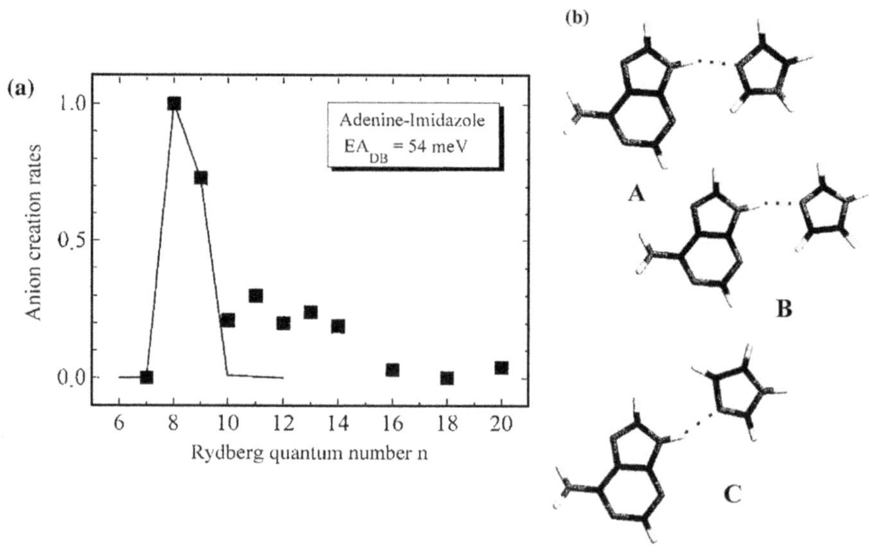

Figure 2.3.4 (a) Experimental Rydberg electron transfer (RET) anion formation rates as a function of the principal quantum number n of Rydberg atoms in collisions with adenine–imidazole complexes. The full line corresponds to the best fit of the experimental data obtained from a curve-crossing model [213]. The electron binding energy EA_{db} is deduced from the peak n-values by means of a semi-empirical formula (see text). (b) Structures of the lowest energy configurations of the adenine–imidazole complex. The respective energies and dipole moments of configurations A, B and C are: 0, 4.1 D; 25 meV, 5.3 D; 40 meV, 6.5 D. Experimentally, configuration A is observed (reproduced with permission from reference [189] ©2000 American Chemical Society).

through establishment of weak bonds to nucleobases (see Section 4.5). Those hairpins are designed with the help of modelling of those weak interactions, present, for example in the adenine–imidazole or adenine–pyrrole complexes. Separately, adenine and imidazole with respective dipole moments of 2.5 and 3.8 D give birth to dipole-bound anions with respective electron binding energies of 11 and 23 meV. According to the different hydrogen bond patterns possibly existing in an adenine–imidazole complex, the geometrical vector sum of the individual molecule dipole moments can take different values. The three lowest energy conformations respectively labelled A, B and C (Figure 2.3.4) have dipole moments, predicted energies and electron binding energies respectively equal to A (0, 4.1 D, 80 ± 20 meV), B (25 meV, 5.3 D, 165 ± 20 meV) and C (40 meV, 6.5 D, 260 ± 50 meV). Experimentally, a single peak is observed in the RET n-dependency at $n_{max} = 8$, corresponding to an electron binding energy of 54 meV according to the semi-empirical above formula. The only experimentally observed configuration of the adenine–imidazole complex is thus the lowest energy configuration A. This structure is coplanar with a hydrogen bond between the N(9)-H group of the five-member ring of adenine and the N(1) atom of imidazole with a H···N distance of 2.1 Å, together with an electrostatic interaction between the N(3) atom of the six-member ring of adenine and the C(2)-H bond located between the N(1) and N(3)-H atoms of imidazole (Figure 2.3.4). In the adenine–pyrrole complex, the only experimentally observed configuration is not the lowest energy one but only the second configuration possessing a sufficient dipole moment for electron binding.

The advantages of the RET method are the following. This method relies on a very mild ionization process avoiding fragmentation processes and leading to anions with binding energies as low as 0.3 meV and up to 200 meV. The structures of the dipole-bound anions and that of their neutral parents are identical. This process, moreover, is isomer selective [218] and it is thus possible to prepare beams of mass- and isomer-selected species. It also allows to record IR spectra of neutral species [44, 50] by IR depletion in a similar fashion to R2PI/IR spectroscopy (see Section 2.1.4.1) without the requirement of any chromophores in contrast with R2PI. Its main drawback is that it can only be applied to molecular species possessing both a dipole moment in the 2.5–5.5 D range and a negative valence electron affinity. For dipole moments above 6 D, dipole-bound anions become similar to valence anions since excess electrons are no longer in diffuse orbitals but rather localize on the positive end of the dipoles. Well-defined peaks in the n-dependencies of the anion creation rates are then no longer observed.

2.3.3.2.1 Mixing of valence and dipole-bound anion states

If the valence electron affinity of a polar molecule is close to zero or positive, the dipole-bound state (DBS), due to the long-range electron–dipole potential acts as a "doorway" for electron attachment [173, 219–224]. Sharp resonances at energies below 3 eV have been observed in DEA to uracil and thymine and interpreted as vibrational Feshbach resonances (see Section 2.3.1). While electron transmission scattering (ETS) experiments only observe resonances due to temporary attachment in anti-bonding π^* orbitals, the interpretation of low-energy DEA measurements show that the vibrational Feshbach resonances are due to temporary electron attachment of an excess electron in low-lying σ^* orbitals [173]. As shown in Figure 2.3.5, there is an excellent overlap between the excess electron orbital in the DBS and the valence state (VS). The corresponding potential curves along the dissociative N_1-H nuclear coordinate thus present a significant avoided crossing. The observed sharp resonances are due to the long-lifetime of the temporary anion non-dissociated bound state U^{-*} since

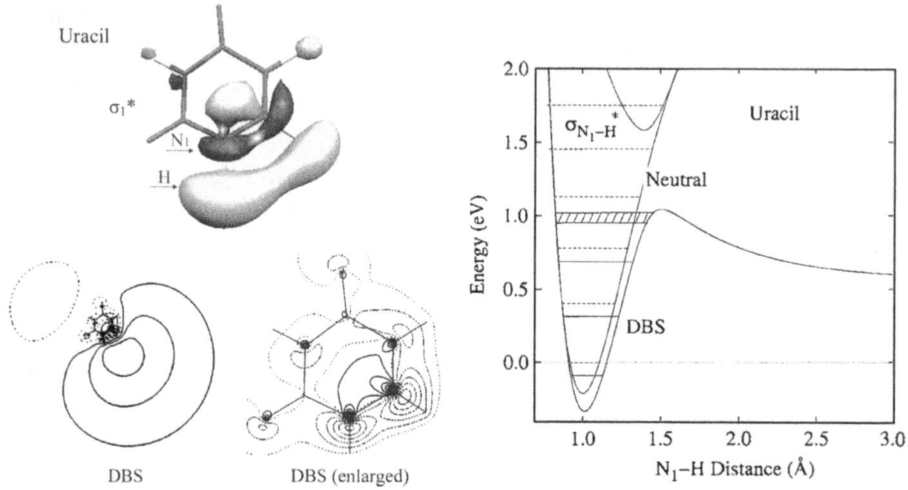

Figure 2.3.5 Left: σ^* orbital of the dipole-bound (DBS) and valence anion states of uracil. Right: Potential energies of the neutral, dipole-bound and valence σ^* anion states of uracil as a function of the N_1-H internuclear distance (reproduced with permission from reference [173] ©2006 American Institute of Physics).

the hydrogen atom liberated in the $U + e \rightarrow U^{-*} \rightarrow [U-H]^- + H$ process must tunnel through the barrier provoked by the avoided crossing (Figure 2.3.5).

2.3.3.3 Negative ion photoelectron spectroscopy

In photoelectron spectroscopy, anions are photo-excited from a bound state $M^-(v)$ into the continuum $M(v') + e$ (Figure 2.3.7). Conservation of energy relates the photon energy $h\nu$ to the kinetic energy EKE of the ejected electron and the electron binding energy EBE in the anion:

$$h\nu = \text{EBE} + \text{EKE}. \tag{2.3.6}$$

Experimentally, a beam of mass-selected anions is crossed with a laser of known frequency ν and the photoelectron kinetic energy spectrum is recorded by means of an electron monochromator that can be, for example either a time-of-flight tube, a hemispherical analyser or a magnetic bottle [185, 225, 226]. As displayed in Figure 2.3.6, the shape of the photoelectron spectrum provides a direct signature of the nature of anions. In the multipole-bound case, the photoelectron spectrum exhibits a narrow peak and the VDE coincides with the EA_{adia} since the geometries of the neutral parent and the anion are identical. In the VS case, the peak of the photoelectron spectrum corresponds to the VDE. However, one must be cautious that the threshold of the photoelectron spectrum does not necessarily correspond to the EA_{adia}. If the geometries of the neutral parent and its anion are too different, the Franck–Condon factor can become too small at the equilibrium internuclear distance of the anion corresponding to the EA_{adia} and the threshold then only represents the value at which the experimental photoelectron signal/noise ratio becomes sufficient.

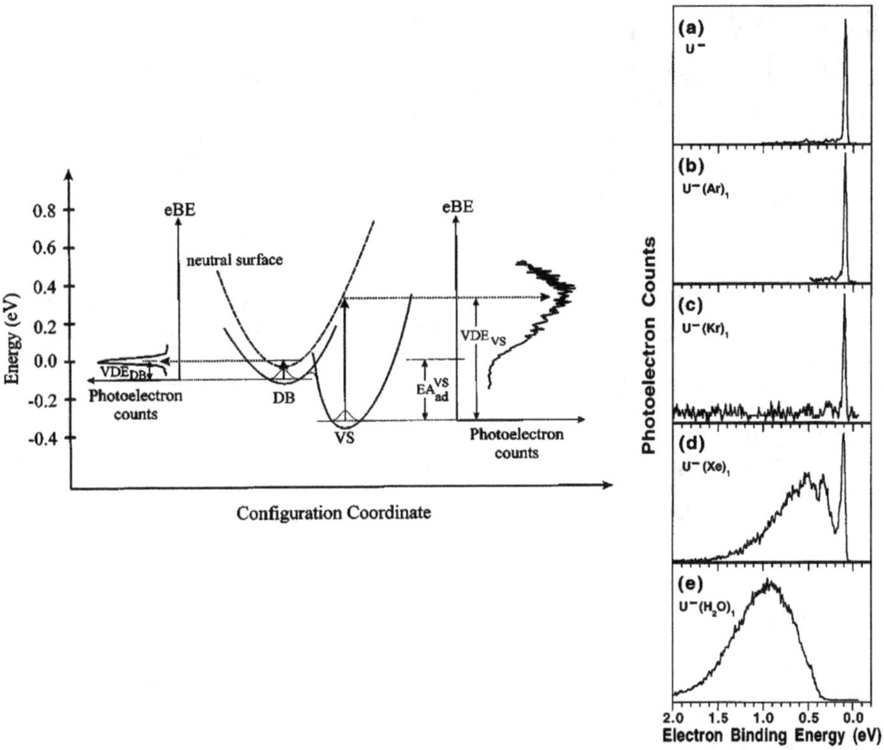

Figure 2.3.6 Left: Schematic energy diagram of negative photoelectron spectroscopy. The valence (VS) and the dipole-bound (DB) anion states of the pyridine tetramer are represented here (reproduced with permission from reference [222] ©1998 American Institute of Physics). Right: Photoelectron spectrum of isolated uracil (a) and uracil solvated by one (b) or two (c) argon atoms displaying a dipole-bound behaviour. The photoelectron spectra (d) and (e) respectively correspond to uracil solvated by a xenon atom or a water molecule and display a valence anion character (reproduced with permission from reference [220] ©1998 American Institute of Physics).

In order to extract as much information as possible from photoelectron experiments, a fit of the experimental spectra can be conducted by assuming Morse potentials for the neutral and anion diatomic-like potential energy curves along a suitable coordinate. Once the Franck–Condon [227] factors are calculated, the parameters that can be determined by adjustment are the VDE as well as the vertical and adiabatic electron affinities (Figure 2.3.8).

2.3.3.4 Time-resolved photoelectron spectroscopy

In TRPES, detached electrons are used as pump–probe signals. This technique is based on the idea that during ionization of neutral clusters or electron detachment from ion clusters, emission of an independent outer electron occurs without simultaneous electronic reorganization of the "core" (cation or neutral). This is called the "molecular orbital" or Koopmans' picture. The final state after the ionization is a continuum state, of which the probability is correlated with the molecular orbital character of the excited state. Therefore, the electronic character

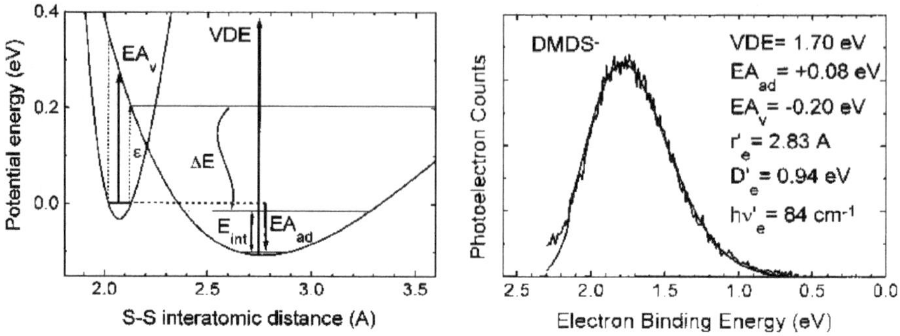

Figure 2.3.7 Left: Energy diagram of negative ion photoelectron spectroscopy of a disulfide-bond model system. The potential energy curve of dimethyldisulfide (DMDS) is plotted as a function of the S–S distance. Rydberg electron transfer (RET, see above) shows that anions are formed from attachment of electrons with 0.2 eV energy (ε). Following stabilization (ΔE) by collision in the photoelectron experiment anion source, anions are left with an amount of internal energy E_{int}. Right: A Franck–Condon analysis assuming an anion temperature of 150 K provides a fit of the experimental PES spectrum leading to the given molecular parameters. D_e, r_e and n'_e are respectively the dissociation energy, equilibrium S–S distance and stretch frequency of the DMDS anion (reproduced with permission from reference [227] ©2001 American Chemical Society).

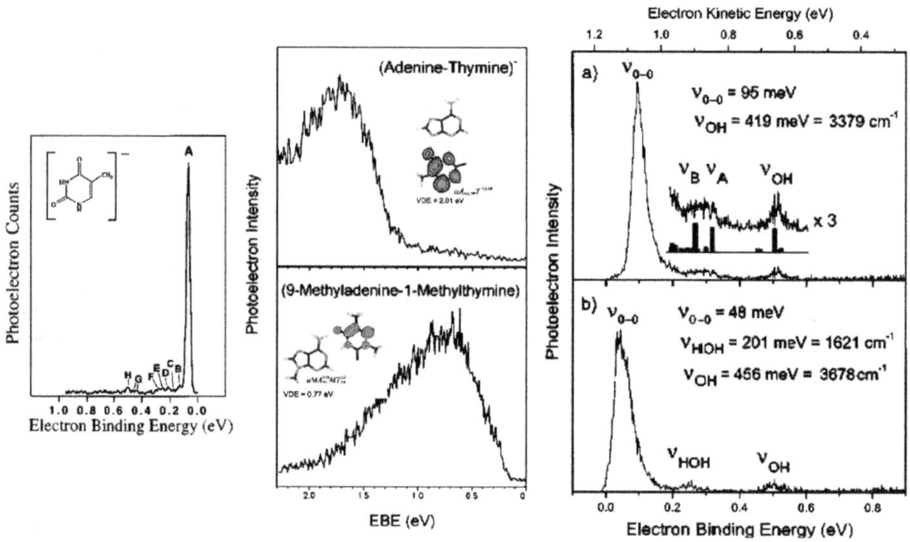

Figure 2.3.8 Left: Photoelectron spectrum of thymine. Middle: Photoelectron spectra of the adenine–thymine (A–T)$^-$ pair anion and its methylated analog, 9-methyladenine-1-methylthymine (MA–MT)$^-$, recorded with 2.54 eV photons. Lowest energy structures of the anions are calculated at the B3LYP/6-31+G** level. Methylation only introduces a very small difference in valence affinities [228]. The strong difference observed here between the A–T and MA–MT pairs arises from the large difference between structures of those pairs. The A–T pair is in both the $aA_{N3}^{N9}T_{O8}^{N3}$ and $aA_{N3N9}T^{N3O8}$ configurations whereas the MA–MT pair is in both the Hoogsteen and Watson Crick configurations (reproduced with permission from reference [226] ©2005 American Institute of Physics). Right: Photoelectron spectra of the glycine molecular anion Gly$^-$ (top) and water dimer anion $(H_2O)_2^-$ (bottom) recorded with 1.165 eV photons [70]. The dominant peak of the Gly$^-$ spectrum ν_{0-0} corresponds to an electron binding energy of 0.095 eV. The calculated and observed free O—H (ν_{O-H}), carbonyl stretching (ν_A) and C—O—H bending (ν_B) vibrations are displayed in the insert. The weak features in the water dimer anion spectrum correspond to the H—O—H bend (ν_{HOH}) and O—H stretching (ν_{O-H}) modes (reproduced with permission from reference [70] ©2005 American Institute of Physics).

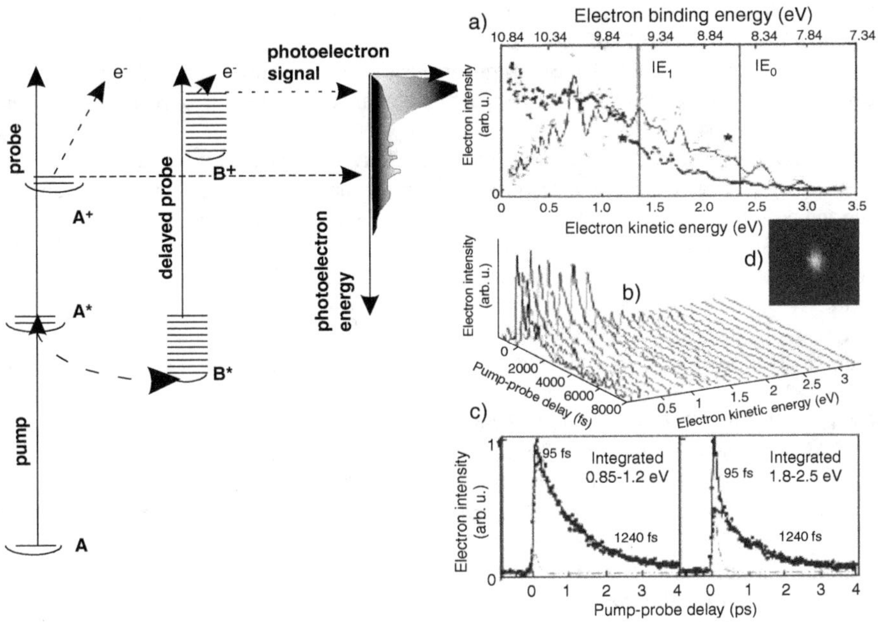

Figure 2.3.9 Left: Time-resolved photoelectron spectroscopy. The studied molecular system is first excited by the probe laser into state A* that further converts into state B* possessing a different electronic structure and thus a different ionization potential. The photoelectron spectrum recorded by exciting with the probe laser varies and is recorded as function of the time-delay between pump and probe (adapted from reference [233]). Right: Time-resolved spectroscopy of 9-methyl-adenine using velocity imaging technique (a) energy distributions of the 9-methyl-adenine photoelectrons corresponding to the first (dots and smoothed curved) and second (dotted line) components of the decay respectively at 95 and 1,240 fs (see Figure 2.2.1). (b) Typical photoelectron velocity imaging spectrum. (c) Transients observed by integration of selected electron energies and corresponding to the first and second components of decay. (d) Photoelectron intensity as a function of the position on the detector. By Abel inversion, the profile of the initial photoelectron velocity distribution can be reconstructed (reproduced with permission from reference [168] ©2006 Royal Society of Chemistry).

as well as the nuclear motion evolution itself can be traced using fs-TRPES. TRPES is particularly well-suited to the study of ultrafast non-adiabatic processes because photoelectron spectroscopy is sensitive to both electronic configurations (molecular orbitals) and vibrational dynamics [229–232].

Experimentally, the radial and angular dependencies of photo-produced electrons are recorded. The radial dependence provides the electron kinetic energy whereas the angular dependence is due to the emission anisotropy. It is also possible to record both photoelectron signals through a velocity map imaging device and photo-produced ions by means of a time-of-flight mass spectrometer [168] (Figure 2.3.9).

REFERENCES

1. Borho N, Suhm MA: Tailor-made aggregates of α-hydroxy esters in supersonic jets. *Physical Chemistry Chemical Physics* 2004, 6:2885–2890.

2. Linder R, Nispel M, Haber T, Kleinermanns K: Gas-phase FT-IR-spectra of natural amino acids. *Chemical Physics Letters* 2005, 409:260–264.
3. Piseri P, Li Bassi A, Milani P: Time-of-flight analysis of neutral cluster beams through detection of charged particles produced by cluster impact on a channeltron. *Review of Scientific Instruments* 1998, 69:1647–1649.
4. Lepère V, Lucas B, Barat M, Fayeton JA, Picard YJ, Jouvet C, Carçabal P, Nielsen IB, Dedonder C, Grégoire G, *et al.*: Photodissociation of protonated tryptophan: towards a full dynamics analysis from time and position resolved detection of all fragments. *Physical Chemistry Chemical Physics* 2007.
5. Fraser GT, Suenram RD, Lovas FJ: Electric-dipole moments of H_2O-formamide and CH_3OH-formamide. *Journal of Molecular Structure* 1988, 189:165–172.
6. Godfrey PD, Brown RD: Shape of glycine. *Journal of the American Chemical Society* 1995, 117:2019–2023.
7. Godfrey PD, Brown RD, Hunter AN: The shape of urea. *Journal of Molecular Structure* 1997, 413:405–414.
8. Godfrey PD, Firth S, Hatherley LD, Brown RD, Pierlot AP: Millimeter-wave spectroscopy of biomolecules – alanine. *Journal of the American Chemical Society* 1993, 115:9687–9691.
9. Godfrey PD, Hatherley LD, Brown RD: The shape of neurotransmitters by millimeter-wave spectroscopy – 2-phenylethylamine. *Journal of the American Chemical Society* 1995, 117: 8204–8210.
10. Alonso JL, Cocinero EJ, Lessari A, Sanz L, Lopez JC: The glycine–water complex. *Angewandte Chemie-International Edition* 2006, 45:3471–3475.
11. Blanco S, Lesarri A, Lopez JC, Alonso JL: The gas-phase structure of alanine. *Journal of the American Chemical Society* 2004, 126:11675–11683.
12. Lesarri A, Cocinero EJ, Lopez JC, Alonso JL: Gas-phase structure of *N,N*-dimethylglycine. *Chemphyschem* 2005, 6:1559–1566.
13. Lesarri A, Mata S, Cocinero EJ, Blanco S, Lopez JC, Alonso JL: The structure of neutral proline. *Angewandte Chemie-International Edition* 2002, 41:4673–4676.
14. Chass GA, Mirasol RS, Setiadi DH, Tang TH, Chin W, Mons M, Dimicoli I, Dognon JP, Viskolcz B, Lovas S, *et al.*: Characterization of the conformational probability of *N*-acetyl-phenylalanyl-NH_2 by RHF, DFT, and MP2 computation and AIM analyses, confirmed by jet-cooled infrared data. *Journal of Physical Chemistry A* 2005, 109:5289–5302.
15. Snyder LE, Lovas FJ, Hollis JM, Friedel DN, Jewell PR, Remijan A, Ilyushin VV, Alekseev EA, Dyubko SF: A rigorous attempt to verify interstellar glycine. *Astrophysical Journal* 2005, 619: 914–930.
16. Aviles Moreno JR, Petitprez D, Huet TR: The conformational flexibility in *N*-phenylformamide: an ab initio approach supported by microwave spectroscopy. *Chemical Physics Letters* 2006, 419: 411–416.
17. Lovas FJ, Lavrich RJ, Plusquellic DF: Methyl torsional state analysis of the jet-cooled microwave spectrum of *N*-acetylglycine. *Journal of Molecular Spectroscopy* 2004, 228:251–257.
18. Ilyushin VV, Alekseev EA, Dyubko SF, Motiyenko RA, Lovas FJ: Millimeter wave spectrum of glycine. *Journal of Molecular Spectroscopy* 2005, 231:15–22.
19. Ilyushin VV, Alekseev EA, Dyubko SF, Kleiners I, Hougen JT: Ground and first torsional states of acetamide. *Journal of Molecular Spectroscopy* 2004, 227:155–169.
20. Pszczolkowski L, Bialkowska-Jaworska E, Kisiel Z: The millimeter wave rotational spectrum of lactic acid. *Journal of Molecular Spectroscopy* 2005, 234:106–112.
21. Ohashi N, Hougen JT, Suenram RD, Lovas FJ, Kawashima Y, Fujitake M, Pyka J: Analysis and fit of the Fourier-transform microwave spectrum of the two-top molecule *N*-methylacetamide. *Journal of Molecular Spectroscopy* 2004, 227:28–42.
22. Caminati W: The rotational spectra of conformers of biomolecules: tryptamine. *Physical Chemistry Chemical Physics* 2004, 6:2806–2809.

23. Connell LL, Corcoran TC, Joireman PW, Felker PM: Conformational analysis of jet-cooled tryptophan analogs by rotational coherence spectroscopy. *Chemical Physics Letters* 1990, 166: 510–516.
24. Felker PM: Rotational coherence spectroscopy: studies of the geometries of large gas-phase species by picosecond time-domain methods. *Journal of Physical Chemistry* 1992, 96: 7844–7857.
25. Jarzeba W, Matylitsky VV, Riehn C, Brutschy B: Rotational coherence spectroscopy of jet-cooled molecules by femtosecond degenerate four-wave mixing: non-rigid symmetric and asymmetric tops. *Chemical Physics Letters* 2003, 368:680–689.
26. Bergner A, Heugen U, Bründermann E, Scwaab G, Hevenith M, Chamberlin DR, Haller EE: New p-Ge THz laser spectrometer for the study of solutions: THz absorption spectroscopy of water. *Review of Scientific Instruments* 2005, 76:063110.
27. Stephenson SK, Saykally RJ: Terahertz laser velocity modulation spectroscopy of ions. *Journal of Molecular Spectroscopy* 2005, 231:145–153.
28. Kutteruf MR, Brown CM, Iwaki LK, Campbell MB, Korter TM, Heilweil J: Terahertz spectroscopy of short-chain polypeptides. *Chemical Physics Letters* 2003, 375:337–343.
29. Korter TM, Plusquellic DF: Continous-wave terahertz spectroscopy of biotin: vibrational anharmonicity in the infrared. *Chemical Physics Letters* 2004, 385:45–51.
30. Hollas JM: *High resolution spectroscopy*: Wiley, 1998.
31. Wang J, Hochstrasser RM: Anharmonicity of amide bonds. *Journal of Physical Chemistry B* New York 2006, 110:3798–3807.
32. Nesbitt DJ, Field RW: Vibrational energy flow in highly excited molecules: role of intramolecular vibrational redistribution. *Journal of Physical Chemistry* 1996, 100:12735–12756.
33. Leitner DM: Vibrational energy transfer in helices. *Physical Review Letters* 2001, 87:1881021–1881024.
34. Cui W, Hu Y, Lifshitz C: Time resolved photo dissociation of small peptide ions – combining laser desorption with ion trap/reflectron TOF mass spectromerty. *European Physical Journal D* 2002, 20:565–571.
35. Windhorn L, Yeston JS, Witte T, Fuss W, Motzkus M, Proch D, Kompa KL, Bradley-Moore C: Getting ahead of IVR: a demonstration of mid-infrared induced dissociation on a sub-statistical time scale. *Journal of Chemical Physics* 2003, 119:641–645.
36. Pouthier V, Falvo C: Relaxation channels of two-vibronic bound states in α-helix proteins. *Physical Review E* 2004, 69:041906–041912.
37. Yamada Y, Ebata T, Kayano M, Mikami N: Picosecond IR-UV pump-probe spectroscopic study of the dynamics of the vibrational relaxation of jet-cooled phenol. I. Intramolecular vibrational energy redistribution of the OH and CH stretching vibrations of bare phenol. *Journal of Chemical Physics* 2004, 120:7400–7409.
38. Price EA, Robertson WH, Diken EG, Weddle GH, Johnson MA: Argon predissociation infrared spectroscopy of the hydroxide-water complex (OH—H_2O). *Chemical Physics Letters* 2002, 366: 412–416.
39. Yamada Y, Mikami N, Ebata T: Real-time detection of doorway states in the intramolecular vibrational energy redistribution of the OH/OD stretch vibration of phenol. *Journal of Chemical Physics* 2004, 121:11530–11534.
40. Valle JJ, Eyler JR, Oomens J, Moore DT, van der Meer AFG, von Helden G, Meijer G, Hendrickson CL, Marshall AG, Blakney GT: Free electron laser-Fourier transform ion cyclotron resonance mass spectrometry facility for obtaining infrared multiphoton dissociation spectra of gaseous ions. *Review of Scientific Instruments* 2005, 76:23103.
41. Dian BC, Longarte A, Winter PR, Zwier TS: The dynamics of conformational isomerization in flexible biomolecules. I. Hole-filling spectroscopy of *N*-acetyl tryptophan methyl amide and *N*-acetyl tryptophan amide. *Journal of Chemical Physics* 2004, 120:133–147.

42. Ebata T, Hashimoto T, Ito T, Inokuchi Y, Altunsu F, Brutschy B, Tarakeshwar P: Hydration profiles of aromatic amino acids: conformations and vibrations of L-phenylalanine-$(H_2O)_n$ clusters. *Physical Chemistry Chemical Physics* 2006, 8:4783–4791.
43. Mons M, Piuzzi F, Dimicoli I, Gorb L, Leszczynski J: Near-UV resonant two-photon ionization spectroscopy of gas phase guanine: evidence for the observation of three rare tautomers. *Journal of Physical Chemistry A* 2006, 110:10921–10924.
44. Lucas B, Lecomte F, Reimann B, Barth HD, Gregoire G, Bouteiller Y, Schermann JP, Desfrancois C: A new infrared spectroscopy technique for structural studies of mass-selected neutral polar complexes without chromophore. *Physical Chemistry Chemical Physics* 2004, 6:2600–2604.
45. Chin W, Compagnon I, Dognon JP, Canuel C, Piuzzi F, Dimicoli I, von Helden G, Meijer G, Mons M: Spectroscopic evidence for gas-phase formation of successive β-turns in a three-residue peptide chain. *Journal of the American Chemical Society* 2005, 127:1388–1389.
46. Choi MY, Dong F, Miller RE: Multiple tautomers of cytosine identified and characterized by infrared laser spectroscopy in helium nanodroplets: probing structure using vibrational transition moment angles. *Philosophical Transactions of the Royal Society of London Series A: Mathematical Physical and Engineering Sciences* 2005, 363:393–412.
47. Choi MS, Miller RE: Four tautomers of isolated guanine from infrared laser spectroscopy in helium droplets. *Journal of the American Chemical Society* 2006, 128:7230–7328.
48. Carcabal P, Jockusch RA, Hunig I, Snoek LC, Kroemer RT, Davis BG, Gamblin DP, Compagnon I, Oomens J, Simons JP: Hydrogen bonding and cooperativity in isolated and hydrated sugars: mannose, galactose, glucose, and lactose. *Journal of the American Chemical Society* 2005, 127:11414–11425.
49. Bludsky O, Chocholousova J, Vacek J, Huisken F, Hobza P: Anharmonic treatment of the lowest-energy conformers of glycine: a theoretical study. *Journal of Chemical Physics* 2000, 113: 4629–4635; Benoit DM: Efficient correlation-corrected vibrational self-consistent field computation of OH-stretch frequencies using a low-scaling algorithm *Journal of Chemical Physics* 2006, 125:244110–244116.
50. Lucas B, Gregoire G, Lecomte F, Reimann B, Schermann JP, Desfrancois C: Infrared spectroscopy of mass-selected neutral molecular systems without chromophore: the formamide monomer and dimer. *Molecular Physics* 2005, 103:1497–1503.
51. Gaigeot MP, Sprik M: Ab initio molecular dynamics computation of the infrared spectrum of aqueous uracil. *Journal of Physical Chemistry B* 2003, 107:10344–10358.
52. Gaigeot MP, Vuilleumier R, Sprik M, Borgis D: Infrared spectroscopy of *N*-methylacetamide revisited by ab initio molecular dynamics simulations. *Journal of Chemical Theory and Computation* 2005, 1:772–789.
53. Gregurick SK, Chaban GM, Gerber RB: Ab initio and improved empirical potentials for the calculation of the anharmonic vibrational states and intramolecular mode coupling of *N*-methylacetamide. *Journal of Physical Chemistry A* 2002, 106:8696–8707.
54. Brauer B, Chaban GM, Gerber RB: Spectroscopically-tested, improved, semi-empirical potentials for biological molecules: calculations for glycine, alanine and proline. *Physical Chemistry Chemical Physics* 2004, 6:2543–2556.
55. Brauer B, Gerber RB, Kabelac M, Hobza P, Bakker JM, Riziq AGA, de Vries MS: Vibrational spectroscopy of the G···C base pair: experiment, harmonic and anharmonic calculations, and the nature of the anharmonic couplings. *Journal of Physical Chemistry A* 2005, 109:6974–6984.
56. Li Q, Wu G, Yu Z: The role of methyl groups in the formation of hydrogen bonds in DMSO-methanol mixtures. *Journal of the American Chemical Society* 2006, 128:1438–1439.
57. Hobza P, Havlas Z: Blue shifting hydrogen bonds. *Chemical Reviews* 2000, 1000:4253–4264.
58. Chocholousova J, Spirko V, Hobza P: First local minimum of the formic acid dimer exhibits simultaneosly red-shifted O—H···O and improper C—H···O hydrogen bonds. *Physical Chemistry Chemical Physics* 2004, 6:37–41.

59. Lu P, Liu GQ, Li JC: Existing problems in theoretical determination of red-shifted or blue-shifted hydrogen bond. *Journal of Molecular Structure (Theochem)* 2005, 723:95–100.
60. John U, Nair KPR: Near overtone spectral investigations of liquid phase imidazole and the blue-shifting hydrogen bonds. *Spectrochimica Acta Part A: Molecular and Biomolecular Spectroscopy* 2006, 63:169–173.
61. Chin W, Piuzzi F, Dimicoli I, Mons M: Probing the competition between secondary structures and local preferences in gas phase isolated peptide backbones. *Physical Chemistry Chemical Physics* 2006, 8:1033–1048.
62. Vass E, Hollosi M, Besson F, Buchet R: Vibrational spectroscopic detection of β- and γ-turns in synthetic and natural peptides and proteins. *Chemical Reviews* 2003, 103:1917–1954.
63. Keutsch FN, Saykally RJ: Water clusters: untangling the mysteries of the liquid, one molecule at a time. *Proceedings of the National Academy of Sciences of United States of America* 2001, 98: 10533–10540.
64. Kobko N, Paraskevas L, del Rio E, Dannenberg JJ: Cooperativity in amide hydrogen bonding chains: implications for protein-folding models. *Journal of the American Chemical Society* 2001, 123:4348–4349.
65. la Cour Jansen T, Dijkstra AG, Watson TM, Hirst JD, Knoester J: Modeling the amide I bands of small peptides. *Journal of Chemical Physics* 2006, 125:044312.
66. Singh BR: *Infrared analysis of peptides and proteins: principles and applications*: Oxford University Press, Oxford 2000.
67. Ramirez FJ, Tunon I, Silla E: Aminoacid chemistry in solution: structural properties and vibrational dynamics of serine using density functional theory and a continuum solvent model. *Chemical Physics* 2004, 303:85–96.
68. Hering HA, Innocent PR, Haris PI: Towards developing a protein infrared spectra databank (PISD) for proteomics research. *Proteomics* 2004, 4:2310–2319.
69. Oomens J, Moore DT, Meijer G, von Helden G: Infrared multiple photon dynamics and spectroscopy of cationic PABA and its dehydroxylated fragment ion. *Physical Chemistry Chemical Physics* 2004, 6:710–718. Oomens J, Sartakov B, Meijer G, von Helden G: Gas-phase infrared multiple photon dissociation spectroscopy of mass-selected molecular ions. *International Journal of Mass Spectrometry* 2006, 254:1–19.
70. Diken EG, Hammer NI, Johnson MA: Preparation and photoelectron spectrum of the glycine molecular anion: assignment to a dipole-bound electron species with a high-dipole moment, non-zwitterionic form of the neutral core. *Journal of Chemical Physics* 2004, 120:9899–9902.
71. Choi MY, Miller RE: Multiple isomers of uracil-water complexes: infrared spectroscopy in helium nanodroplets. *Physical Chemistry Chemical Physics* 2005, 7:3565–3573.
72. Choi MY, Miller RE: Infrared laser spectroscopy of imidazole complexes in helium nanodroplets: monomer, dimer, and binary water complexes. *Journal of Physical Chemistry A* 2006, 110: 9344–9351.
73. Pei L, Zhang JH, Wu C, Kong W: Conformational identification of tryptamine embedded in superfluid droplets using electronic polarization spectroscopy. *Journal of Chemical Physics* 2006, 125: 024305–024308.
74. Scherer JJ, Voelkel N, Rakestraw DJ, Paul JB, Collie CP, Saykally RJ, O'Keefe A: Infrared cavity ringdown laser absorption spectroscopy (IR-CRLAS). *Chemical Physics Letters* 1995, 245:73–280.
75. Scherer JJ, Paul JB, Collier CP, Okeefe A, Rakestraw DJ, Saykally RJ: Cavity ring-down laser spectroscopy: a new ultrasensitive absorption technique. *Spectroscopy* 1996, 11:46–50.
76. Terasaki A, Kondow T, Egashira K: Continuous-wave cavity ring-down spectroscopy applied to solids: properties of a Fabry-Perot cavity containing a transparent substrate. *Journal of the Optical Society of America* 2005, 22:675–686.
77. Shaw AM, Zare RN, Bennet CV, Kolner BH: Bounce-by-bounce cavity ring-down spectroscopy: femtosecond temporal imaging. *ChemPhysChem* 2001, 2:118–121.

78. Casaes RN, Paul JB, McLaughlin RP, Saykally RJ, van Mourik T: Infrared cavity ring-down spectroscopy of jet-cooled nucleotide base clusters and water complexes. *Journal of Physical Chemistry A* 2004, 108:10989–10996.
79. Wysocki VH, Resing KA, Zhang QF, Cheng GL: Mass spectrometry of peptides and proteins. *Methods* 2005, 35:211–222.
80. Bohmse DS, Woodin R, Beauchamp JL: Molecular activation with low-intensity CW infrared laser radiation. Multiphoton dissociation of ions derived from diethyl ether. *Journal of the American Chemical Society* 1979, 101:5503–5512.
81. Lemaire J, Boissel P, Heninger M, Mauclaire G, Bellec G, Mestdagh H, Simon A, Caer SL, Ortega JM, Glotin F, *et al.*: Gas phase infrared spectroscopy of selectively prepared ions. *Physical Review Letters* 2002, 89:273002.1–273002.4.
82. Asmis KR, Brümmer M, Kaposta C, Sambrogio G, von Helden G, Meijer G, Rademann K, Wöste L: Mass-selected infrared photodissociation spectroscopy of $V_4O_{10}^+$. *Physical Chemistry Chemical Physics* 2002, 4:1101–1104.
83. Kapota C, Lemaire J, Maitre P, Ohanessian G: Vibrational signature of charge solvation vs salt bridge isomers of sodiated amino acids in the gas phase. *Journal of the American Chemical Society* 2004, 126:1836–1842.
84. Lucas B, Gregoire G, Lemaire J, Maitre P, Glotin F, Schermann JP, Desfrancois C: Infrared multiphoton dissociation spectroscopy of protonated *N*-acetyl-alanine and alanyl-histidine. *International Journal of Mass Spectrometry* 2005, 243:105–113.
85. Oh HB, Lin C, Hwang HY, Zhai HL, Breuker K, Zabrouskov V, Carpenter BK, McLafferty FW: Infrared photodissociation spectroscopy of electrosprayed ions in a Fourier transform mass spectrometer. *Journal of the American Chemical Society* 2005, 127:4076–4083.
86. Mac Aleese L, Simon A, McMahon TB, Ortega JM, Scuderi D, Lemaire J, Maitre P: Mid-IR spectroscopy of protonated leucine methyl ester performed with an FT-ICR or a Paul type ion-trap. *International Journal of Mass Spectrometry* 2006, 249:14–20.
87. Ohanessian G: personal communication.
88. Li T, Talvenheimo J, Zeni L, Rosenfeld R, Stearns G, Arakawa T: Changes in protein conformation and dynamics upon complex formation of brain-derived neurotrophic factor and its receptor: investigation by isotope-edited Fourier transform IR spectroscopy. *Biopolymers (Biospectroscopy)* 2002, 67:10–19.
89. Bakker JM, Plutzer C, Hunig I, Haber T, Compagnon I, von Helden G, Meijer G, Kleinermanns K: Folding structures of isolated peptides as revealed by gas-phase mid-infrared spectroscopy. *ChemPhysChem* 2005, 6:120–128.
90. Abo-Riziq A, Crews B, Callahan MP, Grace L, de Vries M: Spectroscopy of isolated gramicidin peptides. *Angewandte Chemie-International Edition* 2006, 118:5290–5293.
91. Oomens J, Polfer N, Moore DT, van der Meer L, Marshall AG, Eyler JR, Meijer G, von Helden G: Charge-state resolved mid-infrared spectroscopy of a gas-phase protein. *Physical Chemistry Chemical Physics* 2005, 7:1345–1348.
92. Ideue S, Sakamoto K, Honma K, Clemmer DE: Conformational change of electrosprayed cytochrome c studied by laser-induced fluorescence. *Chemical Physics Letters* 2001, 337:79–84.
93. Chung HS, Tokmakoff A: Visualization and characterization of the infrared active amide I vibrations in proteins. *Journal of Physical Chemistry B* 2006, 110:2888–2898.
94. Valentine SJ, Clemmer DE: H/D exchange levels of shape-resolved cytochrome c conformers in the gas phase. *Journal of the American Chemical Society* 1997, 119:3558–3566.
95. Badman ER, Myung S, Clemmer DE: Evidence for unfolding and refolding of gas-phase cytochrome c ions in a Paul trap. *Journal of the American Society for Mass Spectrometry* 2005, 16:1493–1497.
96. Boyarkin OV, Mercier SR, Kamariotis A, Rizzo TR: Electronic spectroscopy of cold, protonated tryptophan and tyrosine. *Journal of the American Chemical Society* 2006, 128:2816–2817.

97. Kamariotis A, Boyarkin OV, Mercier SR, Beck RD, Bush MF, Williams ER, Rizzo TR: Infrared spectroscopy of hydrated amino acids in the gas phase: protonated and lithiated valine. *Journal of the American Chemical Society* 2006, 128:905–916.
98. Yu C, Irudayaraj J: Spectroscopic characterization of microorganisms by Fourier transform infrared microscopy. *Biopolymers* 2005, 77:368–377.
99. Simons JP, Jockusch RA, Carcabal P, Hung I, Kroemer RT, Macleod NA, Snoek LC: Sugars in the gas phase. Spectroscopy, conformation, hydration, co-operativity and selectivity. *International Reviews in Physical Chemistry* 2005, 24:489–531.
100. Çarçabal P, Patsias T, Hünig I, Liu B, Kaposta C, Snoek LC, Gamblin DP, Davis BG, Simons JP: Spectral signatures and structural motifs in isolated and hydrated monosaccharides: phenyl $\alpha-$ and β-L-fucopyranoside. *Physical Chemistry Chemical Physics* 2006, 8:129–136.
101. Simons JP: Getting into shape? The interaction of molecules of biological interest with molecules of water: from simplicity to complexity. *Comptes Rendus Chimie* 2003, 6:17–31.
102. Zwier TS: Laser probes of conformational isomerization in flexible molecules and complexes. *Journal of Physical Chemistry A* 2006, 110:4133–4150.
103. Kim NJ, Jeong G, Kim YS, Sung J, Kim SK: Resonant two-photon ionization and laser induced fluorescence spectroscopy of jet-cooled adenine. *Journal of Chemical Physics* 2000, 113:10051–10055.
104. Kim NJ, Kang YK, Park YD, Kim SK: Dispersed fluorescence spectroscopy of jet-cooled adenine. *Physical Chemistry Chemical Physics* 2004, 6:2802–2806.
105. Dashtiev M, Zenobi R: Effect of buffer gas on the fluorescence yield of trapped gas-phase ions. *Journal of American Society for Mass Spectrometry* 2006, 17:855–858.
106. Novikov E, Boens N, Hofkens J: New strategies for low light level detection in single molecule spectroscopy. *Chemical Physics Letters* 2001, 338:151–158.
107. Birch DJS: Multiphoton excited fluorescence spectroscopy of biomolecular systems. *Spectrochimica Acta Part A: Molecular and Biomolecular Spectroscopy* 2001, 57:2313–2336.
108. Neusser HJ, Brutschy B, Pratt DW: High resolution spectroscopy in the gas phase – preface. *Chemical Physics* 2002, 283:1–3.
109. Yi JT, Pratt DW: Rotationally resolved electronic spectroscopy of tryptophol in the gas phase. *Physical Chemistry Chemical Physics* 2005, 7:3680–3684.
110. Schmitt M, Böhm M, Ratzer C, Vu C, Kalkman I, Meerts WL: Structural selection by microsolvation: conformational locking of tryptamine. *Journal of the American Chemical Society* 2005, 127:10356–10364.
111. Chervenkov S, Wang PQ, Braun JE, Neusser HJ: Fragmentation and conformation study of ephedrine by low- and high-resolution mass selective UV spectroscopy. *Journal of Chemical Physics* 2004, 121:7169–7174.
112. Rizzo TR, Park YD, Levy DH: Dispersed fluorescence of jet-cooled tryptophan: excited state conformers and intramolecular exciplex formation. *Journal of Chemical Physics* 1986, 85: 6945–6951.
113. Piuzzi F, Mons M, Dimicoli I, Tardivel B, Zhao Q: Ultraviolet spectroscopy and tautomerism of the DNA base guanine and its hydrated form in a supersonic jet. *Chemical Physics* 2001, 270:205–214.
114. Brause R, Fricke H, Gerhards M, Weinkauf R, Kleinermanns K: Double resonance spectroscopy of different conformers of the neurotransmitter amphetamine and its clusters with water. *Chemical Physics* 2006, 327:43–53.
115. Le Barbu-Debus K, Lahmani F, Zehnacker A, Guchhait N: Electronic and infrared spectroscopy of chiral-*cis*-1-amino-indan-2-ol in a supersonic jet. *Chemical Physics Letters* 2006, 422:218–225.
116. Nguyen TV, Korter TM, Pratt DW: Tryptamine in the gas phase. A high resolution laser study of the structural and dynamic properties of its ground and electronically excited states. *Molecular Physics* 2005, 103:1603–1613.

117. Schmitt M, Brause R, Marian CM, Salzmann S, Meerts WL: Electronically excited states of tryptamine and its microhydrated complex. *Journal of Chemical Physics* 2006, 125:124309.1–124309.10.
118. Kuzmenkina EV, Heyes CD, Nienhaus GU: Single-molecule Förster resonance energy transfer of protein dynamics under denaturating conditions. *Proceedings of the National Academy of Sciences of United States of America* 2005, 102:15471–15476.
119. Shaner NC, Steinbach PA, Tsien RY: A guide to choosing fluorescent proteins. *Nature Methods* 2005, 2:905–909.
120. Pfleger KDG, Eidne KA: Illuminating insights into protein-protein interactions using bioluminescence resonance energy transfer (BRET). *Nature methods* 2006, 3:165–174.
121. Khoury JT, Rodriguez-Cruz SE, Parks JH: Pulsed fluorescence measurements of trapped molecular ions with zero background detection. *Journal of the American Society for Mass Spectrometry* 2002, 13:696–708.
122. Iavarone AT, Meinen J, Schulze S, Parks JH: Fluorescence probe of polypeptide conformational dynamics in gas phase and in solution. *International Journal of Mass Spectrometry* 2006, 253:172–180.
123. Danell AS, Parks JH: FRET measurements of trapped oligonucleotide duplexes. *International Journal of Mass Spectrometry* 2003, 229:35–45.
124. Dashtiev M, Azov V, Frankevich V, Scharfenberg L, Zenobi R: Clear evidence of fluorescence energy transfer in gas-phase ions. *Journal of the American Society for Mass Spectrometry* 2005, 16:1481–1487.
125. Olofson M, Kalinin S, Zdunek J, Oliveberg M, Johansson LBA: Tryptophan–BODIPY: a versatile donor–acceptor pair for probing generic changes of intraprotein distances. *Physical Chemistry Chemical Physics* 2006, 8:3130–3140.
126. Danell AS, Parks JH: Fraying and electron autodetachment dynamics of trapped gas phase oligonucleotides. *Journal of the American Society for Mass Spectrometry* 2003, 14: 1330–1339.
127. Rizzo TR, Park YD, Levy DH: A molecular beam of tryptophan. *Journal of the American Chemical Society* 1985, 107:277–278.
128. Cohen R, Brauer B, Nir E, Grace L, de Vries MS: Resonance-enhanced multiphoton ionization spectroscopy of dipeptides. *Journal of Physical Chemistry A* 2000, 104:6351–6355.
129. Nir E, Muller M, Grace LI, de Vries MS: REMPI spectroscopy of cytosine. *Chemical Physics Letters* 2002, 355:59–64.
130. Lee KT, Sung J, Lee KJ, Kim SK, Park YD: Resonant two-photon ionization study of jet-cooled amino acid: L-phenylalanine and its monohydrated complex. *Journal of Chemical Physics* 2002, 116:8251–8254.
131. Brutschy B: The structure of microsolvated benzene derivatives and the role of aromatic constituents. *Chemical Reviews* 2000, 100:3891–3920.
132. Robertson EG, Simons JP: Getting into shape: conformational and supramolecular landscapes in small biomolecules and their hydrated clusters. *Physical Chemistry Chemical Physics* 2001, 3:1–18.
133. Mons M, Dimicoli I, Tardivel B, Piuzzi F, Brenner V, Millie P: Site dependence of the binding energy of water to indole: microscopic approach to the side chain hydration of tryptophan. *Journal of Physical Chemistry A* 1999, 103:9958–9965.
134. Ullrich S, Muller-Dethlefs K: A REMPI and ZEKE spectroscopic study of *trans*-acetanilide-H_2O and comparison to ab initio CASSCF calculations. *Journal of Physical Chemistry A* 2002, 106:9188–9195.
135. Georgiev S, Neusser HJ: Mass analyzed threshold ionization of hydrogen bonded clusters of biological molecules: the 3-methylindole C_6H_6 complex. *Journal of Electron Spectroscopy and Related Phenomena* 2005, 142:207–213.

136. Ullrich S, Tarczay G, Tong X, Dessent CEH, Müller-Derhlefs K: A ZEKE photoelectron spectroscopy and ab initio study of the *cis*- and *trans*-isomers of formanilide: characterizing the cationic amide bond. *Physical Chemistry Chemical Physics* 2001, 3:5450–5458.
137. Snoek LC, Robertson EG, Kroemer RT, Simons JP: Conformational landscapes in amino acids: infrared and ultraviolet ion-dip spectroscopy of phenylalanine in the gas phase. *Chemical Physics Letters* 2000, 321:49–56.
138. Plutzer C, Nir E, de Vries MS, Kleinermanns K: IR-UV double-resonance spectroscopy of the nucleobase adenine. *Physical Chemistry Chemical Physics* 2001, 3:5466–5469.
139. Carney JR, Fedorov AV, Cable JR, Zwier TS: Infrared spectroscopy of H-bonded bridges stretched across the *cis*-amide group: I. Water bridges. *Journal of Physical Chemistry A* 2001, 105:3487–3497.
140. Yahagi T, Fujii A, Ebata T, Mikami N: Infrared spectroscopy of the OH stretching vibrations of jet-cooled salicylic acid and its dimer in S_0 and S_1. *Journal of Physical Chemistry A* 2001, 105: 10673–10680.
141. Nir E, Janzen C, Imhof P, Kleinermanns K, de Vries MS: Pairing of the nucleobase guanine studied by IR-UV double-resonance spectroscopy and ab initio calculations. *Physical Chemistry Chemical Physics* 2002, 4:740–750.
142. Unterberg C, Gerlach A, Schrader T, Gerhards M: Structure of the protected dipeptide Ac-Val-Phe-OMe in the gas phase: towards a β-sheet model system. *Journal of Chemical Physics* 2003, 118:8296–8300.
143. Chin W, Dognon JP, Canuel C, Piuzzi F, Dimicoli I, Mons M, Compagnon I, von Helden G, Meijer G: Secondary structures of short peptide chains in the gas phase: double resonance spectroscopy of protected dipeptides. *Journal of Chemical Physics* 2005, 122:054317.1–054317.8.
144. Chin W, Mons M, Dognon JP, Mirasol R, Chass G, Dimicoli I, Piuzzi F, Butz P, Tardivel B, Compagnon I, *et al.*: The gas-phase dipeptide analogue acetyl-phenylalanyl-amide: a model for the study of side chain/backbone interactions in proteins. *Journal of Physical Chemistry A* 2005, 109:5281–5288.
145. Longarte A, Redondo C, Fernandez JA, Castano F: IR/UV and UV/UV double-resonance of guaiacol and eugenol dimers. *Journal of Chemical Physics* 2005, 122:164304.
146. Bakker JM, Compagnon I, Meijer G, von Helden G, Kabelac M, Hobza P, de Vries MS: The mid-IR absorption spectrum of gas-phase clusters of the nucleobases guanine and cytosine. *Physical Chemistry Chemical Physics* 2004, 6:2810–2815.
147. Nosenko Y, Kunitski M, Thummel RP, Kyrychenko A, Herbich J, Waluk J, Riehn C, Brutschy B: Detection and structural characterization of clusters with ultrashort-lived electronically excited states: IR absorption detected by femtosecond multiphoton ionization. *Journal of the American Chemical Society* 2006, 128:10000–10001.
148. Abo-Riziq A, Bushnell JE, Crews B, Callahan M, Grace L, de Vries MS: Gas phase spectrosopy of the pentapetide FDAVS. *Chemical Physics Letters* 2006, 431:227–230.
149. Dian BC, Longarte A, Mercier S, Evans DA, Wales DJ, Zwier TS: The infrared and ultraviolet spectra of single conformations of methyl-capped dipeptides: *N*-acetyl tryptophan amide and *N*-acetyl tryptophan methyl amide. *Journal of Chemical Physics* 2002, 117:10688–10702.
150. Evans DA, Wales DJ, Dian BC, Zwier TS: The dynamics of conformational isomerization in flexible biomolecules. II. Simulating isomerizations in a supersonic free jet with master equation dynamics. *Journal of Chemical Physics* 2004, 120:148–157.
151. Clarkson JR, Dian BC, Moriggi L, DeFusco A, McCarthy V, Jordan KD, Zwier TS: Direct measurement of the energy thresholds to conformational isomerization in Tryptamine: experiment and theory. *Journal of Chemical Physics* 2005, 122, 214311–214326.
152. Lee KT, Sung J, Lee KJ, Park YD, Kim SK: Conformation-dependent ionization energies of L-phenylalanine. *Angewandte Chemie-International Edition* 2002, 41:4114–4117.

153. Wilson KR, Belau L, Nicolas C, Jimenez-Cruz M, Leone SR, Ahmed M: Direct determination of the ionization energy of histidine with VUV synchrotron radiation. *International Journal of Mass Spectrometry* 2006, 249:155–161.
154. Matsuda Y, Mori M, Masaki M, Fujii A, Mikami N: Infrared spectroscopy of size-selected neutral clusters combined with vacuum-ultraviolet-photoionization mass spectrometry. *Chemical Physics Letters* 2006, 422:378–381.
155. Asplund MC, Zanni MT, Hochstrasser RM: Two-dimensional infrared spectroscopy of peptides by phase-controlled femtosecond vibrational echoes. *Proceedings of the National Academy of Sciences of United States of America* 2000, 97:8219–8224.
156. Zhong D, Pal SK, Zewail AH: Femtosecond studies of protein-DNA binding and dynamics: histone I. *ChemPhysChem* 2001, 2:219–227.
157. Groot ML, van Wilderen LJ, van der Horst MA, van Stokkum IH, Hellingwerf KJ, van Grondelle R: Initial steps of signal generation in photoactive yellow protein revealed with femtosecond mid-infrared spectroscopy. *Biochemistry* 2003, 2003:10054–10059.
158. Zhuang W, Abramavicius D, Hayashi T, Mukamel S: Simulation protocols for coherent femtosecond vibrational spectra of peptides. *Journal of Physical Chemistry A* 2006, 110: 3362–3374.
159. Nilsson L, Halle B: Molecular origin of time-dependent fluorescence shifts in proteins. *Proceedings of the National Academy of Sciences of United States of America* 2005, 102: 13867–13872.
160. Pal SK, Zewail AH: Dynamics of water in biological recognition. *Chemical Reviews* 2004, 104: 2099–2123.
161. Hertel IV, Radloff W: Ultrafast dynamics in isolated molecules and molecular clusters. *Reports on Progress in Physics* 2006, 69:1897–2003.
162. Schneider M, Maksimenka R, Buback FJ, Kitsopoulos T, Lago LR, Fisher I: Photodissociation of thymine. *Physical Chemistry Chemical Physics* 2006, 8:3017–3021.
163. Nix MGD, Devine AL, Cronin B, Ashfold MNR: High resolution photofragment translational spectroscopy of the near UV photolysis of indole: dissociation via the $^1\pi\sigma^*$ state. *Physical Chemistry Chemical Physics* 2006, 8:2610–2618.
164. Ashfold MNR, Cronin B, Devine AL, Dixon RN, Nix MGD: The role of $\pi\sigma^*$ excited states in the photodissociation of heteroaromatic molecules. *Science* 2006, 312:1637–1640.
165. Kang H, Jouvet C, Dedonder-Lardeux C, Martrenchard S, Charriere C, Gregoire G, Desfrancois C, Schermann JP, Barat M, Fayeton JA: Photoinduced processes in protonated tryptamine. *Journal of Chemical Physics* 2005, 122:84307.1–84307.7.
166. Kang H, Dedonder-Lardeux C, Jouvet C, Martrenchard S, Gregoire G, Desfrancois C, Schermann JP, Barat M, Fayeton JA: Photo-induced dissociation of protonated tryptophan $TrpH^+$: a direct dissociation channel in the excited states controls the hydrogen atom loss. *Physical Chemistry Chemical Physics* 2004, 6:2628–2632.
167. Kang H, Dedonder-Lardeux C, Jouvet C, Gregoire G, Desfrancois C, Schermann JP, Barat M, Fayeton JA: Control of bond-cleaving reactions of free protonated tryptophan ion by femtosecond laser pulses. *Journal of Physical Chemistry A* 2005, 109:2417–2420.
168. Canuel C, Elhanine M, Mons M, Piuzzi F, Tardivel B, Dimicoli I: Time-resolved photoelectron and photoion fragmentation spectroscopy study of 9-methyladenine and its hydrates: a contribution to the understanding of the ultrafast decay radiationless decay of excited DNA bases. *Physical Chemistry Chemical Physics* 2006, 8:3978–3987.
169. Grégoire G, Dedonder C, Jouvet C: What is the meaning of lifetime measurements *CNRS HAL* 2006, http://hal.archives-ouvertes.fr/.
170. Scheer AM, Aflatooni K, Gallup GA, Burrow PD: Bond breaking and temporary anion states in uracil and halouracils: Implications for the DNA bases. *Physical Review Letters* 2004, 92: 068102–068105.

171. Abdoul-Carime H, Gohlke S, Illenberger E: Fragmentation of tryptophan by low-energy electrons. *Chemical Physics Letters* 2005, 402:497–502.
172. Abouaf R, Dunet H: Structures in dissociative electron attachment cross-sections in thymine, uracil and halouracils. *European Journal of Physics D* 2005, 35:405–410.
173. Burrow PD, Gallup GA, Scheer AM, Denifl S, Ptasinska S, Mark T, Scheier P: Vibrational Feshbach resonances in uracil and thymine. *Journal of Chemical Physics* 2006, 124:124310.1–124310.7.
174. Abdoul-Carime H, Gohlke S, Illenberger E: Conversion of amino-acids by electrons at sub-excitation energies. *Physical Chemistry Chemical Physics* 2004, 6:161–164.
175. Pshenichnyuk SA, Lomakin GS, Fokin AI, Pshenichnyuk IA: Temperature dependence of the mean autodetachment lifetime of the p-benzoquinone molecular radical anion. *Rapid Communications in Mass Spectrometry* 2006, 20:383–386.
176. Tonzani S, Greene CH: Low energy electron scattering from DNA and RNA bases: shape resonances and radiation damages. *Journal of Chemical Physics* 2005, 124:054312.
177. Ptasinska S, Denifl S, Grill V, Mark TD, Illenberger E, Scheier P: Bond- and site-selective loss of H^- from pyrimidine bases. *Physical Review Letters* 2005, 95:093201–4.
178. Ptasinska S, Denifl S, Grill V, Mark TD, Scheier P, Gohlke S, Huels MA, Illenberger E: Bond-selective H^- ion abstraction from thymine. *Angewandte Chemie-International Edition* 2005, 44:1647–1650.
179. Desfrancois C, Abdoulcarime H, Khelifa N, Schermann JP: From 1/R to $1/R^2$ potentials: electron-exchange between Rydberg atoms and polar molecules. *Physical Review Letters* 1994, 73: 2436–2439.
180. Gutowski M, Skurski P, Boldyrev AI, Simons J, Jordan KD: Contribution of electron correlation to the stability of dipole-bound anionic states. *Physical Review A* 1996, 54: 1906–1909.
181. Desfrancois C, Carles S, Schermann JP: Weakly bound clusters of biological interest. *Chemical Reviews* 2000, 100:3943–3962.
182. Jordan KD, Wang F: Theory of dipole-bound anions. *Annual Review of Physical Chemistry* 2003, 54:367–396.
183. Desfrancois C, Bouteiller Y, Schermann JP, Radisic D, Stokes ST, Bowen KH, Hammer NI, Compton RN: Long-range electron binding to quadrupolar molecules. *Physical Review Letters* 2004, 92:083003–083007.
184. Hendricks JH, Lyapustina SA, deClercq HL, Snodgrass JT, Bowen KH: Dipole bound, nucleic acid base anions studied via negative ion photoelectron spectroscopy. *Biophysical Journal* 1996, 70:SU495.
185. Schiedt J, Weinkauf R, Neumark DM, Schlag EW: Anion spectroscopy of uracil, thymine and the amino-oxo and amino-hydroxy tautomers of cytosine and their water clusters. *Chemical Physics* 1998, 239:511–524.
186. Lee GH, Arnold ST, Eaton JG, Bowen KH: Electronic properties of dipole-bound $(H_2O)_2^-$, $(D_2O)_2^-$, $(H_2O)_2^-$ $Ar_{(n=1,2,3)}$, and $(D_2O)_2^-$ $Ar_{(n=1,2,3)}$ using negative ion photoelectron spectroscopy. *Chemical Physics Letters* 2000, 321:333–337.
187. Hammer NI, Shin JW, Headrick JM, Diken EG, Roscioli JR, Weddle GH, Johnson MA: How do small water clusters bind an excess electron? *Science* 2004, 306:675–679.
188. Abdoul-Carime H, Desfrancois C: Electrons weakly bound to molecules by dipolar, quadrupolar or polarization forces. *European Physical Journal D* 1998, 2:149–156.
189. Carles S, Lecomte F, Schermann JP, Desfrancois C: Gas-phase experimental and theoretical studies of adenine, imidazole, pyrrole, and water non-covalent complexes. *Journal of Physical Chemistry A* 2000, 104:10662–10668.
190. Clary DC, Benoit DM: Diffusion Monte Carlo simulations of the dipole-bound state of the water dimer anion. *Journal of Chemical Physics* 1999, 111:10559–10565.

191. Frigato T, Svozil D, Jungwirth P: Valence-and dipole-bound anions of the thymine-water complex: Ab initio characterization of the potential energy surfaces. *Journal of Physical Chemistry A* 2006, 110:2916–2923.
192. Ervin KM: Experimental techniques in gas-phase ion thermochemistry. *Chemical Reviews* 2001, 101:391–444.
193. Rienstra-Kirakofe JC, Tschumper GS, Schaeffer HF, Nandi S, Ellison GB: Atomic and molecular electron affinities: photoelectron experiments and theoretical computations. *Chemical Reviews* 2002, 102:231–282.
194. Bachorz RA, Rak J, Gutowski M: Stabilization of very rare tautomers of uracil by an excess electron. *Physical Chemistry Chemical Physics* 2005, 7:2116–2125.
195. Bachorz RA, Klopper W, Gutowski M: Coupled-cluster and explicitly-correlated perturbation-theory calculations of the uracil anion. *Journal of Chemical Physics* 2007, 126:085101–085107.
196. Desfrancois C, AbdoulCarime H, Schermann JP: Electron attachment to isolated nucleic acid bases. *Journal of Chemical Physics* 1996, 104:7792–7794.
197. Denifl S, Zappa F, Mähr I, Lecointre J, Probst M, Märk T, Scheier P: Mass spectrometric investigation of anions formed upon free electron attachment to nucleobases molecules and clusters embedded in superfluid helium droplets. *Physical Review Letters* 2006, 97:043201–043204.
198. Sevilla MD, Besler B, Colson AO: Ab-initio molecular-orbital calculations of DNA radical ions.5. Scaling of calculated electron-affinities and ionization-potentials to experimental values. *Journal of Physical Chemistry* 1995, 99:1060–1063.
199. Weselowski SS, Leininger ML, Pentchev PN, Schaeffer HF: Electron affinities of the DNA and RNA bases. *Journal of the American Chemical Society* 2001, 123:4023–4028.
200. Li XF, Cai ZL, Sevilla MD: DFT calculations of the electron affinities of nucleic acid bases: dealing with negative electron affinities. *Journal of Physical Chemistry A* 2002, 106:1596–1603.
201. Oyler NA, Adamowicz L: Theoretical ab-initio calculations of the electron-affinity of thymine. *Chemical Physics Letters* 1994, 219:223–227.
202. Skurski P, Gutowski M, Simons J: How to choose a one-electron basis set to reliably describe a dipole-bound anion. *International Journal of Quantum Chemistry* 2000, 80:1024–1038.
203. Bouteiller Y, Desfrancois C, Schermann JP, Latajka Z, Silvi B: Calculation of electronic affinity and vertical detachment energy of the water dimer complex using the density functional theory. *Journal of Chemical Physics* 1998, 108:7967–7972.
204. Lee GH, Arnold ST, Eaton JG, Sarkas HW, Bowen KH, Ludewigt C, Haberland H: Negative-ion photoelectron spectroscopy of solvated electron cluster anions, $(H_2O)_N^-$ and $(NH_3)_N$. *Zeitschrift Fur Physik D-Atoms Molecules and Clusters* 1991, 20:9–12.
205. Vera DMA, Pierini AB: Species with negative electron affinities and standard DFT methods. *Physical Chemistry Chemical Physics* 2004, 6:2899–2903.
206. Modelli A: Electron attachment and intramolecular electron transfer in unsaturated chloroderivatives. *Physical Chemistry Chemical Physics* 2003, 5:2923–2930.
207. Seydou M, Modelli A, Lucas B, Konate K, Desfrancois C, Schermann JP: Electron attachment to strongly polar clusters – formamide molecule and clusters. *European Physical Journal D* 2005, 35:199–205.
208. Fast PL, Sanchez ML, Corchado JC, Trulhar DG: The Gaussian-2 method with proper dissociation, improved accuracy, and less cost. *Journal of Chemical Physics* 1999, 110:11679–11681.
209. Aflatooni K, Hitt B, Gallup GA, Burrow PD: Temporary ion states of selected amino acids. *Journal of Chemical Physics* 2001, 115:6489–6494.
210. Periquet V, Moreau A, Carles S, Schermann JP, Desfrancois C: Cluster size effects upon anion solvation of *N*-heterocyclic molecules and nucleic acid bases. *Journal of Electron Spectroscopy and Related Phenomena* 2000, 106:141–151.
211. Aflatooni K, Gallup GA, Burrow PD: Electron attachment energies of the DNA bases. *Journal of Physical Chemistry A* 1998, 102:6205–6207.

212. Abdoul-Carime H, Gohlke S, Fischbach E, Scheike J, Illenberger E: Thymine excision from DNA by subexcitation electrons. *Chemical Physics Letters* 2004, 387:267–270.
213. Desfrancois C: Determination of electron-binding energies of ground-state dipole-bound molecular anions. *Physical Review A* 1995, 51:3667–3675.
214. Lecomte F, Lucas B, Gregoire G, Schermann JP, Desfrancois C: Structures of amide-water neutral complexes from dipole-bound anion formation. *European Physical Journal D* 2002, 20: 449–457.
215. White S, Szewczyk JW, Turner JM, Baird EE, Dervan PB: Recognition of the four Watson-Crick pairs in the DNA minor groove by synthetic ligands. *Nature* 1998, 291:468–471.
216. Dervan PB: Molecular recognition of DNA by small molecules. *Bioorganic and Medicinal Chemistry* 2001, 9:2215–2235.
217. Best TP, Edelson BS, Nickols NG, Dervan PB: Nuclear localization of pyrrole-imidazole polyamide-fluorescein conjugates in cell culture. *Proceedings of the National Academy of Sciences of United States of America* 2003, 100:12063–12068.
218. Desfrancois C, Abdoul-Carime H, Schulz CP, Schermann JP: Laser separation of geometrical isomers of weakly-bound molecular complexes. *Science* 1995, 269:1707–1709.
219. Compton RN, Carman HS, Desfrancois C, Abdoul-Carime H, Schermann JP, Hendricks JH, Lyapustina SA, Bowen KH: On the binding of electrons to nitromethane: dipole and valence bound anions. *Journal of Chemical Physics* 1996, 105:3472–3478.
220. Hendricks JH, Lyapustina SA, de Clercq HL, Bowen KH: The dipole bound-to-covalent anion transformation in uracil. *Journal of Chemical Physics* 1998, 108:8–11.
221. Sommerfeld T: Coupling between dipole-bound and valence states: the nitromethane anion. *Physical Chemistry Chemical Physics* 2002, 4:2511–2516.
222. Han SY, Kim JH, Song JK, Kim SK: Simultaneous observation of dipole-bound and valence electron states in pyridine tetramer anion. *Journal of Chemical Physics* 1998, 109:9656–9659.
223. Svozil D, Frigato T, Havlas Z, Jungwirth P: Ab initio electronic structure of thymine anions. *Physical Chemistry Chemical Physics* 2005, 7:840–845.
224. Haranczyk M, Gutowski M: Valence and dipole-bound anions of the most stable tautomers of guanine. *Journal of the American Chemical Society* 2005, 127:699–706.
225. Diken EG, Headrick JM, Johnson MA: Photoelectron spectroscopy of the glycine center $(H_2O)^-_{1,2}$ clusters: sequential hydration shifts and observation of isomers. *Journal of Chemical Physics* 2005, 122:224317.1–224317.6.
226. Radisic D, Bowen KH, Dabkowska I, Storoniak P, Rak J, Gutowski M: AT base pair anions versus (9-methyl-A)(1-methyl-T) base pair anions. *Journal of the American Chemical Society* 2005, 127:6443–6450.
227. Carles S, Lecomte F, Schermann JP, Desfrancois C, Xu S, Nilles JM, Bowen KH, Berges J, Houee-Levin C: Nondissociative electron capture by disulfide bonds. *Journal of Physical Chemistry A* 2001, 105:5622–5626.
228. Desfrancois C, Abdoul-Carime H, Carles S, Periquet V, Schermann JP, Smith DMA, Adamowicz L: Experimental and theoretical ab initio study of the influence of *N*-methylation on the dipole-bound electron affinities of thymine and uracil. *Journal of Chemical Physics* 1999, 110:11876–11883.
229. Lee SH, Tang KC, Chen IC, Schmitt M, Shaffer JP, Schultz T, Underwood JG, Zgierski MZ, Stolow A: Substituent effects in molecular electronic relaxation dynamics via time-resolved photoelectron spectroscopy: $\pi\pi^*$ states in benzenes. *Journal of Physical Chemistry A* 2002, 106:8979–8991.
230. Smits M, de Lange CA, Ullrich S, Schultz T, Schmitt M, Underwood JG, Shaffer JP, Rayner DM, Stolow A: Stable kilohertz rate molecular beam laser ablation sources. *Review of Scientific Instruments* 2003, 74:4812–4817.

231. Ullrich S, Schultz T, Zgierski MZ, Stolow A: Direct observation of electronic relaxation dynamics in adenine via time-resolved photoelectron spectroscopy. *Journal of the American Chemical Society* 2004, 126:2262–2263.
232. Neumark DM: Probing chemical dynamics with negative ions. *Journal of Chemical Physics* 2006, 125:132303.
233. Ullrich S, Schultz T, Zgierski MZ, Stolow A: Electronic relaxation dynamics in DNA and RNA bases studied by time-resolved photoelectron spectroscopy. *Physical Chemistry Chemical Physics* 2004, 6:2796–2801.
234. Scott AP, Radom L: Harmonic vibrational frequencies: an evaluation of Hartree-Fock, Moller-Plesset, quadratic configuration interaction, DFT, and semi-empirical scale factors. *Journal of Chemical Physics* 1996, 100:16502–16513.

– 3 –

Experimental Methods

3.1 BRINGING BIOMOLECULES INTO GAS-PHASE

GENERAL FEATURES

A vast majority of biomolecules have extremely low vapour pressures at room temperature and are thermally fragile. For example, amino acids sublimate and the vapour pressure at their temperature threshold for evaporation ranges from 9×10^{-10} atm (L-Cys) down to 1.8×10^{-10} atm (L-Phe) [1]. Different types of sources have been devised to bring intact neutral [2–5] or charged biomolecules in the gas phase while still avoiding rapid degradation from heating. Some methods, such as electrospray ionization (ESI) and matrix-assisted laser desorption/ionization (MALDI), have been widely used in mass-spectrometry for a long time and recently for spectroscopic studies of ionic species. Other methods have been specially devised in order to produce cold neutral beams suitable for spectroscopy and will be first examined in this chapter. The separation between production of either neutral or charged species is sometimes arbitrary since some sources, such as matrix-assisted laser desorption (MALD), can deliver both.

3.1.1 Production of neutral species from thermal and supersonic expansions

The difficulty for setting molecules of biological interest into gas phase is somewhat dependent upon the sensitivity of the used experimental setup. According to the spectroscopic method and its sensitivity, a more or less large density of neutral species is required. Among molecules, a fraction of them can be simply heated in order to obtain sufficiently high vapour pressure before reaching decomposition. For example, among nucleobases, reasonable partial vapour pressures allowing for some spectroscopic measurements concerning uracil, thymine, adenine and cytosine [6] can be reached while guanine often requires more elaborate methods than simple heating [7–9].

Immediately after vaporization, the molecular temperature is still rather high and different methods are available for cooling molecular systems down to a temperature such that only a restricted number of low-lying conformations are populated. For that purpose, free jet expansions and deposition on helium clusters (see Section 3.1.3) are used for neutrals. Carrier gases (usually helium or argon) can be seeded by the molecular systems of interest either before or after the source nozzle creating the supersonic expansion (Figure 3.1.1).

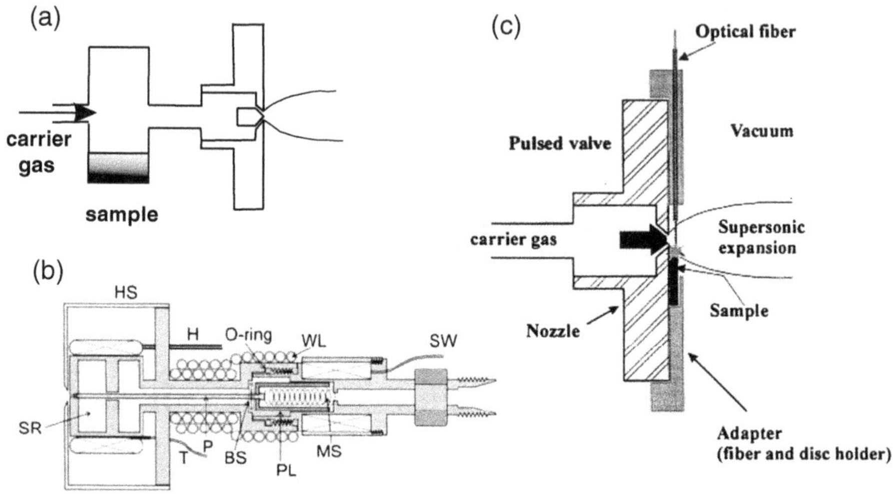

Figure 3.1.1 (a) The carrier gas of a supersonic expansion can be seeded by biomolecular systems in the sample reservoir situated before the supersonic expansion; (b) pratical realization of a hyperthermal source producing pulsed supersonic expansion with a sample reservoir held at 750°C. (HS heat shield, SR sample reservoir, H heater, T thermocouple wire, P stainless steel poppet, BS buffer spring, MS main spring, SW solenoid wire, WL water-cooled line; (c) practical realization of a laser vaporization source combined with a pulsed supersonic expansion. The sample is mixed with graphite powder and pressed into a pellet that is illuminated by the second harmonic of a pulsed YGG laser through a multimode fibre (reproduced with permission from references [8] and [10] ©2000 Elsevier).

Prior to the expansion, one can assume that a Boltzmann population distribution exists among the different conformers corresponding to a temperature typically in between 400 and 550°C. The question is whether this initial distribution is maintained or not during the cooling process induced by the supersonic expansion. The number of collisions per second, z_{coll}, in the supersonic jet depends upon the reservoir density n_0, the nozzle diameter D, the collision cross-section σ with the carrier gas and the local Mach number M (ratio of the beam velocity to the local speed of sound) and the heat capacity ratio $\gamma = C_p/C_v$.

$$z_{\text{coll}} = \sqrt{2} n_0 \sigma \bar{v}_0 \left[1 + \frac{1}{2}(\gamma - 1)M^2 \right]^{-1/2[(\gamma+1)/(\gamma-1)]} \tag{3.1.1}$$

The energy loss ΔE in each collision of a heavy biomolecule of energy E with a light carrier gas atom can be approximated by

$$\Delta E = \lambda(E - E_{\text{ZPE}}) \tag{3.1.2}$$

where E_{ZPE} is the zero-point energy of the lowest energy conformation. The local Mach number M depends upon the ratio x/D of the distance from the nozzle x and D and varies along the beam as

$$M = 3.26\left(\frac{x}{D} - 0.075\right)^{0.67} - 0.61\left(\frac{x}{D} - 0.075\right)^{-0.67} \approx C\left(\frac{x}{D}\right)^{\gamma-1} \tag{3.1.3}$$

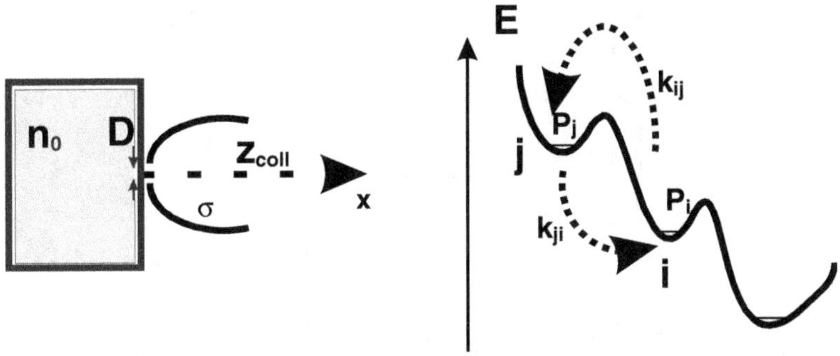

Figure 3.1.2 Left: Parameters of a supersonic expansion used in simulation of the cooling process. Right: Schematic diagram representing minima of a potential energy surface and transition rates between those minima corresponding to different transition states.

If the temperature of the reservoir is T, then the temperature in the beam T varies with distance x from the nozzle as

$$T = \frac{T_0}{(1+(\gamma-1)M^2/2)} \tag{3.1.4}$$

The dynamics of conformational isomerization in a jet expansion has been experimentally [10] and theoretically [11] studied in detail in relation with the systematic exploration of the potential energy surface (PES) (see Sections 1.5 and 2.1.4.4). The change of the occupation probability $P_i(t)$ of a given minimum i of the PES depends upon the occupation probabilities $P_j(t)$ of the other populated minima (Figure 3.1.2) and the rate constants k summed over the different transition states directly connecting minima between themselves.

$$\frac{\mathrm{d}P_i(t)}{\mathrm{d}t} = \sum_{j\neq i}[k_{ij}(E)P_j(t) - k_{ji}(t)P_i(t)] \tag{3.1.5}$$

The cooling process along the supersonic expansion can then be simulated once the different minima and transition state energies have been estimated by a systematic exploration of the molecular PES conducted with a force-field.

An example of the variation of an experimentally observed laser-induced fluorescence (LIF) spectrum recorded at different distances from a beam nozzle is represented in Figure 3.1.3. The variation of the temperature along the expansion (eq. (3.1.4)) can be estimated by comparison between the intensities of the $\nu'' = 0 \rightarrow \nu' = 1$ cold band and the $\nu'' = 1 \rightarrow \nu' = 0$ hot band according to the relation

$$\frac{I_{0\rightarrow 1}}{I_{1\rightarrow 0}} = \exp\left(\frac{-\Delta E}{kT}\right) \tag{3.1.6}$$

In the case of a flexible molecule such as N-acetyl-tryptophan methyl amide, a vibrational temperature down to 10–15°K can be obtained.

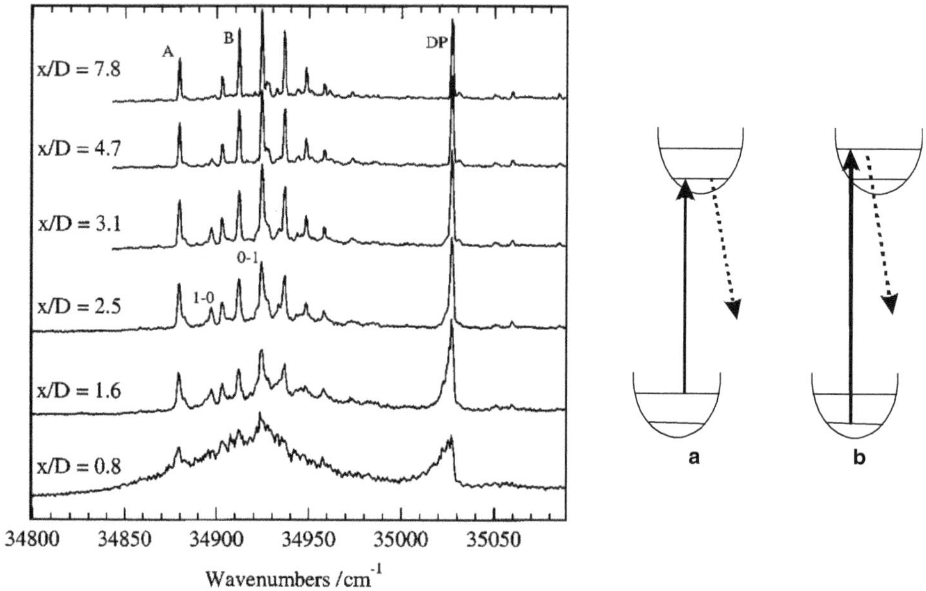

Figure 3.1.3 Variation of the laser-induced fluorescence (LIF) spectrum of *N*-acetyl-tryptophan methyl amide recorded at different distances x from the beam nozzle of diameter D. The temperature is estimated from the ratio of the intensities of the hot (a) and cold (b) bands (reproduced with permission from reference [11] ©2004 American Chemical Society).

A "rule of thumb" has been proposed by J.P. Simons [12]. When barriers between conformers in the PES are larger than 12–15 kJ/mol, the initial Boltzmann population distribution remains unchanged. On the contrary, in presence of energy barriers lower than 5–6 kJ/mol, collisional relaxation brings the molecular systems down to their lowest ZPE conformations.

For experiments requiring long interaction paths between photon beams and molecular systems due to weak absorption in the mid- and far-infrared region, slit nozzles [13–15] or "ragout-type" [16, 17] beams are employed. Very large buffer chambers up to 23 m^3 are then used to feed oversized slit nozzles up to 60 cm long and gas pulses up to 1 s.

3.1.2 Production of neutral species from laser desorption

Thermal degradation can be avoided by means of a very rapid heating of biomolecules. Biomolecules can be either bare [9, 18] or deposited on surfaces such as graphite [19, 20]. They can also be embedded in a carbon [8], cellulose or nicotinic acid [21] matrix. Fast laser irradiation of the surrounding medium introduces a very fast expansion before thermal degradation of the system of interest. This method can be considered as relatively mild since seeded supersonic beams, mostly containing intact parents as well as heterogeneous clusters of matrix molecules and amino acids, can be obtained [22]. Expansion conditions (e.g. backing pressure) must then be carefully controlled to avoid pyrolysis. However, getting non-fragmented molecules can sometimes become a problem. For example, nucleosides can

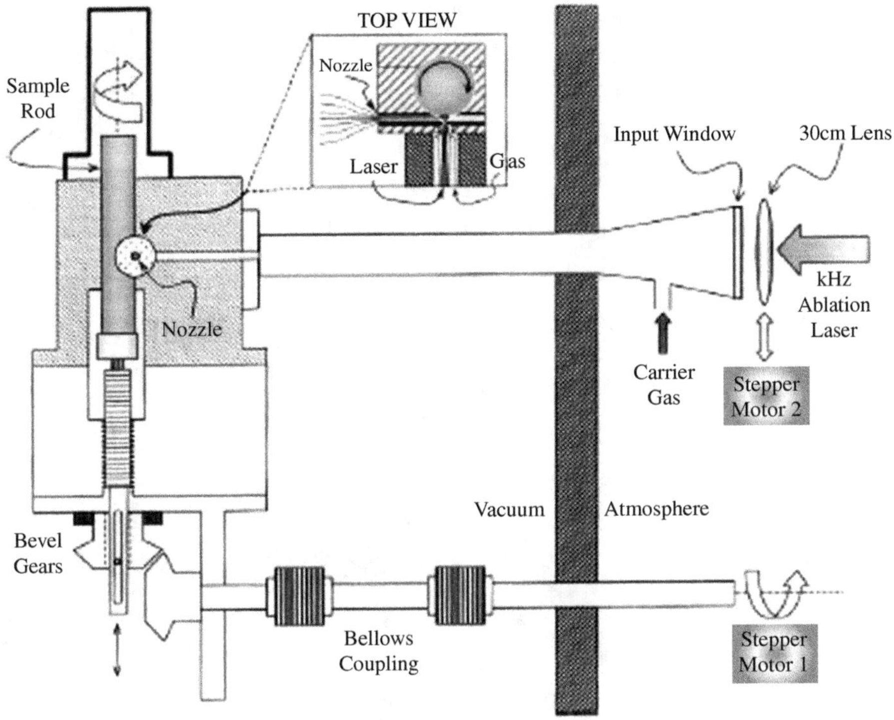

Figure 3.1.4 Schematic representation of a high-repetition rate laser desorption source. The beam of a Nd:YLF laser at 527 nm illuminates a translated rotating cylindrical target of 6 mm diameter (reproduced with permission from reference [24] ©2003 American Institute of Physics).

have different behaviours. Guanosines do not fragment whereas some substituted adenosines do [23].

Most spectroscopic experiments using laser desorption sources are run at very low repetition rates, typically in between 1 and 30 Hz. Some experiments such as photoelectron–photoion coincidence (PEPICO) spectroscopy require less than one particle per laser shot in order to avoid false coincidences and high-repetition rates then be employed for both laser and source. A design of a 1 kHz source used for the generation of a jet-cooled molecular beam of guanine [24] is shown in Figure 3.1.4.

3.1.3 Deposition of neutral species on helium droplets

Molecules, even possessing very low vapour pressures, can be deposited on the surface or embedded inside very small droplets of liquid ^{4}He typically containing 10^3–10^5 atoms (see Section 2.1.3.2.3) [25–31]. Those droplets are produced by cryogenic nozzle expansion. Pressures range from 1 to 100 bars and the nozzles with diameters from 5 to 20 μm are cooled from 10 to 30°K. The equilibrium temperature achieved inside the superfluid droplets has been evaluated to 0.37 ± 0.05°K by fitting the rotational contour of embedded SF_6 [32]. In exceptional cases, very costly ^{3}He can be used to attain temperatures down to

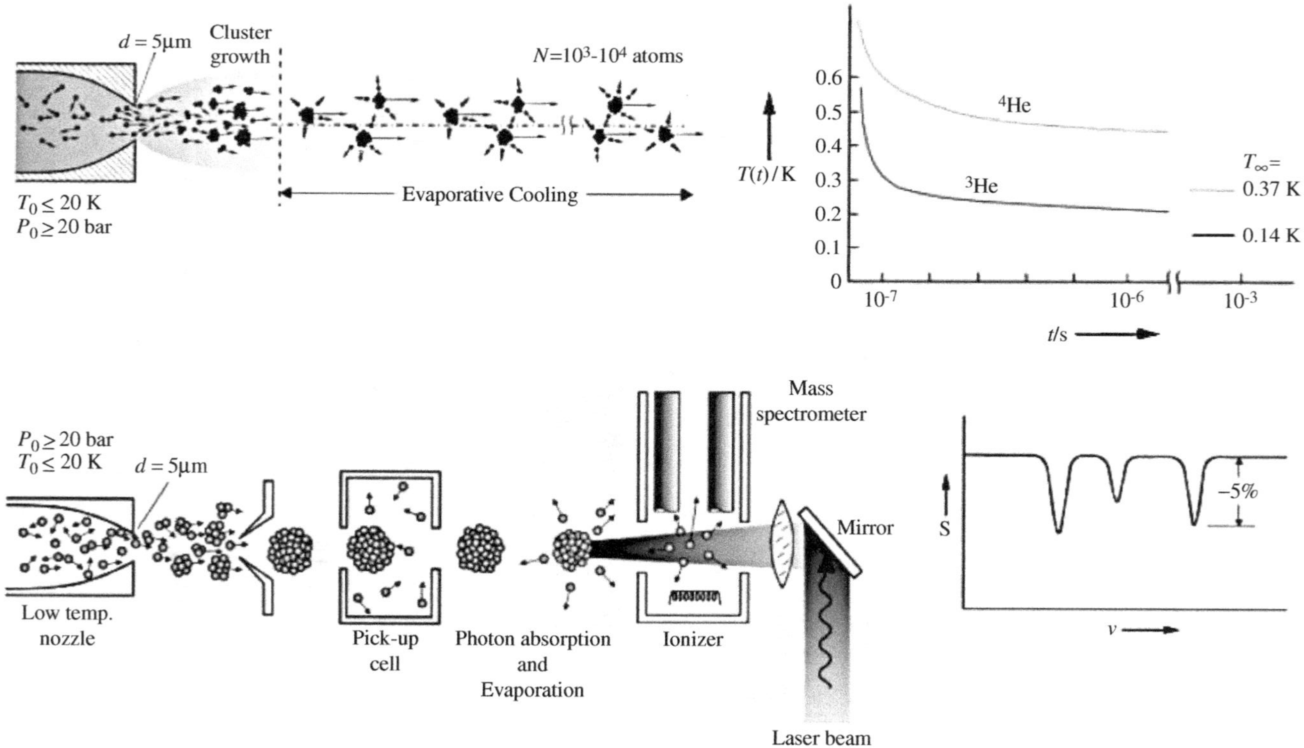

Figure 3.1.5 Top left: Free jet expansion leading to production of helium clusters. Gaseous helium held in a reservoir at a pressure in between 20 and 100 bar and a temperature 5–20°K is expanded through a several micron diameter nozzle. The extensive cooling in the expansion leads to formation of droplets in the cluster growth region. Droplets then evaporate in vacuum and carry on cooling, respectively, down to 0.4 and 0.15°K for ^{4}He and ^{3}He. Right: Calculated dependence of the droplet temperature as function of time for ^{4}He and ^{3}He after they have left the cluster growth region. Bottom: Setup used for pick-up and laser depletion spectroscopy of molecules embedded in helium droplets. The absorption of resonant infrared photons results in evaporation of a large fraction of helium atoms that is detected as a decrease of an ion signal following ionization, mass-selection and detection (reproduced with permission from reference [33] ©2004 Wiley).

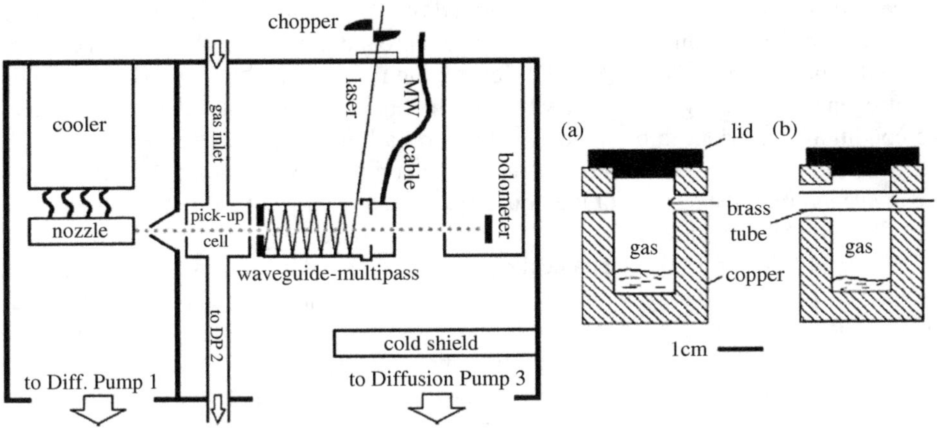

Figure 3.1.6 Left: Typical design of a helium droplet experiment for high-resolution spectroscopy with a conventional pick-up cell (reproduced with permission from reference [25] ©2001 American Institute of Physics). Right: (a) conventional pick-up cell; (b) modified pick-up cell reducing cluster deflections and improving detection of doped clusters (reproduced with permission from reference [34] ©2005 American Institute of Physics).

0.15°K. After collimation, the droplets are seeded with the molecule of interest (dopant) by passing them through a pick-up cell (Figure 3.1.5). If L is the path length through the cell, n the density of the gas in the cell and σ the attachment cross-section, then the ratio between L and the mean free path λ $(= n\sigma L)$ represents the expected number of sticking events. The probablility that a passing cluster will pick-up k molecules (admitting a sticking probability equal to 1) is given by the Poisson distribution

$$P(k) = \frac{\lambda^k e^{-k}}{k!} \tag{3.1.7}$$

The initial kinetic and internal energies of the dopant molecules are rather high since the dopant must be heated for sublimation. Evaporation of each helium atom takes off an energy equivalent to approximately 5 to 7°K. Several hundred He atoms evaporate due to dissipation of kinetic and internal energy of a very small dopant molecule during the pick-up process. Thousands of He atom evaporations are required for large molecules, requiring the use of large droplets. A droplet containing N helium atoms has a very large cross-section equal to $15.5N^{2/3}$ Å, and a dopant vapour pressure in between 10^{-5} and 10^{-4} mbar is sufficient to pick up a single molecule. The pick-up process results in a Poisson distribution of the number of dopants per droplet (Figure 3.1.6).

3.1.4 Liquid and supercritical beams, microjets and liquid droplets

3.1.4.1 Liquid beams

Neutral and ionized biomolecular systems can be directly injected from their natural medium, that is water, into the gas phase. Continuous liquid beams can be produced into vacuum by injection of a liquid through a 10–20 μm aperture in diameter at a pressure in between 10 and 100 bars leading to a flow speed of 10–100 m/s [35, 36]. Although a large amount of water is introduced in vacuum, the background pressure in the vacuum chamber can still be maintained in the 10^{-5}–10^{-6} Torr region by capturing the liquid beam in a liquid nitrogen trap. The high-speed flow renews the liquid beam surface that remains clean. The conditions experienced by molecules lying on a liquid surface have been studied by means of molecular dynamics simulation and experiments with metastable atoms [37]. Due to evaporation, the surface of the liquid beam drops rapidly at a rate of 5°K/mm down to 220°K [38]. Alcohols are easily used as solvents but water may present some difficulties due to a rapid freezing [39]. Mechanical chopping of the ice beam can then be necessary. The studied biomolecular systems introduced in the solvent are ejected from the liquid beam by means of a pulsed nanosecond infrared laser exciting stretching vibrations of the solvent (e.g. the C–O stretch vibration of methanol at 9.66 μm or the O–H stretch vibration of water in the 3,000–3,800 cm^{-1} region) [40–42]. A first explanation of the observation of ion ejection from the liquid beam is given in the reference [43]. After the absorption of pulsed IR photon by the solvent molecules, the vibrational energy relaxation produces within a few picoseconds an extremely rapid heating of the solvent, in the order of 10^{10} °K/s. This raises the pressure up to several hundred bars and brings the solvent in a supercritical state (for water, $T_c = 647$°K, $p_c = 221$ bars, $\rho_c = 322\,kg/m^3$). The dielectric constant then drops suddenly and most ions recombine except few "survivors". Those were highly solvated or far from counter ions, and can then escape into vacuum. By varying the IR laser frequency and power, it is possible to modify its penetration depth in the liquid beam. Fast neutral clusters are also ejected from the beam surface while slow ones are ejected from the bulk region inside the beam [44]. An alternative explanation [45] implies a strong shock wave produced by the fast energy deposition leading to an explosive thermal volume expansion. The liquid is then dispersed into very tiny droplets too small for supporting more than a single charge. In contrast with ESI (see Section 3.1.6), only singly or doubly charged ions are observed. Polydisperse nanoparticles of 150 nm can also be obtained by atomizing solutions of amino acids. A heater and a diffusion dryer are then used to evaporate the solvent. The dry particles are entrained by a nitrogen flow and passed through a skimmer. A copper heater then produces the neutral amino acids in the gas phase [46].

In the case of amino acids, both protonated and negatively charged arginine ions have been observed in the gas phase. It is supposed that following absorption of more than four IR photons, water molecules dissociate into H^+ and OH^- ions which then react with arginine to produce the observed ions [47, 41]. Interestingly, in the case of large biomolecules such as cytochrome *c* or bovine serum albumine desorbed from a liquid beam, it has been shown that the relative intensities of the mass peaks reflect the relative concentration in the solution [45]. Liquid beams have also been used for near-edge X-ray absorption fine structure (NEXFS) studies of aqueous proline and diglycine [48].

Fast liquid beams (100–500 m/s, 1–20 μm diameter) can be created from modified DESI sources (see Section 3.1.6) by using a pulled quartz capillary nozzle. Aqueous solvent systems are then introduced into the nozzle at 300–400 bars. The jet desorption/ionization method (JeDI) allows in-depth analysis of frozen tissue samples [49, 50].

3.1.4.2 Supercritical expansions

Supercritical fluids havc large solvent power. Beams of small molecules such as adenine, caffeine, guanine or vitamin K3 [51, 52] can be generated by seeding carbon dioxide

($T_c = 304°K$, $p_c = 73.8$ bars). The solvent power can be adjusted between that of gas or liquid phase with only moderate changes in pressure. Moreover, carbon dioxide is compatible with ultrahigh vacuum conditions. Supercritical fluids are also used in high-performance liquid chromatography (HPLC). Intermediate-size peptides such as bradykinin or angiotensin can then be dissolved in a CO_2/methanol mobile phase with addition of trifluoroacetic acid to suppress deprotonation of carboxylic groups and to protonate amino groups. The corresponding ions are then electrosprayed (see Section 3.1.6) [53].

3.1.4.3 Liquid droplets

The laser-induced liquid beam ionization/desorption (LILBID) process is very mild and allows the study of non-covalent complexes (see Chapter 4.5) [54, 55]. However, the main drawback of this method is that a large amount of solute is required due to the production of continuous liquid beams while ions are set into gas phase by means of very short-pulsed lasers. An alternative possibility has thus been devised to considerably reduce the consumption by increasing the duty cycle and bringing it close to unity. Microdroplets [56, 57] are introduced into vacuum by means of a piezoelectric driven generator synchronized with an infrared [58] or UV [59] desorption laser. The irradiated droplets are submitted to a shockwave [60] and they explode, leading to a small fraction (10^{-4}) of the dissolved ions they contain being detected in the gas phase. In ESI, droplets created at atmospheric pressure evaporate slowly and produce highly charged and totally desolvated species. In the microdroplet method, droplets evaporate in less than 100 ns. Only low-charge states are observed and some solvent molecules can be retained or removed, if required, by a second evaporation laser shot. The method is also less sensitive than ESI to the presence of salts. The absolute amount of analyte that can be detected from a single droplet is in the attomolar range [58].

This mild method allows the observation of specific non-covalent complexes (see Chapter 4.5) such as the minor groove binder distamycin A bound to the single strand Dickerson dodecamer [61] or cytochrome *c* complexes of bacteria [62]. It also allows the measurement of the stoichiometry of membrane protein complexes. Often, membrane complexes are oligomers and their quaternary structure can be ascertained only by mass-spectrometry (see Section 4.8.1). At very low laser intensities, use of the microdroplet method provides detection of the complete assembly of the non-covalently bound subunits. At higher laser intensities, partial rupture of the non-covalent bonds leads to individual detection of subunits. Protein membranes play a crucial role for cells since they are responsible for communication between the intracellular medium and the cytoplasm. Mass-spectrometric methods encounter problems since those proteins are highly hydrophobic and cannot be solubilized in water or any polar solvent [63, 64]. The necessary use of surface-active detergents strongly reduces the ionization efficiency in electrospray (see Section 3.1.6) and make difficult crystallization in MALDI (see Section 3.1.5). In contrast, the laser-ablated droplet method is much more tolerant to solutions containing buffers and detergents make the method particularly interesting for the study of membrane proteins [62].

3.1.5 Matrix-assisted laser adsorption ionization (MALDI)

3.1.5.1 Ionization processes

For a long time, the most widely used ionization process has been electron impact (EI, see Section 2.3.1). For reproducibility of the mass spectra, the used electron energy was in the range of 50–80 eV, much above ionization energies of molecules, and this exaggerated supplied energy lead to extensive fragmentation. Much milder ionization processes using free electrons leading to electron capture followed by non-dissociative (RET) or specific dissociative. (ECD) processes are, respectively, considered in Section 2.3.2.2 and Chapter 4.7. As EI is limited to low masses, other ionization processes have been devised for different molecular masses. One or several protons can be added to bimolecular systems and reside on the most basic sites. Those protonated sites must be sufficiently separated in order to limit Coulomb repulsion, and it is usually admitted that, for example in peptides, the most basic amino acids (Arg, Lys) must be distant from at least three or four units. However, in some cases, the π-electron cloud of an aromatic side chain offers screening between positive charges and doubly charged peptides can be observed. Removal of protons is another ionization process useful in the case of acidic compounds. In some cases, fragmentation can be minimized by replacing protons with metal cations. For example, Na^+ or K^+ are common adducts for carbohydrates (see Chapter 4.4).

3.1.5.2 MALDI operation

In MALDI, analytes are embedded in a surplus of a matrix of small organic molecules [65, 66] or ultrafine metal powder [67] absorbing the beam of UV lasers usually delivering pulses in the nanosecond range. IR lasers are also used but to a less extent [68]. Following absorption of a laser beam by the matrix, a plume is produced that ejects the analytes into the gas phase. Only a small fraction of ejected material is detected as ions in UV-MALDI, in between 10^{-2} and 10^{-5}, and further mass-analysed. During preparation, the analyte molecules and the matrix are dissolved, for example, in water or in a 1:1 mixture of water and methanol, and deposited as small spots of a few microlitres on a stainless-steel plate, either at atmospheric pressure or under vacuum conditions [69]. In the latter case, vacuum sublimation purifies the matrix compound and an extremely pure layer of small matrix crystals with diameter less than 300 nm is obtained. For analytical purposes, biomolecular solutions are present in between 10^{-5} and 10^{-3} M concentrations. For spectroscopic purposes, larger amounts of material are required since the number of laser shots can be larger than 10^4, for example in order to record a full IR spectrum [70]. Most often, MALDI sources are preferentially associated to time-of-flight (TOF) mass-spectrometers (see Section 3.2.1.1) and the acquisition time of mass spectra can be very short when high-repetition rate lasers are used. For example, a rate of 2 spots per second, each spot being analysed with 800 laser shots, can be obtained allowing a rapid on-line coupling to a liquid chromatograph (LC) [71]. MALDI sources in combination with ion trapping devices can also be used for long spectroscopic scans [72–74] lasting several hours.

The choice of a matrix molecule containing a UV chromophore, such as *trans*-cinnamic acid or 2,5-dihydroxybenzoic acid (DHB), is dictated by the analyte according to the matrix ability to ionize and form adducts [75]. The matrix plays several roles. The first one is isolation to prevent formation of analyte aggregates. In the matrix, analytes are incorporated into the charge state already existing in the solution, in presence of their counterions to ensure neutrality. Once the UV laser photon energy is electronically deposited very locally in the matrix on a timescale of a few nanoseconds, much shorter than the time required for energy redistribution by thermal diffusion, two physical regimes can occur. At low laser fluences, neutral and ionic species are individually set into gas phase in a desorption process. When the laser fluence increases, the matrix layer explodes in an ablation regime and ejects molecular clusters as well as individual species. The modelling of the MALDI process involves a large number of processes occurring on three different timescales. The initial excitation by the UV laser of matrix molecules containing chromophores in their first excited state S_1 is redistributed by hopping on a picosecond timescale to neighbouring molecules. Those excited molecules interact through their aromatic rings and can thus acquire enough energy (energy pooling) to overcome their ionization limit, providing ions and free electrons with a mean free path of 10 nm in the matrix. Those electrons can escape the matrix leaving it with a positively charged surface. The electronic energy is also converted into vibrational energy, and a fast heating, within nanoseconds, leads to ejection of a surface region of the irradiated target and formation of the plume. A large number of reactions take place within microseconds in the plume and produce mostly singly charged ions [76, 77].

The analyte ions have a high initial velocity ν_0 typically in between 300 and 800 m/s, with a considerable spread ($\nu_0 \pm 0.5\nu_0$). In order to improve mass resolution, a technique called time-lag focusing is introduced. A time delay is applied between ionization and extraction of the ions out of the ion source into the drift tube of the analyser. Ions are extracted from the source by an electric pulse after the time delay expires in order to minimize the arrival time distribution of the ions at the detector. When a second time-dependent field is applied in the second extraction region of the TOF, the time lag method becomes mass independent over a broad mass range (Figure 3.1.7).

MALDI can be used for a wide variety of biomolecular systems such as DNA [80], peptides [81], proteins [82] or oligonucleotides [83]. It is also a mild process leading to intact non-covalent complexes (see Chapter 4.5) in the gas phase [78, 84]. When combined with a TOF mass-spectrometer, a high mass accuracy ($<$10 ppm) can be obtained up to very high masses ($>10^6$) [85], with a subfemtomole sensitivity and the possibility to mass-analyse more than a thousand spots in a single run.

3.1.5.3 Bioaerosols

In order to reduce the amount of analyte and increase the sensitivity of MALDI, it is possible to replace matrix spots with typical millimetre diameter by calibrated aerosol droplets with micrometre diameter [86]. Lasers or synchrotron radiation [87] can be used for ionization. In off-line experiments, the matrix and the analyte are premixed in solution prior to aerosolization and the droplets composition reflects that of the solution. In on-line experiments, analyte particles enter a chamber coated with the matrix and particles containing an analyte core with a matrix coating are produced. A detection limit of 14 zeptomoles, corresponding to 8,400 molecules, has been reached with this method.

Amino acids such as glycine, phenylalanine or tryptophan and peptides such as β-carotene or phenylalanine–glycine–glycine can be set into gas phase from aerosol dry nanoparticles that are further thermally vaporized in high vacuum. The resulting vapour that contains neutral molecules can then be ionized by tunable VUV synchrotron radiation with negligible fragmentation. Particles are obtained by atomizing solutions containing 0.5 to 1 g of amino acids or peptides in 1 l of H_2O, methanol or ethanol. The liquid droplets forming the bioaerosols are then entrained in a nitrogen carrier gas, preheated to 50°C and dried at room pressure to form solid particles. The mean size of those particles is typically 100 nm and the concentration ca. 10^7 particles/cm^3. Those particles enter vacuum by passing through a 200 μm diameter hole and five apertures acting as an aerodynamic lens. The focused beam enters the first region of a TOF (see Section 3.2.1.1) mass-spectrometer where they

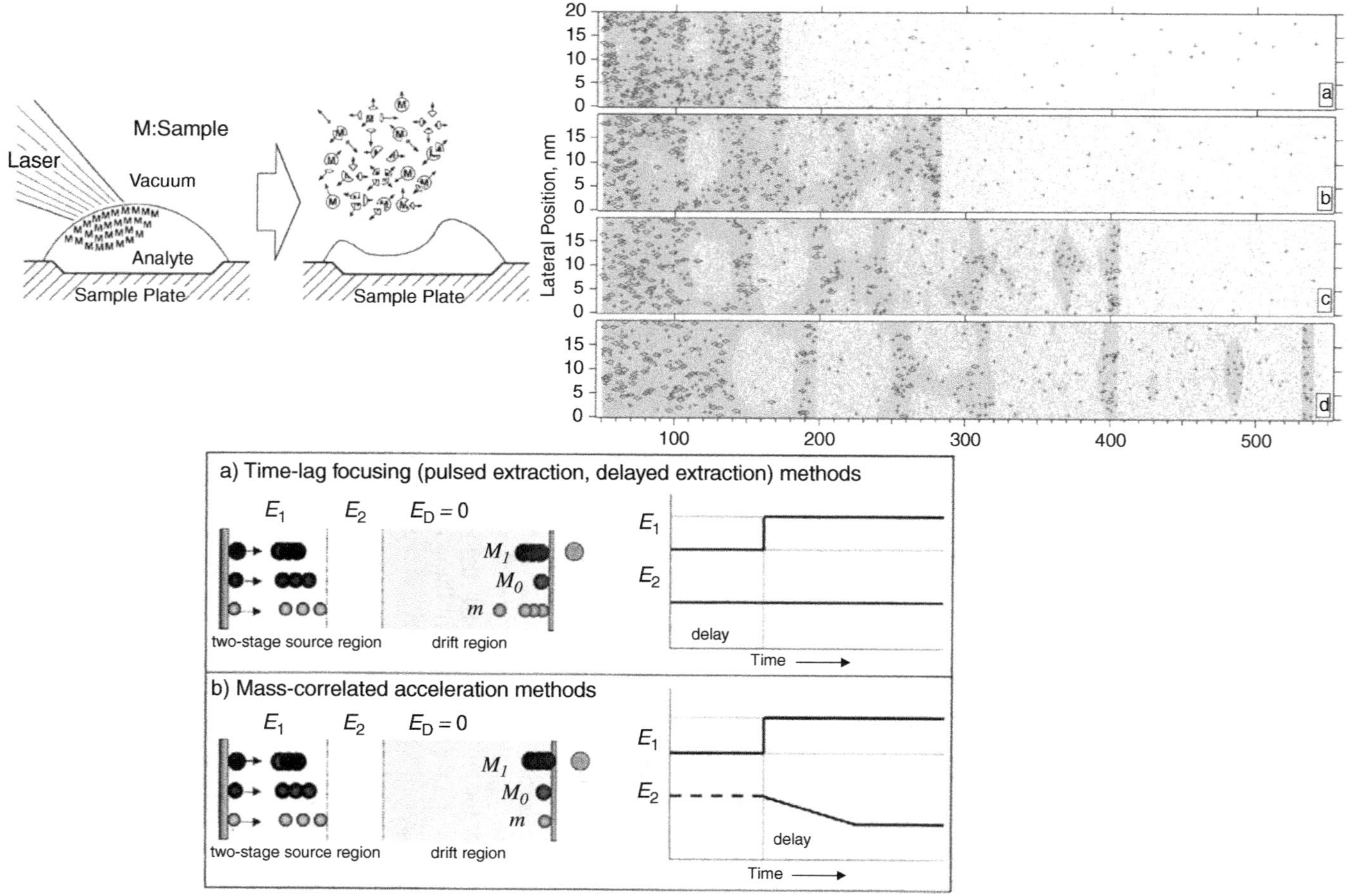

Figure 3.1.7 Top. left: Matrix-assisted laser desorption/ionization (reproduced with permission from reference [67] ©2002 Wiley). Right: Snapshots of a simulation showing ablation and cluster formation comprising 1,022,976 molecules at 500 (a), 750 (b), 1,000 (c) and 1,250 ps after a 30 ps laser shot (reproduced with permission from reference [78] ©2003 American Chemical Society). Neutral molecules are denoted as grey points, positive ions as crosses and negative ions as diamonds. Bottom: Time lag extraction in a MALDI-TOF device with a single pulse (a) or a second time-dependent pulse (b) (reproduced with permission from reference [79] ©2002 American Chemical Society).

hit a 3 mm diameter heated copper tip at a temperature in between 300 and 850°K. The beam particles are thus vaporized and the expanding gas acts as a localized molecular source at the entrance of the TOF setup [88].

3.1.6 Electrospray

The ESI process transfers large unvolatile molecules from liquid phase into the gas phase as highly charged isolated entities [89]. It is a very mild process as demonstrated by the possibility to observe, for example, intact protein quaternary structures (see Chapter 4.8) [90] or non-covalent drug binding sites [91, 92]. A solution containing the analyte is introduced by means of a syringe in a capillary. The application of a high electric field to the capillary accumulates ions attracted by a counter electrode at the surface of the capillary tip [93]. The meniscus at the tip acquires the shape of a so-called Taylor cone that continuously breaks into droplets [56, 94]. Those droplets are enriched in ions (e.g. cations in the case of a positive tip) when the applied electric field exceeds a threshold value in between 700 eV and several keV. Their diameter varies from microns at flow rates in between 1 and 100 μl/min down to 100 nm for nanolitres per minute rates in nanospray [95, 96]. At large flow rates, the spraying process is generally pneumatically assisted by a coaxial dry gas that is possibly heated. The introduction of the ions in the mass-spectrometer uses a dry nitrogen "curtain" or "sheath" gas. A heated capillary interface aids in desolvation and declustering of ions from neutrals. The mist of highly charged droplets initially at atmospheric pressure [57, 97] pass down an electrostatic potential and pressure gradient towards the high vacuum of a mass-analyser. In the nanospray regime (nano-ESI) 1–10 μm capillary exits and nanolitres per minute rates are used without the need of pushing syringe and sheath gas. The electric field at the tip is sufficient for droplet formation and desolvation [98]. A very short distance (mm) can be applied between the tip of the capillary and the entrance of the mass-spectrometer. A high efficiency, defined as the ratio of the flux of ions reaching the ion detector after mass-selection and the flux of ions leaving the tip, of typically 1% and up to 12% can then be reached. Droplets reduce in size by evaporation of the solvent and Coulomb explosion during their passage through the desolvation chamber, which is usually kept at room temperature.

Two mechanisms are mainly invoked for explaining the fate of the charged droplets (Figure 3.1.8). Reviews can be found in references [99, 100]. The droplets undergo a sequence of solvent evaporation and Coulomb explosion cycles. In the "ion evaporation" model (IEM), it is assumed that the high electric field at the surface lifts ions from the solute medium into vacuum when the droplets, following successive evaporations, acquire a radius less than 10 nm. In the "charge-residue" model (CR), droplets sequentially evaporate their solvent molecules. When their radius becomes too small ("Rayleigh limit") [57, 97], their surface charge density provides an electrostatic force overcoming their surface tension. For example, the critical radius for a droplet carrying 200 charges is 12 nm. Finally, a bare charged molecule is left without any solvent. It seems that in nano-ESI where very small droplets are formed, the CR mechanism dominates for large species while the IEM mechanism is favoured in ESI in the case of small ions.

Mass-spectra obtained with electrospray show that after evaporation the charged droplets leave intact sample molecules containing a rather large number of charges [101, 102]. The m/Ze ratio is usually limited to 3,000 although values up to 30,000 can still be observed

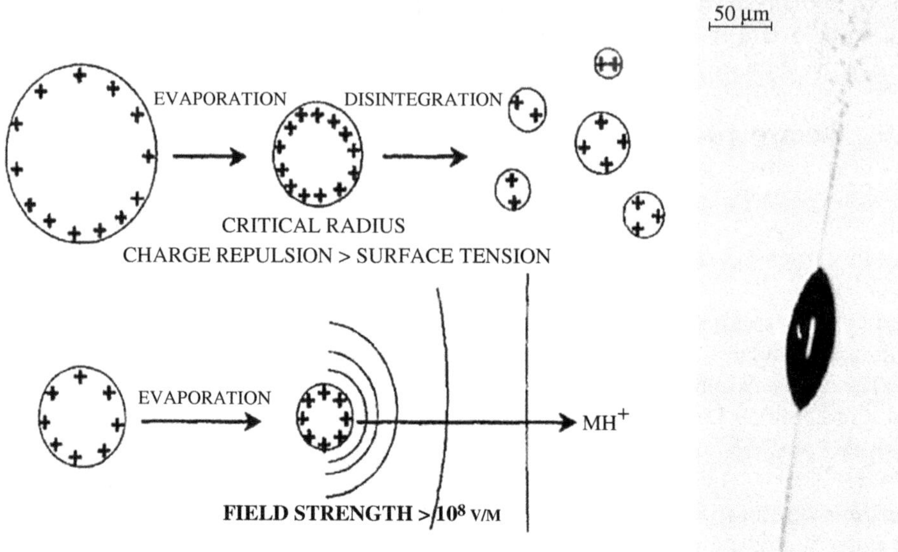

Figure 3.1.8 Left top: "Charged-residue model" (CR). Offspring droplets resulting from a first Rayleigh instability continue to evaporate solvent until they become too small and undergo a second Rayleigh instability and explode. Following a sequence of evaporation-explosion, very small droplets only containing a single solute molecule are produced. Bottom: In the "ion evaporation" model (IEM), before ultimate droplets are formed, the electric field at the droplet surface becomes large enough to expel solute ions (reproduced with permission from reference [89] ©2002 Wiley). Right: Implosion of a glycerol droplet investigated by ultra fast microscopy. In the moment of instability, the deformed droplets have a spindle-like shape from which two highly charged liquid jets are emitted (reproduced with permission from reference [97] ©2005 EDP Sciences).

[103] (see Chapter 4.8). Thus, ESI is well adapted to mass-spectrometers with a limited *m/Ze* range [99, 104].

The great versatility of electrospray allows to combine two ESI sources for producing ions of both signs and studying ion/ion reactions (Figure 3.1.9) [105].

Mass-spectrometry is widely used to obtain structural information of biomolecules through their fragmentation pattern. In the case of MS/MS (see Section 3.2.2), a spatial or temporal sequence of events is required to first isolate parent ions, and then dissociate them by collision-induced dissociation (CID) with a neutral gas and analyse their fragments. In an electrospray source, a collisional exchange occurs between the translational and the internal (vibrational plus rotational) energies of the biomolecular ions. By increasing the applied acceleration voltage leading to their entrance in the mass-spectrometer, it is possible to raise the ion internal energy and thus to increase their fragmentation rates. Above a critical energy, the different fragmentation pathways are characterized by different variations of the dissociation rates as a function of the ion internal energy. This so-called source-CID has been studied in detail and it has been shown that it can be used to test the kinetic stability of non-covalent complexes, providing, for example insights about binding properties of drugs to DNA [107, 108].

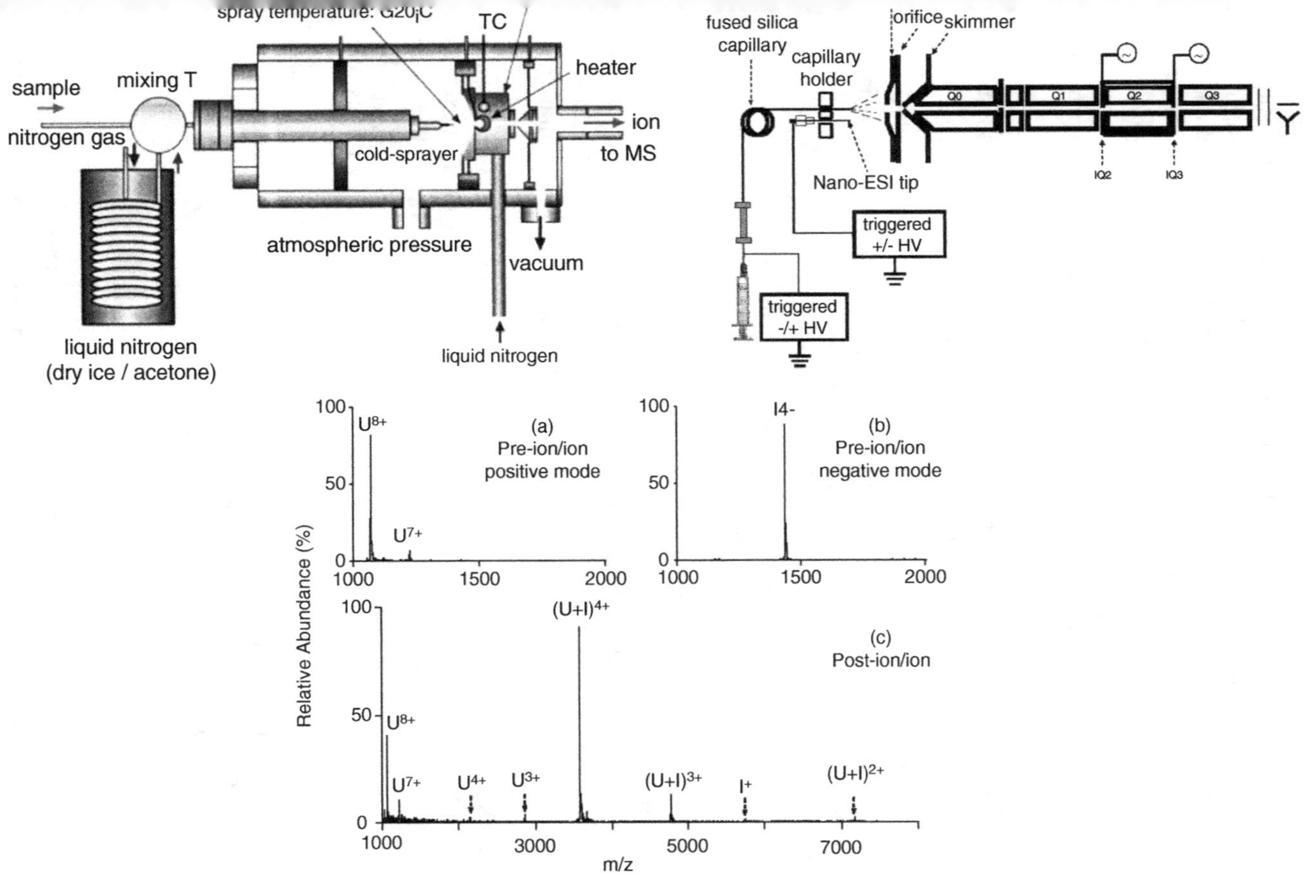

Figure 3.1.9 Top left: Electrospray source with a desolvation chamber held at low temperature (−80–10°C) in order to reduce breaking of non-covalent interactions (reproduced with permission from reference [106] ©2003 Wiley). Right: Pulsed dual electrospray source and triple quadrupole/linear trap tandem mass-spectrometer (see Section 3.2). Each ESI source is pulsed alternatively and ions of opposite signs penetrate separately in Q1 where they are isolated. During mutual storage in Q2 and Q3, ions of opposite charges react. Following ejection of the undesired ions of a given sign, the other ions are mass-analysed. Bottom: Mass-spectra derived from the pulsed dual electrospray source. Positive bovine ubiquitin (U)$[M + 8H]^{8+}$ ions (a) and $[M - 4H]^{4-}$ bovine insulin (I) ions (b) react for 300 ms and preferentially form the protein–protein complex $[U + I]^{4+}$ complex (c) (reproduced with permission from reference [105] ©2005 Elsevier).

Other methods related to electrospray are desorption sonic spray ionization (DeSSI) [109–111] or laser spray [112] using a laser beam in presence of a weak electric field [113] to produce droplets. In atmospheric pressure chemical ionization (APCI), vaporization of the solution containing the analyte is assisted by a strong gas flow at a high temperature (ca. 500°C) in absence of high voltage. The solvent is excited and ionized by a corona discharge and either transfers or removes protons from the analyte, which becomes ionized. In a photo spray, the gaseous mixture is transferred through a heated quartz tube. An added dopant is then ionized by 10 eV photons issued from a Kr lamp and in turn ionizes the analyte [114–117]. Synchrotron radiation can also be used to obtain a far better control of ionization conditions.

3.1.6.1 Desorption electrospray ionization (DESI) and electrospray-assisted laser desorption/ionization (ELDI)

Analyte samples do not necessarily need to be manipulated into solution and further desolved. The desorption electrospray ionization (DESI) allows the production of ions directly in the ambient environment [118–121]. In DESI [111, 121–124], charged droplets of a solvent, produced by an electrospray emitter (Figure 3.1.10), impact at few hundred millimetres per second on a surface and cover an area having a diameter 3–10 times larger than their initial diameter, producing offspring droplets. Ionization of the analyte deposited on the surface takes place by a heterogeneous charge-transfer mechanism and a droplet pick-up mechanism. Ions of both signs can be obtained. For example, DESI mass spectra of protonated alkaloids, polar lipids or carbohydrates have been directly obtained from plant tissues at atmospheric pressure [125]. DESI combines advantages of both MALDI and ESI for protein sequencing in proteomics. ESI provides multiply charged ions more easily identified by MS/MS than the singly-charged ions produced in MALDI. Since it does not require any chromatographic step, it is as fast and easy to automate as MALDI.

ESI and MALDI can be combined for ambient mass-spectrometry in electrospray-assisted laser desorption/ionization (ELDI). In MALDI, the number of desorbed neutral far exceeds the number of ions and post-ionization methods have thus been devised. Among them, one can find electron ionization, REMPI or corona discharge APCI [126, 127]. In ELDI [128], ESI is used to post-ionize neutral proteins desorbed from a metallic plate by means of a laser beam.

The direct analysis in real time method (DART) [120] uses a plasma of helium or nitrogen that possesses high ionization potential, much larger than that of biomolecules. Metastable atoms or molecules M^* are separated from ions and impinge upon the analyte a, that is ionized by the Penning process $M^* + a \rightarrow M + a^+ + e$.

3.1.6.2 Sonic spray

In a sonic spray nebulizer, the sample solution is introduced in a silica capillary fixed in the middle of an orifice drilled in a chamber. Nitrogen introduced into the chamber exits through the orifice with the sample solution and disperses it. There is no need to apply a potential drop between the sample introduction capillary and the quadrupole mass analyser entrance. As in ESI, produced droplets evaporate and form ions. The ion signal intensity is then two orders of magnitude less than in usual ESI. A stable ion yield is obtained with a sample flow rate as low as 1 μl/min [129, 130]. Such an ion source has been used to produce

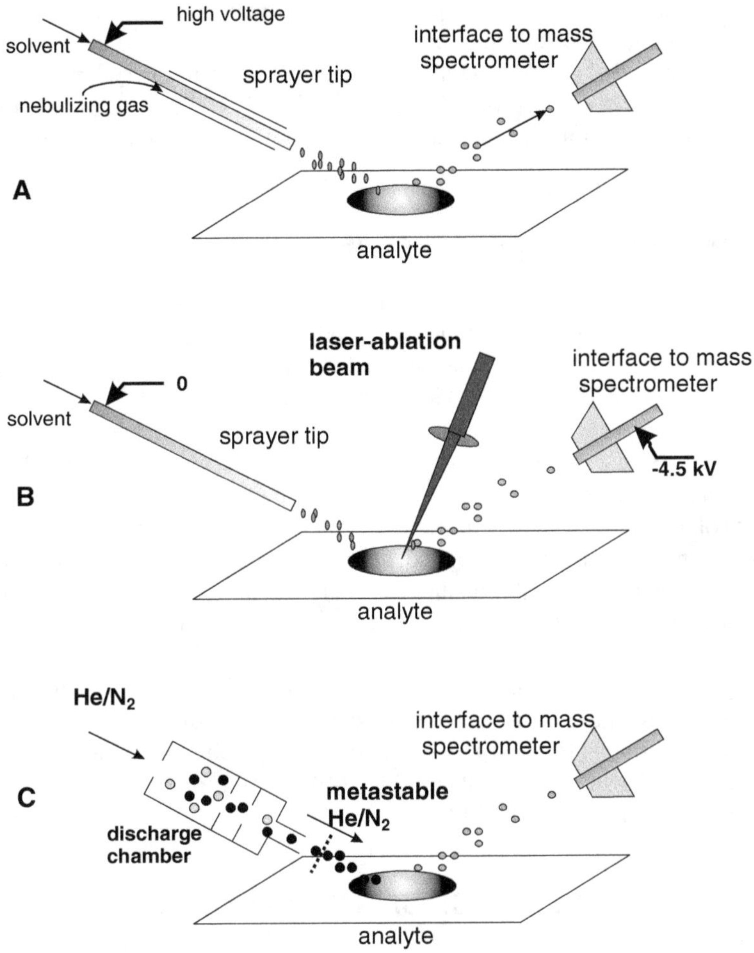

Figure 3.1.10 (A) Schematic view of a desorption electrospray ionization (DESI) source; (B) Schematic view of an electrospray-assisted laser desorption/ionization (ELDI) source. The laser-ablated analyte molecules issued from the MALDI matrix are ionized by the electrospray solvent plume. (C) Schematic view of a direct analysis in real time (DART) source.

$[m\text{Serine} + nH]^{n+}$ serine nanoparticles with m up to 600 and n in between 1 and 10. The chiral selectivity observed in $[8\text{Serine} + H]^{+}$ ions (see Chapter 4.6) is still present in the large nanoparticles [131].

3.1.7 Laser-induced acoustic desorption

An alternate method that can be run in quadrupole trap (Section 3.2.1.2) or Fourier-transform ion cyclotron resonance (FT-ICR) (Section 3.2.1.4) mass-spectrometers is laser-induced acoustic desorption. A high-intensity laser irradiates on one side a thin (10 μm) metal foil or thick silicium (0.5 mm) wafer. Ablation of the metal creates an acoustic shockwave that propagates through the foil and induces gentle desorption on the opposite side. This method can be used, for example, for bringing peptides in an FT-ICR cell and letting them further react to produce radicals [132]. It can also be used for desorbing very large biomolecular systems such as bacteria or cells (see Chapter 4.8) [133]. The advantage with respect to electrospray is then avoiding the production of aerosols when dealing with pathogenic bioparticles.

3.1.8 Production of hydrated species

In order to establish a link between the behaviour of biological species under isolated conditions and in the crowded cellular medium (see Section 5.1), a large number of experiments are devoted to the study of species in presence of a progressive number of water molecules. Those studies are described in Chapter 5 and only sources of hydrated species are considered here. In the case of supersonic expansions, the rare gas carrier (He, Ne or Ar) is seeded by passing over a reservoir containing water, generally at room temperature [6, 134–140].

Hydrated neutral species can also be set in gas phase by dissolving them in water and then freezing drops of the solution at low temperature. A plume of desorbed hydrated clusters is then produced by resonant absorption of ice by 3.1 μm radiation. Neutral tryptophan–(water)$_{N<11}$ clusters have been generated with this method [141]. Tryptophan–(water)$_{N<16}$ clusters can also be obtained from a pick-up source by deposition of tryptophan on a water cluster beam [142].

In electrospray sources, ions can be prepared from a water/methanol solution and different possibilities are offered [143]. One can operate the ESI source under conditions where ions are not totally dehydrated [144–146]. It is also possible to rehydrate ions either in specific ESI sources designed with two chambers [147–149] or by further exposing ions to water vapour, for example, in a drift cell [143].

3.2 MASS-SPECTROMETRY

GENERAL FEATURES

Mass-spectrometry is a fundamental analytical technique that is very widely used for identification, determination of structures and study of the dynamics of biomolecules [150]. Reviews can be found in references [151–156]. Mass-spectrometry is also widely used in pharmaceutical drug discovery [157]. Spectroscopy of isolated biomolecules also takes advantage of the use of mass-spectrometers for a perfect identification of the studied species.

Mass-spectrometry can today provide chemical and spatial information about biological tissues at the sub-cellular resolution [158]. In a mass-spectrometer, the studied molecular system must be introduced and ionized, and then mass-analysed and detected. In this chapter, only the third and fourth steps will be considered. We will first examine the most basic designs that are TOF instruments, linear quadrupoles (Q) and linear ion traps (LIT), three-dimensional quadrupole ion traps (3D-QIT), magnetic sectors and FT-ICR. Those instruments are very often used in hybrid combinations such as triple quadrupoles, Q-TOF or LIT-ICR. We will then briefly describe electrostatic ion storage devices such as the Orbitron, the storage ring and the Zajfman design. For identification and sequencing of biomolecules, tandem mass-spectrometry (MS/MS) preferentially uses ion fragmentation induced by CID (see Section 3.2.2.2). Other methods such as infrared multiphoton dissociation (IRMPD, see Section 2.1.3) and electron capture dissociation (ECD, see Chapter 4.7) are also now important tools.

3.2.1 Mass-spectrometers

3.2.1.1 Time-of-flight instruments

A TOF mass-spectrometer measures the distribution of time intervals it takes for ions of different masses to move from a region where they are created or introduced into a detector. This requires that the starting time at which the ions leave this region is well defined. Ions possessing different *m/z* ratios acquire different velocities in a region where they are submitted to an electric field and then drift in a field-free region. The lightest ions arrive earlier to the detector than the heaviers. This presupposes that ions are either created nearly at the same location within a very short amount of time by means of a pulsed ionization process (e.g. in a MALDI source, see Section 3.1.5) or extracted as brief ion packages from a continuous ion source (e.g. from a ESI source, see Section 3.1.6) by means of rapid electric field switching used as an ion gate after storage. One of the main benefits of TOF mass-spectrometers is their large ion transmission and their ease of use with pulsed ion sources.

3.2.1.1.1 Principle and limitations to resolution

The region where ions are created (or introduced) is situated in between two parallel grids G_1 and G_2, at a distance d_1 away from one another and at respective voltages V_1 and V_2 (typically in between 1 and 30 keV). In a first step, one assumes that the initial ion kinetic energies are null and that all ions are created at the same place. Grids produce an electric field that provide to ions of mass m and electric charge $z = Ze$ (Z is an integer) a kinetic energy $mv^2/2 = z(V_1 - V_2)$. The ions cross the exit grid G_2 with the velocity v and penetrate into a field-free region of length L. An ion TOF t_{flight} from G_2 to reach the detector is equal to $t_{\text{flight}} = L/v = L\sqrt{m/z}\sqrt{1/2(V_1 - V_2)}$ and thus depend upon the ratio of *m/z* (in practice, the relationship between TOF and *m/z* ratio is most often used after calibration with known masses).

Typical ion TOFs are in the microsecond range, and thus a complete mass spectrum is generated within less than a millisecond. Upon impact by an ion, an electron burst is produced in a detector and further amplified by a set of dynodes in electron multipliers or microchannel plates. The electron current out of the detector is then displayed on an oscilloscope or digitized and averaged to increase the signal-to-noise ratio. It is worth noting that if ions are first accelerated to energies larger than a few keV and further neutralized, the obtained neutral particles retain nearly the same kinetic energies and are also detected.

In principle, there is no mass limit and TOF setups can be used for a wide range of masses, from electrons up to molecules of 10^6 Da. The above-described situation is however ideal, and in reality, ions can be produced at different locations and different times with different velocities, thus introducing resolution-limiting factors. A first improvement with two acceleration zones has been introduced by Wiley and McLaren [159]. In the first zone (G_1, G_2), a pulsed voltage (typically during several microseconds and with several tens of V/cm amplitude) is

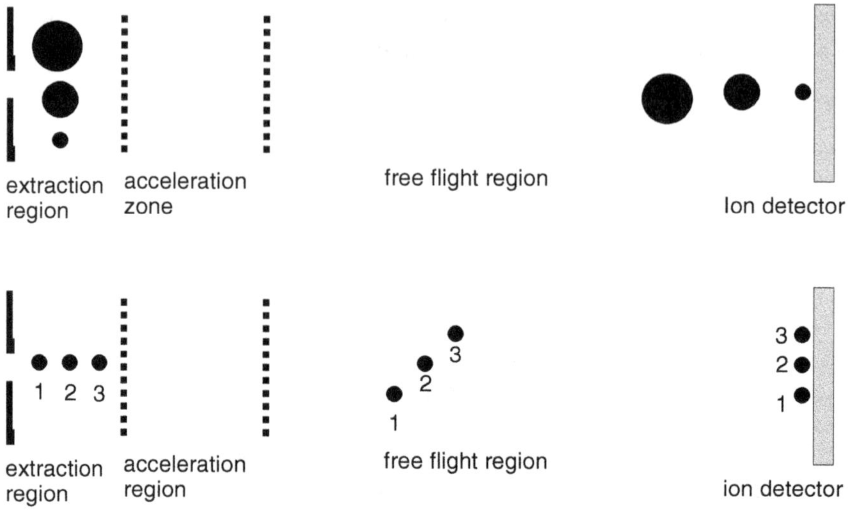

Figure 3.2.1 Principle of a Wiley–McLaren time-of-flight mass-spectrometer. Top: Ions of different *m/z* ratio created at the same positions are detected at different arrival times. Bottom: Ions of the same *m/z* ratio created at different positions are detected at the same time.

applied immediately after the ion creation time. Ions are supposed initially at rest. Those that have been created the farthest from the detector are accelerated for a longer time and thus acquire enough kinetic energy to catch up ions that have been created the closest from the detector. All ions then penetrate into a second region (G2, G3) where they are submitted to a constant accelerating electric field and further drift in a field-free region. By adjusting the lengths of the three regions and the applied electric fields, ions with a given *m/z* ratio can reach the detector nearly independently of the position at which they have been created, ensuring spatial focusing (Figure 3.2.1). This assumes that ions are produced at the different initial locations nearly rigorously at the same time. Narrow ion creation time distributions (typically few nanoseconds) can be achieved by either using short electron pulses created either by a gated electron gun or by a multichannel plate irradiated with a pulsed laser [160] or using a pulsed laser as ionization source [134].

The situation is different when ions are created by laser desorption from a surface, as in MALDI. The ions are then generated within a very small spatial region called the plume (see Section 3.1.5) but with a rather large energy spread and with a creation time distribution. A time delay (typically several hundreds of nanoseconds) is then applied between the laser shot creating the plume and the extraction-acceleration of ions. In the (G_1, G_2) zone, ions are first allowed to drift in absence of electric field. During that amount of time, reactions induced by the laser desorption process terminate. A strong electric field with a short rise-time ($\ll 1\ \mu s$) is then applied, which drives ions in the second acceleration (G_2, G_3) region where they acquire their final velocity. At the onset of acceleration in the (G_1, G_2) zone, ions created with a large initial velocity and/or at an earlier time have travelled farther than the slow ions. Those created at late times thus experience acceleration along a shortest path. This procedure is called "time-lag focusing" or "delayed extraction" and

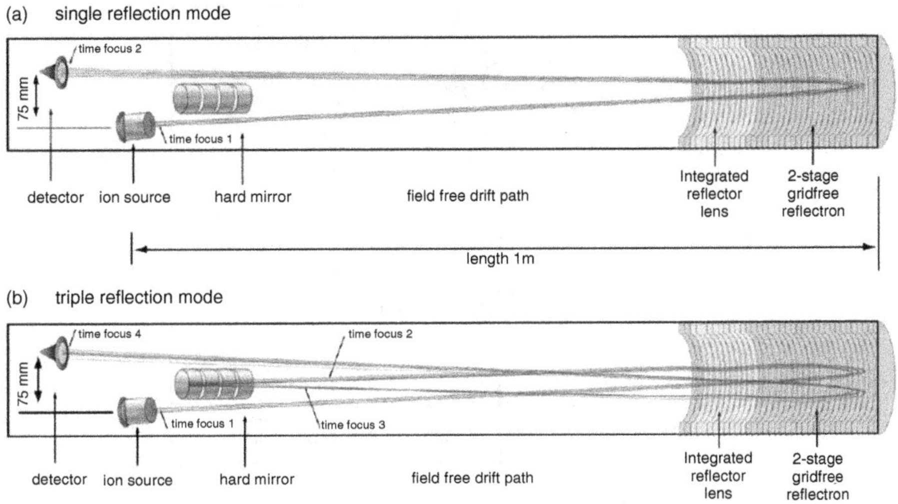

Figure 3.2.2 Ion optical elements of a single-reflection (upper) and a triple-reflection (lower) mode reflectron time-of-flight mass-spectrometer (reproduced with permission from reference [163] ©2006 Elsevier).

enhances the mass resolution. An analogous delay method is used in ZEKE or MATI experiments [161] (see Section 2.1.2.1).

3.2.1.1.2 Reflectrons

A reflectron is an electrostatic device introduced by Mamyrin [162] that compensates for distributions of initial kinetic energies of analysed ions [163]. Two ions with the same *m*/*z* ratio but with different kinetic energies reach the detector in a very narrow time-window. A reflectron is added to a linear Wiley–McLaren TOF setup at the end of the field-free region and acts as a focusing mirror. It comprises two zones. In the first shortest zone, ions are strongly decelerated and then penetrate a longer second zone where they are progressively slowed down. The reflection voltage is set to about 1.1 times the total acceleration voltage applied in the linear TOF and all ions are reflected towards the detector. An ion entering with a large kinetic energy will have a longer travel distance inside the reflectron than an ion with a low kinetic energy and will thus catch it at a focusing distance where the detector will be set. The focusing condition introduced by the Wiley–McLaren linear TOF design in order to compensate the distribution of initially created positions can be separated from the focusing condition introduced by the reflectron in order to compensate the distribution of initial velocities (Figure 3.2.2).

3.2.1.1.3 Time-of-flight analysers with orthogonal extraction

The previously described TOF designs are well suited for mass-analysis of pulsed ion sources. Orthogonal extraction TOF (oTOF) devices are designed for highly efficient mass-analysis of ions extracted from continuous ion beams such as those delivered by ESI sources (see Section 3.1.6). In an oTOF, ions are extracted from a low-energy (5–50 eV) parallel ion beam (Figure 3.2.3). This beam enters the oTOF in an extraction region of length ℓ that is

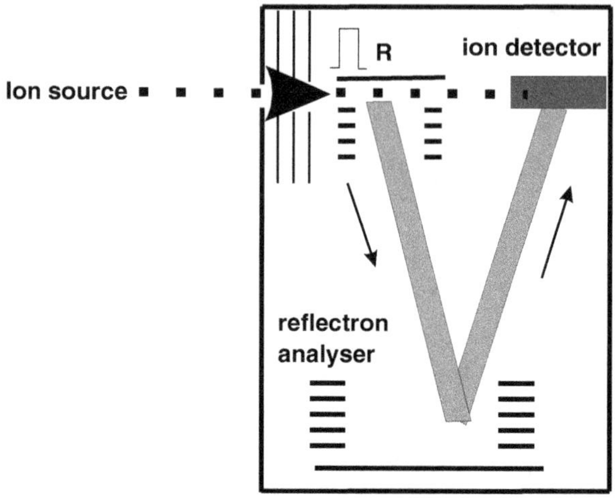

Figure 3.2.3 Schematic of an orthogonal extraction time-of-flight mass-spectrometer. A voltage pulse is applied to the repeller plate R.

first field-free. A steep rise-time (few nanoseconds) pulse of several kiloelectronvolts is then applied between grids parallel to the ion beam and pushes a bunch of ions of length ℓ in the direction orthogonal to the continuous ion beam. The ion bunch is then accelerated by a second zone and enters a drift region, as in a conventional TOF, followed by a reflectron. The ion bunch reaches a detector that must be several centimetres long after covering an overall distance L. While a first ion bunch travels through the oTOF, the extraction region is being refilled and the duty cycle, defined as the ratio ℓ/L, can be rather high. The velocity dispersion along the extraction direction is very small due to the parallelism of the incoming ion beam and the mass resolution is large, typically several thousands and the m/z range can reach 5,000.

Using corkscrew trajectories, ions can be mass-analysed by means of four toroidal electrostatic sectors. A mass resolution of 35,000 can then be achieved for ions with m/z ratios over 300 [164].

3.2.1.1.4 Combined electron and ion time-of-flight setup

Photoionization processes $AB + h\nu \rightarrow AB^+ + e^-$ can be studied by using the technique of photoelectron–photoion coincidence spectroscopy. The light photoelectrons and heavy photoions are, detected after travelling in opposite directions respectively through a short and a long TOF region. If the photon energy is known with great precision (20 meV at 10 eV) and if photoelectrons are detected at nearly zero energy, the internal energy content and the unimolecular dissociation channels of the mass-selected photoions can be followed as a function of the photon energy [165, 166] (see also Sections 2.3.2.4 and 3.2.1.5.2).

3.2.1.2 Ion trapping devices using radio frequency electric fields

Ions can be stored and mass-analysed in devices using radio frequency fields. Let us first consider two-dimensional (2D) configurations and further 3D configurations.

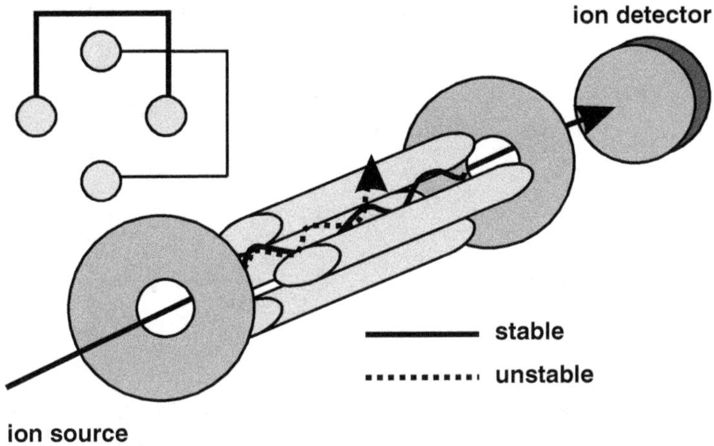

Figure 3.2.4 Stable and unstable trajectories in a quadrupolar mass-spectrometer. Opposite quadrupole rods are connected to the applied AC and DC voltages.

3.2.1.2.1 Linear quadrupoles, linear ion traps and ion guides

Ion motion in multipole electric fields

We here first examine ion motion in linear multipole fields, emphasizing the quadrupole case. A full review can be found in reference [167]. We consider a set of an even number of rods acting as electrodes extending in the z-direction and mounted in a symmetrical configuration in the x–y plane (Figure 3.2.4). An electric potential composed of DC and AC components, $\Psi(t) = \pm(U - V_{RF}\cos\Omega t)$, is applied from alternate rods to ground. This creates a 2D potential, $\Phi(x, y, t) = \Phi(x, y)\Psi(t)$, inside the region surrounded by the rods. The spatial dependency $\Phi(x, y)$ can be expanded as a sum of multipoles $\Phi(x, y) = \Sigma_{N=0}^{\infty} A_N \phi_N(x, y)$ where A_N is the amplitude of the multipole $\phi_N(x, y)$. The lowest order term $\phi_0(x, y) = 1$ is a constant potential. $\phi_1(x, y)$ is the potential of a linear dipole created by two planes of opposite charges. For ion confinement, mass-selection and transmission, we will be interested in the following terms: $\phi_2(x, y)$, $\phi_3(x, y)$ and $\phi_4(x, y)$, which are, respectively, quadrupole, hexapole and octopole potentials. Analytically, those inhomogeneous multipole potentials can be expressed as

$$\phi_2(x, y) = \frac{(x^2 - y^2)}{r_0^2}; \quad \phi_3(x, y) = \frac{(x^3 - 3xy^2)}{r_0^3}; \quad \phi_4(x, y) = \frac{(x^4 - 6x^2y^2 + y^4)}{r_0^4} \tag{3.2.1}$$

Those potentials can be created, respectively, by sets of 4, 6 and 8 parallel rods spaced symmetrically from the central axis by the radius r_0. The force exerted upon an ion with charge $z = Ze$ is $\vec{F} = -z\vec{\nabla}\phi_N(x, y, z, t)$. Let us first consider an important case of a pure quadrupole field. Equations of motion can then be written as

$$-A_2\frac{\partial\phi_2}{\partial x} = \frac{-2x}{r_0^2}(U_{DC} - V_{RF}\cos\Omega t) \quad \text{and} \quad -A_2\frac{\partial\phi_2}{\partial y} = \frac{+2y}{r_0^2}(U_{DC} - V_{RF}\cos\Omega t) \tag{3.2.2}$$

These equations are usually written as a set of dimensionless differential equations

$$\frac{d^2x}{d\xi^2}+(a_x-2q_x\cos 2\xi)x=0 \quad \text{and} \quad \frac{d^2y}{d\xi^2}+(a_x-2q_y\cos 2\xi)y=0 \tag{3.2.3}$$

where

$$a_x=-a_y=\frac{8ZeU_{DC}}{m\Omega^2 r_0^2};\; q_x=-q_y=\frac{4ZeV_{RF}}{m\Omega^2 r_0^2};\; \xi=\frac{\Omega t}{2} \tag{3.2.4}$$

The x and y motions are uncoupled and each equation of motion has the mathematical form of a Mathieu equation. In a plot of a (DC field) parameter as a function of q (AC field), the Mathieu equations have stable solutions for all values of a and q situated inside a region of the (a, q) plane called the stability diagram.

Ions introduced inside the four-rod assembly that creates the quadrupole field will oscillate within the distance 2 r_0 between the rods if the voltages applied to the electrodes are such that a and q parameters are inside the stability diagram. For low values of the applied trapping voltages, ions behave as if they were confined in a well created by an effective electric potential $V_{\text{eff}}(r)=D_{x,y}(r/r_0)^2$. The depth of the well is given by $D_{x,y}=qV_{RF}/4$.

Generally, for any multipole configuration of inhomogeneous oscillating electric field, ions will move to regions of lower electric field but, in contrast with the quadrupole case, there is no stability diagram for motion in hexapole, octopole,… fields [168, 169]. At low values of the trapping voltages, ion trajectories are approximately determined by an effective mechanical potential

$$U_{\text{eff}}(r)=zV_{\text{eff}}(r)=\frac{N^2(Ze)^2V^2}{4m\Omega^2 r_0^2}\left(\frac{r}{r_0}\right)^{2N-2} \tag{3.2.5}$$

where N is the order of the multipole. In a quadrupole, a hexapole and an octopole, the effective potentials are, respectively, proportional to r^2, r^4 and r^6. When the order of the multipole increases, the effective potential becomes flatter and flatter near the central axis and increases more abruptly near the rods (Figure 3.2.5).

3.2.1.2.2 Linear ion traps

In a LIT, ions are axially confined in the (x, y) plane by 2D radio frequency (RF) multipole fields and are furthermore axially confined in the z-direction by applying stopping potentials. LITs can be used to store rather large number of ions prior to mass-analysis in a 3D ion trap or in a FT-ICR cell (see Section 3.2.1.4.). They can also be operated as stand-alone mass-spectrometers. An estimate of the maximum ion density that might be stored in a LIT can be obtained by comparing the repulsive electric field induced by the ionic cloud space charge and the confining electric field due to the effective electric potential $V_{\text{eff}}(r)=U_{\text{eff}}(r)/Ze$. One can assume that ions are confined as a cylindrical cloud with radius r_0. Those ions repel each other by creating a space-charge electric field $|E_{SC}|=n_iZer_0/2\varepsilon_0$ deduced from

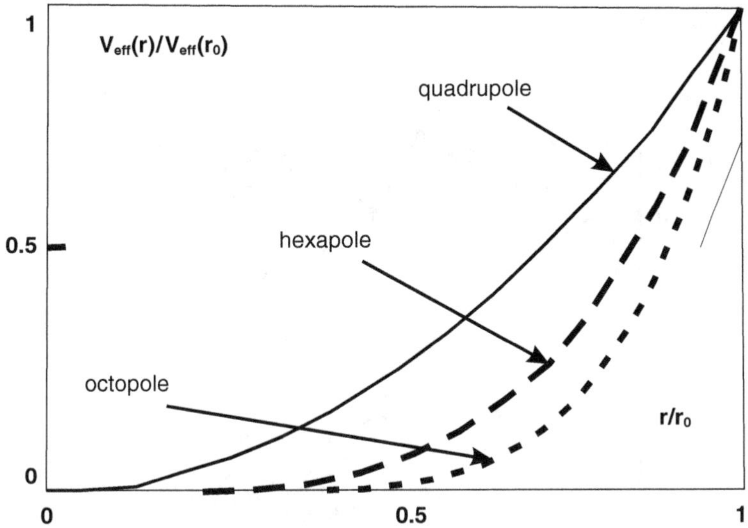

Figure 3.2.5 Comparison between effective potentials for quadrupole, hexapole and octopole fields.

Gauss's law where n_i is the ion density. The confining electric field is obtained from the effective electric potential: $|E_{eff}| = |-\partial V_{eff}(r)/\partial r|_{r=r_0} = qV_{RF}/2r_0 = 2D_{x,y}/r_0$.

The maximum ion density $n_{i,max}$ is reached when the effective electric field is equal to the space-charge induced electric field. It is then equal to $n_{i,max} = 4D_{x,y}\varepsilon_0/Zer_0^2$. In a linear quadrupole ion trap with a radius of few millimetres and an applied RF of 1 MHz, single-charge ion densities of 10^6 ions/mm^3 can be reached. Since LIT can be made rather long, large number of ions can be stored. As compared to 3D traps, LITs do not use an RF field component in the z-direction and thus offer efficiencies for ion injection and extraction that are close to 100%.

3.2.1.2.3 Quadrupole mass analysers

Quadrupole mass analysers take advantage of the 2D stability diagram corresponding to ions with different m/z ratio and plotted in the (V_{RF}, U_{DC}) plane (Figure 3.2.6). In a quadrupole mass analyser, the ratio of the applied DC field to the applied AC field is chosen close to 0.336 (scanning line) and the (x, y) stability region shrinks to a very small triangle at the apex of the stability diagram. For given values of the applied AC and DC fields, the only ions that are transmitted through the quadrupole rods are those with characteristic points inside this small apex triangle. The mass resolution can be changed by varying the apex triangle size.

Ions are introduced with a small velocity component in the axial direction into the set of four rods creating the quadrupole field. Only ions with masses such that a ($\propto U$) and q ($\propto V_{RF}$) parameters are within the small stability region will be confined in the (x, y) plan. Their stable trajectories will then be confined in the neighbourhood of the central axis and those ions will be transmitted through the analyser. By linearly scanning the applied voltages while keeping the ratio U_{DC}/V_{RF} constant (scanning line), ions with increasing m/z

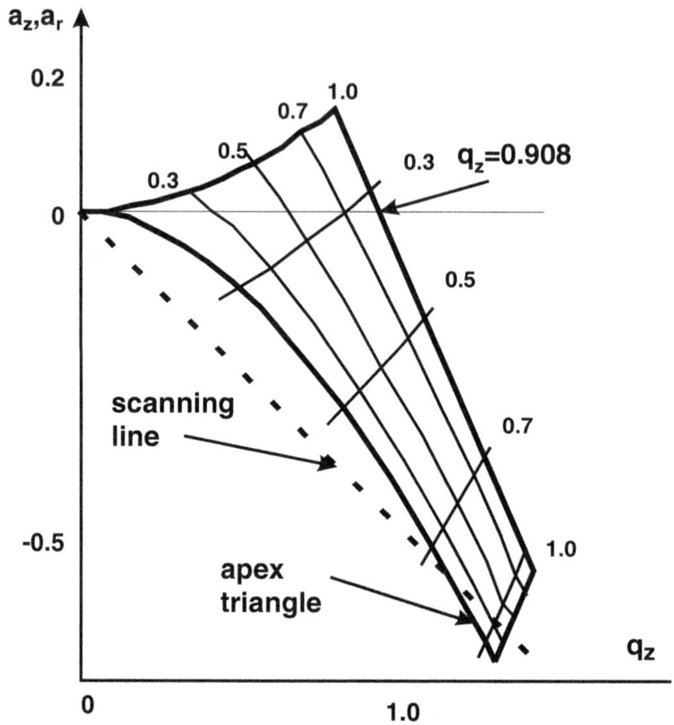

Figure 3.2.6 Stability diagram in the (a, q) or (V_{RF}, U_{DC}) plane. The dotted line is a scanning line ratio keeping the U_{DC}/V_{RF} ratio constant.

ratios will be successively detected at the exit of the quadrupole (Figure 3.2.6). The length and diameter of the rods determine the mass range and the ultimate resolution that can be achieved by the quadrupole assembly. Typical mass ranges that are normally used are around 4,000 Da with a resolution of around 2,000 but the transmission efficiency is usually rather low (typically 10^{-3}) as compared to that of TOF devices.

3.2.1.2.4 Ion guides and collision cells

Linear radio frequency multipole devices can be used as ion guides in order to transport ions from sources to analysers [170]. In that case, no DC component is applied ($a_x = a_y = 0$) and the stability diagram is reduced to a fraction of the q_x and q_y axis. RF-only quadrupoles (q) are then band-pass mass filters. Hexapole and octopoles are also used and have some characteristics more favourable than quadrupoles. When filled by neutral gas at rather high pressure, such filters are used for ion fragmentation in tandem (MS/MS) mass-spectrometry (see Section 3.2.2).

LIT offer a unique possibility for trapping and cooling ions down to very low temperatures ($\approx$10°K). Multipoles up to $N = 22$ are then used to confine ions and let them collide with a helium buffer gas in contact with the trap housing maintained at a few °K [146, 171] (see Section 5.5.3) (Figure 3.2.7).

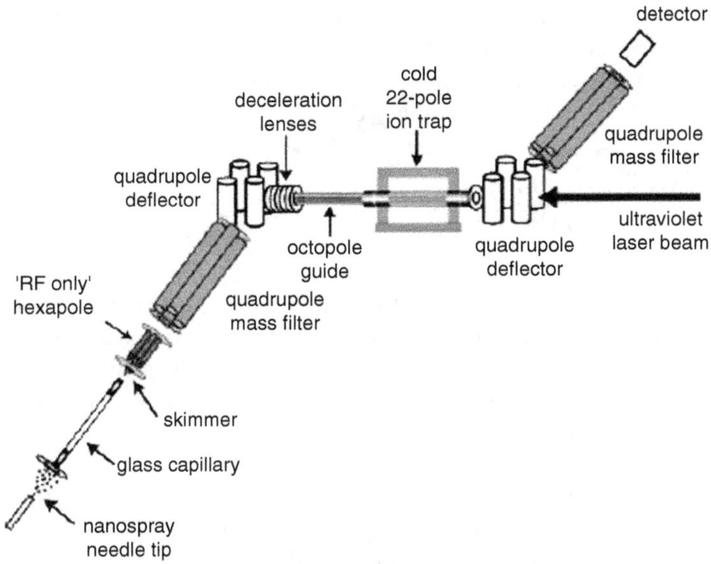

Figure 3.2.7 Tandem mass-spectrometer with a very low temperature 22-pole ion trap. Ions produced by a nanoelectrospray source are collected in an RF-only hexapole and mass-selected by a quadrupole filter. They are then cooled to 10°K and studied by R2PI/IR in the 22-pole trap. After ejection, the fragments are mass-analysed by a second quadrupole mass filter (reproduced with permission from reference [171] ©2006 American Chemical Society).

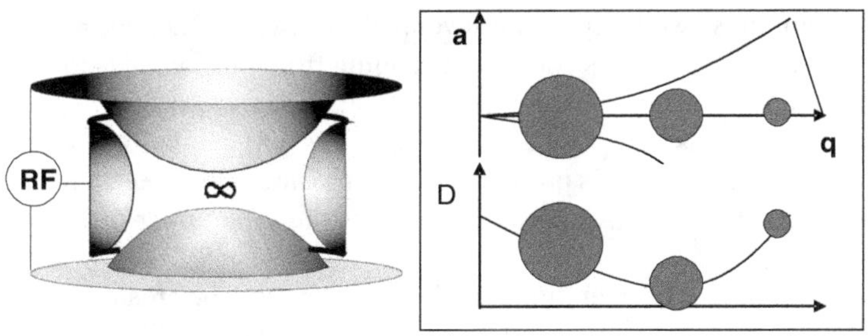

Figure 3.2.8 RF-only quadrupole ion trap. Schematic stability diagram displaying the variation of the potential well D_z as a function of the ion masses (represented by circles).

3.2.1.2.5 Three-dimensional ion traps

In a 3D ion trap (Figure 3.2.8), a 3D quadrupole field is created by applying an electric potential composed of a DC and an AC component, $\Psi(T) = (U_{\text{DC}} - V_{\text{RF}} \cos \Omega t)$, between two grounded hyberbolic electrodes serving as end caps and a ring electrode. The quadrupole field can be expressed in the set of coordinates r and z as

$$\Phi(r,z) = \frac{U_{\text{DC}} + V_{\text{RF}} \cos \Omega t}{r_0^2 + z_0^2}(r^2 - 2z^2 + 2z_0^2) \qquad (3.2.6)$$

The electric field then varies linearly as a function of r and z, creating a restoring force exerted upon the ions towards the centre of the trap. As in the 2D case, the equations of motion can be rewritten as a set of dimensionless Mathieu equations. If $u = r$ or z, those equations are $d^2u/d\xi^2 + (a_u - 2q_u \cos 2\xi)u = 0$ where

$$a_z = -2a_r = -\frac{16ZeU_{\mathrm{DC}}}{m\Omega^2(r_0^2 + 2z_0^2)}; \; q_z = -2q_r = \frac{8ZeV_{\mathrm{RF}}}{m\Omega^2(r_0^2 + 2z_0^2)}; \; \xi = \frac{\Omega t}{2} \tag{3.2.7}$$

The solutions of Mathieu equations in either r or z are stable if the corresponding set of dimensionless variables (a_r, q_r) or (a_z, q_z) are inside the corresponding stability diagram either in the axial $(u = z)$ or radial $(u = r)$ coordinate. Using the single coordinate u, those two stability diagrams can be superposed into a single one. If the applied voltages U_{DC} and V_{RF} are such that the functioning point with coordinates (a_u, q_u) is inside the stability diagram, the ion motion is stable and confined in the trap.

The ion motion has then two components, a "micro" motion at the trapping frequency Ω of the applied RF field and a slow motion called "secular" or "macro" motion at a lower frequency $\omega_u u = r, z$. The relation between the fast micro-motion frequency Ω and the secular motion frequency ω_u is then $\omega_u = (n + \beta_u/2)\Omega$ (where n is an integer). The secular ion trajectories are Lissajous figures due to the combination of an axial oscillation along the z-axis at frequency ω_z and a radial oscillation along r at frequency ω_r. For low values of the RF field amplitude such that $q_u < 0.4$, the micro-motion can be neglected and the macro-motions along r or z are simply oscillations in harmonic wells with respective depths [172] that are equal to $D_z = -(m\Omega^2/8Ze)(a_z + q_z^2/2)z_0^2 = (1/2)D_r$.

Typical values of those well-depths are several electronvolts. 3D traps allow the confinement of ions during time durations commonly ranging from milliseconds to several seconds [173, 174]. Dynamical studies of proteins [175] (see Chapter 4.2) and optical studies of very large systems such as bacteria [176] (see Chapter 4.8) can be then conducted in 3D traps. Using the same arguments as precedently, the maximum ion density that can be stored in a 3D trap is given by $n_{i,\max} = 3D_z\varepsilon_0/Zez_0^2$ and has the same order of magnitude as in a linear trap.

There are several modes of operation for 3D traps. Ions can be created inside the trap by electron bombardment, chemical ionization or electron capture [177], by charge exchange between neutral crossed beams [172] or can be introduced from external sources such as electrospray sources. Ions are then trapped for some and finally ejected through an aperture in one of the trap end caps in order to be detected by a particle detector. In the case of extremely heavy ions, particle detection can be replaced by LIF (see Chapter 4.8). A first possibility is similar to that used in quadrupole mass analysers. The $U_{\mathrm{DC}}/V_{\mathrm{RF}}$ ratio can be set at the apex of the stability diagram for each selected m/z ratio. Although mass selectivity can be very high, this mass-selective mode is not used since it provides unfavourable trapping conditions and a full mass spectrum can only be obtained by repeating several hundred times the procedure for different values of the functioning point. In the second possibility, DC fields are not necessarily used in 3D trap mass analysers. Ions are then trapped with $a_u = 0$ and q_u values comprised in between 0.15 and 0.908 ($q_u = 0.908$ is the limit of the stability diagram if $a_u = 0$). The lowest q_u value corresponds to the largest trapped ion masses. In principle, it could be as small as possible but space-charge effects set the lower limit. Ion densities up to 7×10^6 ions/cm^3 can be reached but then degrade the mass accuracy. The ion density should be limited to 2×10^3 ions/cm^3 if a good mass accuracy is required. Ions with m/z values, such that their q_u values lie outside the stability diagram, are not confined and thus a 3D ion trap can be considered as a band-pass mass filter. The low to high mass range is then approximately a factor of six (0.908/0.15) for low m/z values (< 300). Ions with larger m/z values are more effectively trapped at low q_u values (< 0.15) and m/z values can reach several thousands [178]. The well depths vary with the used functioning point and reach a maximum value for $q_z = 0.8$. This parameter value is the most favourable for ion injection and trapping.

QIT are generally operated with the addition of a helium buffer gas at a pressure of 1 mTorr [179]. This buffer gas has several effects. Collisions with light particles tend to cool and thermalize the trapped heavy ions by decreasing their kinetic energy and focus them at the centre of the trap. The addition of helium also allows fragmentation of the ions inside the trap and thus their analysis by sequential mass-spectrometry (MS/MS) [74, 180, 181]. After their injection or creation inside the trap, ions are accumulated. In order to study the fragmentation of precursor ions of mass m_{p}, those ions must be first isolated. Ions with masses less than m_{p} are ejected by rapidly increasing the V_{RF} voltage value. When an RF field is applied to the trap end-cap electrodes at the secular frequency of an ion, this dipole excitation increases the amplitude of motion and the kinetic energy of the concerned ion. Ions with masses larger than m_{p} are thus resonantly ejected by applying RF voltages over a wide frequency band covering their secular motion frequencies. A small time-delay allows collisions to refocus isolated precursors at the centre of the trap. Those precursors are then submitted to dipole excitation at their secular frequency. They acquire kinetic energy and fragment after several hundreds of collisions with the helium bath (see Section 3.2.2.2). By applying a fast RF voltage and detecting the ejected ions, a fragmentation mass spectrum of the precursor ions is then obtained.

Three-dimensional traps are usually made of very precisely machined hyperbolic electrodes, which ensures as much as possible the exactness of the created quadrupole potential and thus the mass accuracy. However, small imperfections are always present and the mass accuracy is generally not as good as that of other instruments such as TOF or FT-ICR. If a high mass accuracy is not required, cylindrical traps are more easily machined [182] and can be miniaturized for building multiplexed mass-spectrometers allowing the simultaneous record of several mass spectra [183].

3.2.1.3 Magnetic and electric sector mass analysers

Sector-field mass analysers use magnetic and/or electric fields to deflect ions through a curved trajectory onto a detector. Ions of mass m and charge Ze are first accelerated from their source by a potential U which provides them a kinetic energy $K_0 = (L^2/t_{\mathrm{D}}V)(p/760)$ $(273.2/T)$. When entering a sector where a homogeneous magnetic field B is applied, the trajectory of those ions is bent along a circular path of radius r_{M} since they are submitted to both the Lorentz force with amplitude $F_{\mathrm{M}} = ZevB$ and the centripetal force $F_{\mathrm{C}} = mv^2/r_{\mathrm{M}}$. The radius r_{M} is equal to $r_{\mathrm{M}} = (1/B)\sqrt{2mU/Ze}$. In fact, ions with the same m/z ratio and same kinetic energy are emitted by the source over different angles and do not enter the field B at the same point. They do not follow the same trajectory inside the magnetic sector but this sector refocuses them at some distance from the sector exit where a detector is placed behind a slit. This slit defines the mass resolution since ions with different masses are not focused at the same point. Ions of mass m and charge Ze are accelerated by a potential U. They enter a region where a radial electric field E is applied between cylindrical or hemispherical plates. They follow a circular path of radius r_E since they are submitted to the Coulomb force ZeE and to the centripetal force $F_C = mv^2/r_E$. The radius r_E is given by $r_E = 2U/E$. Ions with the same kinetic energy but different masses and emitted with some angular divergence from the source are refocused by the electric sector. Ions entering the electric sector with different kinetic energies are refocused at different points. Electric sectors are thus not mass analysers but with exit slits; they can reduce kinetic energy dispersion.

Using a combination of electric and magnetic sectors [184], also called double focusing, it is possible to focus ions onto nearly a single image point of the source even if they are emitted with slightly different kinetic energies and directions. In the forward geometry, the electric sector is placed first at the exit of the source and defines energy-resolved ion beams before mass-analysis by the magnetic sector. In a reversed geometry, the magnetic sector is placed first and separated from the electric field by a field-free region where collision experiments can be performed. Double focusing devices offer high-resolution, accurate mass determination over a wide mass range but they are large and costly instruments.

3.2.1.4 Fourier-transform ion cyclotron resonance cells

The principle and examples of applications of FT-ICR mass spectroscopy are given in great details by its inventor in reference [185]. Here we only give an outline of the technique. The motion of a single ion in an ion cyclotron resonance (ICR) cell is first recalled and it is then shown how the periodic motions of an ion assembly with different m/z ratios in the time domain is monitored in the frequency domain to obtain a mass spectrum. In an ICR cell, an ion with charge $z = Ze$ and mass m is in presence of a uniform magnetic field B_0 directed along the z-axis. This ion follows a circular path with radius r_c and its velocity component v_{xy} in the (x, y) plane, in absence of collision, is given by the balance between the Lorentz force and the centripetal force $mv_{xy}^2/r_c = ZeB_0v_{xy}$. The cyclotron frequency ν_c is related to the velocity v_{xy} by the relation $v_{xy} = 2\pi r_c \nu_c$ and is thus equal to $\nu_c = (1/2\pi)(ZeB_0/m)$. Ions with the same m/z ratio thus have the same cyclotron frequency, which is independent of their velocity. In absence of any applied external RF field and in a high vacuum, an ion can be trapped for a long time and comes into equilibrium with the surrounding walls of the ICR cell (see Section 3.2.4) at temperature T. It acquires an average thermal energy $m\bar{v}_{xy}^2 = kT$, where k is the Boltzmann constant, and its cyclotron motion radius is given by $r_c = \sqrt{2mT}/ZeB_0$.

For ion detection, a spatially uniform electric field $E(t) = E_0\cos 2\pi\nu_c t$ is first applied in a plane perpendicular to B_0 (for example, the x-axis) by two planar electrodes at the cyclotron frequency ν_c for a time duration T_{excite}. An ion possessing this cyclotron frequency resonantly absorbs energy and its cyclotron radius increases up to a value r_{excite} given by $r_{excite} = E_0T_{excite}/2B_0$, independent of the ion m/z ratio. Ions acquire a kinetic energy $E_{kinetic} = ((Ze)^2 E_0^2(T_{excite})^2)/8m$. Using a numerical example, one can see that the cyclotron radius and the kinetic energy strongly increase after excitation. Let us consider a singly charged ion of mass 1,000 initially thermalized at $T = 300°$K, in a magnetic field of 5 Tesla (typically employed magnetic fields range from 1T obtained with a permanent magnet [74] up to several T [186]

obtained with a superconducting magnet). Its initial cyclotron radius ν_c is 0.14 mm. After applying an electric field of 100 V/m (created by means of two parallel electrodes 2 cm apart with a ± 1 V voltage) for 1 ms, the cyclotron radius becomes $r_{excite} = 1$ cm and the acquired energy becomes $E_{kinetic} = 120$ eV allowing fragmentation in MS/MS (see Section 3.2.2).

Following excitation, an ion with cyclotron frequency ν_c induces an electric charge image and thus an electric current in detection electrodes perpendicular to the *y*-axis (Figure 3.2.9). A first possibility for getting a mass spectrum consists in a very slow sweep over the whole range of ICR excitation frequencies ν_c. Ion currents induced in the detection plates by ions with different *m/z* ratios are recorded as a function of the ICR frequency ν_c. This scanning method is not used because the mass accuracy and resolution are limited. Instead, a wide range of selected *m/z* values are excited by means of stored waveform inverse Fourier-transform (SWIFT) signals. For excitation of each desired *m/z* ratio, the amplitude, duration and frequency of an excitation signal are chosen and then summed over all ν_c values. This signal defined in the frequency domain is then digitally Fourier-transformed in the time domain and applied to the excitation plates. Following this broadband excitation, the whole image currents induced in the detection plates by the different excited ions are simultaneously recorded as a function of time. Fourier-transform provides the ICR ν_c frequency spectrum and thus the stored ion mass spectrum.

The ion motion is only confined in the (x, y) plane by the magnetic field B_0. A static electric field is applied between plates perpendicular to the *z*-axis in order to trap ions along that direction. This introduces an oscillation between those plates and also a radial magnetron rotation at frequencies much lower than the cyclotron frequency ν_c (Figure 3.2.9). The large radius magnetron motion can introduce problems when trapped ions are studied with a strongly focused laser beam over a long period of time [74]. Ions then only interact with the laser beam along a small fraction of their trajectory. The situation is more favourable in a 3D ion trap in presence of a buffer gas since the ion cloud then shrinks at the centre of the trap and does not escape out of the laser beam [74].

Since collisions of the stored ions would damp their cyclotron radius r_{excite} after excitation down to the initial thermal values, ICR cell mass measurements are made under high-vacuum conditions (typically less than 10^{-8} Torr). Due to the fact that frequencies are the most precisely measured physical quantities and that ion frequencies are monitored over a long

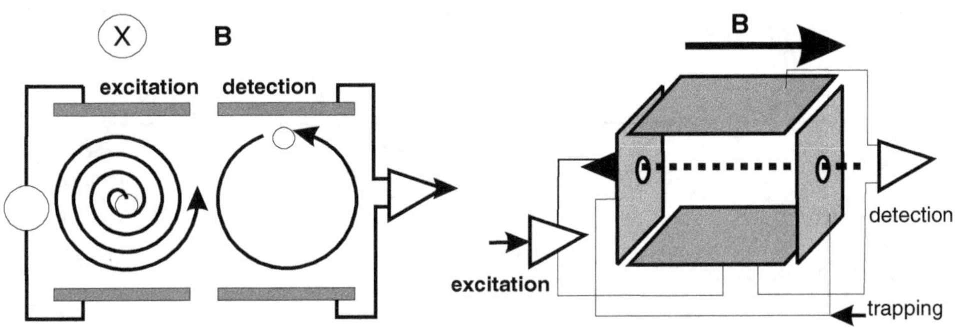

Figure 3.2.9 Principle of ion cyclotron resonance (ICR) mass-spectrometry. The application of a rotating electric field at the ICR frequency of stored ions of a given *m/z* value transforms an incoherent ion cyclotron orbital motion into a coherent and detectable motion.

acquisition time after broadband excitation, ICR mass-spectrometers are very high-resolution and high-accuracy instruments. For studying ion–molecule reactions, ICR cells can be filled with a reagent gas at very low pressures since interaction times can be very long. Typically, with a reagent partial pressure of 10^{-5} Torr and a 10 s reaction time, bimolecular rate constants down to $2 \times 10^{-12}\,cm^{-1}\,mol^{-1}\,sec^{-1}$ can be measured.

3.2.1.5 Electrostatic ion storage devices

The mass-spectrometers that have been presented above are being constructed by industrial manufacturers as "press-button" instruments and routinely used in genomics and proteomics. Electrostatic ion storage devices [187–189] that are now presented are still more or less laboratory instruments but offer high performances and versatility by storing kiloelectronvolt ion beams.

3.2.1.5.1 Ion storage rings

The ion storage ring ELISA experimental setup built at the University of Aarhus is used for biophysics experiments [187]. Ions provided by an electrospray source are steered through an octopole guide into a cylindrical trap for accumulation. They are then accelerated, mass-analysed by a bent magnet and injected in the electrostatic ring where they circulate for very long times (typically 1–100 s). Except in the bent focusing electrostatic elements, ions travel at constant speed in straight sections and make a full turn in approximately 50 μs (Figure 3.2.10). In one of the straight sections, they can be submitted to a copropagating tunable laser beam in the visible range and absorb enough electronic energy to further possibly dissociate into fast neutrals. When this happens in the second straight section, those neutrals are then detected. With the laser kept at the fixed wavelength of 414 nm, the number of neutrals issued from protonated and deprotonated protoporphyrin has been monitored as a function of time providing lifetimes of the corresponding first electronic triplet state [190]. By monitoring the neutral yield as a function of the excitation laser wavelength, the absorption spectrum of protonated and deprotonated chromophore ions such

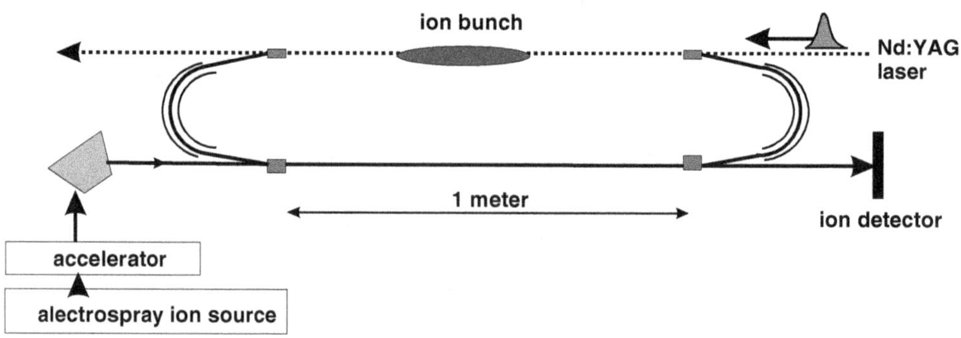

Figure 3.2.10 Principle of electrostatic ion ring spectroscopy. Mass-analysed bunches of ions are injected in an electrostatic ring. A few tens of milliseconds after injection, ions are irradiated with a pulse of tunable UV photons. The accelerated neutral fragments produced in the other straight section of the ring are detected (adapted with permission from reference [196] ©2004 RSC).

as the green fluorescent protein (GFP) and the red fluorescent protein have been recorded [191–194] (see Chapter 4.2). The propensity of non-statistical dissociation of photoexcited protonated and di-protonated dinucleotides has been investigated [195].

Electrostatic ion beam traps

We have seen in Section 3.2.1.1 that reflectrons act as ion mirrors. The design of an electrostatic ion beam trap is similar to an optical Perot–Fabry cavity. If photons are sent in between two parallel mirrors, a system of stationary waves can build up under certain conditions and photons are trapped due to the multiple reflections. The motion of an ion in an electric potential U can be shown to be equivalent to a photon in a material with index of refraction $n = \sqrt{U}$. In the electrostatic ion beam trap designed by Zajfman [189, 197] (Figure 3.2.11), two coaxial sets of grids act as reflectron mirrors. Ions with kiloelectronvolts energies are introduced into or extracted from the trap by opening or closing of the trap end electrodes. With suitable adjustment of voltages, ions oscillate in the trap for several seconds and their periodic motion is detected by means of a pick-up electrode. The Fourier-transform of the induced signal provides the ion mass spectrum. At the turning points in the reflectrons, the ion density strongly increases and Coulomb forces tend to slow down the fastest particles and speed up the slowest [197]. Ions are then no longer moving independently and oscillate in very narrow bunches, leading to a mass resolution better than 10^6. This ion bunching is in fact also observed at very low ion densities in absence of space-charge effect.

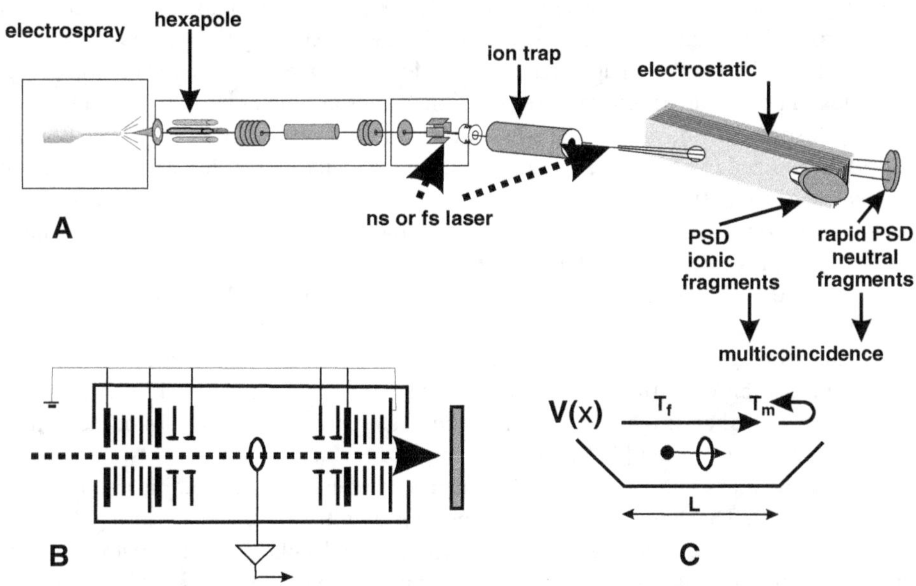

Figure 3.2.11 (A) Zafjman mass-analyser and ion/neutral detection. After dissociation of mass-selected ions induced by a nanosecond laser, neutral and ionic fragments are detected in coincidence between position-sensitive detectors (PSD). The kinematics of dissociation can then be fully reconstructed (by courtesy of B. Lucas and J. Fayeton). (B) Schematic of the Zafjman mass analyser. The ion motion is detected through the current induced in a ring electrode. (C) Potential well in a Zafjman mass-analyser.

3.2.1.5.2 Orbitrap mass-spectrometers

An Orbitrap mass-spectrometer [198, 199] operates without any magnetic or RF voltage but by applying high DC voltages between an inner spindle-like electrode and a coaxial barrel-like electrode (Figure 3.2.12). Stable trajectories consist in an orbital motion between these electrodes and an axial oscillation the frequency of which is independent of the energy. The spatial spread of the injected ions is inversely proportional to the square root of their mass [120]. Fourier-transform of the signals induced in the outer electrode provides the mass spectrum of the trapped ions. An Orbitrap can be used in combination with a LIT [200–202]. The mass accuracy can be close to 2 ppm and the mass resolution larger than 40,000 [201].

3.2.2 Tandem (MS/MS) mass-spectrometry

3.2.2.1 Aim of MS/MS analysis

Once genomes of living species are deciphered, it is possible to predict the corresponding ensemble of proteins that can in principle be produced. However, before becoming biologically active, proteins issued from ribosomes (see Chapter 4.8) undergo post-translational modifications (PTMs). For example, proteins responsible for the packaging of DNA, called histones, are sites of numerous PTM. A single expressed gene can thus lead to a large number of different proteins (away from the one gene → one protein paradigm), and tremendous efforts are devoted in proteomics to identify those proteins [204]. Tandem mass-spectrometry is then a crucial tool for locating those PTMs [205, 206] (see also Chapter 4.7). Proteins are extracted from biological cells and separated on 2D gels according to their mass and charge (neutralization pH). They then appear as more or less separated small spots on gels and can thus be picked up for identification. Two possible routes are then open. In the conventional bottom-up fashion [156], proteins are first digested by the trypsin enzyme and fragment peptides are ionized. The fragments are further fragmented (MS/MS) and very precisely mass analysed [151, 207, 208]. In the more recent top-down mass-analysis, intact proteins are directly sequenced and the PTMs are localized without the digestion step and the reconstruction procedure from identified fragments. It is possible to use the top-down mass-spectrometry for proteins with masses larger than 200 kDa [209] (Scheme 1).

Not only protein identification [156], but also elucidation of carbohydrate (see Chapter 4.4) and oligonucleotide (see Chapter 4.2) structures rely on tandem mass-spectrometry [208, 210]. From a complex mixture, first a mass-analysis (MS_1 in Figure 3.2.13) isolates a precursor ion which is then fragmented. In the second step, the mass spectrum of the obtained fragments (MS_2) is recorded [201]. MS_2 can be used for discrimination between ions possessing the same mass but different chemical structures such as the leucine and isoleucine amino acids [211]. The same procedure can be repeated in a $(MS)_n$ procedure [212].

In order to restrict the enormous number of possible molecular formulas that might correspond to a given molecular mass, highly accurate mass analysers with resolution up to 1 ppm and even up to 0.1 ppm [213] are desirable. TOF instruments can provide 2–5 ppm accuracies while QIT are usually limited to at best 20 ppm. Sub-ppm accuracies are provided by double-focusing magnetic sector instruments as well as FT-ICR cells. In order to improve the versatility and performances such as duty cycles of analytical instruments, the different devices described in Section 3.2.1, that is TOF, Q, LIT or 3D QIT, ICR cells, as well as RF-only

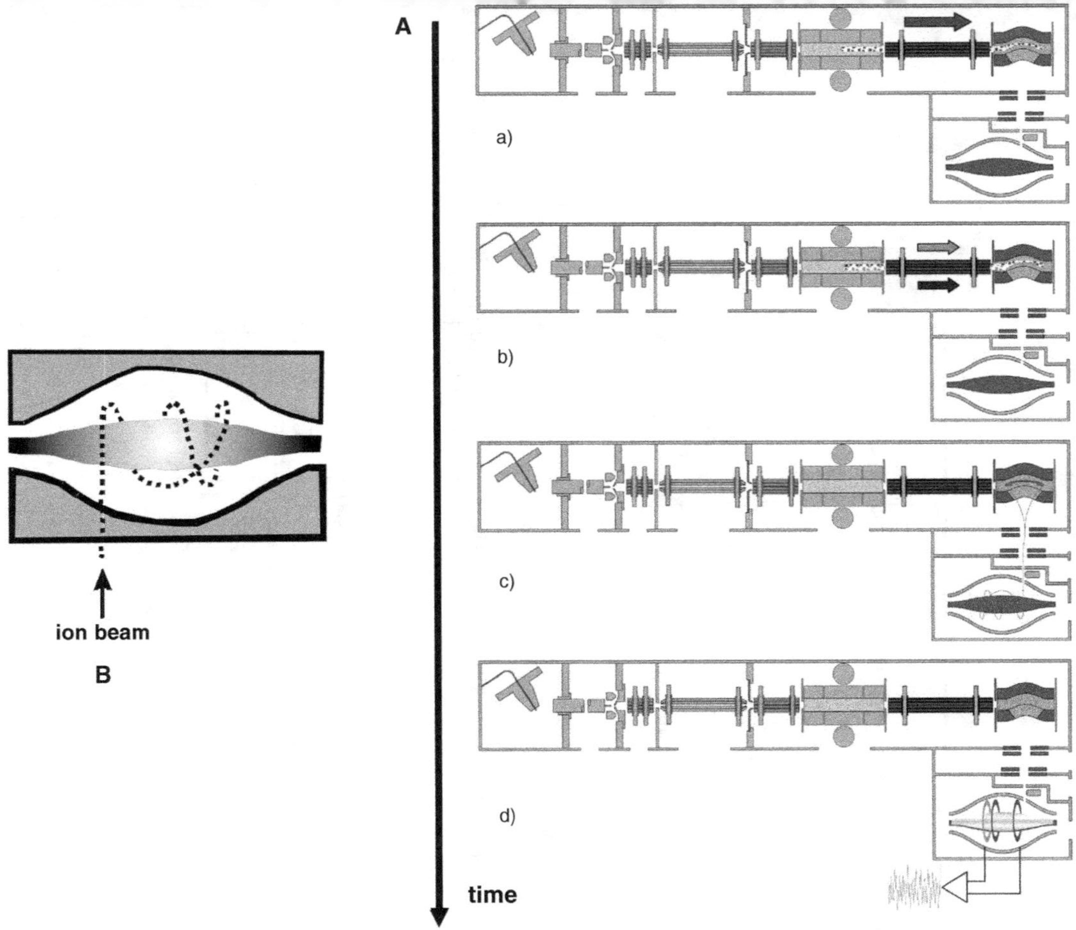

Figure 3.2.12 (A) Combination of a linear ion trap with radial ejection and an Orbitrap. Sequence of events for ejection of selected ions from the LQT into the Orbitrap. (B) Ions selected by the linear ion trap are introduced tangentially in the electrostatic Orbitrap. A voltage is applied between the outer and central electrodes. Ions rotate around the central electrode and their axial motion induces an image current, which is detected and Fourier-transformed (reproduced with permission from reference [200] ©2006 American Chemical Society).

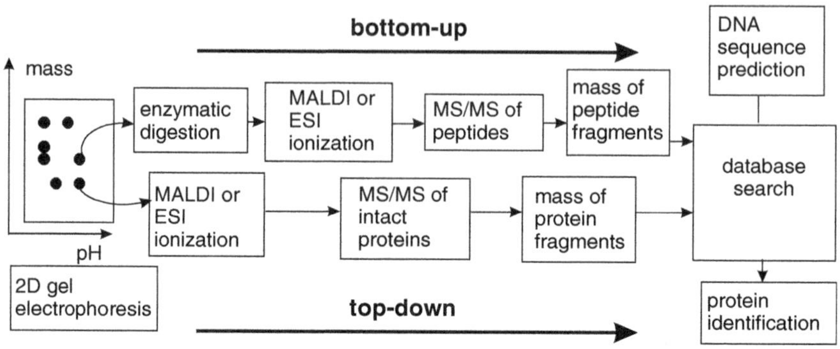

Scheme 1 Principle of bottom-up and top-down mass-spectrometry of proteins.

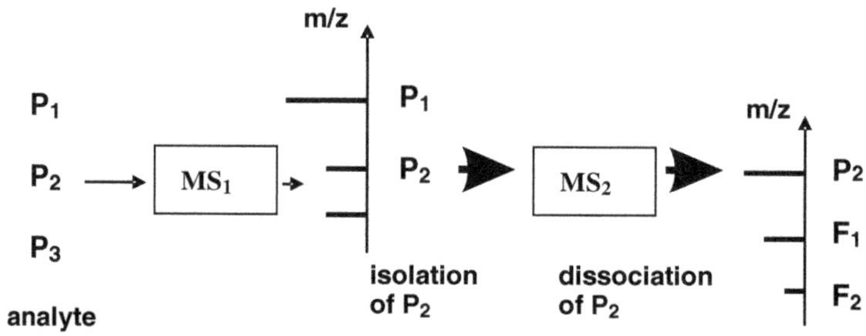

Figure 3.2.13 Principle of MS/MS analysis. The analyte composed of several species is mass-analysed and the P_2 ion is selected (MS_1). P_2 is then dissociated into fragments F_1 and F_2 that are mass-analysed (MS_2).

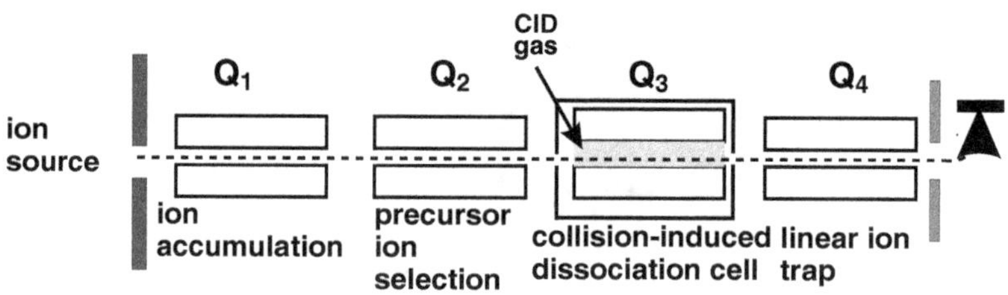

Figure 3.2.14 Quadruple quadrupole analyser. After ion accumulation in first quadrupole mass-spectrometer (Q_1), a given precursor ion is selected (Q_2) and further fragmented by collision-induced dissociation in cell (Q_3). Fragments are then mass-analysed in a linear ion trap (Q_4).

quadrupoles (q), are very often combined for building analytical mass-spectrometers. Those so-called hybrid systems can be divided into tandem-in-space instruments where fragmentation is performed in spatially separated regions (e.g. QqQ, see Figure 3.2.14) or tandem-in-time instruments such as Q-TOF and LIT-FT-ICR combinations where the time sequence isolation–fragmentation–fragment detection is performed in the same spatial region.

In proteomics [204], complex mixtures of proteins issued from biological cells are first separated by 2D gel electrophoresis into spots that are then mass-analysed for identification. Experimental MS/MS spectra are then compared to theoretical spectra of candidate peptide sequences from a comprehensive primary sequence database. By using scoring algorithms, the closest matches between the unknown protein sequence and the already known sequences are searched. If the sequence database does not contain the unknown protein, entries that exhibit the closest homology are pulled out. Data searching programs such as MASCOT [214] mostly use *m/z* information but do not take advantage of the knowledge of relative ion abundances that might be obtained by applying fragmentation pattern rules [215].

In this chapter, we first consider two fragmentation methods using multiple-step excitation of ions, also called slow-heating methods, that is low-energy CID [216] and non-resonant IRMPD [217, 218] which provide similar fragmentation patterns in contrast with ECD [219] (see Chapter 4.7). A fast-heating (single step) activation of ions such as the surface-induced dissociation (SID) method is compared to CID in reference [216]. We further examine models that are being developed to improve the prediction of fragment ion intensity relationships in the very important case of MS/MS spectra of protonated peptides.

3.2.2.2 Collision-induced dissociation (CID)

In CID [216, 220–225], an ion ABC^+, with mass m_{ABC} and velocity v_{ABC}, undergoes a single or multiple collision with rare gas atoms N with mass m_N and velocity v_N. A fraction of the ion kinetic energy is directly or progressively transferred into internal energy of the ion which fragments.

$$ABC^+ + N \rightarrow [ABC^+]^* \rightarrow AB^+ + C + N \qquad (3.2.8)$$

For each individual collision, the maximum possible energy transfer is the centre of mass (CM) collision energy $T_{CM} = (1/2)(m_{ABC}m_N/(m_{ABC} + m_N))(v_{ABC} - v_N)^2$. It increases with mass m_N (argon is more efficient than helium) but monotonically decreases when the size of the polyatomic ion ABC^+ increases. Classical trajectory calculations conducted on polyglycine and poly-alanine ions [226] show that most of the collision energy is transferred into internal energy which goes into vibrations and not into rotations. The energy transfer dynamics in a single collision is mostly influenced by the local region of the peptide which collides with the neutral atom. It thus only moderately depends upon the peptide structure and is practically independent of the peptide size. Among the different internal coordinates, stretches, bends and torsions, the latter are the most specifically excited.

A single collision with a cross-section $\sigma = 1\,\text{nm}^2$ between an ion with a velocity of 1,000 m/s interacting over 1 nm with a neutral atom lasts approximately 1 ps, and the time elapsed between two collisions can typically vary from several microseconds in a high-pressure quadrupole cell to a fraction of second in an ICR cell. The internal energy acquired in a small number of internal modes after a single collision has, thus, plenty of time to redistribute over most of the high-frequency internal modes. One should however remind that equipartition of energy can sometimes only occur after extremely long amounts of time when high-frequency degrees of freedom are strongly uncoupled from low-frequency bath modes [227].

Fragmentation of an ABC^+ ion requires accumulation of an internal energy E_f (typically a few electronvolts) after several collisions, at an activation rate k_{act}, for appearance of an

AB^+ fragment [225]. The ABC^+ ion internal energy is more or less randomly distributed over the very large number of internal degrees of freedom ($s = 3N - 6$ in a system with N atoms) considered as harmonic oscillators with frequencies ν_j. This energy must concentrate into the AB···C reaction coordinate to overpass the dissociation energy barrier (transition state TS with $s - 1$ degrees of freedom and frequencies ν_i^*). It is usual to estimate the fragmentation rate by means of the statistical Rice–Ramsperger–Marcus–Kassel (RRKM) theory and its adaptation called quasi-equilibrium theory (QET). The fragmentation rate $k(E)$ (in s^{-1}) of an ion with internal energy E_{int} is given by

$$k(E) = \left(\frac{E_{int} - E_f}{E_{int}}\right)^{s-1} \frac{\prod_{j=1}^{s} \nu_j}{\prod_{i=1}^{s-1} \nu_i^*} \quad (3.2.9)$$

The number of collisions experienced by an ion depends upon the collision cross-section, the neutral density and the ion velocity. It is given by a Poisson distribution. The situation strongly depends upon the ratio of the collisional activation rate k_{act}, the fragmentation rate $k(E)$ and the experimental time-window allowed for observing fragmentation. The kinetic shift (KS) is defined as the internal energy in excess of the fragmentation energy E_f required to produce detectable dissociation on the timescale of the tandem mass-spectrometer. Tandem-in-space mass-spectrometers are characterized by brief time-windows (1–100 ms) and large rate constants (10^4–$10^6\,s^{-1}$). In tandem-in-time mass-spectrometers, long time windows allow for much efficient fragmentation and KSs can become negligible while smaller rate constants (10^{-1}–$10^2\,s^{-1}$) can then be observed.

3.2.2.3 Peptide fragmentation mechanisms and pathways

There are two major routes to derive fragmentation rules of protonated peptide spectra. The top-down strategy uses statistical interpretation of databases containing large number of MS/MS spectra [129]. The bottom-up strategy [215, 228] is here outlined in the case of low-energy CID. The notion of charge-directed or charge-remote fragmentation is first recalled together with the nomenclature. It is further shown how resonant IRMPD can ascertain protonation sites of peptides in their lowest energy configurations. The mobile proton model is then described and finally an example of quantum and RRKM calculation of fragmentation pathway is given for a model peptide.

Peptide precursor ions submitted to low-energy CID mostly dissociate along the backbone at the amide bonds providing informative sequence ions. Less useful non-sequence ions corresponding to losses of small neutrals such as H_2O or NH_3 will not be considered here. The accepted nomenclature [229] is displayed in Figure.3.2.14. On this schematic diagram, the rupture of amide bonds leads to either b ions or y ions whether the single positive charge is retained on the NH_2 side (N-terminal) or the COOH side (C-terminal). The other neutral moiety is usually not observed although new experiments consider both dissociation channels (see Figure 3.2.11 in Section 3.2.1.5.2). Cleavage of the protonated amide bond is considered in reference [230]. Protonation on the carbonyl xygen leads to an increase in length of the C—N bond whereas protonation on the N—H group has the opposite effect.

In the case of ions with Z protons, each fragment can take away charges and, for example, ion pairs b_n/y_m are observed with $m + n = Z$ (Figure 3.2.15).

A summary of observed cleavages can be presented in a dissociation map where summed b- and y-ion yields are plotted as a function of charge state. An example is shown in Figure 3.2.16 for multiply protonated human haemoglobin α-chain species [174].

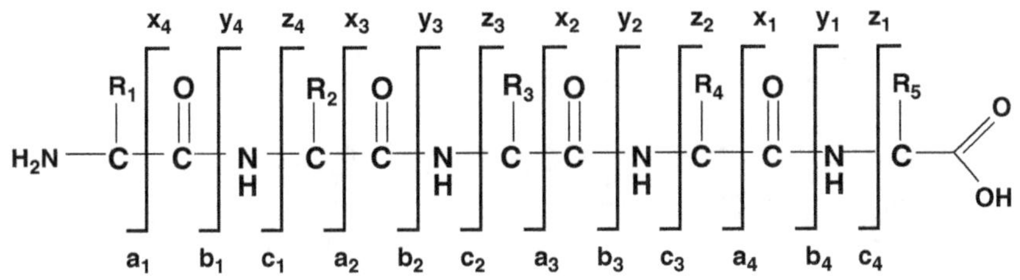

Figure 3.2.15 Accepted nomenclature for ions created from rupture of amide bonds.

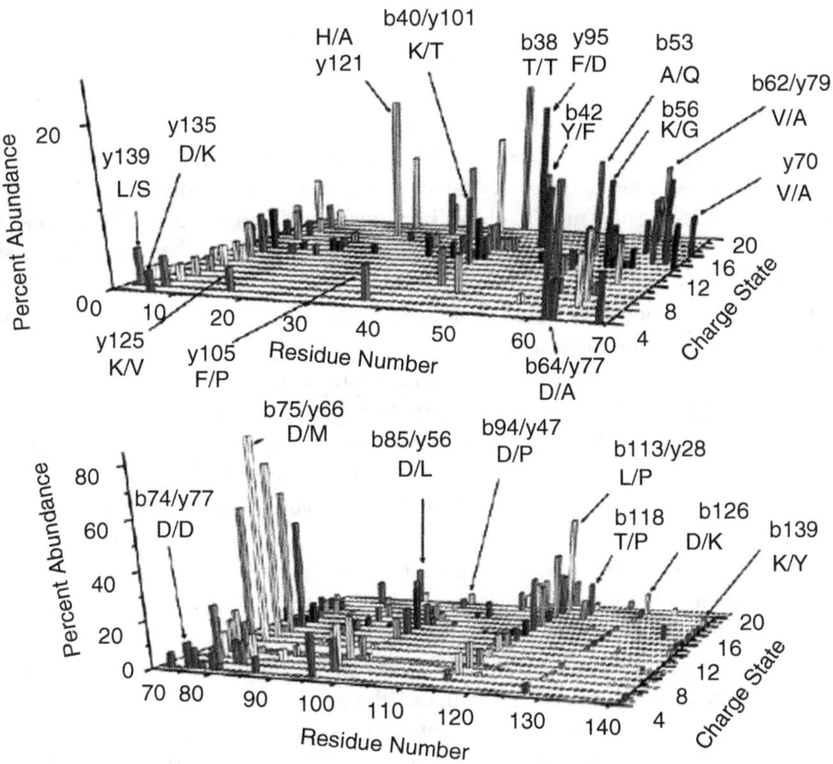

Figure 3.2.16 Dissociation map of haemoglobin α-chain cations obtained from collision-induced dissociation in a quadrupole trap mass-spectrometer. The residues ranges 1–70 and 71–141 are, respectively, presented in the upper and lower figures (reproduced with permission from reference [174] ©2006 Elsevier).

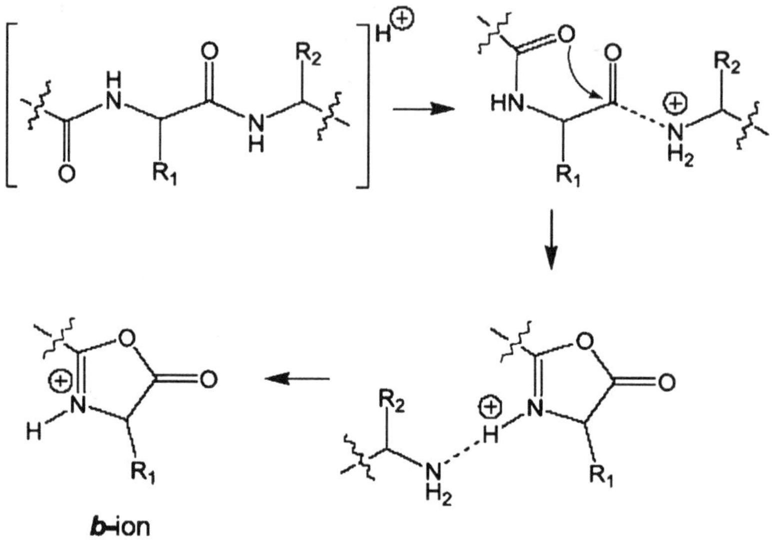

Figure 3.2.17 Fragmentation pathways in collision-induced dissociation of Leu–enkephalin to form b_4^+ fragments (reproduced with permission from reference [240] ©2005 Elsevier).

The knowledge of mass is of course not sufficient for determining the structure of fragments. Modelling provides important clues but quantum calculations [215, 230–236] supported by experimental observations bring definitive answers. A striking example is that of *b* ions for which several different structures have been proposed (Figure 3.2.17). Among them, a linear acylium structure and a cyclic oxazolone structure [237] have been mostly considered. An infrared spectroscopy study of the b_4^+ fragment of Leu–enkephalin [238, 239] (Figure 3.2.18) and a neutralization–re-ionization study [240] have both shown that peptide *b* fragment ions possess cyclic oxazolone structures.

The "mobile proton" model [156, 241, 242] assumes that, upon excitation, a proton added to a peptide can easily migrate along the peptide backbone to various protonation sites prior to fragmentation. This "charge-directed" pathway where cleavage is initiated by a charge that is transferred to the vicinity of the cleavage site cannot occur in peptides containing lysine and arginine, which sequest protons on their strongly basic side chains. This is the case of peptides obtained by tryptic digestion, which contain arginine at their C-terminus. A large excitation energy is then required to remove a proton from those basic sites towards energetically less favourable protonation sites then leading to a "charge-remote" fragmentation [243]. For example, in a singly protonated YYVTIIDAPGHR peptide [156], the proton is immobilized on the arginine (R). Among the very small number of observed fragmentations in the charge-remote pathway, the cleavage of the A–P amide bond belongs to enhanced cleavages at acidic residues (aspartic and glutamic acids) [156]. In the doubly protonated YYVTIIDAPGHR, the second proton is now mobile and CID produces a much richer fragmentation pattern (Figure 3.2.19).

The mobile proton has received confirmation by means of experimental and theoretical approaches. For example, fragmentation of deuterated peptides shows that deuterons, initially on the most favourable sites, are redistributed amongst exchangeable labile hydrogen sites

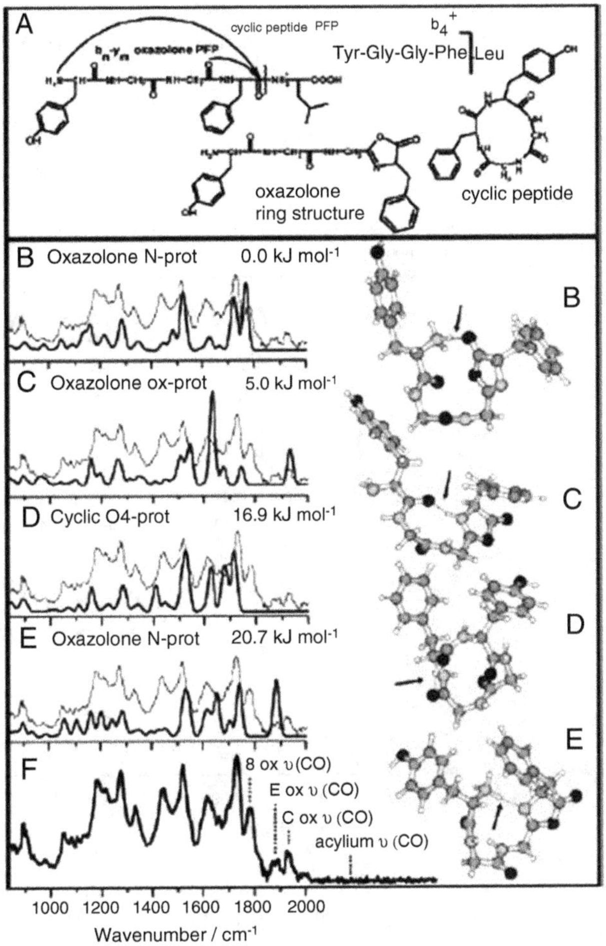

Figure 3.2.18 (A) Fragmentation pathways in collision-induced dissociation of Leu-enkephalin to form b_4^+ fragments (Figure 3.2.16). (B–F) Calculated IR spectra and corresponding structures of N–terminal protonated oxazolone, oxazolone ring protonated oxazolone, cyclized peptide protonated on O4 and N-terminal protonated oxazolone (higher energy structure) are, respectively, represented in B, C, D and E. Arrows indicate proton solvation sites. The experimental IR spectrum obtained from b_4^+ fragment depletion is represented in F and compared to calculated spectra in B, C, D and E (reproduced with permission from reference [238] ©2005 American Chemical Society).

[244]. Isomers O (protonation on the amide oxygen) have stronger amide bonds than neutrals and are also energetically favoured with respect to isomers N (protonation on the amide nitrogen) that contain the weakest amide bonds and are the most likely to fragment [231]. Moreover, protonation of the amide nitrogen makes the carbon of the concerned amide group a likely target for a nucleophilic attack from nearby electron-rich groups, initiating a cascade of rearrangements. The proton migration along the peptide backbone in the charge-directed fragmentation can be followed by infrared spectroscopy [239]. It can also be followed by exploration of PESs, that is determination of minima for different isomers corresponding to the

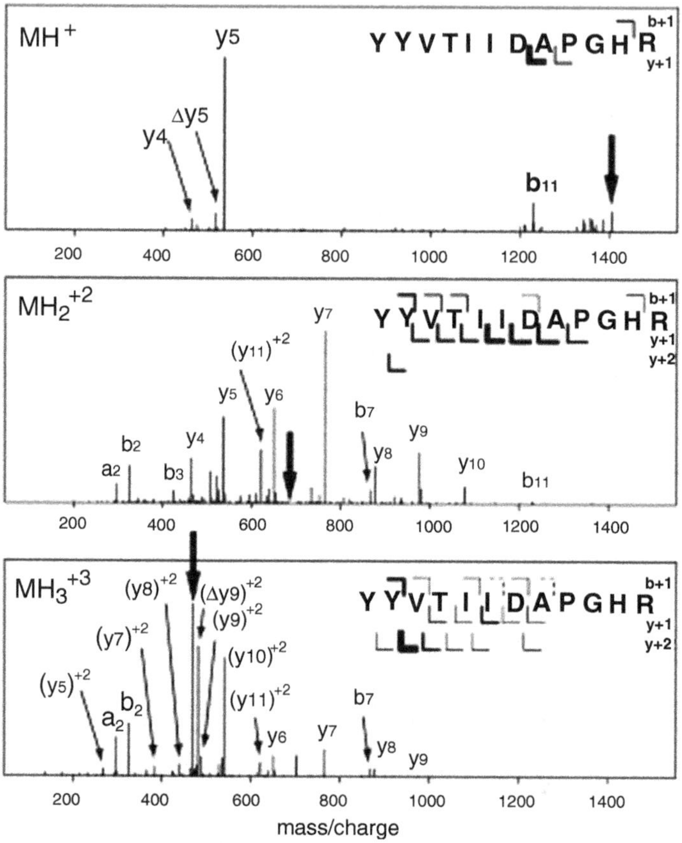

Figure 3.2.19 MS/MS fragmentation pattern of the singly, doubly and triply protonated YYVTI-IDAPGHR peptides observed in a quadrupole ion trap (by courtesy of V. Wysocki)

different protonation sites and transition structures connecting those minima. The proton hopping rate between those minima can be evaluated by RRKM calculations. In the protonated diglycine model peptide, very similar to the protonated dialanine peptide, the lowest energy configurations A1 (taken as the reference energy) and O1 are planar. There is no barrier for proton transfer between A1 and O1 and the proton is delocalized between the N atom of the terminal amino group and the carbonyl O atom (see Chapter 4.2). The same situation exists for the different A—O minima and only A configurations will be represented. For fragmentation, the internal energy must reach 10.5 kJ/mole [245]. Nevertheless, configuration D_1 only lying 4.4 kJ/mole above A_1 is suggested as the fragmenting configuration [231]. The proposed theoretical pathway is depicted in Figure 3.2.20. Two internal rotations with small energy barriers bring the mobile proton to be shared between the amino group and the amide nitrogen atom. The fragmentation rate is governed by the proton transfer $N_{amino} - H \cdots N_{amide} \rightarrow N_{amino} \cdots H \cdots N_{amide} \rightarrow N_{amino} \cdots H - N_{amide}$ which has an RRKM rate constant of 1 μs for an internal energy of 1 eV. This rate is much lower than the RRKM rates of the two internal rotations.

The non-covalent structure of a peptide affects its CID- and photo-fragmentation spectra due to rearrangements of the peptide conformation in the vicinity of a cleavage site [246].

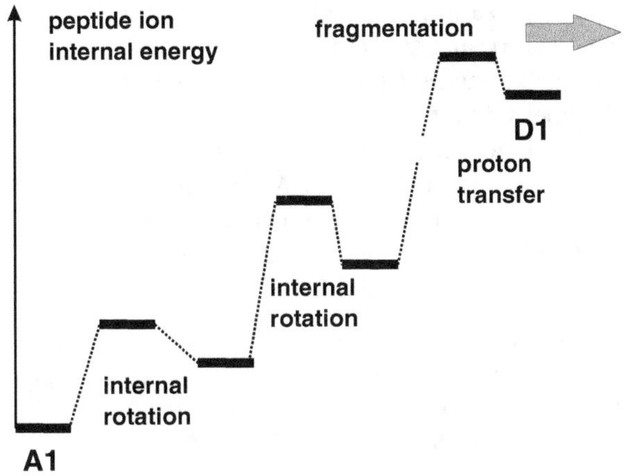

Figure 3.2.20 Proposed fragmentation pathway for the amide bond fragmentation of the protonated diglycine ion.

Only a restricted number of non-covalent bonds are broken near a cleavage site. For example, in a peptide such as melittin, a 26-residue peptide containing 20 hydrogen bonds with a bond strength of 1–1.5 kJ/mole each, rupture of all those bonds would require an excess internal energy of 25 kJ/mole while only 10 kJ/mole is sufficient to break an amide bond. Only a few hydrogen bonds are broken around a cleavage site. This leads to a model [247] in which the different CID fragmentation channels go through different transition states. This model provides complementary information about the gas-phase structure of peptides.

3.3 DETERMINATION OF STRUCTURES OF MASS-SELECTED GAS-PHASE BIOMOLECULAR SYSTEMS

In this chapter, we consider different experimental methods providing structures of mass-selected gas-phase biomolecular systems without use of resonant interactions between those systems and electromagnetic radiation. Similarly to spectroscopic methods that use resonant interactions with electromagnetic fields (see Chapter 2), the methods considered here rely on experimental measurements of structural parameters and their comparison to structural parameters predicted from calculations (see Chapter 1).

3.3.1 Ion-mobility spectrometry

Ion-mobility spectrometry [248] relies on the difference of mobility of molecular ions drifting in an electric field in presence of an inert buffer gas [249–251]. In an ion-mobility cell, ions undergo a very large number of collisions and their collision cross-sections are dependent upon their geometrical structures. If those ions are injected in the cell at the same time, those with the most compact structures arrive before those with more extended structures, and thus ion-mobility spectrometry presents similarities with ion chromatography. By comparing the experimental arrival times to those predicted from collisional process calculations, it is possible to infer structural information [252–261].

Experimentally, a short pulse of ions is injected in a drift cell of length L filled with a buffer gas at pressure p (in Torr) and temperature T. Ions undergo a voltage drop V until they reach the cell exit after an amount of time t_D. The mobility K of an ion is defined as the ratio of the acquired velocity L/t_D to the applied electric field $E = V/L$. Ion mobilities are usually reported as reduced mobilities defined as $K_0 = (L^2/t_D V)(p/760)(273.2/T)$. Pressure p corresponds to a buffer gas number density N, and the parameter determining the ion energy is E/N. At low E/N values, the mobility is independent of the applied electric field and drifting ions do not align along the electric field. The parameters of collisions with the buffer gas must then be averaged over the different orientations.

3.3.1.1 Structure determinations

If σ_{avg} is the average cross-section, m and Ze, respectively, the mass and charge of the studied ions and m_b that of the buffer gas, the mobility is given by

$$K = \frac{\sqrt{18\pi}}{16}\sqrt{\frac{1}{m}+\frac{1}{m_b}}\frac{Ze}{\sqrt{k_B T}\sigma_{avg}}\frac{1}{N} \tag{3.3.1}$$

Ion-mobility measurements provide experimental average cross-sections σ_{avg} that must then be related to ion structures [260, 262, 263] (Figure 3.3.1). A set of possible ion conformations are first generated from quantum or force-field PESs and their systematic exploration (see Section 1.5). In a first approximation, the individual atoms of the molecular ion can be treated as hard spheres. The geometrical cross-section must then be averaged over all possible orientations by projecting the molecular ion onto a randomly chosen plane. A circle is drawn in this plane at the position of each projected atom with the corresponding collision radius. Points in the plane are then randomly picked within a square of area A containing the molecular ion. Those points are considered as hits if they fall inside one or more circles. The cross-section corresponding to this particular projection is then the ratio of the number of hits to the number of tries multiplied by A. The procedure is then repeated and averaged for many randomly selected projections [264].

Evaluation of intrinsic size parameters of individual amino acids [265, 266] and the projection approximation (PA) is employed for rather small (10–200 atoms) systems while the exact hard sphere scattering model (EHSS) is employed for large ones (>1,000 atoms) [267, 268]. Trajectory calculations are used for intermediate (200–1,000 atoms) sizes.

3.3.1.2 Experimental designs

In an ion-mobility setup, ions are produced by either an electrospray [251] or a MALDI source [249, 270, 271]. Ions can then be either directly injected in the drift cell using an ion funnel [272–274] or temporarily trapped [275, 276]. If an ion pulse is injected in a sufficiently short time (typically, several microseconds), the resolution is limited by the width of the arrival time distribution at the exit of the drift cell. If $t_{1/2}$ is the width of the arrival time peak at half-height, the resolving power is given by (Figure 3.3.2)

$$\frac{t_D}{t_{1/2}} = \frac{1}{4}\sqrt{\frac{Ze}{k_B \ln 2}\frac{V}{T}} \tag{3.3.2}$$

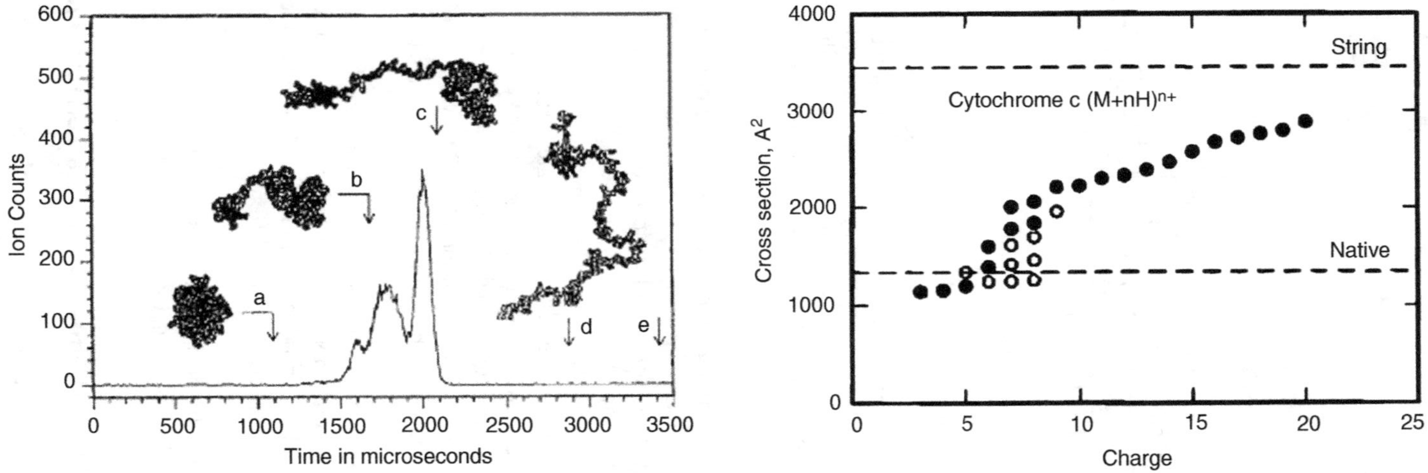

Figure 3.3.1 Left: Distribution of drift times for the +7 charge state of bovine cytochrome *c*. Arrows show the expected drift times of possible conformations of cytochrome *c* (a) native crystal structure, (b) partially unfolded structure, (c) unfolded coil retaining the α helices, (d) random coil without any secondary or tertiary structure and (e) nearly linear conformation [269] Right: Cross-sections determined for the main features resolved in the arrival time distributions of cytochrome *c*. Filled and open points correspond, respectively, to features that dominate at high (collisionally heated and unfolded conformations) and metastable conformations observed at low injection (representing conformations issued from the ESI source) energies. The dash lines show the cross-sections determined using exact hard spheres model, respectively, for the crystal structure native conformation and a fully extended string conformation (reproduced with permission from reference [269] ©1995 and [143] ©1999 American Chemical Society).

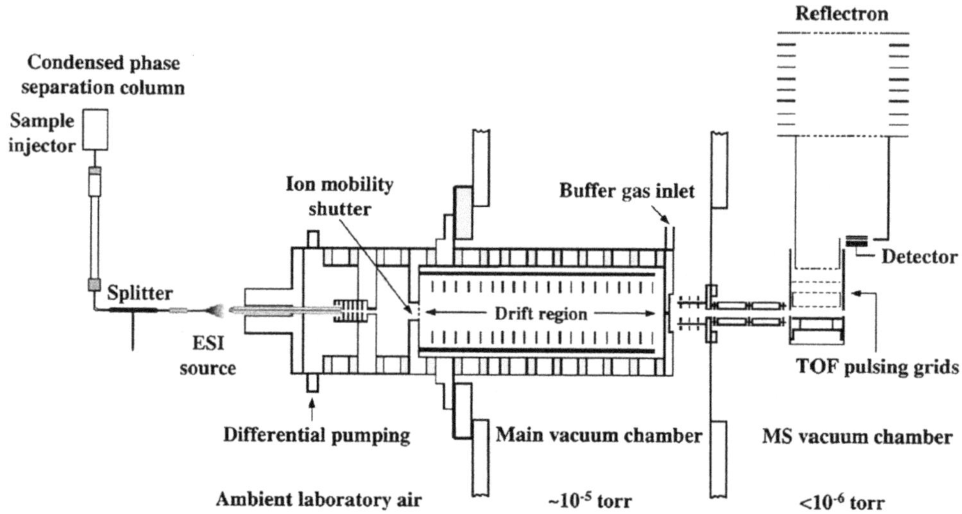

Figure 3.3.2 Ion-mobility cell with an ion drift region for mobility separation and a time-of-flight tube for mass-analysis. Note the different timescales: separation by high-performance liquid chromatography (HPLC) takes ≈45 min. After being electrosprayed and injected in the mobility cell, ions drift for milliseconds and are then mass-analysed in microseconds (reproduced with permission from reference [277] ©2001 Elsevier).

When biomolecular ions are produced from a source at atmospheric pressure (e.g. ESI or DESI) delivering mixtures, it is possible to isolate ions of interest and remove spurious species by using high-field asymmetric waveform ion-mobility spectrometry (FAIMS) [278–280] (Figure 3.3.3). A high-voltage asymmetric waveform, composed of a high-voltage component termed the dispersion voltage (DV) of several kiloelectronvolts and an opposite polarity low voltage component, is applied to a set of electrodes separated from a few millimetres (Figure 3.3.3). These voltages are applied using a high-frequency asymmetric waveform with a null mean value. Ions introduced between two electrodes oscillate as a result of the applied asymmetric waveform and, for a sufficient value of the DV voltage, hit one of the two electrodes. The ion separation is based on the difference between ion mobilities at high and low electric fields.

The geometrical size of very large biomolecular systems in the gas phase can be determined by using gas-phase electrophoretic mobility molecular analysis (GEMMA) [281, 282]. The charges of multiply charged ions produced by electrospray are reduced down to a single charge allowing for an easier separation in a differential mobility analyser. The size of the intact 150S human rhinovirus has been measured in the gas phase as 29.8 ± 3 nm, which is in good agreement with dimensions determined by X-ray crystallography or electron microscopy. The condensed-phase structure thus seems to be preserved in gas phase.

3.3.1.3 Travelling wave-based radio frequency only stacked ring ion guides

Ions travelling through a radio frequency guide undergo multiple collisions that result in a reduction of their axial velocity and thus their transit time. In some cases, it can be interesting

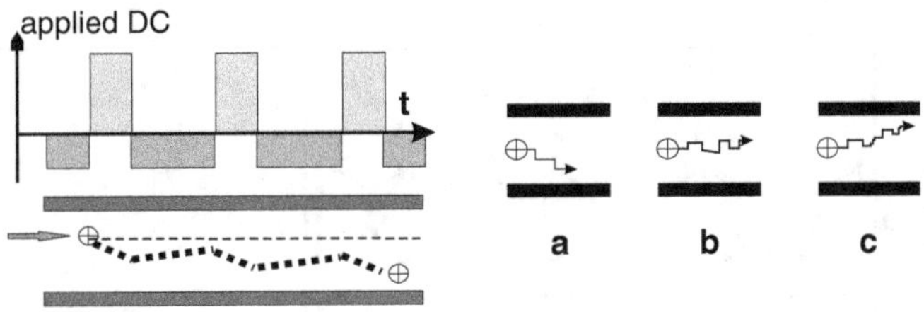

Figure 3.3.3 Principle of the FAIMS method. A mixture of ions drifts into an atmospheric pressure cell between two electrodes. A high-frequency asymmetric waveform with a null mean value is applied as well as a small DC compensating voltage (CV). This voltage is tuned to only allow ions with the right mobility (ion *b* on the right figure) to drift all through the cell while the others (ions *a* and *c*) hit the electrodes. By adding a small continuous DC field, it is possible to compensate the influence of the alternating DV field for ions with a chosen mobility.

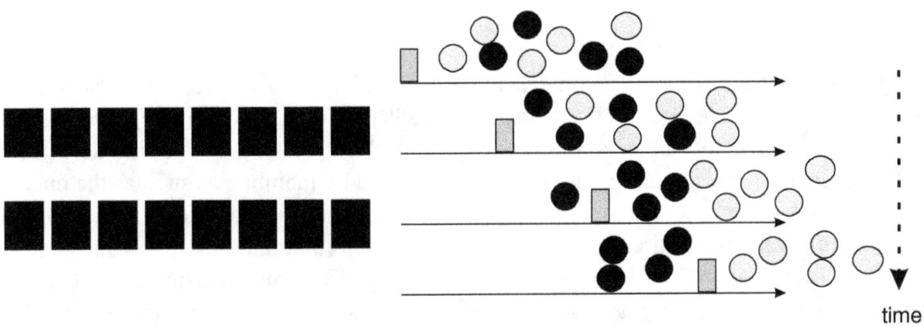

Figure 3.3.4 Travelling wave-based radio frequency-only stacked-ring ion guide. A pulse of ions of a few hundred microseconds is introduced in a set of coaxial rings, and a travelling wave radio frequency voltage pulse is applied. High-mobility ions (grey circles) surf on the wave while low-mobility ions (dark circles) roll over the wave top. For example, gramicidin S ($m/z = 571$) and Leu–enkephalin ($m/z = 556$) can easily be separated.

to keep the guiding properties while transporting ions at rather high pressures (10^{-3}–10^{-1} mbars) with short transit times. This can be achieved by using a set of coaxial ring electrodes, adjacent rings having opposite phases of a radio frequency voltage applied to them [283]. Ions keep up with or "surf" with a travelling wave voltage pulse applied to the electrodes with a typical velocity of 300 m/s. In presence of a buffer gas, the ion mobility plays an important role. High-mobility ions keep up with the wave while low-mobility ions roll over the top of the wave (Figure 3.3.4).

3.3.1.4 Monitoring structural changes

Ion-mobility experiments allow the monitoring of ion structural transitions induced by a change of ion source temperature [284] or that of the drift cell [273]. The amount of time spent in a trap before entering the ion drift cell can also be varied from milliseconds up to nearly a minute, allowing for a rather wide temporal observation range [276, 285]. For example,

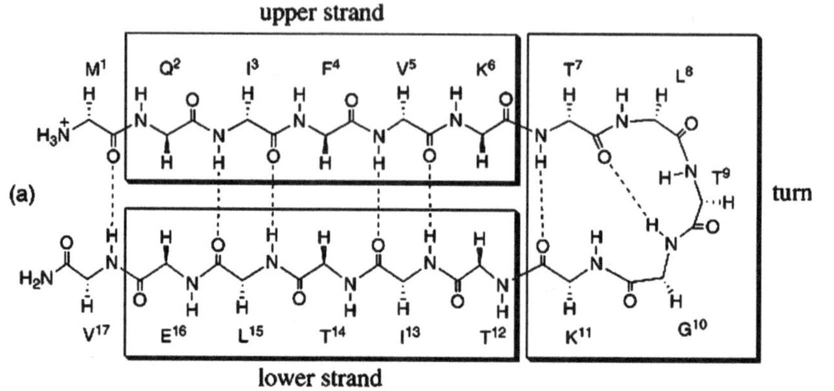

Figure 3.3.5 Structure of the N-terminal β-hairpin (residues 1–17) of ubiquitin (reproduced with permission from reference [301] ©2004 American Chemical Society).

folding and unfolding of cytochrome *c* has been widely investigated by means of ion mobility [143, 253, 276, 286–289], fluorescence [289] or femtosecond [290] spectroscopic methods. Another widely studied system is ubiquitin [285, 291–295], a 76-amino-acid protein found in all eukaryotic (containing organelles surrounded by membranes such as the nucleus) cells. It has 13 basic residues (4 Arg, 7 Lys, 1 His and the N-terminal amino group). Following attachment of several ubiquitin molecules to a given protein (ubiquitilation), this one becomes tagged and is then degraded by a protease [296–298]. This mechanism is involved in a large number of biological processes, including gene silencing. Ubiquitin has also been studied in the gas phase by means of H/D exchange (see Section 3.3.4) [299] and CID (see Section 3.2.2.2) [174, 300]. It is a flexible protein into which a β-hairpin composed of two β-strands connected by a β-turn (see Chapter 4.2) at its N-terminal plays a great role in its folding [301] (Figure 3.3.5).

Ubiquitin ions in the three different charged states (i.e. +5, +7 and +8) are initially formed with relatively compact structures. The +6 charge state remains in rather compact forms for very long periods of time. The +8 charge state is initially partially unfolded and, after an induction period, unfolds to more elongated structures that do not seem to interconvert. The most intriguing state is the +7 charge state (Figure 3.3.6). It initiates from the ESI source in three types of distinguishable compact states labelled A, A′ and A″. After a short delay (35 ms), state A abruptly unfolds to partially folded states B, C and D and then to an elongated state E. The compact conformers A′ seems to be more stable than those of the A type and unfold to elongated forms after a long time. The A″ state unfolds to B, C and D′ states, which still remain partially folded over a long period of time and do not convert to the elongated state E.

3.3.2 Blackbody infrared radiative dissociation (BIRD)

An alternative to CID for fragmentation of biomolecular ions such as peptides or oligonucleotides [302] is absorption of infrared radiation that also leads to an increase of ion internal energies. When resonant absorption of tunable radiation is used, not only fragmentation but

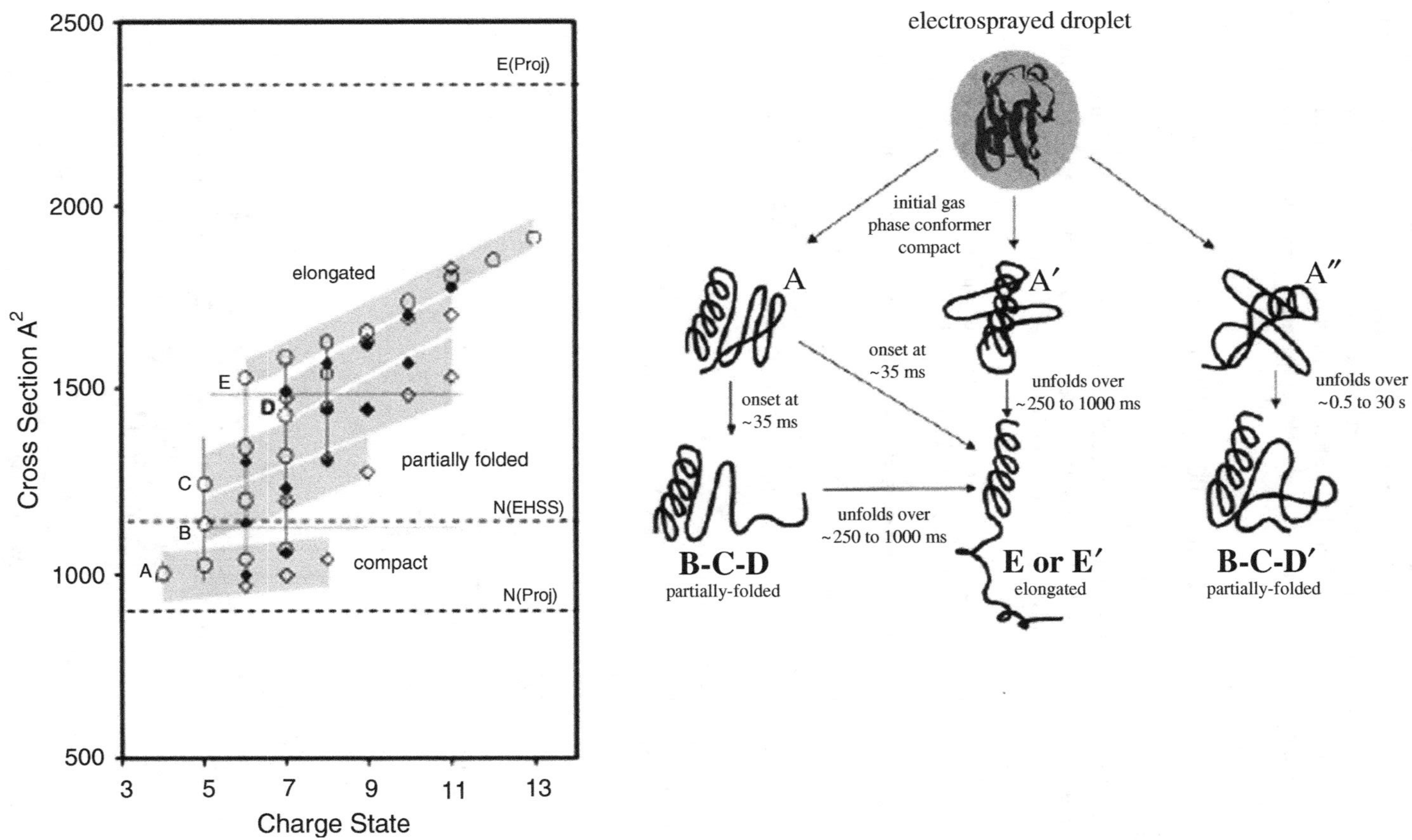

Figure 3.3.6 Left: Cross-sections of the +4 to +13 charge states of ubiquitin ions determined under different experimental conditions with the ion-mobility method. The vertical lines show a range of cross-sections corresponding to unresolved structures in ion-mobility drift time distributions. The five grey regions correspond to five different conformer types (A through E). The dashed lines correspond to cross-sections deduced from the crystal structures and calculated with the EHSS and projection approximation (PA) methods. Right: Schematic of the fate of ubiquitin ions in +7 charge state (reproduced with permission from reference [285] ©2002 American Chemical Society).

also structures of biomolecular ions are obtained (see Section 2.1.3). Here, only dissociation of ions above an activation barrier induced by non-tunable radiation, monitored by in situ mass-spectrometry, is considered. In that case, infrared absorption is not necessarily resonant with $v = 0 \rightarrow v = 1$ vibrational transitions but rather occurs in the quasi-continuum formed by overtones and combination bands of ions at sizable temperature. Several schemes are used for ion excitation: quasi-monochromatic radiation delivered by a fixed-frequency CW laser [218, 303, 304] and blackbody radiation issued from cell walls [305–307] or hot filaments [308]. When the ion temperature varies from 200 to 600°K, the peak wavelength of the blackbody radiation shifts from 6 to 2 μm. The whole infrared absorption region of biomolecules is then spanned (Figure 3.3.7).

BIRD rates are measured as function of temperature and threshold dissociation energies of the ions are then deduced from the temperature dependence of their unimolecular dissociation rate constant k_{frag}. Those threshold dissociation energies are dependent upon ion structures since each of them correspond to a different bonding scheme and thus to a different binding energy. Comparison between prediction of structures and experimental measurements of binding energies then leads to structure determination.

3.3.2.1 Energy exchange between gas-phase ions and their environment

In BIRD experiments, ions are confined in a cell and rapidly exchange their internal energy with the cell walls. The rapid-exchange (REX) condition means that the population of molecules keeps its Boltzmann equilibrium distribution of internal energies in spite of the

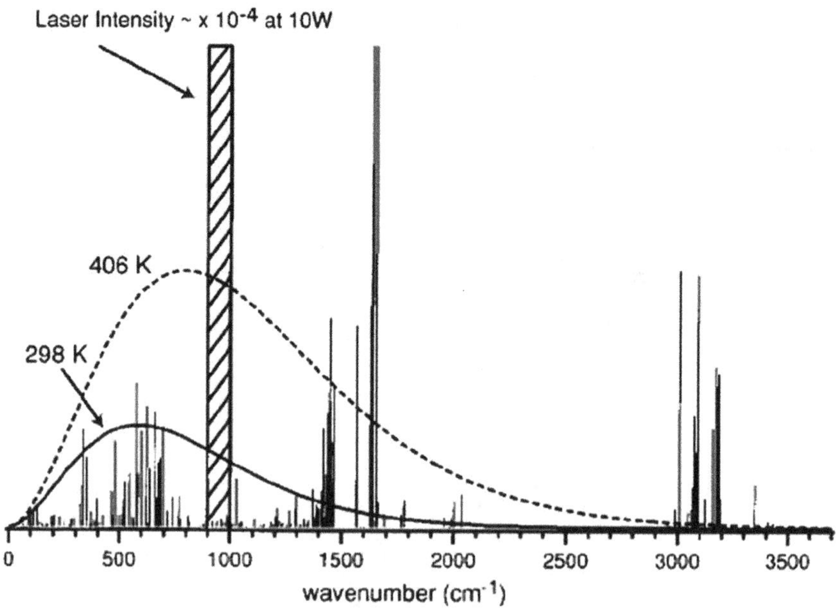

Figure 3.3.7 Planck distributions of blackbody radiation at 298°K and 406°K (magnified by a factor 10^4) and the vibrational absorption spectrum of the protonated peptide leucine–enkephalin (YGGFL). For comparison, the spectral density of a CW laser used for non-resonant IR multiphoton dissociation is displayed. (Reproduced with permission from reference [218] ©2000 American Chemical Society).

tendency of the most energetic molecules to dissociate. A Boltzmann steady-state distribution of internal energies corresponding to the cell wall temperature is then established independently of the initial conditions for creation or introduction of ions into the cell. The dissociation kinetic is considered for this thermal population under the assumption that the fragmentation only depletes the high-energy tail of the Boltzmann distribution. The REX condition is fulfilled if the rate of radiative exchange of energy is sufficiently higher than the fragmentation rate of the biomolecular ions. The number of vibrational degrees of freedom responsible for absorption and emission of radiation, and thus the radiation-exchange rate, increases with molecular size whereas the fragmentation rate is relatively independent of molecular size. Typically, the REX condition (also called large-molecule behaviour) is met for ions with molecular weight larger than 300.

An ion AB^+ can be activated or deactivated by collisions with a rare gas bath C with a rate constant k_{coll} or by absorbing and reemitting infrared photons with a radiative rate constant k_{rad}. The population of the activated ion $[AB^+]^*$ depends upon the competing channels

$$AB^+ \underset{k_{-coll}[C]+k_{rad}}{\overset{k_{coll}[C]+k_{rad}}{\rightleftarrows}} [AB^+]^* \xrightarrow{k_{frag}} A^+ + B \tag{3.3.3}$$

If the unimolecular fragmentation rate constant k_{frag} is smaller than the activation/deactivation rate constants, an equilibrium is established and the ion internal energy distribution is of the Boltzmann-type. In CID, this is achieved by raising pressure until the collision time becomes smaller than the dissociation time. In BIRD, it is strongly desired to remove any collisional process in order to only rely upon absorption/reemission processes induced by the blackbody radiation. The ion temperature is then only determined from the temperature of the cell containing the studied dissociating ions. Experimentally, this imposes the confinement of mass-selected ions at sufficiently low pressures (typically $10^{-7}-10^{-9}$ Torr) for times reaching 100 s. This is achieved in ICR cells taking great care to confine ions in a region where the temperature can be as homogeneous as possible. This temperature is controlled usually in between room temperature and 200°C but possibly down to liquid nitrogen temperature [310]. In the experimental setup displayed in Figure 3.3.8, at pressures below

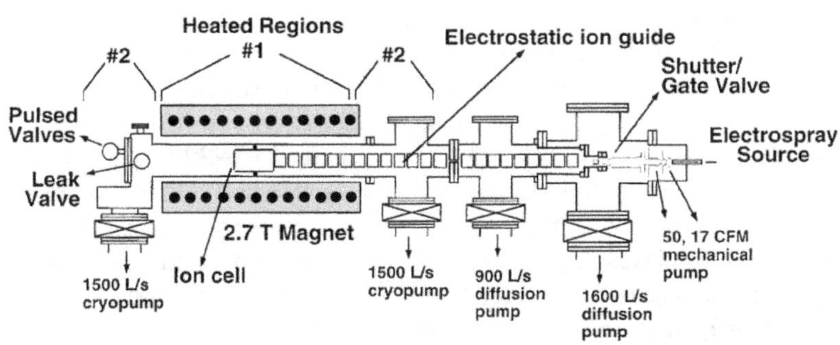

Figure 3.3.8 BIRD experimental setup. Ions issued from an electrospray source are guided into a 2.7 T ICR cell where a vacuum better than 10^{-7} Torr is maintained. The vacuum chamber walls can be heated up to 220°C. Typical observation times for appearance or depletion of ions are in the 10–250 s range (reproduced with permission from reference [306] ©1996 American Chemical Society).

8×10^{-8} Torr, ion dissociation rates are independent of the pressure, indicating that the ions are only activated by blackbody radiation of the vacuum chamber walls.

Experimentally, the abundance of parent (precursor) ions is measured as a function of time for different temperatures in a BIRD experiment or as a function of laser power in an infrared dissociation experiment (Figure 3.3.9). For extracting molecular parameters, the modelling is conducted by dividing the photon energy axis in a large number of small intervals. The time dependences of the ion populations N_i in each interval i are described by a set of coupled linear differential equations

$$\frac{dN_i}{dt} = \sum_{i \neq j} R_{ij} N_j - R_{ji} N_i - k_{\text{frag}} N_i \tag{3.3.4}$$

The radiative energy exchange coefficients R_{ij} can be obtained from calculations of molecular vibrations leading to Einstein coefficients for photon emission and absorption. A best fit of the model and the experimental measurements provides a value of the fragmentation rate that can be expressed with the Arrhenius form $k_{\text{frag}} = Ae^{-E_a/k_B T}$. Typically, for an activation energy of 100 kJ/mole (1 eV), the measurement accuracy is in between 5 and 10%.

3.3.2.2 Examples of BIRD studies

A detailed review of BIRD studies is given in reference [307]. Studies have concerned structural information deduced from measurement of radiation-induced fragmentation rates for peptides [145, 311–313], non-covalent complexes [314], oligonucleotides [314, 315], complexed polysaccharides [316] and proton-bound or metal-bound [317] systems. The role of solvent in molecular recognition [318] and in zwitterion formation [312, 319] has also been considered.

The protonated leucine–enkephalin (YGGFL) peptide fragmentation has been investigated in a BIRD experiment and with multiple photon absorption (IRMPD) of CW CO_2 laser radiation. In both the BIRD and the IRMPD experiment, the logarithm of the parent ion abundance varies linearly as a function of time (after an induction time of .4 s in the IRMPD case). The modelling is more difficult in the laser case, and the accuracy of the determination of the activation energy is only 40%.

The observed variation of the dissociation rate of complexes is not necessarily linear. This can be interpreted as due to the existence of kinetically distinct structures in the ionic complexes. For example, complexes of a protein (P), bovine carbonic anhydrase II (a Zn II metalloenzyme), with carbohydrates (C) ranging in size from mono- to tetra-saccharides have been studied by BIRD in the 60–190°C temperature range [320]. Those complexes are produced during the course of their formation in an electrospray source and are not specific (the interactions responsible for their stability are not present in solution). The dissociation rates of the ionic complexes $(P+C)^{n+}$ ($n = 9$–12) containing a mono- or di-saccharide decrease during the course of the dissociation time-window while those containing a tri- or tetra-saccharide obey first-order dissociation kinetics (Figure 3.3.10). The dissociation proceeds exclusively through loss of the carbohydrate C in its neutral form.

An ionic $(P+C)^{n+}$ complex can adopt different structures S1, S2, … Si over the course of its fragmentation and the time dependence of its abundance A_{PC} can be fitted by a sum of exponentials (Figure 3.3.11)

$$A_{\text{PC}} = c_{\text{S1}} \exp(-k_{\text{S1}} t) + c_{\text{S2}} \exp(-k_{\text{S2}} t) + \cdots + c_{\text{S}i} \exp(-k_{\text{S}i} t) \tag{3.3.5}$$

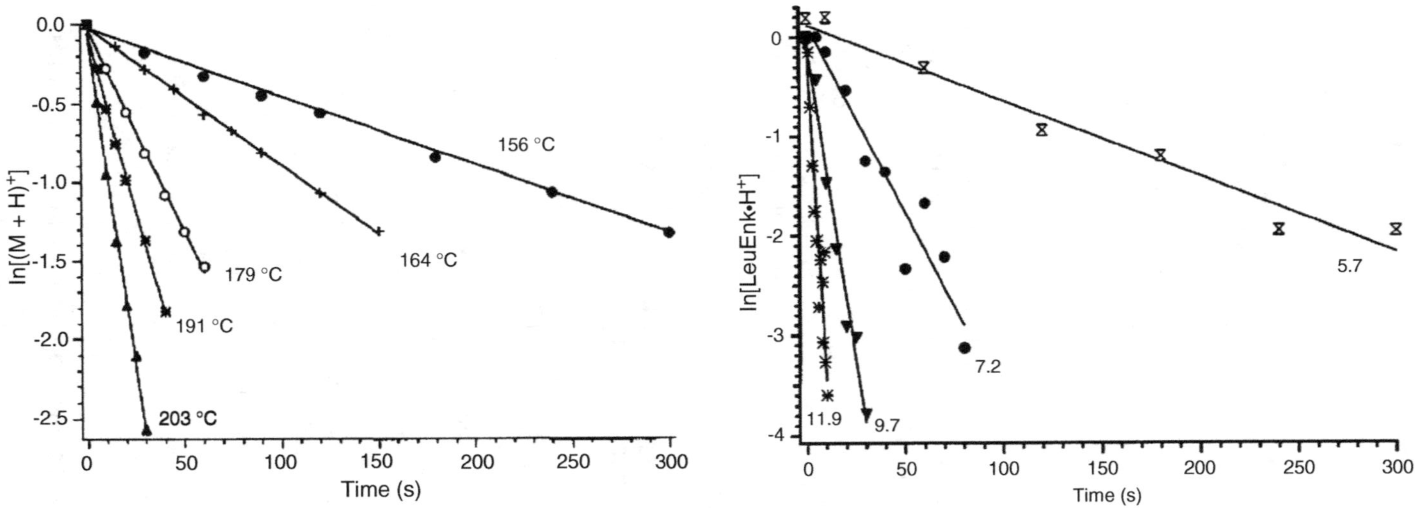

Figure 3.3.9 Left: Dissociation of protonated leucine–enkephalin ions induced by blackbody radiation. The logarithm of the parent ion abundance is plotted as a function of time at different cell temperatures. Right: Dissociation of protonated leucine–enkephalin ions induced by a CW CO_2 laser at different laser powers (in Watt) (reproduced with permission from reference [309] ©2000 American Chemical Society).

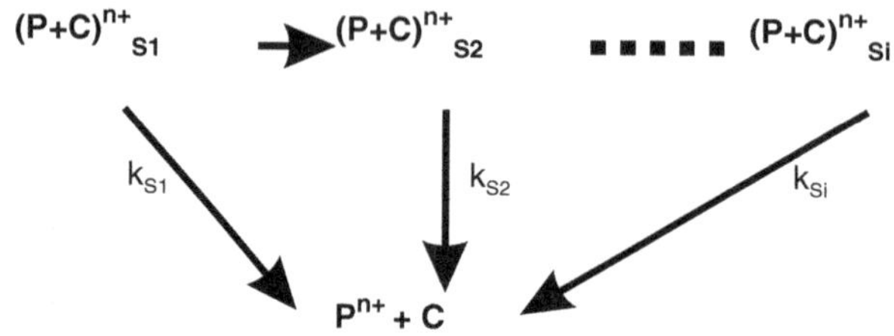

Figure 3.3.10 Dissociation pathways for a protein–ligand complex undergoing conformation changes (S1, S2, ... S*i*) over the course of its fragmentation (adapted with permission from reference [320] ©2005 Elsevier).

Both neutral and ionic intermolecular interactions play a role in the formation of specific and non-specific complexes (see Chapter 4.5). The energetic and kinetic stabilities of the complexes strongly depend upon the formation of hydrogen bonds between the protein and the carbohydrate hydroxyl groups [321]. The non-specific complexes are probably originally produced with different preferred carbohydrate interactions sites and can further undergo structural changes during the course of their dissociation. The observed decrease of the dissociation rates can be interpreted as a relaxation of the complexes towards lower energy states. In those more bound structures, additional or stronger intermolecular interactions are established. By studying carbohydrates with different structures, BIRD studies show the influence of the number of different available OH binding sites.

3.3.3 Determination of geometrical conformations from dipole moment measurements

Building blocks of biomolecules possess sizable dipole moments in the 1–10 D range (1 Debye = 3.336×10^{-30} C/m). For example, peptide dipole moments [322] can be estimated from vector addition of dipoles of peptide bonds (3.46 D for amides) and polar groups: 1.3, 0.7, 2.3 and 1.5 D, respectively, for —N—H, —C═O, —C—O and —O—H. In α-helices, amide dipoles align to form a macro-dipole while they are nearly head-to-tail and almost cancel in β-sheets (see Chapter 4.2). Prebiotic molecules (see Chapter 4.6) possess non-null dipole moments such as those of cyanic acid (2.95 D), formaldehyde (2.33 D), hydrogen sulphide (0.97 D), formic acid (1.42 D) or acetaldehyde (2.75 D). A prebiotic route has been proposed [324] leading from cyanoacetylene (C_3HN, 3.73 D) observed in the Urey–Bradley experiment [323] and water (1.85 D) to formation of cytosine (7.4 D). In the following paragraph, an experimental method relying on direct measurements of total dipole moments from deflection of dipoles in high electric field and leading to structure determination is described. A second method called Rydberg electron transfer (RET) spectroscopy relying on very low electron attachment to polar systems is presented in Section 2.3.2.2 (Figure 3.3.12).

3.3.3.1 Deflection of molecular dipoles in high electric fields

A neutral molecule possessing an electric dipole moment can be deflected in an electric field gradient $\partial F/\partial z$ (Figure 3.3.13) [325–327]. The force is then proportional to the sum of

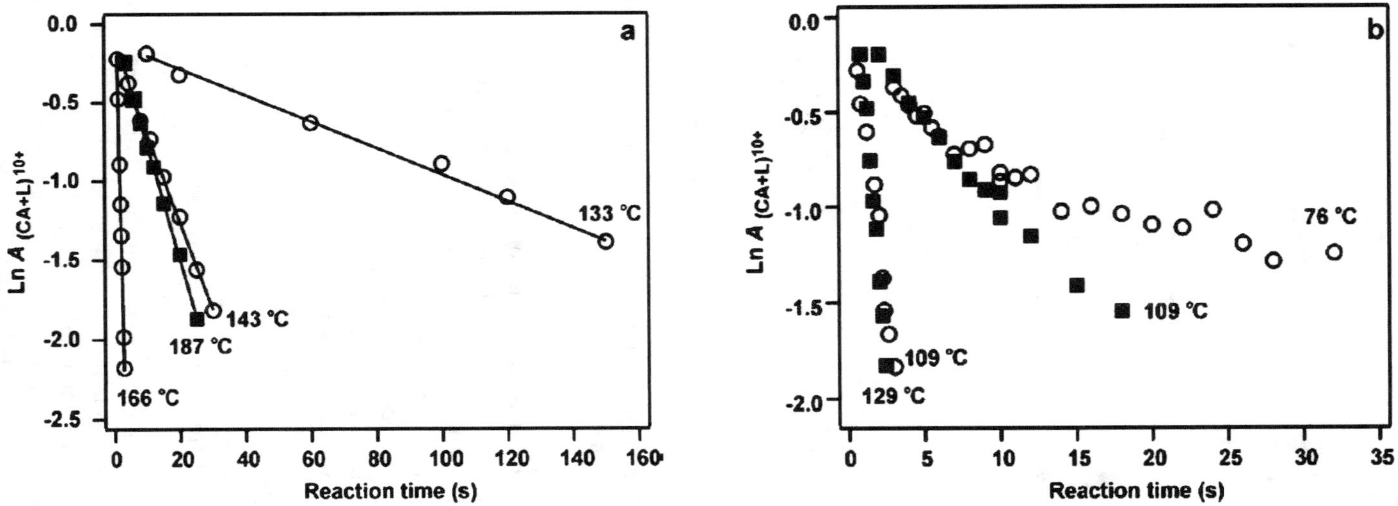

Figure 3.3.11 Plots of the natural logarithm of the normalized abundances of protein–carbohydrate $(P + C)^{10+}$ complex ions as a function of reaction time at the indicated temperatures. P is the bovine carbonic anhydrase. In figure a (left), C is αTal[αAbe]αMan (open circles) and αAbe(2-O-CH_3-αMan)αMan (filled squares). In figure b (right), C is D-Gal (open circles) and αAbe(2-O-CH_3-αMan) (filled squares) (reprinted with permission from reference [320] ©2005 Elsevier).

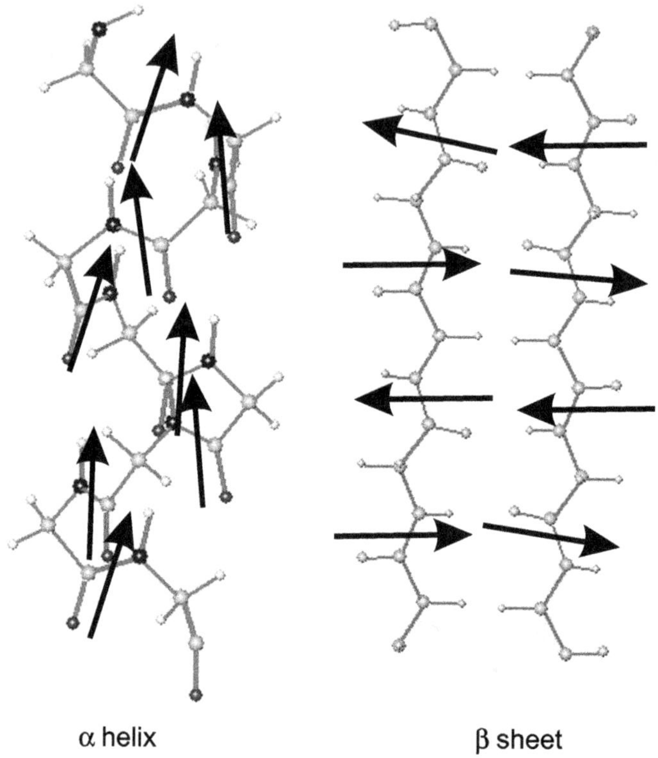

Figure 3.3.12 Amide dipole configurations in an α-helix (left) and a β-sheet (right).

the averaged permanent dipole plus induced dipole component $\langle \mu_z \rangle$ along the electric field gradient direction. The dipole spatial deviation d is given by $d = (K/mv^2)\langle \mu_z \rangle(\partial F/\partial z)$ where m and v are the mass and velocity of the molecule, respectively, and K is a geometrical factor. The deviation of the polar molecule is proportional to the average value of the projection of its dipole along the z-axis of the deflector perpendicular to the beam axis.

Molecular dynamics simulations are used to interpret experimental data [328] (see Chapter 4.2). The electric susceptibility χ is defined by the relation $\langle \mu_z \rangle = \chi F$. It is related to the electric polarizability α and to the average square dipole moment at the nozzle temperature T and, in absence of electric field, by the Langevin–Debye relation $\chi = \alpha + (\langle \mu^2 \rangle_{T,F=0}/3kT)$. The electric susceptibility characterizes the response of the molecular electronic cloud, and the reorientation of polar groups and of the whole molecule in the electric field. In a dipole deflection experiment, the beam profile depends upon the molecular system rigidity. For rigid molecules, this profile is symmetrically broadened by the presence of the electric field in the deflector while it is globally deflected towards high fields in the case of non-rigid systems. The broadening is the result from the distribution of initial orientations of the molecular system before entering the electric deflection region. Electric deflection of dipoles allows the study of intermediate size neutral peptides [329].

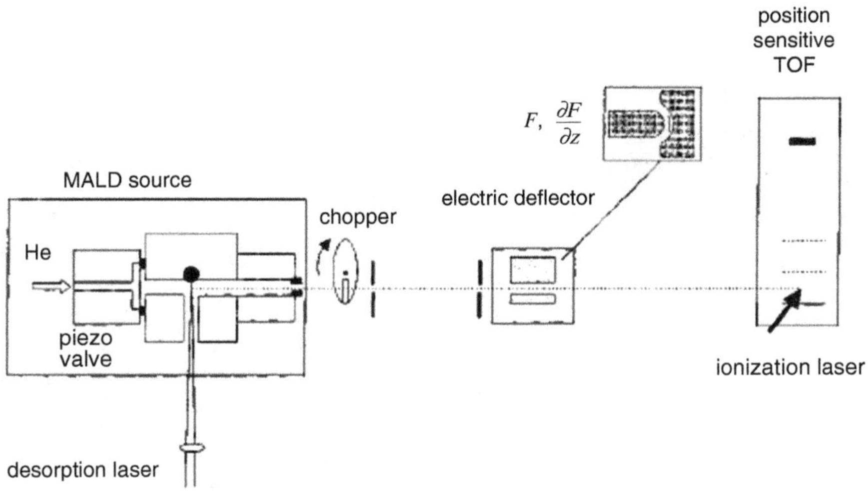

Figure 3.3.13 Measurement of molecular dipole moments by electric deflection. Neutral molecules are laser-desorbed into a supersonic expansion and leave the source through a 5 cm long nozzle. The velocity of the molecules is selected and measured by a mechanical chopper synchronized with the detection laser. The temperature of the nozzle can be adjusted from 85 to 300°K. The beam is tightly collimated by two slits and travels through an electric deflector composed of two wires. In the deflector, the applied electric field (typically 1.5×10^7 V/m) and its gradient are nearly constant over the width of the collimated beam. For detection, 1 m after the deflector, the molecular beam is ionized by a laser in the extraction region of a position-sensitive TOF mass-spectrometer (reproduced with permission from reference [325] ©2002 American Chemical Society).

3.3.4 Hydrogen/deuterium exchange

The gas-phase study of exchange of hydrogen and deuterium atoms can provide information concerning peptide and protein folding or unfolding in absence of solvent and complementary to that obtained by NMR in solution [155, 330]. The folding and unfolding of proteins in solution can also be followed by taking snapshots of H/D exchange and analysis by mass-spectrometric techniques. In this method, molecules of interest, peptides [244, 299, 331–334], nucleic acids [335], proteins [336–338] and protein–ligand complexes [339] are brought in presence of a deuterated solvent [340] (e.g. D_2O, CH_3OD, $(CH_3)_2CO$, ND_3) for a time duration ranging from milliseconds up to hours. More or less rapidly, some hydrogen atoms, called labile or exchangeable hydrogens, are replaced by deuterium atoms. In a mass-spectrometer, such exchanges appear as mass increases (Figure 3.3.14). This method relies on the assumption that hydrogens buried or engaged in strong hydrogen bonds inside biomolecules are not in direct contact with the solvent and thus undergo extremely slow exchange while hydrogens at the surface exchange rapidly. Surface accessibility, which is related to the compactness of conformers, is thus supposed to be one of the primary factors that can be accessed by measuring H/D exchange. In this chapter, we will first consider labile hydrogens [341] with the exchange mechanism and further present examples of H/D exchange studies on gas-phase folding and unfolding of peptides and proteins.

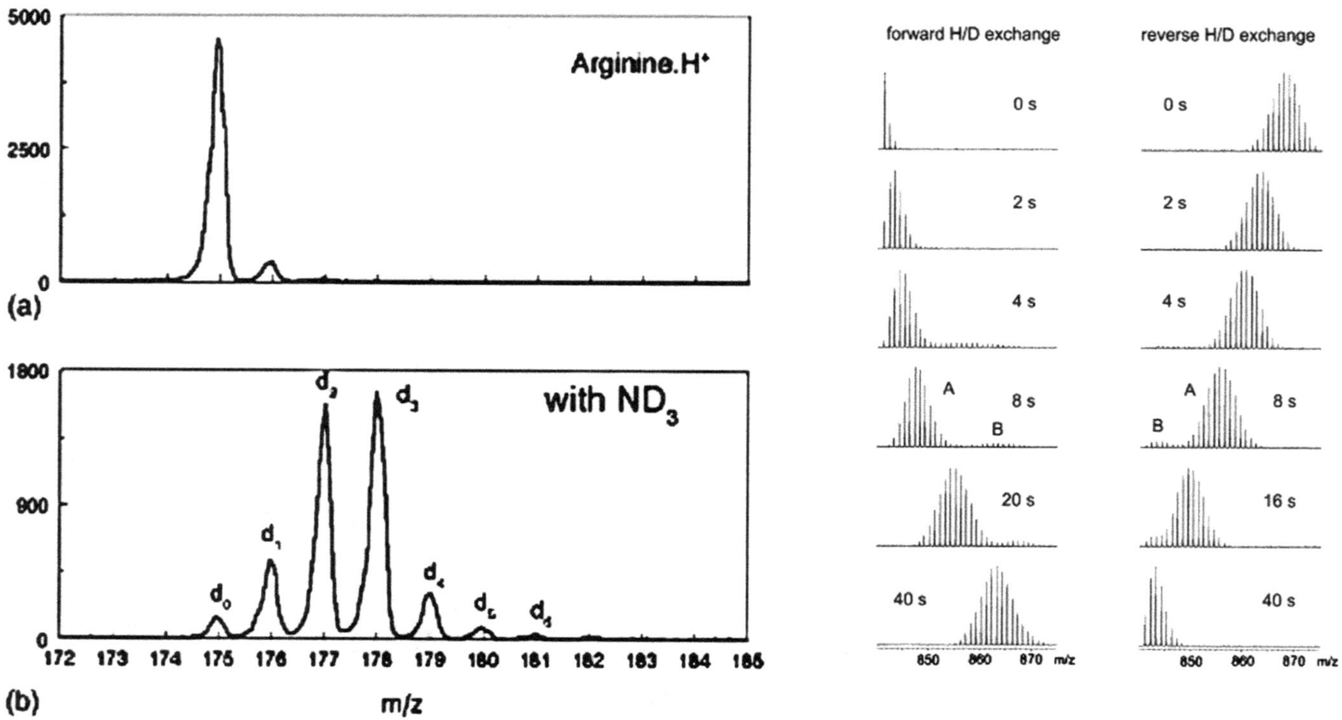

Figure 3.3.14 Left: Mass spectra of the protonated monomer of arginine before (a) and following (b) H/D exchange with ND_3 in a selected-ion flow-tube (SIFT) cell. Right: Forward and reverse H/D exchange of $[(serine)_8H]^+$ ions with CD_3OD and CH_3OH, respectively. The bimodal exchange kinetics is explained by the existence of two conformer populations (reproduced with permission from reference [340] ©2004 and [343] ©2006 Elsevier).

Exchangeable hydrogens of peptides [244] and proteins [336] are those belonging to amine, amide, hydroxyl, carboxyl groups of the main chain and the side chain protons bound to N, O and S atoms of polar groups such as guanidyl in arginine. In nucleic acids, hydrogens of ring imino NHs and of exocyclic amino groups exchange rapidly. On the contrary, hydrogens bound to carbon atoms do not exchange rapidly. Prior to any experiment, it is thus possible to determine the maximum number of exchangeable hydrogens. In the nonapeptide bradykinin, for example [342], which contains four amide bonds and two arginines, there are 17 exchangeable hydrogens given between parentheses in the following sequence: NH_2 (2)—R(4)—P—P—(1)—G—(1)—F—(1)—S(1)—P—(1)—F—(1)—R(4)—CO_2H(1).

Several mechanisms have been considered for H/D exchange in peptides [330, 344, 345]. Molecular dynamics calculations show that a deuterated water molecule undergoing multiple collisions (typically 10^5 to 10^6 in an H/D experiment) samples the entire surface. It can thus find all exchangeable hydrogens on the peptide surface but has no access to those that are buried or engaged in hydrogen bonds. Semi-empirical and ab initio calculations show that barriers for exchange of backbone amide hydrogens on charged remote sites are too high. A relay mechanism is then widely invoked for H/D exchange in the gas phase (Figure 3.3.15) [345–348]. This mechanism relies on total or partial proton transfer taking place in a transient hydrogen-bonded complex $SH^+\cdots RD$ formed between a protonated substrate SH^+ and a deuterating reagent RD

$$SH^+ + RD \leftrightarrow SH^+ \cdots RD \leftrightarrow S \cdots RDH^+ \leftrightarrow SD^+ \cdots RH \leftrightarrow SD^+ + RH \qquad (3.3.6)$$

This requires close proximity between the N-terminus or any protonated amino group in the peptide or protein and an amide carbonyl oxygen. At large separations, proton transfer between the substrate and the reagent is endothermic due to the difference between proton affinities $\Delta PA = PA_S - PA_R$. The formation of hydrogen bonds leading to the intermediate complex is exothermic and can facilitate proton transfer by lowering barriers. H/D exchange is thus dependent upon ΔPA and the molecule flexibility. For example, experiments and calculations [349] show that in the protonated glycine monomer Gly_1H^+, the difference between proton affinities of the N-terminus nitrogen and the acid carbonyl is 56 kJ/mol and

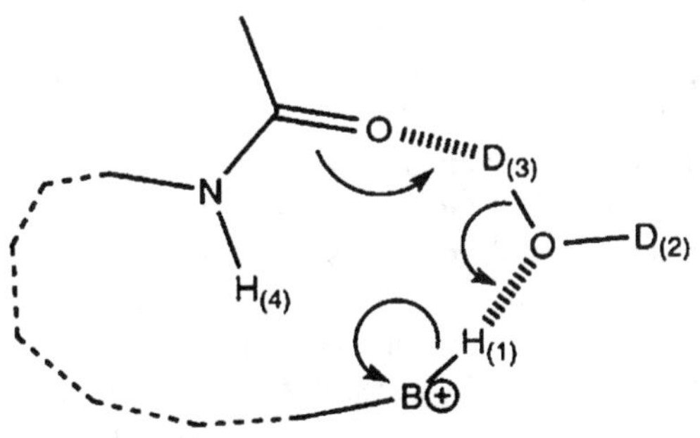

Figure 3.3.15 In this H/D "relay" exchange scheme, the R-O-D(3) reagent establishes hydrogen bonds between a protonated $-S^+-H$ site and a basic site. Following departure of the reagent, the H(1) hydrogen (or any other acidic hydrogen like H(4)) is exchanged by D(3) (reproduced with permission from reference [345] ©1999 Elsevier).

that H/D exchange with D_2O proceeds slowly. On the contrary, ΔPA between the N-terminus nitrogen and the amide oxygen in the Gly_2H^+ oligomer is only 33.8 kJ/mol, and the formation of hydrogen bonds in the intermediate complex between D_2O and Gly_2H^+ is sufficient to allow rapid H/D exchange.

The amino acid sequence is also an important factor. In the $GlyProH^+$ dipeptide, the H/D exchange of the first three hydrogen atoms that belong to the amino group with D_2O takes place rapidly while the exchange of the fourth hydrogen that belongs to the carboxylate group is much slower [350]. In $ProGlyH^+$, the proton remains very strongly bonded to the nitrogen atom of the amino group, thus acquiring a very low mobility towards the other basic sites, and the H/D exchange becomes extremely slow. This influence of the proline residue upon the proton mobility [351] is related to the problem of backbone fragmentation of protonated peptides (IIb2), which shows a preference for cleavage adjacent to proline [352]. Side chain effects are also observed in the case of the aspartic acid [353].

Gas-phase studies have been conducted with H/D exchange that has already taken place between reactants in solution [344, 354] or during the solution to gas-phase process in an electrospray source [334, 355]. H/D exchange can also be performed directly in mass-spectrometers [356, 357] or ion-mobility devices [288]. In solution, NMR studies have shown that H/D exchange is influenced by temperature and pH. In the gas phase, temperature can be tuned by using a CO_2 laser in a FT-ICR cell [331] or by modifying the temperature of an ion-mobility cell [287]. The number of charges can be also modified by adding a high-proton affinity [358] reagent (butylamine) that removes protons.

In H/D exchange experiments, the number of exchanged hydrogens is generally smaller than the number of predicted labile atoms. Let us consider the case of cytochrome *c* with 198 exchangeable hydrogens. In solution, cytochrome *c* adopts its native conformation at pH $\approx$ 7 with 144 exchangeable hydrogens and a less ordered "molten globule" conformation at pH $\approx$ 1 which undergoes larger H/D exchange up to 154 exchangeable hydrogens (see Section 2.1.3.2.6). In an FT-ICR cell, lowering the number of positive charges in cytochrome *c* from 15 down to 7 (similar to a raise of pH) induces a 50% decrease of H/D exchange down to 64 [331]. The most direct interpretation is that the Coulomb repulsion between the positive charges is decreased and favours a partial refolding. Similarly, raising temperature favours unfolding, and thus one would expect that H/D exchange should increase. The number of exchangeable hydrogens has been measured for the 5+ and 9+ charge states of cytochrome *c* [288] in between 300 and 450°K. The ion mobility of those states shows that they are, respectively, compact and extended (respective conformations a and d of Figure 3.3.1) as one would expect from the Coulomb repulsion argument. The compact 5+ and extended 9+ charge states, respectively, exchanges only 53 and 63 hydrogens at 300°K. Surprisingly, above 430°K, the 5+ state exchanges a maximum value of 200 hydrogens (196+4H) and the 9+ state only 190 hydrogens. Simulations that have been conducted provide a possible explanation. They show that at short times, the exchange sites are close to the initial protonated site and become accessible with only small structural changes. At longer times, remote sites become more rapidly accessible in the compact 5+ structure than in the extended 9+ structure due to the presence of tertiary structures. The lack of tertiary folds forbids exchanges across widely separated regions in the extended conformer.

Exchange rates are strongly dependent upon conformations, and the amount of time allowed for H/D exchange in gas phase is thus another important parameter for comparison between structures determined by means of different methods. In ion-mobility measurements, ion conformations can be directly determined within milliseconds after their formation in an electrospray [359] or MALDI source [270] and up to 30 s if ions are first confined in RF quadrupole trap experiments [360]. In an FT-ICR cell, the H/D exchange process can be recorded up to hours. According to the experimental procedures, different charge and temperature dependences of conformers can be observed [295]. Gas-phase H/D exchange in a FT-ICR cell can also be used for spectroscopic study of zwitterion formation (see Section 5.5.3.2).

REFERENCES

1. Gaffney JS, Pierce RC, Friedman L: Mass spectrometer study of evaporation of α-amino acids. *Journal of the American Chemical Society* 1977, 99:4293–4298.
2. Sukhodub L, Yanson IK: Mass spectrometric studies of binding energies for nitrogen bases of nucleic acids *in vacuo*. *Nature* 1976, 264:245–247.
3. Sukhodub L: Interactions and hydration of nucleic acid bases in a vacuum. Experimental study. *Chemical Reviews* 1987, 87:589–606.
4. Rizzo TR, Park YD, Levy DH: A molecular beam of tryptophan. *Journal of the American Chemical Society* 1985, 107:277–278.
5. Peteanu LA, Levy DH: Spectroscopy of complexes of tryptamine and 3-indole propionic acid with various solvents. *Journal of Physical Chemistry* 1988, 92:6554–6561.
6. Kim SK, Lee W, Herschbach DR: Cluster beam chemistry hydration of nucleic acid bases, ionization potentials of hydrated adenine and thymine. *Journal of Physical Chemistry* 1996, 100:7933–7937.
7. Nir E, Grace L, Brauer B, de Vries MS: REMPI spectroscopy of jet-cooled guanine. *Journal of the American Chemical Society* 1999, 121:4896–4897.
8. Piuzzi F, Dimicoli I, Mons M, Tardivel B, Zhao QC: A simple laser vaporization source for thermally fragile molecules coupled to a supersonic expansion: application to the spectroscopy of tryptophan. *Chemical Physics Letters* 2000, 320:282–288; Cho SH, Yoon M, Kim SK: Spectroscopy and energy disposal dynamics of phtalocyanine-Ar (n = 1,2) complexes generated by hyperthermal pulsed nozzle source. *Chemical Physics Letters* 2000, 326:65–72.
9. Saigusa H, Tomioka A, Katayama T, Iwase E: A matrix-free laser desorption method for production of nucleobase clusters and their hydrates. *Chemical Physics Letters* 2006, 418:119–125.
10. Dian BC, Longarte A, Winter PR, Zwier TS: The dynamics of conformational isomerization in flexible biomolecules. I. Hole-filling spectroscopy of N-acetyl tryptophan methyl amide and N-acetyl tryptophan amide. *Journal of Chemical Physics* 2004, 120:133–147.
11. Evans DA, Wales DJ, Dian BC, Zwier TS: The dynamics of conformational isomerization in flexible biomolecules. II. Simulating isomerizations in a supersonic free jet with master equation dynamics. *Journal of Chemical Physics* 2004, 120:148–157.
12. Snoek LC, van Mourik T, Simons JP: Neurotransmitters in the gas phase: a computational and spectroscopic study of noradrenaline. *Molecular Physics* 2003, 101:1239–1248.
13. Viant MR, Feller RS, McLaughlin RP, Saykally RJ: Infrared laser spectroscopy of uracil in a pulsed slit jet. *Chemical Physics* 1995, 103:9502–9505.
14. Liu K, Fellers RS, Viant MR, McLaughlin RP, Brown MG, Saykally RJ: A long path length pulsed slit valve appropriate for high temperature operation: infrared spectroscopy of jet-cooled large water clusters and nucleotide bases. *Review of Scientific Instruments* 1996, 67:410–416.
15. Casaes RN, Paul JB, McLaughlin RP, Saykally RJ, van Mourik T: Infrared cavity ringdown spectroscopy of jet-cooled nucleotide base clusters and water complexes. *Journal of Physical Chemistry A* 2004, 108:10989–10996.
16. Seurre N, Le Barbu-Debus K, Lahmani F, Borho N, Suhm MA, Zehnacker A: Chiral recognition in jet-cooled complexes. *Australian Journal of Chemistry* 2004, 57:1149–1152.

17. Farnik M, Steinbach C, Weimann M, Buck U, Borho N, Suhm MA: Size-selective vibrational spectroscopy of methyl glycolate clusters: comparison with ragout-jet FTIR spectroscopy. *Physical Chemistry Chemical Physics* 2004, 6:4614–4620.
18. Grotemeyer J, Schlag EW: Peptides investigated by laser desorption – multiphoton ionization mass spectrometry. *Organic Mass Spectrometry* 1986, 23:388–396.
19. Nir E, Imhof P, Kleinermanns K, de Vries MS: REMPI spectroscopy of laser desorbed guanosines. *Journal of the American Chemical Society* 2000, 122:8091–8092.
20. Nir E, Kleinermanns K, de Vries MS: Pairing of isolated nucleic-acid bases in the absence of the DNA backbone. *Nature* 2000, 408:949–951.
21. Compagnon I, Antoine R, Rayane D, Broyer M, Dugourd P: Vibration induced electric dipole in a weakly bound molecular complex. *Physical Review Letters* 2002, 89:253001–253005.
22. Meffert A, Grotemeyer J: Dissociative proton transfer in cluster ions: clusters of aromatic carboxylic acids with aminoacids. *International Journal of Mass Spectrometry* 2001, 210:521–530.
23. Nir E, de Vries MS: Fragmentation of laser-desorbed 9-substituted adenines. *International Journal of Mass Spectrometry* 2002, 219:133–138.
24. Smits M, de Lange CA, Ullrich S, Schultz T, Schmitt M, Underwood JG, Shaffer JP, Rayner DM, Stolow A: Stable kilohertz rate molecular beam laser ablation sources. *Review of Scientific Instruments* 2003, 74:4812–4817.
25. Callegari C, Lehmann KK, Schmied R, Scoles G: Helium nanodroplet isolation rovibrational spectroscopy: methods and recent results. *Journal of Chemical Physics* 2001, 115:10090–10110.
26. Choi MY, Miller RE: Multiple isomers of uracil-water complexes: infrared spectroscopy in helium nanodroplets. *Physical Chemistry Chemical Physics* 2005, 7:3565–3573.
27. Denifl S, Zappa F, Mähr I, Lecointre J, Probst M, Märk T, Scheier P: Mass spectrometric investigation of anions formed upon free electron attachment to nucleobases molecules and clusters embedded in superfluid helium droplets. *Physical Review Letters* 2006, 97:043201–043204.
28. Choi MS, Miller RE: Four tautomers of isolated guanine from infrared laser spectroscopy in helium droplets. *Journal of the American Chemical Society* 2006, 128:7230–7328.
29. Huisken F, Werhahn O, Ivanov AY, Krasnokutski SA: The O-H stretching vibrations of glycine trapped in rare gas matrices and helium clusters. *Journal of Chemical Physics* 1999, 111:2978–2984.
30. Lindinger A, Toennies JP, Vilesov AF: High resolution of vibronic spectra of the amino acids tryptophan and tyrosine in 0.38°K cold helium droplets. *Journal of Chemical Physics* 1999, 110:1429–1436.
31. Stiles PL, Nauta K, Miller RE: Dipole moments of molecules solvated in helium nanodroplets. *Physical Review Letters* 2003, 90:135301–135305.
32. Hartmann M, Miller RE, Toennies JP, Vilesov A: Rotationally resolved spectroscopy of SF_6 in liquid helium clusters: a molecular probe of cluster temperature. *Physical Review Letters* 1995, 75:1566–1569.
33. Toennies JP, Vilesov A: Superfluid helium droplets: a uniquely cold nanomatrix for molecules and molecular complexes. *Angewandte Chemie-International Edition* 2004, 43:2622–2648.
34. Moro R, Rabinovitch R, Kresin VV: Pick-up cell for cluster beam experiments. *Review of Scientific Instruments* 2005, 76:056104–056105.
35. Faubel M, Kisters T: Non-equilibrium molecular evaporation of carboxylic acid dimers. *Nature* 1989, 339:527–529.
36. Wattenberg A, Sobott F, Barth HD, Brutschy B: Laser desorption mass spectrometry on liquid beams. *European Mass Spectrometry* 1999, 5:71–76.
37. Oberbrodhage J, Morgner H, Tapia O, Siegbahn HOG: Molecular dynamics simulation of the free surface of liquid formamide. *International Journal of Quantum Chemistry* 1998, 63:1123–1131.
38. Wilson KR, Rude BS, Smith J, Cappa C, Co DT, Schaller RD, Larsson M, Catalano T, Saykally RJ: Investigation of volatile liquid surfaces by synchrotron X-ray spectroscopy of liquid microjets. *Review of Scientific Instruments* 2004, 75:725–736.

39. Kondow T, Mafune F: Stuctures and dynamics of molecules on liquid beam surfaces. *Annual Review of Physical Chemistry* 2000, 51:731–761.
40. Sobott F, Wattenberg A, Kleinekofort W, Pfenninger A, Brutschy B: Laser desorption mass spectrometry on thin liquid jets. *Fresenius Journal of Analytical Chemistry* 1998, 360:745–749.
41. Toyama N, Kohno JY, Mafune F, Kondow T: Solvation structure of arginine in aqueous solution studied by liquid beam technique. *Chemical Physics Letters* 2006, 419:369–373.
42. Charvat A, Stasicki B, Abel B: Product screening of fast reactions in IR-laser-heated liquid water filaments in a vacuum by mass-spectrometry. *Journal of Physical Chemistry A* 2006, 110: 3297–3306.
43. Sobott F, Kleinkofort W, Brutschy B: Cation selectivity of natural and synthetic ionophores probed with laser-induced liquid beam mass spectrometry. *Analytical Chemistry* 1997, 69: 3587–3594.
44. Horimoto N, Kohno J, Mafune F, Kondow T: Ejection mechanism of molecules and neutral clusters from liquid beam under irradiation of IR laser. *Chemical Physics Letters* 2000, 318:536–542.
45. Abel B, Charvat A, Diederichsen U, Faubel M, Girmannc B, Niemeyer J, Zeeck A: Applications, features, and mechanistic aspects of liquid water beam desorption mass spectrometry. *International Journal of Mass Spectrometry* 2005, 243:177–188.
46. Wilson KR, Belau L, Nicolas C, Jimenez-Cruz M, Leone SR, Ahmed M: Direct determination of the ionization energy of histidine with VUV synchrotron radiation. *International Journal of Mass Spectrometry* 2006, 249:155–161.
47. Kohno J, Mafune F, Kondow T: Isolation of adenine salts in the gas phase from a liquid beam of aqueous solutions by IR laser irradiation. *European Journal of Physics D* 2002, 20:339–345.
48. Messer BM, Cappa CD, Smith JD, Drisdell WS, Schwartz CP, Cohen RC, Saykally RJ: Local hydration environments of amino acids and dipeptides studied by X-ray spectroscopy of liquid microjets. *Journal of Physical Chemistry B* 2005, 109:21640–21646.
49. Takats Z, Katona M, Czuczy N, Skoumai R: Jet desorption ionization – a new ionization method for tissue analysis.
50. Wiseman JM, Ifa DR, Song QY, Cooks RG: Tissue imaging at atmospheric pressure using desorption electrospray ionization (DESI). *Angewandte Chemie-International Edition* 2006, 45: 7188–7192.
51. Christen W, Geggier S, Grigorenko S, Rademann K: Pulsed supersonic expansion of nonvolatile solids. *Review of Scientific Instruments* 2004, 75:5048–5049.
52. Christen W, Rademann K, Even U: Efficient cooling in supersonic expansions of supercritical fluids: CO and CO_2. *Journal of Chemical Physics* 2006, 125:174307.
53. Zheng J, Pinkston JD, Zoutendam PH, Taylor LT: Feasibility of supercritical fluid chomatography/ mass spectrometry of polypeptides with up to 40-mers. *Analytical Chemistry* 2006, 78: 1535–1545.
54. Kleinekofort W, Pfenninger A, Plomer T, Griesinger C, Brutschy B: Observation of noncovalent complexes using laser-induced liquid beam ionization desorption. *International Journal of Mass Spectrometry and Ion Processes* 1996, 156:195–202.
55. Wattenberg A, Sobott F, Barth HD, Brutschy B: Studying noncovalent protein complexes in aqueous solution with laser desorption mass spectrometry. *International Journal of Mass Spectrometry* 2000, 203:49–57.
56. Grimm RL, Beauchamp JL: Dynamics of field-induced droplet ionization: time-resolved studies of distortion, jetting, and progeny formation from charged and neutral methanol droplets exposed to strong electric fields. *Journal of Physical Chemistry B* 2005, 109:8244–8250.
57. Duft D, Achtzehn T, Müller R, Huber BA, Leisner T: Rayleigh jets from levitated microdroplets. *Nature* 2003, 421:6919–6922.
58. Morgner N, Barth HD, Brutschy B: A new way to detect noncovalently bonded complexes of biomolecules from liquid micro-droplets by laser mass spectrometry. *Australian Journal of*

Chemistry 2006, 59:109–114. Morgner N, Barth HD, Schmidt TL, Scheffer U, Göbel M, Fucini P, Brutschy B: Detecting specific ligand binding to nucleic acids: a test for ultrasoft laser mass spectrometry *Zeitschrift f. Physikalische Chemie* 2007, 221:689–704.

59. Kohno J, Toyama N, Kondow T: Ion formation to the gas-phase by laser ablation on a droplet beam. *Chemical Physics Letters* 2006, 420:146–150.
60. Dessiaterik Y, Nguyen TV, Baer T, Miller RE: IR vaporization mass spectrometry of aerosol particles with ionic solutions: the problem of ion–ion recombination. *Journal of Physical Chemistry A* 2003, 107:11245–11252.
61. Wellenzohn B, Flader W, Winger RH, Hallbrucker A, Mayer E, Liedl KR: Exocyclic groups in the minor group influence the backbone conformation of DNA. *Nucleic Acids Research* 2001, 29:5036–5043.
62. Morgner N, Barth HD, Brutschy B, Kleinschroth T, Müller W, Ludwig B: A new type of laser mass spectrometry utilized to analyse membrane proteins. *Biochemistry* 2007, to appear.
63. Gabelica V, Lemaire D, Laprevote O, De Pauw E: Kinetics of solvent addition on electrosprayed ions in an electrospray source and in a quadrupole trap. *International Journal of Mass Spectrometry* 2001, 210:113–119.
64. Nonose S, Iwaoka S, Tanaka H, Okai N, Shibakusa T, Fuke K: Structures and reactions of biomolecular ions produced with electrospray ionization. *European Journal of Physics D* 2003, 24:335–338.
65. Karas M, Krüger R: Ion formation in MALDI: the cluster ionization mechanism. *Chemical Reviews* 2003, 103:427–439.
66. Gabelica V, Schulz E, Karas M: Internal energy build-up in matrix-assisted laser desorption/ionization. *Journal of Mass Spectrometry* 2004, 39:579–593.
67. Tanaka K: The origin of macromolecule ionization by laser irradiation (Nobel Lecture). *Angewandte Chemie-International Edition* 2003, 42:3861–3870.
68. Zhang W, Niu S, Chait BT: Exploring infrared wavelength matrix-assisted laser desorption/ionization with delayed-extraction time-of-flight mass spectrometry. *Journal of the American Society of Mass Spectrometry* 1998, 9:879–884.
69. Röhl H, Goethel S, Reihs K: MALDI chips for high-sensitivity and high-throughput peptide analysis by MALDI-TOF MS. *Nature Methods* 2005, 2:478–480.
70. Lucas B, Gregoire G, Lemaire J, Maitre P, Glotin F, Schermann JP, Desfrancois C: Infrared multiphoton dissociation spectroscopy of protonated N-acetyl-alanine and alanyl-histidine. *International Journal of Mass Spectrometry* 2005, 243:105–113.
71. Moskovets E, Preisler J, Chen HS, Rejtar T, Andreev V, Karger BL: High-throughput axial MALDI-TOF MS using a 2 kHz repetition rate laser. *Analytical Chemistry* 2006, 78:912–919.
72. Kapota C, Lemaire J, Maitre P, Ohanessian G: Vibrational signature of charge solvation vs. salt bridge isomers of sodiated amino acids in the gas phase. *Journal of the American Chemical Society* 2004, 126:1836–1842.
73. Lucas B, Gregoire G, Lemaire J, Maitre P, Ortega JM, Rupenyan A, Reimann B, Schermann JP, Desfrancois C: Investigation of the protonation site in the dialanine peptide by infrared multi-photon dissociation spectroscopy. *Physical Chemistry Chemical Physics* 2004, 6:2659–2663.
74. Mac Aleese L, Simon A, McMahon TB, Ortega JM, Scuderi D, Lemaire J, Maitre P: Mid-IR spectroscopy of protonated leucine methyl ester performed with an FT-ICR or a Paul type ion-trap. *International Journal of Mass Spectrometry* 2006, 249:14–20.
75. Cai Y, Jiang Y, Cole RB: Anionic adducts of oligosaccharides by matrix-assisted laser desorption/ionization time-of-flight mass spectrometry. *Analytical Chemistry* 2003, 75:1638–1644.
76. Sadeghi M, Wu X, Vertes A: Conformation changes, complexation, and phase transition in matrix-assisted laser desorption. *Journal of Physical Chemistry B* 2001, 105:2578–2587.
77. Zhigilei LV, Leveugle E, Garrison BJ, Yingling YG, Zeffman MI: Computer simulations of laser abalation of molecular substrates. *Chemical Reviews* 2003, 103:321–347.

78. Knochenmuss R, Zhigelei L: Molecular dynamics model of ultraviolet matrix-assisted laser desorption/ionization including ionization processes. *Journal of Physical Chemistry B* 2005, 109:22947–22957.
79. Kovtoun SV, English RD, Cotter RJ: Mass correlated acceleration in a reflectron MALDI TOF mass spectrometer: an approach for enhanced resolution over a broad mass range. *Journal of the American Society of Mass Spectrometry* 2002, 13:135–143.
80. Gross J, Hahner S, Karas M, Schäfer J, Lützenkirchen F, Nordhoff E: Investigations of metastable decay of DNA under ultraviolet matrix-assisted laser desorption/ionization conditions with post-source decay analysis and hydrogen/deuterium exchange. *Journal of the American Society of Mass Spectrometry* 1998, 9:866–878.
81. Yassin FH, Marynick DS: Computational study of matrix-peptide interactions in MALDI mass spectrometry: interactions of 2,5- and 3,5-dihydroxybenzoic acid with the tripeptide valine-proline-leucine. *Journal of Physical Chemistry A* 2006, 110:3820–3825.
82. Wang CL, Wong CF: Molecular dynamics simulation of laser desorption of a fragment of protein kinase A from two MALDI matrix. *Journal of Physical Chemistry A* 2006, 110:5355–5360.
83. Hsu JF, Chang J, Franz AH: MALDI-TOF and ESI-MS analysis of oligosaccharides labelled with a new multifunctional oligosaccharide tag. *Journal of the American Society of Mass Spectrometry* 2006, 17:194–204.
84. Dreisewerd K: The desorption process in MALDI. *Chemical Reviews* 2003, 103:395–425.
85. Wenzel RJ, Matter U, Schultheis L, Zenobi R: Analysis of megadalton ions using cryodetection MALDI time-of-flight mass spectrometry. *Analytical Chemistry* 2005, 77:4329–4337.
86. Zhang JH, Schubothe K, Li BS, Russell S, Lebrilla CB: Infrared multiphoton dissociation of O-linked mucin-type oligosaccharides. *Analytical Chemistry* 2005, 77:208–214.
87. Mysak ER, Wilson KR, Jimenez-Cruz M, Ahmed M, Baer T: Synchrotron radiation based aerosol time-of-flight mass spectrometry for organic constituents. *Analytical Chemistry* 2005, 77:5963–5960.
88. Wilson KR, Jimenez-Cruz M, Nicolas C, Belau L, Leone SR, Ahmed M: Thermal vaporization of biological nanoparticles: fragment-free vacuum ultraviolet photoionization mass spectra of tryptophan, phenylalanine-glycine-glycine, and β-carotene. *Journal of Physical Chemistry A* 2006, 110:2106–2113.
89. Fenn JB: Electrospray wings for molecular elephants (Nobel Lecture). *Angewandte Chemie-International Edition* 2003, 42:3871–3894.
90. Ruotolo B, Giles K, Campuzano I, Sandercock AM, Bateman RH, Robinson CJ: Evidence for macromolecular protein rings in the absence of bulk water. *Science* 2005, 310:1658–1661.
91. Benkestock K, Edlund PO, Roeraade J: Electrospray ionization mass spectrometry as a tool for determination of drug binding sites to human serum albumin by noncovalent interaction. *Rapid Communications in Mass Spectrometry* 2005, 19:1637–1643.
92. Rosu F, Pirotte S, De Pauw E, Gabelica V: Positive and negative-mode ESI-MS and MS-MS for studying drug-DNA complexes. *International Journal of Mass Spectrometry* 2006, 253: 156–171.
93. Iavarone AT, Williams ER: Mechanism of charging and supercharging molecule in electrospray ionisation. *Journal of the American Chemical Society* 2003, 125:2319–2327.
94. Smith JN, Flagan RC, Beauchamp JL: Droplet evaporation and discharge dynamics in electrospray ionization. *Journal of Physical Chemistry A* 2002, 106:9957–9967.
95. Jurascheck R, Dülcks T, Karas M: Nanoelectrospray – more than just a minimized flow electrospray source. *Journal of the American Society of Mass Spectrometry* 1999, 10:300–308.
96. Valaskovic GA, Utley L, Lee MS, Wu JT: Ultra-low flow nanospray for the normalization of conventional liquid chromatography/mass spectrometry through equimolar response: standard-free quantitaive estimation of metabolite levels in drug discovery. *Rapid Communications in Mass Spectrometry* 2006, 20:1087–1096.

97. Achtzehn T, Müller R, Duft D, Leisner T: The Coulomb instability of charged microdroplets: dynamics and scaling. *European Journal of Physics D* 2005, 34:311–313.
98. Janini GM, Conrade TP, Wilkens KL, Issaq HJ, Veenstra TD: A sheathless nanoflow electrospray interface for on-line capillary electrophoresis mass spectrometry. *Analytical Chemistry* 2003, 75:1615–1619.
99. Gaskell SJ: Electrospray: principles and practice. *Journal of Mass Spectrometry* 1997, 32:677–688.
100. Rohner TC, Lion N, Girault HH: Electrochemical and theoretical aspects of electrospray ionisation. *Physical Chemistry Chemical Physics* 2004, 6:3056–3068.
101. Felitsyn N, Peschke M, Kebarle P: Origin and number of charges observed on multiply-protonated native proteins produced by ESI. *International Journal of Mass Spectrometry* 2002, 219:39–62.
102. Verkerk UH, Kebarle P: Ion-ion and ion-molecule reactions at the surface of proteins produced by nanospray. Information on the number of acidic residues and control of the number of ionized acidic and basic residues. *Journal of the American Society for Mass Spectrometry* 2005, 16:1325–1341.
103. Benesch JLP, Robinson CV: Mass spectrometry of macromolecular assemblies. *Current Opinion in Structural Biology* 2006, 16:245–251.
104. Loo JA: Electrospray ionization mass spectrometry: a technology for studying noncovalent macromolecular complexes. *Journal of Mass Spectrometry* 2000, 200:175–186.
105. Xia Y, Liang X, McLuckey SA: Pulsed dual electrospray ionization for ion/ion reactions. *Journal of the American Society of Mass Spectrometry* 2005, 16:1750–1756.
106. Yamaguchi K: Cold-spray ionization mass spectrometry: principle and applications. *Journal of Mass Spectrometry* 2003, 38:473–490.
107. Rosu F, Gabelica V, Shin-ya K, De Pauw E: Telomestatin-induced stabilization of the human telomeric DNA quadruplex monitored by electrospray mass spectrometry. *Chemical Communications* 2003:2702–2703.
108. Gabelica V, De Pauw E: Internal energy and fragmentation of ions produced in electrospray sources. *Mass Spectrometry Reviews* 2005, 24:566–587.
109. Hirabayashi A, Sakairi M, Koizumi H: Sonic spray ionization method for atmospheric pressure ionization mass spectrometry. *Analytical Chemistry* 1994, 66:4457–4459.
110. Wu L, Cooks RG: Chiral and isomeric analysis by electrospray ionization and sonic spray ionization using the fixed-ligand kinetic method. *European Journal of Mass Spectrometry* 2005, 11:231–242.
111. Haddad R, Sparrapan R, Eberlin MN: Desorption sonic spray ionization for high voltage-free ambient mass spectrometry. *Rapid Communications in Mass Spectrometry* 2006, 20:2901–2905.
112. Shi X, Takamizawa A, Nishimura Y, Hiraoka K, Akashi S: Stability analysis for double-stranded DNA oligomers and their noncovalent complexes with drugs by laser spray. *Journal of Mass Spectrometry* 2006, 41:1086–1095.
113. Takamizawa A, Itoh Y, Osawa R, Iwasaki N, Nishimura Y, Akashi S, Hiraoka K: Selective dissociation of non-covalent bonds in biological molecules by laser spray. *Journal of Mass Spectrometry* 2004, 39:1053–1058.
114. Cai Y, Kingery D, McConnel O, Bach AC: Advantages of atmospheric pressure photoionization mass spectrometry in support of drug discovery. *Rapid Communications in Mass Spectrometry* 2005, 19:1717–1724.
115. Delobel A, Halgand F, Laffranchise-Gosse B, Snijders H, Laprévote O: Characterization of hydrophobic peptides by atmospheric pressure photoionization mass-spectrometry and tandem mass spectrometry. *Analytical Chemistry* 2003, 75:5961–5968.
116. Robb DB, Blades MW: Factors affecting primary ionization in dopant-assisted atmospheric pressure photoionization (DA-APPI) for LC/MS. *Journal of the American Society of Mass Spectrometry* 2006, 17:130–138.

117. Delobel A, Roy S, Touboul D, Gaudin K, Germain DP, Baillet A, Brion F, Prognon P, Chaminade P, Laprévote O: Atmospheric pressure photoionization coupled to porous graphitic carbon liquid chromatography for the analysis of globotriaosylceramides. Application to Fabry disease. *Journal of Mass Spectrometry* 2006, 41:50–58.
118. Takats Z, Wiseman JM, Gologan B, Cooks RG: Mass spectrometry sampling under ambient conditions with desorption electrospray ionization. *Science* 2004, 306:471–473.
119. Takats Z, Wiseman JM, Cooks RG: Ambient mass spectrometry using desorption electrospray ionization (DESI): instrumentation, mechanisms and applications in forensics, chemistry, and biology. *Journal of Mass Spectrometry* 2005, 40:1261–1275.
120. Wu G, Noll RJ, Plass WR, Hu Q, Perry RH, Cooks RG: Ion trajectory simulations of axial ac dipolar excitation in the orbitrap. *International Journal of Mass Spectrometry* 2006, 254:53–62.
121. Mulligan CC, Talaty N, Cooks RG: Desorption electrospray ionization with a portable mass spectrometer: in situ analysis of ambient surfaces. *Chemical Communications* 2006:1709–1711.
122. Laughlin BC, Mulligan CC, Cooks RG: Atmospheric pressure ionization in a miniature mass spectrometer. *Analytical Chemistry* 2005, 77:2928–2939.
123. Kauppila TJ, Wiseman JM, Ketola RA, Kotiaho T, Cooks RG, Kostiainen R: Desorption electrospray ionization mass spectrometry for the analysis of pharmaceuticals and metabolites. *Rapid Communications in Mass Spectrometry* 2006, 20:387–392.
124. Cooks RG, Ouyang Z, Takats Z, Wiseman JM: Ambient mass spectrometry. *Science* 2006, 311:1566–1570.
125. Myung S, Wiseman JM, Valentine SJ, Takats Z, Cooks RG, Clemmer DE: Coupling desorption electrospray ionization with ion mobility/mass spectrometry for analysis of protein structure: evidence for desorption of folded and denatured states. *Journal of Physical Chemistry B* 2006, 110:5045–5051.
126. Santos SS, Knaack L, Metzger JO: Investigation of chemical reactions in solution using API-MS. *International Journal of Mass Spectrometry* 2005, 246:84–104.
127. Cai ZL, Dextraze ME, Cloutier P, Hunting D, Sanche L: Induction of strand breaks by low-energy electrons (8–68 eV) in a self-assembled monolayer of oligonucleotides: Effective cross sections and attenuation lengths. *Journal of Chemical Physics* 2006, 124:024705.
128. Shiea J, Huang MZ, Hsu HJ, Lee CY, Yuan CH, Beech I, Sunner J: Electrospray-assisted laser desorption/ionization mass spectrometry for direct ambient analysis of solids. *Rapid Communications in Mass Spectrometry* 2005, 19:3701–3704.
129. Huang YY, Wysocki VH, Tabb DL, Yates JR: The influence of histidine on cleavage C-terminal to acidic residues in doubly protonated tryptic peptides. *International Journal of Mass Spectrometry* 2002, 219:233–244; Tabb DL, Smith LL, Breci LA, Wysocki VH, Lin D, Yates JR: Statistical characterization of ion trap tandem mass spectra from doubly charged tryptic peptides. *Analytical Chemistry* 2003, 75:1155–1163.
130. Huang MZ, Kojima H, Hirabayashi A, Koizumi H: Sonic spray nebulizer for inductively coupled plasma atomic emission spectrometry. *Analytical Sciences* 1999, 15:265–268.
131. Myung S, Julian RR, Nanita SC, Cooks RG, Clemmer DE: Formation of nanometer-scale serine clusters by sonic spray. *Journal of Physical Chemistry B* 2004, 108:6105–6111.
132. Petzold CJ, Ramirez-Arizmendi LE, Heidbrink JL, Perez J, Kenttamaa HI: Gas-phase reactions of charged phenyl radicals with neutral biomolecules evaporated by laser-induced acoustic desorption. *Journal of the American Society for Mass Spectrometry* 2002, 13:192–194.
133. Peng WP, Yang C, Kang MW, Tzeng YK, Nie Z, Chang HC, Chang W, Chen CH: Laser-induced acoustic desorption mass spectrometry of single bioparticles. *Angewandte Chemie-International Edition* 2006, 45:1423–1426.
134. Piuzzi F, Mons M, Dimicoli I, Tardivel B, Zhao Q: Ultraviolet spectroscopy and tautomerism of the DNA base guanine and its hydrated form in a supersonic jet. *Chemical Physics* 2001, 270:205–214.

135. Lee KT, Sung J, Lee KJ, Kim SK, Park YD: Resonant two-photon ionization study of jet-cooled amino acid: L-phenylalanine and its monohydrated complex. *Journal of Chemical Physics* 2002, 116:8251–8254.
136. Snoek LC, Kroemer RT, Simons JP: A spectroscopic and computational exploration of tryptophan–water cluster structures in the gas phase. *Physical Chemistry Chemical Physics* 2002, 4:2130–2139.
137. Zwier TS: Laser spectroscopy of jet-cooled biomolecules and their water-containing clusters: water bridges and molecular conformation. *Journal of Physical Chemistry A* 2001, 105: 8827–8839.
138. Jockusch RA, Kroemer RT, Talbot FO, Simons JP: Hydrated sugars in the gas phase: spectroscopy and conformation of singly hydrated phenyl β-D-glucopyranoside. *Journal of Physical Chemistry A* 2003, 107:10725–10732.
139. Crews B, Abo-Riziq A, Grace L, Callahan M, Kabelac M, Hobza P, de Vries MS: IR-UV double resonance spectroscopy of guanine–H_2O clusters. *Physical Chemistry Chemical Physics* 2005, 7:3015–3020.
140. Aviles Moreno JR, Demaison J, Huet TR: Conformational flexibility in hydrated sugars: the glycolaldehyde–water complex. *Journal of the American Chemical Society* 2006, 128: 10467–10473.
141. Ziskind M, Mihesan C, Lebrun N, Chazallon B, Focsa C: Hydrated clusters of organic molecules generated by IR OPO resonant desorption of frozen aqueous solutions. *Applied Physics A* 2004, 79:991–995.
142. Moro R, Rabinovitch R, Kresin VV: Amino acid and water molecules adsorbed on water clusters in a beam. *Journal of Chemical Physics* 2005, 123:074301.
143. Jarrold MF: Unfolding, refolding, and hydrations of proteins in the gas phase. *Accounts of Chemical Research* 1999, 32:360–367.
144. Rodriguez-Cruz SE, Klassen JS, Williams ER: Hydration of gas-phase ions formed by electrospray ionization. *Journal of the American Society for Mass Spectrometry* 1999, 10:958–968.
145. Rodriguez-Cruz SE, Klassen JS, Williams ER: Hydration of gas-phase gramicidin S $(M+2H)^+$ ions formed by electrospray: the transition from solution to gas-phase structure. *Journal of the American Chemical Society* 1997, 8:565–568.
146. Kamariotis A, Boyarkin OV, Mercier SR, Beck RD, Bush MF, Williams ER, Rizzo TR: Infrared spectroscopy of hydrated amino acids in the gas phase: protonated and lithiated valine. *Journal of the American Chemical Society* 2006, 128:905–916.
147. Zhan D, Fenn JB: Gas phase hydration of eletrospray ions from small peptides. *International Journal of Mass Spectrometry* 2002, 120:1–10.
148. Nonose S, Iwaoka S, Mori K, Shibata Y, Fuke K: Structure and reactions of hydrated biomolecular cluster ions. *The European Physical Journal D* 2005, 34:315–319.
149. Wincel H: Hydration of gas-phase protonated alkylamines, amino acids and dipeptides produced by electrospray. *International Journal of Mass Spectrometry* 2006, 251:23–31.
150. Domon B, Aebersold R: Mass spectrometry and protein analysis. *Science* 2006, 312:212–217.
151. Bowers MT, Marshall AG, McLafferty FW: Mass spectrometry: recent advances and future directions. *Journal of Physical Chemistry* 1996, 100:12897–12910.
152. McLuckey SA, Wells JM: Mass analysis at the advent of the 21st century. *Chemical Reviews* 2001, 101:571–606.
153. Brenton G, Monoghan J, Ashcroft A: *Advances in mass spectrometry*: Elsevier, Amsterdam, 2004.
154. March RE, Todd JFJ: *Quadrupole ion trap mass spectrometry* 2nd edn: John Wiley, 2005.
155. Kaltashov IA, Eyles SJ: *Mass spectrometry in biophysics*: Wiley, 2005.
156. Wysocki VH, Resing KA, Zhang QF, Cheng GL: Mass spectrometry of peptides and proteins. *Methods* 2005, 35:211–222.
157. Geoghegan KF, Kelly MA: Biochemical applications of mass spectrometry in pharmaceutical drug discovery. *Mass Spectrometry Reviews* 2005, 24:347–366.

158. McDonnell, Piersma SR, Maarten Altelaar AF, Mize TH, Luxembourg SL, Verhaert PDE, van Minnen J, Heeren RMA: Subcellular imaging mass spectrometry of brain tissue. *Journal of Mass Spectrometry* 2005, 40:160–168.
159. Wiley WC, McLaren LH: Time of flight mass spectrometer with improved resolution. *Review of Scientific Instruments* 1955, 26:1150–1157.
160. Schiedt J, Weinkauf R: Efficient and robust anion source, based on a microchannel plate electron source. *Review of Scientific Instruments* 1999, 70:2277–2281.
161. Brutschy B: The structure of microsolvated benzene derivatives and the role of aromatic constituents. *Chemical Reviews* 2000, 100:3891–3920.
162. Mamyrin BA: Laser assisted reflectron time-of-flight mass spectrometry. *International Journal of Mass Spectrometry* 1994, 131:1–19.
163. Scherer S, Altwegg K, Balsiger H, Fischer J, Jäckel A, Korth A, Mildner M, Piazza D, Reme H, Wurz P: A novel principle for an ion mirror design. *International Journal of Mass Spectrometry* 2006, 251:73–81.
164. Satoh T, Tsuno H, Iwanaga M, Kammei Y: The design and characteristic features of a new time-of-flight mass spectrometer with a spiral ion trajectory. *Journal of the American Society of Mass Spectrometry* 2005, 16:1969–1975.
165. Loos J, Schröder D, Zummack W, Schwarz H, Thissen R, Dutuit O: Dissociation behavior of ionized valeramide. Part I: experimental studies. *International Journal of Mass Spectrometry* 2002, 214:105–128.
166. Thissen R, Dutuit O, Alcaraz C, Laprévote O, Serani L, Millié P, Soudan JM: State-selected dissociation of dehydrovincamine alkaloid stereo-isomers. *Physical Chemistry Chemical Physics* 2002, 4:661–667.
167. Douglas DJ, Frank AJ, Mao D: Linear ion traps in mass spectrometry. *Mass Spectrometry Reviews* 2005, 24:1–29.
168. Gerlich D: Molecular ions and nanoparticles in RF and AC traps. *Hyperfine Interactions* 2003, 146/147:293–306.
169. Gerlich D: Applications of RF fields and collisions dynamics in atomic mass spectrometry. *Analytical Atomic Spectrometry* 2004, 19:581–590.
170. Teloy E, Gerlich D: Integral cross-sections for ion-molecule reactions. The guided beam technique. *Chemical Physics* 1974, 4:417–427.
171. Boyarkin OV, Mercier SR, Kamariotis A, Rizzo TR: Electronic spectroscopy of cold, protonated tryptophan and tyrosine. *Journal of the American Chemical Society* 2006, 128:2816–2817.
172. Major FG, Dehmelt HG: Exchange-collision technique for the rf spectroscopy of stored ions. *Physical Review A* 1968, 70:91–107.
173. Black DM, Patyne AH, Glish GL: Determination of cooling rates in a quadrupole ion trap. *Journal of the American Society for Mass Spectrometry* 2006, 17:932–938.
174. Mekecha TT, Amunugama R, McLuckey SA: Ion trap collision-induced dissociation of human hemoglobin α-chain cations. *Journal of the American Society of Mass Spectrometry* 2006, 17: 923–931.
175. Iavarone AT, Parks JH: Conformational change in unsolvated Trp-cage protein probed by fluorescence. *Journal of the American Chemical Society* 2005, 127:8606–8607.
176. Peng WP, Cai Y, Chang HC: Optical methods for mass spectrometry of macroions. *Mass Spectrometry Reviews* 2004, 23:443–465.
177. Carette M, Zerega Y, March RE, Perrier P, André J: Coupling of a Rydberg electron capture ion source with a mass filter. *Rapid Communications in Mass Spectrometry* 2000, 14:979–986.
178. Amanuga R, Hogan JM, Newton KA, McLuckey SA: Whole protein dissociation in a quadrupole ion trap: identification of an *a priori* unknown modified protein. *Analytical Chemistry* 2004, 76:720–727.
179. Gronert S: Quadrupole ion traps studies of fundamental organic reactions. *Mass Spectrometry Reviews* 2005, 24:100–120.

180. Marian C, Nolting D, Weinkauf R: The electronic spectrum of protonated adenine: Theory and experiment. *Physical Chemistry Chemical Physics* 2005, 7:3306–3316.
181. Nolting D, Marian C, Weinkauf R: Protonation effect on the electronic spectrum of tryptophan in the gas phase. *Physical Chemistry Chemical Physics* 2004, 6:2633–2640.
182. Wu G, Cooks RG, Ouyang Z: Geometry optimization for the cylindrical ion trap: field calculations, simulations and experiments. *International Journal of Mass Spectrometry* 2005, 241:119–132.
183. Tabert AM, Griep-Raming J, Guymon AJ, Cooks RG: High-throughput miniature cylindrical ion trap array mass spectrometer. *Analytical Chemistry* 2003, 75:5656–5664.
184. Gluch K, Leubner SM, Echt O, Concina B, Scheier P, Märk TD: High resolution kinetic energy release distribution and dissociation energies for fullerence ions C_n $42<n<90$. *Journal of Chemical Physics* 2004, 121:2137–2143.
185. Marshall AG, Hendrickson CL, Jackson GS: Fourier transform ion cyclotron resonance mass spectrometry: a primer. *Mass Spectrometry Reviews* 1998, 17:1–35.
186. Oomens J, Polfer N, Moore DT, van der Meer L, Marshall AG, Eyler JR, Meijer G, von Helden G: Charge-state resolved mid-infrared spectroscopy of a gas-phase protein. *Physical Chemistry Chemical Physics* 2005, 7:1345–1348.
187. Andersen JU, Andersen LH, Hvelplund P, Lapierre A, Moller SP, Nielsen SB, Pedersen UV, Tomita S: Studies of clusters and biomolecules in ELISA. *Hyperfine Interactions* 2003, 146:283–291.
188. Nielsen IB, Boye-Peronne S, El Ghazaly MOA, Kristensen MB, Nielsen SB, Andersen LH: Absorption spectra of photoactive yellow protein chromophores in vacuum. *Biophysical Journal* 2005, 89:2597–2604.
189. Andersen LH, Herber O, Zajfman D: Physics with electrostatic ring and traps. *Journal of Physics B* 2004, 37:R57–R88.
190. Calvo MR, Andersen JU, Hvelplund P, Nielsen SB, Pedersen UV, Rangama J, Tomita S, Forster JS: Photophysics of protoporphyrin ions in vacuo: triplet-state lifetimes and quantum yields. *Journal of Chemical Physics* 2004, 120:5067–5072.
191. Andersen LH, Bluhme H, Boye S, Jorgensen TJD, Krogh H, Nielsen IB, Nielsen SB, Svendsen A: Experimental studies of the photophysics of gas-phase fluorescent protein chromophores. *Physical Chemistry Chemical Physics* 2004, 6:2617–2627.
192. Andersen LH, Lapierre A, Nielsen SB, Nielsen IB, Pedersen SU, Pedersen UV, Tomita S: Chromophores of the green fluorescent protein studied in the gas phase. *European Physical Journal D* 2002, 20:597–600.
193. Boye S, Krogh H, Nielsen IB, Nielsen SB, Pedersen SU, Pedersen UV, Andersen LH, Bell AF, He X, Tonge PJ: Vibrationally resolved photoabsorption spectroscopy of red fluorescent protein chromophore anions. *Physical Review Letters* 2003, 90.
194. Boye SV, Nielsen IB, Nielsen SB, Krogh H, Lapierre A, Pedersen HB, Pedersen SU, Pedersen UV, Andersen LH: Gas-phase absorption properties of a green fluorescent protein-mutant chromophore: the W7 clone. *Journal of Chemical Physics* 2003, 119:338–345.
195. Worm ES, Anderen IH, Andersen JU, Holm AIS, Hvelplund P, Kadhane U, Nielsen SB, Poully JC, Stochkel K: Statistical versus nonstatistical dissociation processes of dinucleotide ions *in vacuo*. *Physical Chemistry Chemical Physics* 2006, in press.
196. Andersen JU, Cederquist H, Forster JS, Huber BA, Hvelplund P, Jensen J, Liu B, Manil B, Maunoury L, Nielsen SB, *et al.*: Photodissociation of protonated amino acids and peptides in an ion storage ring. Determination of Arrhenius parameters in the high-temperature limit. *Physical Chemistry Chemical Physics* 2004, 6:2676–2681.
197. Pedersen HB, Ring S, Rappaport ML, Rudich Y, Sagi I, Zajfman D: Ion motion synchronization in an ion-trap resonator. *Physical Review Letters* 2001, 87:055001–055004.
198. Hu Q, Noll RJ, Li A, Makarov AA, Cooks RG: The Orbitrap: a new mass spectrometer. *Journal of Mass Spectrometry* 2005, 40:430–443.

199. Hu Q, Makarov AA, Cooks RG, Noll RJ: Resonant ac dipolar excitation for ion motion control in the orbitrap mass analyzer. *Journal of Physical Chemistry A* 2006, 110:2682–2689.
200. Makarov A, Denisov E, Lange O, Horning S: Dynamic range of mass accuracy in LTQ Orbitrap hybrid mass spectrometer. *Journal of the American Society of Mass Spectrometry* 2006, 17: 977–982.
201. Yates JR, Cociorva D, Liao L, Zabrouskov V: Performance of a linear trap-Orbitrap hybrid for peptide analysis. *Analytical Chemistry* 2006, 78:593–500.
202. Peterman SM, Duczak N, Kalgutkar AS, Lame ME, Soglia JR: Application of a linear ion trap/Orbitrap mass spectrometer in metabolite characterization studies: examination of the human liver microsomal metabolism of the non-tricyclic anti-depressant nefadozone using data-dependent accurate mass measurements. *Journal of the American Society for Mass Spectrometry* 2006, 17:363–375.
203. Makarov A, Denisov E, Kholomeev A, Balschun W, Lange O: Performance evaluation of a hybrid linear ion trap/Orbitrap mass spectrometer. *Analytical Chemistry* 2006, 78:213–218.
204. Meng F, Forbes AJ, Miller LM, Kelleher NL: Detection and localization of protein modifications by high-resolution tandem mass-spectrometry. *Mass Spectrometry Reviews* 2005, 24: 126–134.
205. Freitas MA, Sklenar AR, Parthun MR: Application of mass-spectrometry to the identification and quantification of histone post-translational modifications. *Journal of Cellular Biochemistry* 2004, 92:691–700.
206. Krokhin O, Ens W, Standing KG, Wilkins J, Perreault H: Site-specific N-glycosilation analysis: matrix-assisted laser/desorption ionization quadrupole-quadrupole time-of-flight mass spectra signatures for recognition and identification of glycopeptides. *Rapid Communications in Mass Spectrometry* 2004, 18:2020–2030.
207. McLafferty FW, Fridriksson EK, Horn DM, Lewis MA, Zubarev RA: Biochemistry–Biomolecule mass spectrometry. *Science* 1999, 284:1289–1290.
208. McLafferty FW: Tandem mass spectrometric mass analysis of complex biological mixtures. *International Journal of Mass Spectrometry* 2001, 212:81–87.
209. Han X, Jin M, Breuker K, McLafferty FW: Extending top-down mass spectrometry to proteins with masses greater than than 200 kilodaltons. *Science* 2006, 314:109–112.
210. Horn DM, Zubarev RA, McLafferty FW: Automated *de novo* sequencing of proteins by tandem high-resolution mass spectrometry. *Proceedings of the National Academy of Sciences of the United States of America* 2000, 97:10313–10317.
211. Armirotti A, Millo E, Damonte G: How to discriminate between leucine and isoleucine by low energy ESI-trap MS. *Journal of American Society for Mass Spectrometry* 2007, 18:57–63.
212. Eberlin ME: Triple-stage pentaquadrupole (QqQqQ) mass spectrometry and ion/molecule reactions. *Mass Spectrometry Reviews* 1997, 6:113–144.
213. Kim SK, Rodgers RP, Marshall AG: Truly "exact" mass: elemental composition can be determined uniquely from molecular mass measurement at 0.1 mDa accuracy for molecules up to 500 Da. *International Journal of Mass Spectrometry* 2006, 251:260–265.
214. Perkins DL, Pappin DJ, Creasy DM, Cottrel JS: Probability-based protein identification by searching sequence databases using mass spectrometry data. *Electrophoresis* 1999, 20: 3551–3567.
215. Paizs B, Suhai S: Fragmentation pathways of protonated peptides. *Mass Spectrometry Reviews* 2005, 24:508–548.
216. Laskin J, Futrell JH: Collisional activation of peptide ions in FT-ICR mass spectrometry. *Mass Spectrometry Reviews* 2003, 22:158–181.
217. Bomse DS, Woodin RL, Beauchamp JL: Molecular activation with low-intensity CW infrared-laser radiation – multi-photon dissociation of ions derived from diethyl-ether. *Journal of the American Chemical Society* 1979, 101:5503–5512.

218. Jockusch RA, Paech K, Williams ER: Energetics from slow infrared multiphoton dissociation of biomolecules. *Journal of Physical Chemistry A* 2000, 104:3188–3196.
219. Zubarev RA: Electron-capture dissociation tandem mass spectrometry. *Current Opinion in Biotechnology* 2004, 15:12–16.
220. Rogalewicz F, Hoppilliard Y, Ohanessian G: Fragmentation mechanisms of α-amino acids protonated under electrospray ionization: a collisional activation and *ab initio* theoretical study. *International Journal of Mass Spectrometry* 2000, 196:565–590.
221. Rodgers MT, Armentrout PB: Noncovalent metal-ligand bond energies as studied by threshold collision-induced dissociation. *Mass Spectrometry Reviews* 2000, 19:215–247.
222. Rodgers FM, Armentrout PB: Non covalent interactions of nucleic acid bases (uracil, thymine, and adenine) with alkali metal ions. Threshold collision-induced dissociation and theoretical studies. *Journal of the American Chemical Society* 2000, 122:8548–8558.
223. Nesatyy VJ, Laskin J: Dissociation of noncovalent protein complexes by triple quadrupole tandem mass spectrometry: comparison of Monte Carlo simulation and experiment. *International Journal of Mass Spectrometry* 2002, 221:245–262.
224. Laskin J, Denisov E, Futrell JH: Fragmentation energetics of small peptides from multiple-collision activation and surface-induced dissociation in FT-ICR MS. *International Journal of Mass Spectrometry* 2002, 219:189–201.
225. Laskin J, Futrell JH: Activation of large ions in FT-ICR mass spectrometry. *Mass Spectrometry Reviews* 2005, 24:135–167.
226. Meroueh O, Hase WL: Energy transfer pathways in the collisional activation of peptides. *International Journal of Mass Spectrometry* 2000, 201:233–244.
227. Desfrancois C, Schermann JP: Energy transfer in inhomogeneous van der Waals clusters. *Zeitschrift Fur Physik D-Atoms Molecules and Clusters* 1990, 20:185–187.
228. Paizs B, Suhai S: Towards understanding the tandem mass spectra of protonated oligopeptides. 1: mechanism of amide bond cleavage. *Journal of the American Society for Mass Spectrometry* 2004, 15:103–113.
229. Roepstorff P, Fohlman J: Proposal for a common nomenclature for sequence ions in mass spectra of peptides. *Biomedical Mass Spectrometry* 1984, 11:601.
230. Paizs B, Schnolzer M, Warnken U, Suhai S, Harrison AG: Cleavage of the amide bond of protonated dipeptides. *Physical Chemistry Chemical Physics* 2004, 6:2691–2699.
231. Csonka IP, Paizs B, Lendvay G, Suhai S: Proton mobility in protonated peptides: a joint molecular orbital and RRKM study. *Rapid Communications in Mass Spectrometry* 2000, 14:417–431.
232. Csonka IP, Paizs B, Lendvay G, Suhai S: Proton mobility and main fragmentation pathways of protonated lysylglycine. *Rapid Communications in Mass Spectrometry* 2001, 15:1457–1472.
233. Paizs B, Suhai S: Theoretical study of the main fragmentation pathways for protonated glycylglycine. *Rapid Communications in Mass Spectrometry* 2001, 15:651–663.
234. Paizs B, Suhai S: Towards understanding some ion intensity relationships for the tandem mass spectra of protonated peptides. *Rapid Communications in Mass Spectrometry* 2002, 16:1699–1702.
235. Paizs B, Suhai S: Combined quantum chemical and RRKM modeling of the main fragmentation pathways of protonated GGG. II. Formation of b_2, y_1, and y_2 ions. *Rapid Communications in Mass Spectrometry* 2002, 16:375–389.
236. Komaromi I, Somogyi A, Wysocki VH: Proton migration and its effect on the MS fragmentation of N-acetyl OMe proline: MS/MS experiments and *ab initio* and density functional calculations. *International Journal of Mass Spectrometry* 2005, 241:315–323.
237. Griffiths JW, Jonnson AP: Gas-phase conformation can have an influence on peptide fragmentation. *Proteomics* 2001, 1:934–945.
238. Polfer NC, Oomens J, Suhai S, Paizs B: Spectroscopic and theoretical evidence for oxazolone ring formation in collision-induced dissociation of peptides. *Journal of the American Chemical Society* 2005, 127:17154–17155.

239. Polfer N, Oomens J, Suhai S, Paizs B: Infrared spectroscopy and theoretical studies on gas-phase protonated Leu-enkephalin and its fragments: direct experimental evidence for the mobile proton. *Journal of the American Chemical Society* 2007, 129:5887–5897.
240. Chen X, Turecek F: Simple b ions have cyclic oxazolone structures. A neutralization–reionization mass spectrometric and computational study of oxazolone radicals. *Journal of the American Society of Mass Spectrometry* 2005, 16:1941–1956.
241. Wysocki VH, Tsaprailis G, Smith LL, Breci LA: Special feature: commentary – mobile and localized protons: a framework for understanding peptide dissociation. *Journal of Mass Spectrometry* 2000, 35:1399–1406.
242. Kohtani M, Schneider JE, Jones TC, Jarrold MF: The mobile proton in polyalanine peptides. *Journal of the American Chemical Society* 2004, 126:16981–16987.
243. Weinkauf R, Schanen P, Metsala A, Schlag EW, Burgle M, Kessler H: Highly efficient charge transfer in peptide cations in the gas phase: threshold effects and mechanism. *Journal of Physical Chemistry* 1996, 100:18567–18585.
244. Kaltashov IA, Doroshenko VM, Cotter RJ: Gas-phase hydrogen/deuterium exchange reactions of peptide ions in a quadrupole ion trap mass spectrometer. *Proteins-Structure Function and Genetics* 1997, 28:53–58.
245. Klassen JS, Kebarle P: Collision-induced dissociation threshold energies of protonated glycine, glycinamide, and some related small peptides and peptides amino amides. *Journal of the American Chemical Society* 1997, 119:6552–6563.
246. Schlag EW, Selze HL, Schanen P, Weinkauf R: Dissociation kinetics of peptide ions. *Journal of Physical Chemistry A* 2006, 110:8497–8500.
247. Zhang Z, Bordas-Nagy J: Peptide conformation in gas phase probed by collision-induced dissociation and its correlation to conformation in condensed phases. *Journal of the American Society of Mass Spectrometry* 2006, 17:786–794.
248. Eiceman GA, Karpas Z: *Ion mobility spectrometry*. CRC Press, Boca Raton, FL, 2005.
249. von Helden G, Wyttenbach T, Bowers MT: Inclusion of a MALDI ion source in the ion chromatography technique: conformational information on polymer and biomolecular ions. *International Journal of Mass Spectrometry* 1995, 146:349–364.
250. Jarrold MF: Peptides and proteins in the vapor phase. *Annual Review of Physical Chemistry* 2000, 51:179–207.
251. Verbeck GF, Ruotolo B, Sawyer HA, Gillig KJ, Russell DH: A fundamental introduction to ion mobility mass spectrometry applied to the analysis of biomolecules. *Journal of Biomolecular Techniques* 2002, 13:56–61.
252. Wyttenbach T, Witt M, Bowers MT: On the question of salt bridges of cationized amino acids in the gas phase: glycine and arginine. *International Journal of Mass Spectrometry* 1999, 183:243–252.
253. Wong RL, Williams ER, Counterman AE, Clemmer DE: Evaluation of ion mobility spectroscopy for determining charge-solvated versus salt-bridge structures of protonated trimers. *Journal of the American Society for Mass Spectrometry* 2005, 16:1009–1019.
254. Taraszka JA, Counterman AE, Clemmer DE: Large anhydrous polyalanine ions: substitution of Na^+ for H^+ destabilizes folded states. *International Journal of Mass Spectrometry* 2001, 204:87–100.
255. Kinnear BS, Kaleta DT, Kohtani M, Hudgins RR, Jarrold MF: Conformations of unsolvated valine-based peptides. *Journal of the American Chemical Society* 2000, 122:9243–9256.
256. Srebalus CA, Clemmer DE: Assessing intrinsic side chain interactions between i and i+4 residues in solvent free peptides: a combinatorial gas-phase approach. *Journal of Physical Chemistry B* 2003, 107:10566–10579.
257. Barran PE, Roeseke RW, Pawson AJ, Sellar R, Bowers MT, Morgan K, Lu ZL, Tsuda M, Kusakabe T, Millar RP: Evolution of constrained gonadotropin-releasing hormone ligand

conformation and receptor selectivity. *Journal of Biological Chemistry* 2005, 280: 38569–38575.

258. Julian RR, Myung S, Clemmer DE: Do homochiral aggregates have an entropic advantage? *Journal of Physical Chemistry B* 2005, 109:440–444.
259. Sudha R, Kohtani M, Jarrold MF: Non-covalent interactions between unsolvated peptides: helical complexes based on acid-base interactions. *Journal of Physical Chemistry B* 2005, 109: 6442–6447.
260. Barran PE, Polfer NC, Campopiano DJ, Clarke DJ, Langridge-Smith PRR, Langley RJ, Govan JRW, Maxwell A, Dorin JR, Millar RP, *et al.*: Is it biologically relevant to measure the structures of small peptides in the gas-phase? *International Journal of Mass Spectrometry* 2005, 240: 273–284.
261. Baker ES, Bernstein SL, Bowers MT: Structural characterization of G-quadruplexes in deoxyguanosine clusters using ion mobility mass spectrometry. *Journal of the American Society for Mass Spectrometry* 2005, 16:989–997.
262. Wyttenbach T, von Helden G, Batka JJ, Carlat D, Bowers MT: Effect of the long-range potential on ion mobility measurements. *Journal of Mass Spectrometry* 1997, 8:275–282.
263. Clemmer DE, Jarrold MF: Ion mobility measurements and their applications to clusters and biomolecules. *Journal of Mass Spectrometry* 1997, 32:577–592.
264. Jin L, Barran PE, Deakin JA, Lyon M, Uhrin D: Conformation of glycoaminoglycans by ion mobility mass spectrometry and molecular modelling. *Physical Chemistry Chemical Physics* 2005, 7:3464–3471.
265. Shvartsburg AA, Siu KWM, Clemmer DE: Prediction of peptide ion mobilities via a priori calculations from intrinsic size parameters of amino acid residues. *Journal of the American Society for Mass Spectrometry* 2001, 12:885–888.
266. Kim HJ, Kim H, Beegle LW, Johnson PV, Beauchamp JL, Kanik I: Theoretical ion mobility studies of amino acids. *Lunar Planetary Science* 2006, 37:2127.
267. Shvartsburg AA, Jarrold MF: An exact hard-spheres scattering model for the mobilities of polyatomic ions. *Chemical Physics Letters* 1996, 261:86–91.
268. Hudgins RR, Jarrold MF: Helix formation in unsolvated alanine-based peptides: helical monomers and helical dimers. *Journal of the American Chemical Society* 1999, 121: 3494–3501.
269. Clemmer DE, Hudgins RR, Jarrold MF: Naked protein conformations – cytochrome-c in the gas-phase. *Journal of the American Chemical Society* 1995, 117:10141–10142.
270. Woods AS, Fuhrer K, Gonin M, Egan T, Ugarov M, Gillig KJ, Schultz JA: Angiotensin II-acetylcholine noncovalent complexes analyzed with MALDI-ion mobility-TOF MS. *Journal of Biomolecular Techniques* 2003, 14:1–8.
271. Loboda A: Novel ion mobility setup combined with collision cell and time-of-flight mass spectrometer. *Journal of the American Society for Mass Spectrometry* 2006, 17:691–699.
272. Tomachev AV, Kim TK, Udseth HR, Smith RD, Bayley TH, Futrell JH: Simulation-based optimization of the electrodynamic funnel for high sensitivity electrospray ionization mass spectrometry. *International Journal of Mass Spectrometry* 2000, 203:31–47.
273. Wyttenbach T, Kemper PR, Bowers MT: Design of a new electrospray ion mobility mass spectrometer. *International Journal of Mass Spectrometry* 2001, 212:13–23.
274. Julian RR, Mabbett SR, Jarrold MF: Ion funnels for the masses: experiments and simulations with a simplified ion funnel. *Journal of the American Society for Mass Spectrometry* 2005, 16:1708–1712.
275. Hoaglund CS, Valentine SJ, Clemmer DE: An ion trap interface for ESI-ion mobility experiments. *Analytical Chemistry* 1997, 69:4156–4161.
276. Badman ER, Myung S, Clemmer DE: Evidence for unfolding and refolding of gas-phase cytochrome c ions in a Paul trap. *Journal of the American Society for Mass Spectrometry* 2005, 16:1493–1497.

277. Valentine SJ, Kulchania M, Barnes CAS, Clemmer DE: Multidimensional separations of complex peptide mixtures: a combined high-performance liquid chromatography/ion mobility/time-of-flight mass spectrometry approach. *International Journal of Mass Spectrometry* 2001, 212: 97–109.
278. Guevremont R, Purves RW: Atmospheric pressure ion focusing in a high-field asymmetric waveform ion mobility spectrometer. *Review of Scientific Instruments* 1999, 70:1370–1383.
279. Guevremont R, Barnett DA, Purves RW, Viehland LA: Calculation of ion mobilities from electrospray ionization high-field asymmetric waveform ion mobility spectrometry mass spectrometry. *Journal of Chemical Physics* 2001, 114:10270–10277.
280. Venne K, Bonneil E, Eng K, Thibault P: Improvement in peptide detection for proteomics analyses using nanoLC-MS and high field asymmetry waveform ion mobility mass spectrometry (FAIMS). *Analytical Chemistry* 2005, 77:2176–2186.
281. Bacher G, Szymansli WW, Kaufman SL, Zöllner P: Charge-reduced nano electrospray ionization combined with differential mobility analysis of peptides, proteins, glycoproteins, noncovalent protein complexes and viruses. *Journal of Mass Spectrometry* 2001, 36:1038–1052.
282. Loo JA, Berhane B, Kaddis CS, Wooding KM, Xie YM, Kaufman SL, Chernushevich IV: Electrospray ionization mass spectrometry and ion mobility analysis of the 20S proteasome complex. *Journal of the American Society for Mass Spectrometry* 2005, 16:998–1008.
283. Giles K, Pringle SD, Worthington KR, Little D, Wildgoose JL, Bateman RH: Applications of a travelling wave-based radio-frequency only stacked ring ion guide. *Rapid Communications in Mass Spectrometry* 2004, 18:2401–2414.
284. Counterman AE, Clemmer DE: Compact → extended helix transitions of polyalanine in vacuo. *Journal of Physical Chemistry B* 2003, 107:2111–2117.
285. Myung S, Badman ER, Lee YJ, Clemmer DE: Structural transitions of electrosprayed ubiquitin ions stored in an ion trap over similar to 10 ms to 30 s. *Journal of Physical Chemistry A* 2002, 106:9976–9982.
286. Mao Y, Woenckhaus J, Kolafa J, Ratner MA, Jarrold MF: Thermal unfolding of unsolvated cytochrome c: Experiment and molecular dynamics simulations. *Journal of the American Chemical Society* 1999, 121:2712–2721.
287. Valentine SJ, Clemmer DE: Temperature-dependent H/D exchange of compact and elongated cytochrome c ions in the gas phase. *Journal of the American Society for Mass Spectrometry* 2002, 13:506–517.
288. Valentine SJ, Clemmer DE: H/D exchange levels of shape-resolved cytochrome c conformers in the gas phase. *Journal of the American Chemical Society* 1997, 119:3558–3566.
289. Ideue S, Sakamoto K, Honma K, Clemmer DE: Conformational change of electrosprayed cytochrome c studied by laser-induced fluorescence. *Chemical Physics Letters* 2001, 337:79–84.
290. Sahu K, Mondal SK, Ghosh S, Roy D, Sen P, Bhattacharyya K: Femtosecond study of partially folded states of cytochrome c by solvation dynamics. *Journal of Physical Chemistry B* 2006, 110:1056–1062.
291. Valentine SJ, Counterman AE, Clemmer DE: Conformer-dependent proton-transfer reactions of ubiquitin ions. *Journal of the American Society for Mass Spectrometry* 1997, 8:954–961.
292. Badman ER, Hoaglund-Hyzer CS, Clemmer DE: Dissociation of different conformations of ubiquitin ions. *Journal of the American Society for Mass Spectrometry* 2002, 13:719–723.
293. Geller O, Lifshitz C: A fast flow tube study of gas phase H/D exchange of multiply protonated ubiquitin. *Journal of Physical Chemistry A* 2005, 109:2217–2222.
294. Koeniger SL, Merenbloom SI, Clemmer DE: Evidence for many resolvable structures within conformation types of electrosprayed ubiquitin ions. *Journal of Physical Chemistry B* 2006, 110:7017–7021.
295. Robinson EW, Williams ER: Multidimensional separations of ubiquitin conformers in the gas phase: Relating ion cross sections to H/D exchange measurements. *Journal of the American Society for Mass Spectrometry* 2005, 16:1427–1437.

296. Ciechanover A: The ubiquitin proteasome pathway: on protein death and cell life. *EMBO Journal* 1998, 17:7151–7160.
297. Herschko A: The ubiquitin system for protein degradation and some of its roles in the control of the cell division cycle. *Nobel Lecture* 2004.
298. Ciechanover A: From the lysosome to ubiquitin and the proteasome. *Nature Reviews Molecular Cell Biology* 2005, 6:79–86.
299. Freitas MA, Hendrickson CL, Emmet MR, Marshall AG: Gas-phase bovine ubiquitin cation conformations resolved by gas-phase hydrogen/deuterium exchange rate and extent. *International Journal of Mass Spectrometry* 1999, 185:565–575.
300. Xia Y, Liang X, McLuckey SA: Ion trap low-energy beam-type collision-induced dissociation of protonated ubiquitin ions. *Analytical Chemistry* 2006, 78:1218–1227.
301. Huang F, Hudgins RR, Nau WM: Primary and secondary structure dependence of peptide flexibility assessed by fluorescence-based measurements of end-to-end collision rates. *Journal of the American Chemical Society* 2004, 126:16665–16675.
302. Aaserud DJ, Guan Z, Little DF, McLafferty FW: DNA sequencing with blackbody infrared radiative dissociation of electrosprayed ions. *International Journal of Mass Spectrometry* 1997, 167:705–712.
303. Dunbar RC: Kinetics of low-intensity infrared laser photodissociation. The thermal model and application of the Tolman theorem. *Journal of Chemical Physics* 1991, 95:2537–2548.
304. Freitas MA, Hendrickson CL, Marshall AG: Gas phase activation energy for unimolecular dissociation of biomolecular ions determined by Focused Radiation for Gaseous Multiphoton Energy Transfer (FRAGMENT). *Rapid Communications in Mass Spectrometry* 1999, 13:1639–1642.
305. Price WD, Williams ER: Activation of peptide ions by blackbody radiation: factors that lead to dissociation kinetics in the rapid energy exchange limit. *Journal of Physical Chemistry A* 1997, 101:8844–8852.
306. Price WD, Schnier PD, Williams ER: Tandem mass spectrometry of large biomolecule ions by blackbody infrared dissociation. *Analytical Chemistry* 1996, 68:859–866.
307. Dunbar RC: BIRD (blackbody infrared radiative dissociation): evolution, principles, and applications. *Mass Spectrometry Reviews* 2004, 23:127–158.
308. Beyer MK, Fox BS, Reinhard BM, Bondybey VE: Wet electrons and how to dry them. *Journal of Chemical Physics* 2001, 115:9288–9297.
309. Jockusch R, Pasch K, Williams ER: Energetics from slow infrared multiphoton dissociation of biomolecules. *Journal of Physical Chemistry A* 2000, 104:3188–3196.
310. Wong RL, Paech K, Williams ER: Blackbody infrared radiative dissociation.at low temperature: hydration of $X_2^+ (H_2O)_n$, for Mg, Ca. *International Journal of Mass Spectrometry* 2004, 232: 59–66.
311. Jurchen JC, Garcia DE, Williams ER: Gas-phase dissociation pathways of multiply charged peptide clusters. *Journal of the American Society for Mass Spectrometry* 2003, 14:1373–1386.
312. Lemoff AS, Bush MF, Wu CC, Williams ER: Structures and hydration enthalpies of cationized glutamine and structural analogues in the gas phase. *Journal of the American Chemical Society* 2005, 127:10276–10286.
313. Lemoff AS, Wu CC, Bush MF, Williams ER: Binding energies of water to doubly hydrated cationized glutamine and structural analogues in the gas phase. *Journal of Physical Chemistry A* 2006, 110:3662–3669.
314. Williams ER, Schnier PD, Jockusch R: Blackbody infrared radiative dissociation of noncovalent biomolecule complexes. *Abstracts of Papers of the American Chemical Society* 1998, 215: U93–U93.
315. Strittmatter EF, Schnier PD, Klassen JS, Williams ER: Dissociation energies of deoxyribose nucleotide dimer anions measured using blackbody infrared radiative dissociation. *Journal of the American Society for Mass Spectrometry* 1999, 10:1095–1104.

316. Kitova EN, Bundle DR, Klassen JS: Evidence for the preservation of specific intermolecular interactions in gaseous protein-oligosaccharide complexes. *Journal of the American Chemical Society* 2002, 124:9340–9341.
317. Schnier PD, Price WD, Strittmatter EF, Williams ER: Dissociation energetics and mechanisms of leucine enkephalin $(M+H)^+$ and $(2M+X)^+$ ions (X=H, Li, Na, K, and Rb) measured by blackbody infrared radiative dissociation. *Journal of the American Society for Mass Spectrometry* 1997, 8:771–780.
318. Kitova EN, Bundle DR, Klassen JS: Partitioning of solvent effects and intrinsic interactions in biological recognition. *Angewandte Chemie-International Edition* 2004, 43:4183–4186.
319. Lemoff AS, Bush MF, Williams ER: Structures of cationized proline analogues: evidence for the zwitterionic form. *Journal of Physical Chemistry A* 2005, 109:1903–1910.
320. Wang WJ, Kitova EN, Sun JX, Klassen JS: Blackbody infrared radiative dissociation of non-specific protein–carbohydrate complexes produced by nanoelectrospray ionization: The nature of the noncovalent interactions. *Journal of the American Society for Mass Spectrometry* 2005, 16:1583–1594.
321. Kitova EN, Bundle DR, Klassen JS: Thermal dissociation of protein–oligosaccharide complexes in the gas phase: Mapping the intrinsic intermolecular interactions. *Journal of the American Chemical Society* 2002, 124:5902–5913.
322. Kobko N, Paraskevas L, del Rio E, Dannenberg JJ: Cooperativity in amide hydrogen bonding chains: implications for protein-folding models. *Journal of the American Chemical Society* 2001,123:4348–4349.
323. Miller SL, Urey HC: Production of some organic compounds under possible primitive Earth conditions. *Journal of the American Chemical Society* 1955, 77:2351–2361.
324. Shapiro R: Prebiotic cytosine synthesis: a critical analysis and implications for the origin of life. *Proceedings of the National Academy of Sciences of the United States of America* 1999, 96:4396–4401.
325. Antoine R, Compagnon I, Rayane D, Broyer M, Dugourd P, Breaux G, Hagemeister FC, Pippen D, Hudgins RR, Jarrold MF: Electric susceptibility of unsolvated glycine-based peptides. *Journal of the American Chemical Society* 2002, 124:6737–6741.
326. Antoine R, Compagnon I, Rayane D, Broyer M, Dugourd P, Breaux G, Hagemeister FC, Pippen D, Hudgins RR, Jarrold MF: Electric dipole moments and conformations of isolated peptides. *European Physical Journal D* 2002, 20:583–587.
327. Antoine R, Broyer M, Dugourd P, Breaux G, Hagemeister FC, Pippen D, Hudgins RR, Jarrold MF: Direct probing of zwitterion formation in unsolvated peptides. *Journal of the American Chemical Society* 2003, 125:8996–8997.
328. Dugourd P, Antoine R, Abd El Rahim M, Rayane D, Broyer M, Calvo F: Molecular dynamics simulations of molecular beam deflection experiments. *Chemical Physics Letters* 2006, 423:13–16.
329. Poulain P, Antoine R, Broyer M, Dugourd P: Monte Carlo simulations of flexible molecules in a static electric field: electric dipole and conformation. *Chemical Physics Letters* 2005, 401:1–6.
330. Wales TE, Engen JR: Hydrogen exchange mass spectrometry for the analysis of protein dynamics. *Mass Spectrometry Reviews* 2006, 25:158–170.
331. Wood TD, Chorush CD, Wampler FM, Little DP, O'Connor PB, McLafferty FW: Gas-phase folding and unfolding of cytochrome c cations. *Proceedings of the National Academy of Sciences of the United States of America* 1995, 92:2451–2454.
332. Rogalewicz F, Hoppilliard Y, Ohanessian G: Structures and fragmentations of zinc(II) complexes of amino acids in the gas phase. I. Electrosprayed ions which are structurally different from their liquid phase precursors. *International Journal of Mass Spectrometry* 2000, 201: 307–320.

333. Wyttenbach T, Paizs B, Barran P, Breci L, Liu DF, Suhai S, Wysocki VH, Bowers MT: The effect of the initial water of hydration on the energetics, structures, and H/D exchange mechanism of a family of pentapeptides: an experimental and theoretical study. *Journal of the American Chemical Society* 2003, 125:13768–13775.
334. Polfer N, Dunbar RC, Oomens J: Observation of zwitterion formation in the gas-phase H/D-exchange with CH_3OD: solution-phase structures in the gas phase. *Journal of the American Society of Mass Spectrometry* 2007, in press.
335. Kalodimos CG, Biris N, Bonvin AMJ, Levandowski MM, Guennegues M, Boelens R, Kaptein R: Structure and flexibility adaptation on non specific and specific protein-DNA complexes. *Science* 2004, 305:386–389.
336. Englander JJ, Del Mar C, Li W, Kim JS, Starnz DD, Hamuro Y, Woods VL: Protein structure change studied by hydrogen-deuterium exchange, functional labeling, and mass-spectrometry. *Proceedings of the National Academy of Sciences of the United States of America* 2003, 100:7057–7062.
337. Komives EA: Protein-protein interaction dynamics by amide H/D exchange mass spectrometry. *International Journal of Mass Spectrometry* 2005, 240:285–290.
338. Xiao H, Hoerner JK, Eyles SJ, Dobo A, Voigtman E, Melcuk AJ, Kaltashov IA: Mapping protein energy landscapes with amide hydrogen exchange and mass spectrometry: I. A generalized model for a two-state protein and comparison with experiment. *Protein Science* 2005, 14: 543–557.
339. Chalmers MJ, Busby SA, Pascal BD, He YJ, Hendrickson CL, Marshall AG, Griffin PR: Probing protein ligand interactions by automated hydrogen/deuterium exchange mass spectrometry. *Analytical Chemistry* 2006, 78:1005–1014.
340. Lifshitz C: A review of gas-phase H/D exchange experiments: the protonated arginine dimer and bradykinin nonapeptide systems. *International Journal of Mass Spectrometry* 2004, 234:63–70.
341. Jorgensen TJD, Gardsvoll H, Ploug M, Roepstorff P: Intramolecular migration of amide hydrogens in protonated peptides upon collisional activation. *Journal of the American Chemical Society* 2005, 127:2785–2793.
342. Freitas MA, Marshall AG: Rate and extent of gas-phase hydrogen/deuterium exchange of bradykinins evidence for peptide zwitterions in the gas phase. *International Journal of Mass Spectrometry* 1999, 183:221–231.
343. Mazurek U, Engeser M, Lifshitz C: Gas-phase H/D exchange of the protonated serine octamer cluster: "Ion ping pong" of populations A and B. *International Journal of Mass Spectrometry* 2006, 249:473–476.
344. Englander SW, Sosnick TR, Englander JJ, Mayne L: Mechanisms and uses of hydrogen exchange. *Current Opinion in Structural Biology* 1996, 6:18–23.
345. Wyttenbach T, Bowers MT: Gas phase conformations of biological molecules: the hydrogen/deuterium exchange mechanism. *Journal of the American Society for Mass Spectrometry* 1999, 10:9–14.
346. Green MK, Lebrilla CB: Ion-molecule reactions as probes of gas-phase structures of peptides and proteins. *Mass Spectrometry Reviews* 1997, 16:53–71.
347. Green MK, Lebrilla CB: The role of proton-bridged intermediates in promoting hydrogen/deuterium exchange in gas-phase protonated diamines, peptides, and proteins. *International Journal of Mass Spectrometry* 1998, 175:15–26.
348. Evans SE, Lueck N, Marzluff EM: Gas phase hydrogen/deuterium exchange of proteins in an ion trap mass spectrometer. *International Journal of Mass Spectrometry* 2002, 222:175–187.
349. Campbell S, Rodgers MT, Marzluff EM, Beauchamp JL: Deuterium exchange reactions as a probe of biomolecule structure. Fundamental studies of gas phase H/D exchange reactions of protonated glycine oligomers with D_2O, CD_3OD, CD_3CO_2D, and ND_3. *Journal of the American Chemical Society* 1995, 117:12840–12854.

350. Sinnige CG, de Konning LJ, Nibbering NMM: Water molecule assisted proton mobility in gaseous protonated GlyPro and ProGly. *International Journal of Mass Spectrometry* 2000, 196:115–119.
351. Harrison AG, Yalcin T: Proton mobility in protonated amino acids and peptides. *International Journal of Mass Spectrometry* 1997, 165:339–347.
352. Harrison AG, Young AB: Fragmentation reactions of deprotonated peptides containing proline. The proline effect. *Journal of Mass Spectrometry* 2005, 40:1173–1186.
353. Rozman M: Aspartic acid side chain effect – experimental and theoretical insight. *Journal of the American Society for Mass Spectrometry* 2007, 18:121–127.
354. Deng Y, Smith DL: Rate and equilibrium constants for protein unfolding and refolding determined by hydrogen exchange-mass spectrometry. *Analytical Biochemistry* 1999, 276: 150–160.
355. Takats Z, Schlosser G, Vekey K: Hydrogen/deuterium exchange of electrosprayed ions in the atmospheric interface of a commercial triple-quadrupole mass spectrometer. *International Journal of Mass Spectrometry* 2003, 228:729–741.
356. Suckau D, Shi Y, Beu SC, Senko MW, Quinn JP, Wampler FM, McLafferty FW: Coexisting stable conformations of gaseous protein ions. *Proceedings of the National Academy of Sciences of the United States of America* 1993, 90:790–793.
357. Gard E, Green MK, Bregar J, Lebrilla CB: Gas-phase hydrogen/deuterium exchange as a molecular probe for the interaction of methanol and protonated peptides. *Journal of the American Chemical Society of Mass Spectrometry* 1994, 5:623–631.
358. Bouchoux G: Evaluation of the protonation chemistry obtained by the extended kinetic method. *Journal of Mass Spectrometry* 2006, 41:1006–1013.
359. Heck AJR, Jorgensen TJD: Vancomycin in vacuo. *International Journal of Mass Spectrometry* 2004, 236:11–23.
360. Badman ER, Hoaglund-Hyzer CS, Clemmer DE: Monitoring structural changes of proteins in an ion trap over similar to 10–200 ms: Unfolding transitions in cytochrome c ions. *Analytical Chemistry* 2001, 73:6000–6007.

– 4 –

CASE STUDIES

– 4.1 –

Nucleobases, Nucleosides and Oligonucleotides

GENERAL FEATURES

Deoxyribonucleic acid (DNA) carries the genetic information in all cells and in many viruses. A long time elapsed between the discovery of DNA in cells (1869), the first connection of DNA to the inheritance of traits (1944) until the first description of its molecular structure (Figure 4.1.1) by Watson and Crick (1953) [1]. They showed that DNA is constituted of a right-handed helix with two strands that are sequences of nucleotides chosen among four molecules: thymine (T), adenine (A), cytosine (C) and guanine (G) [2, 3]. Those molecules are matched as hydrogen-bonded base pairs, either A—T or G—C, linking the two strands [4–7]. Each base is linked to a sugar forming a nucleoside and in each strand nucleosides are linked by phosphates (Figure 4.1.1). The choice made by Nature of a five-atom ring sugar (pentose) instead of a more common six-atom ring (hexose) such as glucose is explained by the steric problems encountered with the latter choice that would have led

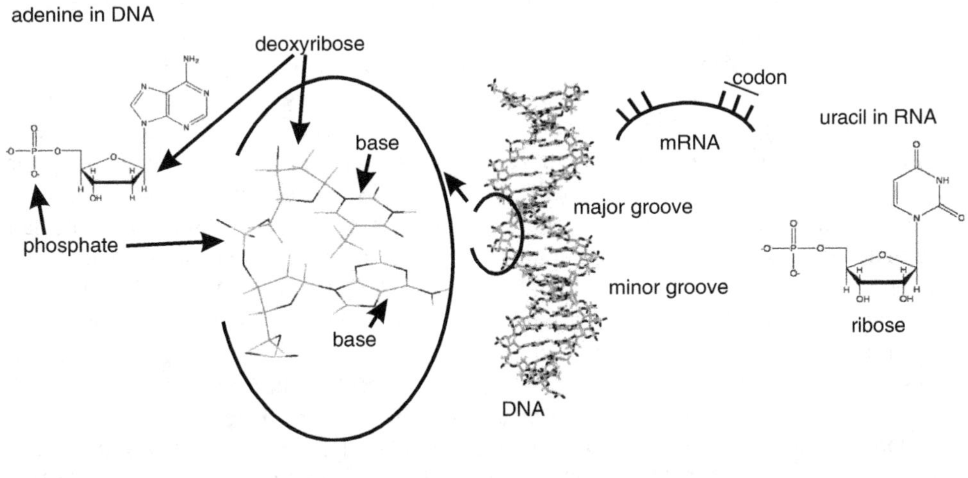

Figure 4.1.1 DNA structure and transcription of DNA into messenger RNA. Note the difference between the deoxyribose (adenine in DNA, left) and the ribose (uracil in RNA, right) in position 2 of the sugars.

to unstable base pairing [8]. In messenger and ribosomal RNA (rRNA) (see Chapter 4.8), thymine is replaced by uracil and bases are linked by a ribose. The collection of bases T, A, C, G which describes a DNA sequence is the primary structure or genome. The base pairing of the two DNA strands is the DNA secondary structure.

DNA sequences adopt different high-order conformations (or tertiary structures) called A, B and Z [9–11]. Form B [12] is the most usual and corresponds to 10 base pairs per turn. Dehydrated DNA often adopts the A form with 11 base pairs per turn. Z DNA is a left-handed helix and methylation of the C bases can induce a B → Z transformation. A full description of DNA taking into account local distorsions can be found in reference [13]. Since the degree of humidity plays a great role, it is not obvious that helices survive in absence of any solvent or crystal packing effect.

Among the whole DNA, sequences of nucleobases constitute genes. The role of the DNA spatial packaging upon gene expression is an active field of research [14]. The term chromatin corresponds to the complex of DNA and proteins and covers a wide level of organization of DNA [15]. Short DNA segments are wound around spools made of histone proteins to form nucleosomes, lowest level of compaction to fit meters of DNA into cell nuclei. The chromatin fibre is made of long arrays of nucleosomes that coil into higher order structures which are still not perfectly understood. The spatial folding of the chromatin fibre is tuned by molecular interactions of the histones that either permit or forbid expression of the genes. An intriguing question is how cells "know" when they must divide. The status of a cell seems to be signalled to chromatin by means of marks such as methyl or acetyl groups on histone tails that are then recognized by proteins that regulate how the genetic information has to be processed.

Expression of the genes constitutes the so-called central dogma of molecular biology. Only a fraction of DNA is transcripted into messenger mRNAs (ribonucleic acid) that are further translated into proteins. Among the 3×10^9 base pairs of human DNA, only a small fraction of 150×10^6 base pairs constitute the genetic information dispersed among expressed sequences of a few thousands base pairs representing the genes. Fourty-seven years after the discovery of the double helix, the Human Genome Project announced the initial sequencing of the human genome [16].

Drug–DNA binding can be studied by means of mass-spectrometry [17–21] and modelling [22] (see Chapter 4.5). In order to turn on or off genes [23], a long-standing therapeutic goal is the search for drugs specifically binding to target DNA sequences [18, 24, 25]. Artificial polyamides containing pyrrole and imidazole establish very specific binding and can, for example distinguish between A—T and T—A pairs [26–28]. Those polyamides are designed using modelling and weak interactions between nucleobases and pyrrole or imidazole can be studied in the gas-phase [29]. Mimicking the great ability of complementary nucleobases to recognize between themselves through Watson–Crick (WC) pairs (A with T, G with C), peptide nucleic acids (PNA) have been synthesized. PNAs are composed of WC pairs linked to a neutral, achiral 2-aminoethyl glycine backbone [30]. PNAs form more stable double helices with DNA than DNA itself and may thus be interesting therapeutic tools [31, 32].

In fact, the so-called human genome is an averaged genome and there are, for each human being, DNA sequences variations, called single nucleotide polymorphisms or SNPs, that occur when a single nucleotide (A, T, C, or G) is altered. For a variation to be considered a SNP, it must occur in at least 1% of the population. SNPs make up ~90% of all human genetic variation and occur every 100–300 bases [33, 34]. SNPs can occur in both coding (gene) and non-coding regions of the genome and they can be studied by mass-spectrometry [35]. Many of them have no effect on cell function, but it is believed that some could predispose people to diseases or influence their response to a drug. Since several codons correspond to a given amino acid (the genetic code is degenerate), SNPs can be considered either as "effective" or "silent". In this last case, although a codon is replaced by another one, there is no change of amino acid and the protein sequence remains intact. At first sight, it might thus seem that silent SNPs should not be discernable. Nevertheless, they can influence protein folding [35] (see Section 4.2.6). This is due to the non-uniform rate of translation of messenger RNAs in the ribosome (see Chapter 4.8). This rate is proportional to the different usages of codons. Consequently, a silent SNP leading to an unusual amino acid will induce a slowing down of translation. Protein folding occurring simultaneously with translation ("co-translational") might thus be affected and lead to a different structure. Other important polymorphisms are insertions or deletion (indels) of nucleotides from genes. Relating SNPs and individualized drugs opens the route for pharmacogenomics and pharmacogenetics [36, 37] (Figure 4.1.2).

The knowledge of the genome can be used to predict all the proteins that a cell is capable of producing by translation of mRNAs but more important is the proteome, that is the actually expressed proteins at a specific time. The link between genes and proteins is not a simple one gene/one protein relationship, due to the existence of post-translational modifications (PTMs), determined by mass-spectrometry (see Section 3.2) [38, 39] and is better described by fuzzy logic [40, 41]. A PTM corresponds to the alteration of the primary structure of a protein after it has been translated and is already folded [42]. A wide range of modifications can take place and they act on individual residues either by cleavage at specific points, deletions, additions or having the side-chains converted or modified. The most

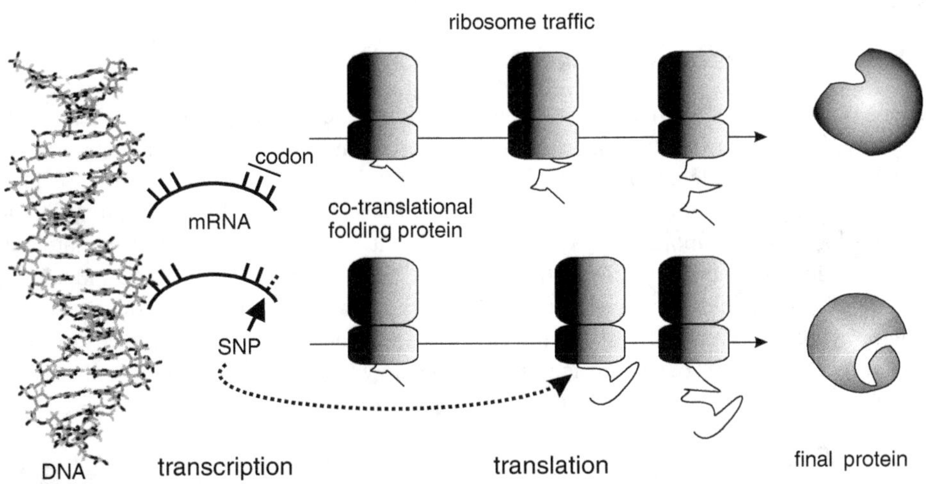

Figure 4.1.2 Schematic representation of translation of a messenger RNA into a folding protein. Each codon is recognized by a transfer RNA (not represented) and an amino acid is added to the nascent protein. This protein folds during the course of translation. If a codon is modified by an SNP, the recognition process can be slowed down and a different protein fold can occur leading to a different final protein structure.

common modification is cleavage of the initial methionine (Met) residue that signals the start of the protein.

DNA replicates by unzipping the double helix. The bases become exposed and both strands act as templates [43]. The structure of DNA in solution is the result of a balance between, on one side, stabilizing non-covalent forces that are the hydrogen bonds linking bases in complementary A—T and G—C pairs and the stacking interactions in between adjacent bases in the same strand as well as hydrophobic forces (see Section 5.3) and, on the other side, destabilizing forces due to the electrostatic repulsion between phosphate groups belonging to each strand [44]. DNA double helices are thus strongly favoured enthalpically by the bonding interactions and are disfavoured entropically by the restrictions imposed to the flexible sugar-phosphate backbone [45].

Structural gas-phase studies presented in the next paragraphs allow the detailed investigation of the different contributions in absence of backbone constraints or hydration effects [46]. Quantum chemistry and spectroscopic studies concerning isolated bases and base pairs mostly focus on hydrogen bonding and stacking interactions [12, 47]. The problem of helix formation in rather short double-strand oligomers that has been first studied by means of X-crystallography and modelling [48] is considered in the gas-phase by means of the ion-mobility (3.3.1) and BIRD (3.3.2) techniques. Finally, mass-spectrometric studies of quadruplex formation in telomers and modelling of RNA are described.

4.1.1 Isolated nucleobases

The pioneer experimental gas-phase studies of nucleobases have been conducted by the groups of Sukhodub [49–50], Herschbach [51] and Levy [52] followed by several groups [53–67].

4.1.1.1 Tautomers of nucleobases

Nucleobases can also exist under other tautomeric forms due to the mobility of their hydrogen atoms [68, 69]. The relative energies of the different tautomers of a given nucleobase are dependent on the used level of theory [70] (Figure 4.1.3).

Tremendously large numbers of tautomers should, in principle, be taken into account. Programs have been elaborated to systematically built those tautomers and calculate their properties allowing to consider respectively 499 and 625 tautomers for guanine and adenine [71]. In fact, one should only keep in mind that only tautomers with reasonable Boltzmann populations are experimentally observed. However, attribution of spectral features to calculated tautomer structures can be difficult as shown below in the case of guanine and phenylalanine (see Section 5.5.3).

4.1.1.1.1 Guanine

Guanine that is mostly in its amino-oxo (N9-H keto) form (Figure 4.1.6) in DNA can also exist in other keto or amino-hydroxy (enol) forms (Figure 4.1.6). LIF, hole-burning R2PI and IR/R2PI (see Section 2.1.4) [72, 73], supported by calculations conducted at the MP2 level [74], have first shown that in a cold supersonic beam, four tautomers are observable. Since in DNA, guanine is attached to the backbone through the N9 atom, it might be tempting

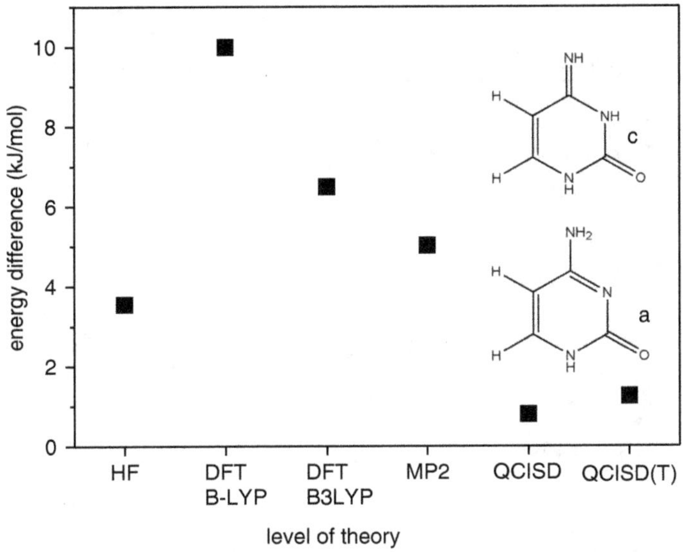

Figure 4.1.3 Influence of the level of theory upon the energy difference between the non-aromatic 2-oxo (a) and the 4-imino (c) cytosine tautomers.

to lock this hydrogen-bonding site by methylation. However, it then turns out that in the corresponding 9-Me-guanine, the keto form which is biologically relevant is not observed but rather the enol form. The problem of assignments of guanine tautomers in R2PI and IR/R2PI has been solved [75] when the results of those studies have been compared with those obtained in helium cluster deposition experiments [76]. The most stable isomers are observed in helium cluster experiments (see Section 3.1.3) because of the extremely rapid cooling process that takes a "snapshot" of the initial distribution obtained during the guanine sublimation process at 350°C. In contrast, REMPI experiments use supersonic expansion cooling (see Section 3.1.1) followed by photoexcitation in the nanosecond range. A tautomer-dependent ultrafast process (see Section 4.1.1.2) taking place in the femto or picosecond range relaxes the most stable tautomers that become unobservable. Only highly excited tautomers are then observed in REMPI experiments. The four tautomers observed by R2PI are thus not those of lowest energy but rather less stable ("rare") ones which stabilities lie in the 12–30 kJ/mol range [75]. In order to evaluate the relative populations of the different tautomers, what really matters are relative free energies that are respectively equal (in kJ/mol) to −1.7, −0.1, 2.5 and 3.5 for the G7K, G9K, G9Ea and G9Eb [76]. A comparison between calculated spectral features of tautomers supposedly populated according to those free energies and experimental spectra proves that indeed helium cluster deposition leads to correct assignments (Figure 4.1.4).

4.1.1.1.2 Cytosine

The amino-oxo and amino-hydroxy tautomers of cytosine have been studied by photoelectron spectroscopy of the corresponding dipole-bound anions [58]. In isolated nucleobases, intramolecular hydrogen bonding does not take place. The situation is quite different in nucleotides.

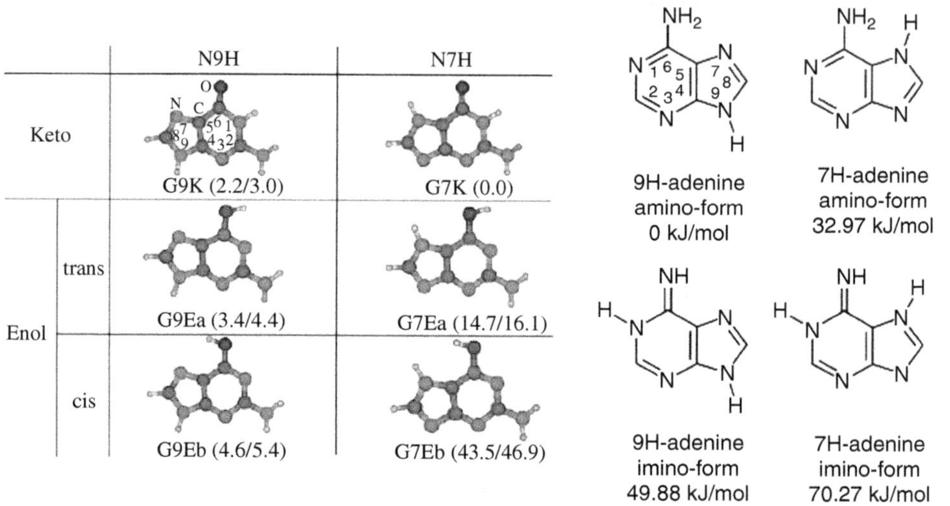

Figure 4.1.4 Left: Most stable guanine tautomers. Energies (in kJ/mol, with and without zero-point energy correction) are calculated at the MP2/aug-cc-pVDZ level and given relative to that of the G7K form. The keto (amino-oxo) and enol (amino-hydroxy) forms are classified according to the NH_2 (amino), C═O(oxo) and hydroxyl (O—H) groups in positions 2 and 6. The enol forms have two rotational orientations a (*trans*) and b (*cis*) of the OH group with respect to the five-member ring (reprinted with permission from reference [76] ©2006 American Chemical Society). Right: Most stable adenine tautomers calculated at the MP2/6-311G(d,p) level (reprinted with permission from reference [77] ©2001 Royal Society of Chemistry).

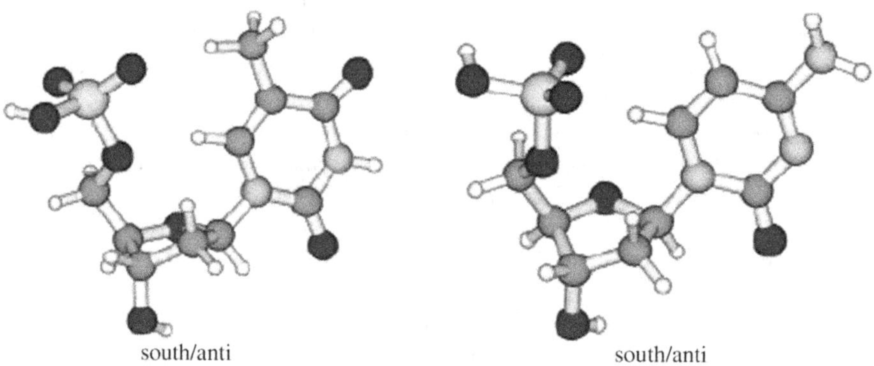

Figure 4.1.5 Most stable conformers of isolated canonical 2′-deoxyribonucleotides thymidine-5′-phosphate (pdT, left) and cytidine-5′-phosphate (pdC, right) calculated at the B3LYP/6-31++G(d,p) level of theory (reprinted with permission from reference [78] ©2006 American Chemical Society).

The different conformers of the canonical 2′-deoxyribonucleotide thymidine-5′-phosphate (pdT) have been considered and their analysis reveals the existence of intramolecular O—O and C—H···O bonds that significantly influence the equilibrium conformations and relative stabilities of the conformers [78] (Figure 4.1.5).

4.1.1.1.3 Adenine

Adenine can in principle exist under 14 different tautomeric forms: four amino tautomers 9H, 7H, 3H and 1H, eight imino tautomers, 1H/7H, 9H, 3H/7H and 9H. Each of these last forms possesses two stereoisomers Z and E. Two zwitterionic forms can also exist. The most stable form is the canonical 9H tautomer (the nitrogen atom N9 is linked to the sugar in adenosine). Tautomer 7H is rather close, in between only 0.2 and 0.4 eV above 9H in isolated adenine and 0.1 eV above 9H in adenine–water clusters. All other tautomers are lying 0.5 eV above tautomer 9H.

4.1.1.2 Excited state behaviour of nucleobases

DNA exhibits a maximum of absorption around 260 nm and a large amount of photo-induced chemical reactions should take place in excited states [79]. In fact, the genetic information is not too sensitive to solar radiation although formation of covalently bound thymine dimers is an important triggering factor of skin lesions [80, 81]. It seems that Evolution has selected nucleobases among molecules and their tautomers among those possessing the fastest deactivation mechanisms [82, 83]. Several mechanisms have been invoked to explain the extremely short lifetimes of the five nucleobases. One of these mechanisms [84] invokes proton transfer along the N—H···N bonds in base pairs (see below). However, this mechanism requires the presence of a double-strand structure that is ubiquitous in DNA but not in RNA. Since RNA may have preceded DNA in the course of Evolution (see Section 4.1.4), other mechanisms invoking non-radiative processes in isolated nucleobases have been also proposed. The fate of nucleobases in their excited states is dominated by relative energies and couplings of the initially excited $\pi\pi^*$ states and the nearby $n\pi^*$ or $\pi\sigma^*$ states (see Section 1.7). A sizeable number of studies have been devoted to calculations of potential energy surfaces in order to interpret non-radiative processes [85–95]. In parallel, femtosecond experiments have probed competing pathways by means of different methods [81, 96–101].

4.1.1.2.1 Neutral adenine

Among nucleobases, adenine has been one of the most widely studied thanks to its UV spectrum into which regions exhibit sharp lines, its restricted number of accessible tautomers and its reasonable ease of vaporization as compared with guanine. It will be the only considered nucleobase in the following overview. An overview of UV absorption of adenine in the first excited state region is provided by fluorescence and R2PI experiments [61, 77, 102]. As displayed in Figure 4.1.6, well-defined spectral lines are observed for excitation energies up to 36,700 cm^{-1} while a broad unresolved spectrum is observed above. The spectral region between 36,000 and 36,700 cm^{-1} can be attributed to the most stable 9H tautomer (see Figure 4.1.4). The observed lifetime of adenine decreases when the amount of energy deposited above the lowest identified origin 0–0 line increases. Excitation near threshold at 277 nm [103] corresponds to a rather long lifetime of 40–50 ps while excitation at 250 nm corresponds to lifetimes of only 110 and 1.5 ps [97]. This can be interpreted as due to the presence of a small energy barrier. Excitation below this barrier leads to sharp lines while excitation above this barrier opens a very fast relaxation channel.

Femtosecond experiments [81, 84, 97, 99, 100, 104] supported by calculations provide a picture of the de-excitation processes. Transients represented in Figure 2.2.1 can be interpreted with the scheme displayed in Figure 4.1.7. Following excitation, the $\pi\pi^*$ L_b state

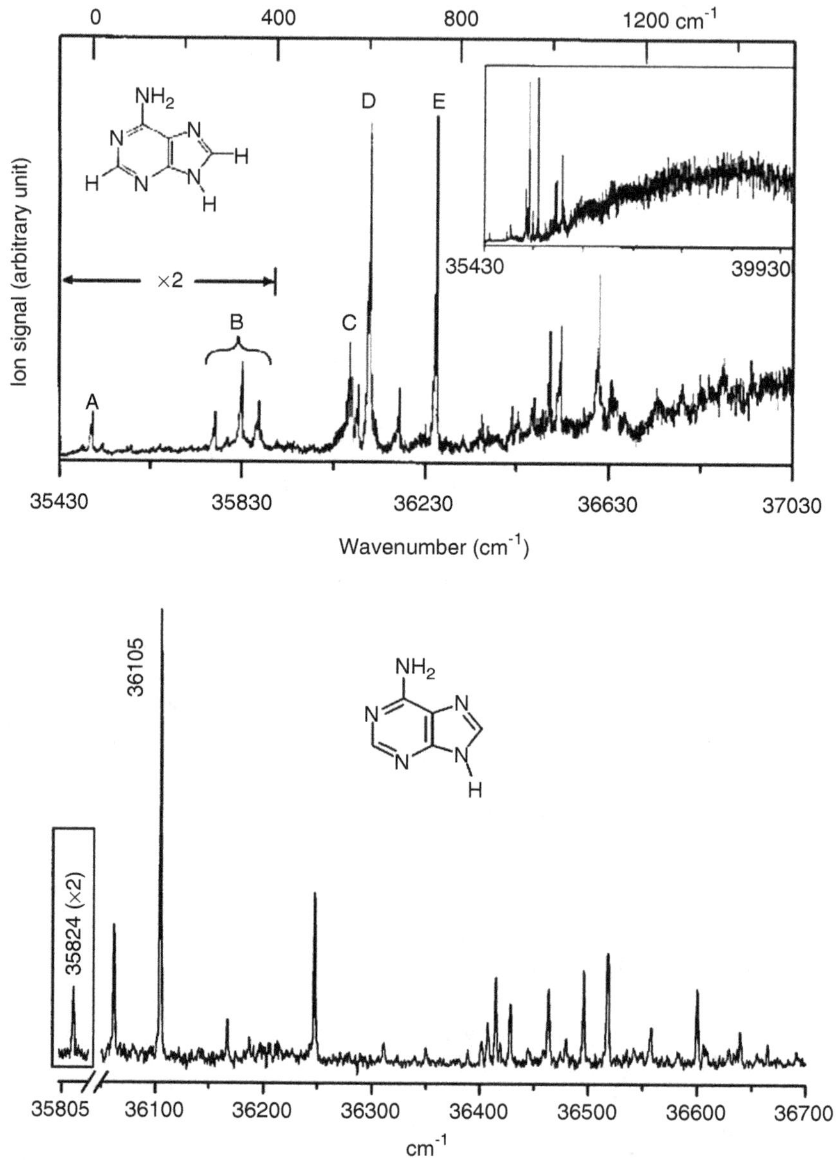

Figure 4.1.6 Top: R2PI spectrum of adenine displaying sharp lines at excitation wavelengths close to the electronic threshold and an unresolved background. The upper scale shows the amount of deposited excess energy (reprinted with permission from reference [61] ©2000 American Institute of Physics). Bottom: R2PI spectra of the 9H tautomer of adenine (reprinted with permission from reference [77] ©2001 Royal Society of Chemistry).

relaxes within 100 fs towards its minimum energy nuclear geometry $(L_b)_{Min}$. This state has a conical intersection $(L_b/n\pi^*)_X$ with a $n\pi^*$ state. Analysis of photoelectron transients show that this $n\pi^*$state introduces a time constant of 1 ps. Two other conical intersections $(L_a/n\pi^*)_X$ and $(L_a/L_b)_X$ are responsible for the existence of competing pathways leading from the initially

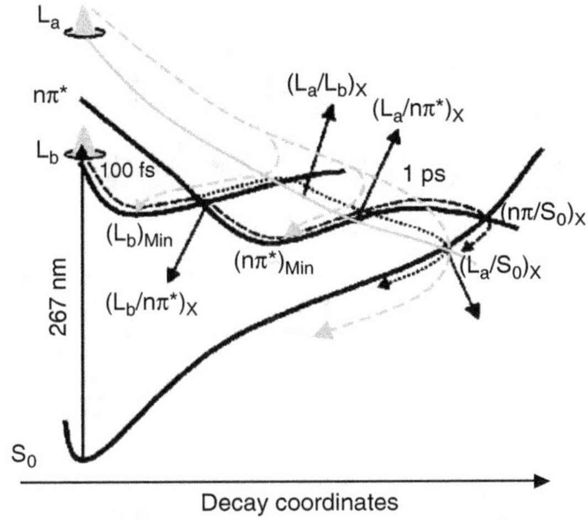

Figure 4.1.7 Qualitative representation of relaxation channels of electronically excited neutral adenine. The decay coordinates involve out-of-plane vibration modes. The photo-excited L_b state relaxes mostly to the $n\pi^*$ state through a small barrier. In turn, this $n\pi^*$ state relaxes to the S_0 ground state through high barriers either directly ($(n\pi^*/S_0)_X$ conical intersection) or through the L_a state $(L_a/S_0)_X$ conical intersection) (reprinted with permission from reference [100] ©2006 Royal Society of Chemistry).

excited L_b state back to ground state S_0. At low excitation energies (corresponding to excitation wavelengths larger than 265 nm), photoexcitation of adenine leads to highly vibrationally excited molecules in ground state S_0 [98]. At larger excitation energies, observation of hydrogen atoms can be attributed to the opening of another channel due to the existence of a higher $\pi\sigma^*$ excited state dissociative along the N—H coordinate (Figure 1.7.3) [105].

4.1.1.2.2 Protonated adenine

Protonation of molecules containing chromophores strongly modifies their electronic spectrum when the protonation site belongs to the chromophore itself. On the contrary, only minor changes are observed when the protonation site is not localized on the chromophore, for example in the case of tryptophan [106] (see Chapter 4.2). In protonated adenine, the most stable tautomer 1H-9H-adenine corresponds to protonation of the nitrogen atom in position 1 [90]. The onset of the electronic spectrum recorded by observation of photo-fragmentation of trapped ions at relatively high temperature (>300°K) presents a slow rise and a broad substructure characteristic of a very short lifetime (Figure 4.7.8). Indeed, calculations suggest an ultrafast barrier-less relaxation of the $n\pi^*$ and $\pi\pi$ states through a conical intersection with the S_0 ground state (Figure 4.1.8).

4.1.1.3 Non-dissociative interactions of nucleobases with low-energy electrons

Canonical nucleobases have large dipole moments and can thus give birth to both valence or multipole-bound anions (see Section 2.3.1 and Figure 4.1.9) [107–110]. The biological importance of electron interactions with DNA in the field of radiation chemistry [111] has prompted a large number of non-dissociative and dissociative (see Section 4.7.1) electron-attachment

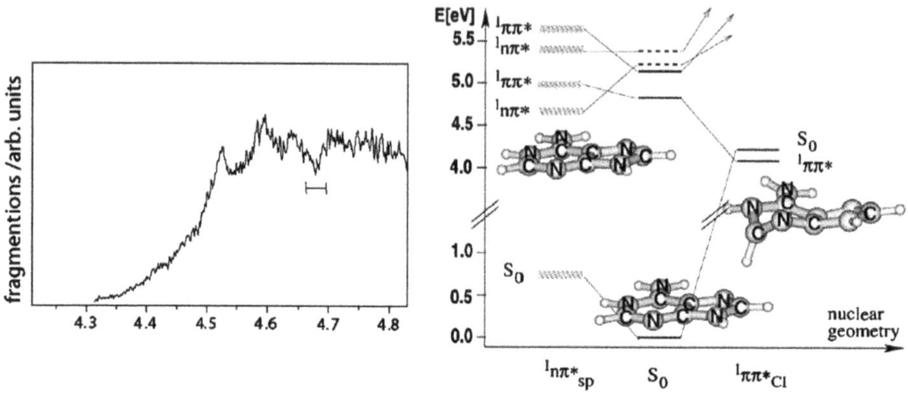

Figure 4.1.8 Left: Fragmentation spectrum of photo-excited protonated adenine ions produced by electrospray and confined in a quadrupole trap. Right: Relaxation pathways of photo-excited protonated adenine calculated by TD-DFT with valence triple zeta plus d,p polarization (TZVP) basis set optimization followed by single-point DFT/MRCI calculations. The singlet $\pi\pi^*$ vertical photoexcitation is calculated at 4.80 eV and the $(\pi\pi^*/S_0)_X$ conical intersection at 4.14 eV (reprinted with permission from reference [90] ©2005 Royal Society of Chemistry).

experimental and theoretical studies. Only non-dissociative attachment is here considered and a detailed review is given in reference [112]. Vertical attachment affinities (Figure 4.1.9) of isolated nucleobases have been determined by electron transmission spectroscopy and are highly negative (Table 4.1.1) [60].

Multipole-bound electron affinities of thymine, adenine and uracil have been rather accurately predicted by L. Adamowicz [113] before the experimental measurements by several groups [58, 107, 108]. The values provided by different experimental techniques are in very good agreement between themselves and with theoretical predictions (Table 4.1.3). Methylation reduces those electron affinities by 40 meV for 1,3-dimethyl-uracil and 45 meV for 1-methyl-thymine [114]. Multipole-bound anions of guanine have not been observed experimentally but it is predicted that the G and GHN7H tautomers support dipole-bound states with respective electron affinities of 65 and 36 meV [115]. Similarly, xanthine (X), a degradation product of nucleobases [116] has a predicted dipole-bound electron affinity of 41 meV [117].

The situation is very different for valence electron affinities. According to the used method and basis sets (see Section 1.2), ab initio predictions of valence electron affinities of the five nucleobases (including uracil) vary considerably (Table 4.1.2) [112, 118, 120, 125, 126]. At the CCSD(T) level of theory, adiabatic valence electron affinities of all canonical nucleobases are negative. However, except cytosine, all of them possess tautomers with positive adiabatic electron affinities [120]. This may give one of the clues for explaining differences between predictions and experimental observations. Valence anions have been experimentally observed in electron attachment to uracil and thymine complexed with rare gas atoms, followed or not by evaporation [107]. A stabilization process against electron auto-detachment seems to be necessary to observe those valence anions. This stabilization process can be brought by either three-body collisions [56], presence of an atomic ionic core [121] or

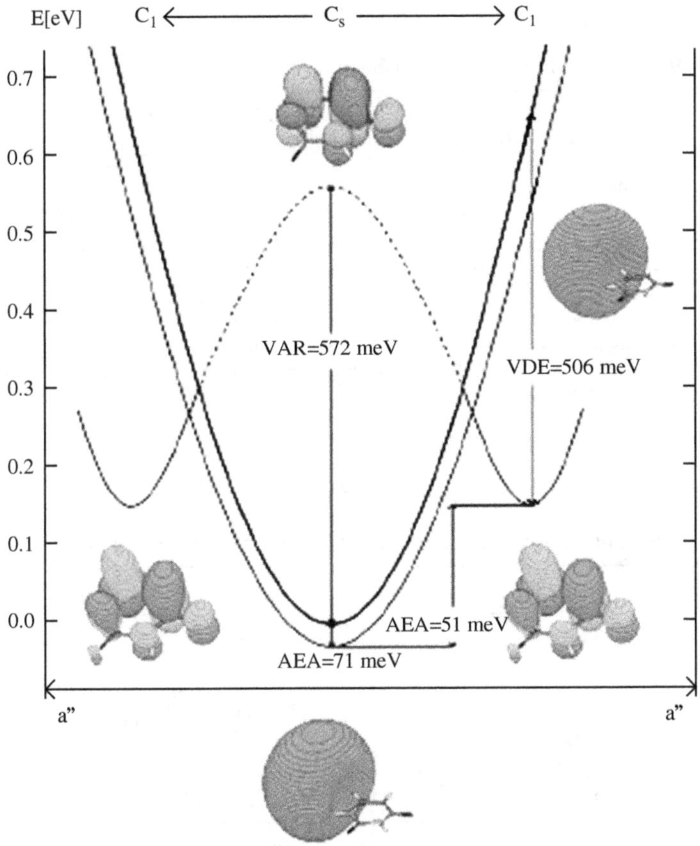

Figure 4.1.9 Uracil potential energy curves and anion excess electron orbitals in the multipole-bound state (planar geometry C_s) and the valence state (puckered geometry C_1) (reprinted with permission from reference [109] ©2005 Royal Society of Chemistry).

Table 4.1.1

Experimental time-constants of transients observed following photoexcitation of nucleobases

Nucleobase	Adenine	Thymine	Uracil	Guanine	Cytosine
τ_1 (fs)	100	105	130	148	160
τ_2 (ps)	1.1	5.1	1.0	0.36	1.86

Note: The extremely short lifetime τ_2 of guanine.

evaporation of helium atoms in the case of free electron attachment to thymine deposited on helium clusters [127].

While valence electron affinities of the bare canonical bases are negative but rather close to zero, those of their radicals are very large, in between 2 and 3 eV [128] as well as those of nucleosides and base pairs that are much more biologically relevant than isolated bases [111, 129].

Table 4.1.2

Valence electron affinities of isolated canonical nucleobases (in eV)

Nucleobase	ETS [60] EA_v	DFT [118] EA_{adia}	DFT [119]	CCSD(T) [120]	RET [121] EA_v	PES [58] EA_{adia}
G	−0.46	−0.75	−0.07	−0.459	–	–
A	−0.54	−0.35	−0.28	−0.038	−0.45	–
C	−0.32	−0.05	0.03	−0.188	−0.55	0.13 ± 0.12
T	−0.29	0.22	0.20	−0.087	−0.30	0.12 ± 0.12
U	−0.22	0.20	–	−0.051	−0.30	–

Note: The electron transmission spectroscopy (ETS) vertical electron affinities are directly obtained from isolated nucleobases. The Rydberg electron transfer (RET) and photoelectron spectroscopy (PES) values are deduced from the solvation method (see Section 2.3.2.3) and are in between vertical and adiabatic electron affinity values.

Table 4.1.3

Dipole-bound electron affinities of isolated nucleobases (in meV)

Nucleobase	Predicted	PES [56]	PES [58]	RET
G	34 [122]			
A	0.6 [113]			11
C			85 ± 8 amino-hydroxy; 230 ± 8 amino-oxo	
T	88 [123]	69 ± 0.007	69	68
U	86 [124]	93 ± 0.007	93	50
X	41 [117]			

Note: Values given in the second column have been predicted without any prior knowledge of the experimental values. For guanine, the predicted value corresponds to the N(9)H amino-oxo tautomer.

Table 4.1.4

Ionization potentials of nucleobases (in eV) obtained from photoionization spectroscopy with synchrotron radiation, mass-analysed threshold spectroscopy (MATI) or calculations

Nucleobase	U	T	C	A	G
IP_{ad}	9.15[a]	8.9178 (10)[b]	8.08[c]	8.20[a]	7.77[c]

Note: The influence of the level of calculation upon prediction can be found in reference [132].
[a]Photoionization spectroscopy with synchrotron radiation [134].
[b]MATI [135].
[c]Calculations [130, 131].

In contrast with electron affinities, ionization potentials can be accurately predicted [130–133] and measured (Table 4.1.3) [134, 135]. The DFT UB3LYP/6-311+G(d,p) value is only 0.113 eV lower than the experimental adiabatic ionization value measured through observation of the $0\text{–}0^+$ band for thymine [135] (Table 4.1.4).

Proton affinities and gas-phase basicities have been calculated at the DFT level up to the QCISD and QCISD(T) levels with large 6-311+(3df,2P) basis sets [136]. For neutral uracil, the di-oxo tautomer is the only populated species. Protonation can, in principle, take place at any C, N or O atom and a large number of tautomers are possible. At any level of theory, the most stable protonated uracil tautomer corresponds to protonation on the N1 atom in the di-hydroxy-pyrimidine form (and not the di-oxo). This tautomer can thus be formed from several neutral tautomers but not from the most stable one. The predicted proton affinity is 858 kJ/mol at the highest calculation level. It must be noted that predicted proton affinities can strongly depend on the level of theory.

Nucleosides and nucleotides have been much less studied in gas-phase than isolated bases [53, 79, 137] and will be considered in Chapter 4.7.

4.1.2 Base pairing: hydrogen-bonded, stacked and wobble pairs

In biological cells, enzymes that synthesize DNA (DNA polymerases) incorporate the correct nucleotide opposite each of the four template bases with the correct pairing configuration called WC base pairing [43]. DNA in crystal or in solution adopts a relatively small number of pair structures. For example, the Hoogsteen geometry [138, 139] is the most favourable for A—T base pairs in solution [140, 141]. In gas-phase, the absence of the double helix backbone and of any solvent surrounding offers a wide variety of possibilities for pairing nucleobases [142]. Base pairing has been first theoretically studied [143] and a first repertoire of 29 possible natural DNA base pairs has been identified [144]. This repertoire has now been extended to unnatural synthesized base pairs [145].

A problem arises during comparisons between quantum calculations and experimental base pair structures determined from X crystallography [146]. Since the calculated hydrogen bond length values in the A—T and G—C WC pairs (N6—O4 and N1—N3 distances in A—T, O6—N4, N1—N3, N2—O2 distances in G—C) cannot be reconciled with crystal values by only considering the isolated bases, the base with their sugars or even the full 5′-ribose-monophosphate. A good agreement is only obtained by simulation of the DNA environment, that is by taking into account counterions Na^+ and water molecules present in the crystal. In other words, great care must be taken when comparing condensed-phase results and "gas-phase" calculations.

Gas-phase experiments, on the contrary, are directly comparable with quantum calculations [147] but face the following problem. Due to the large number of pairing possibilities, the biologically relevant pairs are not necessarily among the most energetically favoured structures and thus may not be experimentally observable. The G—C and G—G pairs are among the most stable possible pairs. The lowest energy configurations of the G—C pair (Figure 4.1.10) correspond to hydrogen-bonded structures and have first been studied by means of a very original and subtle mass-spectrometry experiment [49, 148]. The base pairs were produced in vapours obtained by thermal evaporation from Knudsen cells. Assuming thermal equilibrium, the base pairing enthalpies were then deduced by monitoring the pair decomposition rates as a function of temperature in Arrhenius plots. The ionization process was field-detachment performed by setting at very high-voltages a needle situated in the vapour and then bravely extrapolating the results to zero-field. Those studies provided the only estimates of base pairing enthalpies and are still used by theoreticians to validate their calculations [149]. With widely tunable synchrotron radiation [134] and supersonic beams,

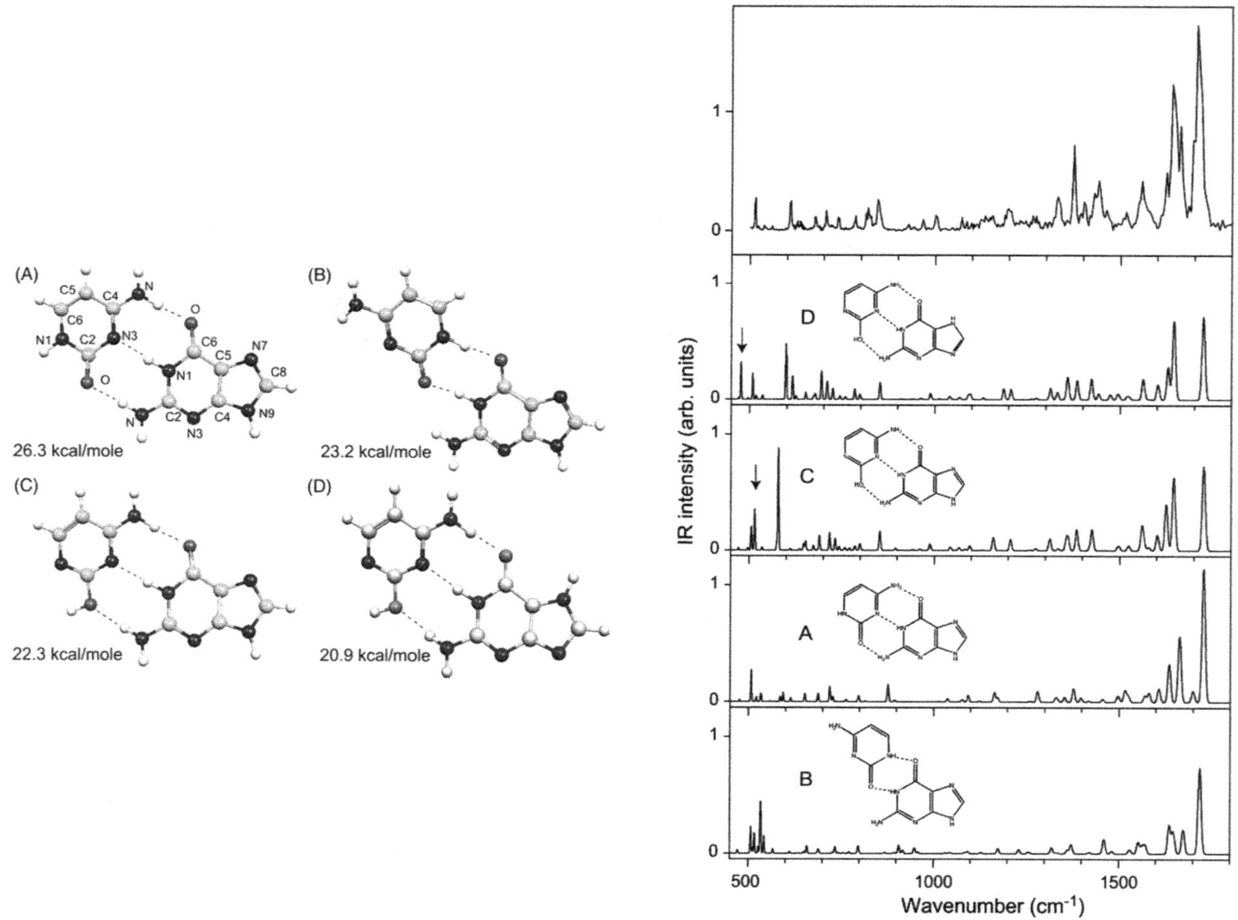

Figure 4.1.10 Left: Lowest energy configurations of the G—C pair. Right: Comparison between the mid-infrared spectrum of jet-cooled G—C pairs obtained from an UV/IR depletion experiment using a free electron laser and theoretical IR spectra of the four most stable configurations of the G—C pair predicted from RI-MP2/cc-pVDZ calculations. The best agreement is obtained for structures C and D and not for the most stable Watson–Crick pair A (reprinted with permission from reference [153] ©2004 Royal Society of Chemistry).

it might be possible to improve the accuracy of those field-detachment experiments and possibly have some isomer selectivity. For example, the ionization potential of guanine $IP_{ad}(G)$ is known (Table 4.1.3) and has some dependence on tautomerization. The measurement of the adiabatic ionization potential $IP_{ad}(G—C)$ of the G—C pair and/or the threshold value $AP(G^+)$ for apparition of G^+ fragments from jet-cooled G—C pairs would then lead to the dissociation energy $D_e(G—C)$ of the G—C pair through the relationship [150].

$$D_e(G—C) = AP(G^+) - IP(G) \tag{4.1.1}$$

Field-detachment mass-spectrometric studies [50] do not provide any information concerning base tautomerization. The answer comes from confrontations between high-level calculations (RI-MP2/cc-pVDZ) and spectroscopic experiments [151–153]. The PES of the G—C pair has been explored from 16 different conformations obtained by pairing the four most stable tautomeric forms of G and C. The most stable structures respectively corresponding to the global and first minima are the nearly planar canonical WC pair A and structure B. They are followed by the non-planar structures C and D respectively lying 16.7 and 36 kJ/mol above A. In structure C, guanine is in its canonical tautomeric form while cytosine is in an enol form. The mid-infrared spectrum of the G—C pair is displayed in Figure 4.1.10. The best agreement between observed and calculated spectra is obtained for structures C and D.

As shown by IR–UV hole-burning spectroscopy [83], the canonical WC configuration of the G—C pair (structure A in Figure 4.1.10) is not only the most energetically favoured but also exhibits a very broad UV absorption ($\approx$400 cm^{-1}). This in contrast with the other observed G—C structure spectra such as that of the very similar WC structure C containing an enol (instead of keto) cytosine tautomer. A possible explanation of this very broad absorption is a rapid internal conversion [154–156] making the biologically relevant structure very stable against degradation by UV radiation during the prebiotic period (see Section 4.6.1).

The A—T pair can exist in a very large number of configurations [147, 157–159]. As experimentally observed [160], the WC pair (structure 1364 in Figure 4.1.12) is somewhat far away ($\approx$0.7 kJ/mol) from being the most stable in gas-phase. The most stable configuration contains the N3 and N9 atoms of A respectively H-bonded to the N1 and O2 atoms of T (structure 3192 in Figure 4.1.11).

When an electron is added to an A—T pair in its most stable gas-phase configuration, the anion cannot be simply described as a T^- anion (T has the largest electron affinity, see Table 4.1.1) solvated by the neutral adenine. More accurately, as shown by photoelectron spectra experiments (see Section 2.3), a barrier-free proton transfer (BFPT) takes place between A and T. In contrast, when both A and T are methylated, the hydrogen bond pattern becomes identical to the WC pattern in the double helix and the BFPT process does not take place [147].

Several other pairing schemes of nucleobases have been investigated theoretically [12, 149, 160–164]. Calculation of base stacking structures requires high-level of theory. The DFT methods provide acceptable predictions for hydrogen-bonded pairs but fail to locate correct minima of the potential energy surface corresponding to stacked structures. Due to the importance of dispersion in that case (see Section 1.1.3.1), an important improvement for interaction energies and geometries comes from its inclusion in DFT (see Section 1.2.4) [165]. The already costly MP2/cc-pVTZ level of theory does not seem to be sufficient since it underestimates intermolecular distances as compared with reference CCSD(T) results

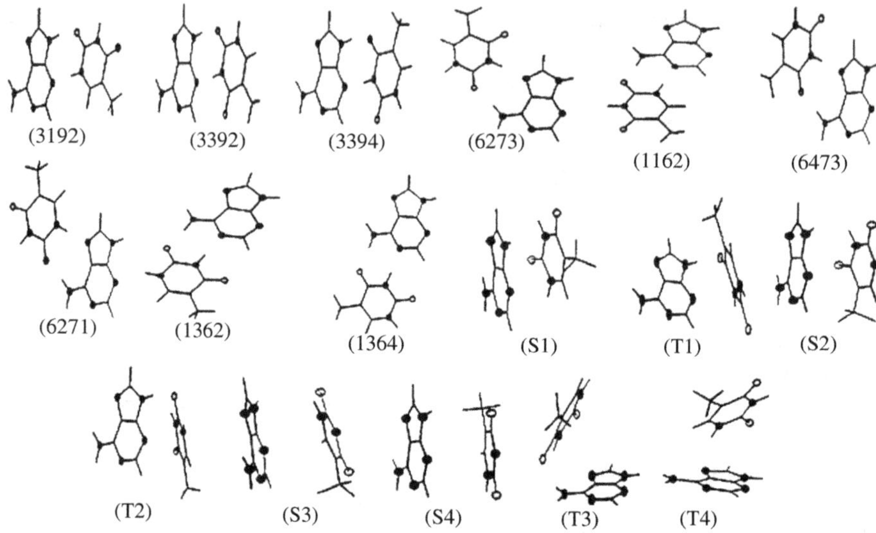

Figure 4.1.11 Different possible structures of the A—T pair. Numbers indicate hydrogen bonds; T and S correspond to T-shape and stacked structures. Note that the Watson–Crick (WC) pair (1364) ranks only ninth (reprinted with permission from reference [157] ©2000 American Chemical Society).

(see Section 1.2.3). Stabilizing energies of stacked pairs are more dependent on the level of calculation than hydrogen-bonded pairs. In particular, keeping the MP2 level but increasing the basis set size from 6-31G* up to complete basis set (CBS) enlarges the stabilization energy by more than 40 kJ/mol.

Experimentally, formation of stacked structures can be imposed in the gas-phase by methylation disfavouring H-bonded structures (Figure 4.1.12) [166, 167] or adding water molecules (see Section 5.4.1.2).

Experimental gas-phase studies of pairing have not been limited to isolated nucleobases. Dissociation kinetics of deprotonated deoxyribose nucleotide dimers (adenosine, cytosine, guanosine and thymidine) have been measured by Blackbody Infrared Radiative Dissociation (BIRD, see Section 3.2.4) [168]. G—C pairs are much more stable than A—T pairs. Similarly, WC base pairs occur in the lowest energy configurations of guanosine–cytosine pairs but not in adenosine–thymidine pairs.

Excited-state dynamics of adenine–adenine, thymine–thymine and mixed adenine–thymine dimers have been investigated by femtosecond pump–probe ionization spectroscopy with excitation wavelengths of 250–272 nm [84]. Similar decay time constants have been found for isolated adenine ($\tau_1 = 110 \pm 30$ fs; $\tau_2 = 900 \pm 100$ fs) and its A_2 dimer ($\tau_1 = 100 \pm 30$ fs; $\tau_2 = 700 \pm 300$ fs) although the contribution of the second component is greatly reduced in A_2. In both cases, time-constant τ_1 corresponds to the decay of the photo-excited singlet $\pi\pi^*$ state coupled to the $n\pi^*$ state. In A_2, the highly polar $\pi\sigma^*$ state of one of the adenines is stabilized through interaction with the other polar adenine and becomes the main de-excitation channel, thus suppressing the contribution of the $n\pi^*$ channel. The similarity of dynamical behaviours of the adenine monomer and adenine in hydrogen-bonded pairs rules out the possibility of intermolecular H-transfer between base pairs observed in model systems [169, 170].

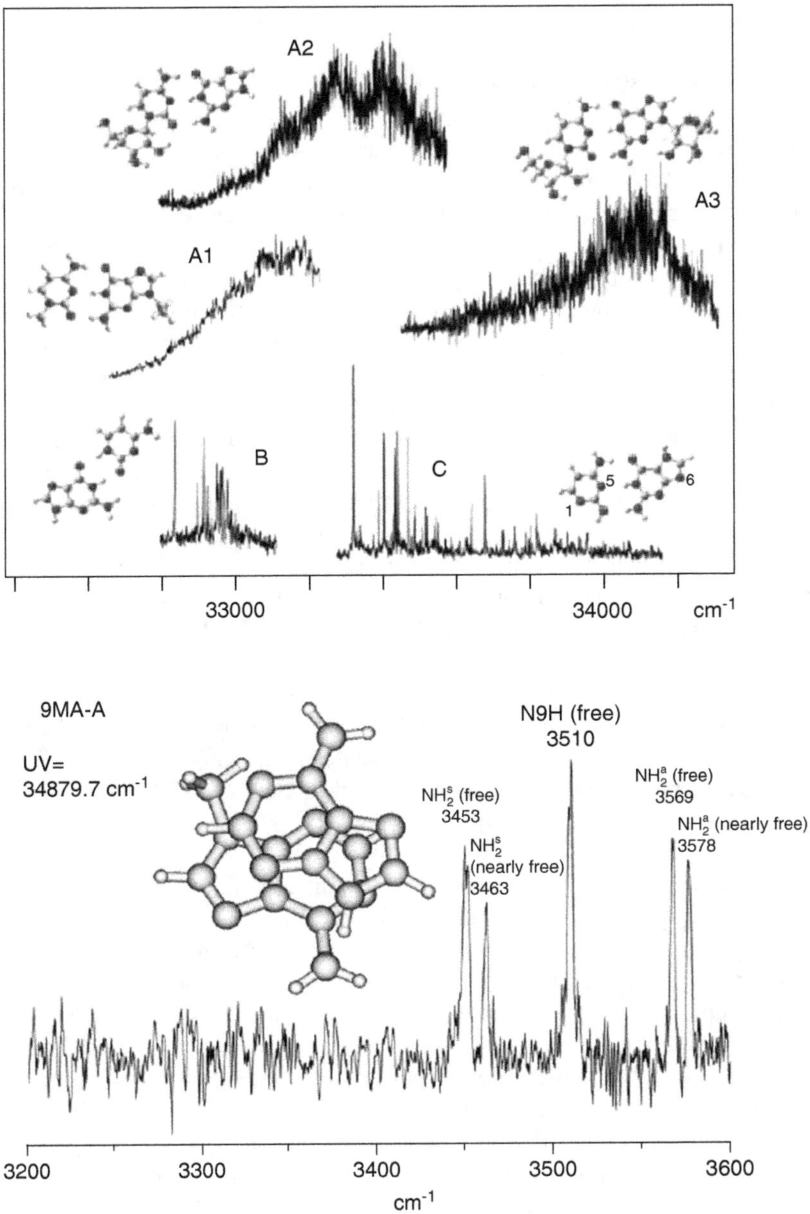

Figure 4.1.12 Top: R2PI spectra of the G—C (B and C), 9-ethylguanine-1-methylcytosine (A1), guanine–cytidine (A2) and guanosine–cytidine (A3). The structures are optimized at the DFTB3LYP/6-311G** level of theory. The comparison between infrared spectra recorded with the UV laser tuned to resonant features in spectra A1, B and C and predicted spectra show that the broad features correspond to the Watson–Crick structure (reprinted with permission from reference [156] ©2005 National Academy of Sciences). Bottom: UV–IR spectrum and stacked structure of the 9-methyladenine–adenine pair (reprinted with permission from reference [166] ©2003 Royal Society of Chemistry).

4.1.3 Oligonucleotides

4.1.3.1 DNA duplexes and quadruplexes

Helical structures of DNA observed in solution are destabilized in the gas-phase due to the absence of hydrophobic interactions (see Section 5.3), counterions and screening of electrostatic repulsive forces between negatively charged phosphate groups. Nevertheless, at least for a limited amount of time, helical structures can remain more or less intact in the gas-phase, the onset of helicity in the gas-phase then being a function of the sequence and length of oligomers. Stability of gas-phase oligonucleotides has been studied by means of blackbody radiation-induced dissociation (BIRD) [168, 171, 172], liquid beams [173], mass-spectrometry [19, 174–178], ion mobility [179–184] and simulations [9, 185–197]. WC base pairing is a factor of stability and has been studied by comparing dimer ions with complementary strands such as $A_7 \cdot T_7^{3-}$or non-complementary strands such as $T_7 \cdot T_7^{3-}$or $A_7 \cdot A_7^{3-}$. The former dimer dissociation energy, measured with BIRD (see3.3.2) is much larger than for the latter dimers. Another proof of the importance of WC hydrogen bonding is given by the observation that an extensive loss of neutral adenine occurs for $A_7 \cdot A_7^{3-}$ and $A_7 \cdot C_7^{3-}$ but not for $A_7 \cdot T_7^{3-}$. Molecular dynamics simulations show that WC base pairing is indeed preserved for $A_7 \cdot T_7^{3-}$ duplex, although the helix structure disappears [171].

Collision-induced dissociation studies (CID, see Section 3.2.2.2) of DNA duplexes have been conducted on non-complementary 16-mer duplexes containing different amounts of strongly bound G—C pairs or duplexes containing the same fraction of G—C pairs and different sequences. Not only specific hydrogen bonds between bases belonging to different strands but also stacking interactions are conserved in the gas-phase [174, 175]. The CID dissociation of three 12-mer A, B and C duplexes respectively containing 33, 67 and 100% of C—G base pairs has been studied [174, 176]. The B duplex, for example contains the following complementary sequence: 5′-CGCAATTCGG-3′. In those DNA duplexes (or in aptamer–drug complexes), the network of hydrogen bonds confers a high stability. When energy is added to those complexes by CID, rupture of this network of non-covalent bonds competes with cleavage of covalent bonds leading to nucleobase losses. The dependences of the rate constants and proposed mechanisms for neutral base loss and unzipping of the duplexes into single strands (non-covalent dissociation) are schematically represented in Figure 4.1.13. A higher G—C pair content corresponds to a larger collision energy required for reaching non-covalent dissociation. Non-covalent dissociation is an entropy-driven process and its rate constant increases more rapidly with energy than covalent base loss which is entropy disfavoured. The competition between those dissociation channels is thus dependent on the amount of time required to increase the internal energy and to overcome the dissociation barrier. If CID is conducted in a quadrupole collision cell (see Section 3.2.1.5) under high-pressure conditions (“fast heating”) with a short time-window (typically within microseconds), the non-covalent channel dominates. On the contrary, if CID is performed in a quadrupole trap within a typical time-window of 3–300 ms (“slow heating” or “low collision energy regime”), the covalent base loss channel is observed.

These studies show that experimental conditions for supplying internal energy can strongly influence the order of appearance of dissociation channels. In the case of an infinitely long time to reach the dissociation energy threshold, the order of appearance would be imposed by the true energy thresholds. When the time-window allowed for monitoring dissociation

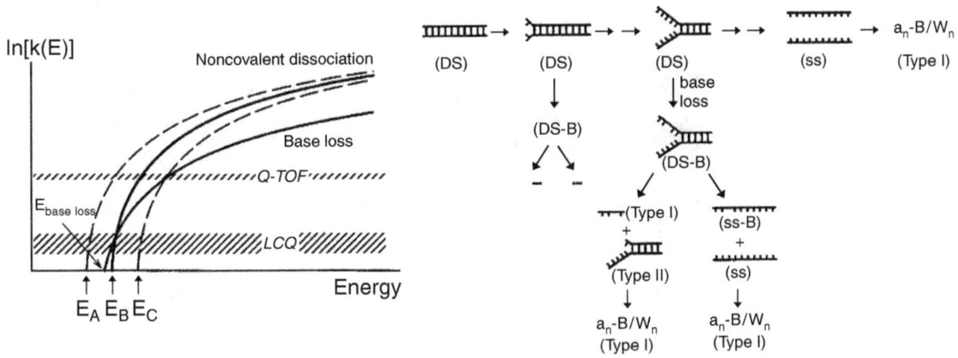

Figure 4.1.13 Left: Schematic dependency of the dissociation rates (in s^{-1}) of DNA duplexes in collision-induced dissociation. The energy thresholds for non-covalent dissociation of three duplexes A, B and C (see text) and covalent loss of a base are represented as well as the experimental time-windows for collisional heating conditions in a quadrupole collision cell (Q-TOF) and a quadrupole ion trap (LCQ)) Right: Proposed mechanism for dissociation of complementary double-stranded (DS) DNA duplexes. During the course of unzipping (preferentially from the terminals), bases can be lost by rupture of covalent bonds leading to DS–B duplexes. In turn, base losses can induce backbone fragmentation or the DS–B duplexes carry on unfolding (reprinted with permission from reference [176] ©2002 Elsevier).

diminishes, apparent energy thresholds ("kinetic shifts") larger than the true ones are observed. These studies also provide insights about the mechanism for non-covalent dissociation of isolated DNA duplex which is a progressive unzipping preferentially starting from terminal positions or in presence of drugs intercalated between nucleobases [18, 19] (see Chapter 4.5).

The onset of helicity has been investigated by ion-mobility measurements and simulation of gas-phase conformations of deprotonated d(C—G)$_n$ (C—G)$_n$ duplexes [184]. They show a dramatic change in structure for $n = 4$ (8-mer length). Below the 8-mer, predictions of force-field simulations (conducted during 2 ns at 300°K) and experimental observations correspond to only globular structures with a very small number of WC G—C pairs (one for the 4-mer). For the 8-mer, both a globular structure (with 3 WC G—C pairs) and a minor component with helical structure (with seven out of eight possible WC G—C pairs) start to be observed as shown in Figure 4.1.14. This helical structure becomes dominant from the 10-mer up to the 18-mer. Gas-phase helical structures are not the usual A, B or Z structures but most probably metastable structures preserving some memory from solution [9, 192] and that have not yet folded as globules. When A—T pairs replace G—C pairs, they are preferentially broken, as can be expected, and they are more easily broken when placed at the ends of the duplexes than in the middle.

The unzipping of oligonucleotide duplexes has also been dynamically observed by means of fluorescence resonance energy transfer (FRET) [198, 199] (see Section 2.1.4.1). Double-stranded oligonucleotides anions were prepared with FRET donor/acceptor (BODIPY-TMR and BODIPY-TR dyes) and electrosprayed from solution. Approximately 7,000 7-charge ions were confined in a variable temperature quadrupole ion trap (see Section 3.1.2.1.2). The dissociation of the duplexes into single strands was followed by monitoring, as a function of temperature, both their laser-induced fluorescence and mass respectively indicates the onset of unzipping of outer pairs and of single strands. The chosen sequences were AATTAATCCGGCCG and its complement, ensuring that A—T pairs followed by

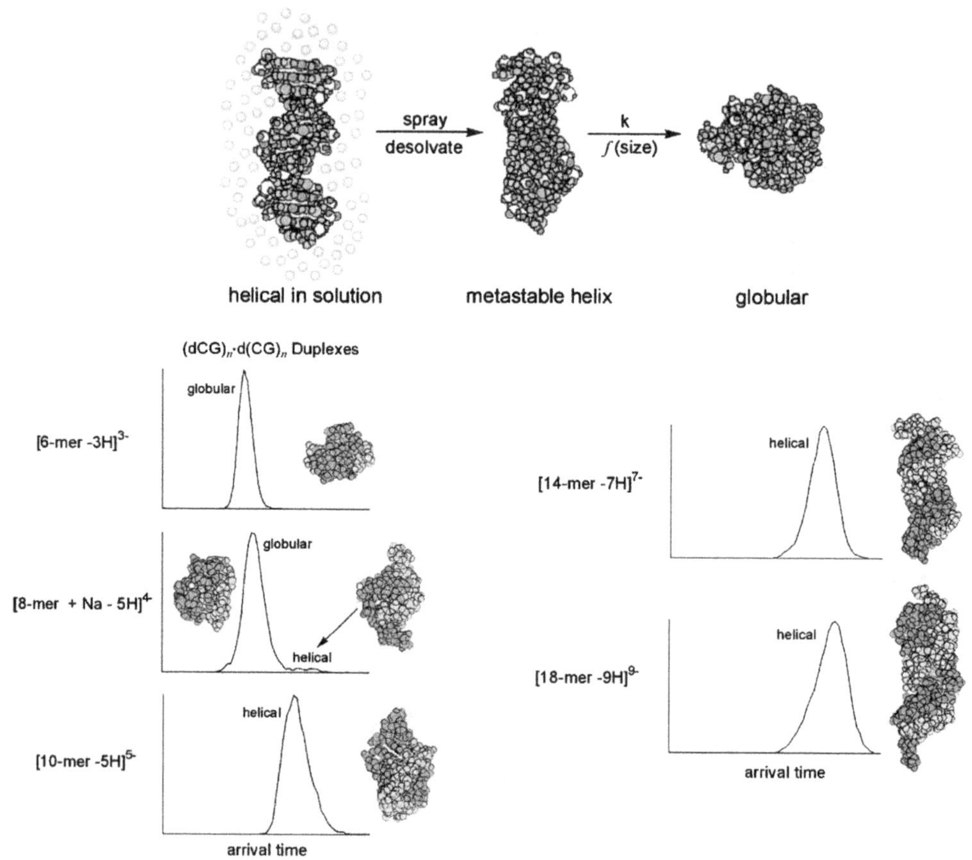

Figure 4.1.14 Top: Proposed sequence of events during desolvation of d(CG)$_n$ (CG)$_n$ duplexes ($n = 4$–9). Initial solution helical structures collapse into the lowest energy globular form. Bottom: Ion-mobility study of the fate of deprotonated d(CG)$_n$ (CG)$_n$ duplexes ($n = 4$–9).generated by electrospray ionization. The arrival time distributions (ATD) (see Section 3.3.1) show that the 6-mer ($n = 3$) (as well as the 4-mer not shown) are globular. Only the 8-mer ($n = 4$) ATD displays two peaks corresponding to a 25% charge in cross-sections. This size corresponds to the onset of helicity. The larger duplexes retain the helical structures they had in solution. They will eventually collapse in their lowest energy globular form but after the time more than the above ATDs which are in the few millisecond range (reprinted with permission from reference [184] ©2005 Elsevier).

strongly bound G—C pairs would lead to long-lived duplexes and that the unzipping would initiate from the outer A—T end where the dyes were attached (Figure 4.1.15). The experimental results were interpreted by a three-state model. In the unzipped state $|1\rangle$, the donor fluorophore D is within 3–5 Å from the acceptor A and the donor does not fluoresce while its fluorescence becomes maximum in the totally unzipped state $|3\rangle$. In the intermediate state $|2\rangle$ where outer pairs start to break, the donor fluorescence depends strongly on the distance $\langle R_2\rangle$ relative to the Förster distance R_0.

There exit other important secondary structures intervening in DNA oligonucleotides such as hairpins [200] or quadruplexes [201]. Ends of chromosomes, called telomers,

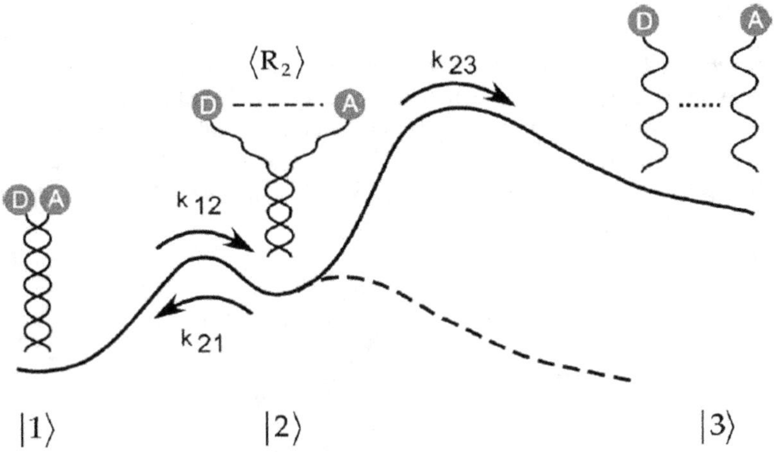

Figure 4.1.15 Schematic potential energy diagram showing the melting transition of AT-rich duplex $|1\rangle$ to single strands $|3\rangle$ through formation of a partially unzipped intermediate state $|2\rangle$ where outer pairs start to break (reprinted with permission from reference [199] ©2003 Elsevier).

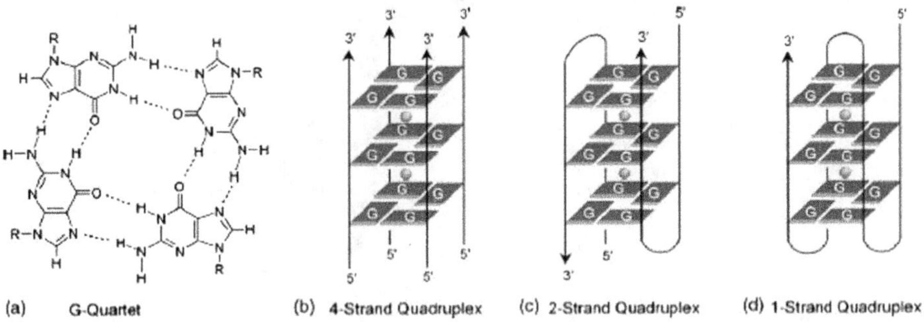

Figure 4.1.16 Guanine quadruplex and folding structure of guanine-rich oligonucleotides [201]; (a) represents the structure of a hydrogen-bonded guanine quartet; (b) is a four-strand quartet held by inter-molecular H-bonds; (c) is a two-strand quadruplex held by intra- and inter-molecular H-bonds; (d) is a folded single strand. These quadruplexes are here stabilized by cations (metals or ammoniums) (reprinted with permission from reference [201] ©2006 Elsevier).

are guanine-rich sequences such as $(TTAGGG)_n$ in human DNA [202]. These extreme parts of DNA are single-strand and form protecting cap structures. For that purpose, tetrads of four guanines assemble through Hoogsteen pairing and stack to form quadruplexes (Figure 4.1.16). While in normal somatic cells (cells that cannot divide or differentiate to produce new generations of offspring), telomer lengths progressively decrease, telomers keep a constant length in cancer cells. They are thus important targets in oncologic pharmacy [203] and mass-spectrometry is used to provide the number of binding sites of drugs per nucleic acid [20, 204–206].

Quadruplexes [11, 207–209] have been investigated by mass-spectrometry [207, 208], ion-mobility [206, 210], IRMPD [211] and modelling [10, 44, 212].

4.1.3.2 Aptamers

Aptamers (*aptus*, "to fit" in Latin) are linear sequences of 15–60 A, T, C, G or U nucleotides designed for binding to specific molecular ligands. The different aptamers can fold into a huge number of possible 3D structures and thus bind to virtually any amino acid, protein, enzyme etc. Their binding properties (affinities) are due to establishment of several precise hydrogen and π-stacking bonds and are thus extremely sensitive [213]. Mass-spectrometry

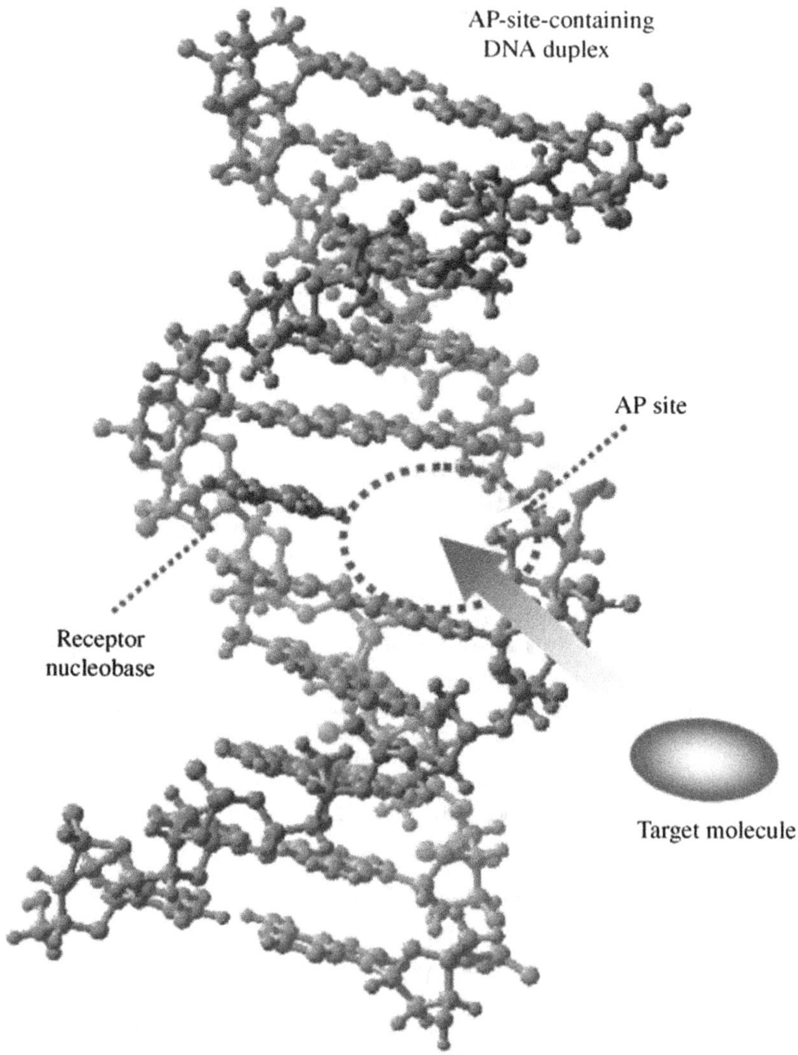

Figure 4.1.17 Binding motif of a DNA duplex as an aptamer for a small ligand. A site lacking a base (abasic site AP) is included in the pale strand and used as an active cavity. The complementary strand is the darker one and contains the receptor nucleobase (reprinted with permission from reference [213] ©2005 Wiley).

is used to determine binding stoechiometry and affinities (see Section 4.5.1) [214, 215] (Figure 4.1.17).

4.1.4 RNA

In this chapter, we recall the roles of the different RNAs and further examine the basic principles of their structures [216]. Investigations of RNA base pairing by means of gas-phase spectroscopy and quantum calculations are considered. The very large ribozymes are considered in Chapter 4.8. According to the central dogma of molecular biology, sequences of nucleobases belonging to a DNA strand and representing genes are transcribed into mRNA. The genes are usually interrupted by non-coding regions known as introns and are also separated by large nucleobase sequences that do not lead to any protein. Genes are transcribed into mRNA. After splicing and removing of introns, each of the messenger RNA codon is recognized by the anti-codon of a transfer RNA (tRNA) leading to the insertion of the corresponding amino acid into a nascent polypeptide (Figure 4.1.18). The mRNA is translated into a protein (that must be further processed by PTMs and folding) by rRNA (see 4.8).

The role of RNA is fundamental and it has been proposed that RNA may have been at the origin of Life since it was able to replicate itself and thus self-sufficient [217]. This hypothesis raises several problems since some RNA components such as cytosine [218] are short-lived and are much more difficult to synthesize under plausible prebiotic conditions than amino acids, for example (see Section 4.6.1). Several types of RNA, other than mRNA, tRNA and rRNA, exist in cells. For example, microRNAs (miRNAs) come from the transcription of genes, are spliced like usual mRNAs but do not code for proteins [219, 220]. Their role is to control messenger RNAs and they are involved in the regulation of approximately 30% of all human genes. The miRNA structure contains an ensemble formed of a double-stranded (ds) region, called stem, and a loop that serves as trigger for an important pathway called RNA interference (RNAi). This stem-loop ensemble is excised by an enzyme and exported to the cytoplasm where another enzyme, called Dicer, processes it into a small interfering RNA (siRNA) containing ~21 nucleotides. The siRNA is then incorporated into the RNA-induced silencing complex (RISC) and guides it to its target messenger RNA by establishing base complementarity (A pairing with U, C with G) [221]. The target mRNA, denounced

5'...ATGGCCTGGACTTCA...3' DNA sense strand
3'...TACCGGACCTGAAGT...5' DNA anti sense strand

transcription of the anti sense strand ↓

5'...AUGGCCUGGACUUCAA...3' **mRNA**

translation of the mRNA ↓

--methionine-alanine-tryptophan-threonine-serine- **peptide**

Figure 4.1.18 Transfer of information between DNA and peptides. The antisense DNA strand is transcribed into messenger RNA (mRNA). After splicing and removing of introns, mRNA codes and is translated into a peptide.

by the siRNA, is then enzymatically cleaved and thus prevented to express the gene as it was supposed to do. miRNAs show many similarities to siRNAs but the double-stranded siRNAs are perfectly self-complementary and complementary to their mRNA targets. miRNAs are in hairpin forms and are not perfectly complementary to the corresponding mRNAs. More than 300 different miRNAs have been identified in humans and are considered as potential targets for drugs [222]. miRNAs undergo covalent modifications such as methylations that can be investigated by mass-spectrometry [222]. The mechanism called post-transcriptional gene silencing (PTGS) is supposed to originate from an ancient defence mechanism against viral dsRNA. A hypothesis concerning the move from an RNA-ruled world to a DNA-ruled world proposes indeed that viruses played a major role (see Chapter 4.8) [223]. PTGS is today widely used for studying gene functions [224] by introducing dsRNA into the cells of an organism. RNAi leads to the sequence specific destruction of the endogenous RNAs that match the dsRNA [225].

RNAs serve as information carriers (mRNA and tRNA), catalysts (rRNA) [226–229] or regulation elements (siRNA or aptamers) [230]. The folding of RNAs, as that of proteins, is hierarchical. The primary structure is the sequence of nucleotides. RNAs contain some of the same nucleobases as DNA (C, A and G) but T is replaced by U. The secondary structure is dependent on the formation of the canonical A—U, C—G and G—U base pairs (Figure 4.1.19). The tertiary structure is the 3D arrangement of atoms and the quaternary structure corresponds to the interaction with proteins or other RNA strands. Once a secondary structure is determined by minimizing the free energy, a tertiary structure can be predicted [231, 232] with some help from prior knowledge of tertiary structures. For example, nucleotides that are distant from one another in the primary sequence can nevertheless pair (Figure 4.1.9) but their mutual distance place constraints upon their 3D arrangement.

DNA can establish very long helices due to the complementarity of nucleobases in its two strands. Only one strand is transcribed into RNA (Figure 4.1.19) and this overall complementarity is lost. Nevertheless, the RNA chain can fold on itself and creates double helices in the A form (stems) separated by single-strand portions. The replacement of deoxyribose in DNA by ribose in RNA allows the formation of 2′-O—H···O4′ hydrogen bonds between neighbouring riboses that stiffen the structures. The RNA structure is made of several types of multipurpose building blocks or motifs [233, 234] such as helices, internal loops, hairpins [235, 236], pseudoknots [13, 232, 237] or kink-turns [234]. Hairpins intervene in various biological functions such as gene expression. They possess a stem containing pairs (e.g. A—G, U—U or G—C) capped by a loop with unpaired nucleosides [10, 238]. Kink-turns are sharp bends of the backbone acting as elbows. They involve —C—G···A—bonds and have been highly conserved through Evolution.

Base pairing plays also an important role in the structure of messenger and ribosomal RNAs between A, C, G, U and other bases such as inosine (I) [239, 240] or pseudouridine (Ψ) [241]. Other pairs other than WC or Hoogsteen AU base pairing patterns are found and are called reverse WC or reverse Hoogsteen. In fact, 40% of RNA pairs are non-WC pairs. Those non-canonical base pairs found in RNA [242–245] have been investigated theoretically [163, 191, 246–248]. Some pairs play a crucial role in the translation of RNA. The mRNA contains a sequence of triplets of base (codons), each characteristic of an amino acid transported by a tRNA. Each tRNA is recognized by its anticodon. The genetic code makes up for disparities in the number of amino acids (20) for codons (64) by using modified base pairs ("wobble pairs") [249] in the first base of the anti-codon. Those pairs are

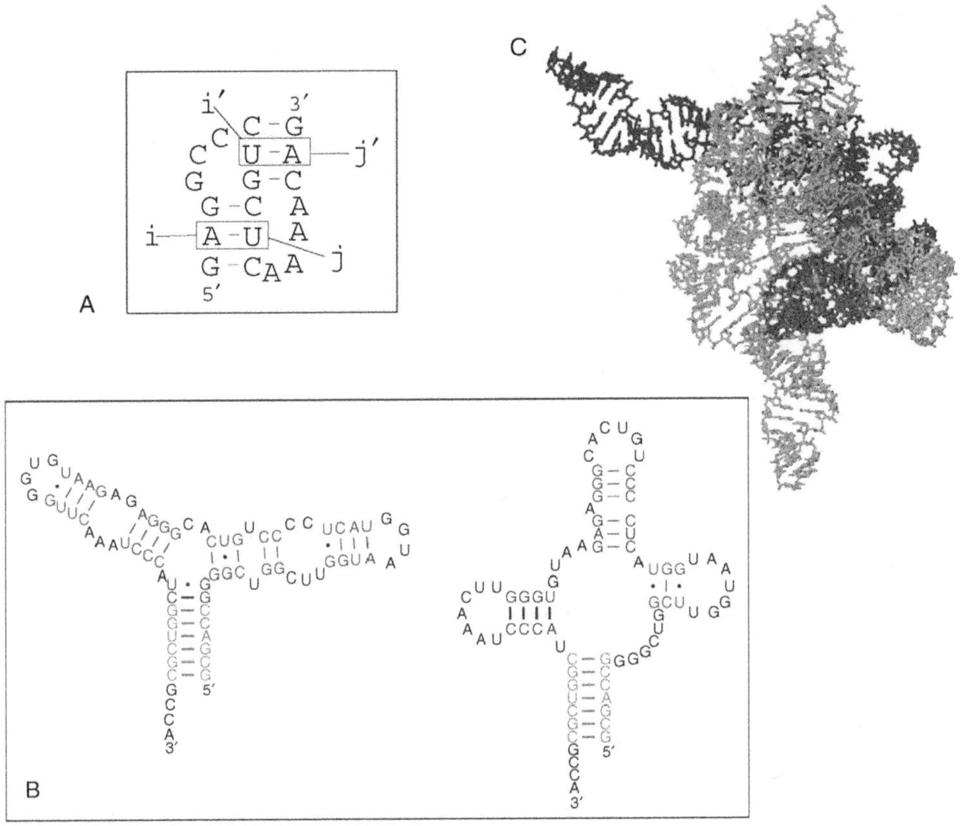

Figure 4.1.19 (A) Example of RNA pseudoknot topology. The illusion of a knot is provided by a spatial organization of base pairs ($i = 2$ and $j = 11$; $i' = 8$ and $j' = 18$). (B) Two predicted lowest free-energy secondary structures of a tRNA. Both exhibit loops separated by helices (stems). (C) Tertiary structure predicted on the basis of a comparative analysis (reprinted with permission from reference [232] ©2006 Elsevier).

G—U, I—U, I—A and I—C. Among them, the wobble G—U pair has attracted most of the attention since it is conserved in nearly all living organisms [250].

For gas-phase experimental studies, mock-up pairs, such as the uracil-2-pyridone pair, respecting the hydrogen bond pattern between neighbouring N—H and C—O groups are considered (Figure 4.1.20) [249, 251–255]. If U—U dimers had been studied, eight different double H-bonded configurations should have been considered since uracil offers four H-bonding sites. Instead, 2-pyridone (2PY) offers only two H-bonding groups. There are thus only three nucleobase dimer 2-pyridone–uracil (2PY—U) mimics (Figure 4.1.20). In the most stable configuration, uracil binds to 2-pyridone via its N1–H donor group. In RNA, uracil is bound to ribose through its N1 atom and this 2PY—U pair is thus called sugar-edge. Since it is not a biologically relevant pair, it is preferable to methylate the uracil N1 atom (methylation on N3 still allows formation of sugar-edge pairs) and then favour formation of WC and wobble pairs. Those wobble pairs play a crucial role in codons (belonging to mRNA) and anticodons

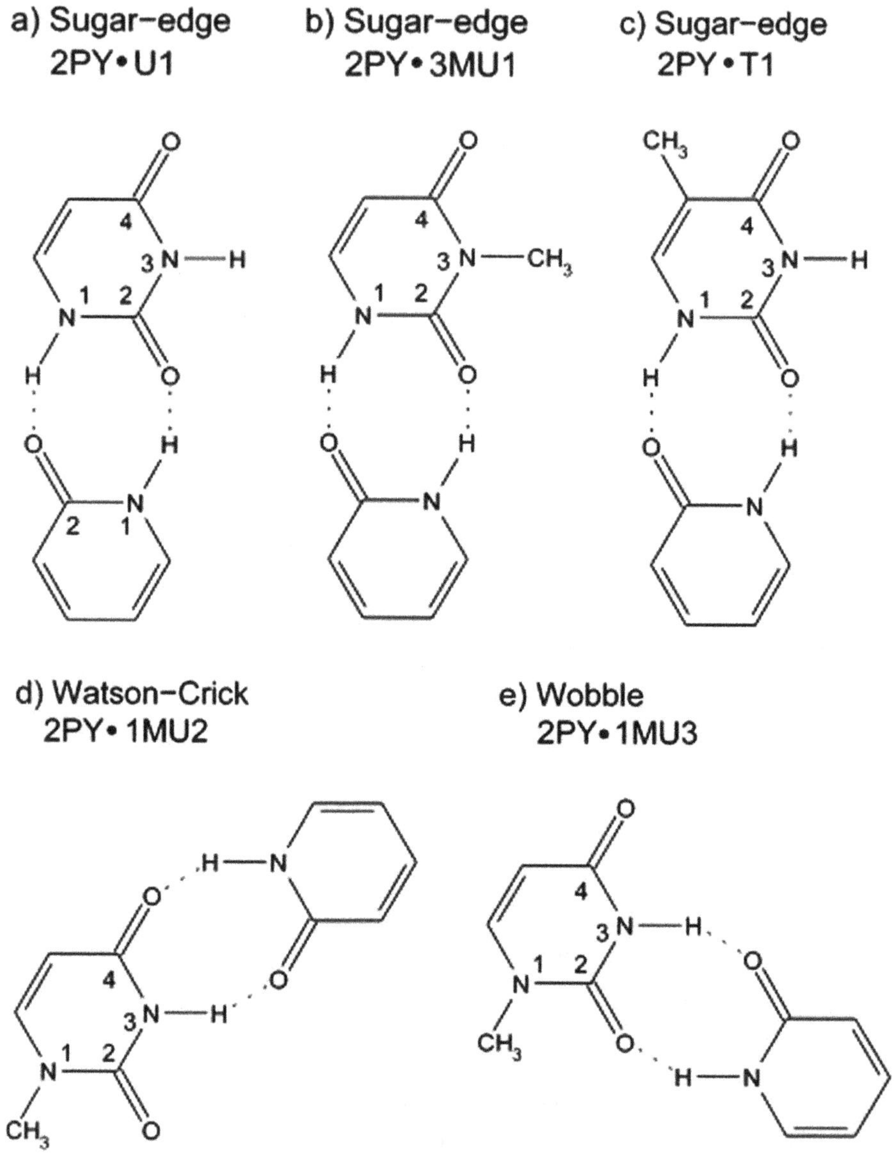

Figure 4.1.20 Structures of the sugar-edge dimers of 2-pyridone-uracil (2PY-U1; a), 2-pyridone-3-methyluracil (2PY-3MU1; b) and 2-pyridone-thymine (2PY-T1; c). Structures of the Watson–Crick (2PY-1MU2) and wobble (2PY-1MU3) pairs are respectively represented in (d) and (e) (reprinted with permission from reference [249] ©2005 American Chemical Society).

(in tRNAs) that establish three hydrogen bonds between themselves. Those bonds must be strong enough to first ensure correct recognition between a codon and the right anticodon at the entrance of the rRNA but not too strong because separation between those entities must take place to allow the incorporation of the amino acid carried by the tRNA.

Table 4.1.5

Calculated ground state dissociation energies D_0 of the different natural or methylated 2-pyridone–uracil pairs of Figure 4.1.20 at the DFT/6-311++G(d,p) level with the PW91 functional (from [249])

Isomer	Sugar-edge		Watson–Crick		Wobble	
	U1	3MU1	U2	1MU2	U3	1MU3
D_0	82.7	81.9	66.0	67.7	55.5	56.5

The different pairs have been experimentally investigated in 2-color R2PI experiments and their binding energies have been calculated [249]. As shown in Table 4.1.5, wobble pairs are much less stable than WC pairs.

REFERENCES

1. Watson JD, Crick FHC: Genetic implications of the structure of deoxyribonucleic acid. *Nature* 1953, 171:965–967.
2. Dickerson RE: Definitions and nomenclature of nucleic acid structure components. *Nucleic Acids Research* 1989, 17:1797–1803.
3. Couch GS, Hendrix DK, Ferrin TE: Nucleic acid visualization with UCSF Chimera. *Nucleic Acids Research* 2006, 34:e29–e34.
4. Tinoco I: Physical chemistry of nucleic acids. *Annual Review of Physical Chemistry* 2002, 53: 1–15.
5. Buyukdagli S, Sanrey M, Joyeux M: Towards more realistic dynamical models for DNA secondary structure. *Chemical Physics Letters* 2006, 419:434–438.
6. Matta CF, Castillo N, Boyd RJ: Extended weak bonding interactions in DNA: π stacking (base–base), base–backbone, and backbone–backbone interactions. *Journal of Physical Chemistry B* 2006, 110:563–578.
7. Wain-Hobson SW: The third bond. *Nature* 2006, 439:539.
8. Egli M, Pallan PS, Pattanayek R, Wilds CJ, Lubini P, Minasov G, Dobler M, Leumann CJ, Eschenmoser A: Crystal structure of homo-DNA and Nature's choice of pentose over hexose in the genetic system. *Journal of the American Chemical Society* 2006, 128:10847–10856.
9. Rueda M, Kalko SG, Luque FJ, Orozco M: The structure and dynamics of DNA in the gas phase. *Journal of the American Chemical Society* 2003, 125:8007–8014.
10. Hernandez B, Baumruk V, Gouyette C, Ghomi M: Thermal stability, structural features, and B-to-Z transition in DNA tetraloop hairpins as determined by optical spectroscopy in $d(CG)_3T_4(CG)_3$ and $d(CG)_3A4(CG)_3$ oligodeoxynucleotides. *Biopolymers* 2005, 78:21–34.
11. Phan AT, Kuryavyi V, Patel DJ: DNA architecture. *Current Opinion in Structural Biology* 2006, 16:288–298.
12. Sponer J, Jurecka P, Marchan I, Luque FJ, Orozco M, Hobza P: Nature of base stacking: reference quantum-chemical stacking energies in ten unique B-DNA base-pair steps. *Chemistry: A European Journal* 2006, 12:2854–2865.
13. Daune M: *Molecular biophysics. Structures in motion:* Oxford University Press, Oxford 1999.
14. Becker PB: A finger on the mark. *Nature* 2006, 442:31–32.
15. Woodcock CL: Chromatin architecture. *Current Opinion in Structural Biology* 2006, 16:213–220.
16. Human G: Initial sequencing and analysis of the human genome. *Nature* 2001, 409:860–921.

17. Hofstadler SA, Griffey RH: Analysis of non-covalent complexes of DNA and RNA by mass spectrometry. *Chemical Reviews* 2001, 101:377–390.
18. Rosu F, Gabelica V, De Pauw E: Determination of affinity, stoechiometry and sequence selectivity of minor groove binder complexes with double-stranded oligodeoxynucleotides by electrospray ionization mass spectrometry. *Nucleic Acids Research* 2002, 30:e82–e91.
19. Gabelica V, De Pauw E, Rosu F: Interaction between antitumor drugs and a double-stranded oligonucleotide studied by electrospray ionization mass spectrometry. *Journal of Mass Spectrometry* 1999, 34:1328–1337.
20. Carrasco C, Rosu F, Gabelica V, Houssier C, De Pauw E, Garbay-Jaureguiberry C, Roques B, Wilson WD, Chaires JB, Waring MJ, *et al.*: Tight binding of the antitumor drug ditercalinium to quadruplex DNA. *ChemBioChem* 2002, 3:1235–1241.
21. Shi X, Takamizawa A, Nishimura Y, Hiraoka K, Akashi S: Stability analysis for double-stranded DNA oligomers and their noncovalent complexes with drugs by laser spray. *Journal of Mass Spectrometry* 2006, 41:1086–1095.
22. Rueda M, Luque FJ, Orozco M: Nature of minor-groove binders: DNA complexes in the gas phase. *Journal of the American Chemical Society* 2005, 127:11690–11698.
23. Dickinson LA, Burnett R, Melander C, Edelson BS, Arora PS, Dervan PB, Gottesfeld JM: Arresting cancer proliferation by small-molecule gene regulation. *Chemistry and Biology* 2004, 11:1583–1594.
24. White S, Szewczyk JW, Turner JM, Baird EE, Dervan PB: Recognition of the four Watson–Crick pairs in the DNA minor groove by synthetic ligands. *Nature* 1998, 291:468–471.
25. Warren CL, Kratochvil NCS, Hauschild KE, Foister S, Brezinski ML, Dervan PB, Phillips GN, Ansari AZ: Defining the sequence-recognition profile of DNA-binding molecules. *Proceedings of the National Academy of Sciences of the United States of America* 2006, 103:867–872.
26. Dervan PB: Molecular recognition of DNA by small molecules. *Bioorganic and Medicinal Chemistry* 2001, 9:2215–2235.
27. Best TP, Edelson BS, Nickols NG, Dervan PB: Nuclear localization of pyrrole–imidazole polyamide–fluorescein conjugates in cell culture. *Proceedings of the National Academy of Sciences of the United States of America* 2003, 100:12063–12068.
28. Doss RM, Marques MM, Foister S, Chenoweth DM, Dervan PB: Programmable oligomers for minor groove DNA recognition. *Journal of the American Chemical Society* 2006, 128: 9074–9079.
29. Carles S, Lecomte F, Schermann JP, Desfrançois C: Gas-phase experimental and theoretical studies of adenine, imidazole, pyrrole and water non-covalent complexes. *Journal of Physical Chemistry A* 2000, 104:10662–10669.
30. Garner P, Huang YP, Dey S: Enhancement of α PN binding affinity and specificity through hydrophobic interactions. *ChemBioChem* 2001, 3:224–226.
31. Baker ES, Hong JW, Gaylord BS, Bazan GC, Bowers MT: PNA/dsDNA complexes: Site specific binding and dsDNA biosensor applications. *Journal of the American Chemical Society* 2006, 128:8484–8492.
32. Herbert HE, Halls MD, Hratchian HP, Raghavachari K: Hydrogen-bonding interactions in peptide nucleic acid and deoxyribonucleic acid: a comparative study. *Journal of Physical Chemistry B* 2006, 110:3336–3343.
33. Group TISMW: A map of human genome sequence variation containing 1.42 million single nucleotide polymorphisms. *Nature* 2001, 409:928–933.
34. Syvanen AC: Toward genome-wide SNP genotyping. *Nature* 2005, 409:S5–S10.
35a. Krebs S, Meugorac I, Seitcher D, Förster M: RNaseCut: a MALDI mass spectrometry-based method for SNP discovery. *Nucleic Acids Research* 2003, 31:e37.
35b. Kimchi-Sarfaty C, Oh JM, Kim IW, Sauna ZE, Calcagno AM, Ambudkar SV, Gottesmann MM: A “silent” polymorphism in the MDR1 gene changes substrate specificity. *Science* 2007, 315:525–528.

36. McCarthy JJ, Hilfiker R: The use of single nucleotide polymorphism maps in pharmacogenomics. *Nature Biotechnology* 2000, 18:505–508.
37. Harold D, MacGregor S, Patterson CE, Hollingworth P, Moore F, Owen MJ, Williams J, O'Donovan M, Passmore P, McIlroy S, *et al.*: A single nucleotide polymorphism in CHAT influences response to acetylcholinesterase inhibitors in Alzheimer's disease. *Pharmacogenetics and Genomics* 2006, 16:75–77.
38. Norhoff E, Kirpekar F, Roepstorff P: Mass spectrometry of nucleic acids. *Mass Spectrometry Reviews* 1996, 15:67–138.
39. Wysocki VH, Resing KA, Zhang QF, Cheng GL: Mass spectrometry of peptides and proteins. *Methods* 2005, 35:211–222.
40. Woolf PJ, Wang YS: A fuzzy logic approach to analyzing gene expression data. *Physiological Genomics* 2000, 3:9–15.
41. Jensen LJ, Gupta R, Blom N, Devos D, Tamanes J, Kesmir C, Nielsen H, Staerfeldt HH, Rapacki K, Workman C, *et al.*: Prediction of human protein function from post-translational modifications and localization features. *Journal of Molecular Biology* 2002, 319:1257–1265.
42. Tsugita A, Miyazaki K, Nabetami T, Nozawa T, Masaharu M, Kawakami T: Application of chemical selective cleavage methods to analyze post-translational modification in proteins. *Proteomics* 2001, 1:1062–1091.
43. Bell SD: Prime-time progress. *Nature* 2006, 439:542–543.
44. Cheatham TE, Kollman PE: Molecular dynamics simulation of nucleic acids. *Annual Review of Physical Chemistry* 2000, 51:435–471.
45. Kool ET: Hydrogen bonding, base stacking, and steric effects in DNA replication. *Annual Review of Physical Chemistry* 2001, 30:1–22.
46. Nir E, Kleinermanns K, de Vries MS: Pairing of isolated nucleic-acid bases in the absence of the DNA backbone. *Nature* 2000, 408:949–951.
47. Hesselmann A, Jansen G, Schütz M: Interaction energy contributions of H-bonded and stacked structures of the AT and GC DNA base pairs from the combined density functional theory and intermolecular perturbation theory approach. *Journal of the American Chemical Society* 2006, 128:11730–11731.
48. Drew HR, Wing RM, Takano T, Broka C, Tanaka S, Ikatura K, Dickerson RE: Structure of a B-DNA dodecamer: conformation and dynamics. *Proceedings of the National Academy of Sciences of the United States of America* 1981, 78:2179–2183.
49. Sukhodub L, Yanson IK: Mass spectrometric studies of binding energies for nitrogen bases of nucleic acids in vacuo. *Nature* 1976, 264:245–247.
50. Sukhodub L: Interactions and hydration of nucleic acid bases in a vacuum. Experimental study. *Chemical Reviews* 1987, 87:589–606.
51. Kim SK, Lee W, Herschbach DR: Cluster beam chemistry: hydration of nucleic acid base; ionization potentials of hydrated adenine and thymine. *Journal of Physical Chemistry* 1996, 100: 7933–7937.
52. Brady BB, Peteanu LA, Levy DH: Cluster beam chemistry: hydration of nucleic acid base; ionization potentials of hydrated adenine and thymine. *Chemical Physics Letters* 1988, 147:538–543.
53. Dey M, Moritz F, Schlag EW: Base pair formation of free nucleobases and mononucleosides in the gas-phase. *Journal of the American Chemical Society* 1994, 116:9211–9215.
54. Desfrancois C, Abdoul-Carime H, Schulz CP, Schermann JP: Laser separation of geometrical isomers of weakly-bound molecular complexes. *Science* 1995, 269:1707–1709.
55. Viant MR, Feller RS, McLaughlin RP, Saykally RJ: Infrared laser spectroscopy of uracil in a pulsed slit jet. *Chemical Physics* 1995, 103:9502–9505.
56. Hendricks JH, Lyapustina SA, deClercq HL, Snodgrass, JT, Bowen KH: Dipole bound, nucleic acid base anions studied via negative ion photoelectron spectroscopy. *Biophysical Journal* 1996, 70:SU495.

57. Nir E, Grace L, Brauer B, de Vries MS: REMPI spectroscopy of jet-cooled guanine. *Journal of the American Chemical Society* 1999, 121:4896–4897.
58. Schiedt J, Weinkauf R, Neumark DM, Schlag EW: Anion spectroscopy of uracil, thymine and the amino-oxo and amino-hydroxy tautomers of cytosine and their water clusters. *Chemical Physics* 1998, 239:511–524.
59. Huels MA, Hahndorf I, Illenberger E, Sanche L: Resonant dissociation of DNA bases by subionization electrons. *Journal of Chemical Physics* 1998, 108:1309–1312.
60. Aflatooni K, Gallup GA, Burrow PD: Electron attachment energies of the DNA bases. *Journal of Physical Chemistry A* 1998, 102:6205–6207.
61. Kim NJ, Jeong G, Kim YS, Sung J, Kim SK: Resonant two-photon ionization and laser induced fluorescence spectroscopy of jet-cooled adenine. *Journal of Chemical Physics* 2000, 113:10051–10055.
62. Weinkauf R, Schermann JP, de Vries MS, Kleinermanns K: Molecular physics of building blocks of life under isolated or defined conditions. *European Physical Journal D* 2002, 20:309–316.
63. Denifl S, Ptasinska S, Cingel M, Matejcik S, Scheier P, Mark TD: Electron attachment to the DNA bases thymine and cytosine. *Chemical Physics Letters* 2003, 377:74–80.
64. Andersen JU, Andersen LH, Hvelplund P, Lapierre A, Moller SP, Nielsen SB, Pedersen UV, Tomita S: Studies of clusters and biomolecules in ELISA. *Hyperfine Interactions* 2003, 146: 283–291.
65. Abdoul-Carime H, Langer J, Huels MA, Illenberger E: Decomposition of purine nucleobases by very low energy electrons. *European Physical Journal D* 2005, 35:399–404.
66. Abouaf R, Dunet H: Structures in dissociative electron attachment cross-sections in thymine, uracil and halouracils. *European Journal of Physics D* 2005, 35:405–410.
67. Ptasinska S, Denifl S, Gohlke S, Scheier P, Illenberger E, Mork TD: Decomposition of thymidine by low-energy electrons: implications for the molecular mechanisms of single-strand breaks in DNA. *Angewandte Chemie-International Edition* 2006, 45:1893–1896.
68. Rueda M, Luque FJ, Orozco M: A theoretical investigation on the effect of remote amino groups in hydrogen bonding of nucleic acids. *Biopolymers* 2001, 61:52–60.
69. Rincon E, Jaque P, Torro-Labbé A: Reaction force analysis of the effect of Mg(II) on the 1,3 intramolecular hydrogen transfer in thymine. *Journal of Physical Chemistry A* 2006, 110: 9478–9485.
70. Piacenza M, Grimme S: Systematic quantum mechanical study of DNA-base tautomers. *Journal of Computational Chemistry* 2004, 25:83–98.
71. Haranczyk, Gutowski M: Combinatorial chemistry based on quantum chemical calculations. Search for the most stable tautomers of anionic nucleic acid bases. *Angewandte Chemie-International Edition* 2005, 44:6585–6588.
72. Nir E, Janzen C, Imhof P, Kleinermanns K, de Vries MS: Guanine tautomerism revealed by UV–UV and IR–UV hole burning spectroscopy. *Journal of Chemical Physics* 2001, 115: 4604–4611.
73. Mons M, Dimicoli I, Piuzzi F, Tardivel B, Elhanine M: Tautomerism of the DNA base guanine and its methylated derivatives as studied by gas-phase infrared and ultraviolet spectroscopy. *Journal of Physical Chemistry A* 2002, 106:5088–5094.
74. Leszczynski J: The potential energy surface of guanine is not flat: an ab initio study with large basis sets and higher order electron correlation contributions. *Journal of Chemical Physics* 1998, 109:2357–2362.
75. Mons M, Piuzzi F, Dimicoli I, Gorb L, Leszczynski J: Near-UV resonant two-photon ionization spectroscopy of gas phase guanine: evidence for the observation of three rare tautomers. *Journal of Physical Chemistry A* 2006, 110:10921–10924.
76. Choi MS, Miller RE: Four tautomers of isolated guanine from infrared laser spectroscopy in helium droplets. *Journal of the American Chemical Society* 2006, 128:7230–7328.

77. Plutzer C, Nir E, de Vries MS, Kleinermanns K: IR-UV double-resonance spectroscopy of the nucleobase adenine. *Physical Chemistry Chemical Physics* 2001, 3:5466–5469.
78. Shishkin OV, Palamarchuk GV, Gorb L, Leszczynski J: Intramolecular hydrogen bonds in canonical 2′-deoxyribonucleotides: an atom-in-molecule study. *Journal of Physical Chemistry B* 2006, 110:4413–4422.
79. Nir E, Kleinermanns K, Grace L, de Vries MS: On the photochemistry of purine nucleobases. *Journal of Physical Chemistry A* 2001, 105:5106–5110.
80. Kim NJ, Kang H, Jeong G, Kim YS, Lee KT, Kim SK: Even-odd alternation in mass spectrum of thymine and uracil clusters: evidence of intra-cluster photodimerization. *Proceedings of the National Academy of Sciences of the United States of America* 2001, 98:4841–4843.
81. Schneider M, Maksimenka R, Buback FJ, Kitsopoulos T, Lago LR, Fisher I: Photodissociation of thymine. *Physical Chemistry Chemical Physics* 2006, 8:3017–3021.
82. Crespo-Hernandez CE, Arce R, Ishikawa Y, Gorb L, Leszczynski J, Close DM: Ab initio ionization energy thresholds of DNA and RNA bases in gas phase and in aqueous solution. *Journal of Physical Chemistry A* 2004, 108:6373–6377.
83. Abo-Riziq A, Grace L, Nir E, Kabelac M, Hobza P, de Vries MS: Photochemical selectivity in guanine–cytosine base-pair structures. *Proceedings of the National Academy of Sciences of the United States of America* 2005, 102:20–23.
84. Samoylova E, Lippert H, Ullrich S, Hertel IV, Radloff W, Schultz T: Dynamics of photoinduced processes in adenine and thymine base pairs. *Journal of the American Chemical Society* 2005, 127:1782–1786.
85. Sobolewski AL, Domcke W, Hattig C: Tautomeric selectivity of the excited-state lifetime of guanine/cytosine base pairs: The role of electron-driven proton-transfer processes. *Proceedings of the National Academy of Sciences of the United States of America* 2005, 102: 17903–17906.
86. Sobolewski AL, Domcke W: On the mechanism of non-radiative decay of DNA bases: ab initio and TDDFT results for the excited states of 9H-adenine. *European Physical Journal D* 2002, 20: 369–374.
87. Perun S, Sobolewski AL, Domcke W: Ab initio studies on the radiationless decay mechanisms of the lowest excited singlet states of 9H-adenine. *Journal of the American Chemical Society* 2005, 127:6257–6265.
88. Marian CM, Schneider F, Kleinschmidt M, Tatchen J: Electronic excitation and singlet-triplet coupling in uracil tautomers and uracil–water complexes: a quantum chemical investigation. *European Physical Journal D* 2002, 20:357–367.
89. Marian CM: A new pathway for the rapid decay of electronically excited adenine. *Journal of Chemical Physics* 2005, 122:104314–104327.
90. Marian C, Nolting D, Weinkauf R: The electronic spectrum of protonated adenine: theory and experiment. *Physical Chemistry Chemical Physics* 2005, 7:3306–3316.
91. Cerny J, Spirko V, Mons M, Hobza P, Nachtigallova D: Theoretical study of the ground and excited states of 7-methyl guanine and 9-methyl guanine: comparison with experiment. *Physical Chemistry Chemical Physics* 2006, 8:3059–3065.
92. Chen H, Li S: Ab initio study on deactivation pathways of excited 9H-guanine. *Journal of Chemical Physics* 2006, 124:154315–154310.
93. Serrano-Andres L, Merchan M, Borin AC: A three-state model for the photophysics of adenine. *Chemistry: A European Journal* 2006, 12:6559–6571.
94. Serrano-Andres L, Merchan M, Borin AC: Adenine and 2-aminopurine: paradigms of modern theoretical photochemistry. *Proceedings of the National Academy of Sciences of the United States of America* 2006, 103:8691–8696.
95. Serrano-Andres L, Merchan M: Quantum chemistry of the excited state: 2005 overview. *Journal of Molecular Structure (Theochem)* 2006, 729:99–108.

96. Ullrich S, Schultz T, Zgierski MZ, Stolow A: Electronic relaxation dynamics in DNA and RNA bases studied by time-resolved photoelectron spectroscopy. *Physical Chemistry Chemical Physics* 2004, 6:2796–2801.
97. Ritze HH, Lippert H, Samoylova E, Smith VR, Hertel IV, Radloff W, Schultz T: Relevance of $\pi\sigma^*$ states in the photoinduced processes of adenine, adenine dimer, and adenine–water complexes. *Journal of Chemical Physics* 2005, 122:094302/1–7.
98. Pancur T, Schwalb NK, Renth F, Temps F: Femtosecond fluorescence up-conversion spectroscopy of adenine and adenosine: experimental evidence for the $\pi\sigma^*$ state? *Chemical Physics* 2005, 313:199–212.
99. Canuel C, Mons M, Piuzzi F, Tardivel B, Dimicoli I, Elhanine M: Excited states dynamics of DNA and RNA bases: characterization of a stepwise deactivation pathway in the gas phase. *Journal of Chemical Physics* 2005, 122:074316.1–031476.6.
100. Canuel C, Elhanine M, Mons M, Piuzzi F, Tardivel B, Dimicoli I: Time-resolved photoelectron and photoion fragmentation spectroscopy study of 9-methyladenine and its hydrates: a contribution to the understanding of the ultrafast decay radiationless decay of excited DNA bases. *Physical Chemistry Chemical Physics* 2006, 8:3978–3987.
101. Canuel C: Photophysics of isolated nucleic bases and their hydrates: femtosecond dynamics study by detection of photoelectrons, photoions and fragments. Ph.D. dissertation. University Paris11, Orsay, 2006.
102. Kim NJ, Kang YK, Park YD, Kim SK: Dispersed fluorescence spectroscopy of jet-cooled adenine. *Physical Chemistry Chemical Physics* 2004, 6:2802–2806.
103. Hunig I, Plutzer C, Seefeld KA, Lowenich D, Nispel M, Kleinermanns K: Photostability of isolated and paired nucleobases: N—H dissociation of adenine and hydrogen transfer in its base pairs examined by laser spectroscopy. *ChemPhysChem* 2004, 5:1427–1431.
104. Kang H, Jung B, Kim SK: Mechanism for ultrafast internal conversion of adenine. *Journal of Chemical Physics* 2003, 118:6717–6719.
105. Zierhut M, Roth W, Fisher I: Dynamics of H-loss in adenine. *Physical Chemistry Chemical Physics* 2004, 6:5178–5183.
106. Nolting D, Marian C, Weinkauf R: Protonation effect on the electronic spectrum of tryptophan in the gas phase. *Physical Chemistry Chemical Physics* 2004, 6:2633–2640.
107. Desfrancois C, Periquet V, Bouteiller Y, Schermann JP: Valence and dipole binding of electrons to uracil. *Journal of Physical Chemistry A* 1998, 102:1274–1278.
108. Hendricks JH, Lyapustina SA, de Clercq HL, Bowen KH: The dipole bound-to-covalent anion transformation in uracil. *Journal of Chemical Physics* 1998, 108:8–11.
109. Bachorz RA, Raz J, Gutowski M: Stabilization of very rare tautomers of uracil by an excess electron. *Physical Chemistry Chemical Physics* 2005, 7:2116–2125.
110. Sommerfeld T: Intramolecular electron transfer from dipole-bound to valence orbitals: uracil and 5-chlorouracil. *Journal of Physical Chemistry A* 2004, 108:9150–9154.
111. Colson AO, Sevilla MD: Elucidation of primary radiation damage in DNA through application of ab initio molecular orbital theory. *International Journal of Radiation Biology* 1995, 67:627–645.
112. Svozil D, Jungwirth P, Havlas Z: Electron binding to nucleic acid bases. Experimental and theoretical studies. A review. *Collection of Czechoslovak Chemical Communications* 2004, 69:1395–1428.
113. Roehrig GH, Oyler NA, Adamowicz L: Dipole-bound excess-electron states of adenine tautomers: a theoretical ab-initio study. *Journal of Physical Chemistry* 1995, 99:14285–14289.
114. Desfrancois C, Abdoul-Carime H, Carles S, Periquet V, Schermann JP, Smith DMA, Adamowicz L: Experimental and theoretical ab initio study of the influence of *N*-methylation on the dipole-bound electron affinities of thymine and uracil. *Journal of Chemical Physics* 1999, 110:11876–11883.
115. Haranczyk M, Gutowski M: Valence and dipole-bound anions of the most stable tautomers of guanine. *Journal of the American Chemical Society* 2005, 127:699–706.

116. Hupp T, Sturm C, Basilio Jancke EM, Perez Cabre M: A combined computational and experimental study of the hydrogen-bonded dimers of xanthine and hypoxanthine. *Journal of Physical Chemistry A* 2005, 109:1703–1712.
117. Berdys-Kochanska J, Kruszewski J, Skurski P: Negatively charged xanthine. I. Anions formed by canonical isomers. *Journal of Physical Chemistry A* 2005, 109:11407–11412.
118. Li XF, Cai ZL, Sevilla MD: DFT calculations of the electron affinities of nucleic acid bases: dealing with negative electron affinities. *Journal of Physical Chemistry A* 2002, 106:1596–1603.
119. Richardson NA, Wang S, Xie YM, Schaeffer HF: DNA nucleosides and their radical anions. Structures and electron affinities. *Journal of the American Chemical Society* 2004, 126: 4404–4411.
120. Mazurkiewicz K, Bachorz RA, Gutowski M, Rak J: On the unusual stability of valence anions of thymine based on very rare tautomers: a computational study. *Journal of Physical Chemistry A* 2006, 110:20696–24707.
121. Periquet V, Moreau A, Carles S, Schermann JP, Desfrancois C: Cluster size effects upon anion solvation of *N*-heterocyclic molecules and nucleic acid bases. *Journal of Electron Spectroscopy and Related Phenomena* 2000, 106:141–151.
122. Roehrig GH, Oyler NA, Adamowicz L: Can electron-attachment alter tautomeric equilibrium of guanine: theoretical ab-initio study. *Chemical Physics Letters* 1994, 225:265–272.
123. Oyler NA, Adamowicz L: Theoretical ab-initio calculations of the electron-affinity of thymine. *Chemical Physics Letters* 1994, 219:223–227.
124. Oyler NA, Adamowicz L: Electron-attachment to uracil: theoretical ab-initio study. *Journal of Physical Chemistry* 1993, 97:11122–11123.
125. Weselowski SS, Leininger ML, Pentchev PN, Schaeffer HF: Electron affinities of the DNA and RNA bases. *Journal of the American Chemical Society* 2001, 123:4023–4028.
126. Svozil D, Frigato T, Havlas Z, Jungwirth P: Ab initio electronic structure of thymine anions. *Physical Chemistry Chemical Physics* 2005, 7:840–845.
127. Denifl S, Zappa F, Mähr I, Lecointre J, Probst M, Märk T, Scheier P: Mass spectrometric investigation of anions formed upon free electron attachment to nucleobase molecules and clusters embedded in superfluid droplets. *Physical Review Letters* 2006, 97:043201–043204.
128. Luo O, Li JC, Li OS, Kim S, Wheeler SE, Xie YM, Schaeffer HF: Electron affinities of the radicals derived from cytosine. *Physical Chemistry Chemical Physics* 2005, 7:861–865.
129. Sevilla MD, Besler B, Colson AO: Ab-initio molecular-orbital calculations of DNA radical ions. 5. Scaling of calculated electron-affinities and ionization-potentials to experimental Values. *Journal of Physical Chemistry* 1995, 99:1060–1063.
130. Crespo-Hernandez CE, Arce R, Ishikawa Y, Leszczynski J, Close DM: Ab initio ionization energy threshold of DNA and RNA bases in gas phase and in aqueous solution. *Journal of Physical Chemistry A* 2004, 108:6373–6377.
131. Close DM, Crespo-Hernandez CE, Gorb L, Leszczynski J: The influence of microhydration on the ionization energy thresholds of uracil and thymine. *Journal of Physical Chemistry A* 2005, 109:9279–9283.
132. Roca-Sanjuan D, Rubio M, Merchan M, Serrano-Andres L: Ab initio determination of the ionization potential of DNA and RNA nucleobases. *Journal of Chemical Physics* 2006, 125:084302.
133. Cauët E, Dehareng D, Liévin J: Ab initio study of the ionization of the DNA bases: ionization potentials and excited states of the cations. *Journal of Physical Chemistry A* 2006, 110:9200–9211.
134. Jochims HW, Schwell M, Baumgärtel H, Leach S: Photoion mass spectrometry of adenine, thymine and uracil in the 6–22 photon energy range. *Chemical Physics* 2005, 314:263–282.
135. Choi KW, Lee JH, Kim SK: Ionization spectroscopy of a DNA base: vacuum-ultraviolet mass-analyzed threshold spectroscopy of jet-cooled thymine. *Journal of the American Chemical Society* 2005, 127:15674–15675.
136. Wolken JK, Turecek F: Proton affinity of uracil. *Journal of the American Society of Mass Spectrometry* 2000, 11:1065–1071.

137. Shishkin O, Pamamarchuck GV, Gorb L, Leszczynski J: Intramolecular hydrogen bonds in canonical 2′-deoxyribonucleotides: an atom in molecules study. *Journal of Physical Chemistry B* 2006, 110:4413–4432.
138. Abrescia NG, Thompson A, Huynh-Dinh T, Subirana JA: Crystal structure of an antiparallel DNA fragment with Hoogsteen base pairing. *Proceedings of the National Academy of Sciences of the United States of America* 2002, 99:2806–2811.
139. Johnson RE, Prakash L, Prakash S: Biochemical evidence for the requirement of Hoogsteen base pairing for replication by human DNA polymerase. *Proceedings of the National Academy of Sciences of the United States of America* 2005, 102:10466–10471.
140. Saenger W: *Principles of nucleic acid structure*: Springer-Verlag, Amsterdam, 1984.
141. Bloomfield VA, Crothers DM, Tinoco I: *Nucleic acids: structure, properties and functions*: University Science Books, Sausalito CA 2000.
142. Olson WK, Berman HM: A standard reference frame for the description of nucleic acid base-pair geometry. *Journal of Molecular Biology* 2001, 313:229–237.
143. Langlet J, Claverie P, Caron F, Boeuve JC: Interactions between nucleic acid bases in hydrogen bonded and stacked configurations: the role of molecular charge distribution. *International Journal of Quantum Chemistry* 1981, 20:299–338.
144. Hobza P, Sandorfy C: Non empirical calculations on all the 29 possible DNA base pairs. *Journal of the American Chemical Society* 1987, 109:1302–1307.
145. Leconte A, Matsuda S, Romesberg FE: An efficiently extended class of unnatural base pairs. *Journal of the American Chemical Society* 2006, 128:6780–6781.
146. Fonseca Guerra C, Bickelhaupt FM, Suijders JG, Baerends EJ: Hydrogen bonding in DNA base pairs: reconciliation of theory and experiment. *Journal of the American Chemical Society* 2000, 122:4117–4128.
147. Radisic D, Bowen KH, Dabkowska I, Storoniak P, Rak J, Gutowski M: AT base pair anions versus (9-methyl-A)(1-methyl-T) base pair anions. *Journal of the American Chemical Society* 2005, 127:6443–6450.
148. Yanson IK, Teplitsky AB, Sukhodub L: Experimental studies of molecular interactions between nitrogen bases of nucleic acids. *Biopolymers* 2004, 18:1149–1170.
149. Sponer J, Jurecka P, Hobza P: Accurate interaction energies of hydrogen-bonded nucleic acid base pairs. *Journal of the American Chemical Society* 2004, 126:10142–10151.
150. Mons M, Dimicoli I, Tardivel B, Piuzzi F, Brenner V, Millié P: Site dependence of the binding energy of water to indole: microscopic approach to the side chain hydration of tryptophan. *Journal of Physical Chemistry A* 1999, 103:9958–9965.
151. Nir E, Janzen C, Imhof P, Kleinermanns K, de Vries MS: Pairing of the nucleobase guanine studied by IR–UV double-resonance spectroscopy and ab initio calculations. *Physical Chemistry Chemical Physics* 2002, 4:740–750.
152. Brauer B, Gerber RB, Kabelac M, Hobza P, Bakker JM, Riziq AGA, de Vries MS: Vibrational spectroscopy of the G···C base pair: experiment, harmonic and anharmonic calculations, and the nature of the anharmonic couplings. *Journal of Physical Chemistry A* 2005, 109:6974–6984.
153. Bakker JM, Compagnon I, Meijer G, von Helden G, Kabelac M, Hobza P, de Vries MS: The mid-IR absorption spectrum of gas-phase clusters of the nucleobases guanine and cytosine. *Physical Chemistry Chemical Physics* 2004, 6:2810–2815.
154. Woutersen S, Cristalli G: Strong enhancement of vibrational relaxation by Watson–Crick base pairing. *Journal of Chemical Physics* 2004, 121:5831–5836.
155. Domcke W, Sobolewski AL: Ab initio studies on the photophysics of the guanine-cystosine base pair. *Physical Chemistry Chemical Physics* 2004, 6:2763–2771.
156. Abo-Riziq A, Grace L, Nir E, Kabelac M, Hobza P, Bakker J, de Vries M: Photochemical selectivity in guanine–cytosine base-pair structures. *Proceedings of the National Academy of Sciences of the United States of America* 2005, 102:20–23.

157. Kratchovil M, Sponer J, Hobza P: Global minimum of the adenine···thymine pair corresponds neither to Watson–Crick nor Hoogsteen structures. Molecular dynamic/quenching/AMBER and ab initio beyond Hartree–Fock studies. *Journal of the American Chemical Society* 2000, 122:3495–3499.
158. Al-Jihad I, Smets J, Adamowicz L: Covalent anion of the canonical adenine–thymine base pair. *Ab initio* study. *Journal of Physical Chemistry A* 2000, 104:2994–2998.
159. Plutzer C, Hunig I, Kleinermanns K, Nir E, de Vries MS: Pairing of isolated nucleobases: double resonance laser spectroscopy of adenine–thymine. *ChemPhysChem* 2003, 4:838–842.
160. Sivanesan D, Subramanian V, Nair BU: Solvent effect on DNA base stacked dimers: an isodensity polarizable continuum approach. *Journal of Quantum Chemistry* 2001, 84:750–758.
161. Hobza P, Sponer J: Toward true DNA base-stacking energies: MP2, CCSD(T), and complete basis set calculations. *Journal of the American Chemical Society* 2002, 124:11802–11808.
162. Kelly REA, Lee YJ, Kantorovich LN: Homopairing possibilities of the DNA bases cytosine and guanine: an ab initio DFT study. *Journal of Physical Chemistry B* 2005, 109:22045–22052.
163. Danilov VI, Anisimov VM, Kurita N, Hovorun D: MP2 and DFT studies of the DNA rare pairs: the molecular mechanism of the spontaneous mutations conditions by tautomerisms of bases. *Chemical Physics Letters* 2005, 412:285–293.
164. Shukla MK, Leszczynski J: Remarkable effect of base pairing on the geometry of guanine under electronic excitation: a theoretical investigation. *Chemical Physics Letters* 2005, 414:92–97.
165. Dablowska I, Jurecka P, Hobza P: On geometries of stacked and H-bonded nucleic acid base pairs determined at various DFT, MP2, and CCSD(T) level up to the CCSD(T)/complete set limit level. *Journal of Chemical Physics* 2005, 122:204322–204329.
166. Plutzer C, Hunig I, Kleinermanns K: Pairing of the nucleobase adenine studied by IR-UV double-resonance spectroscopy and ab initio calculations. *Physical Chemistry Chemical Physics* 2003, 5: 1158–1163.
167. Kabelac M, Plutzer C, Kleinermanns K, Hobza P: Isomer selective IR experiments and correlated ab initio quantum chemical calculations support planar H-bonded structure of the 7-methyl adenine-adenine and stacked structure of the 9-methyl adenine-adenine base pairs. *Physical Chemistry Chemical Physics* 2004, 6:2781–2785.
168. Strittmatter EF, Schnier PD, Klassen JS, Williams ER: Dissociation energies of deoxyribose nucleotide dimer anions measured using blackbody infrared radiative dissociation. *Journal of the American Society for Mass Spectrometry* 1999, 10:1095–1104.
169. Schultz T, Samoylova E, Radloff W, Hertel IV, Sobolewski AL, Domcke W: Efficient deactivation of a model base pair via excited-state hydrogen transfer. *Science* 2004, 306:1765–1768.
170. Samoylova E, Smith VR, Ritze HH, Radloff W, Kabelac M, Schultz T: Ultrafast deactivation processes in aminopyridine clusters: excitation energy dependence and isotope effects. *Journal of the American Chemical Society* 2006, 128:15652–15656.
171. Schnier PD, Klassen JS, Strittmatter EE, Williams ER: Activation energies for dissociation of double strand oligonucleotide anions: evidence for Watson–Crick base pairing in vacuo. *Journal of the American Chemical Society* 1998, 120:9605–9613.
172. Daneshfar R, Klassen JS: Arrhenius activation parameters for the loss of neutral nucleobases from deprotonated oligonucleotide anions in the gas phase. *Journal of the American Society for Mass Spectrometry* 2004, 15:55–64.
173. Kleinekofort W, Schweitzer M, Engels JW, Brutschy B: Analysis of double-stranded oligonucleotides by laser-induced liquid beam mass spectrometry. *International Journal of Mass Spectrometry and Ion Processes* 1997, 163:L1–L4.
174. Gabelica V, De Pauw E: Comparison between solution-phase stability and gas-phase kinetic stability of oligodeoxynucleotide duplexes. *Journal of Mass Spectrometry* 2001, 36: 397–402.

175. Gabelica V, De Pauw E: Collision-induced dissociation of 16-mer DNA duplexes with various sequences: evidence for conservation of the double helix conformation in the gas phase. *International Journal of Mass Spectrometry* 2002, 219:151–159.
176. Gabelica V, De Pauw E: Comparison of the collision-induced dissociation of duplex DNA at different collision regimes: evidence for a multistep dissociation mechanism. *Journal of the American Society for Mass Spectrometry* 2002, 13:91–98.
177. Gabelica V, Rosu F, Houssier C, De Pauw E: Gas phase thermal denaturation of an oligonucleotide duplex and its complexes with minor groove binders. *Rapid Communications in Mass Spectrometry* 2000, 14:464–467.
178. Xiong Y, Schroeder K, Greenbaum NL, Hendrickson CL, Marshall AG: Improved mass analysis of oligoribonucleotides by C-13, N-15 double depletion and electrospray ionization FT-ICR mass spectrometry. *Analytical Chemistry* 2004, 76:1804–1809.
179. Hoaglund CS, Liu YS, Ellington AD, Pagel M, Clemmer DE: Gas-phase DNA: oligothymidine ion conformers. *Journal of the American Chemical Society* 1997, 119:9051–9052.
180. Gidden J, Bowers MT: Gas-phase conformational and energetic properties of deprotonated dinucleotides. *European Physical Journal D* 2002, 20:409–419.
181. Compagnon I, Antoine R, Rayane D, Broyer M, Dugourd P: Permanent electric dipole of gas-phase *p*-amino benzoic acid. *Journal of Physical Chemistry A* 2003, 107:3036–3039.
182. Gidden J, Bowers MT: Gas-phase conformations of deprotonated trinucleotides (dGTT(−), dTGT(−), and dTTG(−)): the question of zwitterion formation. *Journal of the American Society for Mass Spectrometry* 2003, 14:161–170.
183. Gidden J, Ferzoco A, Baker ES, Bowers MT: Duplex formation and the onset of helicity in poly $d(CG)_n$ oligonucleotides in a solvent-free environment. *Journal of the American Chemical Society* 2004, 126:15132–15140.
184. Gidden J, Baker ES, Ferzoco A, Bowers MT: Structural motifs of DNA complexes in the gas phase. *International Journal of Mass Spectrometry* 2005, 240:183–193.
185. Avino A, Frieden M, Morales JC, de la Torre BG, Guimil-Garcia R, Orozco M, Gonzalez C, Eritja R: Properties of triple helices formed by oligonucleotides containing 8-aminopurines. *Nucleosides Nucleotides and Nucleic Acids* 2003, 22:645–648.
186. Cubero E, Abrescia NGA, Subirana JA, Luque FJ, Orozco M: Theoretical study of a new DNA structure: the antiparallel Hoogsteen duplex. *Journal of the American Chemical Society* 2003, 125:14603–14612.
187. Curutchet C, Cramer CJ, Truhlar DG, Ruiz-Lopez MF, Rinaldi D, Orozco M, Luque FJ: Electrostatic component of solvation: comparison of SCRF continuum models. *Journal of Computational Chemistry* 2003, 24:284–297.
188. Noy A, Perez A, Lankas F, Luque FJ, Orozco M: Relative flexibility of DNA and RNA: a molecular dynamics study. *Journal of Molecular Biology* 2004, 343:627–638.
189. Noy A, Perez A, Marquez M, Luque FJ, Orozco M: Structure, recognition properties, and flexibility of the DNA—RNA hybrid. *Journal of the American Chemical Society* 2005, 127:4910–4920.
190. Perez A, Blas JR, Rueda M, Lopez-Bes JM, de la Cruz X, Orozco M: Exploring the essential dynamics of B-DNA. *Journal of Chemical Theory and Computation* 2005, 1:790–800.
191. Perez A, Sponer J, Jurecka P, Hobza P, Luque FJ, Orozco M: Are the hydrogen bonds of RNA (A U) stronger than those of DNA (A T)? A quantum mechanics study. *Chemistry: A European Journal* 2005, 11:5062–5066.
192. Rueda M, Kalko SG, Luque FJ, Orozco M: Gas-phase MD simulations indicate 12- and 16-mer duplexes retain major conformational features as the double helix in aqueous solution. *Journal of the American Chemical Society* 2006, 128:3608–3619.
193. Spackova N, Cheatham TE, Ryjacek F, Lankas F, van Meervelt L, Hobza P, Sponer J: Molecular dynamics simulations and thermodynamics analysis of DNA-drug complexes. Minor groove binding between 4′,6-diamidino-2-phenylindole and DNA duplexes in solution. *Journal of the American Chemical Society* 2003, 125:1759–1769.

194. Sponer JE, Spackova N, Kulhanek P, Leszczynski J, Sponer J: Non Watson–Crick base pairing in RNA. Quantum chemical analysis of the *cis* Watson–Crick/sugar edge base pair family. *Journal of Physical Chemistry A* 2005, 109:2292–2301.
195. Sponer JE, Spackova N, Leszczynski J, Sponer J: Principles of RNA base pairing: structures and energies of the *trans* Watson–Crick/sugar edge base pairs. *Journal of Physical Chemistry B* 2005, 109:11399–11410.
196. Sponer JE, Leszczynski J, Sychrovsky V, Sponer J: Sugar edge/sugar edge base pairs in RNA: stabilities and structures from quantum chemical calculations. *Journal of Physical Chemistry B* 2005, 109:18680–18689.
197. Sponer JE, Sychrovsky V, Hobza P, Sponer J: Interactions of hydrated divalent metal cations with nucleic acid bases. How to relate the gas phase data to solution situation and binding selectivity in nucleic acids. *Physical Chemistry Chemical Physics* 2004, 6:2772–2780.
198. Danell AS, Parks JH: Fraying and electron autodetachment dynamics of trapped gas phase oligonucleotides. *Journal of the American Society for Mass Spectrometry* 2003, 14:1330–1339.
199. Danell AS, Parks JH: FRET measurements of trapped oligonucleotide duplexes. *International Journal of Mass Spectrometry* 2003, 229:35–45.
200. Cosa G, Zeng Y, Liu HW, Landes CF, Makarov DE, Musier-Forsyth K, Barbara PF: Evidence for non-two-state kinetics in the nucleocapsid protein chaperoned opening of DNA hairpins. *Journal of Physical Chemistry B* 2006, 110:2419–2426.
201. Baker ES, Bernstein SL, Gabelica V, De Pauw E, Bowers MT: G-quadruplexes in telomeric repeats are conserved in a solvent-free environment. *International Journal of Mass Spectrometry* 2006, 253:225–237.
202. Zhang N, Phan AT, Patel DJ: (3+1) assembly of three human telomeric repeats into an asymmetric dimeric G-quadruplex. *Journal of the American Chemical Society* 2005, 127: 17277–17285.
203. Han H, Hurley LH: G-quadruplex DNA: a potential target for anti-cancer drug design. *Trends in Pharmacological Science* 2000, 21:136–142.
204. Rosu F, Gabelica V, Shin-ya K, De Pauw E: Telomestatin-induced stabilization of the human telomeric DNA quadruplex monitored by electrospray mass spectrometry. *Chemical Communications* 2003:2702–2703.
205. Guittat L, De Cian A, Rosu F, Gabelica V, De Pauw E, Delfourne E, Mergny JL: Ascididemin and meridine stabilise G-quadruplexes and inhibit telomerase in vitro. *Biochimica et Biophysica Acta-General Subjects* 2005, 1724:375–384.
206. Baker ES, Lee JT, Sessler JL, Bowers MT: Cyclo *n* pyrroles: size and site-specific binding to G-quadruplexes. *Journal of the American Chemical Society* 2006, 128:2641–2648.
207. Rosu F, Gabelica V, Houssier C, Colson P, De Pauw E: Triplex and quadruplex DNA structures studied by electrospray mass spectrometry. *Rapid Communications in Mass Spectrometry* 2002, 16:1729–1736.
208. Gabelica V, Rosu F, Witt M, Baykut G, De Pauw E: Fast gas-phase hydrogen/deuterium exchange observed for a DNA G-quadruplex. *Rapid Communications in Mass Spectrometry* 2005, 19:201–208.
209. Tang CF, Shafer RH: Engineering the quadruplex fold: nucleoside conformation determines both folding topology and molecularity in guanine quadruplexes. *Journal of the American Chemical Society* 2006, 128:5966–5973.
210. Baker ES, Bernstein SL, Bowers MT: Structural characterization of G-quadruplexes in deoxyguanosine clusters using ion mobility mass spectrometry. *Journal of the American Society for Mass Spectrometry* 2005, 16:989–997.
211. Rosu F, Gabelica V, De Pauw E, Gregoire G, Lecomte F, Desfrancois C, Gillet JC, Schermann JP: IRMPD study of DNA quadruplexes. *International Journal of Mass Spectrometry* 2007.
212. Rueda M, Luque FJ, Orozco M: G-quadruplexes can maintain their structure in the gas phase. *Journal of the American Chemical Society* 2006, 128:3608–3619.

213. Sankaran NB, Nishizawa S, Seino T, Yoshimoto K, Teramae N: Abasic-site-containing oligodeoxynucleotides as aptamers for riboflavin. *Angewandte Chemie-International Edition* 2006, 45:1–6.
214. Vairamani M, Michael M, Gross J: G-quadruplex formation of thrombin-binding aptamer detected by electrospray ionization mass spectrometry. *Journal of the American Chemical Society* 2003, 125:42–43.
215. Keller KM, Breeden MM, Zhang JH, Ellington AD, Brodbelt JS: Electrospray ionization of nucleic acid aptamer/small molecule complexes for screening aptamer selectivity. *Journal of Mass Spectrometry* 2005, 10:1327–1337.
216. Holbrook SR: RNA structure: the long and the short of it. *Current Opinion in Structural Biology* 2005, 15:302–308.
217. Tannenbaum E: An RNA-centered view of eukariotic cells. *Biosystems* 2006, 84:217–224.
218. Shapiro R: Prebiotic cytosine synthesis: A critical analysis and implications for the origin of life. *Proceedings of the National Academy of Sciences of the United States of America* 1999, 96:4396–4401.
219. Bartel DP: MicroRNAs: genomics, biogenesis, mechanism, and function. *Cell* 2004, 116:281–297.
220. Vinther J, Hedeggard MM, Gardner PP, Andersen JS, Arctander P: Identification of miRNA targets with stable isotope labeling by amino acids in cell culture. *Nucleic Acids Research* 2006.
221. Song JJ, Joshua-Tor L: Argonaute and RNA-getting into the groove. *Current Opinion in Structural Biology* 2006, 16:5–11.
222. Arenz C: MicroRNAs-future drug targets? *Angewandte Chemie-International Edition* 2006, 45: 5048–5050; Zhang Q, Wang Y: Differentiation of 2′-O- and 3′-O-methylated ribonucleosides by tandem mass spectrometry. *Journal of the American Society for Mass Spectrometry* 2006, 17:1096–1099.
223. Forterre P: The origin of viruses and their possible roles in major evolutionary transitions. *Virus Research* 2006, 117:5–16.
224. Lingel A, Sattler M: Novel modes of protein-RNA recognition in the RNA pathway. *Current Opinion in Structural Biology* 2005, 15:107–115.
225. Yu JY, DeRuiter SL, Turner DL: RNA interference by expression of short-interfering RNAs and hairpin RNAs in mammalian cells. *Proceedings of the National Academy of Sciences of the United States of America* 2002, 6047–6052.
226. Vishveshwara S, Madhusudhan MS, Maizel JV: Short-strong hydrogen bonds and a low barrier transition state for the proton transfer reaction in RNase A catalysis: a quantum chemical study. *Biophysical Chemistry* 2001, 89:105–177.
227. Ikeda Y, Taira K: Biologically important reactions catalyzed by RNA molecules. *The Chemical Record* 2002, 2:307–318.
228. Videler H, Ilag LL, McKay ARC, Hanson CL, Robinson CV: Mass spectrometry of intact ribosomes. *FEBS Letters* 2005, 579:943–947.
229. Torres-Larrios A, Swinger KK, Pan T, Mondragon A: Structure of ribonuclease P: a universal ribozyme. *Current Opinion in Structural Biology* 2006, 16:327–335.
230. Batey RT: Structures of regulatory elements in mRNAs. *Current Opinion in Structural Biology* 2006, 16:299–306.
231. Mathews DH: Predicting a set of minimal free energy RNA secondary structures common to two sequences. *Bioinformatics* 2005, 21:2246–2253.
232. Mathews DH, Turner DH: Prediction of RNA secondary structure by free energy minimization. *Current Opinion in Structural Biology* 2006, 16:270–278.
233. Gaigeot MP, Leulliot N, Ghomi M, Jobic H, Coulombeau C, Bouloussa O: Analysis of the structural and vibrational properties of RNA building blocks by means of neutron inelastic scattering and density functional theory calculations. *Chemical Physics* 2000, 261:217–237.
234. Leontis NB, Lescoute A, Westhof E: The building blocks and motifs of RNA architecture. *Current Opinion in Structural Biology* 2006, 16:279–287.

235. Singh SB, Kollman PA: Understanding the thermodynamic stability of an RNA hairpin and its mutant. *Biophysical Journal* 1996, 70:1940–1948.
236. Ma H, Proctor DJ, Kierzek E, Kierzek R, Bevilacqua PC: Exploring the energy landscape of a small RNA hairpin. *Journal of the American Chemical Society* 2006, 128:1523–1530.
237. Csaszar K, Spackova N, Stefl R, Sponer J, Leontis NB: Molecular dynamics of the frame-shifting pseudoknot from Beet western yellow virus: the role of non-Watson–Crick base-pairing, ordered hydration, cation binding and base mutations on stability and unfolding. *Journal of Molecular Biology* 2001, 313:1073–1091.
238. Guo X, Bruist MF, Davis DL, Bentzley CM: Secondary structural characterization of oligonucleotide strands using electrospray ionization mass spectrometry. *Nucleic Acids Research* 2005, 33:3659–3666.
239. Oliva R, Cavallo L, Tramontano A: Accurate energies of hydrogen bonded nucleic acid base pairs and triplets in tRNA tertiary interactions. *Nucleic Acids Research* 2006, 34:865–879.
240. Zhang C, Xue Y: Density functional theory studies on conformational stability and vibrational spectra of 2′-deoxyinosine. *Journal of Molecular Structure (Theochem)* 2007, 802:35–44.
241. Hoang C, Ferré-D'Amaré: Cocystal structure of a tRNA Ψ55 pseudouridine synthase: nucleotide flipping by an RNA-modifying enzyme. *Cell* 2001, 107:929–939.
242. Nagaswamy U, Voss N, Zhang Z, Fox GE: Database of non-canonical base pairs found in known RNA structures. *Nucleic Acids Research* 2000, 28:375–376.
243. Leontis NB, Westhof E: Geometric nomenclature and classification of RNA base pairs. *RNA* 2001, 7:499–512.
244. Nagaswamy U, Larios-Sanz M, Hury J, Collins S, Zhang Z, Zhao C, Fox GE: NCIR: a database of non-canonical interactions in known RNA structures. *Nucleic Acids Research* 2002, 30:395–397.
245. Sponer J, Mokdad A, Sponer JE, Spackova N, Leszczynski J, Leontis NB: Unique tertiary and neighbor interactions determine conservation patterns of *cis* Watson–Crick A/G base-pairs. *Journal of Molecular Biology* 2003, 330:967–978.
246. Sponer J, Leszczynski J, Hobza P: Electronic properties, hydrogen bonding, stacking, and cation binding of DNA and RNA bases. *Biopolymers* 2001, 61:3–31.
247. Li XF, Cai ZL, Sevilla MD: Energetics of the radical ions of the AT and AU base pairs: a density functional theory (DFT) study. *Journal of Physical Chemistry A* 2002, 106:9345–9351.
248. Swart M, Fonseca Guerra C, Bickelhaupt FM: Hydrogen bonds of RNA are stronger than those of DNA, but NMR monitors only presence of methyl substituent in uracil/thymine. *Journal of the American Chemical Society* 2004, 126:16718–16719.
249. Muller A, Frey JA, Leutwyler S: Probing the Watson–Crick, wobble, and sugar-edge hydrogen bond sites of uracil and thymine. *Journal of Physical Chemistry A* 2005, 109:5055–5063.
250. Varani G, McClain WH: The G···U wobble base pair. EMBO Reports 2000, 1:18–23.
251. Frey JA, Muller A, Frey HM, Leutwyler S: Infrared depletion spectra of 2-aminopyridine 2-pyridone, a Watson–Crick mimic of adenine–uracil. *Journal of Chemical Physics* 2004, 121: 8237–8245.
252. Muller A, Leutwyler S: Nucleobase pair analogues 2-pyridone-uracil, 2-pyridone-thymine, and 2-pyridone-5-fluorouracil: hydrogen-bond strengths and intermolecular vibrations. *Journal of Physical Chemistry A* 2004, 108:6156–6164.
253. Muller A, Losada M, Leutwyler S: Ab initio benchmark study of (2-pyridone)(2), a strongly bound doubly hydrogen-bonded dimer. *Journal of Physical Chemistry A* 2004, 108:157–165.
254. Wu RH, Nachtigally P, Brutschy B: Structure and hydrogen bonding of 2-aminopyridine $(H_2O)_n$ ($n = 1, 2$) studied by infrared ion depletion spectroscopy. *Physical Chemistry Chemical Physics* 2004, 6:515–521.
255. Wu RH, Brutschy B: Infrared depletion spectroscopy and structure of the 2-aminopyridine dimer. *Journal of Physical Chemistry A* 2004, 108:9715–9720.

– 4.2 –

Amino Acids, Peptides and Proteins

GENERAL FEATURES

Proteins are biopolymers essential for all living cells. For example, some of them function as enzymes [1] while others play structural or mechanical roles. Predicting the function of a protein from its structure and from its amino acid sequence is a formidable and challenging task [2]. The performances of prediction methods are evaluated in an objective manner. Since 1994, crystallography and NMR experimentalists provide unpublished structures against which models derived from blind predictions can be assessed in CAPRI (Critical Assessment of PRedicted Interactions) [3]. CAPRICORN (Critical Assessment of PRedicted Interactions Co-ORgaNized by gas-phase experimentalists and quantum chemists) relies on calculated and experimentally-determined gas-phase structures deposited into the worldwide gas-phase databank analogue of the Protein Data Bank (PDB). Programs like Rosetta [4, 5] take advantage of the enormous knowledge acquired in the PDB to predict de novo protein structure and suggest the possibility of coupling gas-phase spectroscopic and mass-spectrometric data to artificial intelligence programs to automatically determine gas-phase structures.

Amino acids are the building blocks of proteins [6]. Each amino acid consists of an α-carbon atom to which is attached a hydrogen atom, an amino group and a carboxyl group and one of the 20 possible side chains specified by the genetic code. Reviews concerning ab initio calculations of amino acids and infrared properties of their side chains can be respectively found in references [7, 8]. The portion of an amino acid which remains after losing a water molecule when it is linked to another amino acid is called a residue. In some cases, the amino and the carboxyl groups are not attached to the same α-carbon atom. This occurs, for example in the natural β-amino acid, β-alanine and in the β-peptide carnosine [9].

Naturally occurring amino acids that are incorporated into proteins are, for the most part, L-isomers [10–13] (see Chapter 4.6). By elimination of a water molecule, two amino acids can form a dipeptide and their link is a —CO—NH— peptide bond (Figure 4.2.1).

The peptide bond is planar and the C—N distance is equal to 1.325 Å. According to the properties of their side-chains, amino acids can engage into different bonds which are fundamental for recognition of other biomolecules. Amino acids can be classified into different classes such as polar (hydrophilic) and non-polar (hydrophobic), charged and uncharged species [14]. In the hierarchy of structures, the sequence of amino acids is called the primary

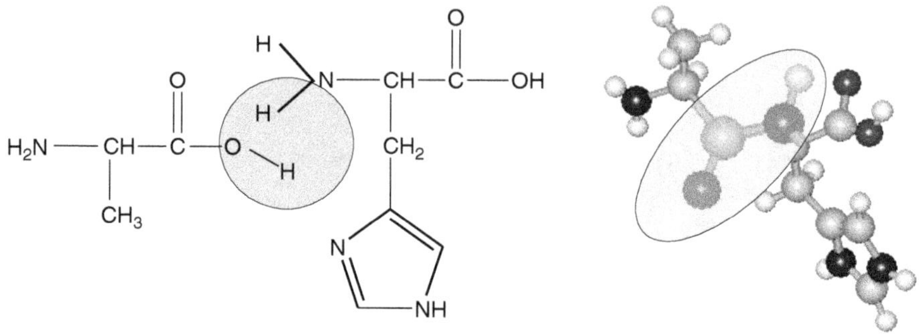

Figure 4.2.1 Left: Formation of a peptide bond by elimination of water from alanine and histidine. Right: The alanine–histidine (Ala–His) peptide.

structure and can be determined by mass-spectrometry [15, 16] (see Section 3.2.2). The backbone of a peptide contains one C=O and one N—H group per amino acid which can establish hydrogen bonds, leading to very stable structures, called secondary structures [17]. For example, an α-helix is formed by H-bondings of the C=O groups of amino acids i with the N—H group of amino acids $i+4$ [18–22]. Long polypeptide chains fold into secondary structures satisfying the hydrogen-bonding possibilities of their backbones (see Section 4.2.2). In turn, these structures pack together into compact structures called tertiary structures. H-bonding and ionic interactions between side-chains influence the folding process in the gas-phase (see Section 4.2.6). In solution, hydrophobic forces (see Section 5.3) are also very important and tend to bury hydrophobic side chains into the centre of the molecule [23]. Tertiary structures are further stabilized by oxidation of cysteine —S—H groups which form —S—S— bonds [24, 25]. Finally, proteins can assemble into larger systems composed of several large subunits linked by non-covalent bonds and forming quaternary structures (see Chapter 4.8) [3, 26–29].

An important paradigm of structural biology is that the structure of a protein is intimately related to its function [30]. A tremendous effort is thus been devoted since many years to theoretical prediction of protein structures [2, 31] and dynamics and their experimental determination in condensed-phase [32]. We will see in Section 4.2.6 that these studies can now be performed in the gas-phase. It is worth noting that a large fraction of proteins are unstructured but still bioactive [33–35].

Gas-phase protein structures have been studied mostly by means of ion-mobility [36–40] (see Section 3.2.3), BIRD [41] or H/D exchange [39, 40] (see Section 3.2.6) but spectroscopic studies are still very scarce [42] and till now mostly been devoted to rather small systems [43].

In proteins, a sizable fraction of amino acids possess ionizable side chains. Under physiological conditions, basic groups (Arg, Lys and His) are generally protonated and acidic groups (Asp, Glu) are deprotonated [23]. In bulk water, neutral amino acids and peptides switch from the neutral form ($H_2N\cdots COOH$) to the zwitterionic form ($H_3N^+\cdots COO^-$) [44] and a large number of theoretical and experimental studies have been devoted to the determination of the number of water molecules required for zwitterion formation in the gas-phase (see Section 5.4).

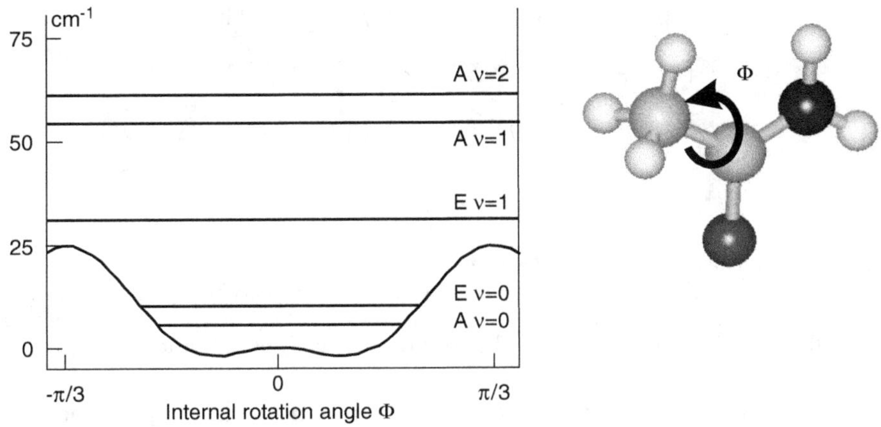

Figure 4.2.2 The first three torsional levels of acetamide (CH_3—C(=O)—NH_2). The internal rotation of the methyl group is nearly free due to the very low barrier of only 25 cm^{-1}.

In this chapter, we will consider frequency- and time-resolved spectroscopic studies of neutral and charged peptide-bond models, amino acids and small peptides. We will further examine structural studies of larger peptides by means of ion mobility, measurement of dipole moments and BIRD methods (see Section 3.2.2). Protein folding can now be followed experimentally in the gas-phase and by simulations during microseconds. Fluorescent proteins can be introduced in living cells and followed in real time. Gas-phase studies can elucidate the structures of their chromophores. Finally, electron spectroscopy of amino acids will be briefly considered.

4.2.1 Peptide bond models

The formamide molecule and its methylated derivatives [45–52] are models of the peptide bond [53, 54]. Their neutral [55] or charged [56–58] clusters have been studied in the gas-phase. The cluster structures [59] are dominated by the competition between hydrogen bonding between C=O and N—H amide groups [60] and dipole–dipole interaction as in peptides [61] due to the large dipole moments of amides (ca. 3.8 D). The *trans* (C=O and N—H bonds in opposition) configuration is energetically favoured with respect to the *cis* configuration [45], in peptides due to steric conflicts between substituents adjacent to the α-carbons [62]. This situation is reversed for the peptide bond models in presence of water (see Section 5.5). Proline represents an exception since the *cis* and *trans* configurations are rather close in energy [63–65]. From a spectroscopic point of view, peptide models offer the opportunity to study in-depth large-amplitude motions. In non-rigid or "floppy" molecules, some atoms execute motions far from their equilibrium position which are difficult to describe theoretically [47] (Figure 4.2.2).

Substituted amides [66] are also considered for studying the conformational flexibility of the peptide bond. In *N*-phenylformamide [67], for example, microwave experiments show that the *trans* conformer has a planar geometry while the *cis* conformer is non-planar.

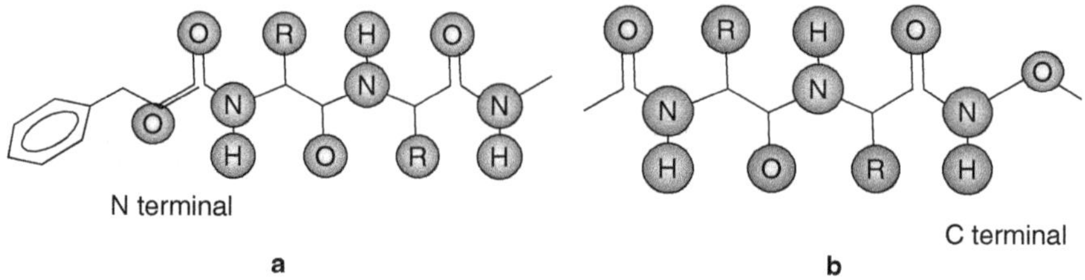

Figure 4.2.3 Examples of peptide chains with different chemical protections. Peptide a is capped on the N-terminal with a benzyl carboxy group (Z-peptide) that prevents hydrogen bonding with the C-terminus and acts as a UV chromophore. Peptide b is capped on the C-terminal with a methyl ester (adapted with permission from reference [77] ©2006 Royal Society of Chemistry).

4.2.2 Spectroscopic studies of neutral amino acids and peptides

The amino NH_2 (N-terminal) and carboxylic COOH (C-terminal) groups of natural peptides can interact and form specific bonds which do not exist inside peptide chains. For example, the lowest energy configuration of glycine corresponds to a bifurcated hydrogen bond $NH_2 \rightarrow O{=}C$ while the slightly less stable conformer corresponds to a $NH_2 \rightarrow O{-}H$ bond (see Figure 1.5.5). In structural studies of short peptide chains aiming to mimic configurations found in proteins, it is interesting to forbid those interactions between the N- and the C-terminal that would not exist if those peptides where imbedded inside a protein. This is achieved by substitution. For example, one of the H atoms of the N-terminal can be replaced by an acetyl group Ac ($CH_3{-}C{=}O$) and the carboxylic group of the C-terminal can be replaced by an amide group (NH_2) leading to "protected" (or "capped") peptides [68–76] (Figure 4.2.3).

In a peptide, amino acids are linked by peptide bonds made of C, O, N and H atoms lying in the same plane. The link between amino acid i and amino acid $i + 1$ creates a plane containing the O_i, C_i, N_{i+1} and H_{i+1} atoms. The α-carbon C_i^{α} is thus situated in between two planes and the set of couples of rotations, defined by the dihedral angles ϕ and ψ, respectively around the $N_i{-}C_i^{\alpha}$ and $C_i^{\alpha}{-}C_i$ bonds, characterizes the conformation of the peptide backbone (Figure 4.2.4). Possible steric hindrances between side chains and the peptide backbone lead to the existence of allowed and forbidden regions in the (ϕ, ψ) planes which are represented on a 2D-Ramachandran diagram [78–80]. When the angles describing the side chain configurations are introduced, 4D-Ramachandran diagrams can display the full topology of the potential energy surface (PES) [81].

4.2.2.1 Competition between local conformational preferences and secondary structures

Structures of large peptides and proteins result from the competition between local conformational preferences involving adjacent units and secondary structures due to interactions between distant sites. Since a long time secondary structures have been identified in large peptides and proteins by means of crystallography and NMR [79, 82]. Local hydrogen bonded structures in small neutral amino acids and peptides have only been recently experimentally studied in the gas-phase by means of microwave spectroscopy (see Section 2.1.2) [83–89] and visible and/or IR spectroscopy (see Section 2.1.4) [70, 90–105] and theoretically [106–113] and reviewed in reference [77].

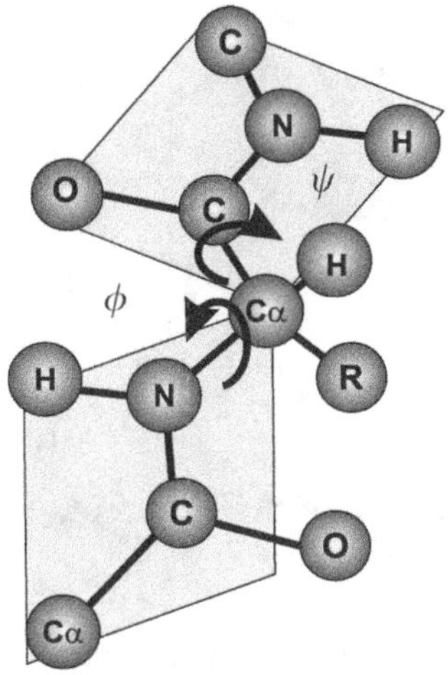

Figure 4.2.4 Definition of dihedral angles ϕ and ψ of the peptide backbone.

Single-residue protected peptides such as Ac—Phe—NH_2 or Ac—Ala—NH_2 contain two peptide links and the study of their different possible structures provides a test of local conformational preferences. In principle, the Ramachandran diagram predicts nine possible conformations but only three most stable conformations, respectively labeled β_L, γ_L and γ_D, are imposed by intramolecular interactions. The possible hydrogen bonds which can establish between C=O and N—H groups are either in between adjacent groups of the same peptide bond (C_5 corresponding to β_L) or in between groups belonging to separate peptide bonds (C_7 corresponding to γ_L and γ_D). The conformation β_L is extended and is equivalent to the secondary structure called β strand in larger peptides. Conformations γ_L and γ_D are folded and correspond to the secondary structure called γ turn in larger peptides. Turns are secondary structures responsible for the compactness of proteins [114]. Turns involve 3(γ), 4(β) and 5(α) residues that respectively form 7-, 10- and 13-membered hydrogen bonded rings (Figure 4.2.6). These turn structures are recognized by universal cell-surface polypeptides called G-protein-coupled receptors [115], most of them being olfactory or smell, taste, vision receptors.

In a non-aromatic amino acid, the peptide backbone hydrogen bonding is stronger in the C_7 conformation than in the C_5 conformation and the folded conformation γ_L is the most stable (see Figure 4.2.5). On the contrary, in an aromatic amino acid, such as Ac—Phe—NH_2, the presence of a strong interaction between the terminal NH ($i+1$) and the aromatic $\pi(i)$ side chain electron cloud favours the β_L form.

Turns are common secondary structure motifs in proteins [114, 116]. The most frequent turns are not the γ turns (previously described) but are β turns characterized by C=O(i) ← N—H ($i+3$) hydrogen bonds implying residues in a C_{10} cycle (Figure 4.2.7). C_{10} cycles related to secondary structures enter into competition with C_7 cycles corresponding to local conformations already in small systems such as protected dipeptides. A systematic series of IR-REMPI studies [77] (see Section 2.1.3) conducted on Ac—Phe—Xxx—NH_2 and Ac—Xxx—Phe—NH_2 protected peptides (alkyl residues Xxx = Gly, Ala, Val, Pro) show that the majority of gas-phase conformers adopt a succession of local conformational preferences

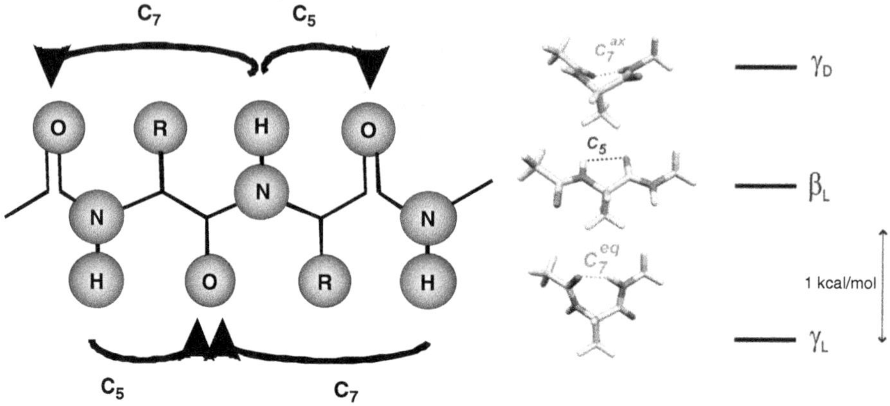

Figure 4.2.5 Left: Possible intramolecular hydrogen bonds in a protected dipeptide. Right: Relative energies of the three conformers γ_L, β_L and γ_D of the protected Ac—Ala—NH—CH_3 peptide (reprinted with permission from reference [77] ©2006 Royal Society of Chemistry).

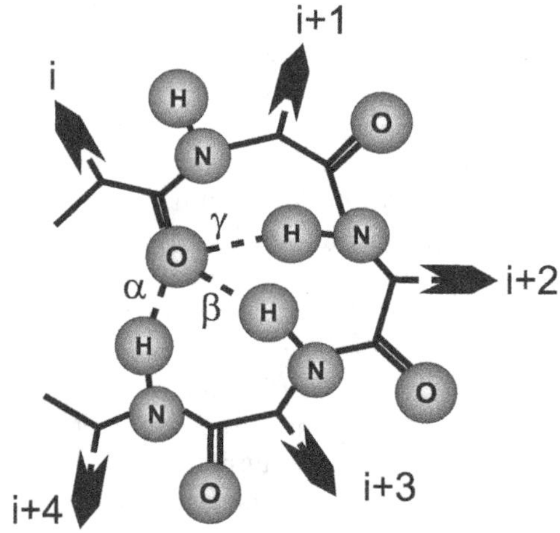

Figure 4.2.6 Intramolecular hydrogen bonds leading to the different peptides turns.

such as $\beta_L - \gamma$ or $\gamma - \gamma$ that are the most stable but β-turns with weaker hydrogen bonds are also observed as minor conformers. It turns out that the distinction between those different conformations requires a rather good spectral resolution (see Section 2.1.3) that cannot be achieved in solution.

The shortest systems in which helical structures can be observed are protected tripeptides (containing four amide bonds). As shown in Figure 4.2.7, several possibilities are opened among which formation of a 3_{10} helix (formation of two C_{10} bonds) or an α-helix (formation of a C_{13} bond between the CO_{Ac} and the NH_2 groups). For infinite chains of homopeptides only composed from Gly or Ala residues, theory predicts that the α-helix is the most stable form but in finite size peptides and proteins, a 3_{10} helix is often found at the

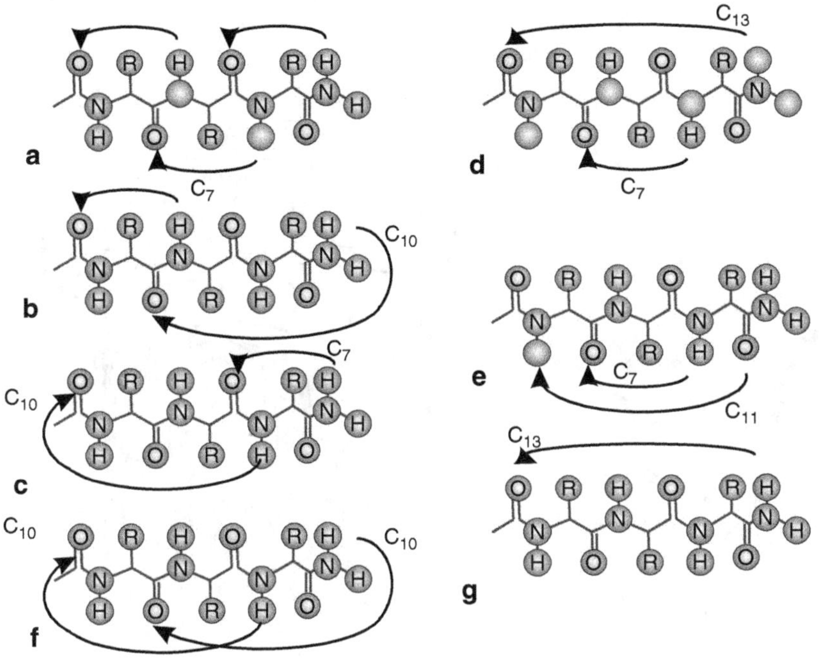

Figure 4.2.7 Possible hydrogen-bonding patterns in protected tripeptides. (a) Successive local preferences (β- or γ-turns); (b, c) combination of a local preference with a β-turn; (d, e) central γ-turn; (f) combination of two β-turns with or without a 3_{10} helix structure; (g) first turn of an α-helix.

C-terminal end of a α-helix [6, 117]. IR–UV studies of protected tripeptides containing one Phe and two Ala residues show that 3_{10} helix formation competes with the formation of other structures such as 2_7 ribbons ($C_7{-}C_7{-}C_7$) [118]. Up to now, IR–UV spectroscopic studies of peptides provide the most conformer selective and precise data but the necessary presence of an aromatic chromophore often introduces perturbation in the structures and is only a step towards the formidable task of a comprehensive investigation of the whole landscape of small peptides. On the other hand, general rules tend to appear when a correlation is established between experimentally determined spectral shifts and calculated hydrogen bond patterns [118] (see Section 2.1.3). It is important to notice that most high-resolution spectroscopic studies are conducted at very low temperatures and calculations consider the corresponding lowest energy conformations, in contrast with biological conditions which are governed by lowest free energy conformations. As shown in Figure 4.2.8, lowest-energy or free-energy conformations are not necessarily the same.

4.2.2.2 α-helices and β-sheets

Two very important secondary structures of proteins are α-helices and β-sheets [22, 120]. In a α-helix [121], carbonyl group of peptide bond (i) establishes an intramolecular hydrogen bond to the N—H group of peptide bond ($i+4$) (Figure 4.2.9). All hydrogen bonds point in the same direction and the result is the building of a large dipole moment [19] (see Section 4.2.2.3). In solution [122] and in the gas-phase, some amino acids such as Ala

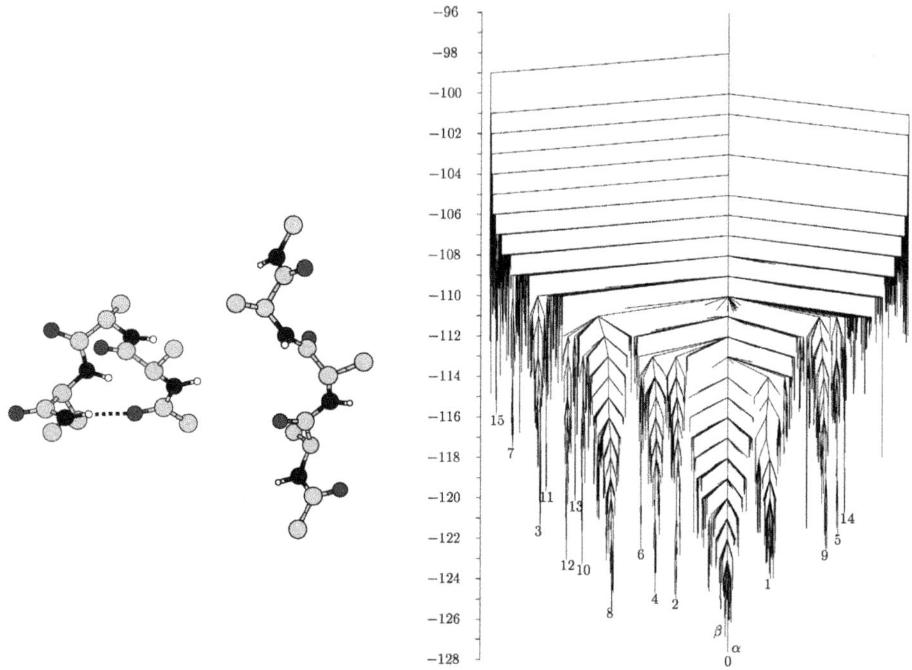

Figure 4.2.8 Left: Structures of the lowest energy (α-helix) and lowest free energy (β-strand) minima of the protected Ac $(Ala)_3$–NHMe tetrapeptide. Right: Potential energy disconnectivity diagram of the same peptide. The α and β lowest energy structures are those displayed on the left figure (reprinted with permission from reference [119] ©2003 American Institute of Physics).

[123–126], Glu [127], Leu and Met present a high propensity for helix formation while Pro [128], Gly [129], Tyr and Ser do not. In a β-sheet, intermolecular hydrogen bonding mostly takes place between C═O and N—H groups belonging to adjacent strands either parallel or anti-parallel. β-sheets are pleated and C_α carbons alternate slightly below and above a plane. Parallel β-sheets can contain hydrophobic residues on both side of this plane while anti-parallel β-sheets contain hydrophobic residues on a single side. In the last case, the aromatic rings of the hydrophobic residues do not interact favourably [130].

Two β-strands linked by a short loop constitute a β-hairpin [79, 131] that represents the simplest model of an anti-parallel β-sheet [132]. Substituting one or several amino acids in a α-helix can stabilize or destabilize the helix form. This has been systematically studied in the capped peptide acetyl—$(Ala)_{17}$—NH_2. Among the 17 alanine amino acids, the 10th has been substituted by amino acid X (X = Gly, Leu, Val, Phe, Ser and Pro) [76]. This choice is dictated by the following reasons. Gly, Leu and Val have alkyl chains and the β-CH can form C—H···O hydrogen bonds (see Section 1.1.2). The OH group of Ser may form an H-bond. Pro does not possess an N—H group and has an important influence upon peptide structures. The respective energies of the α-helices and β-strand structures of the acetyl—$(Ala)_9$—X—$(Ala)_7$—NH_2 peptide have been calculated using a ONIOM (DFT + AM1) (see Section 1.6). The relative stabilities of the β-strands depend upon the steric distortions imposed at the substitution site, upon the cooperative H-bond chain involving C_5 interactions

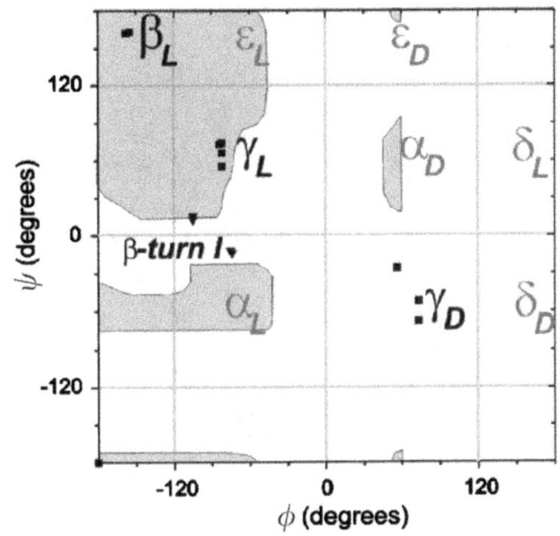

Figure 4.2.9 Ramachandran diagram of the local conformational landscape of a peptide backbone (definition of dihedral angles ϕ and ψ is shown in Figure 4.2.4). Shaded area corresponds to sterically favoured conformations (except for glycine). Dihedral angles corresponding to geometries of γ_L, β_L and γ_D conformations of the protected Ac—Ala—NH—CH_3 peptide (Figure 4.2.5) and a typical β-turn are shown (reprinted with permission from reference [77] ©2006 Royal Society of Chemistry).

and upon establishment of new C—H···O H-bonds. Substitution of Ala by Pro destabilizes the α-helix (49 kJ/mol) more than the β-strand (31 kJ/mol) that becomes the most energetically favoured structure. On the contrary, substitution Ala → Gly stabilizes both α-helix and β-strand but stabilizes β-strand more.

The backbone structure and the intermolecular hydrogen-bond pattern observed in parallel β-sheets are respectively encountered in capped peptide monomers [133] and dimers [68, 70, 134–137] and have been studied by IR/R2PI (Figures 4.2.10 and 4.2.11). In the acetyl—Val—Tyr(Me)—NHMe tripeptide, the N—H and C═O groups are nearly parallel to each other. In the N—H stretching region, the three observed vibrational bonds are situated above 3,400 cm^{-1} (Figure 4.2.10) and are thus not red-shifted by H-bonding (see Section 2.1.3). A similar conclusion can be drawn from the C═O stretch region around 1,700 cm^{-1}. This tripeptide is thus a model of β-sheet when it is included in a hydrogen-bonded cluster with other peptides.

The intermolecular H-bond pattern of the protected acetyl—Phe—OMe dimer is characteristic of a parallel β-sheet (β_L). According to the Phe side chain orientation, several possibilities are offered. The presence of a line at 1,205 cm^{-1} in the experimental spectrum and absent in the predicted spectrum of the $[\beta_L(a)-\beta_L(a)]$ configuration shows that the $[\beta_L(g+)-\beta_L(g+)]$ configuration is energetically favoured and observed.

4.2.3 Spectroscopic studies of charged amino acids and peptides

Structures of isolated positively charged amino acids have been determined by means of UV spectroscopy coupled to modelling and compared to those of the corresponding neutral species [92, 138–141]. Protonated species [142] present different tautomers according

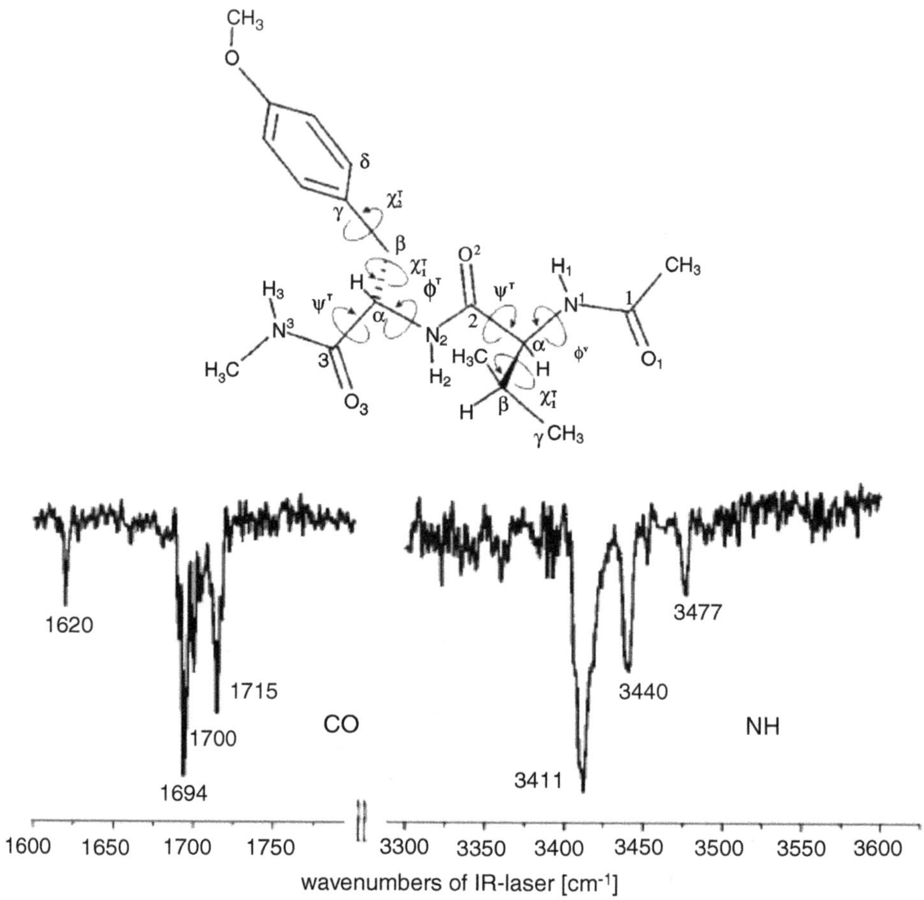

Figure 4.2.10 Top: Structure of the acetyl—Val—Tyr(Me)—NHMe tripeptide. Angles ϕ^V, ψ^V, ϕ^T and ψ^T define the backbone. Angles χ^V, χ_1^T and χ_2^T define the side chain orientation. In a β-sheet (β_L) related structure, angles ϕ and ψ respectively fall in the regions $-180° < \phi < -120°$ (or $180° < \phi < 120°$) and $-180° < \psi < -120°$ (or $120° < \psi < 180°$) of a Ramachandran diagram (see upper left of Figure 4.2.9). The lowest energy configuration of the tripeptide corresponds to $\phi^V = -129°$, $\phi^T = -157°$, $\psi^V = 160°$, $\psi^T = 172°$. Bottom: IR/R2PI spectrum of the acetyl—Val—Tyr(Me)—NHMe tripeptide (reprinted with permission from reference [133] ©2004 Royal Society of Chemistry).

to the protonation site and each tautomer possesses different conformers. In tryptophan, among the three possible protonation sites: the amino group, the indole nitrogen and carbon 3 of the indole ring, the former leads to the most stable tautomer and the lowest energy conformer in the ground state S_0 is stabilized by the positive charge-indole ring π interaction (see Section 1.1). Following UV excitation, the π-electron transfers to the protonated amino group and induces fragmentation with the loss of a hydrogen atom [143] (see Sections 1.7 and 4.2.5).

In peptides, many protonation sites are available. Resonant infrared multiphoton dissociation spectroscopy (resonant IRMPD, see Section 2.1.3.2.5) allows the determination of those protonation sites through comparison between experimentally observed resonant IRMPD

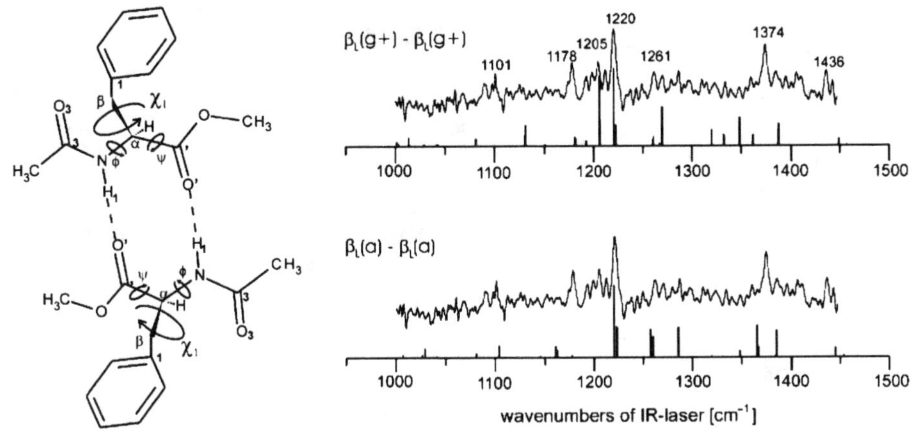

Figure 4.2.11 Left: Structure of the protected acetyl—Phe—OMe dimer. The N—H···C—O hydrogen-bonding pattern is similar to that encountered in a parallel β-sheet. Angle χ_1 define the orientations of the side chains. Right: Experimental and calculated IR/R2PI spectra of the (acetyl—Phe—O—Me)$_2$ dimer in the fingerprint region of 1,000 to 1,450 cm^{-1}. Configurations g+ (*gauche* plus) and a (*anti*) of the monomer respectively correspond to $\chi = 60°$ and 180° (reprinted with permission from reference [137] ©2006 Royal Society of Chemistry).

fragmentation spectra and calculated infrared absorption spectra for the different tautomers. For example, in the Ala—Ala H$^+$ peptide [145], the proton is mostly located on the amino terminal group. As it can be expected from basicity arguments, in the protected *N*-acetyl Ala H$^+$ peptide, the proton becomes located on the carbonyl while it is mostly sequestered on the histidine side chain in the Ala—His H$^+$ peptide [146]. Two different points of view can be adopted for analysing experimental results. One can first calculate the different minima of the peptide PES (see Section 1.5) and then assume that the different structures are populated according to a Boltzmann distribution corresponding to the experimental temperature. This static description can be replaced by a dynamical picture. Instead of considering an ensemble of observable conformers, one can follow the fate of a given peptide as a function of time. As we will now show in the case of the Ala—Ala H$^+$ peptide, at a sufficiently high temperature corresponding to a biological situation, a given peptide does not possess a well-defined structure but rather continuously explore a set of possible conformations that can even correspond to different protonation sites.

A systematic conformational search of the PES of Ala—Ala H$^+$ provides several lowest energy conformations identified according to the respective protonation sites. The different conformers of peptides only possessing protonation sites with rather similar basicities such as terminal amino groups, amide oxygens and nitrogens, are respectively labelled A, O and N [147]. The lowest energy configuration A1 is separated from the second configuration A2 situated 7 kJ/mol above A1 by an energy barrier of 11.5 kJ/mol. Those two configurations correspond to a rotation of the C-terminal around the N—C$_3$ bond. A third configuration called O1, corresponding to a proton jump from the N-terminal to the carbonyl group, is situated 9.7 kJ/mol above A1 with an energy barrier of 14.3 kJ/mol (Figure 4.2.13).

In a static point of view valid at very low temperatures (Figure 4.2.12), one can interpret the observed experimental spectra by assuming that only configurations A1 and A2 are

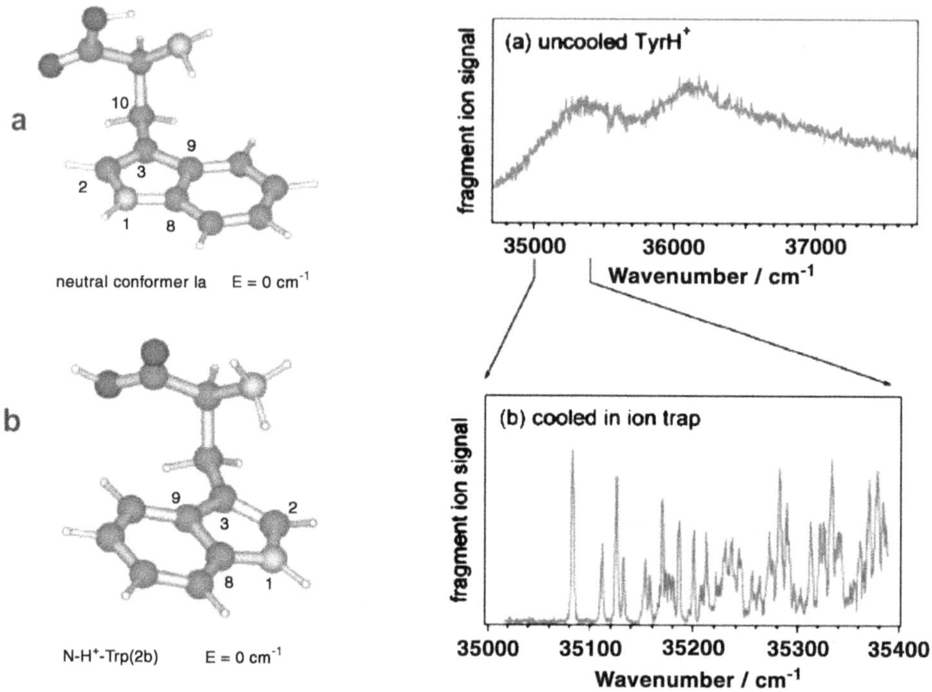

Figure 4.2.12 Left: Calculated structures of the lowest energy neutral (a) and protonated (b) conformers of tryptophan (reprinted with permission from reference [139] ©2004 Royal Society of Chemistry) [139]. Right: Electronic spectrum of protonated tyrosine obtained by UV photoexcitation and monitoring of the fragment corresponding the loss of NH_3. $TyrH^+$ ions are confined in a 22-pole trap (see Section 3.2.1.2) and either un-cooled (a) or cooled down to 10°K (b) (reprinted with permission from reference [144] ©2006 American Chemical Society).

populated according to a Boltzmann distribution at the experimental temperature of 300°K. In a Car-Parrinello simulation (see Section 1.5.5.3) run at 300°K, the picture is totally different. As shown in Figure 4.2.14, there is no longer any free energy barrier between configurations A1 and A2 and the time evolution of the C_2—N—C_3—C_4 dihedral angle Φ corresponds to a continuous isomerization between the A1 and A2 conformers around the N—C_3 bond (Figure 4.2.14). At 300°K, the angle Φ is no longer constrained at values of $-162°$ or $-74°$ respectively corresponding to conformers A_1 and A_2 but rather oscillates in between those two values at a frequency around 1 THz.

One can also observe the proton jumps from the NH_3^+ to the C=O site (Figure 4.2.15). The distance d_{O-H_1} between the hydrogen atom of the amide group which is closest from the oxygen of the carbonyl group makes jumps between 1.6 and 1 Å corresponding to proton transfer between the amino and carbonyl groups. This chemical event cannot be studied by means of classical simulations.

4.2.3.1 β-peptides

Usual amino acids are α-amino acids in which the amino and carboxylic groups are directly bonded to the C_α carbon, itself bonded to the side chain. In β-amino acids, the amino group

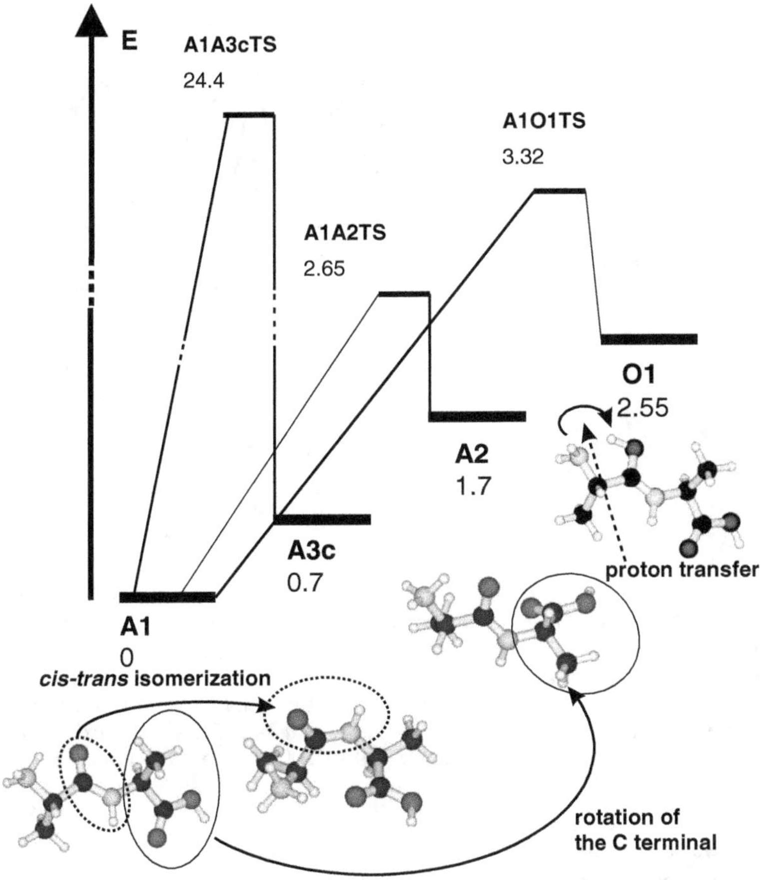

Figure 4.2.13 Potential energy landscape of the protonated dialanine peptide. Four lowest energy configurations of the gas-phase protonated dialanine peptide. The top structures are respectively A1 (left) and A2 (right). The bottom structures are O1 (left) and a low-lying structure *cis* configuration (right) but separated from A1 by a very large energy barrier of 100 kJ/mol.

is bonded to the C_β carbon, itself bonded to the C_α carbon. In nature, only the β-alanine amino acid exists and β-peptides containing β-amino acids are often used to evade resistance to antibiotics. Four different secondary structure motifs are generally encountered in β-peptides. The elongated (E) and spiral (S) structures do not require internal hydrogen bonds while the zigzag (Z) and helical (H) structures do (Figure 4.2.16). The structures of the homo-conformers of the HCO—(β-Ala)$_{1\text{–}6}$—NH_2 peptides have been systematically studied at the RHF/3-21G level [149, 150]. When α- and β-peptides are compared [151], α-peptides have a lower number of homo-conformers possessing relative energies closer to each other (Figure 4.2.17).

4.2.3.2 Cyclic peptides

Cell receptors can recognize very specifically amino acid sequences present at the surface of proteins. When excised from a protein, a peptide containing such a recognized sequence can

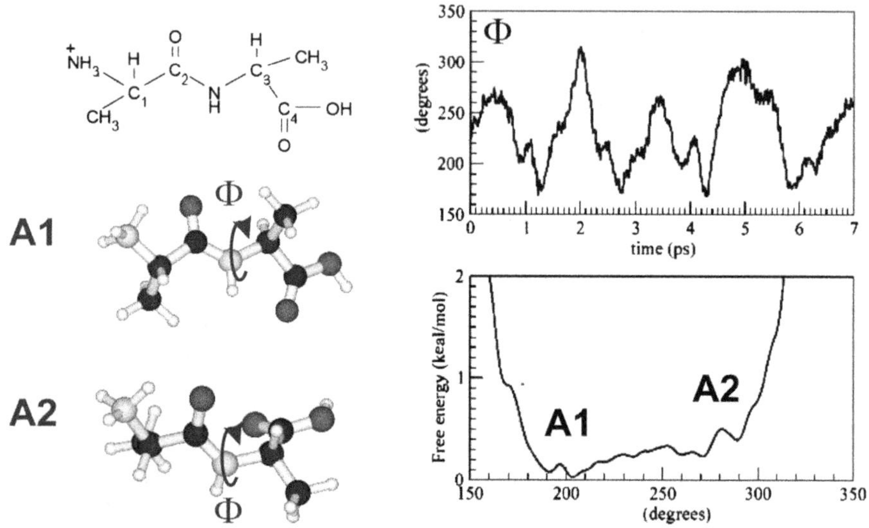

Figure 4.2.14 Left: The two lowest conformers *trans* A1 and A2 of the dialanine cation protonated at the N-terminal. Those conformers respectively correspond to values of −162° and −74° of the rotation angle Φ around the N—C bond. Right top: Dynamics of isomerization of the dialanine peptide at 300°K. The variation of the C_2—N—C_3—C_4 dihedral angle Φ corresponds to isomerization between the two lowest energy configurations A1 and A2. Bottom: Free energy profile along angle Φ shows that two minima respectively at $\phi = 198°$ (A1) and $\phi = 284°$ (A2) still exist but there is no barrier (reprinted with permission from reference [148] ©2006 American Chemical Society).

adopt several conformations that can be quite different from the bioactive structure imposed by the protein backbone. For development of drugs, peptides containing a bioactive sequence identified as an attractive pharmaceutical target are conformationally constrained by means of cyclization (elimination of a water molecule between amino acids) [30, 152–154]. For example, the arginine—glycine—aspartic acid (RGD) sequence is a tripeptide motif recognized by proteins called integrins. When integrin receptor sites interact with the RGD motif, they establish a link between the intracellular medium and the extracellular matrix [155–157]. In the gas-phase, RGD tripeptide and the cyclic Arg—Gly—Asp—D-Phe—Val pentapeptide [158] are both protonated on the guanidinium side chain of Arg [159] (Figure 4.2.18). The RGD sequence in the cyclized pentapeptide is then in a conformation close to that it adopts in an integrin protein such as the protein deposited in the PDB (entry 1ttf) [160].

It is interesting to compare the structures of an unprotected peptide such as the RGD sequence when it is totally isolated or when it is included in a protein such as the dendroaspin. In this neurotoxin homologue (PDB entry 1DRS), the flexible RGD loop structure [161] corresponds to residues 43–45 and is located on the surface, maintained by disulphide bridges [162]. Studies conducted with resonant infrared multiphoton dissociation (IRMPD) spectroscopy (see Section 2.1.3.2.5) show that this loop structure is an intrinsic structural property that is conserved in the gas-phase. The lowest energy configurations CS1 and CS2 of RGD belong to the so-called charge-solvated (CS) family and correspond to the

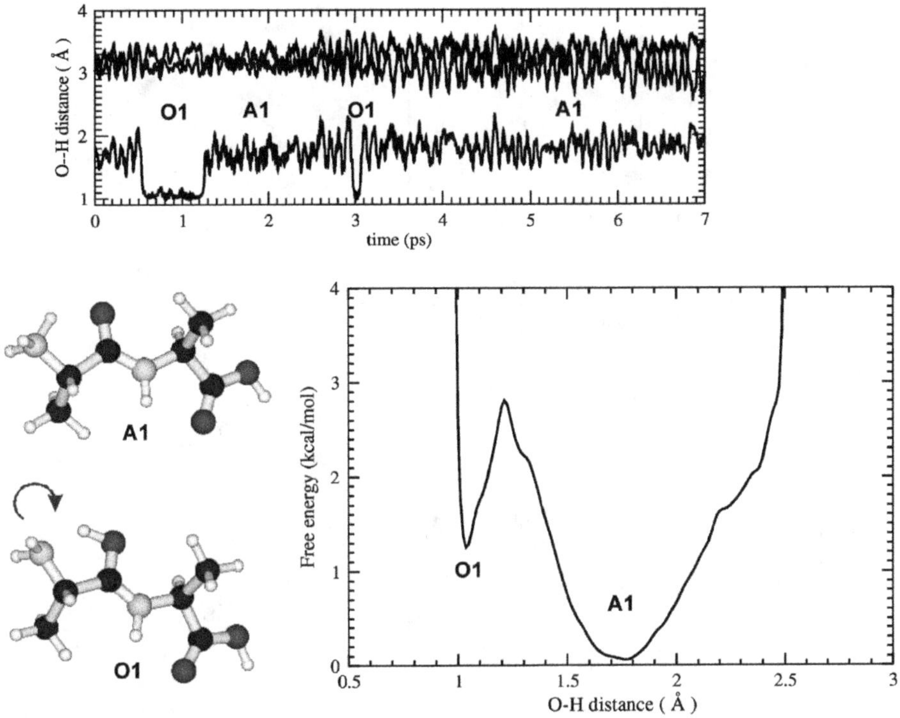

Figure 4.2.15 Proton transfer in the dialanine peptide studied by Car and Parrinello molecular dynamics. Left: Time evolution of the carbonyl oxygen—H distances. Time evolution of the $O_2 \cdots H$ distances. The two upper traces correspond to the two hydrogen atoms of the protonated amino group further away from the carbonyl oxygen atom. The lower trace corresponds to the hydrogen atom of the protonated amino group closest to the carbonyl O_2 oxygen atom. When the $O_2 \cdots H$ distance becomes very short (1 Å), the proton has temporarily jumped from the N-terminal to the carbonyl. Right: Free-energy profile along the $O_2 \cdots H$ distance. A free energy barrier of 11 kJ/mol separates the A1 and O1 minima (reprinted with permission from reference [148] ©2006 American Chemical Society).

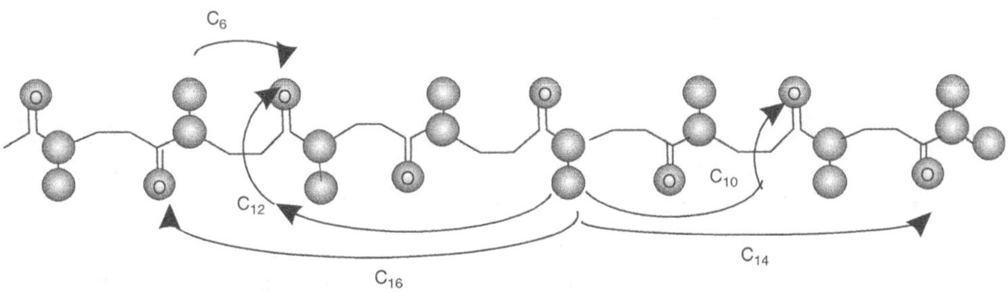

Figure 4.2.16 Typical hydrogen bonds of the different secondary structures of β-peptides.

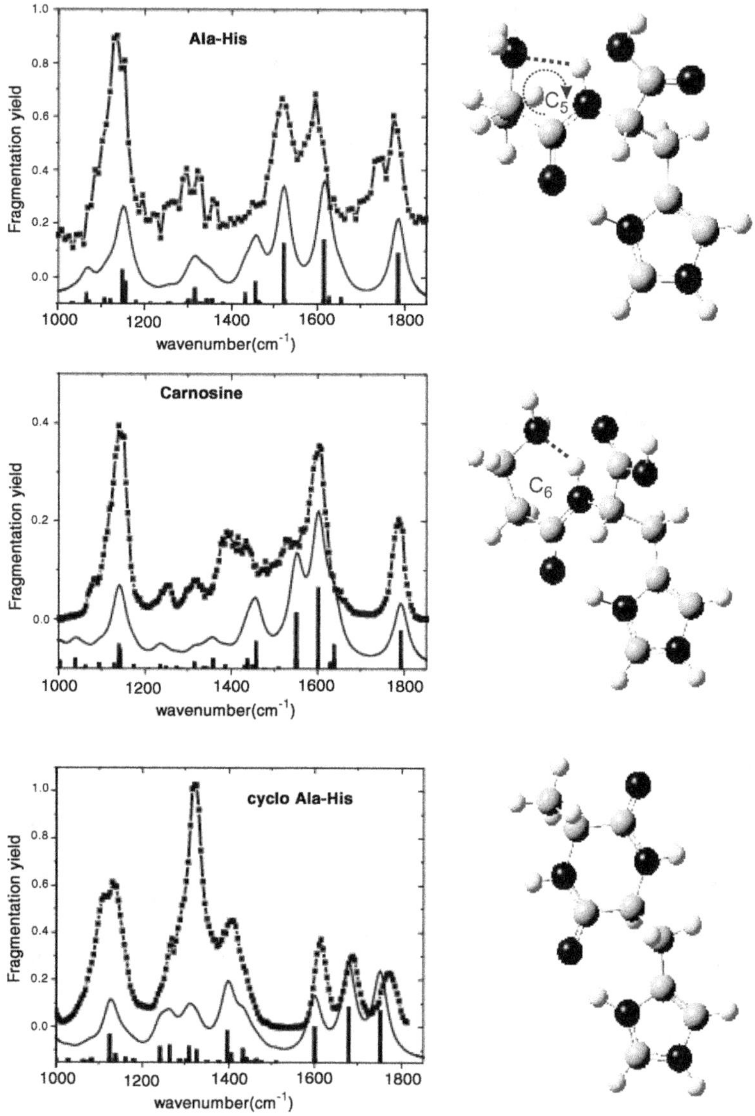

Figure 4.2.17 IRMPD spectra and lowest energy structures of the protonated alanine–histidine peptide, the corresponding β-peptide (carnosine) and the cyclized form. The protonation site remains on the hisdine in the three cases. Large differences are observed in the amide I region (ca. 1,750 cm^{-1}) corresponding to different hydrogen-bonding patterns of the amide carbonyl group. In the cyclic peptide, the amide bond is in *cis* conformation.

protonated very basic guanidine group forming hydrogen bonds with the first carbonyl group of R and the neutral C-terminal and/or side chain carboxyl groups of D. The crucial difference between these CS1 and CS2 structures is that only the CS2 conformation is biologically relevant. In the RGD unprotected peptide, the C-terminal is only free in the CS2 conformation

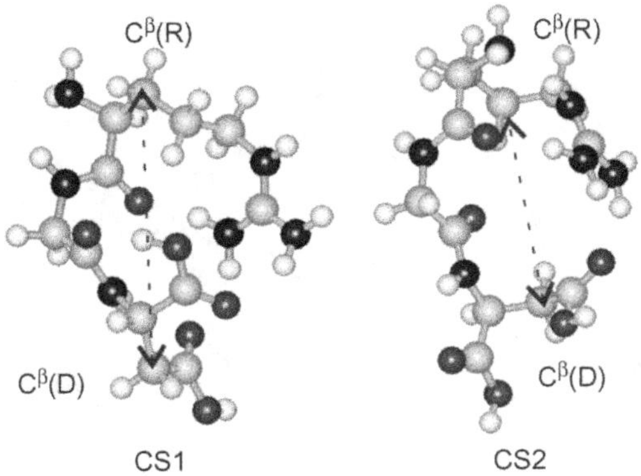

Figure 4.2.18 Lowest energy structures of the unprotected protonated RGD peptide. Arrows indicate the distance between carbon atoms involved in the evaluation of the qualitative structure–activity relationship (QSAR) of RGD containing peptides.

and the side-chain carboxylic group of D is then hydrogen-bonded to R. On the contrary, the carboxylic group of the aspartic acid that would be covalently linked in a peptide bond in a larger peptide is hydrogen bonded in the CS1 conformation

The relationship between the structure and the activity of RGD-containing peptides has been evaluated quantitatively (QSAR) [163] in a platelet aggregation assay study. The RGD structures were then determined by NMR. The observed variety of conformations of the RGD sequence seems to be partially responsible for its ability to be recognized by integrins. The respective positions of the two major recognition sites, that is the charged side chains of R and D are the retained QSAR criteria. Those positions are defined as the distance between the respective C^{β} atoms of R and D and the pseudo-dihedral defining the R and D side chain orientation. This angle is given by the respective position of the C^{ζ} and C^{a} atoms of R and the E^{a} and C^{γ} atoms of D. The prerequisite for bioactivity is that the pseudo-dihedral is comprised in between $-45°$ and $+45°$. Only the biologically relevant gas-phase conformation CS2 satisfies this condition with a dihedral angle value of 21° as compared to $-62.8°$ in CS1. The distance between the respective C^{β} atoms of R and D in the gas-phase is equal to nearly 6.96 Å in configuration C2 and falls in the middle of the distances measured in the studied compounds of reference [163] that vary in between 4.4 and 9 Å. This study shows that small biomolecules can retain in gas-phase their bioactive conformation as demonstrated in larger systems [164].

4.2.4 Non-spectroscopic determinations of charged peptide gas-phase structures

4.2.4.1 Electric deflection of dipoles

Electrostatic forces have long ranges and play a great role in secondary structures of proteins [165]. In α-helices (Figure 4.2.19), for example the alignment of individual backbone amide dipole moments and their cooperativity lead to large macro-dipoles which, in turn, induce large internal electric fields [61]. A variety of other interactions also act to stabilize α-helices: C_{13} structure formation tendencies of constituent amino acids, capping interactions at the

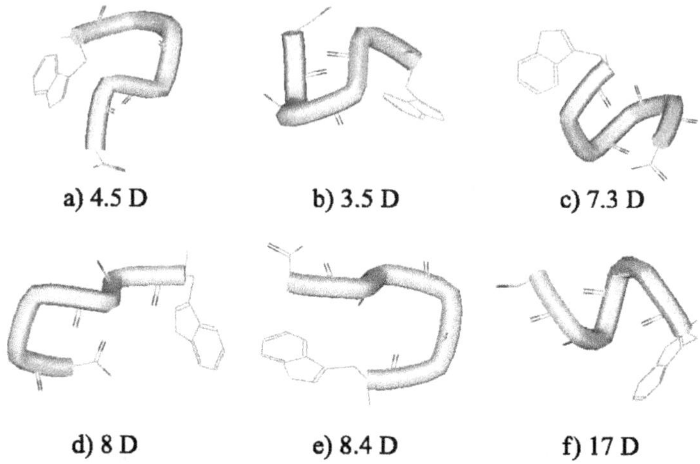

Figure 4.2.19 Dipole moments of different conformations of the WG_5 peptide. Note the large dipole moment value of 17 D corresponding to α-helix conformation (f) (reprinted with permission from reference [167] ©2002 EDP Sciences).

N- and C-terminals, solvent environment [120] and side chain interactions [117, 166]. Gas-phase studies eliminate the influence of the solvent.

In dipole moment measurement studies (see Section 3.3.3) conducted on small neutral peptides containing glycine, the very floppy molecules can convert between different conformations and the very large dipole moment corresponding to the formation of a α-helix is not observed [167] (Figure 4.2.19).

The average dipole moments of protected Ac—Trp—Ala_n—NH_2 peptides have been determined by electric deflection measurements with $n = 3, 5, 10, 13$ and 15 [168].

The average dipoles determined from the susceptibility measurements are shown in Figure 4.2.21. The 300°K dipole moments increase with the peptide size n, at peak $n = 10$ and then decrease. At 500°K, the dipole moments systematically increase with n. Simulations with the Amber force-field and parallel tempering (see Section 1.5) show that four main types of conformations are adopted by the neutral peptides: α-helices, β-sheets, compact random-looking globules and random coils (unfolded structures appearing at high temperatures). For n above 10, the lowest energy structures are α-helices but the situation is strongly modified at room temperature and above when entropy is taken into account. At 100°K, the lowest free energy conformations are still α-helices. At 300°K, the free energy map is displayed in Figure 4.2.20 and shows that there is a broad minimum of the free energy around $N_{\text{Beta}} \approx 3$ and $N_{\text{Alpha}} \approx 2$ corresponding to globules. There is also a small minimum (c) with $N_{\text{Beta}} \approx 9$ corresponding to β-sheets [135, 136]. β-hairpins [169] and S-shaped β-sheets (respectively conformations b and c in Figure 4.2.20) are also present in the vicinity of this β-sheet region.

The comparison between measured and predicted dipole moments is shown in Figure 4.2.21. Polyalanine peptides adopt α-helix structures in solution while the protonated species adopt globular conformations in the gas-phase [129]. Here, considered predicted structures of gas-phase neutral peptides, the aligned dipoles of α-helices simply add and thus strongly increase with size n, in contrast with the low values of the measured dipoles.

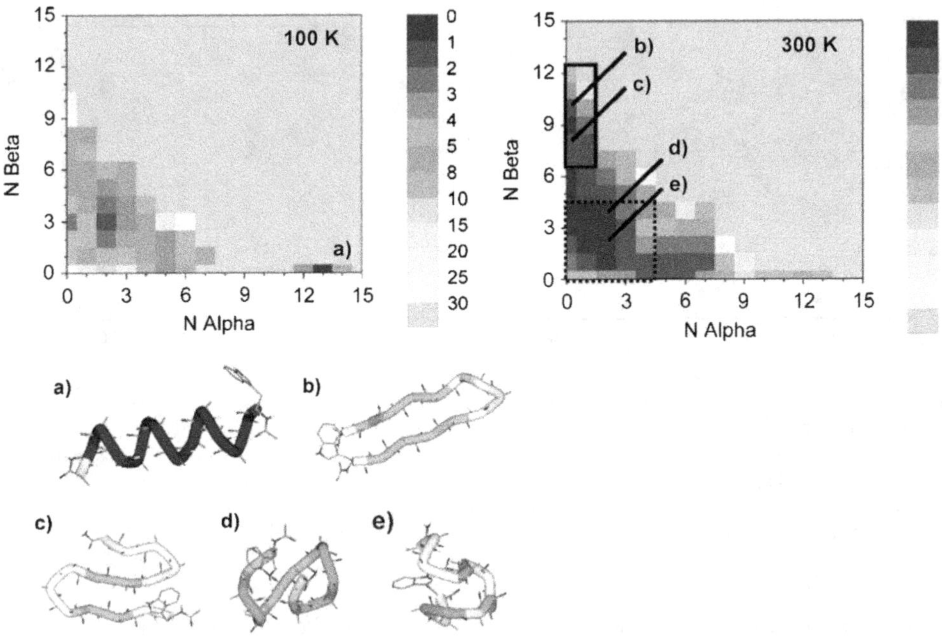

Figure 4.2.20 Top: Free-energy map of neutral Ac—Trp—Ala$_n$—NH$_2$ peptides at 300°K. The horizontal and vertical axes respectively correspond to the number of amino acids in α-helix and β-sheet conformations. The lowest free energy region is (e). The boxes show the region used to predict the average dipole values for β-sheets (full line) and globules (dashed lines). Bottom: Representative helical (a), β-sheet (b and c) and globule (d and e) conformations obtained from simulations. The tryptophan has been added to the alanines to allow simple two-photon ionization for detection (see Section 2.1.2.1) (reprinted with permission from reference [168] ©2005 American Chemical Society).

When considering the largest peptides ($n = 13$ and 15), the best agreement is obtained for β-sheets and β-hairpins conformations

4.2.4.2 Ion-mobility spectrometry

We have seen above that the different amino acids present different propensities to form α-helices in solution and in the gas-phase. Among amino acids, Ala has the highest propensity to form helices in aqueous solution, followed by Leu, Met, Phe, Glu, Gln, Lys, Arg, His whereas glycine exhibits one of the lowest. The helix-breaker amino acids are Pro, Gly, Ser, Asn, and Asp. As demonstrated by ion-mobility studies, the situation is different in the gas-phase. The relative ordering of helix propensities in solution is Ala > Leu > Val > Gly while it is Leu > Ala > Val > Gly in the gas-phase [170].

Protonated poly-Ala and poly-Gly peptides both adopt compact globular conformations [129]. Poly-Ala peptides do not adopt helix conformation because their protonation site is at the N-terminus which is also the positive end of the helix macro-dipole. This repulsive electrostatic interaction is reinforced by the absence of solvent. The electronegative C=O groups of the peptide backbone then wrap around the protonated N-terminal to solvate the proton positive charge.

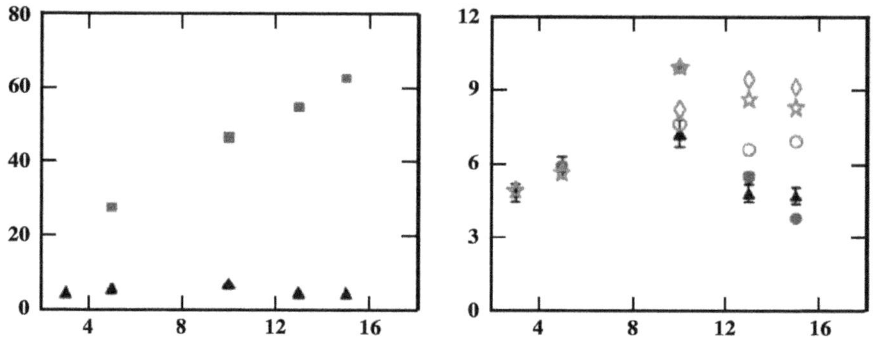

Figure 4.2.21 Comparison between measured and predicted dipole moments as a function of the Ac—Trp—Ala_n—NH_2 peptide size (horizontal axis n) at 300°K. The filled triangles are experimental values (vertical axis in Debye). The predicted α-helices and β-hairpins values are respectively represented by filled squares and filled red circles. The average over the entire free energy map, over the β-sheet region and the globule region are respectively represented by stars, circles and diamonds (reprinted with permission from reference [168] ©2005 American Chemical Society).

In order to reverse this electrostatic situation, the protonation site can be set at the C-terminus by protecting the N-terminus by acetylation (Ac) and even better by adding the strongly basic Lys (K) at the C-terminus that now captures the proton [38, 171, 172]. The helical conformation is then observed in Ac—A_nH^+ and Ac—A_nKH^+ peptides while the Ac—KH^+—A_n peptides are gobular for $n < 13$ since the electrostatic helix-stabilizing factor is then absent. When Gly amino acids replace Ala in protected mixed Ac—$Ala_nGly_xA_mH^+$ peptide, a helix disruption is observed if a long block of Gly residues is added (Figure 4.2.22). For example, more than three Gly residues are required to disrupt helix formation in a 15-residue peptide [173]. This disruption is not due to a local effect and occurs thanks to a global effect on the relative energies of the helix and globular conformations. Similar studies have been conducted on helix formation in poly-valine [170] and in poly-proline peptides [128].

Peptide charges can be located by means of comparisons between ion-mobility measurements and simulations [175]. The location of protons on amino acids in position i inside a peptide strongly influences the folding properties. For example, in $(Ala_n + 3H)^{3+}$ ($n = 24$–41) peptides, ion mobility arrival time distributions (see Section 3.3.1) show that two conformations respectively corresponding to hinged helix-coils and extended α-helices are formed. Those conformations do not interconvert at 300°K during the 10–40 ms experimental timescales while the compact structures unfold and adopt extended helical structures at 360°K.

Classical molecular dynamic simulations show that if the net charge is mostly distributed on the C-terminal side of the peptides ($(\Sigma i)/3 > n/2$), extended α-helices are formed. In contrast, if the net charge is mostly distributed on the N-terminal side of the peptides ($(\Sigma i)/3 < n/2$), compact structures identified as hinged helix-coils are preferentially formed [176]. The peptide is then half-folded and the proton closest from the N-terminal in the coil region stabilizes the helix region of the C-terminal. Simulations have been run with charged peptides satisfying the $(\Sigma i)/3 > n/2$ condition (for example with protons on amide bonds in positions 19, 29 and 34 or 16, 30 and 36) but initially in the hinged helix-coil configuration (protons in positions $i = 1$, 8 and 16) (Figure 4.2.23). Those simulations show that the key steps for the

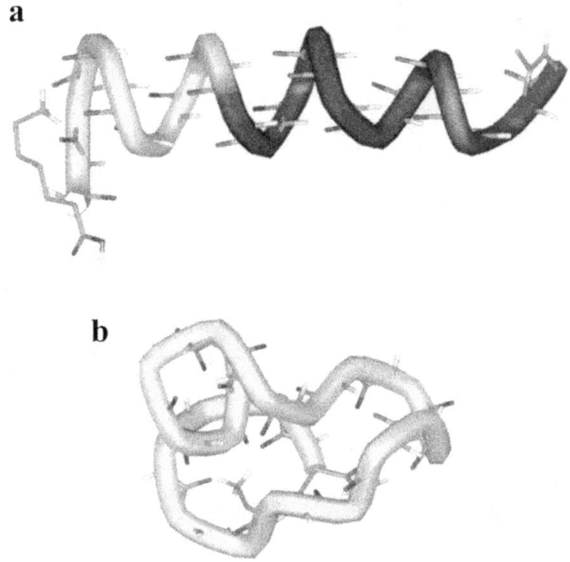

Figure 4.2.22 Structures of gas-phase protonated Gly—LysH$^+$ peptides generated by classical molecular dynamics simulations. (a) Shows a helix and (b) a globular conformation. The helix structure was generated from a α-helix (i, $i+4$ hydrogen bonds, 3.6 residues per turn) and partially relaxed to a π-helix (i, $i+5$, 4.4 residues per turn) hydrogen bonds (reprinted with permission from reference [174] ©2005 American Chemical Society).

compact (hinged helix-coil) → extended helix transition are intramolecular proton transfer (charge repulsion) and separation of the coil and helical regions.

4.2.5 Excited state behaviour of amino acids and peptides

Among the 20 natural amino acids, tryptophan, tyrosine and phenylalanine possess aromatic side chain UV chromophores and have thus been most widely studied. In particular, tryptophan absorbs at the longest wavelengths and its fluorescence dominates that of proteins [177–180]. This fluorescence is very sensitive to the environment and is used as a probe of protein conformation [177, 178, 181], for example in early cancer detection [182]. In solution [183], the tryptophan fluorescence lifetime varies considerably with pH, from 19 ns at pH 11 to a bi-exponential decay of 0.5 and 3.1 ns at pH 7. This bi-exponential decay may correspond to a large number of possible conformers and environment situations (see Section 2.1.4) and not simply to the existence of two excited states. At pH 11, the amino group is neutral and becomes protonated in the zwitterionic form at pH 7. Both fluorescence lifetime and fluorescence yield drop rapidly at low pH indicating the opening of ultra fast de-excitation pathways when the amino group becomes protonated.

Tryptophan [90, 184–187], tyrosine [91, 188–191] and phenylalanine [93, 98, 105, 112, 140, 141, 192, 193] have been studied in R2PI experiments. The nanosecond lifetimes of neutral tryptophan excited states correspond to single exponential decay slightly varying from one conformer to another [194].

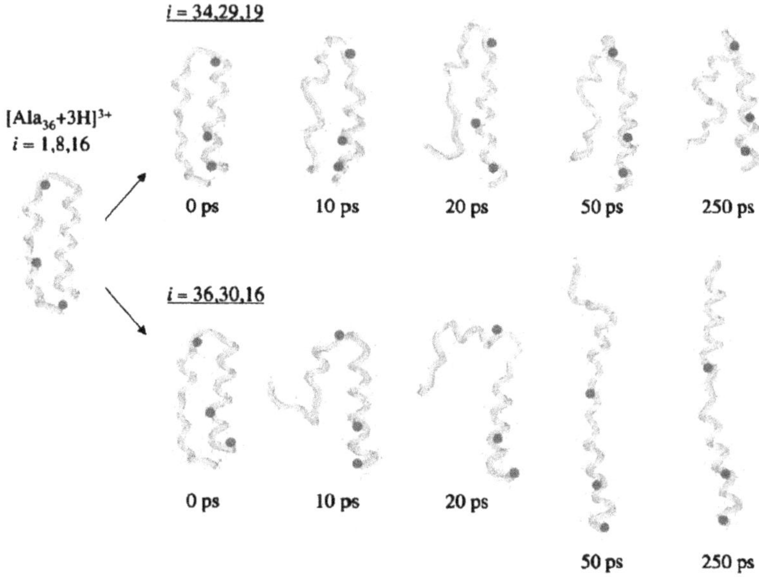

Figure 4.2.23 Structures of gas-phase protonated $(Ala_{36} + 3H)^{3+}$ peptides generated by classical molecular dynamics simulations. The initial hinged helix-coil conformation is imposed with protons in positions $i = 1$, 8 and 16. In the upper simulation, all charges are set in the helix region of the C-terminal with negligible electrostatic interactions and the separation of the coil and helical regions is very slow. In the bottom simulation, a proton is kept at position 16 in the hinge region and its repulsion with protons at positions 30 and 36 induces a fast separation of the coil and helix regions followed by extended helix formation (reprinted with permission from reference [176] ©2003 American Chemical Society).

4.2.5.1 Protonated tryptophan and tryptophan-containing peptides

Although structures and electronic spectra of protonated and neutral tryptophan are rather similar (Figure 4.2.24) [139], excited states of protonated tryptophan and protonated tryptophan-containing peptides exhibit 1,000-fold shorter lifetimes [143, 195–199] than neutrals. The lowest energy conformers of neutral aromatic amino acids correspond, as in glycine (see Figure 1.5.5), to establishment of hydrogen bonds between OH and NH_2 groups, either $OH \rightarrow NH_2$ (in Trp) or $NH_2 \rightarrow OH$ (in Phe) [200]. In tryptophan, several protonation sites are available at the amino group, the indole nitrogen or carbon 3. The lowest energy conformer of protonated tryptophan corresponds to protonation on the NH_2 group. The $OH \rightarrow NH_2$ bond is then broken and replaced by a strong interaction between the NH_3^+ and the π-electron cloud of the indole side chain. This interaction remains identical in the ground state S_0 and the excited state S_1. The energy difference between S_0 and S_1 is thus not modified and the electronic spectra of the neutral [92] and protonated [139] tryptophan are in the same energy range [139]. In contrast, the dynamical behaviours of the neutral and the protonated species are extremely different due to the existence of the $\pi\sigma^*$ de-excitation channel in the protonated species. Following photoexcitation, an electron transfer takes place from a π orbital of the aromatic chromophore to a σ^* orbital of the protonated

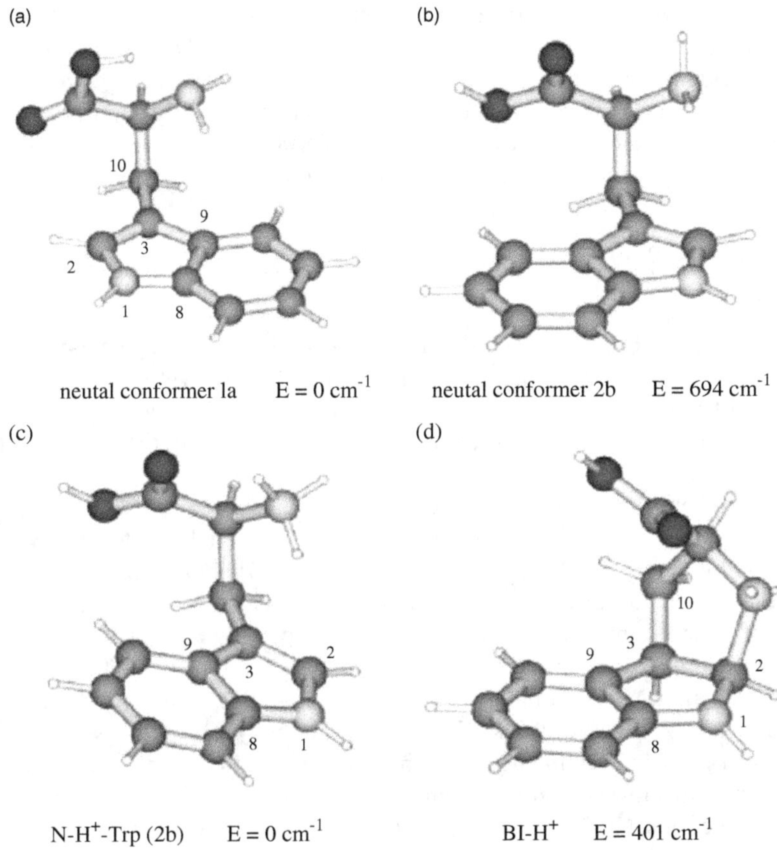

Figure 4.2.24 Top: Calculated structures of the lowest energy conformer of neutral tryptophan (a) and the neutral precursor from which the lowest energy of protonated tryptophan can be formed (b). Bottom: Structures of the lowest energy conformer of tryptophan protonated at the N-terminal (c) and in position 3 of the indole chromophore (d). Calculations have been conducted with combined density functional theory and multi-reference configuration interaction (DFT/MRCI) (reprinted with permission from reference [139] ©2004 Royal Society of Chemistry).

amino group. The NH_3^+ is neutralized and becomes a hypervalent R—NH_3 group (a similar situation can occur in electron capture dissociation, see Section 4.7.2). This group is unstable and the protonated amino acid state switches from the initial $\pi\pi^*$ state to a $\pi\sigma^*$ state which energy decreases when the N—H bond is stretched. As shown in Figure 1.7.3, the $\pi\sigma^*$ state has a conical intersection (see Section 1.7) with the S_0 ground state and a hydrogen loss competes with internal conversion leading to highly vibrationally amino acids that can further fragment. The branching ratio between these two pathways is 50%.

Photo-induced dissociation (PID) of tryptophan leads to observation of fragments different from those obtained in collision-induced dissociation (CID, see Section 3.2.2.2). Usually, only charged fragments are observed but it is now possible to obtain full information from a single dissociation event. The detection system must then allow the spatial and time localization of both neutral and charged fragments, so that the kinematics of the photo-fragmentation event can be fully re-constructed [201]. Following photo-dissociation of a fast (5 keV) ABC^+ ion, neutral B and C fragments and A^+ ions deflected by an electrostatic analyser are detected in coincidence by fast position-sensitive detectors. The analysis of two-dimension position-correlation diagrams allows the distinction between simultaneous or sequential fragmentation events. For example, it is possible to ascertain the order of losses in CID and PID in the following sequential fragmentation pattern of the protonated tryptophan ($TrpH^+$) ion leading to the immonium ion

$$(\mathrm{TrpH}^+)\times(205)\xrightarrow[\text{neutral H}_2\text{O loss}]{} m_{\mathrm{ion}}=187\xrightarrow[\text{neutral CO loss}]{} m_{\mathrm{ion}}=159 \quad (4.2.1)$$

The interesting result of the ion-neutral coincidence study is the fragmentation channel leading to the $m = 132$ ion. The fragmentation path had been previously attributed to the sequential loss of three neutral CO, H_2O and HCN fragments [202] while the correlation diagram unambiguously demonstrates that this ion is associated with the emission of a single $m = 73$ neutral fragment.

The dynamical behaviour of protonated tryptophan-containing peptides has been investigated in cases such as Trp—$LeuH^+$ [197] where tryptophan is on the protonated N-terminal side or not as in Leu—$TrpH^+$ [199], Lys—Trp—$LysH^+$ [203] or Gly—Trp—$GlyH^+$ [204] (the corresponding neutrals are considered in reference [104]). A first result is that CID and PID lead to different fragmentation patterns. In Gly—Trp—Gly H^+, for example the ratio between a_2 and b_2 fragment ion production is equal to 0.1 in CID and 4.5 in PID. y_2 fragment ions are observed in PID and not in CID. A second result is that extremely short decays of $\pi\pi^*$ excited states (550 fs) are observed for Gly—Trp—$GlyH^+$, as for bare tryptophan, indicating that an efficient coupling between $\pi\pi^*$ and $\pi\sigma^*$ is still present. This can be understood if one admits that lowest energy structures of Gly—Trp—$GlyH^+$ also involve a strong interaction between the NH_3^+ terminal and the delocalized aromatic π electronic cloud. The main difference is the disappearance of the "long" time constant observed in bare $TrpH^+$ that is attributed to the hindering of H-loss along the $\pi\sigma^*$ state due to a caging effect.

4.2.5.2 Comparison between excited state behaviours of tryptophan, tyrosine and phenylalanine

The electronic spectrum of cold protonated tyrosine shows well-resolved features (see Figure 4.2.12), with the origin band located at 4.35 eV. The vertical excitation energy of protonated tyrosine is equal to 4.87 and 4.69 eV at the CC2 level respectively with standard SV(P) split-valence double-zeta Gaussian basis set with polarization functions on heavy atoms and the correlation-consistent Dunning–Hay basis set of double-zeta quality augmented with diffuse functions (aug-cc-pVDZ) used for carbon and hydrogen atoms and nitrogen and oxygen atoms. This overestimate in between 0.5 and 0.3 eV of the CC2 method can be compared to the overestimate displayed of Figure 4.1.8 for TD-DFT and density functional theory and multi-reference configuration interaction (DFT/MRCI) calculations.

Dynamical behaviours and lifetimes of excited states of protonated tryptophan, tyrosine and phenylalanine are strongly different. As shown in Figure 4.2.12, the broad and sharp lines observed in the electronic spectra of tryptophan and tyrosine correspond respectively to very short (380 fs for Trp) and intermediate (22.3 ps for Tyr) lifetimes of the $\pi\pi^*$ photoexcited states. The fragmentation of phenylalanine is too low for any measurement of the decay [143]. This can be understood by comparing the energy gap and thus the coupling between the $\pi\pi^*$

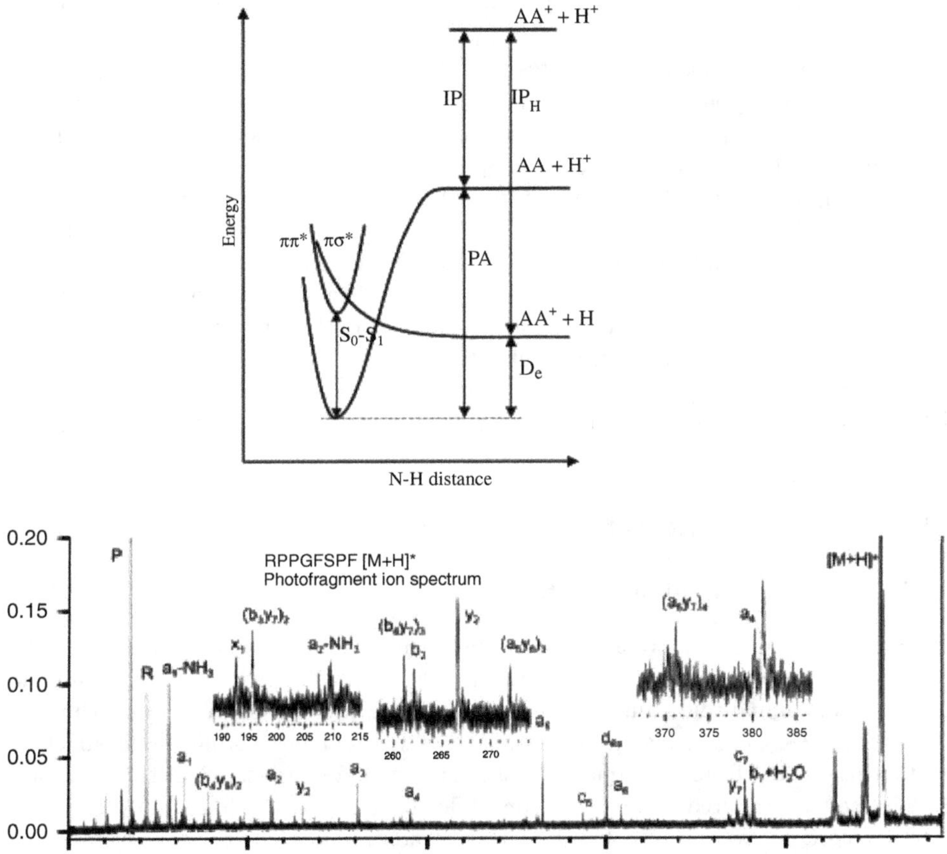

Figure 4.2.25 Top: Schematic potential energy curves of protonated amino acids. The estimation of the dissociation energy D_e of a hydrogen atom provides information concerning the crossing between the $\pi\pi^*$ and $\pi\sigma^*$ states (reprinted with permission from reference [143] ©2005 Royal Society of Chemistry). Bottom: Ion mass spectrum of the protonated 1–8 fragment of bradykinin photo-fragmented by 193 nm photons (reprinted with permission from reference [206] ©2006 American Chemical Society).

and $\pi\sigma^*$ states which increases from tryptophan to tyrosine and phenylalanine. This gap can be estimated by assuming that the dependence of the dissociative $\pi\sigma^*$ state as a function of the N—H distance is identical for the three amino acids. This $\pi\sigma^*$ state is localized on one of the amino group hydrogen atoms and is quite unaffected by any substitution of the aromatic chromophore group. The variation of the respective H-atom dissociation energies D_e then provides an estimate of the variation of the coupling between the $\pi\pi^*$ and $\pi\sigma^*$ states. The values of D_e can be estimated from the respective amino acid proton affinities PA, ionization potentials IP and the ionization potential of the hydrogen atom IP_H (13.6 eV) (Figure 4.2.25)

$$D_e = \mathrm{PA} + \mathrm{IP} - \mathrm{IP_H} \tag{4.2.2}$$

The values of D_e for Trp, Tyr and Phe are respectively equal to 3.4, 4.0 and 4.5 eV. The very strong coupling of the $\pi\pi^*$ and $\pi\sigma^*$ states in tryptophan becomes weak in tyrosine and negligible in phenylalanine in the vicinity of the electronic S_0—S_1 band origin. The observed difference between tryptophan and tyrosine remains in peptides such as Gly—Trp—Gly and Gly—Tyr—Gly [204]. Other photo-dissociation studies have been conducted on Leu—Leu—Leu—Tyr [205] and Gly—Leu—Gly [204] peptides.

Fragmentation patterns and reactions of large photoexcited peptide ions produced either by electrospray [207] or MALDI [206, 208, 209] has been the subject of several studies. Peptide ions are intrinsic chromophores in the VUV region and can absorb 157 or 193 nm laser radiation not only at their aromatic side chains but also at their amide bonds. The de-excitation of photoexcited states is followed by internal conversion leading to ground state vibrationally excited ions. Following statistical redistribution of this internal energy among the very large number of degrees of freedom, fragmentation takes place over a wide range of dissociation times and information concerning sequence of events is then lost owing to the large time interval (tens of microseconds) before mass-analysis of product ions.

4.2.6 Protein folding in the gas-phase

4.2.6.1 The protein folding problem

Protein function and structure are deeply related one to another and a tremendous effort has thus been devoted to the problem of protein folding, prediction of folded structures and functions from amino acid sequences [210, 211] and de novo design of proteins with given structures [212, 213]. It is also generally admitted that proteins sharing similar functions, due to their descent from a common ancestral protein, have similar folds. The function of proteins may have altered during evolution while keeping the same fold. It also seems that there are more or less universal folding laws instead of a unique scenario per protein. An important goal of protein folding studies is the structural basis for the rational design of drugs to therapeutically prevent misfoldings and loss-of-function disorders. In a cell, proteins are synthesized at a rate of approximately 50 amino acids per second [79] and most of them must get a three-dimensional structure. Structured proteins tend to adopt a unique set of very nearby conformations called their native state. Anfinsen suggested in 1973 that the native state structure of a protein is already encoded and uniquely determined by the amino acid sequence. A protein of 150 amino acids has around 10^{300} different conformations among which the native one must be found. The "Blind watcher" paradox states that the protein function cannot be reached from random sequences [214]. In 1968, Levinthal [215] has shown that protein folding cannot be the result of a random search of the native state since it would require a gigantic amount of time. Instead, experiments by Hagen [216] show that proteins fold within relatively short times in between from microseconds up to hundreds of seconds. Levinthal thus suggested that the folding is kinetically accelerated by simultaneous formation of small structures [215]. In 1975, Anfinsen proposed a thermodynamical control of folding into which the native form possesses the lowest free enthalpy [217].

For a large number of proteins, correct folding requires the assistance of other proteins called chaperones [218]. Those proteins are very large and, for example the GroEL chaperonin has a molecular mass of 800 kDa and consists of two heptameric rings stacked back to back, each of these rings containing a large central cavity [219]. When bound to another co-chaperonin, GroES, the GroEL forms a cavity called "Anfinsen" cage where the newborn protein is protected against aggregation or degradation [220]. Conformational changes induced by ligand binding ("induced-fit") [35, 115] have been studied in the gas-phase by tandem (see Section 3.2.2) [221, 222], H/D exchange [223] (see Section 3.3.4) or ion mobility [164] (see Section 3.3.1) mass-spectrometry.

4.2.6.2 Modelling of protein folding

The variations of enthalpy and entropy between the denatured state and the native state are large but the variation of free energy between those states is most often small (typically −5 to −15 kcal/mol) and is comparable to the establishment of a few hydrogen bonds. The decrease of enthalpy favourable to folding is compensated by a loss of entropy due to the passage from an extended form to a compact form (Figure 4.2.26). The enthalpic stabilizing terms comprise van der Waals and electrostatic forces (H-bonds and salt-bridges) as well as the hydrophobic effect and establishment of disulphide bonds. The enthalpy term decreases due to the loss of interactions between the denaturated form and the solvent ("desolvation effet").

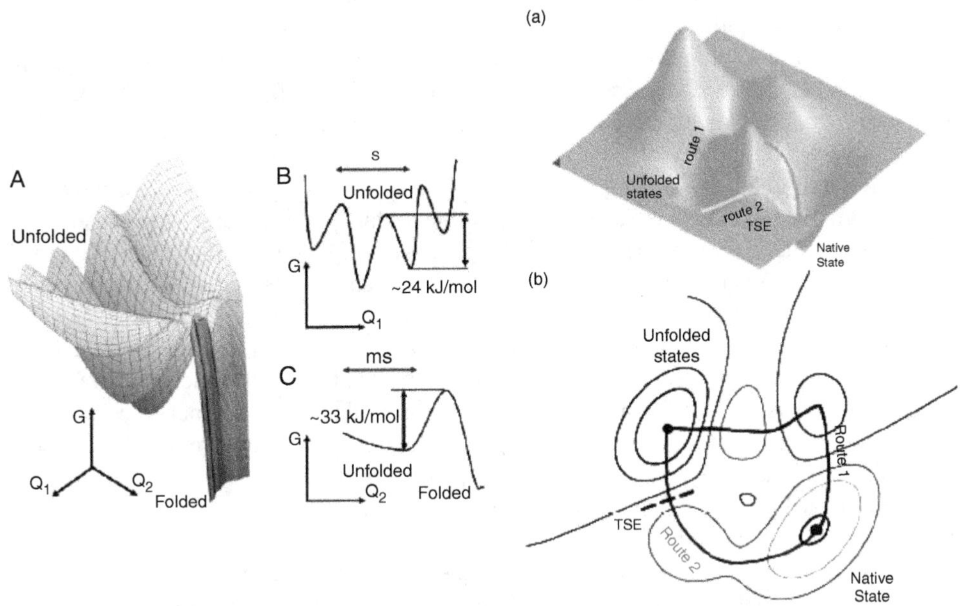

Figure 4.2.26 Left: Free energy landscape of RNase H under denaturating conditions explored by means of FRET. Transitions among unfolded sub-states (coordinate Q_1) occur on the second timescale and correspond to deep potential wells. The barrier between unfolded and folded states (coordinate Q_2) can be approached on a sub-millisecond timescale (reproduced with permission from reference [224] ©2005 National Academy of Sciences). Right: Three-dimensional representation (a) and contour (b) of the free energy landscape of a hypothetical protein. The folding proceeds via two routes 1 and 2. Route 2 connects the unfolded and the native state via a smaller free energy barrier (dashed line) but route 1 can, under certain conditions such as salt concentration and pH, be preferred. A local minimum corresponding to a stable folding intermediate appearing on route 1 then appears (b). This may indicate that condensed-phase and gas-phase folding scenario are not necessarily similar (reprinted with permission from reference [225] ©2006 Elsevier).

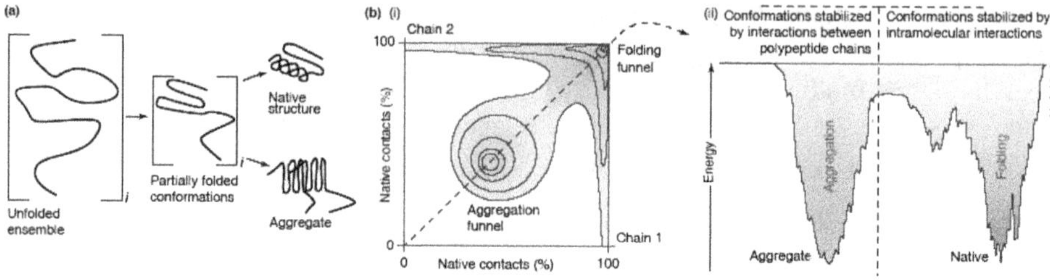

Figure 4.2.27 Schematic representation of the folding funnel leading to the native state. In fact, very often folding to the native state enters into competition with misfolding and aggregation; A pool of identical polypeptide chains, initially in random conformations, partially fold and can either continue to fold towards the native structure or misfold, start to associate with other polypeptide chains and aggregate (reprinted with permission from reference [227] ©2004 Elsevier).

The folding process is often represented by the folding "funnel" diagram [226, 227] (Figure 4.2.27). The funnel describes the progressive decrease of the accessible space dimensionality. Its width represents entropy and its depth energy.

The holes in the folding "funnel" diagram are local metastable minima and the bumps are transition states. The kinetics is represented by the slope of the curve: a larger slope corresponds to a faster kinetics. First, the folding is under kinetic control and lead to a "molten globule" where the majority of the native secondary structures are present but the tertiary structure is not reached. The folding towards the native state is then much slower. In agreement with the Levinthal paradox, there are several possible folding pathways and in agreement with Anfinsen principle, the most native state is the most stable. Under the influence of thermal fluctuations, the protein can move from one sub-state to another. However, large conformational transitions (i.e. large relative movements of change in secondary structures) can also occur within proteins. An important example is rearrangement of peptides and proteins into amyloid-type structures and fibres rich in β-sheets, leading to dysfunctions [228, 229]. In Alzheimer's disease [230], amyloid plaques accumulate between nerve cells [231]. Amyloid is the general term for protein fragments that are normally produced and eliminated. β-amyloids are fragments snipped from precursor proteins (APP) and that accumulates to form hard and insoluble plaques. The major component of these plaques is a 39- to 43-residue β-amyloid peptide Aβ whose predominant structure in the fibrils is a β-sheet. β-sheets, structures and polymerization [232] of β-amyloid peptides into amyloid fibrils thus constitute an active field of research in modelling [229, 233–235] and experiments [236–238].

From a pharmaceutical point of view, finding drugs [231, 239] that target a common motif shared by several proteins with different structures and functions involved in the Alzheimer's disease is a crucial goal. The concept of "one drug, one target" used in drug design is here replaced by the concept of "one-drug-multiple receptors" leading to "promiscuous drugs". This therapeutic research takes advantage of advances in modelling and its validation through experiments as well as chemo-informatics [240] calculations. Experimental gas-phase studies on model peptides by means of IR/R2PI [133] and the real Aβ peptide by means of mass-spectrometry [241] and ion-mobility [236, 238] are also devoted to the problem of β-sheet and amyloid β-peptide formation. For large proteins folding and fluctuation processes in the vicinity of the native state are being studied with coarse-grain models. Each α-carbon of the protein is replaced by a node in an elastic lattice [242, 243]. For small molecules such as the Try-cage protein, it is possible to use an all-atom representation and study folding process by running simulations with force-fields [244, 245].

4.2.6.3 Modelling and gas-phase studies of the Trp-cage protein

The problem of protein folding is now accessible to all-atom modelling and gas-phase experiments since the optimized design of a "miniprotein" by the group of Andersen [246]. The 20 amino acid sequence of the tryptophan-cage protein (Trp-cage, 1L2Y in the PDB data bank) is NLYIQWLKDGGPSSGRPPPS. The residues one to nine form a α-helix, residues 10 to 15 a 3_{10} helix and tryptophan is caged by the proline-rich C-terminal. Aspartic acid 9 and arginine 16 are involved in a stabilizing salt-bridge interaction. The modelling of the structure and folding properties of this protein [225, 244, 247–251] has reached the point where it is now possible to design mutants with improved stability and faster folding (1 μs) in condensed-phase [252] than the parent (4 μs) [216] (Figure 4.2.28).

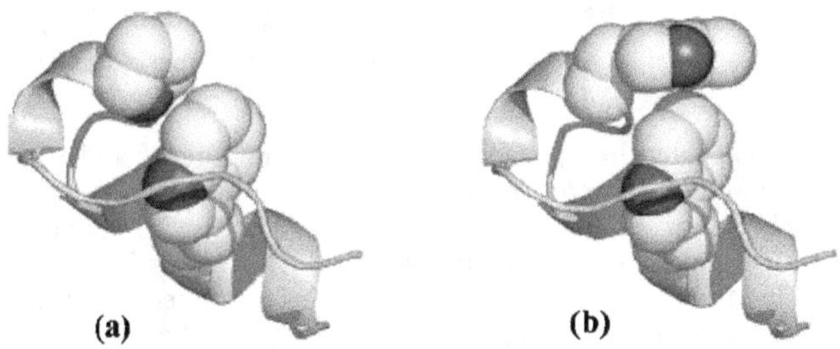

Figure 4.2.28 The original Trp-cage protein (a) and a more stable and ultra fast folding mutant (b) (reprinted with permission from reference [252] ©2006 American Chemical Society).

Electron-capture dissociation (ECD, see Section 4.7.2) [253] and fluorescence-energy transfer (FRET, see Section 3.5.5) [254] gas-phase experiments have been performed on trapped Trp-cage protein cations. ECD shows that although protonation sites in the gas-phase are different from those in solution, most of the structural features of the protein are conserved (Figure 4.2.29). A fluorescent dye (BODIPY TMR) [255] has been covalently attached to Trp-cage ions [254]. Those ions issued from an ESI source are trapped in presence of a buffer gas and the dye fluorescence is recorded as a function of the temperature. At low temperatures, the ions are in the native structure and the tryptophan is caged by the three prolines. As the temperature is raised, the protein unfolds and the tryptophan is no longer protected from quenching collisions with the dye (when the interaction distance becomes close to 5 Å). The fluorescence lifetimes of the dye, recorded as a function of the temperature for the tagged 2+ and 3+ protonated ions (Figure 4.2.29), are directly correlated with the miniprotein conformational changes. Fits of the data by a fluorescent model taking into account protein unfolding (with only the native and the unfolded states) and the quenching collisions provide enthalpy and entropy changes for the 3+ charge state that are respectively 1.5 and 1.2 larger than values determined in solution. In absence of solvent, the strength of the salt-bridge is increased. Hydrogen bond donor and acceptor groups that were interacting with the solvent can now form new intramolecular bonds. The larger decrease of the fluorescence as function of temperature observed for the 3+ charge state, as compared to the 2+ state, can be interpreted as due to the larger unfolding imposed by Coulomb repulsion in the higher charge state.

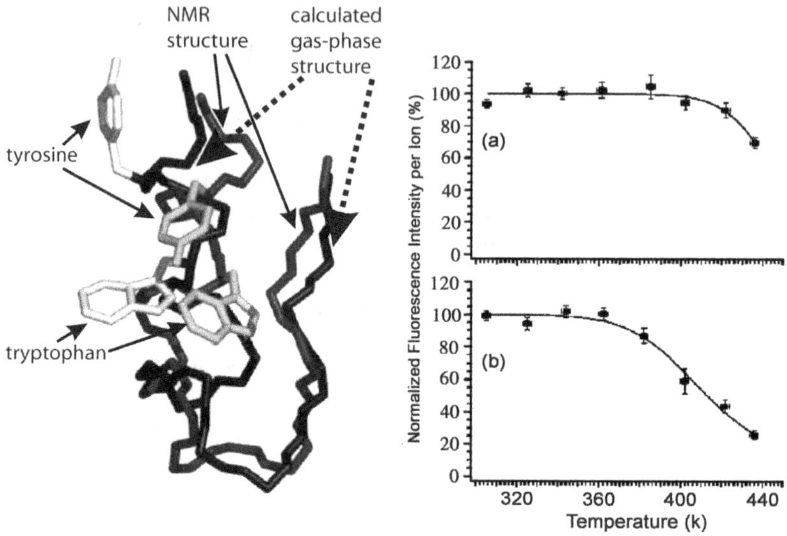

Figure 4.2.29 Left: Calculated lowest energy gas-phase structure of all-L Trp-cage dications at 275°K compared to the native solution phase conformation determined by NMR. The positions of the critical Tyr and Trp side chains are indicated by arrows (reprinted with permission from reference [253] ©2005 Elsevier). Right: Normalized fluorescence intensity per ion versus temperature for $(M+2H)^{2+}$ and $(M+3+)^{3+}$ Trp-cage-(BODIPY TMR) ions (reprinted with permission from reference [254] ©2005 American Chemical Society).

4.2.7 Fluorescent proteins

Fluorescent proteins are today very widely used in many biological fields since they allow living cell imaging [256]. The corresponding mRNAs are mechanically injected in a cell that translates them into the desired proteins. Once produced in cells, it is then possible to photoexcite those proteins in situ and follow their fate by means of fluorescence. The first used fluorescent protein has been the green fluorescent protein (GFP). It is a relatively small protein consisting of 238 amino acids and its natural role in the *Aequorea victoria* jellyfish where it was originally found is the conversion of blue light into green light [257]. In the GFP protein, the chromophore is embedded into the barrel-like protein made of 11 sheets and is rather well protected from the environment. As compared to the isolated chromophore situation, the environment introduces structural modifications and locally modifies any externally applied electromagnetic field. Gas-phase studies have been devoted to the comparison between the absorption spectrum of a model of the GFP chromophore, 4-hydroxybenzylidene-2,3-dimethyl-imidazolidone, and the absorption spectrum of the GFP itself (Figure 4.2.30). Other fluorescence spectral ranges in the red, blue or yellow have been also obtained by mutants [258, 259].

The gas-phase experiments have been conducted with the ELISA electrostatic heavy-ion storage ring (see Section 3.2.1.5) [260, 261]. The protonation state of the GFP chromophore plays an important role and could be changed by modifying the electrospray conditions in the ion sources. Anions were created by removing a proton from the

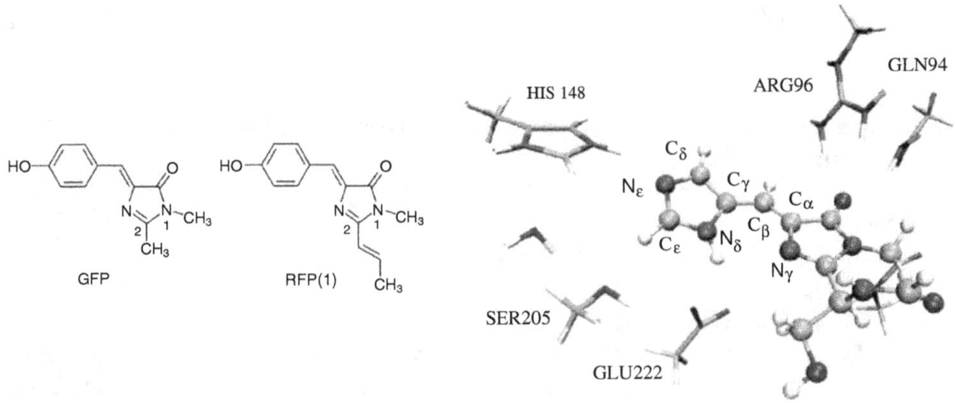

Figure 4.2.30 Left: Structures of the neutral gas-phase green fluorescent protein (GFP) and red fluorescent protein (RFP1) model chromophores (reprinted with permission from reference [261] ©2004 Royal Society of Chemistry). Right: Structure of the blue fluorescent protein chromophore (ball and sticks) and its closest amino acid residues in the protein (reprinted with permission from reference [259] ©2005 American Chemical Society).

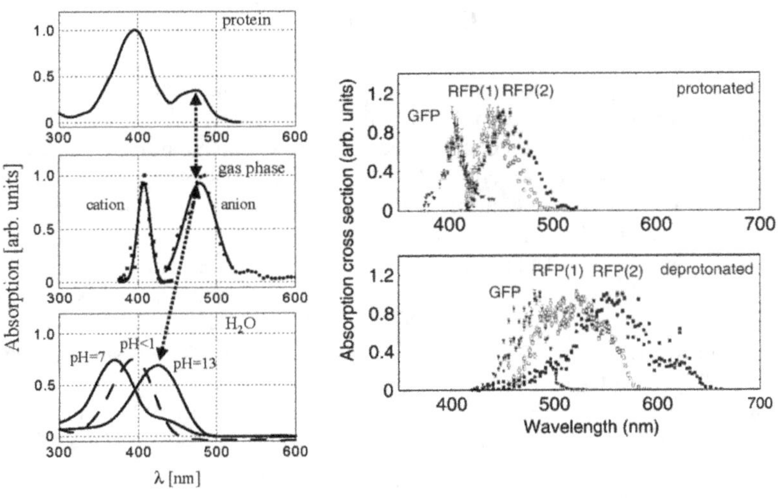

Figure 4.2.31 Left: Absorption spectra of the jellyfish GFP (top), of the gas-phase GFP-model chromophore in its protonated and deprotonated forms (middle) and of the GFP-model chromophore in aqueous solutions at different pHs (bottom) (reprinted with permission from reference [264] ©2002 EDP Sciences). Right: Absorption cross-sections of the protonated and deprotonated forms of the Green Fluorescent Protein (GFP) and Red Fluorescent Protein (RFP) model chromophores (reprinted with permission from reference [261] ©2004 Royal Society of Chemistry).

hydroxyl group (Figure 4.2.30) and cations by attaching a proton to the imidazolidone ring. The gas-phase absorption spectra and photodynamical [262, 263] properties of those different forms of the chromophore have been compared to that of the protein. The anion form absorption spectrum peak is almost identical to one of the two absorption minima of the protein showing that the protein shields the chromophore from its surrounding (Figure 4.2.31).

4.2.8 Electron spectroscopy of peptide bond models and amino acids

The peptide bond models [56, 57, 59], amino acids [265] and their side chains [266, 267] possess rather large permanent dipole moments and can thus attach electrons either in diffuse orbitals (multipole-bound anions) or in valence orbitals [268]. Gas-phase free-electron transmission spectra (ETS) of Gly, Ala, Phe, Pro and Cys [269] show that short-lived anions are formed by temporary attachment into the π^* empty orbitals of the COOH group of those amino acids. The corresponding vertical electron affinities (AE_{vert}) are respectively equal to 1.93, 1.80, 0.87, 0.68, 1.91 and 0.98 eV. Following electron attachment with a resonant character, dissociation takes place and the main fragments correspond to the loss of a hydrogen atom from the anion. For example, in proline this resonance is narrow and takes place around 1.2 eV [270].

REFERENCES

1. Agarwal PK: Role of protein dynamics in reaction rates enhancement by enzymes. *Journal of the American Chemical Society* 2006, 122:15248–15256.
2. Watson JD, Laskowsli RA, Thornton JM: Predicting protein function from sequence and structural data. *Current Opinion in Structural Biology* 2005, 15:275–284.
3. Janin J, Levitt M: Theory and simulation. Accuracy and reliability in modelling proteins and complexes. *Current Opinion in Structural Biology* 2006, 16:139–141.
4. Simons KT, Kooperberg C, Huang E, Baker D: Assembly of protein tertiary structures from fragments with similar local sequences using simulated annealing and Bayesian scoring functions. *Journal of Molecular Biology* 1997, 268:209–225.
5. Rohl CA, Strauss CE, Misura KM, Baker D: Protein structure prediction using Rosetta. *Methods Enzymology* 2004, 383:66–93.
6. Jalkanen KJ, Elstner M, Suhai S: Aminoacids and small peptides as building blocks for proteins: comparative theoretical and spectroscopic studies. *Journal of Molecular Structure* 2004, 675: 61–77.
7. Csazar AG, Perczel A: Ab initio characterization of building units in peptides and proteins. *Progress in Biophysics and Molecular Biology* 1999, 71:243–309.
8. Barth A: The infrared absorption of amino acid side chains. *Progress in Biophysics Molecular Biology* 2000, 74:141–173.
9. Cheng RP, Gellman SH, DeGrado WF: β-peptides: from structure to function. *Chemical Reviews* 2001, 101:3219–3232.
10. Zhang D, Tao WA, Cooks RG: Chiral resolution of D- and L-aminoacids by tandem mass-spectrometry of Ni (II)-bound trimeric complexes. *International Journal of Mass Spectrometry* 2001, 204:159–169.
11. Schalley CA, Weis P: Unusually stable magic number clusters of serine with a surprising preference for homochirality. *International Journal of Mass Spectrometry* 2002, 221:9–19.
12. Nanita SC, Takats Z, Cooks RG: Chiral enrichment of serine via formation, dissociation, and soft-landing of octameric cluster ions. *Journal of the American Society for Mass Spectrometry* 2004, 15:1360–1365.

13. Concina B, Hvelplund P, Nielsen AB, Nielsen SB, Rangama J: Formation and stability of charged amino acid clusters and the role of chirality. *Journal of the American Society for Mass Spectrometry* 2006, 17:275–279.
14. Pommié C, Levadoux S, Sabatier R, Lefranc G, Lefranc MP: IMGT standardized criteria for statistical analysis of immunoglobulin V-region aminoacid properties. *Journal of Molecular Recognition* 2004, 17:17–32.
15. Reid GE, Tichy SE, Perez J, O'Hair RAJ, Simpson RJ, Kenttamaa HI: N-terminal derivatization and fragmentation of neutral peptides via ion-molecule reactions with acylium ions: towards a gas-phase Eldman degradation? *Journal of the American Chemical Society* 2001, 123:1184–1192.
16. Wysocki VH, Resing KA, Zhang QF, Cheng GL: Mass spectrometry of peptides and proteins. *Methods* 2005, 35:211–222.
17. Pagel K, Wagner SC, Samedov K, von Berlepsch H, Böttcher C, Koksch B: Random coils, β-sheet ribbons, and α-helical fibers: one peptide adopting three different secondary structures at will. *Journal of the American Chemical Society* 2006, 126:2196–2197.
18. Jarrold MF: Unfolding, refolding, and hydration of proteins in the gas phase. *Accounts of Chemical Research* 1999, 32:360–367.
19. Jarrold MF: Peptides and proteins in the vapor phase. *Annual Review of Physical Chemistry* 2000, 51:179–207.
20. Kohtani M, Jones TC, Schneider JE, Jarrold MF: Extreme stability of an unsolvated α-helix. *Journal of the American Chemical Society* 2004, 126:7420–7421.
21. Kinnear BS, Jarrold MF: Helix formation in unsolvated peptides: side chain entropy is not the determining factor. *Journal of the American Chemical Society* 2001, 123:7907–7908.
22. Daidone I, D'Abramo M, Di Nola A, Amadei A: Theoretical characterization of α-helix and β-hairpin folding kinetics. *Journal of the American Chemical Society* 2005, 127:14825–14832.
23. Kim J, Mao J, Gunner MR: Are acidic and basic groups in buried proteins predicted to be ionized. *Journal of Molecular Biology* 2005, 348:1283–1298.
24. Valentine SJ, Anderson JG, Ellington AD, Clemmer DE: Disulfide-intact and -reduced lysozyme in the gas phase: Conformations and pathways of folding and unfolding. *Journal of Physical Chemistry B* 1997, 101:3891–3900.
25. Carles S, Lecomte F, Schermann JP, Desfrancois C, Xu S, Nilles JM, Bowen KH, Berges J, Houee-Levin C: Non-dissociative electron capture by disulfide bonds. *Journal of Physical Chemistry A* 2001, 105:5622–5626.
26. Elcock AH, Sept D, McCammon JA: Computer simulation of protein-protein interactions. *Journal of Physical Chemistry B* 2001, 105:1504–1518.
27. Sobott F, Robinson CV: Protein complexes gain momentum. *Current Opinion in Structural Biology* 2002, 12:729–734.
28. Sobott F, McCammon MG, Robinson CV: Gas-phase dissociation pathways of a tetrameric protein complex. *International Journal of Mass Spectrometry* 2003, 230:193–200.
29. Sobott F, McCammon MG, Hernandez H, Robinson CV: The flight of macromolecular complexes in a mass spectrometer. *Philosophical Transactions of the Royal Society of London Series A, Mathematical Physical and Engineering Sciences* 2005, 363:379–389.
30. Becker OM, Levy Y, Ravitz O: Flexibility, conformation space and bioactivity. *Journal of Physical Chemistry B* 2000, 104:2123–2135.
31. Misura KM, Baker D: Progress and challenges in high-resolution refinement of protein structure models. *Proteins-Structure Function and Bioinformatics* 2005, 59:15–29.
32. Lindorff-Larsen K, Best RB, DeProsto MA, Dobson CM, Vendrelusco M: Simultaneous determination of protein structure and dynamics. *Nature* 2005, 433:128–132.
33. Gunasekaran K, Tsai CJ, Kumar S, Zanuy D, Nussinov R: Extended disordered proteins: targeting function with less scaffold. *Trends in Biochemical Science* 2003, 28:81–85.
34. Tompa P: The interplay between structure and function in intrinsically unstructured proteins. *FEBS Letters* 2005, 13:1375–1394.

35. Dyson HJ, Wright PE: Intrinsically unstructured proteins and their functions. *Nature Reviews Molecular Cell Biology* 2005, 6:197–211.
36. Shelimov KB, Clemmer DE, Hudgins RR, Jarrold MF: Protein structure in vacuo: gas-phase confirmations of BPTI and cytochrome c. *Journal of the American Chemical Society* 1997, 119:2240–2248.
37. Mao Y, Ratner MA, Jarrold MF: One water molecule stiffens a protein. *Journal of the American Chemical Society* 2000, 122:2950–2951.
38. Kohtani M, Schneider JE, Jones TC, Jarrold MF: The mobile proton in polyalanine peptides. *Journal of the American Chemical Society* 2004, 126:16981–16987.
39. Valentine SJ, Clemmer DE: Temperature-dependent H/D exchange of compact and elongated cytochrome c ions in the gas phase. *Journal of the American Society for Mass Spectrometry* 2002, 13:506–517.
40. Valentine SJ, Clemmer DE: H/D exchange levels of shape-resolved cytochrome c conformers in the gas phase. *Journal of the American Chemical Society* 1997, 119:3558–3566.
41. Dunbar RC: BIRD (blackbody infrared radiative dissociation): evolution, principles, and applications. *Mass Spectrometry Reviews* 2004, 23:127–158.
42. Oomens J, Polfer N, Moore DT, van der Meer L, Marshall AG, Eyler JR, Meijer G, von Helden G: Charge-state resolved mid-infrared spectroscopy of a gas-phase protein. *Physical Chemistry Chemical Physics* 2005, 7:1345–1348.
43. Abo-Riziq A, Crews B, Callahan MP, Grace L, de Vries M: Spectroscopy of isolated gramicidin peptides. *Angewandte Chemie-International Edition* 2006, 118:5290–5293.
44. Sangarron W, Ruangpomvisuti V: Conformational study of cationic, zwitterionic, anionic species of aspartic acid, water-added forms and their protonation. A DFT method. *Journal of Molecular Structure* 2005, 758:179–185.
45. Sugarawa Y, Hamada Y, Hirakawa AY, Tsuboi M: Occurrence of the *cis*-peptide in gaseous N-methylformamide as revealed by rotational structure in its infrared spectrum. *Chemical Physics Letters* 1979, 67:186–188.
46. Lavrich RJ, Plusquellic DF, Suenram RD, Fraser GT, Walker ARH, Lovas FJ, Suenram RD, Tubergen MJ: Experimental model studies of the peptide bond: identification of the C7 conformation of the alanine dipeptide analog N-acetyl-alanine N′-methylamide from torsion-rotation interactions. *Journal of Chemical Physics* 2003, 118:1253–1265.
47. Ilyushin VV, Alekseev EA, Dyubko SF, Kleiners I, Hougen JT: Ground and first torsional states of acetamide. *Journal of Molecular Spectroscopy* 2004, 227:155–169.
48. Cordeiro MAM, Santana WP, Cusinato R, Cordeiro JMM: Monte-Carlo investigations of intermolecular interactions in water-amide structures. *Journal of Molecular Structure* 2005, 759:159–164.
49. Wu Q, Zhang H, Zhou Z, Lu J, Zhao G: Study of the hydrogen bonding interaction of 1,1 complexes of serine with formamide using density functional theory. *Journal of Molecular Structure* 2005, 757:9–18.
50. Ohashi N, Hougen JT, Suenram RD, Lovas FJ, Kawashima Y, Fujitake M, Pyka J: Analysis and fit of the Fourier-transform microwave spectrum of the two-top molecule N-methylacetamide. *Journal of Molecular Spectroscopy* 2004, 227:28–42.
51. Fujitake M, Kubota Y, Ohashi N: Fourier transform microwave spectroscopy of N, N-dimethylacetamide. *Journal of Molecular Spectroscopy* 2006, 236:97–109.
52. Bounouar M, Scheurer C: Reducing the vibrational coupling network in N-methylacetamide as a model for ab initio infrared spectra computations of peptides. *Chemical Physics* 2006, 323:87–101.
53. Headley AD: A theoretical analysis of substituted formamide conformers. *Journal of Molecular Structure* 2002, 589:423–429.
54. Cuevas G, Renugopalakrishnan V, Madrid G, Hagler AT: Density function studies of peptides. Part I Vibrational frequencies including isotopic effects and NMR chemical shifts of

N-methylacetamide, a peptide model, from density function and MP2 calculations. *Physical Chemistry Chemical Physics* 2002, 4:1490–1499.

55. Desfrancois C, Periquet V, Carles S, Schermann JP, Adamowicz L: Neutral and negatively-charged formamide, N-methylformamide and dimethylformamide clusters. *Chemical Physics* 1998, 239:475–483.
56. Maeyama T, Negishi Y, Tsukuda T, Yagi I, Mikami N: Electron localization in negatively charged formamide clusters studied by photodetachment spectroscopy. *Physical Chemistry Chemical Physics* 2006, 8:827–833.
57. Maeyama T, Mikami N: Characteristic distributions of negatively charged N-monosubstituted amide clusters generated by electron attachment in supersonic expansions. *Physical Chemistry Chemical Physics* 2004, 6:2725–2731.
58. Chang HC, Wu CC, Kuo JL: Recent advances in understanding the structures of medium-sized protonated water clusters. *International Reviews in Physical Chemistry* 2005, 24: 553–578.
59. Seydou M, Modelli A, Lucas B, Konate K, Desfrancois C, Schermann JP: Electron attachment to strongly polar clusters – formamide molecule and clusters. *European Physical Journal D* 2005, 35:199–205.
60. Bleckmann P, Breiteenbach P, Dickhut KU, Keller D, Schwittek C: Intermolecular potentials and force constants from ab initio energies – application to the N—H···O═C hydrogen bonds in formamide dimers. *Fresenius Journal of Analytical Chemistry* 1997, 359:115–120.
61. Kobko N, Paraskevas L, del Rio E, Dannenberg JJ: Cooperativity in amide hydrogen bonding chains: implications for protein-folding models. *Journal of the American Chemical Society* 2001, 123:4348–4349.
62. Dugave C, Demange L: *Cis–trans* isomerization of organic molecules and biomolecules: implications and applications. *Chemical Reviews* 2003, 103:2475–2532.
63. Ramek M, Kelterer AM, Nikolic S: Ab initio and molecular mechanics conformational analysis of neutral L-proline. *International Journal of Quantum Chemistry* 1997, 65:1033–1045.
64. Marino T, Russo N, Tocci E, Toscano M: Density-functional computations of proton affinity and gas-phase basicity of proline. *Journal of Mass Spectrometry* 2001, 36:301–305.
65. Stepanian SG, Reva ID, Radchenko ED, Adamowicz L: Conformers of non-ionized proline. Matrix isolation infrared and post-Hartree–Fock ab initio studies. *Journal of Physical Chemistry A* 2001, 105:10664–10672.
66. Christiansen JJ: The microwave spectrum of cyanoformamide. *Journal of Molecular Spectroscopy* 2005, 231:131–136.
67. Aviles Moreno JR, Petitprez D, Huet TR: The conformational flexibility in N-phenylformamide: an ab initio approach supported by microwave spectroscopy. *Chemical Physics Letters* 2006, 419:411–416.
68. Unterberg C, Gerlach A, Schrader T, Gerhards M: Structure of the protected dipeptide Ac-Val-Phe-OMe in the gas phase: towards a β-sheet model system. *Journal of Chemical Physics* 2003, 118:8296–8300.
69. Gerlach A, Unterberg C, Fricke H, Gerhards M: Structures of Ac-Trp-OMe and its dimer (Ac-Trp-OMe)(2) in the gas phase: influence of a polar group in the side-chain. *Molecular Physics* 2005, 103:1521–1529.
70. Gerhards M, Unterberg C, Gerlach A, Jansen A: β-sheet model systems in the gas phase: structures and vibrations of Ac-Phe-NHMe and its dimer $(Ac\text{-}Phe\text{-}NHMe)_2$. *Physical Chemistry Chemical Physics* 2004, 6:2682–2690.
71. Chin W, Mons M, Dognon JP, Mirasol R, Chass G, Dimicoli I, Piuzzi F, Butz P, Tardivel B, Compagnon I, *et al.*: The gas-phase dipeptide analogue acetyl-phenylalanyl-amide: a model for the study of side chain/backbone interactions in proteins. *Journal of Physical Chemistry A* 2005, 109:5281–5288.

72. Masman MF, Zamora MA, Rodriguez AM, Fidanza NG, Peruchena NM, Enriz RD, Csizmadia IG: Exploration of the full conformational space of N-acetyl-L-glutamate-N-methylamide. *European Physical Journal D* 2002, 20:531–542.
73. Klipfel MW, Zamora MA, Rodriguez AM, Fidanza NG, Enriz RD, Csizmadia IG: Exploration of the full conformational space of N-acetyl-L-glutamine-N-methylamide. An ab initio and density functional theory study. *Journal of Physical Chemistry A* 2003, 107:5079–5091.
74. Enriz RD, Morales ME, Baldoni HE: Conformational study of *cis* and *trans* N-formyl-N-methyl-L-glycine-N′-amide and N-acetyl-N-methyl-L-glycine-N′-methylamide. *Journal of Molecular Structure* 2005, 731:177–185.
75. Watson TM, Hirst JD: Vibrational analysis of capped (Leu) enkephalin. *Physical Chemistry Chemical Physics* 2004, 6:2580–2587.
76. Wieczorek R, Dannenberg JJ: The energetic and structural effects of single amino acid substitutions upon capped α-helical peptides containing 17 aminoacid residues. An ONIOM DFT/AM1 study. *Journal of the American Chemical Society* 2005, 127:17216–17223.
77. Chin W, Piuzzi F, Dimicoli I, Mons M: Probing the competition between secondary structures and local preferences in gas phase isolated peptide backbones. *Physical Chemistry Chemical Physics* 2006, 8:1033–1048.
78. Ramachandran GN, Sassiekharan V: Conformation of polypeptides and proteins. *Advances in Protein Chemistry* 1968, 28:283–437.
79. Daune M: *Molecular biophysics. Structures in motion*: Oxford University Press, Oxford, 1999.
80. Rai R, Aravinda S, Kanagarajadurai K, Raghthama S, Shamala N, Balaram P: Diproline templates as folding nuclei in designed peptides. Conformational analysis of synthetic peptide helices containing amino terminal Pro-Pro segments. *Journal of the American Chemical Society* 2006, 128:7916–7928.
81. Koo JCP, Chass GA, Perczel A, Farkas O, Torday LL, Varro A, Papp JG, Csizmadia IG: Exploration of the four-dimensional-conformational potential energy hypersurface of N-acetyl-L-aspartic acid N′-methylamide with its internally hydrogen bonded side-chain orientation. *Journal of Physical Chemistry A* 2002, 106:6999–7009.
82. Vass E, Hollosi M, Besson F, Buchet R: Vibrational spectroscopic detection of β- and γ-turns in synthetic and natural peptides and proteins. *Chemical Reviews* 2003, 103:1917–1954.
83. Godfrey PD, Brown RD: Shape of glycine. *Journal of the American Chemical Society* 1995, 117:2019–2023.
84. Lovas FJ, Lavrich RJ, Plusquellic DF: Methyl torsional state analysis of the jet-cooled microwave spectrum of N-acetylglycine. *Journal of Molecular Spectroscopy* 2004, 228:251–257.
85. Ilyushin VV, Alekseev EA, Dyubko SF, Motiyenko RA, Lovas FJ: Millimeter wave spectrum of glycine. *Journal of Molecular Spectroscopy* 2005, 231:15–22.
86. Sanz ME, Cortijo V, Caminati W, Lopez JC, Alonso JL: The conformers of phenylglycine. *Chemistry – A European Journal* 2006, 12:2564–2570.
87. Lesarri A, Mata S, Cocinero EJ, Blanco S, Lopez JC, Alonso JL: The structure of neutral proline. *Angewandte Chemie-International Edition* 2002, 41:4673–4676.
88. Lesarri A, Cocinero EJ, Lopez JC, Alonso JL: The shape of neutral valine. *Angewandte Chemie-International Edition* 2004, 43:605–610.
89. Blanco S, Lesarri A, Lopez JC, Alonso JL: The gas-phase structure of alanine. *Journal of the American Chemical Society* 2004, 126:11675–11683.
90. Rizzo TR, Park YD, Levy DH: A molecular beam of tryptophan. *Journal of the American Chemical Society* 1985, 107:277–278.
91. Lindinger A, Toennies JP, Vilesov AF: High resolution of vibronic spectra of the amino acids tryptophan and tyrosine in 0.38°K cold helium droplets. *Journal of Chemical Physics* 1999, 110:1429–1436.
92. Snoek LC, Kroemer RT, Hockridge MR, Simons JP: Conformational landscapes of aromatic amino acids in the gas phase: infrared and ultraviolet ion dip spectroscopy of tryptophan. *Physical Chemistry Chemical Physics* 2001, 3:1819–1826.

93. Lee KT, Sung J, Lee KJ, Kim SK, Park YD: Resonant two-photon ionization study of jet-cooled amino acid: L-phenylalanine and its monohydrated complex. *Journal of Chemical Physics* 2002, 116:8251–8254.
94. Lee KT, Sung J, Lee KJ, Park YD, Kim SK: Conformation-dependent ionization energies of L-phenylalanine. *Angewandte Chemie-International Edition* 2002, 41:4114–4117.
95. Dian BC, Longarte A, Mercier S, Evans DA, Wales DJ, Zwier TS: The infrared and ultraviolet spectra of single conformations of methyl-capped dipeptides: N-acetyl tryptophan amide and N-acetyl tryptophan methyl amide. *Journal of Chemical Physics* 2002, 117: 10688–10702.
96. Reha D, Valdes H, Vondrasek J, Hobza P, Abu-Riziq A, Crews B, de Vries MS: Structure and IR spectrum of phenylalanyl-glycyl-glycine tripetide in the gas-phase: IR/UV experiments, *ab initio* quantum chemical calculations, and molecular dynamic simulations. *Chemistry – A European Journal* 2005, 11:6803–6817.
97. Cohen R, Brauer B, Nir E, Grace L, de Vries MS: Resonance-enhanced multiphoton ionization spectroscopy of dipeptides. *Journal of Physical Chemistry A* 2000, 104:6351–6355.
98. Lee YH, Jung JW, Kim B, Butz P, Snoek LC, Kroemer RT, Simons JP: Alanyl side chain folding in phenylalanine: conformational assignments through ultraviolet rotational band contour analysis. *Journal of Physical Chemistry A* 2004, 108:69–73.
99. Chin W, Compagnon I, Dognon JP, Canuel C, Piuzzi F, Dimicoli I, von Helden G, Meijer G, Mons M: Spectroscopic evidence for gas-phase formation of successive β-turns in a three-residue peptide chain. *Journal of the American Chemical Society* 2005, 127:1388–1389.
100. Compagnon I, Oomens J, Bakker J, Meijer G, von Helden G: Vibrational spectroscopy of a non-aromatic amino acid-based model peptide: identification of the γ-turn motif of the peptide backbone. *Physical Chemistry Chemical Physics* 2005, 7:13–15.
101. Bakker JM, Plutzer C, Hunig I, Haber T, Compagnon I, von Helden G, Meijer G, Kleinermanns K: Folding structures of isolated peptides as revealed by gas-phase mid-infrared spectroscopy. *ChemPhysChem* 2005, 6:120–128.
102. Linder R, Nispel M, Haber T, Kleinermanns K: Gas-phase FT-IR-spectra of natural amino acids. *Chemical Physics Letters* 2005, 409:260–264.
103. Hunig I, Seefeld KA, Kleinermanns K: REMPI and UV-UV double resonance spectroscopy of tryptophan ethylester and the dipeptides tryptophan-serine, glycine-tryptophan and proline-tryptophan. *Chemical Physics Letters* 2003, 369:173–179.
104. Hunig I, Kleinermanns K: Conformers of the peptides glycine-tryptophan, tryptophan-glycine and tryptophan-glycine-glycine as revealed by double resonance laser spectroscopy. *Physical Chemistry Chemical Physics* 2004, 6:2650–2658.
105. Ebata T, Hashimoto T, Ito T, Inokuchi Y, Altunsu F, Brutschy B, Tarakeshwar P: Hydration profiles of aromatic aminoacids: conformations and vibrations of L-phenylalanine-$(H_2O)_n$ clusters. *Physical Chemistry Chemical Physics* 2006, 8:4783–4791.
106. Improta R, Barone V: Assessing the reliability of density functional methods in the conformational study of polypeptides: the treatment of intra-residue nonbonding interactions. *Journal of Computational Chemistry* 2004, 25:1233–1341.
107. MacKerell AD, Feig M, Brooks CL: Extending the treatment of backbone energetics in protein force fields: Limitations of gas-phase quantum mechanics in reproducing protein conformational distributions in molecular dynamics simulations. *Journal of Computational Chemistry* 2004, 25:1400–1415.
108. MacKerell AD, Feig M, Brooks CL: Improved treatment of the protein backbone in empirical force fields. *Journal of the American Chemical Society* 2004, 126:698–699.
109. Chass GA, Mirasol RS, Setiadi DH, Tang TH, Chin W, Mons M, Dimicoli I, Dognon JP, Viskolcz B, Lovas S, *et al.*: Characterization of the conformational probability of N-acetyl-phenylalanyl-NH2 by RHF, DFT, and MP2 computation and AIM analyses, confirmed by jet-cooled infrared data. *Journal of Physical Chemistry A* 2005, 109:5289–5302.

110. Bissety K, Corcho FJ, Canto J, Kruger HG, Perez JJ: Simulated annealing study of the pentacyclo-undecane cage amino acid tripeptides of the type Ac-X-Y-Z-NHMe. *Journal of Molecular Structure* 2005, 759:145–167.
111. Owen MC, Komaroni I, Murphy RF, Lovas S: The local preference of C_α-radicals in proteins. *Journal of Molecular Structure* 2005, 759:117–124.
112. Huang Z, Yu W, Lin Z: Exploration of the full conformational landscapes of gaseous aromatic aminoacid phenylalanine: an ab initio study. *Journal of Molecular Structure* 2006, 758:195–202.
113. Horvath V, Varga Z, Kovacs A: Substituent effects on long-range interactions in the β-sheet structure of oligopeptides. *Journal of Molecular Structure* 2005, 755:247–251.
114. Gunasekaran K, Gomathi L, Ramakrishnan C, Chandrasekhar J, Balaram P: Conformational interconversions in peptide β-turns: analysis of turns in proteins and computational estimates of barriers. *Journal of Molecular Biology* 1998, 284:1505–1516.
115. Tyndall JD, Pfeiffer B, Abbenante G, Fairlie DP: Over one hundred peptide-activated G protein-coupled receptors recognize ligands with turn structure. *Chemical Reviews* 2005, 105:793–826.
116. Fairlie D: Small molecules that mimic components of bioactive surfaces. *Australian Journal of Chemistry* 2004, 57:855–857.
117. Iqbalsyah TM, Doig AJ: Pairwise coupling in an Arg-Phe-Met triplet stabilizes α-helical peptide via shared rotamer preferences. *Journal of the American Chemical Society* 2005, 127: 5002–5003.
118. Chin W, Piuzzi F, Dognon JP, Dimicoli L, Tardivel B, Mons M: Gas phase formation of a 3_{10}-helix in a three-residue peptide chain: role of side chain-backbone interactions as evidenced by IR-UV double resonance experiments. *Journal of the American Chemical Society* 2005, 127:11900–11901.
119. Evans DA, Wales DJ: Free energy landscapes of model peptides and proteins. *Journal of Chemical Physics* 2003, 118:3891–3897.
120. Mayer B, Klein CT: Influence of solvation on the helix-forming tendency of non polar amino acids. *Journal of Molecular Structure: THEOCHEM* 2000, 532:213–236.
121. Shepherd NE, Hoang HN, Abbenante G, Fairlie D: Single turn α-helices with exceptional stability in water. *Journal of the American Chemical Society* 2005, 127:2974–2983.
122. Mikhonin AV, Myshakina NS, Bykov SV, Asher SA: UV resonance Raman determination of polyproline II, extended 2.5_1 -helix, and β-sheet ψ angle energy landscape in poly-L-L-lysine and poly-L-glutamic acid. *Journal of the American Chemical Society* 2005, 127:7712–7720.
123. Kohtani M, Jarrold MF: Water molecule adsorption on short alanine peptides: how short is the shortest gas-phase alanine-based helix? *Journal of the American Chemical Society* 2004, 126:8454–8458.
124. Kohtani M, Jarrold MF: The initial steps in the hydration of unsolvated peptides: water molecule adsorption on alanine-based helices and globules. *Journal of the American Chemical Society* 2002, 124:11148–11158.
125. Kaleta DT, Jarrold MF: Disrupting helix formation in unsolvated peptides. *Journal of Physical Chemistry B* 2001, 105:4436–4440.
126. Hudgins RR, Ratner MA, Jarrold MF: Design of helices that are stable in vacuo. *Journal of the American Chemical Society* 1998, 120:12974–12975.
127. Kimura T, Takahashi S, Akiyama S, Uzawa T, Ishimori K, Morishima I: Direct observation of the multistep helix formation of poly-glutamic acids. *Journal of the American Chemical Society* 2002, 124:11596–11597.
128. Counterman AE, Clemmer DE: Anhydrous polyproline helices and globules. *Journal of Physical Chemistry B* 2004, 108:4885–4898.
129. Hudgins RR, Mao Y, Ratner MA, Jarrold MF: Conformations of Gly_nH^+ and Ala_nH^+ peptides in the gas phase. *Biophysical Journal* 1999, 76:1591–1597.
130. Chung DM, Dou Y, Baldi P, Nowick JS: The absence of favorable aromatic interactions between β-sheet peptides. *Journal of the American Chemical Society* 2005, 127:9998–9999.

131. Hughes RM, Waters ML: Model systems for β hairpins and β sheets. *Current Opinion in Structural Biology* 2006, 16:514–524.
132. Maynard AJ, Sharman GJ, Searle MS: Origin of β-hairpin stability in solution: structural and thermodynamics analysis of the folding of a model peptide supports hydrophobic stabilization in water. *Journal of the American Chemical Society* 1998, 120:1996–2007.
133. Fricke H, Gerlach A, Unterberg C, Rzepecki P, Schrader T, Gerhards M: Structure of the tripeptide model Ac-Val-Tyr(Me)-NHMe and its cluster with water investigated by IR/UV double resonance spectroscopy. *Physical Chemistry Chemical Physics* 2004, 6:4636–4641.
134. Unterberg C, Gerlach A, Schrader T, Gerhards M: Clusters of a protected amino acid with pyrazole derivatives: β-sheet model systems in the gas phase. *European Physical Journal D* 2002, 20:543–550.
135. Gerhards M, Unterberg C, Gerlach A: Structure of a β-sheet model system in the gas phase: analysis of the C=O stretching vibrations. *Physical Chemistry Chemical Physics* 2002, 4: 5563–5565.
136. Gerhards M, Unterberg C: Structures of the protected amino acid Ac-Phe-OMe and its dimer: a β-sheet model system in the gas phase. *Physical Chemistry Chemical Physics* 2002, 4:1760–1765.
137. Fricke H, Gerlach A, Gerhards M: Structure of a β-sheet model system in the gas phase: analysis of the fingerprint region up to 10 μm. *Physical Chemistry Chemical Physics* 2006, 8:1660–1662.
138. Weinkauf R, Schanen P, Yang D, Soukara S, Schlag EW: Elementary processes in peptides: electron mobility and dissociation in peptide cations in the gas phase. *Journal of Physical Chemistry* 1995, 99:11255–11265.
139. Nolting D, Marian C, Weinkauf R: Protonation effect on the electronic spectrum of tryptophan in the gas phase. *Physical Chemistry Chemical Physics* 2004, 6:2633–2640.
140. Snoek LC, Robertson EG, Kroemer RT, Simons JP: Conformational landscapes in amino acids: infrared and ultraviolet ion-dip spectroscopy of phenylalanine in the gas phase. *Chemical Physics Letters* 2000, 321:49–56.
141. Lee KT, Sung JH, Lee KJ, Kim SK, Park YD: Conformation-dependent ionization of L-phenylalanine: structures and energetics of cationic conformers. *Chemical Physics Letters* 2003, 368:262–268.
142. Maupin CM, Wong KF, Soudackov AV, Kim S, Voth GA: A multistate empirical valence bond description of protonable amino acids. *Journal of Physical Chemistry A* 2006, 110:631–639.
143. Kang H, Jouvet C, Dedonder-Lardeux C, Martrenchard S, Gregoire G, Desfrancois C, Schermann JP, Barat M, Fayeton JA: Ultrafast deactivation mechanisms of protonated aromatic amino acids following UV excitation. *Physical Chemistry Chemical Physics* 2005, 7:394–398.
144. Boyarkin OV, Mercier SR, Kamariotis A, Rizzo TR: Electronic spectroscopy of cold, protonated tryptophan and tyrosine. *Journal of the American Chemical Society* 2006, 128:2816–2817.
145. Lucas B, Gregoire G, Lemaire J, Maitre P, Ortega JM, Rupenyan A, Reimann B, Schermann JP, Desfrancois C: Investigation of the protonation site in the dialanine peptide by infrared multiphoton dissociation spectroscopy. *Physical Chemistry Chemical Physics* 2004, 6:2659–2663.
146. Lucas B, Gregoire G, Lemaire J, Maitre P, Glotin F, Schermann JP, Desfrancois C: Infrared multiphoton dissociation spectroscopy of protonated N-acetyl-alanine and alanyl-histidine. *International Journal of Mass Spectrometry* 2005, 243:105–113.
147. Paizs B, Csonka IP, Lendvay G, Suhai S: Proton mobility in protonated glycylglycine and N-formylglycylglycinamide: a combined quantum chemical and RKKM study. *Rapid Communications in Mass Spectrometry* 2001, 15:637–650.
148. Marinica DC, Gregoire G, Desfrancois C, Schermann JP, Borgis D, Gaigeot MP: Ab initio molecular dynamics of protonated dialanine and comparison to infrared multiphoton dissociation experiments. *Journal of Physical Chemistry A* 2006, 110:8802–8810.
149. Beke T, Csizmadia IG, Perczel A: On the flexibility of β-peptides. *Journal of Computational Chemistry* 2004, 25:285–307.

150. Beke T, Somlal C, Perczel A: Toward a rational design of β-peptide structures. *Journal of Computational Chemistry* 2006, 27:20–38.
151. Gillet JC, Grégoire G, Schermann JP, Desfrancois C: Resonant infrared multiphoton dissociation spectra and structures of alanine-histidine, cyclic alanine-histidine and carnosine. To be published.
152. Bredenbeck J, Elbing J, Sieg A, Schrader T, Zinth W, Renner C, Behrendt R, Moroder L, Wachtveitl J, Hamm P: Picosecond conformational transition and equilibration of a cyclic peptide. *Proceedings of the National Academy of Sciences of the United States of America* 2003, 1000:6452–6457.
153. Wiedemann S, Metsala A, Nolting D, Weinkauf R: The dipeptide cyclic(glycyltryptophanyl) in the gas phase: a concerted action of density functional calculations, S-0-S-1 two-photon ionization, spectral UV/UV hole burning and laser photoelectron spectroscopy. *Physical Chemistry Chemical Physics* 2004, 6:2641–2649.
154. Kriz Z, Carlsen PHJ, Koca J: Conformational features of linear and cyclic enkephalins: a computational study. *Journal of Molecular Structure* 2001, 540:231–250.
155. Xiong JP, Stehle T, Diefenbach B, Zhang RG, Dunker R, Scott DL, Joachimiak A, Goodman SL, Arnaout MA: Crystal structure of the extracellular segment of integrin $\alpha V\beta$ 3. *Science* 2001, 294:339–345.
156. Gottschalk KE, Kessler H: The structures of integrins and integrin-ligand complexes: implications for drug design and signal transduction. *Angewandte Chemie-International Edition* 2002, 41:3767–3774.
157. Suvire FD, Rodriguez AM, Mak ML, Papp JG, Enriz RD: Binding mechanism of RGD and its mimetics to receptor GPIIb/IIIa. A theoretical study. *Journal of Molecular Structure: THEOCHEM* 2001, 540:257–270.
158. Gregoire G, Gaigeot MP, Lemaire J, Desfrançois C, Schermann JP: Resonant infrared multiphoton dissociation spectroscopy of gas-phase protonated peptides using free electron laser radiation and quantum chemistry calculations. *Physical Chemistry Chemical Physics* 2007, 9:3082–3097.
159. Csonka IP, Paizs B, Suhai S: Modeling of the gas-phase ion chemistry of protonated arginine. *Journal of Mass Spectrometry* 2004, 39:1025–1035.
160. Harvey AL, Baron TS, Boyd M, Campbell J: The three-dimensional structure of the tenth type III module of fibronectin: an insight into RGD-mediated interactions. *Cell* 1992, 71:671–678.
161. Shiu JH, Chen CY, Chang LS, Chen YC, Lo YH, Liu YC, Chuang WJ: Solution structure of gamma-bungarotoxin: the functional significance of amino acid residues flanking the RGD motif in integrin binding. *Proteins-Structure Function and Bioinformatics* 2004, 57:839–849.
162. Monleon D, Esteve V, Kovacs H, Calvete JJ, Celda B: Conformation and concerted dynamics of the integrin-binding site and the C-terminal region of echistatin revealed by homonuclear NMR. *Biochemical Journal* 2005, 387:57–66.
163. Kostidis S, Stavrakoudis A, Biris N, Tsoukatos D, Sakarellos C, Tsikaris V: The relative orientation of the Arg and Asp side chains defined by a pseudodihedral angle as a key criterion for evaluating the structure-activity relationship of RGD peptides. *Journal of Peptide Science* 2004, 10:494–509.
164. Barran PE, Roeseke RW, Pawson AJ, Sellar R, Bowers MT, Morgan K, Lu ZL, Tsuda M, Kusakabe T, Millar RP: Evolution of constrained gonadotropin-releasing hormone ligand conformation and receptor selectivity. *Journal of Biological Chemistry* 2005, 280:38569–38575.
165. Tyndall JD, Nall T, Fairlie DP: Proteases universally recognize β strands in their active sites. *Chemical Reviews* 2005, 105:973–979.
166. Ananda K, Vasudev PG, Sengupta AM, Raja KMP, Shamata N, Balaram P: Polypeptide helices in hybrid peptide sequences. *Journal of the American Chemical Society* 2005, 127:16668–16674.
167. Antoine R, Compagnon I, Rayane D, Broyer M, Dugourd P, Breaux G, Hagemeister FC, Pippen D, Hudgins RR, Jarrold MF: Electric dipole moments and conformations of isolated peptides. *European Physical Journal D* 2002, 20:583–587.

168. Dugourd P, Antoine R, Breaux G, Broyer M, Jarrold MF: Entropic stabilization of isolated β-sheets. *Journal of the American Chemical Society* 2005, 127:4675–4679.
169. Higo J, Galzitskaya OV, Ono S, Nakamura H: Energy landscape of a β-hairpin peptide in explicit water studied by multicanonical dynamics. *Chemical Physics Letters* 2001, 337:169–175.
170. Kinnear BS, Kaleta DT, Kohtani M, Hudgins RR, Jarrold MF: Conformations of unsolvated valine-based peptides. *Journal of the American Chemical Society* 2000, 122:9243–9256.
171. Hudgins RR, Jarrold MF: Helix formation in unsolvated alanine-based peptides: helical monomers and helical dimers. *Journal of the American Chemical Society* 1999, 121:3494–3501.
172. Kohtani M, Jones TC, Sudha R, Jarrold MF: Proton transfer conformational changes and melting in designed peptides in the gas-phase. *Journal of the American Chemical Society* 2006, 128: 7193–7197.
173. Kaleta DT, Jarrold MF: Non-covalent interactions between unsolvated peptides: dissociation of helical and globular peptide complexes. *Journal of Physical Chemistry B* 2003, 107:14529–14536.
174. Hudgins RR, Jarrold MF: Conformations of unsolvated glycine-based peptides. *Journal of Physical Chemistry B* 2000, 104:2154–2158.
175. Hill HH, Hill CH, Asbury GH, Wu C, Matz LM, Ichiye T: Charge location on gas phase peptides. *International Journal of Mass Spectrometry* 2002, 219:23–37.
176. Counterman AE, Clemmer DE: Compact → extended helix transitions of polyalanine in vacuo. *Journal of Physical Chemistry B* 2003, 107:2111–2117.
177. Vivian JT, Callis PR: Mechanisms of tryptophan fluorescence shifts in proteins. *Biophysical Journal* 2001, 80:2093–2109.
178. Callis PR, Liu T: Quantitative predictions of fluorescence quantum yields for tryptophan in proteins. *Journal of Physical Chemistry* 2004 B, 108:4248–4259.
179. Chen JJ, Flaugh SL, Callis PR, King J: Mechanism of the highly efficient quenching of tryptophan fluorescence in human gamma D-crystallin. *Biochemistry* 2006, 45:11552–11563.
180. Callis PR, Liu TQ: Quantitative prediction of fluorescence quantum yields for tryptophan in proteins. *Journal of Physical Chemistry* 2004 B, 108:4248–4259.
181. Engelborghs Y: The analysis of time resolved protein fluorescence in multi-tryptophan proteins. *Spectrochimica Acta Part A – Molecular and Biomolecular Spectroscopy* 2001, 57: 2255–2270.
182. Palmer GM, Keely PJ, Breslin TM, Ramanujan N: Autofluorescence spectroscopy of normal and malignant human breast cell lines. *Photochemistry and Photobiology* 2003, 78:462–469.
183. Lakowicz JR: *Principles of fluorescence spectroscopy*, edn 2nd: Kluwer Academic/Plenum, New York, 1999.
184. Rizzo TR, Park YD, Levy DH: Dispersed fluorescence of jet-cooled tryptophan: excited state conformers and intramolecular exciplex formation. *Journal of Chemical Physics* 1986, 85:6945–6951.
185. Snoek LC, Kroemer RT, Simons JP: A spectroscopic and computational exploration of tryptophan-water cluster structures in the gas phase. *Physical Chemistry Chemical Physics* 2002, 4: 2130–2139.
186. Piuzzi F, Dimicoli I, Mons M, Tardivel B, Zhao QC: A simple laser vaporization source for thermally fragile molecules coupled to a supersonic expansion: application to the spectroscopy of tryptophan. *Chemical Physics Letters* 2000, 320:282–288.
187. Carcabal P, Kroemer RT, Snoek LC, Simons JP, Bakker JM, Compagnon I, Meijer G, von Helden G: Hydrated complexes of tryptophan: ion dip infrared spectroscopy in the 'molecular fingerprint' region, 100–2,000 cm^{-1}. *Physical Chemistry Chemical Physics* 2004, 6:4546–4552.
188. Grace LI, Cohen R, Dunn TM, Lubman DM, de Vries MS: The R2PI spectroscopy of tyrosine: a vibronic analysis. *Journal of Molecular Spectroscopy* 2002, 215:204–219.
189. Samuels AC, Jensen JO, Hameka HF: Theoretical studies of the fluorescence and phosphorescence of tyrosine. *Journal of Molecular Structure: THEOCHEM* 1998, 454:25–30.

190. Smolyar A, Wong CF: Theoretical studies of the spectroscopic properties of tryptamine, tryptophan and tyrosine. *Journal of Molecular Structure: THEOCHEM* 1999, 488:51–67.
191. Toroz D, Van Mourik T: The structure of the gas-phase tyrosine-glycine dipeptide. *Molecular Physics* 2006, 104:559–570.
192. Polfer NC, Oomens J, Moore DT, von Helden G, Meijer G, Dunbar RC: Infrared spectroscopy of phenylalanine Ag(I) and Zn(II) complexes in the gas phase. *Journal of the American Chemical Society* 2006, 128:517–525.
193. Hashimoto T, Takasu Y, Yamada Y, Ebata T: Anomalous conformer dependent S_1 lifetime of L-phenylalanine. *Chemical Physics Letters* 2006, 421:227–231.
194. Philips LA, Webb SP, Martinez SJ, Fleming GR, Levy DH: Time-resolved spectroscopy of tryptophan conformers in a supersonic jet. *Journal of the American Chemical Society* 1988, 110:1352–1355.
195. Kang H, Dedonder-Lardeux C, Jouvet C, Martrenchard S, Gregoire G, Desfrancois C, Schermann JP, Barat M, Fayeton JA: Photo-induced dissociation of protonated tryptophan $TrpH^+$: a direct dissociation channel in the excited states controls the hydrogen atom loss. *Physical Chemistry Chemical Physics* 2004, 6:2628–2632.
196. Kang H, Dedonder-Lardeux C, Jouvet C, Gregoire G, Desfrancois C, Schermann JP, Barat M, Fayeton JA: Control of bond-cleaving reactions of free protonated tryptophan ion by femtosecond laser pulses. *Journal of Physical Chemistry A* 2005, 109:2417–2420.
197. Gregoire G, Kang H, Dedonder-Lardeux C, Jouvet C, Desfrancois C, Onidas D, Lepere V, Fayeton JA: Statistical vs. non-statistical deactivation pathways in the UV photo-fragmentation of protonated tryptophan-leucine dipeptide. *Physical Chemistry Chemical Physics* 2006, 8: 122–128.
198. Antoine R, Broyer M, Chamot-Rooke J, Dedonder C, Desfrancois C, Dugourd P, Gregoire G, Jouvet C, Onidas D, Poulain P, *et al.*: Comparison of the fragmentation pattern induced by collisions, laser excitation and electron capture. Influence of the initial excitation. *Rapid Communications in Mass Spectrometry* 2006, 20:1648–1652.
199. Nolting D, Schultz T, Hertel IV, Weinkauf R: Excited state dynamics and fragmentation channels of the protonated dipeptide H_2N-Leu-Trp-COOH. *Physical Chemistry Chemical Physics* 2006, 8:5247–5254.
200. Robertson EG, Simons JP: Getting into shape: conformational and supramolecular landscapes in small biomolecules and their hydrated clusters. *Physical Chemistry Chemical Physics* 2001, 3:1–18.
201. Lepère V, Lucas B, Barat M, Fayeton JA, Picard YJ, Jouvet C, Carçabal P, Nielsen IB, Dedonder C, Grégoire G, *et al.*: Photodissociation of protonated tryptophan: towards a full dynamics analysis from time and position resolved detection of all fragments. *Physical Chemistry Chemical Physics* 2007.
202. El Aribi H, Orlova G, Hopkinson AC, Siu KWM: Gas-phase fragmentation reactions of protonated aromatic amino acids: concomitant and consecutive neutral eliminations and radical cation formations. *Journal of Physical Chemistry A* 2004, 108:3844–3853.
203. Andersen JU, Cederquist H, Forster JS, Huber BA, Hvelplund P, Jensen J, Liu B, Manil B, Maunoury L, Nielsen SB, *et al.*: Photodissociation of protonated amino acids and peptides in an ion storage ring. Determination of Arrhenius parameters in the high-temperature limit. *Physical Chemistry Chemical Physics* 2004, 6:2676–2681.
204. Grégoire G, Dedonder C, Jouvet C, Desfrancois C, Fayeton JA: Ultrafast excited state dynamics in protonated GWG and GYG tripeptides. *Physical Chemistry Chemical Physics* 2006, 8:1–5.
205. Cui W, Hu Y, Lifshitz C: Time resolved photo dissociation of small peptide ions – combining laser desorption with ion trap/reflectron TOF mass spectromerty. *European Physical Journal D* 2002, 20:565–571.

206. Morgan JW, Russell DH: Comparative studies of 193-nm photodissociation and TOF-TOFMS analysis of bradykinin analogues: the effects of charge site(s) and fragmentation timescales. *Journal of the American Society of Mass Spectrometry* 2006, 17:721–729.
207. Nonose S, Iwaoka S, Tanaka H, Okai N, Shibakusa T, Fuke K: Structures and reactions of biomolecular ions produced with electrospray ionization. *European Journal of Physics D* 2003, 24:335–338.
208. Thompson MS, Cui W, Reilly JP: Fragmentation of singly charged peptide ions by photodissociation at 157 nm. *Angewandte Chemie-International Edition* 2004, 43:4791–4794.
209. Cui W, Thompson MS, Reilly JP: Pathways of peptide ion fragmentation induced by vacuum ultraviolet light. *Journal of the American Society for Mass Spectrometry* 2005, 16: 1384–1398.
210. Watson JD, Laskowsli RA, Thornton JM: Predicting protein function from sequence and structural data. *Current Opinion in Structural Biology* 2005, 15:275–284.
211. Skolnick J: In quest of an empirical potential for protein structure prediction. *Current Opinion in Structural Biology* 2006, 16:166–171.
212. Pandya MJ, Cerasoli E, Joseph A, Stoneman RG, Waite E, Woolfson DN: Sequence and structural duality: designing peptides to adopt two stable conformations. *Journal of the American Chemical Society* 2004, 126:17016–17024.
213. Tashiro S, Tominga M, Yamaguchi Y, Kato K, Fujita M: Folding a de novo desiged peptide into an α-helix through hydrophobic binding by a bowl-shaped host. *Angewandte Chemie-International Edition* 2005, 44:1–5.
214. Kaltashov IA, Eyles SJ: *Mass spectrometry in biophysics*: Wiley, Hoboken, NJ 2005.
215. Levinthal C: Are they pathways to protein folding? *Journal De Chimie Physique et de Physico-Chimie Biologique* 1968, 65:44–45.
216. Qiu LL, Pabit SA, Roitberg AE, Hagen SJ: Similar and faster: the 20 residues Trp-cage protein folds in 4 microseconds. *Journal of the American Chemical Society* 2002, 124:12952–12953.
217. Anfinsen CB, Scheraga HA: Experimental and theoretical aspects of protein folding. *Advances in Protein Chemistry* 1975, 29:205–300.
218. McLaughlin SH, Sobott F, Yaol ZP, Zhang W, Nielsen PR, Grossmann JG, Laue ED, Robinson CV, Jackson SE: The co-chaperone p23 arrests the Hsp90 ATPase cycle to trap client proteins. *Journal of Molecular Biology* 2006, 356:746–758.
219. Rostom AA, Robinson CJ: Detection of the intact GroEL chaperonin assembly by mass spectrometry. *Journal of the American Chemical Society* 1999, 121:4716–4719.
220. Perham M, Chen M, Ma J, Wittung-Stafshede P: Unfolding of heptameric co-chaperonin protein follows "Fly casting" mechanism: observation of transient non-native heptamer. *Journal of the American Chemical Society* 2005, 127:16402–16403.
221. Cederkvist FH, Zamfir AD, Bahrke S, Eijsink VGH, Sorlie M: Identification of a high-affinity-binding oligosaccharide by nanoelectrospray quadrupole time-of-flight tandem mass spectrometry of a non-covalent enzyme-ligand complex. *Angewandte Chemie-International Edition* 2006, 45:2429–2434.
222. Breuker K: The study of protein-ligand interactions by mass spectrometry – a personal view. *International Journal of Mass Spectrometry* 2004, 239:33–41.
223. Chalmers MJ, Busby SA, Pascal BD, He YJ, Hendrickson CL, Marshall AG, Griffin PR: Probing protein ligand interactions by automated hydrogen/deuterium exchange mass spectrometry. *Analytical Chemistry* 2006, 78:1005–1014.
224. Kuzmenkina EV, Heyes CD, Nienhaus GU: Single-molecule Förster resonance energy transfer of protein dynamics under denaturating conditions. *Proceedings of the National Academy of Sciences of the United States of America* 2005, 102:15471–15476.
225. Dokholyan V: Studies of folding and misfolding using simplified models. *Current Opinion in Structural Biology* 2006, 16:79–85.

226. Chan HS, Dill KA: Protein folding in the landscape perspective: chevron plots and non-Arrhenius kinetics. *Proteins-Structure Function Genetics* 1998, 30:2–33.
227. Clark PL: Protein folding in the cell: reshaping the folding funnel. *Trends in Biochemical Sciences* 2004, 29:527–534.
228. Pastor MT, Estera-Chopo A, Lopez de la Paz M: Design of model systems for amyloid formation: lessons for prediction and inhibition. *Current Opinion in Structural Biology* 2005, 15:57–63.
229. Dokholian NV: Studies of folding and misfolding using simplified models. *Current Opinion in Structural Biology* 2005, 16:1–7.
230. Alzheimer A: Über eine eigenartige Erkrankung der Hirnrinde. *Allg Zeitschr Psychiatr Psychiatr-Gerichtl Med.* 1907:146–148.
231. Stephenson VC, Heyding RA, Weaver DF: The "promiscuous concept" with applications to Alzheimer's disease. *FEBS Letters* 2005, 579:1338–1342.
232. Jang S, Shin S: Amyloid-peptide oligomerization in silico: dimer and trimer. *Journal of Physical Chemistry B* 2006, 110:1955–1958.
233. Fawzi NL, Chubukov V, Clark LA, Brown S, Head-Gordon T: Influence of denatured and intermediate states of folding on protein aggregation. *Protein Science* 2005, 14:993–1003.
234. Marianayagam NJ, Fawzi NL, Head-Gordon T: Protein folding by distributed computing and the denatured state ensemble. *Proceedings of the National Academy of Sciences of the United States of America* 2005, 102:16684–16689.
235. Luttmann E, Fels G: All-atom dynamics studies of the full-length β-amyloid peptides. *Chemical Physics Letters* 2006, 323:138–147.
236. Bernstein SL, Wyttenbach T, Baumketnert A, Shea JE, Bitan G, Teplow DB, Bowers MT: Amyloid β-protein: monomer structure and early aggregation states of A β 42 and its Pro(19) alloform. *Journal of the American Chemical Society* 2005, 127:2075–2084.
237. Baumketner A, Bernstein SL, Wyttenbach T, Lazo ND, Teplow DB, Bowers MT, Shea JE: Structure of the 21-30 fragment of amyloid β-protein. *Protein Science* 2006, 15:1239–1247.
238. Baumketner A, Bernstein SL, Wyttenbach T, Bitan G, Teplow DB, Bowers MT, Shea JE: Amyloid β-protein monomer structure: a computational and experimental study. *Protein Science* 2006, 15:420–428.
239. Cheng X, van Breemen RB: Mass spectrometry-based screening for inhibitors of β-amyloid protein aggregation. *Analytical Chemistry* 2005, 77:7012–7015.
240. Hopkins AL, Mason JS, Overington JP: Can we rationally design promiscuous drugs? *Current Opinion in Structural Biology* 2006, 16:1–10.
241. Bazoti FN, Tsarbopoulos A, Markides KE, Bergquist J: Study of the non-covalent interaction between amyloid-β-peptide and melatonin using electrospray mass spectrometry. *Journal of Mass Spectrometry* 2005, 40:182–192.
242. Navizet I, Cailliez F, Lavery R: Probing protein mechanics: residue-level properties and their use in defining domains. *Biophysical Journal* 2004, 87:1426–1435.
243. Doruker P, Jernigan RL, Navizet I, Hernandez R: Important fluctuation dynamics of large protein structures of large protein structures are preserved upon coarse-grained renormalization. *International Journal of Quantum Chemistry* 2002, 90:822–837.
244. Simmerling C, Strockbine B, Rothberg AE: All-atom structure prediction and folding simulations of a stable protein. *Journal of the American Chemical Society* 2002, 124:11258–11259.
245. Seshasayee ASN: High-temperature unfolding of a Trp-cage mini-protein: a molecular dynamics simulation study. *Theoretical Biology and Medical Modelling* 2005, 2:1–5.
246. Neidigh JW, Fesinmeyer RM, Andersen NH: Designing a 20-residue protein. *Nature Structural & Molecular Biology* 2002, 9:425–430.
247. Snow CD, Zagrovic B, Bande VS: The Trp-cage: folding kinetics and unfolded state topology via molecular dynamics simulations. *Journal of the American Chemical Society* 2002, 124: 14548–14549.

248. Pitera J, Swope W: Understanding folding and design: replica-exchange simulations of "Trp-cage" miniproteins. *Proceedings of the National Academy of Sciences of the United States of America* 2003, 100:7587–7592.
249. Zhou R: Trp-cage: folding free energy landscape in explicit water. *Proceedings of the National Academy of Sciences of the United States of America* 2003, 100:13280–13285.
250. Schug A, Wenzel W: All-atom folding of the Trp-cage protein with an adaptative parallel tempering method. *Europhysics Letters* 2004, 67:307–313.
251. Urbanc B, Cruz L, Ding F, Sammond D, Khare S, Buldyrev SV, Stanley HE, Dokholian NV: Molecular dynamics simulation of amyloid β dimer formation. *Biophysical Journal* 2004, 87:2310–2321.
252. Bunagan MR, Yang X, Saven JG, Gal F: Ultrafast folding of a computationally designed Trp-cage mutant Trp2-cage. *Journal of Physical Chemistry B* 2006, 110:3750–3763.
253. Patriksson A, Adams C, Kjeldsen F, Raber J, van der Spoel D, Zubarev RA: Prediction of N-C_α bond cleavage frequencies in electron capture dissociation of Trp-cage dications by force-field molecular dynamics simulations. *International Journal of Mass Spectrometry* 2006, 248: 124–135.
254. Iavarone AT, Parks JH: Conformational change in unsolvated Trp-cage protein probed by fluorescence. *Journal of the American Chemical Society* 2005, 127:8606–8607.
255. Olofson M, Kalinin S, Zdunek J, Oliveberg M, Johansson LBA: Tryptophan–BODIPY: a versatile donor–acceptor pair for probing generic changes of intraprotein distances. *Physical Chemistry Chemical Physics* 2006, 8:3130–3140.
256. Duden R: Live cell imaging: the 'green revolution' continues apace. *Trends in Cell Biology* 2002, 12:548.
257. Zimmer M: Green fluorescent protein (GFP): applications, structure, and related photophysical behavior. *Chemical Reviews* 2002, 102:759–781.
258. Boye S, Krogh H, Nielsen IB, Nielsen SB, Pedersen SU, Pedersen UV, Andersen LH, Bell AF, He X, Tonge PJ: Vibrationally resolved photoabsorption spectroscopy of red fluorescent protein chromophore anions. *Physical Review Letters* 2003, 90:11803–11806.
259. Lopez X, Marques MA, Castro A, Rubio A: Optical absorption of the blue fluorescent protein: a first-principle study. *Journal of the American Chemical Society* 2005, 2005:12329–12337.
260. Andersen LH, Herber O, Zajfman D: Physics with electrostatic ring and traps. *Journal of Physics B* 2004, 37:R57–R88.
261. Andersen LH, Bluhme H, Boye S, Jorgensen TJD, Krogh H, Nielsen IB, Nielsen SB, Svendsen A: Experimental studies of the photophysics of gas-phase fluorescent protein chromophores. *Physical Chemistry Chemical Physics* 2004, 6:2617–2627.
262. Usman A, Mohammed OF, Nibbering NMM, Dong J, Solntsev KM, Tolbert LM: Excited-state structure of the green fluorescent chromophore. *Journal of the American Chemical Society* 2005, 127:11214–11215.
263. Altoe P, Bernardi F, Garavelli M, Orlandi G, Negri F: Solvent effects on the vibrational activity and photodynamics of the green fluorescent protein chromophore: a quantum-chemical study. *Journal of the American Chemical Society* 2005, 127:3952–3963.
264. Andersen LH, Lapierre A, Nielsen SB, Nielsen IB, Pedersen SU, Pedersen UV, Tomita S: Chromophores of the green fluorescent protein studied in the gas phase. *European Physical Journal D* 2002, 20:597–600.
265. Diken EG, Hammer NI, Johnson MA: Preparation and photoelectron spectrum of the glycine molecular anion: assignment to a dipole-bound electron species with a high-dipole moment, non-zwitterionic form of the neutral core. *Journal of Chemical Physics* 2004, 120: 9899–9902.
266. Carles S, Desfrancois C, Schermann JP, Smith DMA, Adamowicz L: Structures and electron affinities of indole-(water)$_N$ clusters. *Journal of Chemical Physics* 2000, 112:3726–3734.

267. Carles S, Lecomte F, Schermann JP, Desfrançois C: Gas-phase experimental and theoretical studies of adenine, imidazole, pyrrole and water non-covalent complexes. *Journal of Physical Chemistry A* 2000, 104:10662–10669.
268. Sanche L: Low energy electron-driven damage in biomolecules. *European Physical Journal D* 2005, 35:367–390.
269. Aflatooni K, Hitt B, Gallup GA, Burrow PD: Temporary ion states of selected amino acids. *Journal of Chemical Physics* 2001, 115:6489–6494.
270. Abdoul-Carime H, Illenberger E: Fragmentation of proline induced by slow electrons. *Chemical Physics Letters* 2004, 397:309–313.

– 4.3 –

Sugars

GENERAL FEATURES

Sugar chains are the most abundant and structurally diverse biomolecules. They play crucial roles in recognition processes and are polymers composed of monosaccharide building blocks. While proteins and oligonucleotides are linear, sugars, also named carbohydrates or glycans, can form linear or branched structures because linkages through glycosidic bonds (see below) can occur at a number of positions along the cycle [1, 2] (Figure 3.1.1). Their conformation space is extraordinarily rich. A rather small sugar such as a hexasaccharide can already adopt nearly 10^{12} possible isomeric conformations. Moreover, glycan polymers can contain not only monosaccharides but also other groups. Specific biological functions are fulfilled when, for example some of their hydroxyl groups are phosphorylated, sulphated or methylated [3, 4]. Sugars are very hydrophilic, thanks to their many OH groups, and are very often linked to proteins which then become more hydrophilic. The corresponding glycoproteins are thus very often found exposed at the surface of cells to the aqueous intercellular medium. The attachment of the sugar to the protein occurs as a covalent linkage to the hydroxyl group of the side-chains of serine or threonine (O-link) or to the amino group in the side-chain of asparagine (N-link).

In this chapter, we will first consider spectroscopic gas-phase studies of neutral saccharides which have been reviewed in depth by Simons [5]. Modelling approaches will be briefly examined and we will then describe ion-mobility and mass-spectrometric studies of oligosaccharides and glycoaminoglycans. A detailed introduction to mass-spectrometry of carbohydrates can be found in reference [6].

4.3.1 Spectroscopic determination of monosaccharide structures

Monosaccharides are chiral [7] and most naturally occurring sugars are D isomers (except fucose [8]). Their general formula is $(CH_2O)_n$ where n is in between 4 and 9. In their linear form, they are rather flexible due to easy rotations about each of the single C—C bonds. Monosaccharides possess either an aldehyde (e.g. glucose) or a keto (e.g. fructose) group which reacts with an OH group leading to a cyclic form (Figure 4.3.1). In this internal cyclization, the carbonyl group reacts with one of the hydroxyl groups from the other end of the molecule. Depending on the point of attack, the resulting rings can contain five or six atoms, one of which is oxygen (four and seven-membered ring sugars are rather sparse).

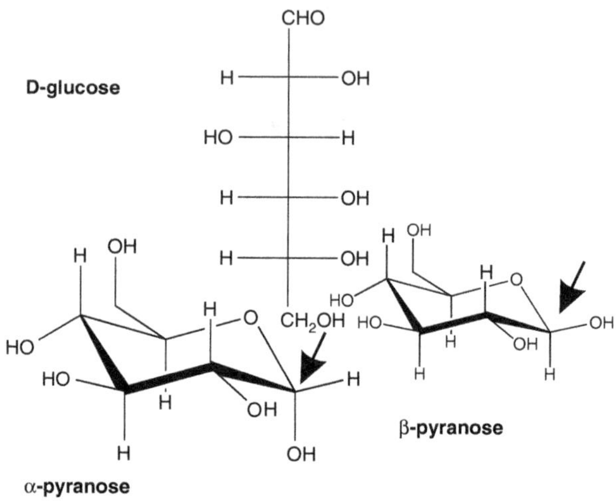

Figure 4.3.1 Cyclization produces an asymmetric centre with two stereoisomers called α and β whether the OH group at the anomeric carbon indicated by the arrow is below or above the ring. For IR/R2PI experiments, a chromophore is linked to the anomeric carbon.

The closing of a linear monosaccharide to make a ring creates a new chiral centre called the anomeric carbon (designated as C1). The two corresponding possibilities that are designated with either an α or β prefix. A comprehensive databank of sugars is available on the website http://www.cermav.cnrs.fr/glyco3D/index.php.

Due to the large number of hydroxyl groups that have a high freedom of rotation and can engage into the creation of various hydrogen bonds, prediction of the hydrogen-bonding network is a difficult task. Usually, those hydrogen bonds are present to either neighbouring carbohydrate molecules, glycoproteins, or surrounding water molecules.

In solution, NMR is the favourite mean of investigation of sugars. However, a major difficulty in the determination of the conformation of an oligosaccharide from NMR data is the flexibility of the carbohydrates, and especially of the glycosidic links. When multiple conformations are present in solution, NMR data will represent a time-averaged conformation. Gas-phase studies conducted at very low temperatures in supersonic expansions [9] (see Section 3.1.2) overcome this problem and provide well-defined identification. Natural sugars can be studied by microwave spectroscopy [10] but do not possess chromophores in the visible/UV region. For R2PI spectroscopy, a benzyl or phenyl group must be covalently attached on the anomeric carbon [5, 8, 11–15] (Figure 4.3.1). Simulations show that this group then plays the role of a spectator and thus does not perturb the gas-phase conformations [8, 16].

The multiple OH groups in the monosaccharide rings are responsible for the glycosidic linkages leading to the variety of linear or branched structures. They can adopt several conformations in the nomenclature (Figure 4.3.2) of which is given in detail in references [5, 12]. The three ring OH groups that create a cooperative hydrogen-bonding chain OH2 → OH3 → OH4 are linked in a clockwise sense (when viewed with the ring oxygen located at the top right) and indicated by letter c (cc would state for counterclockwise). G and T refer to the conformation of the exocyclic hydroxymethyl group represented by the dihedral angle O6–C6–C5–O5 which can adopt Gauche (G+ or G−) or Trans (T) orientations.

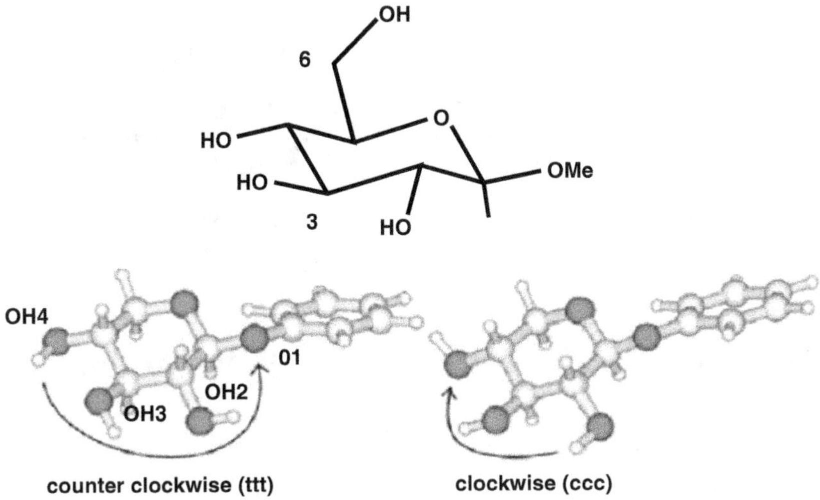

Figure 4.3.2 Top: Orientation of the exocyclic hydroxyl groups. The monosaccharide represented here is methyl β-D-glucopyranoside. Bottom (left and right): Hydroxy group network of β-phenylxyloside with counterclockwise (ttt) and clockwise (ccc) orientation (reproduced with permission from reference [17] ©2005 Royal Society of Chemistry).

The terminal OH6 group is defined by the dihedral angle H6–O6–C6–C5 which can adopt g^+, g^- and t orientations.

The vibrational stretch modes associated to the OH groups are spectrally shifted by their engagement into intra- and inter-molecular hydrogen bonds (see Section 2.1.3). The measurement of those spectral shifts combined with a conformational search and predictions of the vibrational spectra corresponding to the different conformers (B3LYP geometries further optimized at the MP2/6-311++G** level) allows the assignment of the different structures. For example, the infrared hole-burning spectra of three conformers of phenyl-α-D-mannopyranoside is displayed in Figure 4.3.3. The assignment of the lowest energy conformer A is cGg^+. The signature of the cooperativity of hydrogen bonds (see Section 2.1.3.1) is a general shift of the IR spectra towards lower wave numbers.

4.3.2 Glycosidic linkage

Oligo-(2–10 sugar blocks) and poly-(>10 blocks) saccharides are formed by elimination of water molecules from the anomeric OH group of one monosaccharide and any of the other OH groups of another sugar block. Formation of a bond, called glycosidic linkage, from the anomeric carbon C1 prevents the ring opening [18]. Monosaccharides can be linked to produce oligomeric structures through glycosidic bond formation. Water is eliminated between the anomeric hydroxyl and any one of the hydroxyls of a second monosaccharide or oligosaccharide. A disaccharide is shown in Figure 4.3.4. The glycosidic linkage consists two bonds, the glycosidic C1—O and the aglycone O—C4 ones and the oxygen atom is the locus of much of carbohydrate flexibility described by changes in the internal torsion angles. There are two possible linkages, α and β, characterized by the rotation

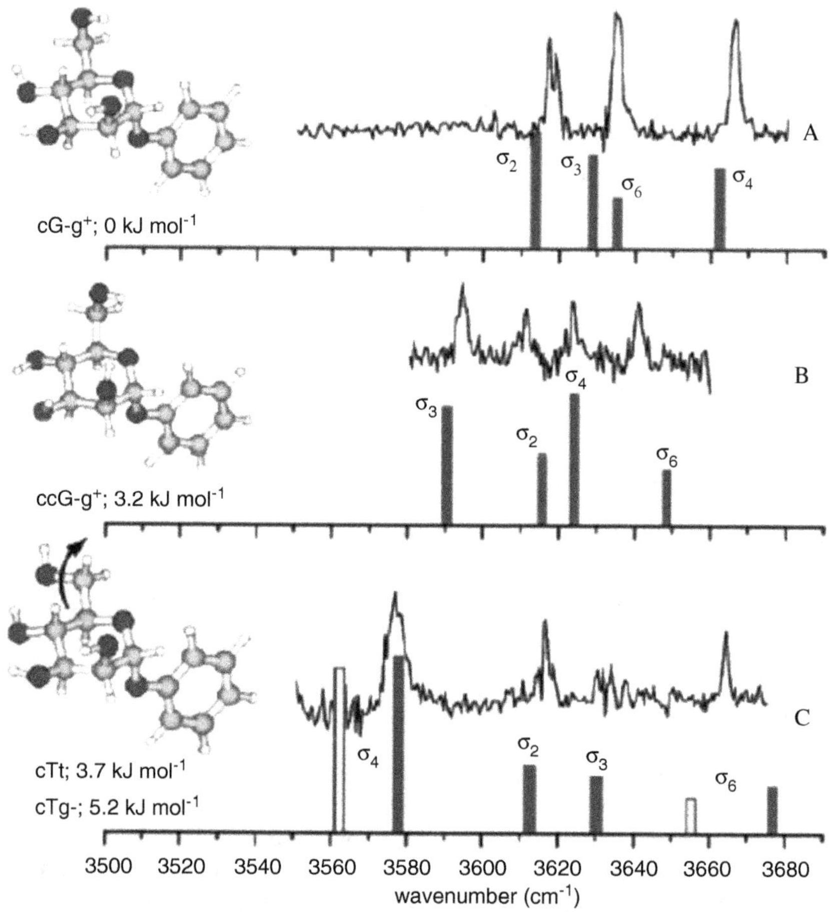

Figure 4.3.3 Experimental IR – hole-burning spectra of the three observed conformers of phenyl-α-D-mannopyranoside together with their calculated structure. For conformer C, the full and empty stick respectively correspond to the cTt and cTg− (displayed structure) which only differ from the torsion around the C—C bond depicted by the arrow (reproduced with permission from reference [16] ©2005 American Chemical Society).

angles between the glycosidic oxygen and its two neighbours – ϕ and ψ (Figure 4.3.4). β-1,4 linkages favour straight chains (e.g. cellulose). The carbohydrates do not move around freely but rather alternate between several preferred conformations. In a large carbohydrate, torsion angles may change while the biomolecule maintains its overall shape due to correlated internal motions.

Even in absence of water molecules, the OH groups of both monosaccharides tend to interact cooperatively by forming uninterrupted hydrogen bond networks and rigidify the glycosidic linkage (see Section 5.6). The biological implication of disaccharide rigid structures observed at very low temperature in the gas-phase, as compared with more flexible structures observed in solution are considered in Section 5.6.

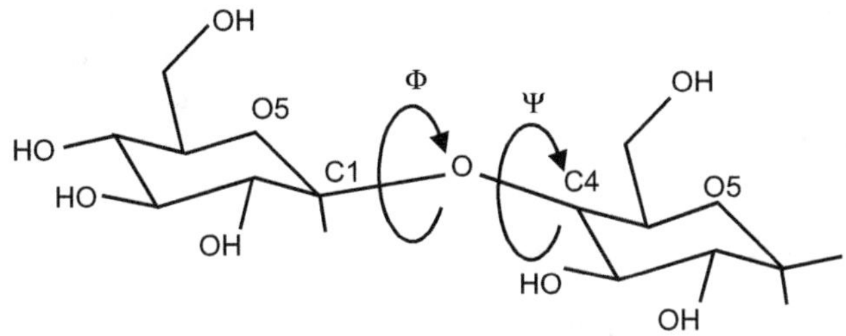

Figure 4.3.4 The disaccharide maltose represented here, a cleavage product of starch, has an α(1 → 4) glycosidic linkage between the C1 hydroxyl of one glucose and the C4 hydroxyl of a second glucose. Maltose is the α anomer, because the O at C1 points down from the ring.

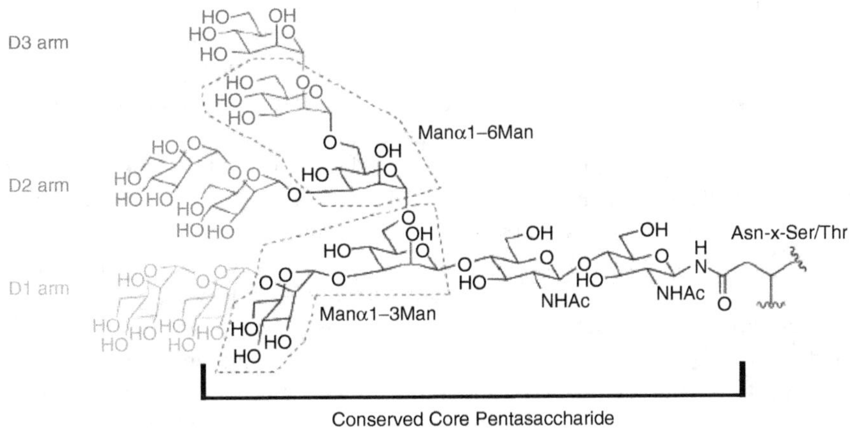

Figure 4.3.5 Recurrent N-glycan pentasaccharide motif of glycoproteins including the N-asparagine linkage [19]. Note the mannose α(1,3) mannose and mannose α(1,6) mannose linkages in the D1 and D3 arms (reproduced with permission from reference [19] ©2006 American Chemical Society).

Due to the extremely large number of possibilities for linking the different monosaccharides, one might expect that gigantic number of structures should be observed in biomolecules. In fact, some structures have been selected through Evolution and play a crucial role. Glycosylation is omnipresent as a post-translational modification of proteins (PTM, see Section 3.2.2.1). Glycosidic linkages which are naturally occurring can be classified in two principal groups, N- and O-glycosides [3] (see above). In vivo, N-glycosylation can only occur at asparagine amino acid side-chains embedded in a consensus tripeptide sequence Asn-Xxx-Ser/Thr, Xxx being any amino acid except proline [3]. The carbohydrate motifs of those N-glycans are well conserved around a central pentasaccharide core (Figure 4.3.5). Two terminal arms of this pentasaccharide contain mannose linkages that have been spectroscopically investigated [19] with a phenyl UV chromophore tag on one of the mannose. The geometry of structures and calculations of vibrational spectra for assignment of structures was performed by using the ONIOM method (see Section 1.6). The IR/R2PI spectrum shown in Figure 4.3.5 displays a red-shift down to 3,520 cm^{-1} characteristic of the stretching of an OH group of one sugar ring (here OH2) hydrogen bonded to an oxygen atom (here O6′) belonging to the other sugar ring. This inter-ring hydrogen bond is the equivalent of hydrogen bonds established in peptides between different residues that lead to secondary structures (see Section 4.2.2).

4.3.3 Modelling of carbohydrates

Carbohydrates can assume an enormous variety of spatial arrangements around the glycosidic linkages and polysaccharides models aim to predict the relative abundance of the various conformations. A number of force-fields (see Section 1.3) have been parameterized for carbohydrates [2, 20–23]. The torsional dihedral potential associated with the rotations of the hydroxyl and anomeric groups have been obtained by fitting the corresponding energy profiles obtained from quantum chemistry calculation of monosaccharides and have been further validated from solution NMR data. Unfortunately, there still exists no consensus force-field acceptable for monosaccharides and glycosidic linkages in polysaccharides [22] and quantum calculations become rapidly extremely costly even for small oligosaccharides. Modelling of infrared spectra for rapid identification of carbohydrates [24] thus only uses the semi-empirical low-cost CNDO/2 method (see Section 1.4).

For higher level of calculations, the optimal basis sets suitable for density functional theory (DFT) calculations of carbohydrates are discussed in reference [25]. The introduction of diffuse functions on heavy atoms, B3LYP/6-31+G(d,p) or B3LYP/6-311+G(d,p) is necessary but introduction of diffuse functions on hydrogen atoms is not necessarily required [26]. Using this DFT level of calculation for optimizing structures of conformers, the calculation of relative conformer populations of gas-phase monosaccharides has been conducted with a fully anharmonic quantum mechanical treatment of torsional modes. β-D-galactose, for example has six torsional internal coordinates and its equilibrium conformer populations have been calculated by means of the torsional path integral Monte Carlo (TPIMC) method (see Section 1.5.3) [27]. The high-level calculations were then only used to locate conformational minima and evaluate their corresponding harmonic frequencies. The low-frequency torsional modes correspond to shallow potential wells and are thus strongly anharmonic. Their treatment was conducted with the affordable AMBER force-field and then an anharmonicity correction factor was introduced. An interesting result is that there is a very low mixing of the torsional internal coordinates with stretches or bends in the composition of normal modes.

4.3.3.1 Oligosaccharides and polysaccharides

The three-dimensional gas-phase structures of oligosaccharides have been determined by means of ion-mobility measurements interpreted by means of molecular dynamics calculations [4, 28–30]. Those studies show that sodium ions interact differently in solution and in the gas-phase [28]. In solution, hydrated Na^+ ions play the role of counterions delocalized around the oligosaccharides. In the gas-phase, the Na^+ ions play a stabilizing structural role by coordinating to several ring oxygen atoms in cations or localizing near negatively charged groups in anions.

Oligosaccharides are often sulphated and constitute an important class of bioactive molecules [31]. For example, heparin and its derivatives share a linear carbohydrate skeleton of a repeating disaccharide unit. Single-charged disaccharide and double-charged tetrasaccharide negative ions respectively containing three and six Na^+ ions have been studied by ion-mobility and modelling with the AMBER force-field parameterized for carbohydrates [4]. Again, sodium ions play an important structural role in stabilizing the gas-phase conformations of the tetrasaccharide ions. The appearance of compact structures is due to the folding of the saccharide chain. Thanks to the flexibility of the glycosidic linkages, outer units are brought together and become bridged by sodium cations.

Oligosaccharides and polysaccharides can be mass-analysed by means of different techniques [6, 32–37]. According to the method used for providing internal energy, different fragmentation patterns are obtained [36]. The nomenclature of the collision-induced dissociation (CID) fragmentation pattern [38] is shown in Figure 4.3.6. CID fragmentation patterns are sensitive to oligosaccharide sequences, branching and stereochemistry and can thus be used (with rather large difficulties) to differentiate structural differences. Glycosidic

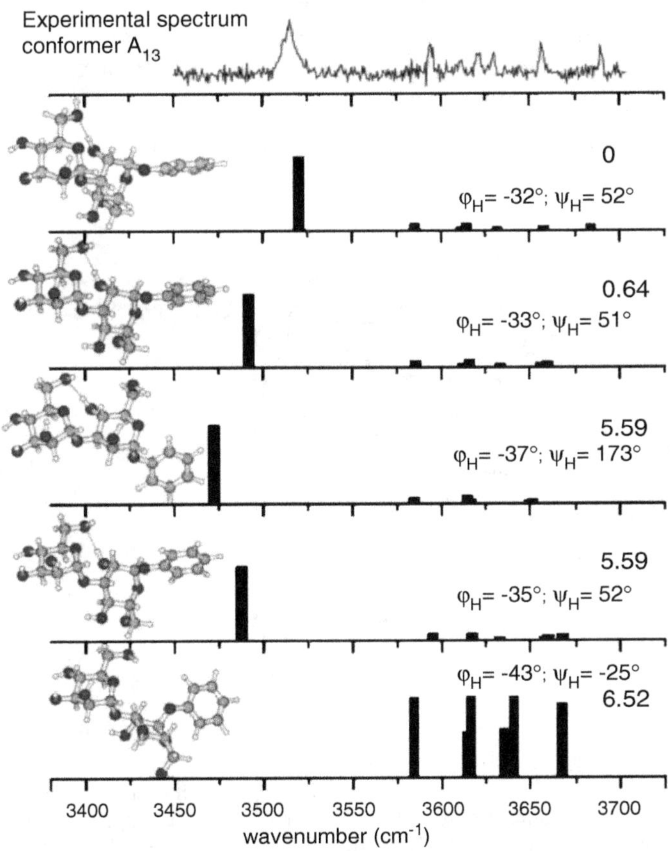

Figure 4.3.6 IR/R2PI spectrum of mannose α(1,3) phenyl-tagged mannose (φ) and calculated IR spectra of its five lowest energy conformers. Relative energies are in kJ/mol. The glycosidic bond angles φ and ψ are defined in Figure 4.3.3 (reproduced with permission from reference [19] ©2006 American Chemical Society).

bond cleavages (b-y or c-z ions) alone however do not provide full information. For example, two cyclic hexoses can be linked in one of the five ways through different hydroxyl groups, in contrast to amino acids which can only form a single amide bond (Figure 4.3.7).

Some detailed structural information can however be obtained from CID experiments. For example [39, 40], in the case of carbohydrates containing acetyl groups on all their carbons, the loss of acetic acid from the anomeric carbon (C1) depends on the anomeric configuration (α or β). This stereo-specificity is understood with the help of quantum calculations

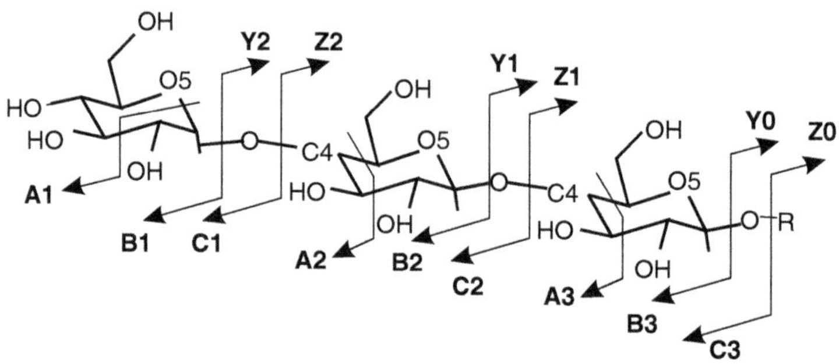

Figure 4.3.7 Nomenclature of oligosaccharide fragmentation.

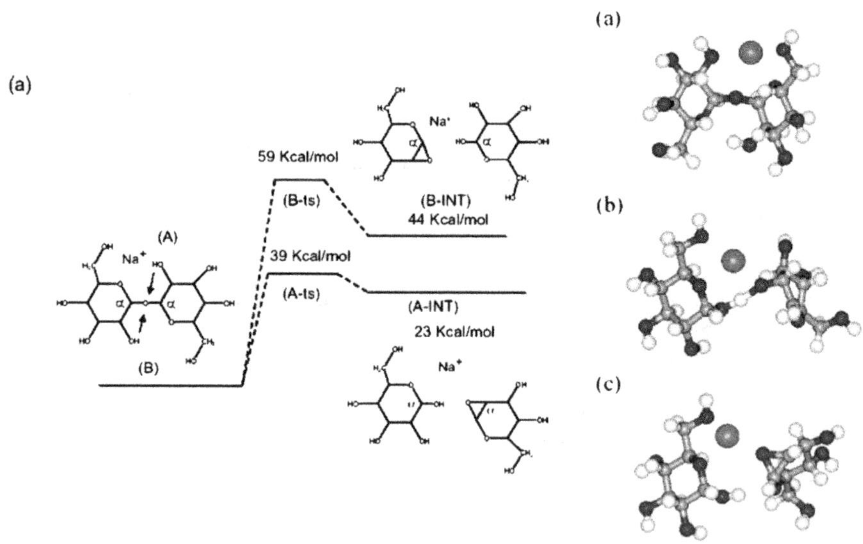

Figure 4.3.8 Left: Potential energy surface along the reaction path for the α-glycosyl bond cleavage of (trehalose + Na)$^+$ ions. Fragmentation of the initial state proceeds through the passage over the transition state (TS) followed by formation of the product intermediate INT. The corresponding geometries are shown in the right figure. In the most favourable path A, the observed charged dissociation product is the (Glc + Na)$^+$ b ion. In the case of β. β-glycosyl bond cleavage (not represented), path B becomes the most favourable and a y fragment ion is observed. Right: Stable complex structures of (a) (α,α trehalose + Na)$^+$, (b) (β,β trehalose + Na)$^+$ and (c) (α,β trehalose + Na)$^+$ ions complexed with a sodium atom (reproduced with permission from reference [42] ©2006 American Chemical Society).

by considering the respective reactivities of the different isomers and is attributed to the relative positions of the acetyl groups on C1 and its neighbour C2.

Glycosidic bond cleavage in CID experiments [41] is widely used and the corresponding mass-spectra allow the identification of the sequence of sugars. However, the presence of α and β configurations at the anomeric position leads to the presence of α,α,α,β and β,β isomers. The abundance of y and b fragments arising from those isomers in mass-spectra of sodiated ions $(\text{trehalose} + \text{Na})^+$ are different since α-glycosyl bonds cleave more easily than β-glycosyl bonds as shown by calculations of the reaction path of glycosyl bond cleavage [42] (Figure 4.3.8). The trehaloses consist of two D-glucopyranoses (Glc1-1Glc) linked at the anomeric C1 position. The fragmentation mechanism was determined by using H/D exchange (see Section 3.3.4). Two paths A and B are possible according to the transferred proton allowing the formation of fragments that comes from an H/D exchangeable hydrogen atom at one of the C2 hydroxy groups of each moiety.

REFERENCES

1. Dwek RA: Glycobiology: towards understanding the function of sugars. *Chemical Reviews* 1996, 96:683–720.
2. Imberty A, Perez S: Structure, conformation and dynamics of bioactive oligosaccharides. *Chemical Reviews* 2000, 100:4567–4588.
3. Seitz O: Glycopeptide synthesis and the effects of glycosylation on protein structure and activity. *ChemBioChem* 2000, 1:214–246.
4. Jin L, Barran PE, Deakin JA, Lyon M, Uhrin D: Conformation of glycoaminoglycans by ion mobility mass spectrometry and molecular modelling. *Physical Chemistry Chemical Physics* 2005, 7:3464–3471.
5. Simons JP, Jockusch RA, Carcabal P, Hung I, Kroemer RT, Macleod NA, Snoek LC: Sugars in the gas phase. Spectroscopy, conformation, hydration, co-operativity and selectivity. *International Reviews in Physical Chemistry* 2005, 24:489–531.
6. Harvey DJ: Matrix-assisted laser desorption/ionization mass spectrometry of carbohydrates. *Mass Spectrometry Reviews* 1999, 18:349–451.
7. Madhusudanan KP: Tandem mass spectra of ammonium adducts of monosaccharides: differentiation of diastereoisomers. *Journal of Mass Spectrometry* 2006, 41:1096–1104.
8. Carcabal P, Patsias T, Hunig I, Liu B, Kaposta C, Snoek LC, Gamblin DP, Davis BG, Simons JP: Spectral signatures and structural motifs in isolated and hydrated monosaccharides: phenyl α- and β-L-fucopyranoside. *Physical Chemistry Chemical Physics* 2006, 8:129–136.
9. Robertson EG, Simons JP: Getting into shape: conformational and supramolecular landscapes in small biomolecules and their hydrated clusters. *Physical Chemistry Chemical Physics* 2001, 3:1–18.
10. Aviles Moreno JR, Petitprez D, Huet TR: The conformational flexibility in N-phenylformamide: an ab initio approach supported by microwave spectroscopy. *Chemical Physics Letters* 2006, 419:411–416.
11. Talbot FO, Simons JP: Sugars in the gas phase: the spectroscopy and structure of jet-cooled phenyl β-D-glucopyranoside. *Physical Chemistry Chemical Physics* 2002, 4:3562–3565.
12. Macleod NA, Johannessen C, Hecht I, Barron LD, Simons JP: From the gas phase to aqueous solution: vibrational spectroscopy, Raman optical activity and conformational structure of carbohydrates. *International Journal of Mass Spectrometry* 2006, 253:193–200.
13. Jockusch RA, Talbot FO, Simons JP: Sugars in the gas phase – Part 2: the spectroscopy and structure of jet-cooled phenyl β-D-galactopyranoside. *Physical Chemistry Chemical Physics* 2003, 5:1502–1507.

14. Jockusch RA, Talbot FO, Asano N, Fleet GW, Simons JP: Gas-phase structure and conformation of the glycosidase and ceramide glucosyltransferase inhibitor N-benzyl deoxynojirimycin. *Physical Chemistry Chemical Physics* 2004, 6:5283–5287.
15. Jockusch RA, Kroemer RT, Talbot FO, Simons JP: Hydrated sugars in the gas phase: spectroscopy and conformation of singly hydrated phenyl β-D-glucopyranoside. *Journal of Physical Chemistry A* 2003, 107:10725–10732.
16. Carcabal P, Jockusch RA, Hunig I, Snoek LC, Kroemer RT, Davis BG, Gamblin DP, Compagnon I, Oomens J, Simons JP: Hydrogen bonding and cooperativity in isolated and hydrated sugars: mannose, galactose, glucose, and lactose. *Journal of the American Chemical Society* 2005, 127:11414–11425.
17. Hunig I, Painter AJ, Jockusch RA, Carcabal P, Marzluff EM, Snoek LC, Gamblin DP, Davis BG, Simons JP: Adding water to sugar: a spectroscopic and computational study of α- and β-phenylxyloside in the gas phase. *Physical Chemistry Chemical Physics* 2005, 7:2474–2480.
18. Jockusch RA, Kroemer RT, Talbot FO, Snoek LC, Carcabal P, Simons JP, Havenith M, Bakker JM, Compagnon I, Meijer G, *et al.*: Probing the glycosidic linkage: UV and IR ion-dip spectroscopy of a lactoside. *Journal of the American Chemical Society* 2004, 126:5709–5714.
19. Carcabal P, Hunig I, Gamblin DP, Liu B, Jockusch RA, Kroemer RT, Snoek LC, Fairbanks AJ, Davis BG, Simons JP: Building up key segments of N-glycans in the gas phase: intrinsic structural preferences of the α(1,3) and α(1,6) dimannosides. *Journal of the American Chemical Society* 2006, 128:1976–1981.
20. Damm W, Frontera A, Tirado-Rives J, Jorgensen WL: OPLS all-atom force field for carbohydrates. *Journal of Computational Chemistry* 1997, 18:1965–1970.
21. Perez S, Imberty A, Engelsen SB, Gruza J, Mazeau K, Jimenez-Barbero J, Poveda A, Espinosa JF, van Eyck BP, Johnson G, *et al.*: A comparison and chemometric analysis of several molecular mechanics force fields and parameter sets applied to carbohydrates. *Carbohydrate Research* 1998, 314:141–155.
22. MacKerell AD: Empirical force fields for biological macromolecules: overview and issues. *Journal of Computational Chemistry* 2004, 25:1584–1604.
23. Lins RD, Hüneberger PH: A new GROMOS force field for hexopyranose-based carbohydrates. *Journal of Computational Chemistry* 2005, 26:1400–1412.
24. Zhbankov RG, Korolevich MV, Derendyaev BG, Piottukh-Peletsky: Structural similarity and modeling of infrared spectra of molecules of organic compounds. *Journal of Molecular Structure* 2005, 144:937–946.
25. Csonka IP: Proper basis set for quantum mechanical studies of potential energy surfaces of carbohydrates. *Journal of Molecular Structure (Theochem)* 2002, 584:1–4.
26. Miura N, Taniguchi T, Monde K, Nishimura SI: A theoretical study of α- and β-D-glucopyranose conformations by the density functional theory. *Chemical Physics Letters* 2006, 419:326–332.
27. Sturdy YK, Skylaris CK, Clary DC: Torsional anharmonicity in the conformational analysis of β-D-galactose. *Journal of Physical Chemistry B* 2006, 110:3485–3492.
28. Lee S, Wyttenbach T, Bowers MT: Gas phase structures of sodiated oligosaccharides by ion mobility ion chromatography methods. *International Journal of Mass Spectrometry* 1997, 167: 605–614.
29. Liu YS, Clemmer DE: Characterizing oligosaccharides using injected-ion mobility mass spectrometry. *Analytical Chemistry* 1997, 69:2504–2509.
30. Leavell MD, Gaucher SP, Leary JA, Taraszka JA, Clemmer DE: Conformational studies of Zn-ligand-hexose diastereomers using ion mobility measurements and density functional theory calculations. *Journal of the American Society for Mass Spectrometry* 2002, 13:284–293.
31. Descroix S, Varenne A, Goasdoue N, Abian J, Carrascal M, Daniel RM, Gareil P: Non-aqueous capillary electrophoresis of the positional isomers of a sulfated monosaccharide. *Journal of Chromatography A* 2003, 987:427–376.

32. Perreault H, Costello CE: Stereochemial effects on the mass spectrometric behavior of native and derivatized trisaccharide isomers: comparisons with result from molecular modeling. *Journal of Mass Spectrometry* 1999, 34:184–197.
33. Ojima N, Masuda K, Tanaka K, Nishimura O: Analysis of neutral oligosaccharides for structural characterization by matrix-assisted laser desorption/ionization quadrupole ion trap time-of-flight mass spectrometry. *Journal of Mass Spectrometry* 2005, 40:380–388.
34. Que AH, Mechref Y, Huang YP, Taraszka JA, Clemmer DE, Novotny MV: Coupling capillary electrochromatography with electrospray Fourier transform mass spectrometry for characterizing complex oligosaccharide pools. *Analytical Chemistry* 2003, 75:1684–1690.
35. Taylor VF, Marh RE, Longerich HP, Stadey CJ: A mass spectrometric study of glucose, sucrose and fructose in an inductively coupled plasma and electrospray ionization. *International Journal of Mass Spectrometry* 2005, 243:71–84.
36. Park YD, Lebrilla CB: Application of Fourier-transform ion cyclotron resonance mass spectrometry to oligosaccharides. *Mass Spectrometry Reviews* 2005, 24:232–264.
37. March RE, Stadey CJ: A tandem mass spectrometric study of saccharides at high mass resolution. *Rapid Communications in Mass Spectrometry* 2005, 19:805–812.
38. Zaia J: Mass spectrometry of oligosaccharides. *Mass Spectrometry Reviews* 2004, 23:161–227.
39. Denekamp C, Sandlers Y: Anomeric distinction and oxonium ion formation in acetylated glycosides. *Journal of Mass Spectrometry* 2005, 40:765–771.
40. Harvey DJ: Collision-induced fragmentation of negative ions from N-linked glycans derivatized with 2-aminobenzoic acid. *Journal of Mass Spectrometry* 2005, 40:642–653.
41. Vrkic AK, O'Hair RAJ: Using non-covalent complexes to direct the fragmentation of glycosidic bonds in the gas phase. *Journal of the American Society for Mass Spectrometry* 2004, 15:715–724.
42. Yamagaki T, Fukui K, Tachibana K: Analysis of glycosyl bond cleavage and related isotope effects in collision-induced dissociation quadrupole-time-of-flight mass-spectrometry of isomeric trehaloses. *Analytical Chemistry* 2006, 78:1015–1022.

– 4.4 –

Neuromolecules

GENERAL FEATURES

Neurons communicate across synapses by means of neurotransmitters [1] produced within neurons at the nerve terminal. Following stimulation, the neuromolecules are released on the pre-synaptic side. After crossing, they bind to their specific receptors and trigger the opening or closing of ion channels leading to excitatory or inhibitory post-synaptic potentials. Neurotransmitters are small molecular systems such as amino acids, for example aspartic acid [2], derivatives of amino acids, for example noradrenaline [3, 4] (Figure 4.4.1) or small peptides, for example met-enkephalin YGGFM [5–9]. Those small entities can be studied by means of spectroscopy and quantum calculations. On the contrary, receptors are large assemblies of polypeptides (with typical masses of 500 kDa) containing a centre hole and protruding on both side of the post-synaptic membrane [10]. Receptors can only be studied with force-field simulations [11]. Larger peptides known as neuropeptides [12] are a set of messenger molecules involved in cell signalling and neuromodulation. They are produced by enzymatic cleavage of protein precursor molecules called pro-hormones. Their search and the characterization of their post-translational modifications (PTMs) have been performed by means of mass-spectrometry [12].

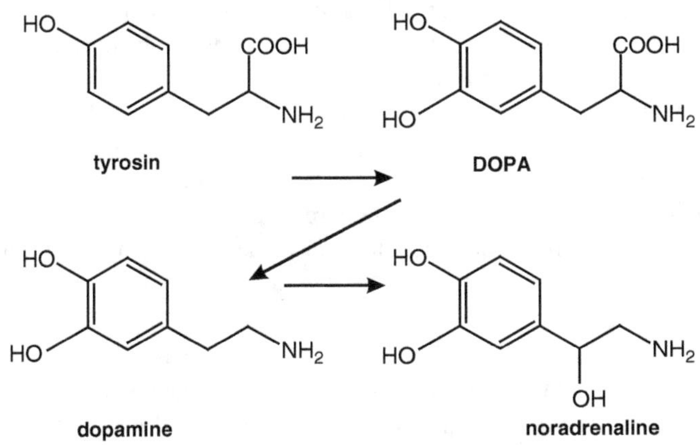

Figure 4.4.1 Biosynthetic pathway from the tyrosine amino acid to the noradrenaline neurotransmitter.

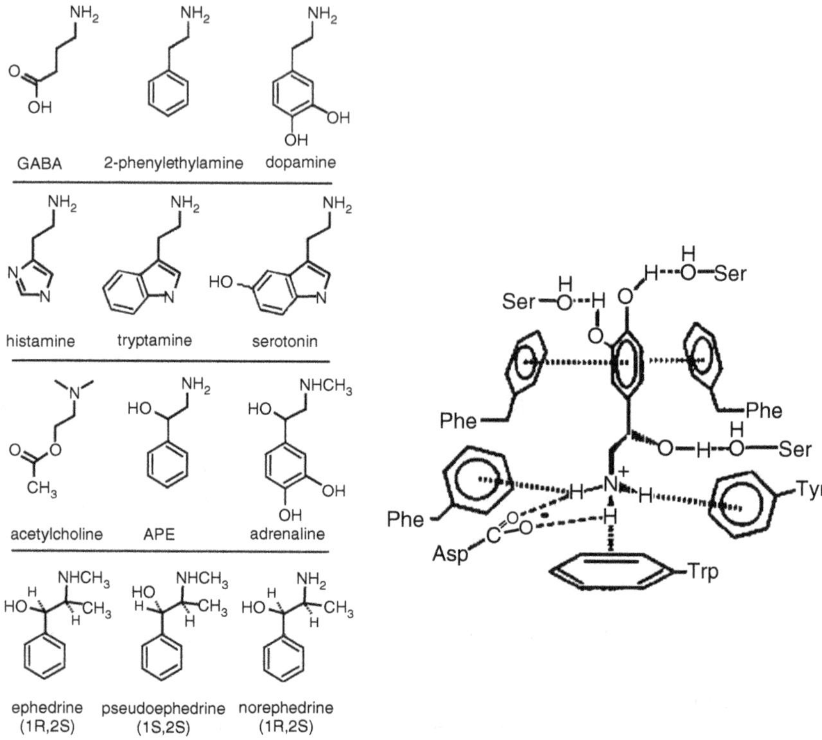

Figure 4.4.2 Left: Structures of a series of neurotransmitters (reprinted by permission from reference [31] ©2001 RCS). Right: Non-covalent interactions between noradrenaline and the β-adrenergic receptor (reproduced by permission from reference [13] ©2001 Wiley).

The search for pharmaceutical ligands binding to those receptors and promoting (agonists) or inhibiting (antagonists) [1, 13–16] can benefit from the understanding of the flexibility of neurotransmitters by exploring their energy landscape either in the gas-phase or in solution. For example, β-blockers [17] bind to adrenergic receptor sites [18] and inhibit the effects of catecholamines (Figure 4.4.2) such as adrenaline or noradrenaline [4].

In this chapter, we first consider R2PI/IR spectroscopic studies of neutral neurotransmitters containing a UV chromophore such as *p*-methoxyphenylethylamine (MPEA) [19], ephedrine and pseudoephedrine [20] and amphetamine [21]. Different investigations of protonated systems are examined. We then turn our attention towards modelling of neurotransmitters and non-covalent binding to their receptors emphasizing the case of acetylcholine, and two of its agonists, nicotine and muscarine, all lacking chromophores.

4.4.1 Neutral species

Among molecules involved in brain functioning, some of them, such as tyramine [22, 23] which is formed from the breakdown of proteins in aged food and known as a "migraine trigger" and γ-amino-*n*-butyric acid (GABA) [24, 25] have been studied by microwave spectroscopy. A sizable fraction of neurotransmitters, the melatonin neurohormone [26],

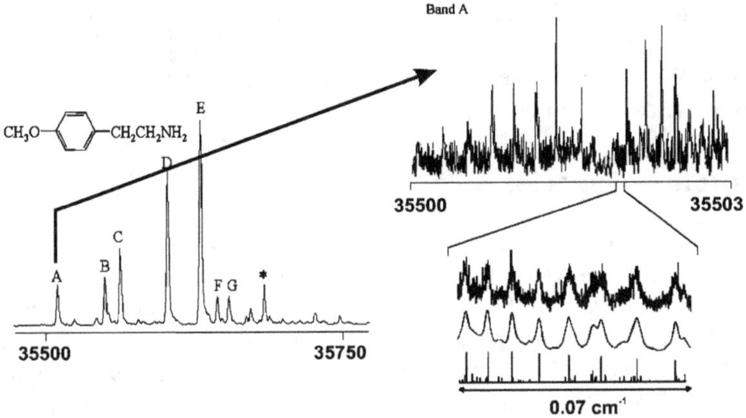

Figure 4.4.3 Left: Vibrationally resolved fluorescence excitation spectrum of gas-phase *p*-methoxyphenylethylamine cooled in a supersonic expansion. Bands A–G are the $S_1 \leftarrow S_0$ electronic origins of seven identified conformers (the asterisk corresponds to a water complex of MPEA). Right: Rotationally resolved fluorescence excitation spectrum of band A of MPEA. The bottom part shows a section of the R-branch with two simulations with and without de-convoluted line-shape. Individual rovibronic lines are fit to a Voigt line-shape resulting from a Gaussian Doppler-broadened component of 18 MHz and a Lorentzian lifetime-broadened component of 50 MHz corresponding to a fluorescence lifetime of 3 ns (reproduced by permission from reference [19] ©2002 Royal Society of Chemistry).

tryptamine [27–30] (see Section 2.1.4.4) or the 2-phenoxy ethylamine β-blocker [17] contain chromophores such as indole or catechol. A comprehensive presentation can be found in reference [31]. Different gas-phase conformations have been studied, together with their hydrated complexes (see Section 5.7) by means of rotational coherent spectroscopy (RCS) [32], rotationally resolved fluorescence excitation [19, 33] and IR–UV depletion spectroscopy [31, 34, 35], mass-spectrometry [36] and modelling [3, 37–39].

The *p*-MPEA substituted aromatic molecule, as many other neurotransmitters, has a flexible structure with an alkyl-amine tail. Some of its conformers can bind to ion channel receptors and open them while some others do not. The attribution of MPEA $S_1 \leftarrow S_0$ electronic spectrum is somewhat difficult [40] and has been solved by means of rotationally resolved fluorescence excitation [19] (Figure 4.4.3). Seven bands are observed in the low-resolution fluorescence excitation (Figure 4.4.3). Each band is resolved into well-defined P and R branches. The fits of rovibrational spectra with rotational constants calculated at the MP2/6-31G** level, as displayed in Figure 4.4.3 for band A, are characteristic for each band and allow their unambiguous assignment to conformer structures.

Ephedrine (1R,2S) and pseudoephedrine (1S,2S) constitute a diastereoisomeric pair (see Chapter 4.6). Their structure is characteristic of most neurotransmitters since they also possess a flexible side-chain, here ethylamine, which can adopt a more or less large number of conformations. All conformers of bare neutral ephedrine and pseudoephedrine feature an intra-molecular hydrogen bond between the alcohol OH group and the nitrogen atom [20]. Bare pseudoephedrine displays a more varied conformational landscape than ephedrine. Interactions depend on both the side-chain conformation and the configuration of the groups around the chiral centre. This chiral discrimination (see Chapter 4.6) turns out to be crucial for the binding of ephedrine and pseudoephedrine to the active sites of their receptors. While both molecules mimic the effects of having an adrenaline rush, ephedrine is active as a vasopressor whereas pseudoephedrine is a decongestant.

4.4.2 Protonated species

In the aqueous physiological environment of a cell (pH = 7.4), all ligands possessing basic groups, such as amines, are protonated [14, 35]. Methods such as MS/MS mass-spectrometry [36] and resonant-IRMPD are then immediately applicable. When neurotransmitters do not possess a UV chromophore, the messenger method (see Section 2.1.3.2.2) can then be applied in IR/R2PI studies. For that purpose, the ethanolamine has been complexed in a supersonic expansion with phenol [35]. In absence of IR irradiation, the large proton affinity of ethanolamine induces the following internal proton transfer:

$$[C_6H_5OH\cdots Et]^{+}_{E} \rightarrow [C_6H_5OH\cdots EtH^{+}]_{E>E_{\text{threshold}}} \rightarrow C_6H_5O + Et\,H^{+}. \qquad (4.4.1)$$

Observation of spontaneous dissociation of the complex takes place if the internal energy is above an energy threshold. In presence of resonant irradiation of the non-dissociated complex, an increase of the protonated ethanolamine complex provides a signature of infrared absorption of ethanolamine assuming a negligible influence of the weakly bound phenoxy radical partner (Figure 4.4.4).

$$[C_6H_5O\cdots Et\,H^{+}]_{E<E_{\text{threshold}}} + h\nu(\text{IR}) \rightarrow C_6H_5O + Et\,H^{+}. \qquad (4.4.2)$$

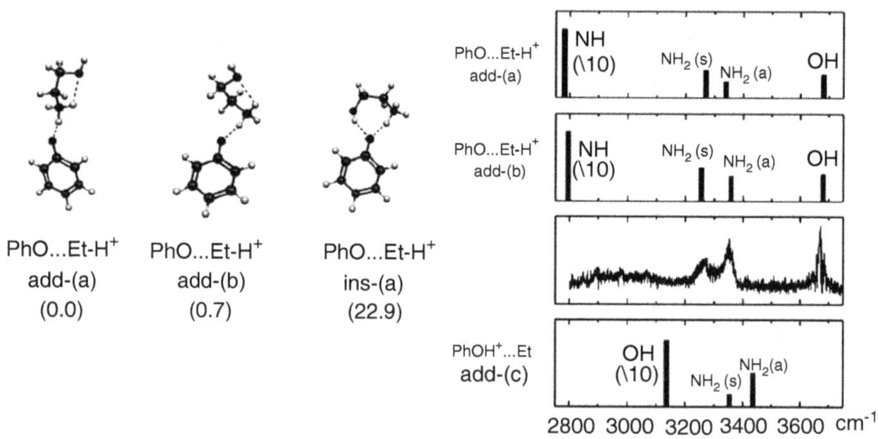

Figure 4.4.4 Left: Computed lowest energy structures of protonated complexes of ethanolamine and phenol calculated at the B3LYP/6-31+G* level (relative energies are in kJ/mol). Right: Experimental infrared photo-dissociation spectrum of the protonated phenol–ethanolamine complex and calculated vibrational spectra of the three lowest energy conformers (reproduced by permission from reference [35] ©2004 Royal Society of Chemistry).

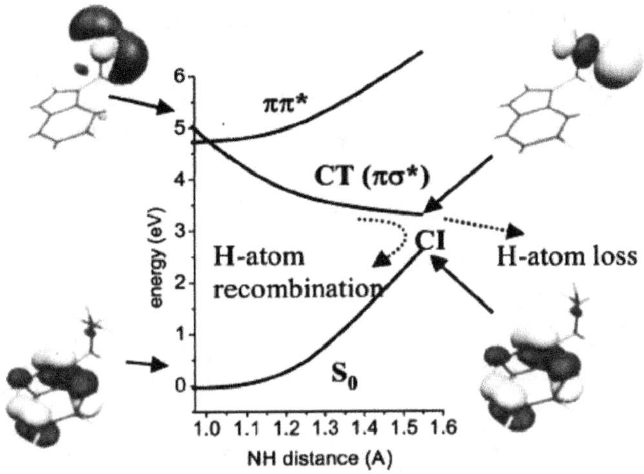

Figure 4.4.5 Electronic states of protonated tryptamine along the amino group N—H dissociation coordinate calculated at the TD-DFT B3LYP/6-31G** level (reproduced by permission from reference [29] ©2005 American Institute of Physics).

The electronic excited state dynamics of protonated tryptamine (TrypH$^+$) has been studied by means of photo-induced dissociation method on a femtosecond timescale [29]. TrypH$^+$ ions are initially excited in their $\pi\pi^*$ state by a pump laser beam at 266 nm (4.66 eV) and further fragment. The time evolution is followed through absorption of a probe photon at 800 nm (1.55 eV). For each fragmentation channel, the intensity of the ion signal is recorded as a function of the pump–probe delay and monitors the fragmentation efficiency between the initial $\pi\pi^*$ state and the higher excited states reached with the probe. Calculations performed with the TD-DFT (see Section 1.7) method show that the active electron initially in a $\pi\pi^*$ orbital on the indole ring becomes localized in a charge-transfer (CT) $\pi\sigma^*$ diffuse orbital around the protonated amino group. When the N—H coordinate of the amino group is stretched, this $\pi\sigma^*$ diffuse orbital collapses into a σ^* antibonding orbital. During the evolution along the N—H coordinate, the $\pi\sigma^*$ orbital has a conical intersection with the initial state S_0. The excited ions can then either dissociate with the loss of a hydrogen atom or undergo internal conversion (IC) and become highly vibrationally excited (Figure 4.4.5). The hot ions can then further fragment and the whole excited state evolution can be followed by combining probe excitation and mass-spectrometric detection of both neutral and charged fragments in coincidence (see Figure 3.2.11).

4.4.3 Modelling of neuromolecules

The great flexibility of neurotransmitters [3, 4, 37–39, 41] and related molecules such as tryptamine leads to large number of conformers with rather similar geometries. A search of conformers can be first performed by means of systematic exploration of the different dihedral angles (see Section 1.5) at a low level such as AM1 or PM3 but a moderate level of theory such as SCF/6-31G* is much more reliable [37]. The obtained minimum energy structures are then optimized with DFT BLYP or B3LYP with a 6-31+G(d) or a aug-cc-pVDZ

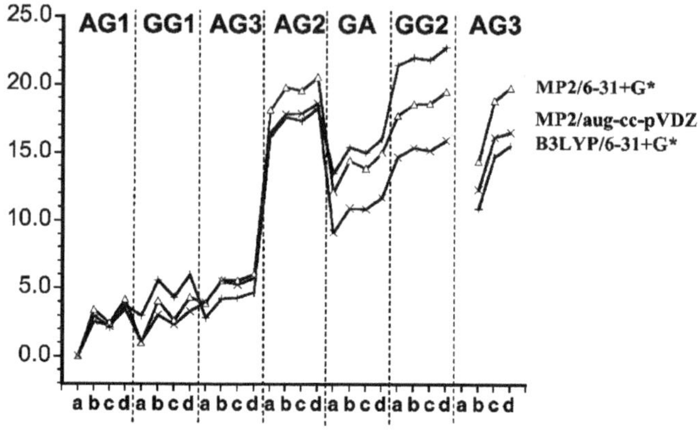

Figure 4.4.6 Relative energies (in kJ/mol) of 28 conformers of pseudoadrenaline, including ZPE corrections. Note that the order obtained from DFT and MP2 calculations can be different (e.g. conformers AG2) (reproduced by permission from reference [39] ©2005 Royal Society of Chemistry).

basis set (see Section 1.2.5). A comparison between the different levels of theory in the case of noradrenaline [39] is displayed in Figure 4.4.6.

In order to describe as best as possible the conformational landscape, not only minima but also energy barriers are needed. Transition state structures can be searched with a synchronous transit algorithm and optimized at the B3LYP/6-31+G(d) level. The next step is getting the correct order of conformer energies and as accurate as possible values of those energies by accurately taking into account dispersion that is ignored in DFT. MP2 calculations with a large basis set being too costly, either RI-MP2/aug-cc-pVDZ and CCSD(T) single-point calculations of the B3LYP/6-31+G(d) structures [28] or MP2 calculations with the compact interaction-optimized DZPi basis set have been performed [3].

Stimulated emission pumping population transfer spectroscopy (SEP-PTS) [28, 42] (see Section 2.1.4.4) has been used to experimentally explore the potential energy surface (PES) of tryptamine represented in Figure 4.4.7.

Enkephalins are small endogeneous neuropentapeptides that bind to the opioid receptors in a similar fashion as morphine [43–45]. They have the general sequence Tyr-Gly-Gly-Phe-Xaa where Xaa is either Leu or Met. They are found in the brain, the spinal cord and the gut. In contrast with the rigid morphine, enkephalins are flexible. The conformations and infrared spectra of these agonist peptides have been theoretically studied. The exploration step of the PES [7] can be performed by means of the CHARMM force-field [5] or the low-cost self-consistent-charge density functional based tight-binding (SCC-DFTB) method [9]. This method is an approximate DFT approach but, in contrast with semi-empirical methods, no fitting of experimental data are required. It allows the modelling of systems with hundreds of atoms with nearly the same accuracy as conventional DFT.

4.4.3.1 Neuroreceptor–drug complexes

Modelling [46, 47] and mass-spectrometry [18, 48–51] are among the tools used for understanding receptor binding site interactions as well as the search for new drugs mimicking natural molecules [14, 18, 51]. A review of predictive tools of the binding affinity of small

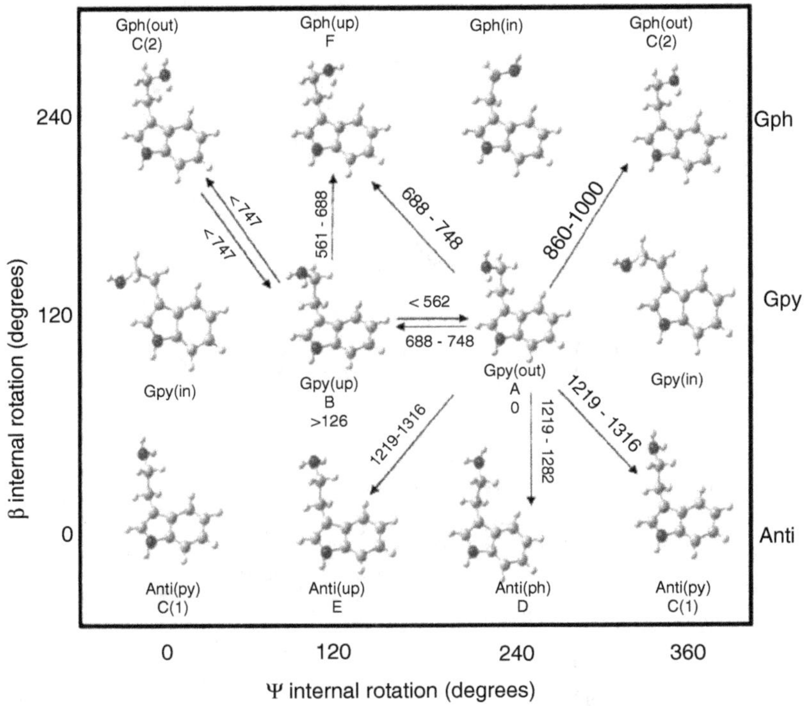

Figure 4.4.7 Schematic PES of tryptamine along two flexible coordinates ψ (C_α-amino nitrogen angle) and β (indole ring C—C_β angle) coordinates. The experimentally observed conformational isomerization paths are shown by arrows. The numbers associated with those arrows correspond to the lower and upper bounds of the isomerization barriers (reproduced by permission from reference [29] ©2005 American Institute of Physics).

molecule ligands to macromolecular receptors can be found in reference [52]. The best-studied ligand-gated ion channel is the nicotinic acetylcholine receptor [11] which belongs to a large family of receptors for serotonin [37, 53], GABA [24, 25] and glycine.

Neurotransmitters such as acetylcholine or its two agonists, muscarine and nicotine [54] do not contain any UV chromophore. Gas-phase structural studies of their ionized species can be conducted by means of resonant IRMPD (see Section 2.1.3.2.5). Those three molecules contain amines which have to fit in the hydrophobic pocket of the acetylcholine receptor [55] itself containing aromatic residues. Those receptors are divided into nicotinic and muscarinic receptors according to the agonist action. Acetylcholine and muscarine contain a quaternary amine group [14] whereas nicotine contains a tertiary amine. Experimental infrared fragmentation spectra of acetylcholine, nicotine and muscarine cations have been obtained by means of IRMPD in the fingerprint (800–2,000 cm^{-1}) region [56]. Four conformers of the acetylcholine cations are observable, each characterized by a different set of the C—O and C=O stretch frequencies corresponding to a characteristic hydrogen bonding pattern of those groups with the positively charged quaternary amine group.

The case of nicotine [57] is interesting because the most energetically favourable protonation site in the gas-phase is the pyridinic-ring nitrogen atom whereas it has been proposed

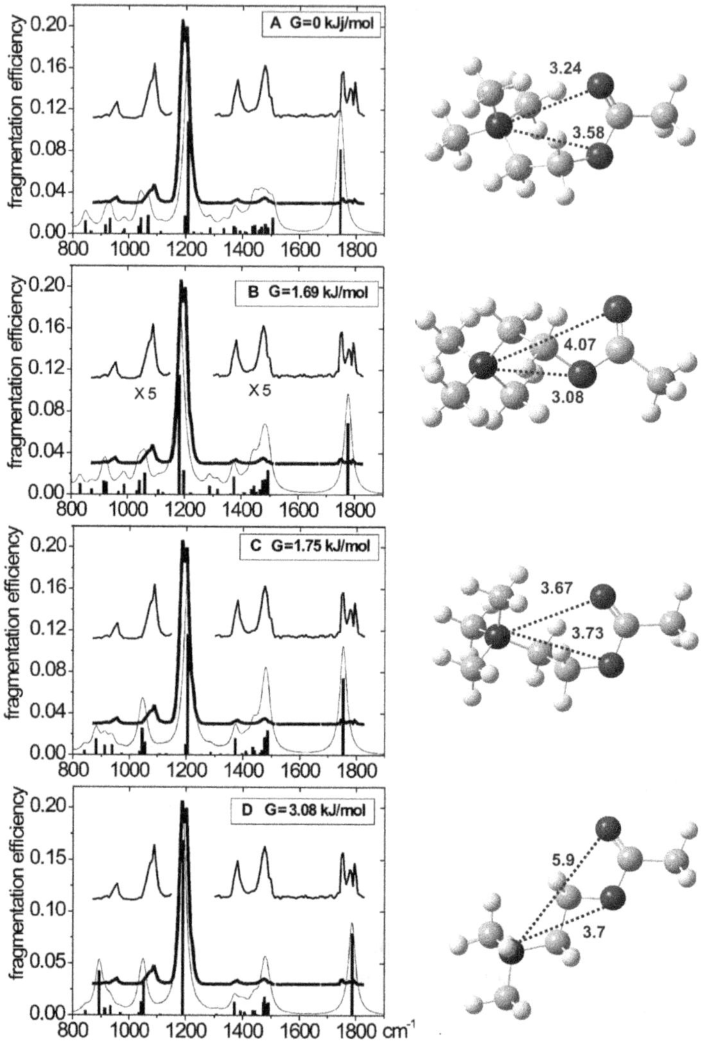

Figure 4.4.8 Left: Infrared multiphoton dissociation spectrum of protonated acetylcholine. The dashed line and the solid line respectively correspond to the experimental IRMPD spectrum and the IR absorption spectrum of the lowest energy configuration calculated at the B3LYP/6-31G* level. Right: Lowest energy configurations of the gas-phase acetylcholine neurotransmitter calculated at the MP2/6-31G*. The different observed shifts of the C—O (ca. 1,150 cm^{-1}) and C═O (ca. 1,780 cm^{-1}) lines correspond to different distances between these groups and the charged quaternary nitrogen atom.

that the bioactive form corresponds to the pyrrolidine-ring nitrogen [58]. The energy difference between those two protonation sites is rather small but the gas-phase barrier is rather large (Figure 4.4.8). This barrier is probably much lower for nicotine in an aqueous environment [54] (Figure 4.4.9).

The recognition process between the acetylcholine receptor and its ligands involves a positively charged amine (cation–π) interaction on one side [55]. It has been proposed that

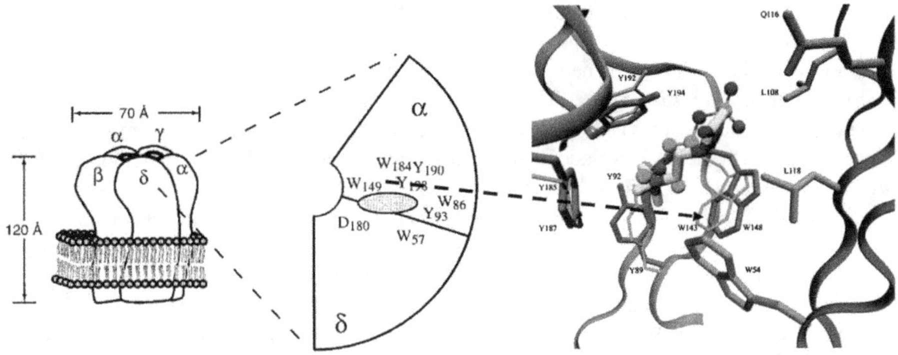

Figure 4.4.9 Left: The nicotine acetylcholine receptor (nAChR) is a ligand-gated ion channel. Among the four tryptophans (W), the α-Trp-149 (designated by the arrow) is identified as the major cation–π interaction site of the quaternary ammonium group of acetylcholine and nicotine (reprinted by permission from reference [55] ©1998 National Academy of Sciences). Right: Acetylcholine inside the nicotinic acetylcholine receptor conserved binding pocket at crystallographic resolution (reproduced by permission from reference [11] ©2006 American Chemical Society).

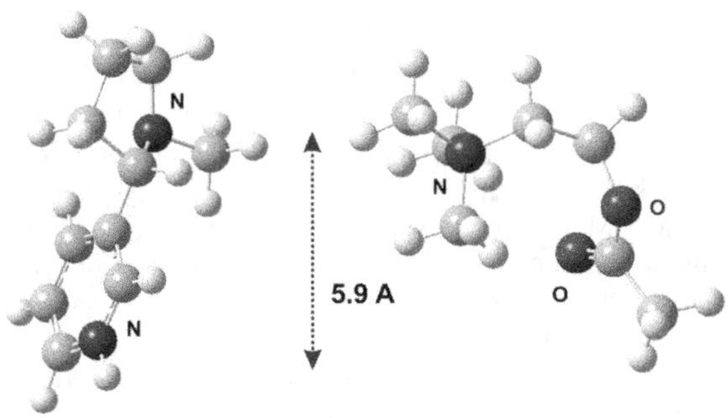

Figure 4.4.10 Protonated nicotine (left) and acetylcholine (right) are two very different molecules but nevertheless interact with the same neuroreceptors. A possible explanation is that the recognized groups of each molecule are at the same distance (adapted by permission from reference [59] ©2003 American Chemical Society).

the carboxyl oxygen is involved in the nicotinic interaction while the ester oxygen is needed for muscarine. An intriguing question is why acetylcholine interacts with the same receptors of either muscarine and nicotine that are quite different molecules. A possible answer is provided by the modelling comparisons between those molecules [54, 59, 60]. Acetylcholine has a muscarinic or a nicotinic activity. It is then proposed that the ester group of acetylcholine acts as the ring of muscarine or its carbonyl oxygen acts as the pyridinic nitrogen of nicotine. The receptors thus seem to electrostatically interact with the cation head on one side and either the negatively charged ester or the carbonyl groups on the other side. What then matters is only the distances between those charged groups called pharmacophoric groups (Figure 4.4.10). Understanding of these interactions provides clues for the design of drugs fighting tobacco addiction [61].

REFERENCES

1. Carlsson A: A half-century of neurotransmitter research: impact on neurology and psychiatry (Nobel lecture). *ChemBioChem* 2001, 2:484–493.
2. Sangarron W, Ruangpomvisuti V: Conformational study of cationic, zwitterionic, anionic species of aspartic acid, water-added forms and their protonation. A DFT method. *Journal of Molecular Structure* 2005, 758:179–185.
3. Van Mourik T, Fruchtl HA: The potential energy landscape of noradrenaline: an electronic structure study. *Molecular Physics* 2005, 103:1641–1654.
4. Snoek LC, Van Mourik T, Simons JP: Neurotransmitters in the gas phase: a computational and spectroscopic study of noradrenaline. *Molecular Physics* 2003, 101:1239–1248.
5. Watson TM, Hirst JD: Vibrational analysis of capped (Leu) enkephalin. *Physical Chemistry Chemical Physics* 2004, 6:2590–2587.
6. Schnier PD, Price WD, Strittmatter EF, Williams ER: Dissociation energetics and mechanisms of leucine enkephalin $(M+H)^+$and $(2M+X)^+$ions (X=H, Li, Na, K, and Rb) measured by blackbody infrared radiative dissociation. *Journal of the American Society for Mass Spectrometry* 1997, 8:771–780.
7. Kriz Z, Carlsen PHJ, Koca J: Conformational features of linear and cyclic enkephalins: a computational study. *Journal of Molecular Structure* 2001, 540:231–250.
8. Evans DA, Wales DJ: The free energy landscape and dynamics of met-enkephalin. *Journal of Chemical Physics* 2003, 119:9947–9955.
9. Abdali S, Niehaus TA, Jalkanen KJ, Cao X, Nafie LA, Frauenheim T, Suhai S, Bohr H: Vibrational absorption spectra, DFT and SCC-DFTB conformational study and analysis of [Leu]enkephalin. *Physical Chemistry Chemical Physics* 2003, 5:1295–1300.
10. Unwin N: Refined structure of the nicotinic acetylcholine receptor at 4Å resolution. *Journal of Molecular Biology* 2004, 346:967–989.
11. Zhang DQ, Gullingsrud J, McCammon JA: Potentials of mean force for acetylcholine unbinding from the α7 nicotinic acetylcholine receptor ligand-binding domain. *Journal of the American Chemical Society* 2006, 128:3019–3026.
12. Hummon AB, Amare A, Sweedler JV: Discovering new invertebrate neuropeptides using mass spectrometry. *Chemical Reviews* 2006, 25:77–98; Verhaert PD, Prieto Canaway MC, Pekar TM, Miller K. Neuropeptide imaging on an LTQ with vMALDI source: the complete "all-in-one" peptidome analysis. *International Journal of Mass Spectrometry* 2007, 260:177–184.
13. Herm M, Molt O, Schrader T: Towards synthetic adrenaline receptors-shape selective adrenaline recognition in water. *Angewandte Chemie-International Edition* 2001, 40:3148–3151.
14. Woods AS: The mighty arginine, the stable quaternary amines, the powerful aromatics and the aggressive phosphate: their role in the noncovalent minuet. *Journal of Proteome Research* 2004, 3:478–484.
15. Baker SR, Boot J, Brunays M, Dobson D, Green R, Hayhurst L, Keenan M, Wallace L: High affinity ligands for the 7 nicotinic receptor that show no reactivity with the 5-HT3 receptor to bind to their receptors. *Bioorganic Medicinal Chemical Letters* 2005, 15:4727–4730.
16. Chalmers MJ, Busby SA, Pascal BD, He YJ, Hendrickson CL, Marshall AG, Griffin PR: Probing protein ligand interactions by automated hydrogen/deuterium exchange mass spectrometry. *Analytical Chemistry* 2006, 78:1005–1014.
17. Macleod NA, Simons JP: β-blocker conformations in the gas phase: 2-phenoxy ethylamine, its hydrated clusters and 3-phenoxy propanolamine. *Physical Chemistry Chemical Physics* 2004, 6:2878–2884.
18. Ramström O, Yu C, Mosbach K: Chiral recognition in adrenergic receptor binding mimics prepared by molecular imprinting. *Journal of Molecular Recognition* 1996, 9:691–696.

19. Yi JT, Robertson EG, Pratt DW: Conformational analysis by laser spectroscopy in the gas-phase: p-methoxyphenetylamine (a neurotransmitter). *Physical Chemistry Chemical Physics* 2002, 4:5244–5248.
20. Butz P, Kroemer RT, Macleod NA, Simons JP: Conformational preferences of neurotransmitters: ephedrine and its diastereoisomer, pseudoephedrine. *Journal of Physical Chemistry A* 2001, 105:544–551.
21. Brause R, Fricke H, Gerhards M, Weinkauf R, Kleinermanns K: Double resonance spectroscopy of different conformers of the neurotransmitter amphetamine and its clusters with water. *Chemical Physics* 2006, 327:43–53.
22. Melandri S, Maris A: Intramolecular hydrogen bonds and conformational properties of biogenic amines: a free-jet microwave study of tyramine. *Physical Chemistry Chemical Physics.* 2004, 6:2863–2866.
23. Caminati W: The rotational spectra of conformers of biomolecules: tryptamine. *Physical Chemistry Chemical Physics* 2004, 6:2806–2809.
24. Lorenzini ML, Bruno-Blanch L, Estin GL: Structural and electronic factors associated with the activity in the GABA-A system. *Journal of Molecular Structure (Theochem)* 1998, 454:1–16.
25. Cocinero EJ, Lesarri A, Sanz ME, Lopez JC, Alonso JL: Conformations of α-aminobutyric acid in the gas phase. *ChemPhysChem* 2006, 7:1481–1487.
26. Zwier TS: Laser probes of conformational isomerization in flexible molecules and complexes. *Journal of Physical Chemistry A* 2006, 110:4133–4150.
27. Peteanu LA, Levy DH: Spectroscopy of complexes of tryptamine and 3-indole propionic acid with various solvents. *Journal of Physical Chemistry* 1988, 92:6554–6561.
28. Clarkson JR, Dian BC, Moriggi L, DeFusco A, McCarthy V, Jordan KD, Zwier TS: Direct measurement of the energy thresholds to conformational isomerization in tryptamine: experiment and theory. *Journal of Chemical Physics* 2005, 122:214311–214326.
29. Kang H, Jouvet C, Dedonder-Lardeux C, Martrenchard S, Charriere C, Gregoire G, Desfrancois C, Schermann JP, Barat M, Fayeton JA: Photoinduced processes in protonated tryptamine. *Journal of Chemical Physics* 2005, 122:84307–84311.
30. Nguyen TV, Pratt DW: Permanent electric dipole moments of four tryptamine conformers in the gas phase: a new diagnostic of structure and dynamics. *Journal of Chemical Physics* 2006, 124:054317–054323.
31. Robertson EG, Simons JP: Getting into shape: conformational and supramolecular landscapes in small biomolecules and their hydrated clusters. *Physical Chemistry Chemical Physics* 2001, 3:1–18.
32. Connell LL, Corcoran TC, Joireman PW, Felker PM: Conformational analysis of jet-cooled tryptophan analogs by rotational coherence spectroscopy. *Chemical Physics Letters* 1990, 166:510–516.
33. Nguyen TV, Korter TM, Pratt DW: Tryptamine in the gas phase. A high resolution laser study of the structural and dynamic properties of its ground and electronically excited states *Molecular Physics* 2005, 103:2453–2453.
34. Snoek LC, van Mourik T, Carcabal P, Simons JP: Neurotransmitters in the gas phase: hydrated noradrenaline. *Physical Chemistry Chemical Physics* 2003, 5:4519–4526.
35. Macleod NA, Simons JP: Neurotransmitters in the gas phase: infrared spectroscopy and structure of protonated ethanolamine. *Physical Chemistry Chemical Physics* 2004, 6:2821–2826.
36. Rogalewicz F, Bourcier S, Hoppilliard Y: Decomposition of protonated noradrenaline and normetanephrine assisted by NH_2 migration studied by electrospray tandem mass spectrometry and molecular orbital calculations. *Rapid Communications in Mass Spectrometry* 2005, 6:2821–2826.
37. van Mourik T, Emson LEV: A theoretical study of the conformational landscape of serotonin. *Physical Chemistry Chemical Physics* 2002, 4:5863–5871.

38. van Mourik T: On the relative stability of two noradrenaline conformers. *Chemical Physics Letters* 2005, 414:364–368.
39. van Mourik T: The shape of neurotransmitters in the gas phase: a theoretical study of adrenaline, pseudoadrenaline, and hydrated adrenaline. *Physical Chemistry Chemical Physics* 2004, 6:2827–2837.
40. Robertson EG, Simons JP, Mons M: Comment on "structural and vibrational assignment of p-methoxyphenethylamine conformers". *Journal of Physical Chemistry A* 2001, 105:9990–9992.
41. Carcabal P, Snoek LC, Van Mourik T: A computational and spectroscopic study of the gas-phase conformers of adrenaline. *Molecular Physics* 2005, 103:1633–1639.
42. Evans DA, Wales DJ, Dian BC, Zwier TS: The dynamics of conformational isomerization in flexible biomolecules. II. Simulating isomerizations in a supersonic free jet with master equation dynamics. *Journal of Chemical Physics* 2004, 120:148–157.
43. Kogan A, Ustyuzhanin P, Reuben BG, Lifshitz C: Hydrogen/deuterium exchange of monomers and dimers of leucine enkephalin. *International Journal of Mass Spectrometry* 2002, 213:1–4.
44. Ishikawa K, Nakamura T, Koga Y: Cross-checking of electrospray ionization mass spectrometry and computer simulation for the evaluation of the interaction strength of non-covalently bound enkephalins in solution. *Journal of Mass Spectrometry* 2001, 36:937–942.
45. Black DM, Patyne AH, Glish GL: Determination of cooling rates in a quadrupole ion trap. *Journal of the American Society for Mass Spectrometry* 2006, 17:932–938.
46. Aarons L, Karlsson MO, Mentré F, Rombout F, Steiner JL, van Peer A: Role of modelling and simulation in Phase I drug development. *European Journal of Pharmaceutical Sciences* 2001, 13:115–122.
47. Zsoldos Z, Szabo I, Szabo Z, Johnson AP: Software tools for structure based rational drug design. *Journal of Molecular Structure (Theochem)* 2003, 666:659–665.
48. Ciruela F, Burgueno J, Casado V, Canals M, Marcellino D, Goldberg SR, Bader M, Fuxe K, Agnati LF, Lluis C, *et al.*: Techniques for the study of receptor heteromerization. Direct epitope–epitope electrostatic interactions between adenosine A_{2A} and dopamine D_2 receptors. *Analytical Chemistry* 2004, 76:5354–5363.
49. Lee MS: *Integrated strategies for drug discovery using mass spectrometry*: Wiley-Interscience, Hoboken, NJ 2005.
50. Geoghehan KF, Kelly MA: Biochemical applications of mass spectrometry in pharmaceutical drug discovery. *Mass Spectrometry Reviews* 2005, 24:347–366.
51. Leonard S: Consequences of low levels of nicotinic acetylcholine receptors in schizophrenia for drug development. *Drug Development Research* 2003, 60:127–136.
52. Gohlke H, Klebe G: Approaches to the description and prediction of the binding affinity of small-molecule ligands to macromolecular receptors. *Angewandte Chemie-International Edition* 2002, 41:2644–2676.
53. Pisterzi LF, Almeida DRP, Chass GA, Torday LL, Papp JG, Varro A, Csizmadia IG: Density functional molecular computations on protonated serotonin in the gas phase and various solvent media. *Chemical Physics Letters* 2002, 365:542–551.
54. Munoz-Caro C, Nino A, Mora M, Reyes S, Melendez FJ, Castro ME: Conformational population distribution of acetylcholine, nicotine and muscarine in vacuum and solution. *Journal of Molecular Structure (Theochem)* 2005, 726:115–124.
55. Zhong W, Gallivan JP, Zhang Y, Li L, Lester HA, Dougherty DA: A cation-pi binding site in the nicotinic receptor. *Proceedings of National Academy of Sciences of United States of America* 1998, 95:12088–12093.
56. Seydou M, Grégoire G, Schermann JP, Desfrancois C: Infrared spectra of isolated and aqueous acetylcholine, muscarine and nicotine in preparation.
57. Lukas RJ, Ke L, Bencherif M, Eisenhour C: Regulation by nicotine of its on receptor. *Drug Development Research* 1996, 38:136–148.

58. Graton J, van Mourik T, Price SL: Interference between the hydrogen bonds to the two rings of nicotine. *Journal of the American Chemical Society* 2003, 125:5988–5997.
59. Galvez-Ruano E, Iriepa-Canalda I, Morreale A, Lipkowitz KB: A computational model of the nicotinic acetylcholine binding site. *Journal of Computer-Aided Molecular Design* 1999, 13: 57–68.
60. Graton J, Berthelot M, Gal JF, Girard S, Laurence C, Lebreton J, Le Questel JY, Maria PC, Naus P: Site of protonation of nicotine and nornicotine in the gas phase: pyridine or pyrrolidine nitrogen? *Journal of the American Chemical Society* 2002, 124:10552–10562.
61. Wang HYJ, Taggl AE, Meinald J, Wise RA, Woods AS: Study of the interaction of chlorisondamine and chlorisondamine analogues with the epitope of the α-2 neuronal acetylcholine nicotinic receptor subunit. *Journal of Proteome Research* 2005, 4:532–539.

– 4.5 –

Non-Covalent Complexes

GENERAL FEATURES

In a very large number of biological processes, biomolecules interact [1] and cooperate [2] through formation of non-covalent complexes [3]. Those non-covalent complexes can be investigated through modelling, spectroscopy and mass-spectrometry. For example, understanding the role of antibiotics [4] and measuring the strength of their interactions as well as designing inhibitors [5, 6] that forbid those interactions take advantage of mass-spectrometric techniques [7–14]. A very large number of complexes have now been investigated in the gas-phase such as DNA duplexes [15, 16], metal–nucleic acids [17–22], nucleic acid–ligands [23], metal–peptides [24–43], protein–protein [9, 44], protein–nucleic acid [45] and protein–DNA or RNA [3, 46] protein–ligand [47–51], peptide–sugars [52], protein–carbohydrate [53], hormones [54] or enzyme–ligand complexes [55, 56].

This chapter is restricted to rather small non-covalent complexes. Non-covalent assemblies of large systems are considered in Chapter 4.8. Mass-spectrometry of non-covalent complexes with some emphasis on drug interactions will be first examined and will be followed by modelling and spectroscopy topics.

4.5.1 Mass-spectrometry of non-covalent complexes

Non-covalent complexes form in solution and mass-spectrometric techniques must be capable of bringing them as intact gas-phase ions and the internal energy transferred to the complexes in the course of ionization must be low enough to prevent their dissociation. Electrospray (see Section 3.1.6) is the most widely used method [57–65] but other techniques such as MALDI [66–70] (see Section 3.1.5), liquid beams [71, 72] (see Section 3.1.4) or liquid droplets [73] are also valuable.

Once mass-spectra of non-covalent complexes are obtained, different studies can be conducted such as H/D exchange [74, 75] (see Section 3.3.4), BIRD [53] (see Section 3.3.2), collision-induced dissociation [10, 76] (CID, see Section 3.2.2.2), infrared multiphoton dissociation [13, 77] (IRMPD, see Section 2.1.3.2.5) or electron capture dissociation [78, 79] (see Section 4.7.2). Although the structural information concerning complexes provided by mass-spectrometry is limited [80] as compared with NMR or X-ray crystallography, its great advantages are that it can provide the exact complex stoechiometry and requires much less material. It is usually admitted that after electrospray ionization [12, 55, 74] or desorption from liquids [72, 73], non-covalent complexes can retain memory of their solution-phase structure. By means of comparison between solution and gas-phase data, it is then possible to investigate the respective influence of the different interactions responsible for specificity. Those interactions are mainly ionic, aromatic and van der Waals, hydrophobic interactions and hydrogen bonds. An example of the role of hydrophobicity in the gas-phase in formation of non-covalent complexes [60] is described in Section 5.3. During the change from the liquid to the gaseous solvent-less state, the stability of the non-covalent complexes (see Chapter 4.1 in the case of DNA duplexes) can become quite different. Electrostatic interactions are usually strengthened due to the decrease of the dielectric constant while hydrophobic interactions decrease and eventually vanish [60]. It may be important to note that if complexes are formed from monomeric molecules, unexpected oligomers can be observed due to formation of non-specific aggregates if too high solution concentrations are used. This non-specific aggregation can be checked and reduced by reduction of the solution concentration.

Mass-spectrometry is widely used for screening drugs [75], for example compounds binding to DNA [81–83] or inhibitors of β-amyloid protein aggregation [84, 85]. The mass-spectrum of a receptor alone presents peaks corresponding to the different charge states (1:0 peaks, case A of Figure 4.5.1). In presence of a specific ligand, peaks corresponding to the receptor–ligand complexes appear while 1:0 peaks have vanished (1:1 peaks,

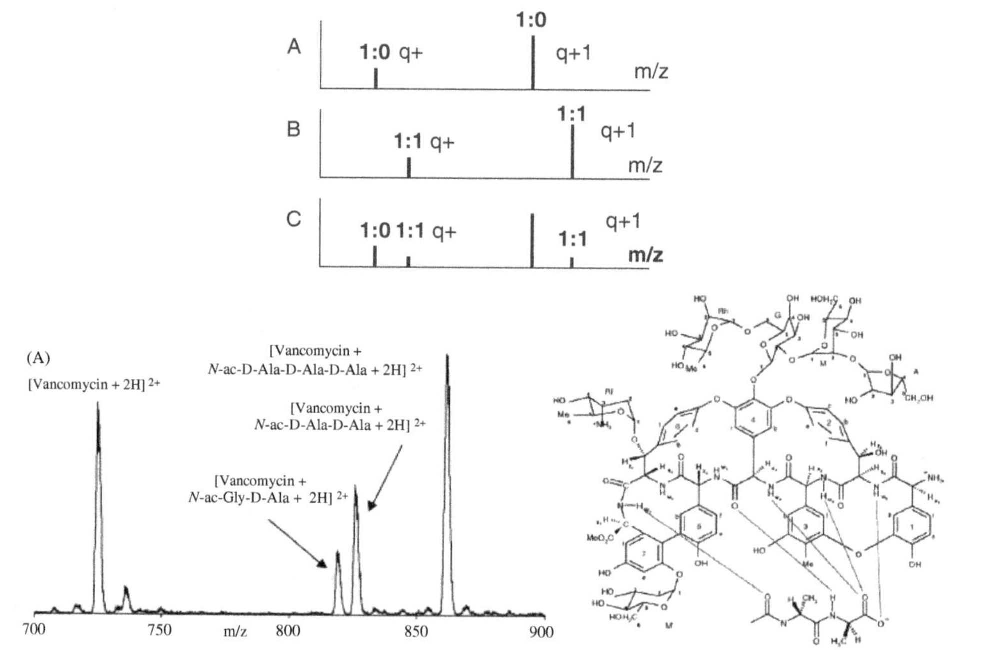

Figure 4.5.1 Top: Schematic mass-spectra displaying specific recognition between a receptor alone (A), bound to one of its specific target ligands (B) or one of its non-specific target ligands (C). Bottom left: Mass-spectrum of the vancomycin antibiotic-D-alanyl-D-alanine receptor complex. Right: Chemical structure of ristocetin and specific interactions with the D-alanyl-D-alanine terminus of its pentapeptide receptor (reproduced with permission from reference [4] ©2004 Elsevier).

case B). The gas-phase relative abundances of the 1 : 0, 1:1 and 1 : 2 complexes can be measured as function of the ligand concentration in solution. The 1 : 0 peak then decreases linearly while the 1:1 peak increases linearly, the sum of both being nearly constant. The 1 : 2 complex only appears when the 1 : 0 has nearly disappeared, showing that the 1:1 and 1 : 2 ligand complexes form sequentially. When the ligand binding is not specific, 1 : 0 and 1:1 peaks coexist in the mass-spectra (case C).

As an example of mass-spectrometric observation of specific recognition, let us consider non-covalent complexes concerning aptamers [23] (see Chapter 4.1). Aptamers are tailor-made unnatural strands of DNA or RNA possessing disordered conformations when isolated in solution but undergo a rapid folding when binding to their recognized ligands (see Chapter 4.1). For example, the RNA aptamer 5′-GGGACUUGGUUUAGGUAAUGAGUCCCU-3′ specifically binds tobramycin, an aminoglycoside antibiotic used to treat eye bacterial infections. Negatively charged aptamer (RNA) mass-peaks that appear in absence of ligands (A) disappear when tobramycin is added while the 1:1 complex RNA–tobramycin complex appears (B). This typical situation with disappearance of isolated reactant peaks and appearance of complex peaks is also obtained with another antibiotic, bekanamicin (B), but not with gentamicin (C). This can be rationalized by examination of the NMR structure of the aptamer (PDB entry 2TOB) which shows that the aptamer folds with a deep pocket for ligand binding. The specificity of the aptamer for tobramycin and bekanamycin is then explained by the size of the pocket and the network of hydrogen bonds between the protonated amino groups of the ligands and the backbone phosphates of the aptamer which strongly favour binding of tobramycin and bekanamicin but precludes binding of gentamicin. Another example is provided by non-covalent interactions between glycopeptide antibiotics such as vancomycin and ristocetin [4] with a bacterial cell pentapeptide (Figure 4.5.1). Those antibiotics specifically target the D-alanyl-D-alanine terminus of the pentapeptide and surprisingly do not interact with the more natural stereoisomeric L-alanyl-L-alanine.

As another example of mass-spectrometry drug-screening study of structure and binding energies, let us consider DNA duplex/drug complexes produced by electrospray and isolated in gas-phase (see Section 3.3.1). Mass-spectrometry is widely used for screening drugs, for example compounds binding to DNA [81–83] or inhibitors of β-amyloid protein aggregation [84].

MS/MS mass-spectrometry (see Section 3.2.2) provides information concerning the energetic of interactions between non-covalent complex partners [86]. However, in large complexes, internal energies and densities required for application of statistical model of dissociation are not accurately known. It is then only possible to proceed through comparisons. The investigated series of dissociating complexes must have received the same experimental amount of internal energy under identical experimental conditions and must have identical masses. Dissociation mechanisms must also be identical. Dissociation rates are then compared in plots of so-called breakdown curves, that is graphs of the percentage of intact parent complex ion as a function of relative collision energy. For different drug families such as intercalators and minor groove binders, the dissociation patterns of the negatively charged 1:1 DNA duplex: drug complexes differ from drug to drug. The possible pathways correspond to a loss of neutral drug, loss of a negatively charged drug or positively charged drug remaining bound to a single strand (e.g. minor groove binders such as netropsin). Subtle differences between drug chemical compositions can strongly modify

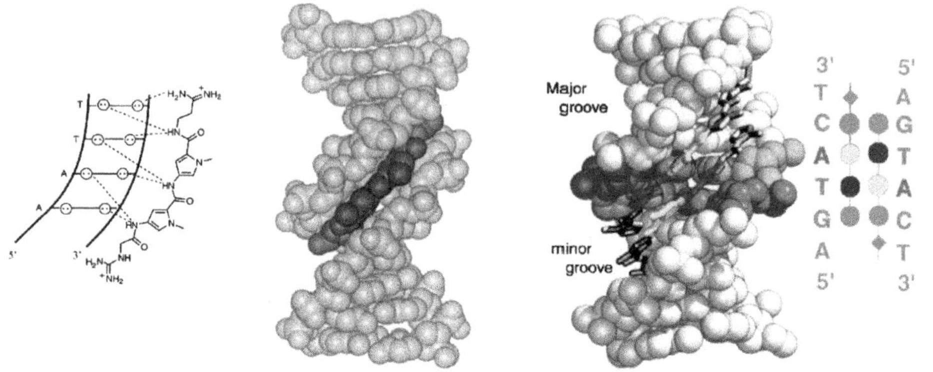

Figure 4.5.2 Left: Binding of netropsin in the minor groove of DNA. Right: Artificial polyamide (imidazole–hydroxypyrrole–pyrrole) hairpin bound in the minor groove of 5′-CCAGTACTGG-3′ (reproduced with permission from reference [90] ©2001 Elsevier).

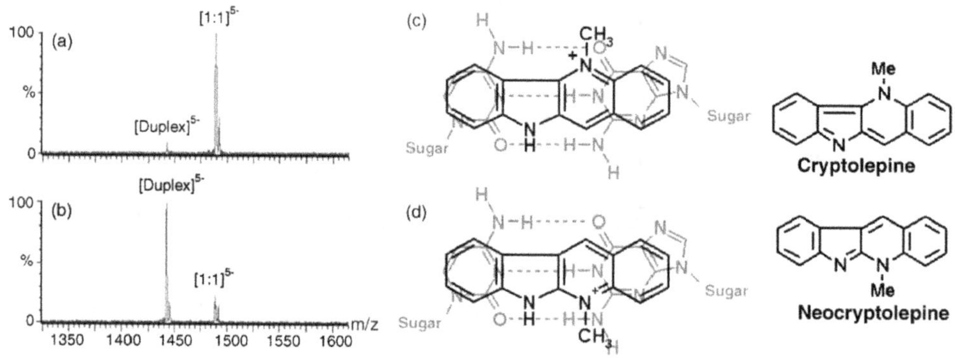

Figure 4.5.3 MS/MS spectra of the $[1:1]^{5-}$ complex of a specific Dickerson duplex with cryptolepine (a) and of the $[1:1]^{5-}$ complex of a non-specific Dickerson duplex with cryptolepine neocryptolepine (b). The superposition of the chemical structures of a G–C pair (in grey) and the drugs (in black). Note the steric hindrance between the methyl group of neocryptolepine (d) and the amino group of guanine (reproduced with permission from reference [81] ©2006 Elsevier).

their binding properties. The same situation arises for artificial gene repressors targeting precise gene sequences [87–91] (see Section 2.3.2.2) (Figure 4.5.2).

Cryptolepine and neocryptolepine are two DNA intercalators that only differ by the position of a methyl group. In cryptolepine, the N—H and N—CH_3 groups are on opposite sides of the molecule and each group can interact with a carbonyl group of the base pair (Figure 4.5.3). In neocryptolepine, the N—H and N—CH_3 groups are on the same side and a single interaction is left. Moreover, the N—CH_3 group has a steric hindrance with one of the amino group of the base pair. The Dickerson duplex/neocryptolepine is thus less stable and dissociates more rapidly than the Dickerson duplex/cryptolepine in an MS/MS mass-spectrometry experiment [81].

4.5.2 Modelling and spectroscopy of non-covalent complexes

Modelling of non-covalent complexes [92] not only allows a detailed description of the interactions that govern their formation in known systems such as DNA–netropsin but it also plays a role in the search and evaluation of new drugs [93] and in the design of molecules such as peptides [6] that will bind specific targets, for example the HIV-1 reverse transcriptase, a critical step in multi-therapies of AIDS [94]. In this chapter, after considering some examples of nucleobase/amino acid and DNA/drug complex calculations, we will present examples of spectroscopic and spectrometric studies of small and medium size metal–ligand complexes.

Recognition of DNA sequences by proteins is a crucial step in gene expression [95, 96]. Purines (A and G) interact preferentially with only a few amino acids (Gln, Arg, Asn and Lys) while pyrimidines (T and C) exhibit a much broader choice. Proton transfer in nucleobase–amino acid complexes has been considered a level for the interpretation of photoelectron spectroscopy experiments [97–101] (see Section 4.7.1.5).

We have seen in the preceding chapter that small or medium size molecules can bind to DNA through minor groove binding or intercalation between base pairs. In the latter case, intercalators bind through a combination of non-covalent stacking, van der Waals and hydrogen bonding [102] and even covalent bonding involving drug side-chains. Most intercalators possess several polar functional groups and can be possibly charged. Electrostatic interactions dominate but in the case of aromatic groups, dispersion and polarization play a great role. DFT calculations are not satisfying in that respect and the MP2 level is required. The obtained reliable results for only A—T or G—C pairs complexed with intercalator drugs such as daunomycin or the 4,6′-diaminide-2-phenylindole (DAPI) [103] can then further be used for parameterization of force-fields such as AMBER (see Section 1.3) that can be used for larger scale molecular dynamics simulations [104]. Simulations have been conducted up to 3.3 μs in the case of Dickerson duplexes not covalently bound to minor groove binders such as the naturally occurring antibiotic netropsin or DAPI [105].

A large number of quantum calculation and studies have been devoted to the prediction of complex structures involving metal ions due to their crucial role in biochemical processes [106]. Metals are able to accept electrons and act as general acids. The shell structure of the metals has an important role. For example, filled-shell metals such as zinc ($3d^{10}$) behave differently from partially filled transition metals [107]. In contrast with protons, metals present the ability to interact with several ligands. Metal ions are required in respiration or enzymatic activity [108]. Half of all proteins contain metal ions that can stabilize their folded structures. Metal–peptide interactions have been investigated at the DFT or CBS-QB3 [41, 109] and at the MP2 level [110, 111]. Metal ions also interact with nucleobases [18, 112] and their complex structure computations can be conducted up to the CCSD(T) level [22, 113].

In infrared photo-dissociation experiments of metal–amino acid and metal–peptide complex ions, the metal ion can be used a messenger [114] (see Section 2.1.3.2.2). The appearance of the K^+ ion from excited potassium-tagged amino acids (Tyr or Phe) or peptide (bradikinin 1-5 fragment Arg-Pro-Pro-Gly-Phe or Leu-enkephalin) complexes provides a signature of the infrared absorption of the weakly bound complexes [37]. For interpreting the obtained infrared photo-dissociation spectra, the PES of the complexes has first been explored with a search engine [115]. The lowest energy structures correspond to either

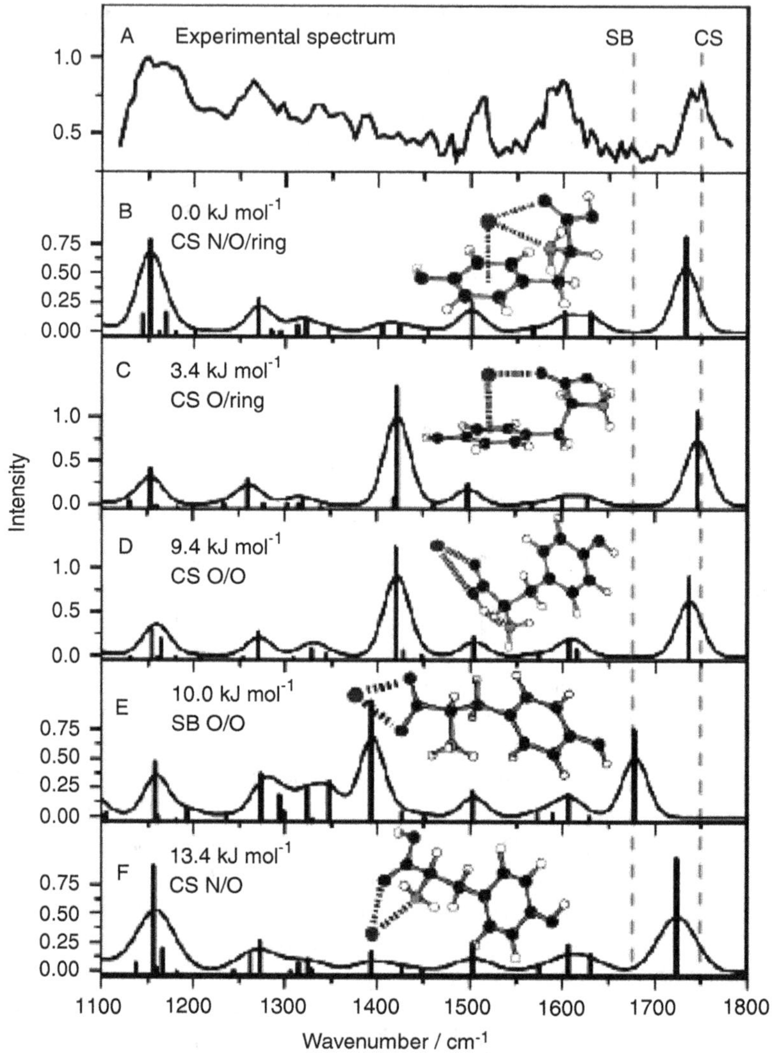

Figure 4.5.4 Experimental photo-dissociation spectra of the [tyrosine+K]$^+$ ion in the IR fingerprint region obtained with a free electron laser. The spectra of different conformers have been simulated by means of calculations at the B3LYP/6-31G(d) level. In the observed lowest energy (B) configuration, the K$^+$ ion is solvated in an N/O/ring structure. A similar situation occurs for the [phenylalanine+K]$^+$ ion (reproduced with permission from reference [37] ©2005 American Chemical Society).

charge-solvated (CS) or salt-bridge (SB) conformations. In the CS structures, the amino acid and peptide terminals are neutral whereas a proton is transferred from the C-terminal to the N-terminal in the zwitterion SB structures. The IR spectra simulations have been conducted at the DFT B3LYP/6-31G(d) level and are presented in Figures 4.5.4 and 4.5.5. In the case of the aromatic amino acids Phe and Tyr, the lowest energy configuration

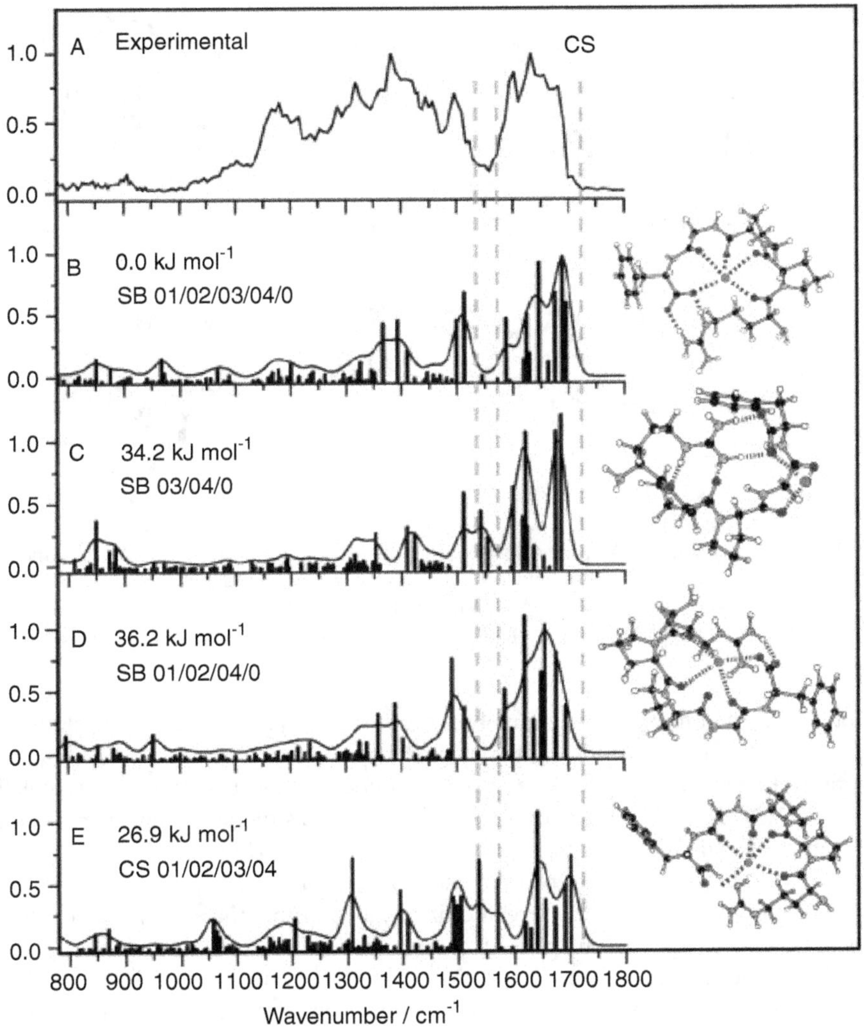

Figure 4.5.5 Experimental infrared photo-dissociation spectra of the [Arg-Pro-Pro-Gly-Phe+K]$^+$ ion in the IR fingerprint region. In the salt-bridge (SB) lowest energy configuration (B), the K^+ ion is bound to five C═O groups by electrostatic interactions (reproduced with permission from reference [37] ©2005 American Chemical Society).

(N/O/ring) corresponds to the K^+ cation bound to the amino nitrogen (N), the carbonyl (O) and to the side-chain phenyl ring (ring) by a cation–π interaction (see Section 1.1.1). The first zwitterion (SB) configuration lies 10 kJ/mol above and is not observed.

In the lowest energy SB conformation of the [Arg-Pro-Pro-Gly-Phe+K]$^+$complex ion, electrostatic interactions dominate and the K^+ ion is bound to the negatively charged deprotonated carboxylate group and the protonated guanidine group (a similar situation is encountered in the protonated arginine–glycine–aspartic acid (RGD) peptide [116].

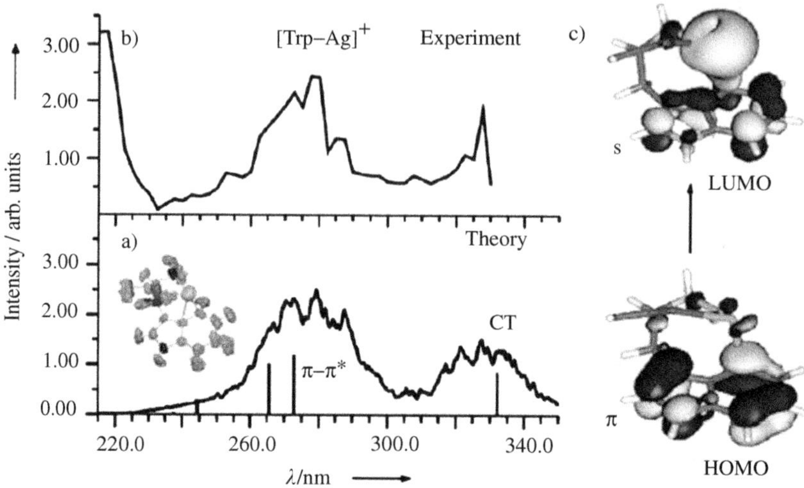

Figure 4.5.6 Experimental UV photo-dissociation spectra of the $[Trp+Ag]^+$ ion confined in a quadrupole trap. The broadening of the absorption lines of the most stable charge-solvated structure has been simulated by means of molecular dynamics in the electronic ground state and simultaneously involving transitions to excited states (reproduced with permission from reference [40] ©2006 Wiley).

Cation–π interactions have also been observed in the UV photo-dissociation study of the tryptophan–Ag^+complex [40]. The $[Trp+Ag]^+$ complex ion structures were obtained by DFT optimization of structures obtained from snapshots of molecular dynamics calculations (see Section 1.3). The UV spectra corresponding to these structures were then calculated with the time-dependent density functional theory (TD-DFT) (Figure 4.5.6).

Haemoglobin and myoglobin are fundamental oxygen transport proteins that contain haemes, assemblies of heterocyclic organic rings called porphyrins and metals (iron). The time evolution of gas-phase photo-excited metalloporphyrins has been investigated by means of photoelectron femtosecond spectroscopy [107] (see Section 2.3.2.4). Following photoexcitation of the ππ macrocycle of the porphyrin, an ultrafast charge transfer (100 fs) is observed towards the d orbital of a metal (different from zinc) through a conical intersection and is further followed by a longer transfer (1 ps) back to an empty orbital of the porphyrin.

In the case of larger systems such as metal–peptide or metal–oligosaccharide [117] (see Chapter 4.4) complexes, ion mobility measurements coupled to molecular dynamics simulation with force-fields provide information on metal binding sites and recognition mechanisms. The specific binding of the oxytoxin peptide hormone (OT) [118] with its receptor is strongly influenced by the presence of doubly charged ions such as Zn^{2+} or Cu^{2+} [54]. This nonapeptide (Cys-Tyr-Ile-Gln-Asn-Cys-Pro-Leu-Gly) has a single basic site at the N-terminus and an aminated C-terminus. Its charge-state under biological conditions is thus $(OT+H)^+$. Its secondary structure is characterized by a disulphide bridge between the two Cys residues (Figure 4.5.7). Specific recognition of OT by its receptor is governed by interactions between the side-chains of Ile 3, Gln4 and Asn5. In the gas-phase or in solution, the N-terminus charge of $(OT+H)^+$ is solvated by the peptide itself and is

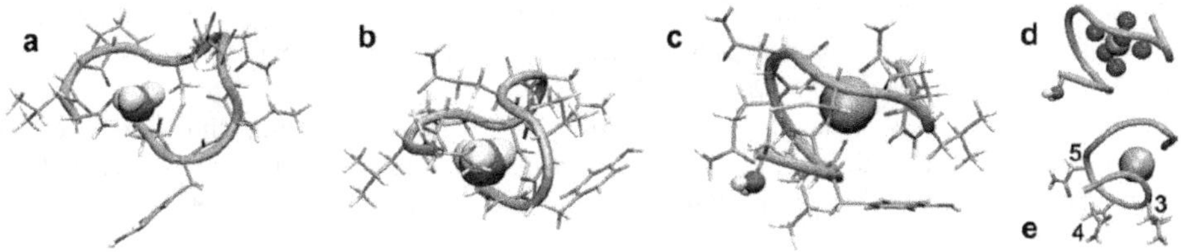

Figure 4.5.7 Examples of minimum energy structures of the [oxytoxin+H]$^+$ and [oxytoxin+Zn]$^{2+}$ ion complexes generated by molecular dynamics calculations and optimized at the DFT level. (a) and (b) respectively represent structures of [OT+H]$^+$ in explicit water solvation and without water. The calculated structure of [OT+Zn]$^{2+}$ is displayed in (c) and figures (d) and (e) respectively emphasize the coordination of Zn^{2+} with the six C=O backbone groups and the side-chain orientation of residues Ile3, Gln4 and Asn5 (reproduced with permission from reference [54] ©2005 American Chemical Society).

thus shielded from interaction with the receptor. The three residues responsible for specific recognition point in different directions and cannot form an efficient interaction network.

In presence of a Zn^{2+} ion, the (OT+H)$^+$ mass-spectrum peak disappears and only the (OT+Zn)$^{2+}$ remains present. Molecular dynamics simulations conducted with the CHARMM force-field and optimized at the DFT level show that the Zn^{2+} ion forms a nearly perfect octahedral complex through binding with six of the backbone C=O groups. This lines up the side-chains of the crucial residues 3, 4 and 5 that now form a nearly planar recognition surface and frees the positively charge N-terminus. Complexation of oxytoxin with an ion thus induces considerable structural changes and strongly favours specific recognition with the receptor.

REFERENCES

1. Di Tullio A, Reale S, De Angelico F: Molecular recognition by mass spectrometry. *Journal of Mass Spectrometry* 2005, 40:845–865.
2. Rogniaux H, Sanglier S, Strupat K, Roitel O, Ball V, Tristsch D, Branlant G, van Dorsslaer A: Mass spectrometry as a novel approach to probe cooperativity in multimeric enzymatic systems. *Analytica Biochemistry* 2001, 291:48–61.
3. Hofstadler SA, Griffey RH: Analysis of non covalent complexes of DNA and RNA by mass spectrometry. *Chemical Reviews* 2001, 101:377–390.
4. Heck AJR, Jorgensen TJD: Vancomycin in vacuo. *International Journal of Mass Spectrometry* 2004, 236:11–23.
5. Geoghegan KF, Kelly MA: Biochemical applications of mass spectrometry in pharmaceutical drug discovery. *Mass Spectrometry Reviews* 2005, 24:347–366.
6. Zhao L, Chmielewski J: Inhibiting protein–protein interactions using designed molecules. *Current Opinion in Structural Biology* 2005, 15:31–34.
7. Loo A, Xhe JX, Cody WL: Higher order structure in the gas phase reflects solution structure. *Journal of the American Chemical Society* 1998, 120:4542–4543.
8. Pramanik BN, Bartner PL, Mirza UA, Liu YH, Ganguly AK: Electrospray ionization mass spectrometry for the study of non-covalent complexes: an emerging technology. *Journal of Mass Spectrometry* 1998, 33:911–920.
9. Miranker AD: Protein complexes and analysis of their assembly by mass spectrometry. *Current Opinion in Structural Biology* 2000, 10:601–610.

10. Nesatyy VJ, Laskin J: Dissociation of noncovalent protein complexes by triple quadrupole tandem mass spectrometry: comparison of Monte Carlo simulation and experiment. *International Journal of Mass Spectrometry* 2002, 221:245–262.
11. Sobott F, Robinson CV: Characterising electrosprayed biomolecules using tandem-MS: the noncovalent GroEL chaperonin assembly. *International Journal of Mass Spectrometry* 2004, 236:25–32.
12. Sobott F, Robinson CV: Protein complexes gain momentum. *Current Opinion in Structural Biology* 2002, 12:729–734.
13. Schafer M, Schmuck C, Geiger L, Chalmers MJ, Hendrickson CL, Marshall AG: Structurally related non-covalent complexes examined by quadrupole ion trap (QIT) MS2 and infrared multiphoton dissociation Fourier transform ion cyclotron resonance mass spectrometry IRMPD-FT-ICR MS: evidence for salt-bridge structures in the gas phase. *International Journal of Mass Spectrometry* 2004, 237:33–45.
14. Benesch JLP, Robinson CV: Mass spectrometry of macromolecular assemblies. *Current Opinion in Structural Biology* 2006, 16:245–251.
15. Gabelica V, De Pauw E: Comparison between solution-phase stability and gas-phase kinetic stability of oligodeoxynucleotide duplexes. *Journal of Mass Spectrometry* 2001, 36:397–402.
16. Gabelica V, De Pauw E: Comparison of the collision-induced dissociation of duplex DNA at different collision regimes: evidence for a multistep dissociation mechanism. *Journal of the American Society for Mass Spectrometry* 2002, 13:91–98.
17. Pedersen DB, Zgierski MZ, Denommee S, Simard B: Photoinduced charge-transfer dehydrogenation in a gas-phase metal–DNA base complex: al-cytosine. *Journal of the American Chemical Society* 2002, 124:6686–6692.
18. Baker ES, Gidden J, Ferzoco A, Bowers MT: Sodium stabilization of dinucleotide multiplexes in the gas phase. *Physical Chemistry Chemical Physics* 2004, 6:2786–2795.
19. Yang ZB, Rodgers MT: Influence of methylation on the properties of uracil and its noncovalent interactions with alkali metal ions: threshold collision-induced dissociation and theoretical studies. *International Journal of Mass Spectrometry* 2005, 241:225–242.
20. Yang ZB, Rodgers MT: Influence of halogenation on the properties of uracil and its noncovalent interactions with alkali metal ions. Threshold collision-induced dissociation and theoretical studies. *Journal of the American Chemical Society* 2004, 126:16217–16226.
21. Lamsabhi AM, Alcami M, Mo O, Yanez M, Tortajada J: Gas-phase depronation of uracil–Cu^{2+} and thiouracil–Cu^{2+} complexes. *Journal of Physical Chemistry A* 2006, 110:1943–1950.
22. Prado MAS, Garcia E, Martins JBL: Theoretical study of cytosine–Mg complex. *Chemical Physics Letters* 2006, 418:264–267.
23. Keller KM, Breeden MM, Zhang JH, Ellington AD, Brodbelt JS: Electrospray ionization of nucleic acid aptamer/small molecule complexes for screening aptamer selectivity. *Journal of Mass Spectrometry* 2005, 10:1327–1337.
24. Rodgers MT, Armentrout PB: Noncovalent metal-ligand bond energies as studied by threshold collision-induced dissociation. *Mass Spectrometry Reviews* 2000, 19:215–247.
25. Rodgers FM, Armentrout PB: Non covalent interactions of nucleic acid bases (uracil, thymine, and adenine) with alkali metal ions. Threshold collision-induced dissociation and theoretical studies. *Journal of the American Chemical Society* 2000, 122:8548–8558.
26. Taraszka JA, Li JW, Clemmer DE: Metal-mediated peptide ion conformations in the gas phase. *Journal of Physical Chemistry B* 2000, 104:4545–4551.
27. Talley JM, Cerda BA, Ohanessian G, Wesdemiotis C: Alkali metal ion binding to amino acids versus their methyl esters: affinity trends and structural changes in the gas phase. *Chemistry: A European Journal* 2002, 8:1377–1388.
28. Rogalewicz F, Hoppilliard Y, Ohanessian G: Structures and fragmentations of zinc(II) complexes of amino acids in the gas phase-IV. Solvent effect on the structure of electrosprayed ions. *International Journal of Mass Spectrometry* 2003, 227:439–451.

29. Rogalewicz F, Hoppilliard Y, Ohanessian G: Structures and fragmentations of zinc(II) complexes of amino acids in the gas phase. III. Rearrangement versus desolvation in the electrospray formation of the glycine–zinc complex. *International Journal of Mass Spectrometry* 2001, 206:45–52.
30. Gapeev A, Dunbar RC: Na^+ affinities of gas-phase amino acids by ligand exchange equilibrium. *International Journal of Mass Spectrometry* 2003, 228:825–839.
31. Woods KK, Lan T, McLaughlin LW, Williams LD: The role of minor groove functional groups in DNA hydration. *Nucleic Acids Research* 2003, 31:1536–1540.
32. Lemoff AS, Wu CC, Bush MF, Williams ER: Binding energies of water to doubly hydrated cationized glutamine and structural analogues in the gas phase. *Journal of Physical Chemistry A* 2006, 110:3662–3669.
33. Lemoff AS, Williams ER: Binding energies of water to lithiated valine: formation of solution-phase structure in vacuo. *Journal of the American Society for Mass Spectrometry* 2004, 15: 1014–1024.
34. Lemoff AS, Bush MF, Wu CC, Williams ER: Structures and hydration enthalpies of cationized glutamine and structural analogues in the gas phase. *Journal of the American Chemical Society* 2005, 127:10276–10286.
35. Lemoff AS, Bush MF, Williams ER: Structures of cationized proline analogues: evidence for the zwitterionic form. *Journal of Physical Chemistry A* 2005, 109:1903–1910.
36. Lemoff AS, Bush MF, Williams ER: Binding energies of water to sodiated valine and structural isomers in the gas phase: the effect of proton affinity on zwitterion stability. *Journal of the American Chemical Society* 2003, 125:13576–13584.
37. Polfer NC, Paizs B, Snoek LC, Compagnon I, Suhai S, Meijer G, von Helden G, Oomens J: Infrared fingerprint spectroscopy and theoretical studies of potassium ion tagged amino acids and peptides in the gas phase. *Journal of the American Chemical Society* 2005, 127:8571–8579.
38. Kozlowski H, Kowalik-Jankowska T, Jezowska-Bojczuk M: Chemical and biological aspects of Cu^{2+} interactions with peptides and aminoglycosides. *Coordination Chemistry Reviews* 2005, 249:2323–2334.
39. Kapota C, Ohanessian G: The low energy tautomers and conformers of the dipeptides HisGly and GlyHis and of their sodium ion complexes in the gas phase. *Physical Chemistry Chemical Physics* 2005, 7:3744–3755.
40. Antoine R, Tabarin T, Broyer M, Dugourd P, Mitric R, Bonacic-Koutecky V: Optical properties of gas-phase tryptophan-silver cations: charge transfer from the indole ring to the silver atom. *ChemPhysChem* 2006, 7:524–528.
41. Remko M, Rode BM: Effect of metal ions (Li^+, Na^+, K^+, Ca^+, Ni^+, Cu^+, and Zn^+) and water coordination on the structure of glycine and zwitterionic glycine. *Journal of Physical Chemistry A* 2006, 1960–1967.
42. Wang P, Polce MJ, Bleiholder C, Paizs B, Wesdemiotis C: Sructural characterization of peptides via tandem mass spectrometry of their dilithiated monocations. *International Journal of Mass Spectrometry* 2006, 249:45–59.
43. Fung YME, Liu H, Chan TW: Electron capture dissociation of peptides metalated with alkaline-earth metal ions. *Journal of the American Society of Mass Spectrometry* 2006, 17:757–771.
44. Heck AJR, van den Heuvel RHH: Investigation of intact protein complexes by mass spectrometry. *Mass Spectrometry Reviews* 2004, 23:368–389.
45. Aggarwal AK, Doudna JA: Protein–nucleic acid interactions: unlocking mysteries old and new. *Current Opinion in Structural Biology* 2005, 15:65–67.
46. Luo Z, Li YM, Quiang W, Zhao YF, Abe H, Nemoto T, Quin XR, Nakanishi H: Detection of specific noncovalent interaction of peptide with DNA by MALDI-TOF. *Journal of the American Society of Mass Spectrometry* 2004, 15:28–31; Cheng AC, Chen WW, Fuhrmann CN, Frankel, AD: Recognition of nucleic acid bases and base-pairs by hydrogen bonding to amino acid side chains. *Journal of Molecular Biology* 2003, 327:781–796.

47. Bligh SAW, Haley T, Lowe PN: Measurement of dissociation constants of inhibitors binding to Src SH2 domain protein by non-covalent electrospray ionization mass spectrometry. *Journal of Molecular Recognition* 2003, 16:139–147.
48. Raha K, van der Vaart AJ, Riley KE, Peters MB, Westerhoff LM, Kim H, Merz KM: Pairwise decomposition of residue interaction energies using semiempirical quantum mechanical methods in studies of protein–ligand interaction. *Journal of the American Chemical Society* 2005, 127:6583–6594.
49. Breuker K: The study of protein–ligand interactions by mass spectrometry: a personal view. *International Journal of Mass Spectrometry* 2004, 239:33–41.
50. Breuker K: New mass spectrometric methods for the quantification of protein–ligand binding in solution. *Angewandte Chemie-International Edition* 2004, 43:22–25.
51. Lengqvist J, Mata de Urquiza A, Perlmann T, Sjovall J, Griffits WJ: Specificity of receptor–ligand interactions and their effect on dimerisation as observed by electrospray mass spectrometry: bile acids form stable adducts to the RXR. *Journal of Mass Spectrometry* 2005, 40:1448–1461; Armstrong KA, Tidor B, Cheng A. Optimal charges in lead progression: a structure-based neuramidase case study. *Journal of Medicinal Chemistry* 2006, 49:2470–2477.
52. von Seggern CE, Cotter RJ: Study of peptide-sugar non-covalent complexes by infrared atmospheric pressure matrix-assisted laser desorption/ionization. *Journal of Mass Spectrometry* 2004, 39:736–742.
53. Wang WJ, Kitova EN, Sun JX, Klassen JS: Blackbody infrared radiative dissociation of nonspecific protein–carbohydrate complexes produced by nanoelectrospray ionization: the nature of the noncovalent interactions. *Journal of the American Society for Mass Spectrometry* 2005, 16:1583–1594.
54. Liu DF, Seuthe AB, Ehrler OT, Zhang XH, Wyttenbach T, Hsu JF, Bowers MT: Oxytocin-receptor binding: why divalent metals are essential. *Journal of the American Chemical Society* 2005, 127:2024–2025.
55. Nesatyy VJ: Gas-phase-binding of non-covalent protein complexes between bovine pancreatic trypsin inhibitor and its target enzymes studied by electrospray ionization tandem mass spectrometry. *Journal of Mass Spectrometry* 2001, 36:950–959.
56. Cederkvist FH, Zamfir AD, Bahrke S, Eijsink VGH, Sorlie M: Identification of a high-affinity-binding oligosaccharide by nanoelectrospray quadrupole time-of-flight tandem mass spectrometry of a non-covalent enzyme-ligand complex. *Angewandte Chemie-International Edition* 2006, 45:2429–2434.
57. Loo A: Studying noncovalent protein complexes by electrospray ionization mass spectrometry. *Mass Spectrometry Reviews* 1997, 16:1–23.
58. Sannes-Lowery KA, Mei HY, Loo JA: Studying aminoglycoside antibiotic binding to HIV-1 TAR RNA by electrospray ionization mass spectrometry. *International Journal of Mass Spectrometry* 1999, 193:115–122.
59. Bacher G, Szymansli WW, Kaufman SL, Zöllner P: Charge-reduced nano electrospray ionization combined with differential mobility analysis of peptides, proteins, glycoproteins, noncovalent protein complexes and viruses. *Journal of Mass Spectrometry* 2001, 36:1038–1052.
60. Zhan D, Fenn JB: Gas phase hydration of electrospray ions from small peptides. *International Journal of Mass Spectrometry* 2002, 120:1–10.
61. Catalina MI, de Mol NJ, Fischer JE, Heck AJR: Probing factors affecting gas-phase stabilities of noncovalent complexes formed by peptides bound to the Grb2 SH2 domain protein. *Physical Chemistry Chemical Physics* 2004, 6:2572–2579.
62. Takamizawa A, Itoh Y, Osawa R, Iwasaki N, Nishimura Y, Akashi S, Hiraoka K: Selective dissociation of non-covalent bonds in biological molecules by laser spray. *Journal of Mass Spectrometry* 2004, 39:1053–1058.
63. Catalina MI, van den Heuvel RHH, van Duijn E, Heck AJR: Decharging of globular proteins and protein complexes in electrospray. *Chemistry: A European Journal* 2005, 11:960–968.

64. Gabelica V, De Pauw E: Internal energy and fragmentation of ions produced in electrospray sources. *Mass Spectrometry Reviews* 2005, 24:566–587.
65. Di Marco VB, Giorgio Bombi G: Elecrospray mass spectrometry (ESI-MS) in the study of metal–ligand solution equilibria. *Mass Spectrometry Reviews* 2006, 25:347–379.
66. Friess SD, Daniel JM, Hartmann R, Zenobi R: Mass spectrometric noncovalent probing of amino acids in peptides and proteins. *International Journal of Mass Spectrometry* 2002, 219:269–281.
67. Gabelica V, Schulz E, Karas M: Internal energy build-up in matrix-assisted laser desorption/ionization. *Journal of Mass Spectrometry* 2004, 39:579–593.
68. Friess SD, Daniel JM, Zenobi R: Probing the surface accessibility of proteins with non covalent receptors and MALDI mass spectrometry. *Physical Chemistry Chemical Physics* 2004, 6:2664–2675.
69. Woods AS, Fuhrer K, Gonin M, Egan T, Ugarov M, Gillig KJ, Schultz JA: Angiotensin II-acetylcholine noncovalent complexes analyzed with MALDI-ion mobility-TOF MS. *Journal of Biomolecular Techniques* 2003, 14:1–8.
70. Ohara K, Smietana MV, J.J.: Characterization of specific non covalent complexes between guanadinium derivatives and single-stranded DNA by MALDI. *Journal of the American Society for Mass Spectrometry* 2006, 17:283–291.
71. Kleinekofort W, Pfenninger A, Plomer T, Griesinger C, Brutschy B: Observation of noncovalent complexes using laser-induced liquid beam ionization desorption. *International Journal of Mass Spectrometry and Ion Processes* 1996, 156:195–202.
72. Wattenberg A, Sobott F, Barth HD, Brutschy B: Studying noncovalent protein complexes in aqueous solution with laser desorption mass spectrometry. *International Journal of Mass Spectrometry* 2000, 203:49–57.
73. Morgner N, Barth HD, Brutschy B: A new way to detect noncovalently bonded complexes of biomolecules from liquid micro-droplets by laser mass spectrometry. *Australian Journal of Chemistry* 2006, 59:109–114.
74. Lorenz SA, Maziarz EP, Wood TD: Using solution phase hydrogen/deuterium (H/D) exchange to determine the origin of non-covalent complexes observed by electrospray ionization mass spectrometry: in solution or in vacuo. *Journal of the American Society of Mass Spectrometry* 2001, 12:795–804.
75. Zhu MM, Chitta R, Gross M: PLIMSTEX: a novel mass spectrometric method for the quantification of protein–ligand interactions in solution. *International Journal of Mass Spectrometry* 2005, 240:213–220.
76. Laskin J, Futrell JH: Entropy is the major driving force for fragmentation of proteins and protein–ligand complexes in the gas phase. *Journal of Physical Chemistry A* 2003, 107:5836–5839.
77. Schafer M, Schmuck C, Heil M, Cooper HJ, Hendrickson CL, Chalmers MJ, Marshall AG: Determination of the activation energy for unimolecular dissociation of a non-covalent gas-phase peptide: substrate complex by infrared multiphoton dissociation Fourier transform ion cyclotron resonance mass spectrometry. *Journal of the American Society for Mass Spectrometry* 2003, 14:1282–1289.
78. Haselmann KF, Jorgensen TJD, Budnik BA, Jensen F, Zubarev RA: Electron capture dissociation of weakly bound polypeptide polycationic complexes. *Rapid Communications in Mass Spectrometry* 2002, 16:2260–2265.
79. Fagerquist CK, Hudgins RR, Emmett MR, Hakansson K, Marshall AG: An antibiotic linked to peptides and proteins is released by electron capture dissociation Fourier transform ion cyclotron resonance mass spectrometry. *Journal of the American Society for Mass Spectrometry* 2003, 14:302–310.
80. Shields SJ, Oyeyemi O, Lightstone FC, Balhorn R: Mass spectrometry and non-covalent protein–ligand complexes: conformation of binding sites and changes in tertiary structure. *Journal of the American Society of Mass Spectrometry* 2003, 14:460–470.

81. Rosu F, Pirotte S, De Pauw E, Gabelica V: Positive and negative-mode ESI-MS and MS-MS for studying drug-DNA complexes. *International Journal of Mass Spectrometry* 2006, 253:156–171.
82. Gabelica V, Rosu F, Houssier C, De Pauw E: Gas phase thermal denaturation of an oligonucleotide duplex and its complexes with minor groove binders. *Rapid Communications in Mass Spectrometry* 2000, 14:464–467.
83. Carrasco C, Rosu F, Gabelica V, Houssier C, De Pauw E, Garbay-Jaureguiberry C, Roques B, Wilson WD, Chaires JB, Waring MJ: Tight binding of the antitumor drug ditercalinium to quadruplex DNA. *ChemBioChem* 2002, 3:1235–1241.
84. Cheng X, van Breemen RB: Mass spectrometry-based screening for inhibitors of β-amyloid protein aggregation. *Analytical Chemistry* 2005, 77:7012–7015.
85. Bazoti FN, Tsarbopoulos A, Markides KE, Bergquist J: Study of the non-covalent interaction between amyloid-β-peptide and melatonin using electrospray mass spectrometry. *Journal of Mass Spectrometry* 2005, 40:182–192.
86. Lifshitz C, Laskin J: *Principles of mass spectrometry applied to biomolecules*: Wiley-VCH Hoboken, NJ 2006.
87. White S, Szewczyk JW, Turner JM, Baird EE, Dervan PB: Recognition of the four Watson–Crick pairs in the DNA minor groove by synthetic ligands. *Nature* 1998, 291:468–471.
88. Warren CL, Kratochvil NCS, Hauschild KE, Foister S, Brezinski ML, Dervan PB, Phillips GN, Ansari AZ: Defining the sequence-recognition profile of DNA-binding molecules. *Proceedings of the National Academy of Sciences of the United States of America* 2006, 103:867–872.
89. Doss RM, Marques MA, Foister S, Dervan PB: DNA minor-groove recognition by 3-methylthiophene/pyrrole pair. *Chemistry and Biodiversity* 2004, 1:886–899.
90. Dervan PB: Molecular recognition of DNA by small molecules. *Bioorganic and Medicinal Chemistry* 2001, 9:2215–2235.
91. Chenoweth DM, Popowski JA, Marques MA, Dervan PB: Programmable oligomers targeting 5′-GGGG-3′ in the minor groove of DNA and NF-B binding inhibition. *Bioorganic and Medicinal Chemistry* 2007, 15:759–770.
92. Janin J, Levitt M: Theory and simulation. Accuracy and reliability in modelling proteins and complexes. *Current Opinion in Structural Biology* 2006, 16:139–141.
93. Gohlke H, Klebe G: Approaches to the description and prediction of the binding affinity of small-molecule ligands to macromolecular receptors. *Angewandte Chemie-International Edition* 2002, 41:2644–2676.
94. Campbell W, Kleiman L, Baranyi L, Li ZQ, Khorchid A, Fujita E, Okada N, Okada H: A novel genetic algorithm for designing mimetic peptides that interfere with the function of a target molecule. *Microbiology Immunology* 2002, 46:211–215.
95. Sarai A, Kono H: Protein–DNA recognition patterns and predictions. *Annual Review of Physical Chemistry* 2005, 34:379–398.
96. Yang W, Van Duyne WY: Protein–nucleic acid interactions: from Argonaute to X(PF). *Current Opinion in Structural Biology* 2006, 16:1–4.
97. Gutowski M, Dabkowska I, Rak J, Xu S, Nilles JM, Radisic D, Bowen KH: Barrier-free intermolecular proton transfer in the uracil–glycine complex induced by excess electron attachment. *European Physical Journal D* 2002, 20:431–439.
98. Dabkowska I, Rak J, Gutowski M, Nilles JM, Stokes ST, Radisic D, Bowen KH: Barrier-free proton transfer in anionic complex of thymine with glycine. *Physical Chemistry Chemical Physics* 2004, 6:4351–4357.
99. Dabkowska I, Rak J, Gutowski M, Nilles JM, Stokes ST, Bowen KH: Barrier-free intermolecular proton transfer induced by excess electron attachment to the complex of alanine with uracil. *Journal of Chemical Physics* 2004, 120:6064–6071.
100. Radisic D, Bowen KH, Dabkowska I, Storoniak P, Rak J, Gutowski M: AT base pair anions versus (9-methyl-A)(1-methyl-T) base pair anions. *Journal of the American Chemical Society* 2005, 127:6443–6450.

101. Dabkowska I, Gutowski M, Rak J: Interaction with glycine increases stability of a mutagenic tautomer of uracil. A density functional theory study. *Journal of the American Chemical Society* 2005, 127:2238–2248.
102. Dolenc J, Borstnick U, Hodoscek M, Koller J, Janezic D: An ab initio QM/MM study of the conformational stability of complexes formed by netropsin and DNA. The importance of van der Waals interactions and hydrogen bonding. *Journal of Molecular Structure (Theochem)* 2005, 718:77–85.
103. Reha D, Kabelac M, Ryjacek F, Sponer J, Sponer JE, Elstner M, Suhai S, Hobza P: Intercalators. 1. Nature of stacking interactions between intercalators (ethidium, daunomycin, ellipticine, and 4′,6′-diaminide-2-phenylindole) and DNA base pairs. Ab initio quantum chemical, density functional theory, and empirical potential study. *Journal of the American Chemical Society* 2003, 125:5581–5581.
104. Gorfe AA, Caflish A, Jelasarov I: The role of flexibility and hydration on the sequence-specific DNA recognition by the Tn916 integrase protein: a molecular dynamics analysis. *Journal of Molecular Recognition* 2004, 17:120–131.
105. Rueda M, Luque FJ, Orozco M: Nature of minor-groove binders: DNA complexes in the gas phase. *Journal of the American Chemical Society* 2005, 127:11690–11698.
106. Dudev T, Lim C: Principles governing Mg, Ca, and Zn binding and selectivity in proteins. *Chemical Reviews* 2003, 103:773–787.
107. Sorgues S, Poisson L, Raffael K, Krim L, Soep B, Shafizadeh N: Femtosecond electronic relaxation of excited metalloporphyrins in the gas phase. *Journal of Chemical Physics* 2006, 124:114302–114310.
108. Deerfield DW, Carter CW, Pedersen LG: Models for protein-zinc ion binding sites. II The catalytic sites. *International Journal of Quantum Chemistry* 2001, 83:150–165.
109. Montgomery JA, Frisch MJ, Ochterski JW, Peterson GAJ: A complete basis set model chemistry. VI. Use of density functional geometries and frequencies. *Journal of Chemical Physics* 1999, 110:2822–2827.
110. Rogalewicz F, Louazel G, Hoppilliard Y, Ohanessian G: Structures and fragmentations of electrosprayed Zn(II) complexes of carboxylic acids in the gas phase. Isomerisation versus desolvation during the last desolvation step. *International Journal of Mass Spectrometry* 2003, 228:779–795.
111. Moision RM, Armentrout PB: An experimental and theoretical dissection of potassium cation/glycine interactions. *Physical Chemistry Chemical Physical* 2004, 6:2588–2599.
112. Zhao Y, Zhou L: Computational study of hydrogen-bonded complexes of the pyrimidinic bases of RNA and L-leucine. *Journal of Molecular Structure (Theochem)* 2005, 726:31–37.
113. Matsui T, Shigeta Y, Hirao K: Influence of Pt complex binding on the guanine–cytosine pair: a theoretical study. *Chemical Physics Letters* 2006, 423:331–334.
114. Kapota C, Lemaire J, Maitre P, Ohanessian G: Vibrational signature of charge solvation vs. salt bridge isomers of sodiated amino acids in the gas phase. *Journal of the American Chemical Society* 2004, 126:1836–1842.
115. Paizs B, Suhai S, Hargittai B, Hruby V, Somogyi A: Ab initio and MS/MS studies on protonated peptides containing basic and acidic amino acid residues I. Solvated proton vs. salt-bridged structures and the cleavage of the terminal amide bond of protonated RD-NH2. *International Journal of Mass Spectrometry* 2002, 219:203–232.
116. Grégoire G, Gaigeot MP, Marinica DC, Lemaire J, Schermann JP, Desfrancois C: Gas-phase protonated peptides studied with IRMPD spectroscopy experiments and quantum molecular dynamics simulations conducted at 300K. *Physical Chemistry Chemical Physics* 2007.
117. Jin L, Barran PE, Deakin JA, Lyon M, Uhrin D: Conformation of glycoaminoglycans by ion mobility mass spectrometry and molecular modelling. *Physical Chemistry Chemical Physics* 2005, 7:3464–3471.
118. Kleinnijenhuis AJ, Mihalca R, Heeren RMA, Heck AJR: Atypical behavior in the electron capture induced dissociation of biologically relevant transition metal complexes of the peptide hormone oxytocin. *International Journal of Mass Spectrometry* 2006, 253:217–224.

– 4.6 –

Chiral Molecular Systems

In this chapter, we first briefly examine some hypothesis concerning the origin of chirality of molecules constituting the building blocks of Life and their precursors in the gas-phase. We then consider the influence of chirality upon non-covalent complex structures as well as spectroscopic and mass-spectrometric differentiation between chiral species.

4.6.1 Origin of chirality and prebiotic molecules

The origin of the homochirality of biological molecules, that is the use in living organisms of only left-handed or L-amino acids and right-handed or D-sugars, is still the subject of debates since the discovery of the relationship between optical rotation and molecular asymmetry by Louis Pasteur in 1848. Biotic theories suppose that Life started from equal mixtures of enantiomers (racemates) and that Evolution favoured the use of specific enantiomers whereas abiotic theories presume that homochirality came from chemical evolution and asymmetry preceded the birth of Life [1–5]. One of the arguments in favour of the latter theories is that Life seems to be result of highly stereoselective asymmetric reactions between chiral molecules like amino acids, peptides, DNA and RNA components as well as sugars. Some abiotic theories invoke an accidental deviation from the racemate that would have been further amplified exponentially by an autocatalytic process [6].

Building blocks of Life: amino acids [7], nucleobases [8], sugars [9] most probably derived from elementary molecules such as water, ammonia, hydrogen cyanide, formaldehyde, formamide, formic acid, urea, etc. [10]. Although the Miller–Urey experiment [11] does not constitute a proof, it provides clues about possible prebiotic scenarios [12]. In this experiment, a mixture of gases (methane, ammonia, hydrogen, carbon monoxide, nitrogen ...) representing the supposed primitive atmosphere has been admitted and driven in a closed circuit in presence of boiling water and subjected to an electrical discharge. After a few days of continuous operation, the solution was found to contain aldehydes and amino acids (both left-handed and right-handed).

Since this first demonstration, production of amino acids from simple molecules observed in interstellar molecular clouds or atmospheres of certain planets has been observed in a variety of experimental conditions comprising, for example proton bombardment simulating the action of cosmic rays [13], UV photolysis. In this last case, the UV photochemical stability of the prebiotic molecules under the intense short-wavelength irradiation either in Space or on Earth has probably played an important selection role. This has been studied in great detail in the case of the nucleobase pairing in absence of DNA backbone [8] (see Section 4.1.2). Recent laboratory experiments [14], calculations [15], spectroscopy [16] as well as mass-spectrometric measurements of meteorite content [1] and possible observations in the interstellar medium [7] have shown that precursors and building blocks of biomolecules can indeed be produced in Space. For example, slow electron (0.5–20 eV) bombardment of formic or acetic acid molecules deposited on films with ammonia at 25°K simulating interstellar ices lead to the formation of glycine [17].

Irradiation by linearly polarized light from neutron or faint stars and planet surface or parity violation in weak interactions [18] are among the large number of hypothesis considered for the origin of chirality of molecules produced in Space or on Earth [19]. Enantiomeric excesses in solid leucine have been induced in a laboratory experiment by VUV asymmetric photochemical reactions under realistic interstellar and circumstellar conditions [20]. This supports the assumption that enantioselective photolysis may have occurred on ice grains irradiated by circularly polarized UV light. A recent attractive possibility is based on the observation of asymmetric amplification leading to enantiomeric excesses from racemic mixture in separation between L- and D-proline in ternary mixtures with DMSO (see Section 5.2) [20]. Of course, different hypothesis [2] such as inclusion polymerization in chiral clays, action of pressure on racemic mixtures or asymmetric catalysis on inorganic surfaces have been invoked for differentiation in condensed-phase but are out of the scope of this brief review.

4.6.2 Chiral molecules and complexes

In this chapter, the possibility of direct distinction between isolated enantiomers through their individual physical properties [21] is discussed first. In practice, it turns out that discrimination between chiral isomers requires an external agent which is chiral (a right or left glove establishes a strong distinction between left and right hands). In a mixture of two chiral species A and B with respective enantiomeric forms R_A or S_A and R_B or S_B, homochiral (R_A–R_B or S_A–S_B) or heterochiral (R_A–S_B or R_B–S_A) complexes (diastereoisomers) can be formed. Enantioselective reactions proceed through the temporary formation of diastereoisomers. Gas-phase studies of the spectroscopic properties of diastereoisomer complexes [22–24] and diastereoisomeric molecules [25] offer the opportunity to investigate the intra and intermolecular interactions responsible for chiral recognition and intervening in drug design. If a drug is chiral (almost all drugs derived from natural products are chiral), its enantiomers do not interact similarly with the receptor and can thus have critically different effects [26]. Combining mass-spectroscopy and molecular modelling will be considered in Section 4.6.3 with emphasis on chiral differentiation between L- and D-amino acids.

4.6.2.1 Spectroscopic differentiation between chiral species

4.6.2.1.1 Spectroscopic differentiation between enantiomers

Among the different possible explanations of the origin of the symmetry breaking between left- and right-handed biomolecules, deterministic abiotic theories consider the intrinsic energy difference between enantiomers as due to parity violation in electroweak forces. Electrons and nucleons are held together by Coulomb interactions (exchange of photons as vectors) which are strictly identical in two mirror-image species. However, electrons and nucleons are also sensitive to electroweak interactions [27] (exchange of Z-bosons as vectors) that do not conserve parity. This has led to the prediction of a very small difference between electronic energies of enantiomers of a given chiral molecule. This may in turn induce extremely minute differences between vibrational frequencies and Boltzmann populations [18, 28]. Life relying on self-replicating processes, the very small initial differences may have been further amplified and lead to sizable enantiomeric discrimination. A comprehensive review can be found in reference [29]. Theoretical calculations have been performed for amino acids and sugars and, for example predicted that L-alanine should have a lower energy than D-alanine by some 10^{-16} kJ/mol [30]. Similarly, α-helix and β-sheets of L-amino acids should have lower energies than those of D-amino acids. In principle, such predictions should be easier to verify on isolated molecules, in absence of any environmental spurious effect. However, experimental tests conducted on camphor and CHFClBr with respective spectroscopic resolutions of 10^{-8} and 10^{-13} provided negative results indicating that this effect, even for different isotopic species is still too small for being observable in molecules only containing C, N, O and H atoms. Since parity-violation effects increase with the atomic number Z as Z^3, introducing iodine [31] or rhenium [32] atoms as chiral centres might be a possibility for observation of those effects.

4.6.2.1.2 Spectroscopic differentiation between diastereoisomeric species

Enantioselective reactions result from the formation of intermediate diastereoisomers with different intermolecular interactions. Those chiral complexes that are transient in solution become stable in cold supersonic beams and can be studied by laser-induced fluorescence [23], REMPI spectroscopy [22, 25, 33–36] and Fourier-transform infrared spectroscopy [23, 24, 37]. Chiral recognition rests on the difference between specific interactions leading to different stabilities between homo and heterochiral complexes that can be predicted [4]. Determination of binding energies of complexes is one of the most severe tests of structure prediction [38] and is very useful for interpretation of experimental observations. Even in the case of neutral dimers, experimental determination of binding energies turns out to be difficult and can be performed with two-colour R2PI measurements if one of the constituents (here A representing either R_A or S_A) contains a UV chromophore (see Section 2.2) and is complexed by B (either R_B or S_B) [34]. A first photon ν_1 selectively excites one of the diastereoisomers A···B and the second photon ν_2 ionizes it (Figure 4.6.1). The binding energy $D_{A\cdots B}$ of A···B is the difference between the apparition potential $AP[(A\cdots B)^+ \rightarrow A^+ + B]$ of fragment A^+ minus the adiabatic ionization potential $IP_{ad}(A)$ of bare A. For example, in the case of the two diastereoisomeric complexes of *R*-1-phenylethanol with *R*- and *S*-butan-2-ol, the homochiral species is more stable than the hetero-chiral complex by 300 ± 175 meV.

4.6.2.1.3 Influence of chirality upon conformational preferences of short peptides

We have seen in Chapter 4.3 that UV/IR spectroscopy is very powerful for studying chiral molecules such as neurotransmitters [35, 39]. Homo and heterodipeptides containing phenylalanine as the UV chromophore and valine have also been investigated by means of IR/R2PI spectroscopy [25]. The R2PI spectra of the hetero-dipeptides L-Val-L-Phe (LL-VF) and D-Val-L-Phe (DL-VF) are different. UV–UV double resonance spectra show the presence of three conformers for LL-VF and four in the case of DL-VF. Absorption lines observed for one diastereoisomer can be absent in the other. The most stable structures of LL-VF and DL-VF, calculated at the B3LYP/6-311G** level and confirmed by IR/R2PI spectroscopy, have a hydrogen bond between the hydroxyl of the C-terminus and the C=O group of the peptide bond. A similar situation is encountered in the case of the homodipeptides where two conformers are observed for LL-FF and one for LD-FF.

Even in presence of the very simple methyl side-chain of alanine, dipeptides containing phenylalanine adopt different conformations according to the chirality of their components. This has been demonstrated in IR/R2PI studies of *N*-acetyl-Xxx-L-Phe-NH_2 dipeptides where Xxx = L-Ala (LL, 1a), D-Ala (DL, 1b) or α-aminoisobutyric acid, Aib (2) [40]. In dipeptide (2) containing the achiral α-aminoisobutyric acid, two spectral features, respectively B and B′, are interpreted with the help of structure calculations conducted at the LMP2/6-31G(d)// B3LYP/6-31+G(d) level as signatures of β-turns (Figure 4.6.2) characterized by a C_{10} H bond. The introduction of a chiral centre strikingly modifies the situation. In the LL dipeptide (1a), two lowest energy conformers are observed and respectively correspond to the type I β-turn (B) and an extended C_7–C_7 H-bond structure. In the DL peptide (1b), only the type II′ β-turn (B) appears. This chiral discrimination is the result of different steric hindrances. In the L–L and DL peptides, these steric hindrances are due to close contacts between one of the

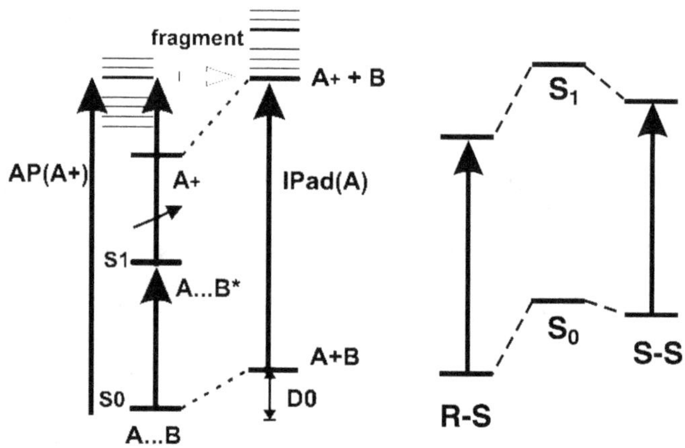

Figure 4.6.1 Left: Principle of the measurement of the binding energy of neutral dimers by means of two-colour R2PI measurements (adapted with permission from reference [34] ©2000 Royal Society of Chemistry). Right: $S_0 \leftarrow S_1$ electronic transitions of hetero-(*R–S* or *S–R*) and homo-(*S–S* or *R–R*) diastereoisomers.

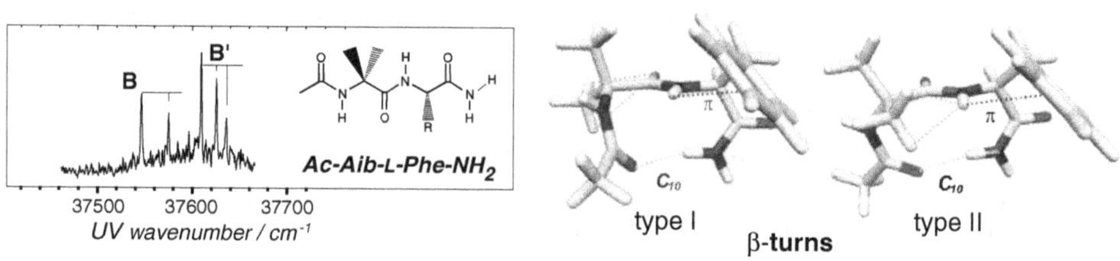

Figure 4.6.2 Left: R2PI spectrum of the capped α-aminoisobutyric acid (Aib)-L-Phe-NH_2 peptide. The sharp spectral features B and B′ in the electronic spectrum respectively correspond to the type I and type II′ β-turns displayed on the right side. Features B and B′ respectively appear in the spectra of the LL and DL dipeptides (by courtesy of M. Mons).

methyl groups and either the N—H or the C=O group of the amide bond. In the Aib-containing dipeptide, the two methyl groups on the Cα atom interact.

An important result is thus that the change of a single chiral centre in a peptide can be spectroscopically detectable. This is also possible with mass-spectrometric methods as shown below.

4.6.3 Mass-spectrometric chiral discrimination between amino acids

Discrimination between enantiomers (with identical *m/z*) by mass-spectrometry is important in chemistry, biology and pharmacy and can be achieved by means of a large variety of methods [41–54].

4.6.3.1 The three-point interaction model

A chiral discrimination mechanism, called the "three-point interaction model", rests on the combination of hydrogen bonding, electrostatic, van der Waals and dispersion interactions in a chiral complex. This model has been derived from condensed-phase studies and can also be applied to gas-phase chiral recognition [43]. We illustrate here this model in the case of discrimination between protonated L- and D-amino acids. Guest amino acids establish with their hosts two attractive interactions involving their protonated amino group and carboxylic group (Figure 4.6.3). According to the amino acid, the interaction of the side-chain can be

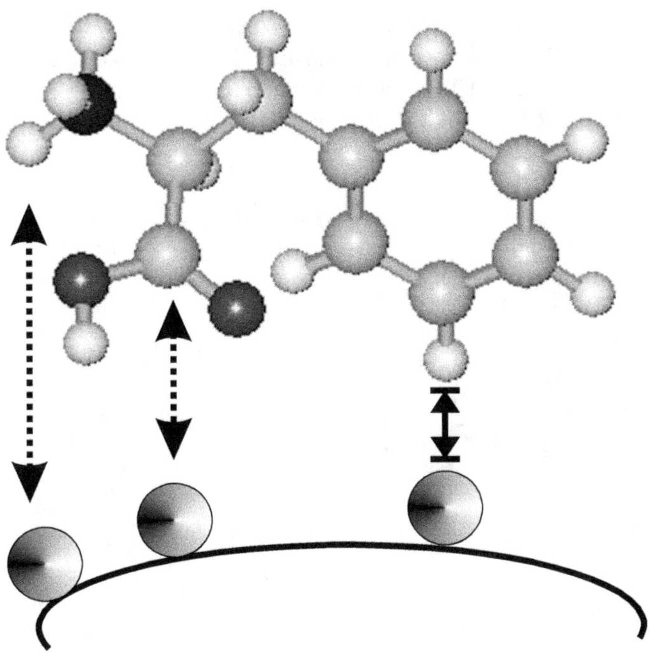

Figure 4.6.3 Illustration of the "three-point" model in the case of protonated phenylalanine. In the gas-phase, optimal enantioselectivity is obtained with two attractive and one repulsive interactions (adapted with permission from reference [43] ©2001 American Chemical Society).

attractive or repulsive. Enantioselectivity is optimal in presence of two attractive and one repulsive interaction. This is the case of guest Phe and Tyr amino acids where the aromatic phenyl group introduces steric hindrance. When their D and L species are in presence of host linear oligosaccharides, host–guest complexes with different structures are produced.

Chiral discrimination is measured by comparison between the guest-exchange reaction rates k_L/k_D of L and D species. In presence of a gaseous alkyl amine, the amino acids transfer their proton to this amine and are lost as neutral species. Values of k_L/k_D up to 4.6 have been observed for Phe complexed with maltoheptaose. Enantioselectivity is strongly dependent on the relative size of the guest and the binding cavity of the host. When the size of the guest is optimal, each enantiomer finds a distinct interaction that yields its characteristic rate of exchange different relative to its mirror image. The shorter linear oligosaccharides, maltohexaose and maltopentaose provide less selectivity than maltoheptaose since they cannot as easily wrap up the guest and the third interaction is less effective (Figure 4.6.4).

Other methods for amino acid chiral discrimination have been investigated. For example, mass-selected trimers of nickel atoms complexed with an analyte and a reference amino acid with known chirality have been submitted to CID and fragmented into diastereoisomers. The branching ratio is sensitive to the chirality of the analyte [46, 55]. Complexation by a transition metal (CuII and Fe) allows diastereomeric differentiation between pentapeptides in CID experiments [54]. Differentiation is then due to interactions sensitive to the stereochemistry of D or L-leucine and D or L-alanine leading to different fragmentation pathways.

4.6.3.2 D-amino acids in a Trp-cage protein

Electron capture dissociation (ECD, see Chapter 4.7) can distinguish the presence of D-amino acids in proteins [56, 57]. This has been demonstrated in a comparative study between the ECD mass-spectra of the natural Trp-cage protein (see Section 4.2.6) and its D-KSS variant NLYIQWL**K**GGP**SS**SGRPPPS into which amino acids in bold have been substituted

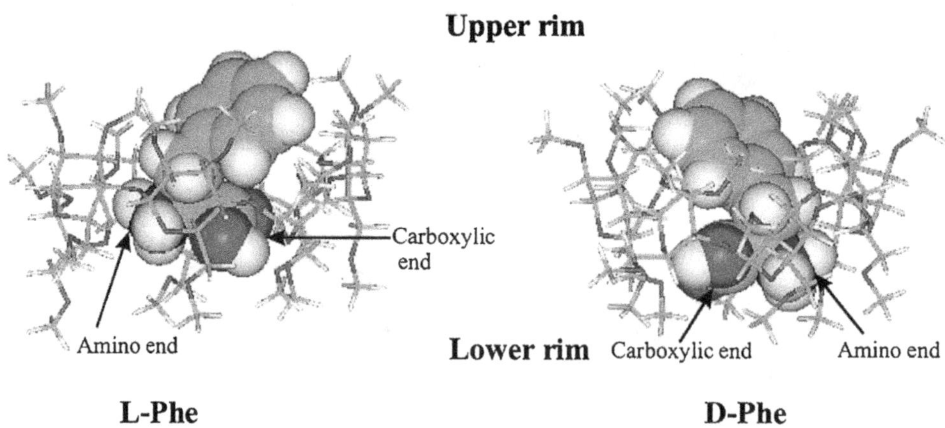

Figure 4.6.4 Lowest energy structures of enantiomeric [Hep:Phe+H]$^+$ complexes of phenylalanine (space filling) with maltoheptaose (Hep, ball and stick). The orientation of the phenyl group strongly differs in the L- and D-enantiomer complexes (reprinted with permission from reference [43] ©2001 American Chemical Society).

for D-amino acids. The used ECD feature is that ECD produces N–Cα bond cleavages (Figure 3.2.14, c and z fragments) near the carbonyl oxygen groups participating into $-NH_3^+\cdots O{=}C$ proton solvation but not when the carbonyl oxygen groups participate into neutral —N—H···O═C hydrogen bonding. In contrast with CID and IRMPD (see Sections 3.2.2.2 and 2.1.3.2.5), ECD fragment abundances are strongly determined by the local sequence. For the doubly charged native Trp-cage ion, nearly no backbone z fragments are observed. This shows that the backbone carbonyls are involved in an extensive neutral hydrogen bond network. On the contrary, the ECD mass-spectrum of the doubly charged D-KSS variant ion presents a large number of z ions, principally the z fragment with the C-terminal adjacent to the D-lysine residue (**K**) in the ion central part. This indicates that an unfolding that may have taken place in the solution from which ions have been electrosprayed, has profoundly modified the hydrogen bonding of the D-KSS due to the D-amino acid substitutions.

4.6.3.3 Chiral preference in serine cluster formation

It has been experimentally observed by several groups that unusually stable protonated octamers of the amino acid serine present a strong preference for homochirality [6, 48, 58–69]. In contrast, negatively and positively charged arginine clusters form stable trimers but without any chiral selectivity. This is attributed to formation of the stable structure that only involves a two-point contact (electrostatic interaction between the guanidinium side-chain and the carboxylate) between pairs of arginine (Figure 4.6.5).

The explanation of the preference of serine octamers for homochirality is not obvious and experimental measurements aiming to provide clues have been performed. CID experiments

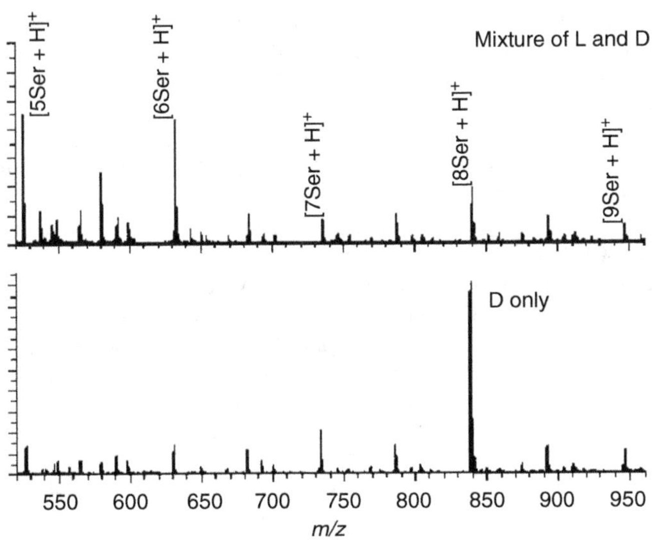

Figure 4.6.5 Mass spectra of protonated serine clusters. The upper spectrum has been obtained by electrospray ionization of an acidic solution containing a racemic mixture of L- and D-serine. The lower spectrum corresponds to a sprayed enantiomerically pure serine solution and displays a magic peak for [8 serine+H]$^+$ ions (reprinted with permission from reference [69] ©2006 Elsevier).

have demonstrated that a protonated octamer (8 serine+H)$^+$ ion preferentially fragments into a protonated hexamer and a neutral dimer. Tetramers and hexamers are more abundant than trimers, pentamers and heptamers. These observations are in favour of formation of zwitterionic dimers as building blocks [6]. For mixed clusters of serine and threonine, a strong chiral preference is also observed. On the contrary, formation of (10 serine+2H)$^{2+}$ ions is strongly favoured when sampled from a racemic solution. This may explain why crystals grown from racemic serine solutions do not spontaneously resolve in homochiral crystals as one might expect from the octamer preference for homochirality. The octamer preference may be counterbalanced by the presence of (10 serine+2H)$^{2+}$ clusters during the growth [65].

REFERENCES

1. Cronin JR, Pizzarello S: Enantiomeric excesses in meteoritic amino acids. *Science* 1997, 275:951–955.
2. Cataldo F, Brucato JR, Keheyan Y: Chirality in prebiotic molecules and the phenomenon of photo- and radio-racemization. *Journal of Physics: Light, Dust and Chemical Evolution* 2005, 6:139–148.
3. Plankenmeister K, Reiner H, Rode BM: Catalytic effects of glycine on prebiotic divaline and diproline formation. *Peptides* 2005, 6:1109–1112.
4. Lu S, Zhu Z, Gong X, Dong X: Theoretical study of helical structure and chiral discrimination of cysteine dimer. *Journal of Molecular Structure (Theochem)* 2005, 717:121–125.
5. Cline DB: Supernova antineutrino interactions cause chiral symmetry breaking and possibly homochiral biomaterials for life. *Chirality* 2005, 17:S234–S239.
6. Schalley CA, Weis P: Unusually stable magic number clusters of serine with a surprising preference for homochirality. *International Journal of Mass Spectrometry* 2002, 221:9–19.
7. Snyder LE, Lovas FJ, Hollis JM, Friedel DN, Jewell PR, Remijan A, Ilyushin VV, Alekseev EA, Dyubko SF: A rigorous attempt to verify interstellar glycine. *Astrophysical Journal* 2005, 619:914–930.
8. Abo-Riziq A, Grace L, Nir E, Kabelac M, Hobza P, de Vries MS: Photochemical selectivity in guanine–cytosine base-pair structures. *Proceedings of the National Academy of Sciences of the United States of America* 2005, 102:20–23.
9. Hollis JM, Lovas FJ, Jewell PR: Interstellar glycolaldehyde: the first sugar. *Astrophysical Journal* 2000, 540:L107–L110.
10. Orgel LE: Prebiotic chemistry and the origin of the RNA world. Critical Reviews in Biochemistry and Molecular Biology 2004, 39:99–123.
11. Miller SL, Urey HC: Production of some organic compounds under possible primitive Earth conditions. *Journal of the American Chemical Society* 1955, 77:2351–2361.
12. Shapiro R: Prebiotic cytosine synthesis: a critical analysis and implications for the origin of life. *Proceedings of the National Academy of Sciences of the United States of America* 1999, 96:4396–4401.
13. Hudson RL, Moore MH, Cook A: IR characterization and radiation chemistry of glycolaldehyde and ethylene glycol ices. *Advances in Space Research* 2005, 36:184–189.
14. Bernstein MP, Sandford SA, Allamandola LJ, Gillette JS, Clemett SJ, Zare RN: UV irradiation of polycyclic aromatic hydrocarbons in ices: production of alcohols, quinones, and ethers. *Science* 1999, 283:1135–1138.
15. Ruette MF, Martorell G, Rodriguez LS: Quantum-chemical modeling of interstellar grain prebiotic chemistry catalytic synthesis of glycine and alanine on the surface of a polycyclic aromatic hydrocarbon flake. *Astrophysical Journal* 2004, 601:L59–L62.

16. Bernstein MP, Baumschlicher CW, Sandford SA: The infrared spectrum of matrix isolated aminoacetonitrile, a precursor to the amino acid glycine. *Advances in Space Research* 2004, 33:40–43.
17. Lafosse A, Bertin M, Domaracka A, Pliszka D, Illenberger E, Azria R: Glycine synthesis at 25 K induced by low-energy electron irradiation of condensed NH_3–CH_3COOD (1:1) mixture. *Physical Chemistry Chemical Physics* 2006, 8:5564–5568.
18. Quack M: How important is parity violation for molecular and biomolecular chirality. *Angewandte Chemie-International Edition* 2002, 41:4618–4630.
19. Pestunova O, Simonov A, Snytnikov V, Stoyanovsky V, Parmon V: Putative mechanism of the sugar formation on prebiotic Earth initiated by UV radiation. *Advances in Space Research* 2005, 36:214–219.
20. Meierhenrich UJ, Nahon L, Alcaraz C, Bredehöft JH, Hoffmann SV, Barbier B, Brack A: Asymmetric vacuum UV photolysis of the amino acid leucine in the solid state. *Angewandte Chemie-International Edition* 2005, 44:5630–5634.
21. Boon G, van Alsenoy CV, De Proft F, Bultinck P, Geerlings P: Molecular quantum similarity of enantiomers of amino acids: a case study. *Journal of Molecular Structure (Theochem)* 2005, 727:49–56.
22. Scuderi D, Pladini A, Satta M, Catone D, Filippi S, Lagana A, Piccirillo S, Speranza M, Giardini A: Gas-phase complexes: non-covalent interactions and stereospecificity. *International Journal of Mass Spectrometry* 2000, 198:137–163.
23. Seurre N, Le Barbu-Debus K, Lahmani F, Borho N, Suhm MA, Zehnacker A: Chiral recognition in jet-cooled complexes. *Australian Journal of Chemistry* 2004, 57:1149–1152.
24. Seurre N, Barbu-Debus KL, Lahmani F, Zehnacker A, Borho N, Suhm MA: Chiral recognition between lactic acid derivatives and an aromatic alcohol in a supersonic expansion: electronic and vibrational spectroscopy. *Physical Chemistry Chemical Physics* 2006, 8:1007–1016.
25. Abo-Riziq AG, Bushnell JE, Crews B, Callahan MP, Grace L, De Vries MS: Discrimination between diastereoisomeric dipeptides by IR–UV double resonance spectroscopy and ab initio calculations. *International Journal of Quantum Chemistry* 2005, 105:437–445.
26. Richards A, McCague R: The impact of chiral technology on the pharmaceutical industry. *Chemistry and Industry* 1997, 11:422–425.
27. Berger R, Quack M: Electroweak quantum chemistry of alanine: parity violation in gas and condensed phases. *ChemPhysChem* 2000, 1:56–59.
28. Lethokov V: Effect of parity violation upon the vibrational spectrum of chiral molecules. *Physics Letters A* 1975, 53:275–276.
29. Bonner W: Parity and the evolution of biomolecular homochirality. *Chirality* 2000, 12:114–126.
30. Laerdahl JK, Wesendrup R, Schwerdtfeger P: D- or L-alanine: that is the question. *ChemPhysChem* 2000, 1:60.
31. Soulard P, Asselin P, Cuisset A, Aviles Moreno JR, Huet TR, Petitprez D, Demaison J, Freedman TB, Cao X, Nafie LA, *et al.*: Chlorofluoroiodomethane as a potential candidate for parity violation measurements. *Physical Chemistry Chemical Physics* 2006, 8:79–92.
32. Ziskind M, Daussy C, Marrel T, Chardonnet C: Improved sensitivity in the search for a parity-violating energy difference in the vibrational spectrum of the enantiomers of CHFClBr. *European Physical Journal D* 2002, 20:219–225.
33. Le Barbu K, Zehnacker A, Lahmani F, Mons M, Piuzzi F, Dimicoli I: Spectroscopic studies of enantiomeric discrimination in jet-cooled chiral complexes. *Chirality* 2001, 13:715–721.
34. Mons M, Piuzzi F, Dimicoli I, Zehnacker A, Lahmani F: Binding energy of hydrogen-bonded complexes of the chiral molecule 1-phenylethanol, as studied by 2C-R2PI: comparison between diastereoisomeric complexes with butan-2-ol and the singly hydrated complex. *Physical Chemistry Chemical Physics* 2000, 2:5065–5070.
35. Butz P, Kroemer RT, Macleod NA, Simons JP: Hydration of neurotransmitters: a spectroscopic and computational study of ephedrine and its diastereoisomer pseudoephedrine. *Physical Chemistry Chemical Physics* 2002, 4:3566–3574.

36. Speranza M, Piccirillo S, Rondino F, Paladini A, Giardini A, Filippi S, Catone D: Chiral recognition by mass-resolved spectroscopy. *Mass Spectrometry Reviews* 2005, 24:588–610.
37. Borho N, Häber T, Suhm MA: Chiral self-recognition in the gas-phase: the case of glycidol dimers. *Physical Chemistry Chemical Physics* 2001, 3:1945–1948.
38. Liu Sz, Zhou Zy, Gong Xl, Dong Xl: Theoretical study of helical structure and chiral discrimination of cysteine dimer. *Journal of Molecular Structure (Theochem)* 2005, 717:121–125.
39. Macleod NA, Butz P, Simons JP, Grant GH, Baker CM, Tranter GE: Structure, electronic circular dichroism and Raman optical activity in the gas phase and in solution: a computational and experimental investigation. *Physical Chemistry Chemical Physics* 2005, 7:1432–1440.
40. Brenner V, Piuzzi F, Dimicoli I, Tardivel B, Mons M: Chirality-controlled formation of β-turn secondary structures in short peptide chains: gas phase experiment vs. quantum chemistry. *Angewandte Chemie-International Edition* 2007, 36:214–219.
41. Grigorean G, Ramirez J, Ahn SH, Lebrilla CB: A mass spectrometry method for the determination of enantiomeric excess in mixtures of D,L-amino acids. *Analytical Chemistry* 2000, 72:4275–4281.
42. Grigorean G, Lebrilla CB: Enantiomeric analysis of pharmaceutical compounds by ion/molecule reactions. *Analytical Chemistry* 2001, 73:1684–1691.
43. Ahn S, Ramirez J, Grigorean G, Lebrilla CB: Chiral recognition in gas-phase cyclodextrin: amino acid complexes – Is the three point interaction still valid in the gas phase? *Journal of the American Society for Mass Spectrometry* 2001, 12:278–287.
44. Gal JF, Stone M, Lebrilla CB: Chiral recognition of non-natural α-amino acids. *International Journal of Mass Spectrometry* 2003, 222:259–267.
45. Cheng Y, Hercules DM: Measurement of chiral complexes of cyclodextrine and amino acids by electrospray ionization time-of-flight mass spectrometry. *Journal of Mass Spectrometry* 2001, 36:834–836.
46. Zhang D, Tao WA, Cooks RG: Chiral resolution of D- and L-aminoacids by tandem mass-spectrometry of Ni (II)-bound trimeric complexes. *International Journal of Mass Spectrometry* 2001, 204:159–169.
47. Wu L, Cooks RG: Chiral and isomeric analysis by electrospray ionization and sonic spray ionization using the fixed-ligand kinetic method. *European Journal of Mass Spectrometry* 2005, 11:231–242.
48. Nanita SC, Takats Z, Cooks RG: Chiral enrichment of serine via formation, dissociation, and soft-landing of octameric cluster ions. *Journal of the American Society for Mass Spectrometry* 2004, 15:1360–1365.
49. Cooks RG, Zhang DX, Koch KJ, Gozzo FC, Eberlin MN: Chiroselective self-directed octamerization of serine: implications for homochirogenesis. *Analytical Chemistry* 2001, 73:3646–3655.
50. Bagheri H, Chen H, Cooks RG: Chiral recognition by proton transfer reactions with optically active amines and alcohols. *Chemical Communications* 2004, 23:2740–2741.
51. Mancel V, Sellier N, Lesage D, Fournier F, Tabet JC: Gas phase enantiomeric distinction of (R)- and (S)-aromatic hydroxy esters by negative ion chemical ionization mass spectrometry using a chiral reagent gas. *International Journal of Mass Spectrometry* 2004, 237:185–195.
52. Mezzache S, Pepe C, Karoyan P, Fournier F, Tabet JC: Proton affinity of diastereoisomers of modified prolines using the kinetic method and density functional theory calculations: role of the *cis/trans* substituent on the endo/exo ring conformation. *Rapid Communications in Mass Spectrometry* 2005, 19:2279–2283.
53. Madhusudanan KP: Tandem mass spectra of ammonium adducts of monosaccharides: differentiation of diastereoisomers. *Journal of Mass Spectrometry* 2006, 41:1096–1104.
54. Lagarrigue M, Bossée A, Afonso C, Fournier F, Bellier B, Tabet JC: Diastereoisomeric differentiation of peptides with CuII and FeII complexation in an ion trap mass spectrometer. *Journal of Mass Spectrometry* 2006, 41:1073–1085.
55. Hu HJ, Guo YL: Chiral recognition of borneol by association with zinc (II) and L-tryptophan in the gas-phase. *Analytica Chimica Acta* 2003, 482:1–7.

56. Adams CM, Zubarev RA: Distinguishing and quantifying peptides and proteins containing D-amino acids by tandem mass spectrometry. *Analytical Chemistry* 2005, 77:4571–4580.
57. Adams CM, Kjeldsen F, Zubarev RA, Budnik BA, Haselmann KF: Electron capture dissociation distinguishes a single D-amino acid in a protein and probes the tertiary structure. *Journal of the American Society for Mass Spectrometry* 2004, 15:1087–1098.
58. Hodyss R, Julian RR, Beauchamp JL: Spontaneous chiral separation in non covalent molecular clusters. *Chirality* 2001, 13:703–706.
59. Takats Z, Cooks RG: Thermal formation of serine octamer ions. *Chemical Communications* 2004, 444–445.
60. Nanita SC, Cooks RG: Negatively charged halide adducts of homochiral serine octamers. *Journal of Physical Chemistry B* 2005, 109:4748–4753.
61. Myung S, Julian RR, Nanita SC, Cooks RG, Clemmer DE: Formation of nanometer-scale serine clusters by sonic spray. *Journal of Physical Chemistry B* 2004, 108:6105–6111.
62. Mazurek U, Geller O, Lifshitz C, McFarland MA, Marshall AG, Reuben BG: Protonated serine octamer cluster: structure elucidation by gas-phase H/D exchange reactions. *Journal of Physical Chemistry A* 2005, 109:2107–2112.
63. Mazurek U, Engeser M, Lifshitz C: Gas-phase H/D exchange of the protonated serine octamer cluster: "Ion ping pong" of populations A and B. *International Journal of Mass Spectrometry* 2006, 249:473–476.
64. Mazurek U: Letter: Some more aspects of formation and stability of the protonated serine octamer cluster. *European Journal of Mass Spectrometry* 2006, 12:63–69.
65. Julian RR, Myung S, Clemmer DE: Spontaneous anti-resolution in heterochiral clusters of serine. *Journal of the American Chemical Society* 2004, 126:4110–4111.
66. Julian RR, Hodyss R, Kinnear B, Jarrold MF, Beauchamp JL: Nanocrystalline aggregation of serine detected by electrospray ionization mass spectrometry: origin of the stable homochiral gas-phase serine octamer. *Journal of Physical Chemistry B* 2002, 106:1219–1228.
67. Counterman AE, Clemmer DE: Magic number clusters of serine in the gas phase. *Journal of Physical Chemistry B* 2001, 105:8092–8096.
68. Julian RR, Myung S, Clemmer DE: Do homochiral aggregates have an entropic advantage? *Journal of Physical Chemistry B* 2005, 109:440–444.
69. Concina B, Hvelplund P, Nielsen AB, Bronsted Nielsen S, Rangama J: Formation and stability of charged amino acid clusters and the role of chirality. *Journal of the American Society for Mass Spectrometry* 2006, 17:275–279.

– 4.7 –

Low-Energy Electron–Molecule Interactions

GENERAL FEATURES

Electron ionization (EI) has since a long time been used as a convenient tool for ionizing neutral biomolecules in mass-spectrometry. Ionization potentials of most molecules lie in between 7 and 15 eV (see Section 3.4). Immediately above the corresponding ionization thresholds, ionization efficiencies are very low but increase rapidly with electron energy and exhibit a plateau, typically around 70 eV, allowing for reproducibility and comparisons of ion mass-spectra. Total cross-sections for ionization can come up to the geometrical cross-section of the molecule, in the 10^{-16}–$10^{-15}\,cm^2$ range. The situation is very different for electrons in the low-energy range which can be captured by molecular systems (neutral molecules, neutral complexes, cations…) leading to the formation of short- or long-lived temporary anions with cross-sections that can exceed the geometrical cross-section by several orders of magnitude. Interestingly, some of the low-energy electron capture events can lead to bond-selective processes. In this chapter, we will examine two kinds of dissociative electron capture processes, one concerning DNA building blocks in radiation chemistry and the other concerning multiply-charged peptides in mass-spectrometry. Non-dissociative electron capture processes have been considered in Section 2.3.

4.7.1 Reactions initiated by electrons in DNA and its building blocks

4.7.1.1 Ionizing radiations and the cellular medium

When high-energy radiations penetrate into living cells [1–3], their primary interactions essentially remove electrons from the cell components. Radiolysis of the majority component, water, in the intracellular fluid induces damages to the genetic information stored in DNA via production of intermediate species such as $^{\bullet}$H and $^{\bullet}$OH radicals and also of secondary electrons. The latter are the most abundant since a 1 MeV particle typically creates around 4×10^4 electrons. Different types of DNA damages can occur: (a) in a base alteration, chemical properties of a nucleobase (A, T, G or C) are abnormally modified from the form found in undamaged DNA; (b) a nucleobase can be deleted from a nucleotide; (c) the chemical properties of a deoxyribose sugar can be abnormally modified; (d) the covalent bond between the deoxyribose sugar unit and the phosphate group can be broken (strand break). A single strand break (SSB) occurs when one of the two DNA backbones is broken. If two breaks occur within a short distance (typically 10 nucleobases), the damage is called a double strand break (DSB).

The genetic information is coded by triplets of nucleotides (codons). In cases a and b, a base sequence is changed (e.g. GCA is changed to GAA) and a point mutation is said to occur. A missense mutation occurs when a change in base sequence converts a codon for one amino acid to a codon for a different amino acid leading to a defective protein. There are very efficient repair mechanisms [4] if the elementary damage sites are more than 10 base pairs apart but permanent mutations can however take place if the damage sites are too close.

Direct damages concern reactive processes occurring directly in DNA. Indirect damages result from the deposition of energy in water and other biomolecules surrounding DNA and most of them are due to the attack of oxidative species [5]. Two thirds of DNA damages are indirect and one third is direct.

4.7.1.2 Oxidative damages

Reactions of the hydroxyl radical with nucleobases have been experimentally studied in the gas-phase by means of neutralization–re-ionization mass-spectrometry [6] and modelling [7]. Those reactions produce adducts and hydrogen abstraction. Oxidative attack by the $^{*}OH$ radical on the deoxyribose moiety also lead to the release of free bases from DNA, generating strand breaks with various sugar modifications and simple deletions of nucleobases. Ionizing radiations are thus classified as potent carcinogens but one cannot ignore that oxidants, such as hydrogen peroxide (see Section 5.1.2.3) and the hydroxyl radical $^{\bullet}OH$, can be also be endogenous (Figure 4.7.1). They may cause cancer in individuals where no obvious exposure to chemical or physical agents known to be carcinogenic has occurred. They also intervene in mutagenesis and aging.

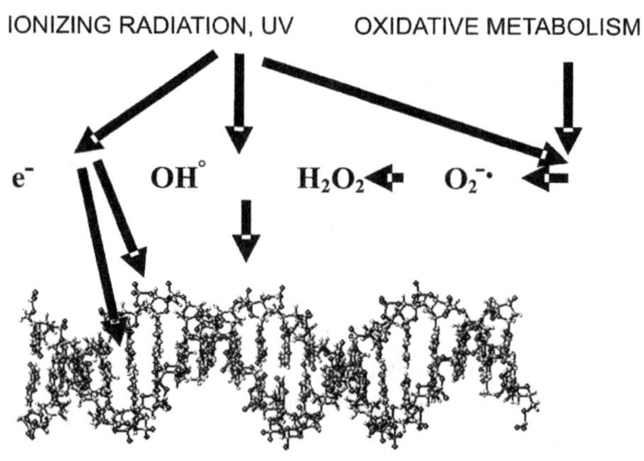

Figure 4.7.1 High-energy ionizing (HEI) [3] and UV [8] radiations can directly attack nucleobases and induce lesions. HEI can also interact with water producing low-energy electrons and $OH^{\bullet}$ radicals that in turn attack nucleobases [9, 10] and deoxyribose [11] (indirect effects). The oxidative metabolism leads to reactive oxygen species modifying bases and sugar [5, 12].

4.7.1.2.1 Radiation damages induced by low-energy electrons to plasmid DNA

Direct damages are mainly produced by the abundant secondary electrons that have low energies with an energy distribution mainly below 15 eV. Those electrons thermalize through inelastic collisions over distances in between 1 and 10 nm. This defines volumes, called spurs, where the initial energy of the ionizing radiations is finally deposited [13]. In the very dense aqueous cellular medium, electrons are rapidly solvated and one must consider both initial free electrons and hydrated electrons (see Section 5.2.1.2). The traditional point of view has been for many years that toxic damages induced by secondary electrons take place above the onset of ionization or upon solvation. Solvated electrons turn out to be much less reactive than free electrons [14] and it thus seems that free electrons (before solvation) are responsible for most DNA damages. This has been strongly suggested by recent studies devoted to interactions of free electrons with plasmid DNA deposited on surfaces under ultrahigh conditions. Those surface studies which further prompted a rather large number of gas-phase studies have challenged the traditional point of view by demonstrating that, even in absence of any environment effect, strand breaks are produced in DNA at energies well below ionization and electronic excitation, down to nearly zero electron energy [2, 15, 16]. In those experiments, supercoiled plasmid DNAs (3187 base pairs) still containing their structural water are deposited on a tantalum sheet and submitted to electron bombardment at different electron energies. By releasing the strained energy, SSB and DSBs respectively transform the supercoiled DNA into a circular or linear configuration. These different forms of

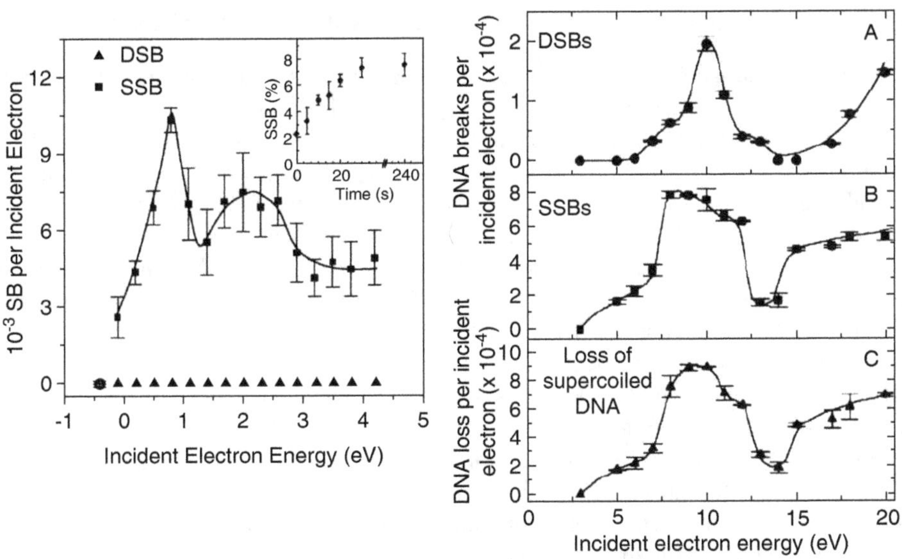

Figure 4.7.2 Quantum yields, per incident electron, for induction of single strand breaks (SSB) and double strand breaks (DSB) in supercoiled DNA. At very low electron energies, only SSB take place (reprinted with permission from reference [16] ©2004 American Institute of Physics). The DSB yield presents a peak around 10 eV (reprinted with permission from reference [15] ©2000 American Society for the Advancement of Science).

DNA are then separated by gel electrophoresis and the percentage of each form is quantified by fluorescence. Very sensitive detection of SSBs and DSBs becomes possible owing to the large amplification factor resulting from the topological changes of the plasmids. While nearly no DSB is observed, up to 1×10^{-2} SSB per incident electron can be observed. The dependence of the SSB yield as a function of energy presents rather sharp peaks around 0.8 and 2.2 eV (Figure 4.7.2) as well as a broader feature around 10 eV. This has been interpreted [2] as due to formation of resonance anionic states. Formation of transient anions by temporary capture of low-energy electrons is an intermediate step occurring very rapidly (within 10^{-13} s). It is then followed by decay into dissociative electronically excited states or dissociation into stable anions plus neutral radicals. Dissociation of those transient anions within DNA can occur either by direct electron capture at the location of breaks or by electron transfer to another subunit. In order to test those hypothesis, experimental gas-phase studies of dissociative electron attachment (DEA) to elementary DNA components (nucleobase, sugar) and quantum modelling have been performed by different groups and are presented in the following paragraph.

Experiments have been performed under single collision condition between monochromatized electron beams and isolated nucleobases, desoxyribose and nucleosides (bases plus sugar). In those experiments, absolute cross-sections are obtained for the DEA process:

$$\mathrm{A-B} + e \rightarrow (\mathrm{A-B})^{-*} \rightarrow \mathrm{A}^{-} + \mathrm{B} \quad \text{or} \quad \mathrm{A} + \mathrm{B}^{-}$$

by measuring the different fragment anion yields as function of the incident electron energy. Since desorbed molecules can be hot, electron attachment can take place to vibrationally excited species. The use of deposition on helium clusters eliminates this problem [17].

4.7.1.3 Dissociative electron attachment to isolated nucleobases

At nearly zero energy, electron attachment to thymine leads to formation of dipole-bound anions with a very low electron binding energy [18–21]. Those anions are not biologically relevant since they cannot be created in the very dense DNA environment. Moreover, except during transcription, nucleobases are paired and A—T or C—G pairs lead to valence anions [22]. Only valence anions will now be considered. In collisions between free electrons and thymine clusters deposited on cold helium clusters (see Section 3.1.3), it has been shown that around 0 eV and in presence of a stabilizing body (see Section 4.1.1.3), intact thymine anions are produced [17]. At free electron energies above 0.5 eV, DEA takes place and has been studied in the case of uracil [23, 24], thymine [25], adenine [26] and cytosine [27, 28]. Three different reactive channels are then observed respectively leading to the loss of one (observation of $(\mathrm{T{-}H})^-$ ions) or two (observation of $(\mathrm{T{-}2H})^-$ ions) hydrogen atoms and H^- anions. Those three channels present bond selectivity as a function of electron energy [29]. The energy threshold for removing an H atom from an atomic site X of thymine is equal to the binding energy $D_{\mathrm{X-H}}$ of the H atom to atom X minus the adiabatic electron affinity of the resulting neutral $\mathrm{EA_{ad}(X)}$. As demonstrated by using thymine partially deuterated on the different carbon atoms and the methyl site, cleavage of a single N—H bond at the nitrogen N1 and N3 sites leading to $(\mathrm{T{-}H})^-$ and H^- anions is a preferential energetic pathway whereas DEA of C—H bonds requires higher electron energies. Calculations at the G2 level [27] of the different energy thresholds confirm that the most energetically favoured removal of an H atom is from the N1 site (attachment site of thymine in DNA). Discrimination between the possible N sites as a function of electron energy is possible through a combination of different methylation

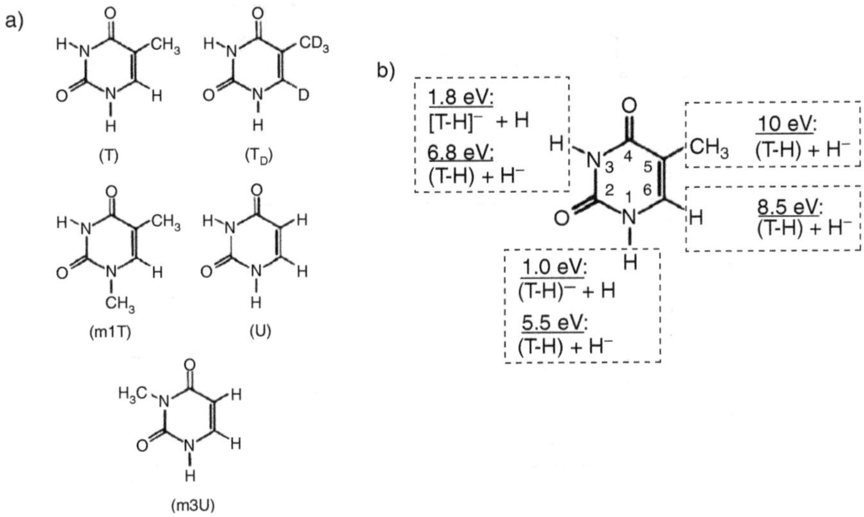

Figure 4.7.3 (a) The different deuterated and methylated thymine molecules and uracil used for the investigation of site and bond-selective dissociative electron attachment (DEA) to thymine (T). (b) Experimental evidence for bond-selective (C—H or N—H bond cleavage) and site-selective DEA to T (N3—H compared with N1H and C6—H compared with CH_2—H) as a function of incident electron energies. Peak values of the electron energies inducing DEA to T according to the different sites and observed product anions (reprinted with permission from reference [3] ©2006 Wiley).

(m1T or m3T in Figure 4.7.3a) and deuteration at the different N positions. For example, at electron energies above 1 eV, the $(m1T—H)^{-1}$ ion yield becomes null whereas the $(m3T—H)^{-1}$ ion yield still exist around 1.8 eV [10, 25]. At higher energies, four different DEA channels corresponding to selective dissociation at the N1, N3, C6 and methyl sites are identified and respectively peak at 5.5, 6.6, 8.5 and 10 eV (Figure 4.7.3b) (Figure 4.7.4).

4.7.1.4 Dissociative electron attachment to desoxyribose, thymidine and phosphates

The sugar links between nucleobases and phosphate groups play an active part in initial processes leading to SSBs induced by low-energy electrons. Low-energy electron attachment to the deoxyribose sugar has been investigated [11] as well as direct electron attachment to phosphate groups [30]. 2-desoxy-D-ribose has a furanose ring and links a nucleobase to a phosphate group in a DNA strand [31]. In contrast to electron attachment to nucleobases where the dehydrogenation channel has the largest absolute cross-section, loss of two water molecules is the most important reaction channel. Moreover, the electron energy threshold for dissociative attachment is close to zero-energy and the sugar seems to be very sensitive to electron interactions. DEA processes to furan, tetrahydrofuran and fructose have been compared [32]. As far as electron damages induced by low energy electrons are concerned, tetrahydrofuran cannot be used as surrogate to model deoxyribose in DNA (Figure 4.7.5).

It is interesting to compare the DEA energy dependencies obtained separately for thymine and deoxyribose to results concerning electron attachment to thymidine, Td, corresponding

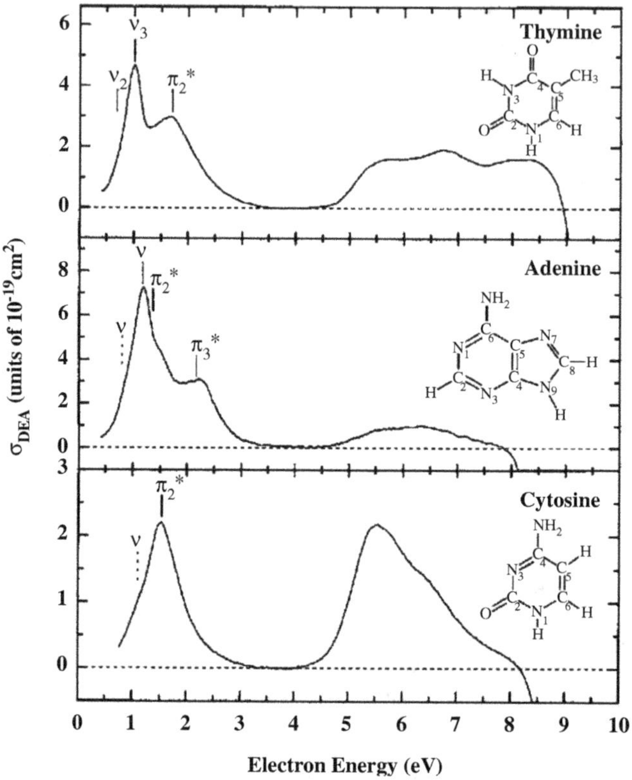

Figure 4.7.4 Total dissociative electron attachment cross-sections for thymine, adenine and cytosine, as function of electron energy. The solid vertical lines correspond to the energies of π^* shape resonances determined by electron transmission spectroscopy (see Section 2.3). Vertical lines labelled ν correspond to vibrational Feshbach resonances (reprinted with permission from reference [28] ©2006 American Institute of Physics).

to the thyminyl radical (thymine minus H in N1) linked to desoxyribose via a N1—C glycosidic bond [9, 33]. The three following DEA processes are in competition according to which fragment takes away the excess negative charge [9] (Figure 4.7.6).

$$\begin{aligned} \text{Td} + e^- \rightarrow \text{Td}^{*-} &\rightarrow (\text{T}-\text{H})^-_{N1} + (\text{2–deoxyribose–OH})^{\bullet} \quad \text{(a)} \\ &\rightarrow (\text{T}-\text{H})^{\bullet}_{N1} + (\text{2–deoxyribose–OH})^- \quad \text{(b)} \\ &\rightarrow (\text{dR—2H})^- \quad \text{(c)} \end{aligned}$$

At low-electron energies, thymidine captures electrons through formation of transient anions (resonances) at 1.2 and 1.8 eV. The 1.2 eV resonance corresponds to the production of the $(\text{dR—2H})^-$ (115 atomic mass units, a.m.u.) ion by capture of an electron in a σ^* orbital localized on the sugar moiety leading to the rupture of the glycosidic N1—C1 bond. The resonance at 1.8 eV corresponds to production of the $(\text{T—H})^-$ (125 a.m.u.) ion by capture of an electron in a π^* orbital localized on the thymine base and leads to rupture of the N3–H bond.

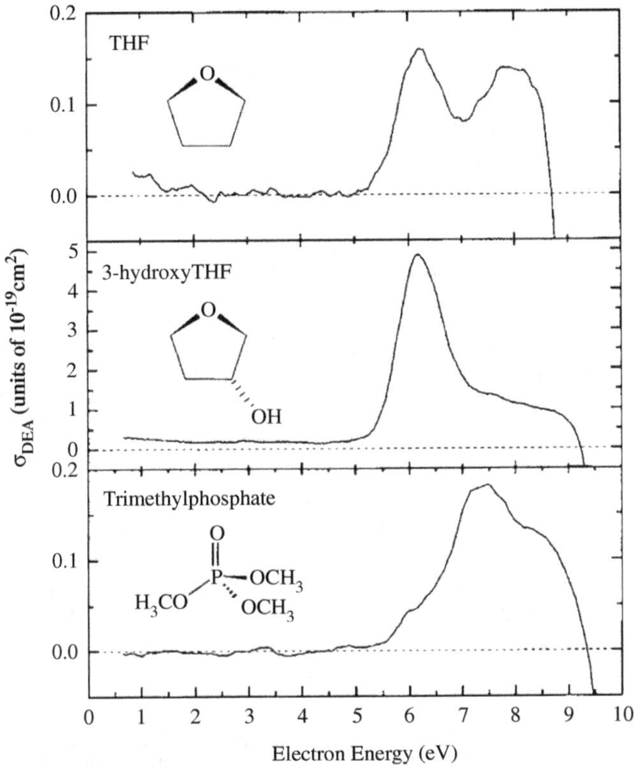

Figure 4.7.5 Total dissociative electron attachment cross-sections of tetrahydrofuran (THF), 3-hydroxytetrahydrofuran (3-hydroxyTHF) and trimethylphosphate (reprinted with permission from reference [28] ©2006 American Institute of Physics).

Results presented in Figure 4.7.5 seem to indicate that the σ^* resonance is not coupled to the π^* resonance since the $(T—H)^-$ and $(Td—H)^-$ ion yield dependencies are quite different from the $(dR—2H)^-$ yield dependency. This coupling between σ^* and π^* resonances presents similarities with the coupling between $\pi\pi^*$ and $\pi\sigma^*$ states in dissociation of photo-excited protonated peptides [34, 35] (see 2.1.4). The general case of coupling between σ^* and π^* resonances has been examined in reference [36] and will be considered in the following chapter.

Alternative experimental explanations to the observation of SSBs induced by very low-energy electrons are provided by measurements of electron attachment to sodium dihydrogen-phosphate films [37] or to a molecule analogous to the DNA phosphate group, dibutyl-phosphate (a neutral phosphate group capped by two neutral butyls) [30]. DEA occurs through resonances observed at 1 and 8 eV. Those resonances lead to direct cleavage of the C—O and P—O bonds without any prior attachment to nucleobases. However, direct electron attachment to the negatively charged phosphate group raises the problem of the neutralization of this ionic group by counterions that are always present in the vicinity of DNA.

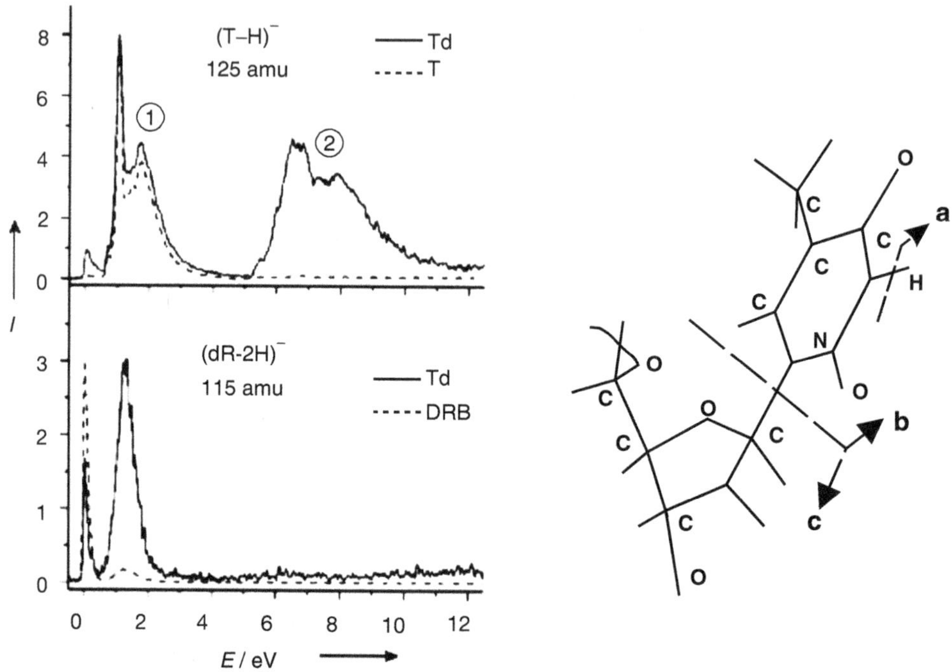

Figure 4.7.6 Left: Comparison between negative ion formation from DEA to thymidine (Td) and its isolated building blocks thymine (T) and 2-deoxyribose (DRB). The DEA cross-sections are in units of $10^{-22}\,m^2$ (reprinted with permission from reference[33] ©2006 Wiley). Right: Fragmentation scheme of thymidine.

4.7.1.5 Modelling of electron-induced DNA strand breaks

Radiation- or chemically induced damages to DNA have been already modelled before the experimental observation of strand breaks induced by electrons with energies well below 4 eV [6]. This experimental observation has triggered several theoretical studies [14, 38–40] that are reviewed in reference [2]. A first possibility invoked for explaining the breaking of sugar-phosphate O—C bond is the direct capture of incident electrons either at the 3′ or 5′ site (C3—O or C5-bond). A model system comprising two desoxyriboses linked by a phosphate has been considered by Sevilla [14]. The neutral and anion sugar-phosphate potential energy curves have been computed as function of the bond length $R_{C—O}$ stretched from the equilibrium distance (1.4 Å) up to large distances (2.8 Å). The rather small calculated energy-barrier of ca. 0.4 eV that must be surmounted for inducing the anion bond rupture either at 3′ or 5′ indicates that very low incident electron energies are sufficient to induce SSBs. However, electron affinities of the nucleobases (see Section 2.3.2) are higher than those of the phosphate group neutralized by a counterion. A model system of DNA single strand obtained by excising a fragment of DNA and comprising a nucleobase (cytosine and thymine which possess the lowest π^* orbitals), a dexoyribose sugar and a phosphate group have been considered. In this model, electron attachment can take place either into the π^* orbital of the nucleobase [41] or into a P—O π^* orbital of the phosphate unit [42]. In order to avoid spurious electron attachment sites, hydrogen atoms artificially cap (see Section 1.6) the —C—O and $-OPO_3^-$

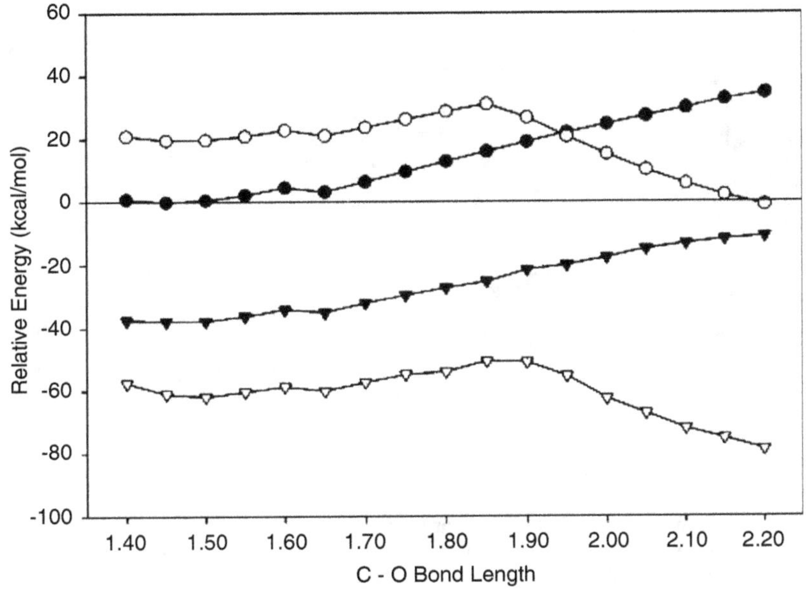

Figure 4.7.7 Electron-driven rupture of the C—O bond linking the sugar and the phosphate groups of the cytosine nucleotide. The upper figure represents the curve-crossing between the potential energy curves (in kcal/mol) of the unsolvated neutral (○) and anionic (●) species as a function of the C—O bond length (in Å). Around the value of 1.864 Å, the electron density shifts from the π^* orbital of cytosine to the sugar and phosphate region. The lower part corresponds to the solvated neutral (▼) and anionic systems (reprinted with permission from reference [41] ©2002 American Chemical Society).

radicals formed when the bonds are cut in DNA and the $-OPO_3^-$ anion is neutralized. The dry DNA that has been used in the surface experiments [15] still contained a mean value of 2.5 water molecules per nucleobase and solvation effects have been taken into account by using the polarization continuum model (Section 5.1.4) with different dielectric constants up to that of bulk water ($\varepsilon = 78$).

For each sugar-phosphate bond length $R_{C\text{-}O}$ value, the diabatic energy curves of the cytosine nucleobase (π^* orbital) and the phosphate-sugar (σ^* orbital) anion have been calculated. When the $R_{C—O}$ distance becomes larger than 1.9 Å, the electron affinity of the phosphate group becomes large and the excess charge migrates from the π^* orbital into which it was captured towards the σ^* orbital, then inducing the SSB (Figure 4.7.7).

The similarities between the switch from an initial π^* (following electron attachment) or $\pi\pi^*$ (following photoexcitation) orbital localized on an aromatic moiety to σ^* or $\pi\sigma^*$ orbitals leading to an electron-induced or photo-induced fragmentation along a dissociative C—O or N—H coordinate are illustrated in Figures 4.7.7 and 4.7.8.

4.7.1.6 Radiosensitisers for tumour therapy: electron attachment to halouracil

High-energy radiation damages to DNA can be lethal but radiations are also used in tumour therapy. The problem is then that the beneficial exposure of the cancerous domains must be

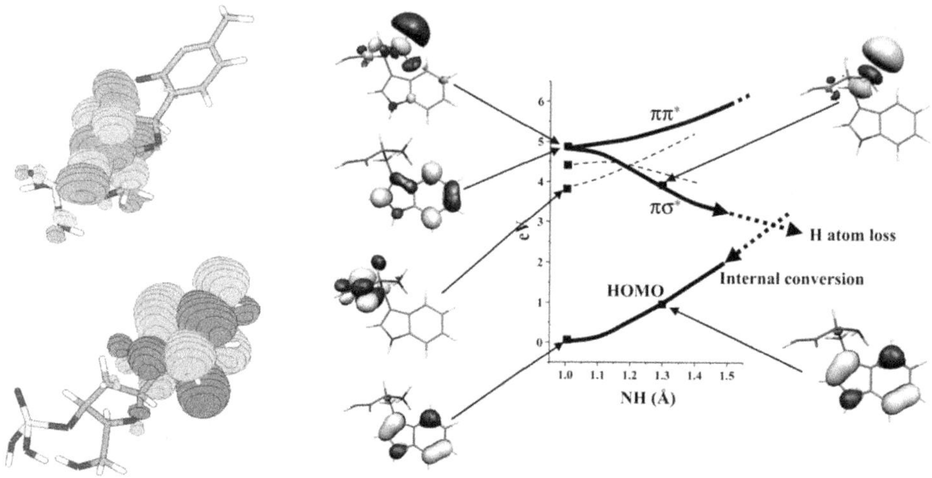

Figure 4.7.8 Left: Singly occupied molecular orbital (SOMO) of the solvated cytosine nucleotide anion represented at the equilibrium R_{CO} distance (top). Following electron capture, the excess electron is then localized in the π^* orbital of cytosine. At a large value R_{CO} distance (bottom) after the C–O bond rupture, the excess electron resides between the sugar and the phosphate groups (reprinted with permission from reference [41]©2002 American Chemical Society). Right: For comparison of interactions between π^* and σ^* orbitals on one side and $\pi\pi^*$ and $\pi\sigma^*$ orbitals on the other, the potential energy curves of the protonated tryptophan cation are displayed as a function of the dissociative coordinate R_{NH} leading to hydrogen atom loss. Orbitals of the ground and excited states are represented on the left side. On the right, the HOMO and LUMO orbitals are represented at the R_{NH} distance of 1.3 Å. Note the switch between the initial $\pi\pi^*$ orbital localized on the indole chromophore and the $\pi\sigma^*$ orbital localized on the amino group (reprinted with permission from reference [43] ©2004 Royal Chemical Society).

harmless for healthy tissues. One of the possibilities for achieving selectivity is to rely on the differences of oxygenation between normal tissues and tumours. Theses difference arise from the fact that irradiation involves free radicals that react rapidly with oxygen to form superoxide radicals (hydrated oxygen anions, see Section 5.1.2.3) which in turn inhibit reduction in normal tissues. Another possibility also involving anions is the replacement of the methyl group of thymine by bromine. Bromouracil (BrU) strongly enhances the sensibility of cells to irradiation. In order to get insights about the mechanisms of the combination of chemo- and radiation chemistry, gas-phase studies of collisions between BrU or 5-bromouridine and low-energy electrons have been conducted. Cross-sections for production of radicals that can cause damage to DNA (mostly Br^- but also BrU^-) through DEA to halouracil are very large in the 10^{-14} cm^2 range, typically one or two order of magnitude larger than for thymine [44, 45]. Once introduced in the modelling of the complex cascade of reactions, those results deduced from gas-phase experiments may improve the strategies of radiotherapy.

4.7.1.7 Electron-induced proton transfer between nucleobases and peptides

DNA establishes a large number of specific interactions with proteins such as repressors, RNA polymerase, transcription factors etc. and the corresponding sites can be found in databases (e.g. http://arep.med.harvard.edu/dpinteract/). The presence of radiation-induced low-energy electrons can induce intermolecular tautomerization through the weak bonds established at contacts between nucleobases and protein recognition motifs. The mechanism is a barrier-less proton transfer from the proton donor towards the proton-accepting nucleobase (see Section 4.5.2). A model system has been studied by photoelectron spectroscopy (see Section 2.3.2.3) of the thymine–glycine complex anion [46]. Calculations show that the most stable neutral complexes correspond to the carbonyl and hydroxyl groups of glycine interacting with the proton donor and acceptor groups of thymine, creating two strong hydrogen bonds. Upon electron capture, the negative charge is first localized in the thymine π^* orbital. A barrier-less proton transfer then takes place from the carboxylic group of glycine to the O8 of thymine. The result is the stabilization of the anion excess electron by the transferred proton which is observed in the PES spectra.

4.7.2 Electron capture dissociation (ECD) of polypeptide cations

ECD is a widely used fragmentation technique in mass-spectrometry. Multiply charged cations are confined in presence of low-energy electrons. The electron capture leads to excited radicals that further fragment along pathways different from those encountered in usual MS/MS techniques such as CID or IRMPD.

$$[\mathrm{M}+n\mathrm{H}]^{n+} + \mathrm{e} \rightarrow [\mathrm{M}+n\mathrm{H}]^{(n-1)+**} \rightarrow \text{fragments}.$$

The exact mechanism of ECD seems to be still an object of debates [47] (in agreement with the heuristic observation "if something is understood, it is generally useless; if it turns out to be useful, most probably no one understands how it works") but nevertheless this technique allows the investigation of labile biomolecules. For example, one of the applications is the identification of post-translational modifications (PTMs) [48–51] such as glycolisation, carboxylation or phosphorylation. Other possible applications are, for example direct distinction between amino acids leucine and isoleucine with the same mass but different carbon positions in their side-chains [52] or discrimination between the presence of L- or D amino acids [53] in proteins (see Chapter 4.6). A comparable dissociation process exists for anions and is called electron detachment dissociation and is considered in references [54–56]. Electron detachment from anions can also be induced by photons [57]. In this chapter, we will first focus on the mass-spectrometry point of view and we will further examine modelling of ECD.

4.7.2.1 Mass-spectrometry considerations

Fragmentation of multiply charged ions is a very important tool for "top down" mass-spectrometric sequencing [58, 59] (see Section 3.2.2). Production of fragments can be performed either by slowly increasing internal energy or suddenly providing activation. In the former case, for example in collision-induced dissociation (CID) (see Section 3.2.2.2) or infrared multiphoton dissociation (IRMPD) (see Section 2.1.3.2.5), vibrational energy is fed into the ions at a much slower rate ("slow heating") than required for energy randomization among all degrees of freedom. Fast cleavage of the weakest bonds, most generally backbone amide bonds, then leads to predominantly b- and y-type ions (Figure 4.7.8). On the contrary, interactions between electrons and biomolecules responsible for ECD [60–64] bring amounts of energy larger than dissociation energies on extremely short timescales, much shorter than nuclear motions.

ECD has been reviewed in references [55, 59, 65]. Experimentally, ECD is most efficient at electron energies below 0.2 eV [52]. Thermal electrons extremely rapidly acquire energy when submitted to the radiofrequency (RF) fields present in instruments such as linear 2D, 3D

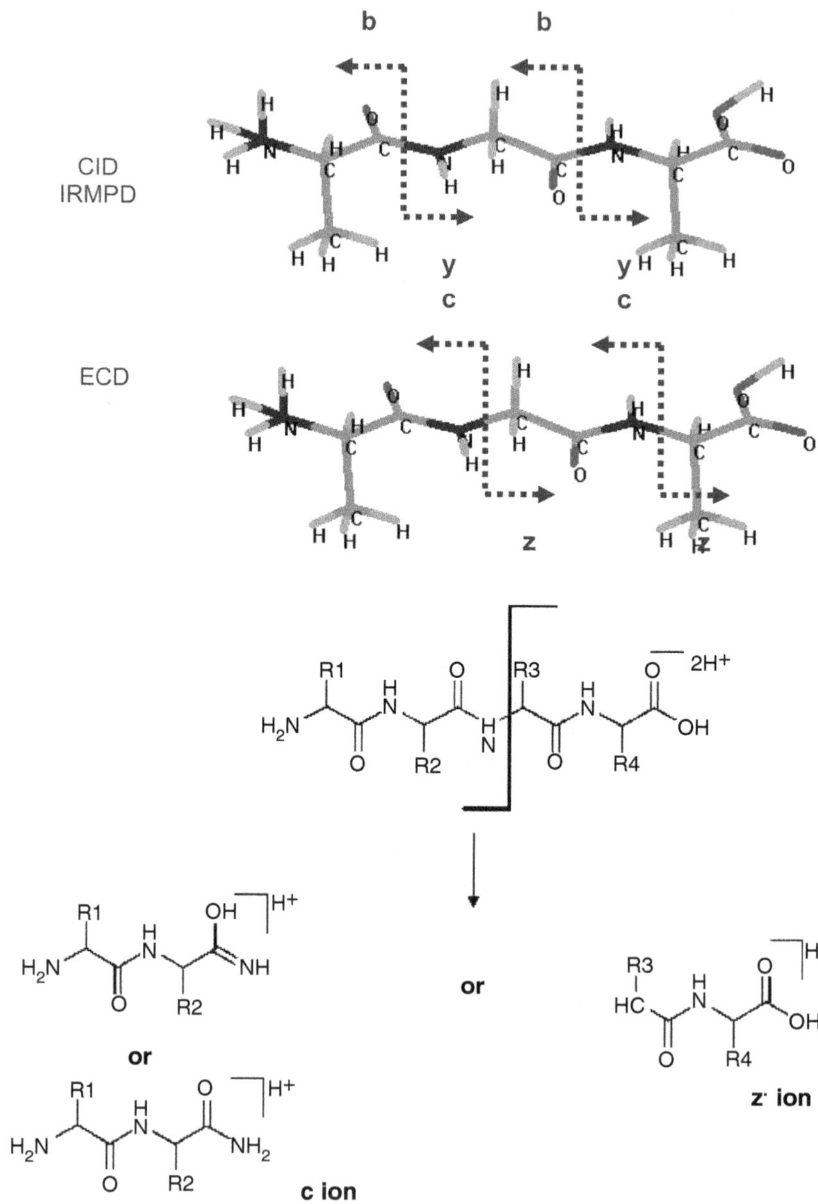

Figure 4.7.9 Comparison between fragmentation patterns observed in CID or IRMPD and ECD.

or time-of-flight quadrupole ion traps. ECD is thus mostly employed in FT-ICR and hybrid linear trap/FT-ICR mass spectrometers [66–76] (see Section 3.2.1.4). In order to overcome this limitation, an alternative scheme has been used in an RF linear quadrupole trap (LQT) [77]. First, multiply protonated peptide precursors to be sequenced are injected in a LQT. In a second step, negatively charged reactant ions (e.g. anthracene anions) are injected in the LQT and transfer their electrons to the peptide cations in ion–ion reactions. This initiates fragmentations leading to *c* and *z* ions as in ECD. A proposed alternative to the use of free electrons which energy is difficult to control is the use of metastable [78] or very highly excited Rydberg atoms [78]. Another scheme can be used in ion trapping devices [79] (see Section 3.2.1.2). The voltages applied to the trap are digitally controlled and can thus be dropped to zero during specific time frames lasting a few microseconds then allowing injection of decelerated electrons.

As in the preceding case of DEA to DNA building blocks (see Section 4.7.1.3), DEA takes place through temporary formation of electron–biomolecule complexes. Very different cleavage patterns, as compared with slow heating methods, are observed and correspond to selective rupture of single bonds. ECD gives rise mainly to c and $z^{\bullet}$ or $c^{\bullet}$ and z fragments due to the fast breaking of the backbone N—C_{α} bonds (although b ions can sometimes be also observed [80]). ECD also lead to selective cleavage of the rather weak disulphide bonds [81–87]. In addition to protons carrying the charges in *c* and *z* ions, an additional proton is abstracted from the precursor peptide. A larger percentage of backbone cleavages occur in ECD than in CID. For example, 75 of the 103 possible cleavages take place in cytochrome c allowing unambiguous protein identification and de novo sequencing [55].

One of the key advantages of ECD is its ability to randomly cleave peptide backbones while still preserving PTMs [88, 89]. The following example illustrates one of the possible uses of ECD. The mass-spectrum of the mono-phosphorylated peptide 48-FQ[pS]EEQQQT-EDELQDK-63 generated by digestion of β-casein is displayed in Figure 4.7.10. Triply charged ions have been obtained by electrospray, mass-selected and then submitted to fragmentation by ECD. Identification of fragments through precise determination of their masses show that the N-terminal fragments c_3—c_{15} contain the intact phosphorylated serine residue [pS]. The phosphoryl PTM is thus localized on the serine residue as confirmed by identification of the C-terminal *z* fragments.

ECD and IRMPD can provide complementary structural information. In a peptide containing a mannose–glycose–fucose glycan modification, IRMPD cleaves each glycosidic bond but leaves an intact backbone. The position of the peptide modification is then provided by ECD. The respective advantages of CID and ECD have been discussed in reference [55]. CID is the most sensitive but is not the best for sequencing because cleavages are site-specific and not as randomly distributed all along backbones as in ECD. CID can also raise problems in the case of peptides containing multiple arginine residues or PTMs.

4.7.2.2 Mechanisms of electron capture dissociation

The experimental observation of very different fragmentation patterns between CID and IRMPD on one hand and ECD on the other hand is attributed to a fundamental difference between the respective mechanisms. While CID and IRMPD are becoming reasonably well understood in terms of statistical (ergodic) redistribution of the excess vibrational energy brought by collisions or photon absorption, different explanations of the N—C_{α} [47, 63, 90, 91] and S–S [92, 93] bond rupture specificity of ECD have been proposed. In a multiply charged cation, the most favourable protonation sites are the basic side-chains of lysine, arginine and histidine [94] and the N-terminal. An incident electron is attracted by one of the protonated sites and attaches into a diffuse Rydberg-like orbital, forming a hypervalent radical [86]. The energy brought by this electron attachment is very large (several eV). The consequence of the electron capture is the formation of an excited transient reduced ion which can further fragment. A common fragment is $[M + (n-1)H]^{(n-1)+}$ corresponding to the loss of a hydrogen atom, as also observed in photo-dissociation processes (see Sections 2.1.4 and 4.7.1.5) [34, 95]. In both cases, non-ergodic fragmentation patterns are observed since strong bonds are dissociated in the presence of other weaker bonds.

A first considered explanation of the ECD fragmentation process has been the "hot H atom" model [65](Figure 4.7.11). In this model, the fast released H atom can attack the oxygen atom of an adjacent carbonyl group or disulphide bond forming radicals which then rearrange and dissociate. This model has been ruled out for the following reasons. Fast H atoms do not induce N–C_α breakings but rather bounce off [87, 96]. Synthetic peptides containing S–S bonds and charged with alkali atoms instead of protons also exhibit disulphide bond breaking without invoking the transfer of a fast H atom [92].

Other mechanisms leading to N—C_α or S—S cleavages have been investigated by means of ab initio quantum calculations of model systems, *N*-methylacetamide (NMA) [90], dimethyldisulphide (DMDS) [86] and a glycine hexamer mimic [63], in presence of a nearby positive charge (NH_4^+ or guanidinium). Isolated amide bond containing molecules

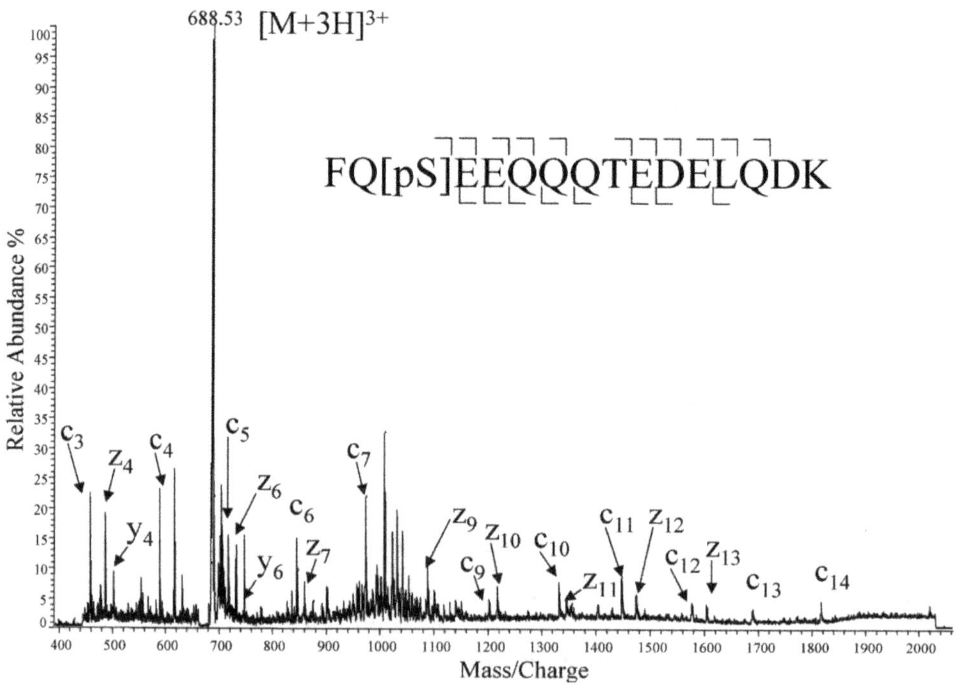

Figure 4.7.10 Electron capture dissociation mass-spectrum of the mono-phosphorylated peptide 48-FQ[pS]EEQQQTEDELQDK-63 (reprinted with permission from reference [79] ©2006 American Chemical Society).

[97] have negative valence electron affinities and small positive dipole-bound affinities. In the high-level quantum calculations of reference [90], the amide group of a peptide ion is modelled by NMA and electron attachment to a R—NH_3^+ group is modelled by the hypervalent methylammonium radical CH_3—NH_3. The relative energies of the (NMA...CH_3—NH_3) complex have then be computed. In presence of a positive charge carried by any remote protonated amino acid side-chain (the considered ions are multiply charged), the valence affinity of the considered amide group becomes positive. Following electron capture, direct bond cleavage is still endothermic but there are several long-lived excited states in which the amide C═O group acquires a very large proton affinity. A key point of this model is that the charge-stabilized anion radical is a superbase and its proton affinity is large enough to allow abstraction of a proton from any amino acid side-chain. A labile radical is then formed and further dissociates. This model thus provides an explanation of why ECD-induced bond cleavages occur preferentially in the vicinity of the most basic groups.

The role of the peptide conformation has been considered in a model system where an ammonium NH_4^+ is imbedded in a glycine hexamer mimic [63]. The H atoms of the NH_4^+ form hydrogen bonds to alternating carbonyl oxygen similarly a charged lysine side-chain would bind in a α-helical structure. When an electron is added, its orbital is not localized on NH_4 but rather delocalized on the three H-bonded carbonyls. After optimization, an H transfer takes place towards one of these carbonyls initiating one of the three N—C_α cleavages allowed by the helical structure and leaves a hydrogen-bonded neutral NH_3. This mechanism invoking the influence of conformation has common features with that observed in photoexcitation [43] although in ECD, the electronic energy is brought by the electron

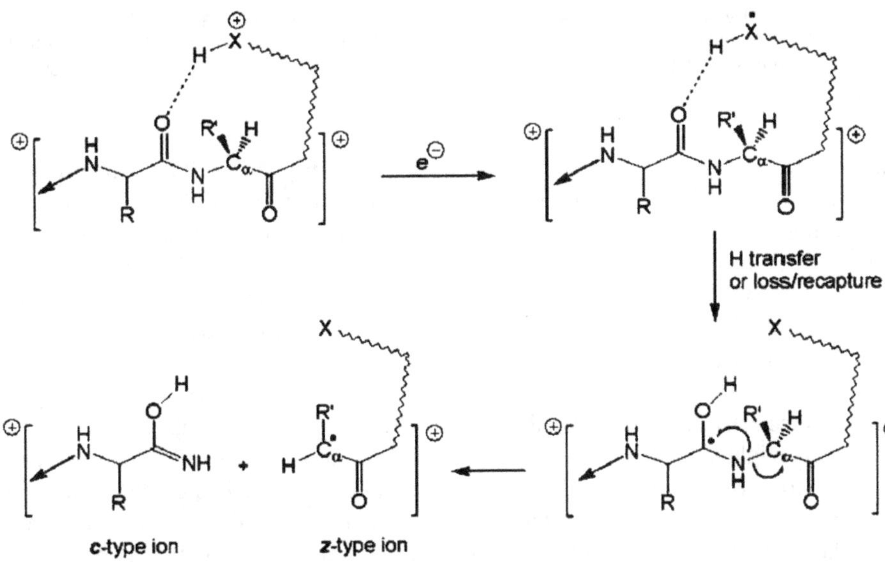

Figure 4.7.11 Proposed mechanism for ECD. Electron attachment takes place at one of the protonated sites (Lys, Arg, His side-chain or N-terminal) of a multiply protonated peptide. A radical intermediate is formed and further dissociate leading to a "hot" hydrogen atom. This H atom is captured by a nearby C═O group yielding an intermediate that cleaves on the adjacent N—C_α bond (reprinted with permission from reference [90] ©2005 Elsevier).

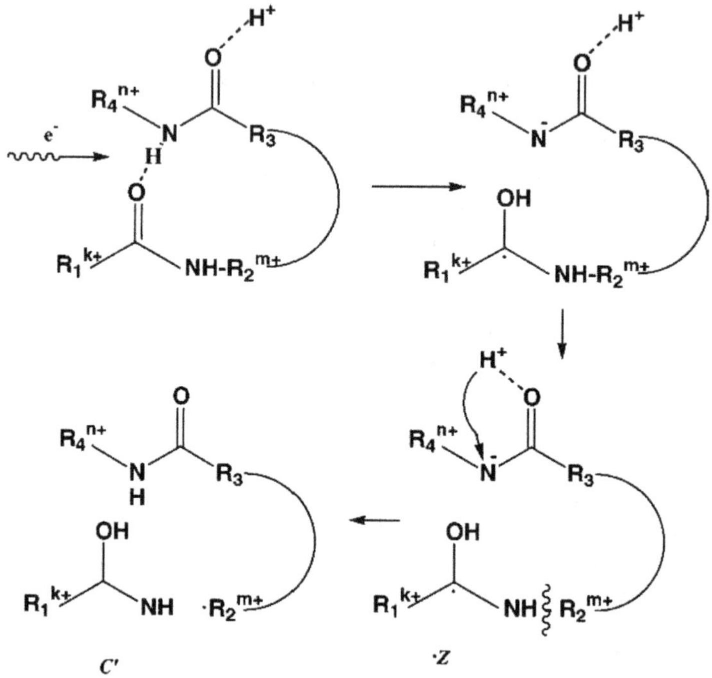

Figure 4.7.12 In a multiply charged peptide, a —N—H...O═C— hydrogen bond is formed between backbone amide groups and captures a low-energy electron. A hydrogen atom is transferred to the C═O group. The nascent negatively charged N atom is neutralized by the most weakly bound adjacent proton and the N—C_α bond cleavage takes place (reprinted with permission from reference [47] ©2006 Elsevier).

capture and the number of electrons increases by one. In photoexcitation of protonated tryptophan, for example an electron is promoted from a π orbital of the indole side-chain to a π^* orbital. This electron can have a large transfer probability towards the protonated side only if the distance between the indole ring and the protonated side is not too large. Thus the geometry of the tryptophan conformer can thus be more or less favourable for the dissociation along the N—H and C_α—N coordinates (see Figure 4.7.7). Loss of H atoms and C_α—N bond cleavages are then observed as in ECD.

For rupture of disulphide bonds in ECD, DMDS has been considered as a model system. The vertical electron affinity of isolated DMDS is strongly negative and its adiabatic electron affinity is only slightly positive (0.2–0.3 eV) [81]. Calculations [86] predict that the presence of a nearby positive charge strongly stabilizes the anion and vertical electron attachment becomes exothermic. Direct attachment of an electron then persists during a time sufficient to allow the stretching of S—S bond until rupture with retention of a hydrogen atom.

Another model supported by the correlation between experimental observations of ECD of Trp-cage protein dications (see Section 4.2.6) and molecular dynamics simulations has been proposed [47]. It is supposed that the low-energy electrons are captured on neutral hydrogen bonds established between backbone N—H and C═O groups such as those encountered in α-helices and β-sheet structures (see Section 4.2.2.2). Negatively charged sites are then formed at the N atoms and simultaneously H atoms are transferred to the C═O groups. The abstraction of a proton by a superbase radical anion in the model of reference [90] is here replaced by an intramolecular proton transfer.

$$\text{—N—H...O═C—} + e \rightarrow \text{—N}^- + {}^{\circ}\text{C(OH)—}$$

The resulting labile radical then fragments as in the mechanism depicted in Figure 4.7.9.

The frequencies of the hydrogen-bond sites (typically 15 H-bonds in a Trp-cage ion) of four Trp-cage stereoisomers obtained by substitution of L- to D-amino acids (see Chapter 4.6) (all-L, D-Tyr^3, D-Gln^5 and D-Leu^7) have been determined through molecular dynamics simulations. A strong correlation between these frequencies and the experimentally observed N—C_α bond cleavages supports this ECD mechanism (Figure 4.7.11) that explains the relative abundances of the observed *z* ions and predict the changes in mass-spectra induced by the chiral substitutions (Figure 4.7.12).

REFERENCES

1. Touati A, Hervé du Penhoat MA, Bailly-Despiney J, Gobert F, Champion C, Fayard B, Abel F, L'Hoir A, Moulin J, Sabatier L, *et al.*: Biological implication of atomic collisions at the molecular level. *Nuclear Instruments and Methods B* 1997, 132:276–279.
2. Sanche L: Low energy electron-driven damage in biomolecules. *European Physical Journal D* 2005, 35:367–390.
3. Swiderek P: Fundamental processes in radiation damage of DNA. *Angewandte Chemie-International Edition* 2006, 45:4056–4059.
4. Stivers J: Inroduction: DNA damage and repair. *Chemical Reviews* 2006, 106:213–214.
5. Kamiya H: Mutagenic potential of damaged nucleic acids produced by reactive oxygen/nitrogen species: approaches using synthetic oligonucleotides and nucleotides. *Nucleic Acids Research* 2003, 31:517–531.
6. Turecek F: Modeling nucleobase radicals in the mass-spectrometer. *Journal of Mass Spectrometry* 1998, 33:779–795.
7. Ji YJ, Xia YY, Zhao MW, Huang BD, Li F: Theoretical study of the OH· reaction with cytosine. *Journal of Molecular Structure (Theochem)* 2005, 723:123–129.
8. Choi KW, Lee JH, Kim SK: Ionization spectroscopy of a DNA base: vacuum-ultraviolet mass-analyzed threshold spectroscopy of jet-cooled thymine. *Journal of the American Chemical Society* 2005, 127:15674–15675.
9. Abdoul-Carime H, Gohlke S, Fischbach E, Scheike J, Illenberger E: Thymine excision from DNA by subexcitation electrons. *Chemical Physics Letters* 2004, 387:267–270.
10. Ptasinska S, Denifl S, Scheier P, Illenberger E, Mark TD: Bond- and site-selective loss of H atoms from nucleobases by very-low-energy electrons (<3 eV). *Angewandte Chemie-International Edition* 2005, 44:6941–6943.
11. Ptasinska S, Denifl S, Scheier P, Mark TD: Inelastic electron interaction (attachment/ionization) with deoxyribose. *Journal of Chemical Physics* 2004, 120:8505–8511.
12. Deby-Dupont G, Deby C, Lamy M: Actual data concerning oxygen toxicity. *Reanimation* 2002, 11:28–39.
13. Sanche L: Nanoscopic aspects of radiobiological damage: fragmentation induced by secondary low-energy electrons. *Mass Spectrometry Reviews* 2002, 21:349–369.
14. Li XF, Sevilla MD, Sanche L: Density functional theory studies of electron interaction with DNA: can zero eV electrons induce strand breaks? *Journal of the American Chemical Society* 2003, 125: 13668–13669.

15. Boudaïffa B, Cloutier P, Hunting D, Huels MA, Sanche L: Resonant formation of DNA strand breaks by low-energy (3–20 eV) electrons. *Science* 2000, 287:1658–1660.
16. Martin F, Burrow PD, Cai ZL, Cloutier P, Hunting D, Sanche L: DNA strand breaks induced by 0–4 eV electrons: the role of shape resonances. *Physical Review Letters* 2004, 93:068101–068105.
17. Denifl S, Zappa F, Mähr I, Lecointre J, Probst M, Märk TD, Scheier P: Mass spectrometric investigation of anions formed upon free electron attachment to nucleobase molecules and clusters embedded in superfluid helium droplets. *Physical Review Letters* 2006, 97:043201–043204.
18. Oyler NA, Adamowicz L: Theoretical ab-initio calculations of the electron-affinity of thymine. *Chemical Physics Letters* 1994, 219:223–227.
19. Desfrancois C, Periquet V, Bouteiller Y, Schermann JP: Valence and dipole binding of electrons to uracil. *Journal of Physical Chemistry A* 1998, 102:1274–1278.
20. Desfrancois C, Abdoul-Carime H, Schermann JP: Electron attachment to isolated nucleic acid bases. *Journal of Chemical Physics* 1996, 104:7792–7794.
21. Hendricks JH, Lyapustina SA, de Clercq HL, Bowen KH: The dipole bound-to-covalent anion transformation in uracil. *Journal of Chemical Physics* 1998, 108:8–11.
22. Desfrancois C, Abdoul-Carime H, Schulz CP, Schermann JP: Laser separation of geometrical isomers of weakly-bound molecular complexes. *Science* 1995, 269:1707–1709.
23. Hanel G, Gstir B, Denifl S, Scheier P, Probst M, Farizon B, Farizon M, Illenberger E, Mark TD: Electron attachment to uracil: effective destruction at sub-excitation energies. *Physical Review Letters* 2003, 90:188104–188108.
24. Aflatooni K, Scheer AM, Burrow PD: Dissociative electron attachment in uracil: total anion yield. *Chemical Physics Letters* 2005, 408:426–428.
25. Ptasinska S, Denifl S, Mroz B, Probst M, Grill V, Illenberger E, Scheier P, Mark TD: Bond selective dissociative electron attachment to thymine. *Journal of Chemical Physics* 2005, 123: 124302–124310.
26. Huber D, Beikircher M, Denifl S, Zappa F, Matejcik S, Bacher A, Grill V, Märk T, Scheier P: High resolution dissociative electron attachment to gas phase adenine. *Journal of Chemical Physics* 2006, 125:084304.
27. Denifl S, Ptasinska S, Probst M, Hrusak J, Scheier P, Mark TD: Electron attachment to the gas-phase DNA bases cytosine and thymine. *Journal of Physical Chemistry A* 2004, 108:6562–6569.
28. Aflatooni K, Scheer AM, Burrow PD: Total dissociative electron attachment cross sections for molecular constituents of DNA. *Journal of Chemical Physics* 2006, 125:054301.
29. Abdoul-Carime H, Gohlke S, Illenberger E: Site-specific dissociation of DNA bases by slow electrons at early stages of irradiation. *Physical Review Letters* 2004, 92:168103–168107.
30. König C, Kopyra J, Bald I, Illenberger E: Dissociative attachment to phosphoric acid esters: the direct mechanism for single strand breaks in DNA. *Physical Review Letters* 2006, 97:018105–018109.
31. Alvarado F, Bari S, Hoekstra R, Schlathölter T: Quantification of ion-induced molecular fragmentation of isolated 2-deoxy-D-ribose molecules. *Physical Chemistry Chemical Physics* 2006, 8:1922–1928.
32. Sulzer P, Ptasinska S, Zappa F, Mielewska B, Milosavljevic AR, Scheier P, Märk T, Bald I, Huels MA, Illenberger E: Dissociative electron attachment to furan, tetrahydrofran, and fructose. *Journal of Chemical Physics* 2006, 125:044304.
33. Ptasinska S, Denifl S, Gohlke S, Scheier P, Illenberger E, Mork TD: Decomposition of thymidine by low-energy electrons: implications for the molecular mechanisms of single-strand breaks in DNA. *Angewandte Chemie-International Edition* 2006, 45:1893-1896.
34. Domcke W, Sobolewski AL: *Ab initio* studies on the photophysics of the guanine–cytosine base pair. *Physical Chemistry Chemical Physics.* 2004, 6:2763–2771.
35. Kang H, Jouvet C, Dedonder-Lardeux C, Martrenchard S, Gregoire G, Desfrancois C, Schermann JP, Barat M, Fayeton JA: Ultrafast deactivation mechanisms of protonated aromatic amino acids following UV excitation. *Physical Chemistry Chemical Physics* 2005, 7:394–398.

36. Burrow PD, Gallup GA: Remote bond breaking by interacting temporary anion states. *Journal of Chemical Physics* 2006, 125:154309.
37. Pan X, Sanche L: Dissociative electron attachment to DNA basic constituents: the phosphate group. *Chemical Physics Letters* 2006, 421:404–408.
38. Berdys J, Skurski P, Simons J: Damage to model DNA fragments by 0.25–1.0 eV electrons attached to a thymine π^* orbital. *Journal of Physical Chemistry B* 2004, 108:5800–5805.
39. Berdys J, Anusiewicz I, Skurski P, Simons J: Theoretical study of damage to DNA by 0.2–1.5 eV electrons attached to cytosine. *Journal of Physical Chemistry A* 2004, 108:2999–3005.
40. Dabkowska I, Rak J, Gutowski M: DNA strand breaks induced by concerted interaction of H radical and low-energy electrons. A computational study on the nucleotide of cytosine. *European Journal of Physics D* 2005, 35:424–436.
41. Barrios R, Skurski P, Simons J: Mechanism for damage to DNA by low-energy electrons. *Journal of Physical Chemistry B* 2002, 106:7991–7994.
42. Berdys J, Anusiewicz I, Skurski P, Simons J: Damage to model DNA fragments from very low-energy ($<$1 eV) electrons. *Journal of the American Chemical Society* 2004, 126:6441–6447.
43. Kang H, Dedonder-Lardeux C, Jouvet C, Martrenchard S, Gregoire G, Desfrancois C, Schermann JP, Barat M, Fayeton JA: Photo-induced dissociation of protonated tryptophan $TrpH^+$: a direct dissociation channel in the excited states controls the hydrogen atom loss. *Physical Chemistry Chemical Physics* 2004, 6:2628–2632.
44. Abouaf R, Pommier J, Dunet H: Negative ions in thymine and 5-bromouracil produced by low-energy electrons. *International Journal of Mass Spectrometry* 2003, 226:397–403.
45. Abdoul-Carime H, Limao-Vieira P, Gohlke S, Petrushko I, Mason NJ, Illenberger E: Sensitization of 5-bromouridine by slow electrons. *Chemical Physics Letters* 2004, 393:442–447.
46. Dabkowska I, Rak J, Gutowski M, Nilles JM, Stokes ST, Radisic D, Bowen KH: Barrier-free proton transfer in anionic complex of thymine with glycine. *Physical Chemistry Chemical Physics* 2004, 6:4351–4357.
47. Patriksson A, Adams C, Kjeldsen F, Raber J, van der Spoel D, Zubarev RA: Prediction of N-C-α bond cleavage frequencies in electron capture dissociation of Trp-cage dications by force-field molecular dynamics simulations. *International Journal of Mass Spectrometry* 2006, 248:124–135.
48. Kelleher RL, Zubarev RA, Bush K, Furie B, Furie BC, McLafferty FW, Walsh CT: Localization of labile posttranslational modifications by electron capture dissociation: the case of γ-carboxyglutamic acid. *Analytical Chemistry* 1999, 71:4250–4253.
49. Kjeldsen F, Haselmann KF, Budnik BA, Sorensen ES, Zubarev RA: Complete characterization of post-translational modification sites in the bovine milk protein PP3 by tandem mass spectrometry with electron capture dissociation as the last stage. *Analytical Chemistry* 2003, 75:2355–2361.
50. Mirgorodskaya E, Roepstorff P, Zubarev RA: Localization of O-glycosylation sites in peptides by electron capture dissociation in a Fourier-transform mass spectrometer. *Analytical Chemistry* 1999, 71:4431–4436.
51. Renfrow MB, Cooper HJ, Tomana M, Kulhavy R, Hiki Y, Toma K, Emmett MR, Mestecky J, Marshall AG, Novak J: Determination of aberrant O-glycosylation in the IgA1 hinge region by electron capture dissociation Fourier-transform-ion cyclotron resonance mass spectrometry. *Journal of Biological Chemistry* 2005, 280:19136–19145.
52. Kjeldsen F, Haselmann KF, Sorensen ES, Zubarev RA: Distinguishing of Ile/Leu amino acid residues in the PP3 protein by (hot) electron capture dissociation in Fourier-transform ion cyclotron resonance mass spectrometry. *Analytical Chemistry* 2003, 75:1267–1274.
53. Adams CM, Kjeldsen F, Zubarev RA, Budnik BA, Haselmann KF: Electron capture dissociation distinguishes a single D-amino acid in a protein and probes the tertiary structure. *Journal of the American Society for Mass Spectrometry* 2004, 15:1087–1098.
54. Budnik BA, Haselmann KF, Zubarev RA: Electron detachment dissociation of peptide di-anions: an electron-hole recombination phenomenon. *Chemical Physics Letters* 2001, 342:299–302.

55. Cooper HJ, Hakansson K, Marshall AG: The role of electron capture dissociation in biomolecular analysis. *Mass Spectrometry Reviews* 2005, 24:201–222.
56. Kalli A, Hakansson K: Preferential cleavage of S–S and C–S bonds in electron detachment and infrared multiphoton dissociation of disulfide-linked peptide anions. *International Journal of Mass Spectrometry* 2007, 263:71–81 .
57. Gabelica V, Tabarin T, Antoine R, Rosu F, Compagnon I, Broyer M, De Pauw E, Dugourd P: Electron photodetachment dissociation of DNA polyanions in a quadrupole ion trap mass spectrometer. *Analytical Chemistry* 2006, 78:6564–6572.
58. Zhai H, Han XM, Breuker K, McLafferty FW: Consecutive ion activation for top down mass spectrometry: improved protein sequencing by nozzle-skimmer dissociation. *Analytical Chemistry* 2005, 77:5777–5784.
59. Wysocki VH, Resing KA, Zhang QF, Cheng GL: Mass spectrometry of peptides and proteins. *Methods* 2005, 35:211–222.
60. Zubarev RA, Kelleher NL, McLafferty FW: Electron capture dissociation of multiply charged protein cations. A non-ergodic process. *Journal of the American Chemical Society* 1998, 120: 3265–3266.
61. Zubarev RA: Electron-capture dissociation tandem mass spectrometry. *Current Opinion in Biotechnology* 2004, 15:12–16.
62. McLafferty FW, Horn DM, Breuker K, Ge Y, Lewis MA, Cerda B, Zubarev RA, Carpenter BK: Electron capture dissociation of gaseous multiply charged ions by Fourier-transform ion cyclotron resonance. *Journal of the American Society for Mass Spectrometry* 2001, 12:245–249.
63. Breuker K, Oh HB, Lin C, Carpenter BK, McLafferty FW: Non-ergodic and conformational control of the electron capture dissociation of protein cations. *Proceedings of the National Academy of Sciences of the United States of America* 2004, 101:14011–14016.
64. Breuker K, McLafferty FW: The thermal unfolding of native cytochrome c in the transition from solution to gas phase probed by native electron capture dissociation. *Angewandte Chemie-International Edition* 2005, 44:4911–4914.
65. Zubarev RA, Haselmann KF, Budnik B, Kjeldsen F, Jensen F: Towards an understanding of the mechanism of electron-capture dissociation: a historical perspective and modern ideas. *European Journal of Mass Spectrometry* 2002, 8:337–349.
66. Fagerquist CK, Hudgins RR, Emmett MR, Hakansson K, Marshall AG: An antibiotic linked to peptides and proteins is released by electron capture dissociation Fourier transform ion cyclotron resonance mass spectrometry. *Journal of the American Society for Mass Spectrometry* 2003, 14:302–310.
67. Cooper HJ, Hudgins RR, Marshall AG: Electron capture dissociation Fourier transform ion cyclotron resonance mass spectrometry of cyclo-depsipeptides, branched peptides, and ε-peptides. *International Journal of Mass Spectrometry* 2004, 234:23–35.
68. Cooper HJ, Hudgins RR, Hakansson K, Marshall AG: Secondary fragmentation of linear peptides in electron capture dissociation. *International Journal of Mass Spectrometry* 2003, 228: 723–728.
69. Cooper HJ, Hudgins RR, Hakansson K, Marshall AG: Characterization of amino acid side chain losses in electron capture dissociation. *Journal of the American Society for Mass Spectrometry* 2002, 13:241–249.
70. Yim YH, Kim B, Ahn S, So HY, Lee S: Evaluation of the internal temperature of a 8.6 kDa protein cation exposed to a hot dispenser cathode employed in electron capture dissociation mass spectrometry. *Rapid Communications in Mass Spectrometry* 2006, 20:1918–1924.
71. Mihalca R, van der Burgt YEM, McDonnell LA, Duursma MC, Cerjak I, Heck AJR, Heeren RMA: Combined infrared multiphoton dissociation and electron-capture dissociation using colinear and overlapping beams in Fourier-transform ion cyclotron resonance mass spectrometry. *Rapid Communications in Mass Spectrometry* 2006, 20:1838–1844.

72. Kjeldsen F, Savitski MM, Adams C, Zubarev RA: Determination of the location of positive charges in gas-phase polypeptide polycations by tandem mass spectrometry. *International Journal of Mass Spectrometry* 2006, 252:204–212.
73. Fung YME, Liu H, Chan TW: Electron capture dissociation of peptides metalated with alkaline-earth metal ions. *Journal of the American Society of Mass Spectrometry* 2006, 17:757–771.
74. Anusiewicz W, Berdys-Kochanska J, Simons J: Electron attachment step in electron capture dissociation (ECD) and electron transfer dissociation (ETD). *Journal of Physical Chemistry A* 2005, 109:5801–5813.
75. Antoine R, Broyer M, Chamot-Rooke J, Dedonder C, Desfrancois C, Dugourd P, Gregoire G, Jouvet C, Onidas D, Poulain P: Comparison of the fragmentation pattern induced by collisions, laser excitation and electron capture. Influence of the initial excitation. *Rapid Communications in Mass Spectrometry* 2006, 20:1648–1652.
76. O'Connor PB, Lin C, Cournoyer JC, Pittman JL, Belyayev M, Budnik B: Long-lived electron capture dissociation product ions experience radical migration via hydrogen abstraction. *Journal of the American Society for Mass Spectrometry* 2006, 17:576–585.
77. Syka JEP: Peptide and protein sequence analysis by electron transfer dissociation mass spectrometry. *Proceedings of National Academic of Sciences of the United States of America* 2004, 101:9528–9533.
78. Misharin AS, Silivra OA, Kjekdsen F, Zubarev R: Dissociation of peptide ions by fast atom bombardment in a quadrupole ion trap *Rapid Communications in Mass Spectrometry* 2005, 19:2163–2171; Carette M, Zerega Y, March RE, Perrier P, André J: Coupling of a Rydberg electron capture ion source with a mass filter. *Rapid Communications in Mass Spectrometry* 2000, 14:979–986.
79. Ding L, Brancia FL: Electron capture dissociation in a digital ion trap mass spectrometer. *Analytical Chemistry* 2006, 78:1995–2000.
80. Cooper HJ: Investigation of the presence of b ions in electron capture dissociation mass spectra. *Journal of the American Society for Mass Spectrometry* 2005, 16:1932–1940.
81. Carles S, Lecomte F, Schermann JP, Desfrancois C, Xu S, Nilles JM, Bowen KH, Berges J, Houée-Levin C: Non-dissociative electron capture by disulfide bonds. *Journal of Physical Chemistry A* 2001, 105:5622–5626.
82. Fung YME, Kjeldsen F, Silivra OA, Chan TWD, Zubarev RA: Facile disulfide bond cleavage in gaseous peptide and protein cations by ultraviolet photodissociation at 157 nm. *Angewandte Chemie-International Edition* 2005, 44:6399–6403.
83. Bergès J, Richards G, Rauk A, Houée-Levin C: QM/MM study of electron addition on protein disulfide bonds. *Chemical Physics Letters* 2006, 421:63–67.
84. Modelli A, Jones D: Temporary anion states and dissociative electron attachment in diphenyl disulfide. *Journal of Physical Chemistry A* 2006, 110:10219–10224.
85. Zubarev RA, Kruger NA, Fridriksson EK, Lewis MA, Horn DM, Carpenter BK, McLafferty FW: Electron capture dissociation of gaseous multiply-charged proteins is favored at disulfide bonds and other sites of high hydrogen atom affinity. *Journal of the American Chemical Society* 1999, 121:2857–2862.
86. Sawicka A, Skurski P, Hudgins RR, Simons J: Model calculations relevant to disulfide bond cleavage via electron capture influenced by positively charged groups. *Journal of Physical Chemistry B* 2003, 107:13505–13511.
87. Uggerud E: Electron capture dissociation of the disulfide bond-a quantum chemical model study. *International Journal of Mass Spectrometry* 2005, 235:45–50.
88. Stensballe A, Jensen ON, Olsen JV, Haselmann KF, Zubarev RA: Electron capture dissociation of singly and multiply phosphorylated peptides. *Rapid Communications in Mass Spectrometry* 2000, 14:1793–1800.
89. Cournoyer JC, Lin C, O'Connor PB: Detecting deamidation products in proteins by electron capture dissociation. *Analytical Chemistry* 2006, 78:1264–1271.

90. Systrad EA, Turecek F: Toward a general mechanism of electron capture dissociation. *Journal of the American Society of Mass Spectrometry* 2005, 16:208–224.
91. Anusiewicz I, Berdys-Kochanska J, Skurski P, Simons J: Simulating electron transfer attachment to a positively charged model peptide. *Journal of Physical Chemistry A* 2006, 110:1261–1266.
92. Sobczyk M, Skurski P, Simons J: Dissociative low-energy electron attachment to the C–S bond of H_3CSCH_3 influenced by Coulomb stabilization. *Advances in Quantum Chemistry* 2005, 48:239–251.
93. Bakken V, Helgader T, Uggerud E: Models of fragmentations induced by electron attachment to protonated peptides. *European Journal of Mass Spectrometry* 2004, 10:625–638.
94. Sterner JL: Apparent proton affinities of highly charged peptide ions. *Journal of the American Society of Mass Spectrometry* 1999, 10:403–491.
95. Kang H, Dedonder-Lardeux C, Jouvet C, Gregoire G, Desfrancois C, Schermann JP, Barat M, Fayeton JA: Control of bond-cleaving reactions of free protonated tryptophan ion by femtosecond laser pulses. *Journal of Physical Chemistry A* 2005, 109:2417–2420.
96. Demirev PA: Generation of hydrogen radicals for reactivity studies in Fourier-transform ion cyclotron resonance mass spectrometry. *Rapid Communications in Mass Spectrometry* 2000, 14: 777–781.
97. Seydou M, Modelli A, Lucas B, Konate K, Desfrancois C, Schermann JP: Electron attachment to strongly polar clusters: formamide molecule and clusters. *European Physical Journal D* 2005, 35:199–205; Maeyama T, Negishi Y, Tsukuda T, Yagi I, Mikami N: Electron localization in negatively charged formamide clusters studied by photodetachment spectroscopy. *Physical Chemistry Chemical Physics* 2006, 8:827–833.

– 4.8 –

Large Biomolecular Systems in the Gas-Phase

GENERAL FEATURES

Some of the experimental methods and instruments developed for the investigation of small- or intermediate-size systems can be extended to biological entities such as viruses, cells or bacteria still isolated in the gas-phase. In the reverse way, techniques devised for macroscopic systems can be downsized and the borders between microscopic, mesoscopic and macroscopic worlds slowly vanish. Meanwhile, modelling approaches extending complex structure studies and moderate-size molecular dynamics are adapted to weakly bound assemblies of biopolymers.

In this chapter, we examine experimental gas-phase and modelling studies involving large biomolecular systems. After some considerations about problems encountered in mass-spectrometry of large mass systems, we will consider examples of studies of monomeric or multimeric assemblies such as adenosine triphosphate (ATP)-synthase and cytochrome complexes, chaperon proteins and ribosomes and then consider viruses and bacteria.

4.8.1 Mass spectrometry and ion mobility studies of large assemblies

Gas-phase and solution-phase properties of protein complexes studied by mass-spectrometry are reviewed in reference [1]. Practical uses of mass-spectrometry for the detection of micro-organisms are reviewed in references [2, 3]. Although some studies have been performed up to masses in excess of 10^8 Da [4, 5], experimental methods devised for the study of intermediate-size biomolecules such as peptides or oligonucleotides cannot usually be extended in a straightforward fashion to the investigation of large protein complexes and intact microorganisms such as viruses and bacteria. Among problems encountered for extending the available mass range, one can note in particular the detection of signals issued from secondary emission-based detectors and the observed square-root relationship between the average number of charges and the mass of electrosprayed ions. Electron multipliers such as micro-channel plates (MCP) are widely used detectors in mass spectrometers and have an efficiency of ca. 80% for ions with m/z ratios up to 10^4 at impact energy of 20 keV but the efficiency drops down to ca. 3% at 5 keV. One would need an acceleration voltage of 10^6 V to accelerate an ion with a *m/z* ratio up of 10^5 to the velocity of 10^4 m/s required to be above the detection threshold of MCPs [6]. Daly detectors overcome this problem [7]. Heavy ions are attracted by a 20–30 kV voltage onto a polished metal surface and expel secondary electrons. Those electrons are repelled and hit a scintillator emitting photons that are detected by a photomultiplier. Other types of detectors have thus been devised for large biomolecular systems such as energy-sensitive (cryogenic bolometers [8]), charge-sensitive [9] and optical detectors reviewed in reference [6]. In presence of a large number of charges, mass-spectra can become extremely difficult or even impossible to disentangle [4, 10]. The use of MALDI (see Section 3.1.5) is a first possibility to overcome this problem since MALDI produces ions only carrying one or two charges and this proves to work for bacteria [11]. However, interactions between the matrix and weakly bound non-covalent complexes such as viruses [12] or edifices lacking rigid walls such as cells [13] can endanger those fragile entities. Another possibility is the use of electrospray ionization with a mean to strongly reduce the number of charges, for example with a radioactive source [5]. The use of a supersonic nebulizing gas ("electrosonic spray ionization") also reduces the extent of charging during the formation of ions in an electrospray source [14, 15].

4.8.1.1 ATP-synthase

Very hydrophobic biomolecules such as membrane proteins represent challenges for mass-spectrometry [16]. Detergents are required as solubilizing agents and prevent them from aggregation in solution. Problems are then encountered when classical methods such as MALDI or ESI (see Section 3.1) are used. Laser-induced liquid bead ion desorption (LILBID) (see Section 3.1.4) is more tolerant to the presence of detergents and moreover provides narrow distributions of relatively low-charge states. This has allowed the determination of the quaternary structure of subunits of ATP-synthase enzymes [17, 18]. ATP-synthases are found in membranes of mitochondria, chloroplasts and bacteria. Their crucial role is production of the "biological fuel" ATP from adenosine diphosphate (ADP) in the reaction ADP + P → ATP. The functioning of an ATP-synthase is similar to a motor and includes a rotor and a stator. The synthesis of ATP arises from conversion of the gradient of H^+ or Na^+ ions across a membrane. A part of the rotor, called F_0, is embedded in the membrane and is a multimeric assembly of *c* subunits (Figure 4.8.1). Those *c* subunits form a ring around a central pore and their number *n* is equal to the number of ion binding sites. A full rotation of the rotor gates *n* ions across the membrane and leads to production of three ATP molecules. This number *n* is thus an important characteristic and varies between 10 and 15 from one living creature to another. The LILBID method allows a relatively rapid (within a day) determination of this number using a precise determination of the stoichiometry of *c* rings [19].

4.8.1.2 Cytochrome bc_1 complex

The proton flow used for synthesizing ATP is produced by a series of trans-membrane electron carriers forming the electron transport chain in mitochondria and photosynthetic bacteria. Four membrane complexes (I–IV) and two freely diffusible electron carriers, ubiquinone (Q) and cytochrome *c*, are involved (Figure 4.8.1). Complex III (cytochrome *c* reductase complex or cytochrome bc_1 complex) transfers electrons to two molecules of cytochrome *c* and moves two protons across the membrane. Four electrons are removed from four cytochrome *c* molecules [20–22] by complex IV (cytochrome *c* oxidase complex) and are

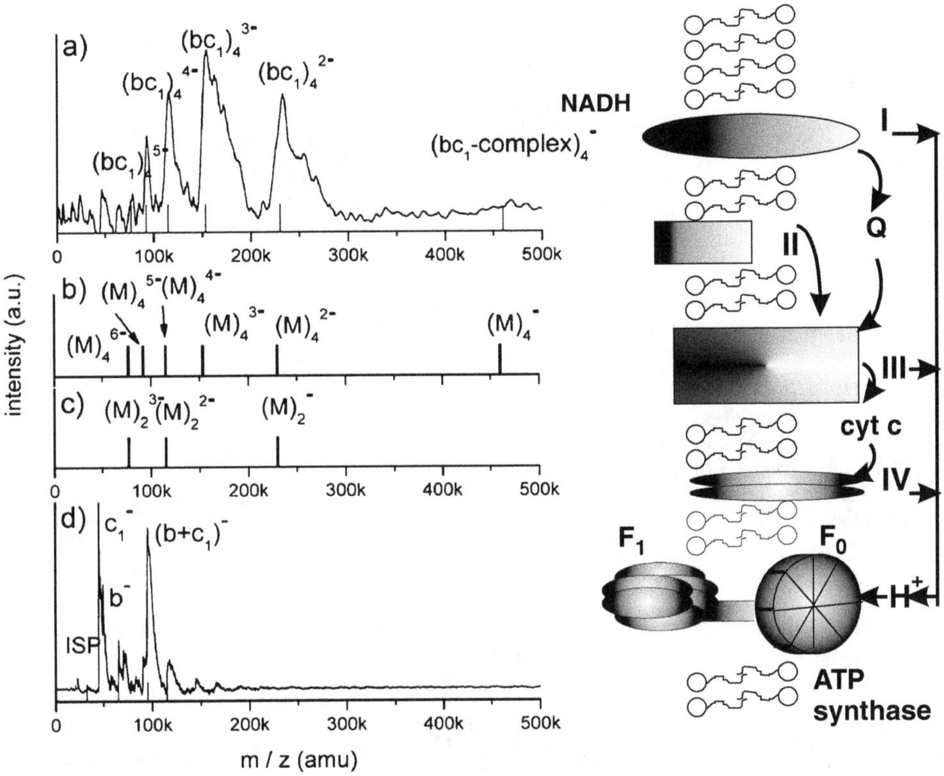

Figure 4.8.1 Left: (a) Experimental laser-induced liquid bead ion desorption mass spectrum of a bacterial bc_1 complex at low desorption laser intensity. (b and c) Respective theoretical charge distributions of the bc_1 complex as a tetramer or a dimer. The bc_1 complex quaternary structure is tetrameric. (d) Experimental LILBID mass spectrum at a larger desorption laser intensity. The subunits of the bc_1 complex appear individually (by courtesy of N. Morgner and B. Brutschy). Right: Simplified scheme of the respiratory chain displaying the bc_1 complex (III) and the ATP-synthase.

used with protons to reduce molecular oxygen to water. The LILBILD method allows the use of an amount of protein present in one single droplet in the femto- to atto-mole region. With this technique, the quaternary structure of the cytochrome bc_1 complex of the *Paracoccus denitrificans* soil bacterium has been determined [23]. The cytochrome bc_1 complex is composed of cytochrome *b* and cytochrome c_1 subunits with haem *c* groups and an iron–sulphur protein (Rieske). The highly hydrophobic cytochrome *b* proteins span the whole membrane and transfer electrons from its outer face to its inner face.

4.8.1.3 Proteasomes

Ion mobility techniques can be extended to very large masses [5, 24–26]. The obtained information is then not as precise and detailed as in usual ion mobility methods (see Section 3.3.1) but can nevertheless complement mass-spectrometry methods. Proteasomes have, for example been studied by gas-phase electrophoretic mobility molecular analysis (GEMMA; see Section 3.3.1). A proteasome [27] is a catalytic complex expressed in the nucleus and

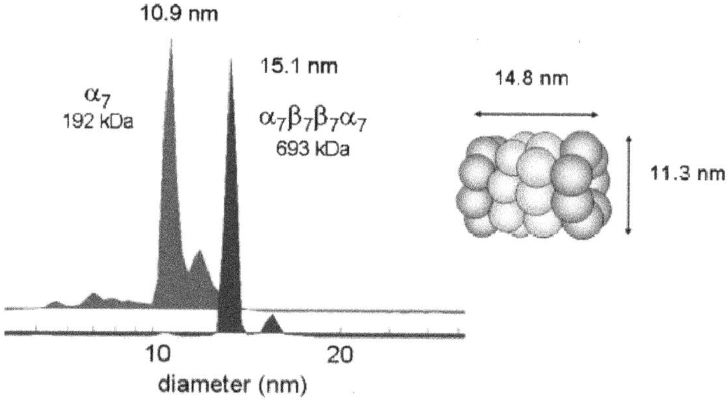

Figure 4.8.2 Gas-phase electrophoretic mobility molecular analysis (GEMMA) measurement of an α_7 subunit and $\alpha_7\beta_7\beta_7\alpha_7$ proteasome complex diameters. The figure on the right displays the complex architecture and dimensions obtained from X-radio crystallography (reproduced by permission from reference [24] ©2005 Elsevier).

cytoplasm of eukaryotic cells and entities such as archeons (intermediates between eukaryotes and bacteria). Its role is to recycle some damaged and misfolded proteins and complement that of the ubiquitin-pathway (see Section 3.1.1). The studied 20S proteasome from the *Methanosarcina thermophilia* archeon is a 693 kDa assembly of 28 α- and β-subunits (see Figure 4.8.2). The outer α-rings contain each of the seven subunits and are not involved in the catalytic process. The β-rings harbour three active centres. The mobility measurements of the bare $\alpha_7\beta_7\beta_7\alpha_7$ proteasome provide a diameter of 15.1 nm comparable to that obtained from X-crystallography indicating that the gas-phase and solution complex architectures are nearly identical. When an inhibitor of protease activity such as lactacystin ($m = 210$ Da) is added, the mass increases from 2,940 Da corresponding to covalent binding of lactacystin to each of the 14 β-subunits. Only a small diameter increase of 0.1 nm is observed. If a normally degraded molecule such as apo-myoglobin ($m = 16{,}951$ Da) is incubated with the now inhibited 20S proteasome+lactacystin complex, a mass increase of ca. 4.4 molecules of apo-myoglobin is measured and the complex becomes larger with a diameter of 15.7 nm (an "apo" enzyme is separated from its non-protein component called prosthetic group while it is bound in the "holo" form). This shows that a distribution of apo-myoglobins becomes trapped in the inhibited proteasome. The proteasome is an important target and this method opens possibilities for drug screening.

4.8.1.4 Ribosomes

Decoding of the genetic code and protein synthesis take place in a large assembly called the ribosome consisting of a catalytic sequence of RNA, called ribosyme, and proteins [28]. The overall architecture of ribosomes has been rather well conserved through evolution and is today better known in the prokaryotic case. In eukaryotes, ribosome masses are estimated ~4.5 MDa while they are only 2.3 MDa for prokaryotes. The incoming messenger RNAs are decoded by the transfer RNA species in a substructure called 30S unit (S stands for

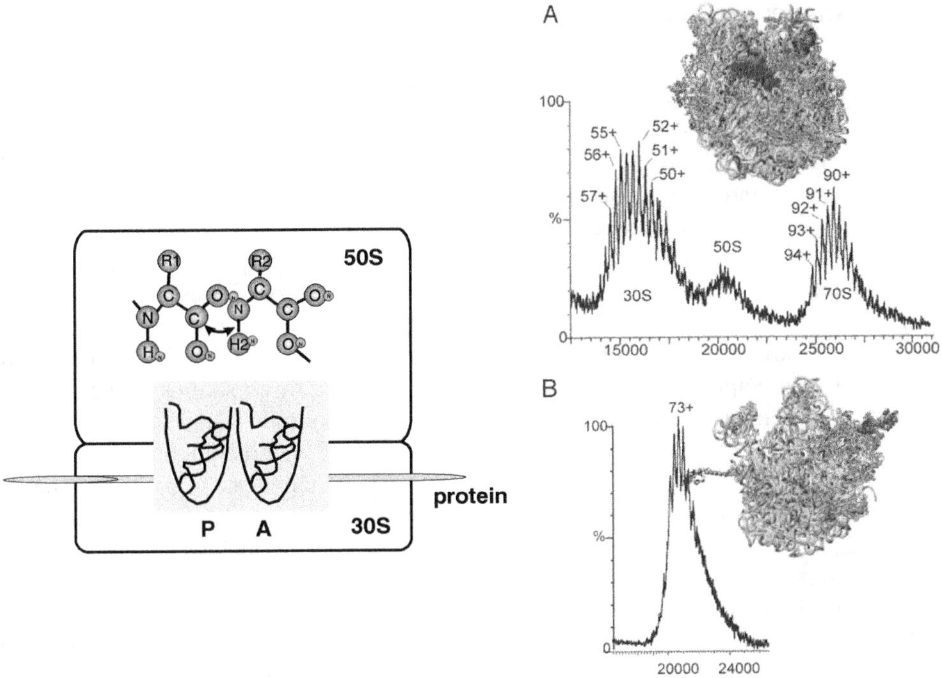

Figure 4.8.3 Left: Schematic representation of protein synthesis in the ribosome. Incoming messenger RNAs are decoded by transfer RNAs at the 30S subunit. Peptide bond formation between tRNAs is catalyzed in the 50S subunit. It involves nucleophilic attack by a nitrogen atom held at the A site on a carbon atom held at the P site (adapted with permission from reference [29] ©2001 Wiley). Right: Mass-spectra and structures of the 70S (A) and 50S (B) subunits of the ribosome of a thermophilic bacterium (*Thermus thermophilus*) (reproduced with permission from reference [33] ©2003 National Academy of Sciences).

sedimentation) while formation of peptide bonds is catalyzed in the larger 50S unit (information about the active site mechanism can be found in references [29, 30]).

Mass-spectrometric studies of intact ribosomes performed by the group of Robinson are rather challenging due to the non-covalent complex heterogeneity making difficult to maintain interactions between RNA and proteins [31–33]. Under mild nano-electrospray conditions, a single ribosome is contained in a droplet. Following droplet evaporation, intact ribosomes and some observed dissociation products are mass-analysed. Determination of the spontaneously released proteins during the mass analysis and collision-induced dissociation (CID; see Section 3.2.2.2) in a MS/MS set up allow the mapping of interactions between the different components (Figure 4.8.3). It is then assumed that the rate of dissociation is related to the interaction surface, that is the lowest area of interaction corresponding to the experimentally observed most weakly bound and released species. This approach has allowed complementary deciphering of a ribosome structure. A vital region for protein synthesis called the stalk complex was scrutinized. This complex that was first supposed to be pentameric turned out to be a heptamer $(L12)_6$ L10 with phosphorylated L12 proteins.

4.8.1.5 Chaperone proteins

When cells are submitted to stresses such as heat (heat-shock proteins, HSPs), some proteins called chaperones assist the folding of proteins and avoid strong modifications of protein assemblies. A system widely studied in the gas-phase is the GroEL/GroES complex of *Escherichia coli* [34–37]. GroES is a tetradecameric complex while its cochaperon GroES is a single-ring heptamer that binds to GroEL in the presence of ATP or ADP [16]. Both are constructed from dimeric building blocks. The cavity of GroES is large enough to accommodate and assist the refolding of the bacteriophage capsid protein gp23 that can be followed with mass-spectrometry through monitoring of the stoichiometry of substrate binding [4].

Another interesting example of the use of mass spectrometry for large fragile assemblies is the study of the quaternary structure of α-crystallins [38]. The eye-crystalline proteins are prevented from precipitation, otherwise they lead to cataract, by crystallins acting as chaperones that are heteromultimeric proteins. α-Crystallin is not only found in the eye where it constitutes as much as 50% of its dry weight and inhibits the precipitation of denatured proteins, but is also expressed in other tissues [39]. αB-crystallins are non-covalent multimeric assemblies with no definite stoichiometry and molecular masses between 300 kDa and 1 MDa [40]. This wide mass range corresponds to poly-disperse oligomers. Electrospray mass spectrometry [38] shows that when an oligomer containing n subunits and with a total charge q is submitted to CID (see Section 3.2.2.2) it dissociates into monomers and $n-1$, $n-2$ and $n-3$ subunits. This study indicates that oligomers can be as small as 10-mers up to large 40-mers and that dominant species are composed of 28 subunits. The quaternary structure of these poly-disperse proteins is affected by phosphorylation that disrupts the dimeric interfaces. The αB-crystallins then no longer preferentially adopt oligomer structures with an even number of subunits (Figure 4.8.4).

4.8.1.6 Viruses

Viruses are parasites of the three cell-based domains of life: bacteria, archea and eukaryotes. They cannot reproduce by themselves and share an overall structure consisting of a protein

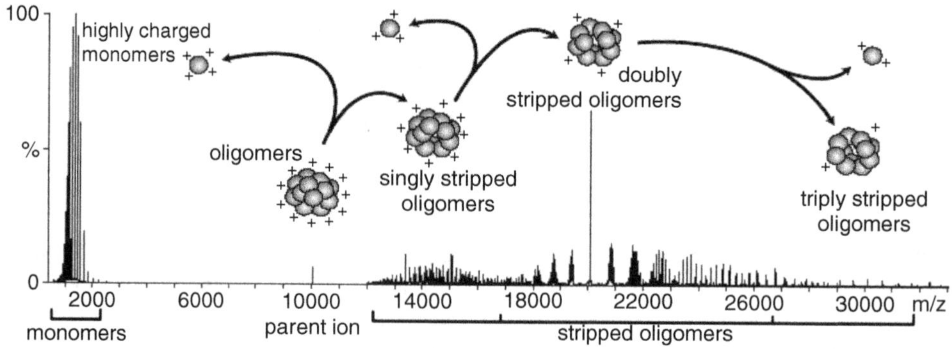

Figure 4.8.4 Collision-induced dissociation (CID) mass spectrum of αB-crystallin. The ejection of highly charged monomers results in the formation of low-charge "stripped" oligomers appearing at higher *m/z* values than their parent oligomers (reproduced with permission from reference [38] ©2003 National Academy of Sciences).

coat (the capsid) enclosing a nucleoprotein filament including either DNA or RNA. The size of the corresponding genome varies widely from only 4×10^4 base pairs up to 8×10^5 base pairs for the Mimivirus [41]. The origin and place of viruses in evolution is still controversial but they may have originated before the last universal cellular ancestor (LUCA) from which the three cellular domains diverged [42]. Understanding how a large number of protein subunits (from 60 to thousands) assemble into complete and reproducible structures is intellectually challenging but also fundamental for fighting viruses. For example, the entry of the HIV virus in cells is better understood than its assembling after replication in infected cells [43].

Tandem mass spectrometry (MS/MS) (see Section 3.2.2) works for disassembling and analysis of large non-covalent assemblies of proteins represented by virus capsids [44, 45]. H/D exchange (see Section 3.3.4) is also used for obtaining detailed information on inter-subunit interfaces. For example, the stability of the dimer interface has been monitored during the polymerization of the HIV-1 capsid [46, 47].

Intact viruses can be set into the gas-phase [5, 9, 10, 41, 48, 49] and still retain their biological function [48]. This was demonstrated by mass-selecting tobacco mosaic viruses (TMVs) and then capturing them on a glycerol-coated plate. The collected TMVs were then inoculated in tobacco cultivars which further developed lesions characteristic of infection [41].

The diameter of a "small" virus, bacteriophage MS2, with a molecular weight of 3.6 MDa has been measured by differential ion mobility and found equal to 27.5 nm similar to that of human rhinoviruses responsible for common-cold infection [50]. Larger viruses like bacteriophages T2 and T4 have double-stranded DNA genomes and molecular masses higher than 1 GDa [5]. Their length greater than 200 nm is larger than the diameter of most droplets produced by the used electrospray source with a 40 μm diameter tip and their measured diameter is close to 87 nm.

4.8.2 Modelling of oligomeric assemblies

Modelling of the assembling of those oligomeric proteins includes several approaches. The protein monomers can be first considered as almost already folded before their association. A rigid-body docking "lock-and-key", a flexible recognition "induced-fit" picture can then be proposed as well as the recognition of pre-selected conformers. Another approach consists in considering the enhanced speed of assembling between unfolded polymers with respect to folded ones. This accelerated assembling [51] of unfolded species is then attributed to their greater capture ratio for recognition of other unfolded polypeptides and is described by a fly-casting model [52] illustrated in Figure 4.8.5.

4.8.2.1 Viruses and their capsid assembling

Viruses borrow from cells their machineries for translating their own genetic code. A large number of theoretical studies have been devoted to the functioning of the involved different steps. In retroviruses such as HIV, the genetic code lies in RNA. RNA–protein interactions govern the reverse transcription and transport of the viral RNA. Molecular dynamics calculations conducted with the Amber force-field, for example have considered the binding of a crucial transcription protein (Tat) and the trans-activation response element RNA (TAR) [53]. Once the poly-protein ($Pr160^{gag\text{-}pol}$) representing the translation of the HIV

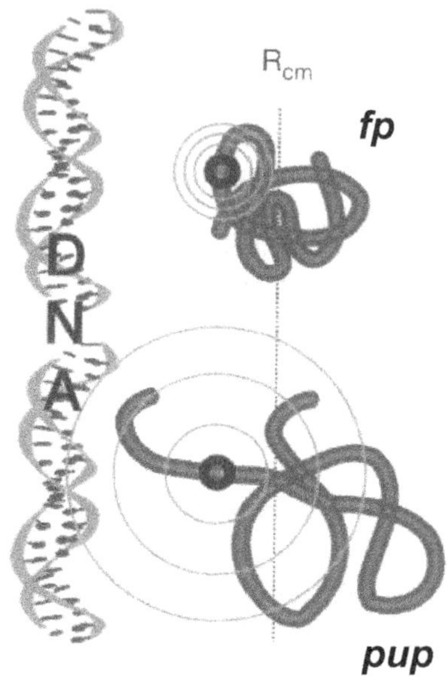

Figure 4.8.5 In this figure, a partially unfolded protein (pup) aiming to recognize a DNA sequence is compared to a folded protein (fp). At a large approach distance, the pup structure has already established initial weak interactions while the fp structure remains out of range because of its much smaller fluctuations. The increased capture radius allows the unfolded protein to find more easily and faster its specific binding site (reproduced with permission from reference [38] ©2000 National Academy of Sciences).

genetic code is performed, a key step on the road towards a matured HIV is the production of its final proteins from which it will be reconstructed. The HIV protease catalyzes its own release from the poly-protein and then proceeds to the hydrolysis of certain bonds leading to the final proteins [54]. Modelling of the HIV protease cleavage is important for designing drugs that can inhibit this enzyme. Force-field-based molecular dynamics fail in describing the conformational properties of the cleavage site. The three-dimensional structure of the HIV-1 protease is represented in Figure 4.8.6 and consists of two identical subunits. The cleavage site is at the border between those subunits and consists of an aspartic acid catalytic dyad, Asp_{25} and $Asp_{25'}$. A dynamical picture appears to be necessary to understand not only the catalytic mechanism but also the delicate role of mutations that enhance this enzyme/substrate activity [55, 56]. Classical molecular dynamics simulations first show that the active site structure is not rigid but that a low frequency motion of the whole protein takes place on a nanosecond timescale. Ab initio molecular dynamics (see Section 1.5.5.3) then point out that the mechanism is a concerted double proton transfer among the Asp dyad, a catalytic water molecule and the substrate that takes place at very short substrate/Asp dyad distances. This allows to understand the role of drug-resistant mutations that modify the flexibility of the protease and thus its ability to work as an "elaborate nutcracker" that brings the substrate into a favourable position for its cleavage.

Viral capsids represent the non-covalent quaternary association of protein subunits [57]. The role of the self-assembling viral capsids [58, 59] is to offer packaging and protection

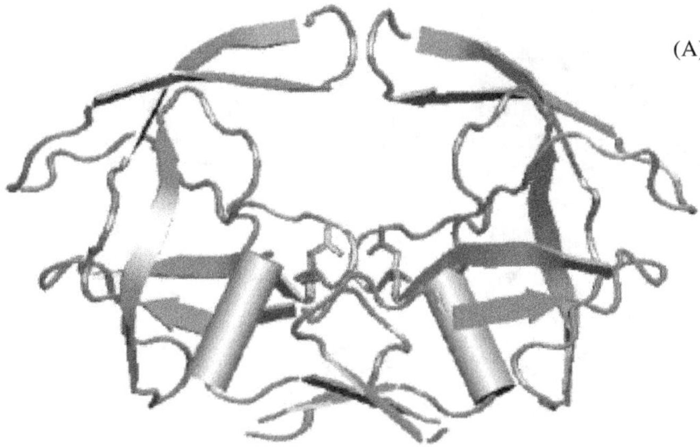

(A)

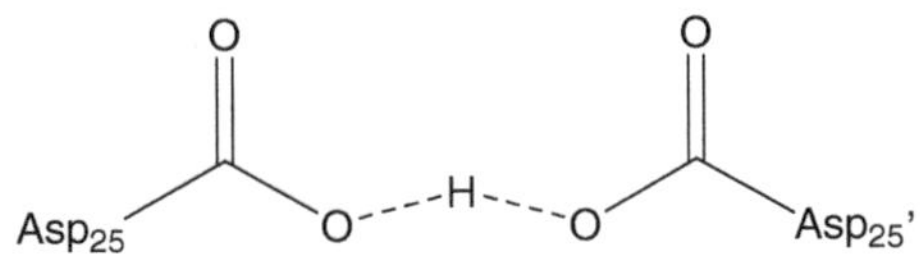

(B)

Figure 4.8.6 (A) Three-dimensional structure of the HIV-1 protease that cleaves the expressed viral polyprotein into functional proteins essential for the virus assembling. This enzyme is composed of two identical subunits of 99 amino acids. Flexible flaps acting as hinges allow the substrate to access the active site. This active site is characterized by the two catalytic triad sequences Asp–Thr–Gly, each belonging to a subunit. (B) A low-barrier hydrogen bond (see Section 1.1.2) is formed between the carboxyl carbons of the aspartic acids in the free enzyme (reproduced with permission from reference [55] ©2003 American Chemical Society).

to the genetic message and deliver it to host cells [43]. Capsids [44, 59] are supramolecular assemblies of a large number, up to several hundreds, of monomeric protein subunits called capsomers. In order to build very organized overall structures, those monomers fold into well-defined tertiary structures leading to attractions between complementary regions of nearby subunits. During its assembling from its elementary subunits and its life, a virus undergoes several transformations from a pro-capsid which is an intermediate metastable state to the mature capsid. Concerted subunit motions are required for these architectural reorganizations and are difficult to follow by means of the unique knowledge of the high-resolution overall three-dimensional structures. The combination of proteolysis and mass-spectrometry then can provide time-resolved information [45].

The formation of the capsid from the pro-capsid also involves transformations of the subunits. In an icosahedral RNA tetravirus [49], each primary α-subunit (1–644 residues) is cleaved into β (1–570) and γ (571–644) secondary subunits. This cleavage occurs on a timescale of hours in solution and can be followed by means of mass-spectrometry. Complementary information on structural transformation can be obtained by following the modification of the Fourier-transform infrared spectrum in the region of the amide I band (see Section 4.2.2) which shows that the pro-capsid is mostly formed of α-helices while the capsid contains more β-sheets.

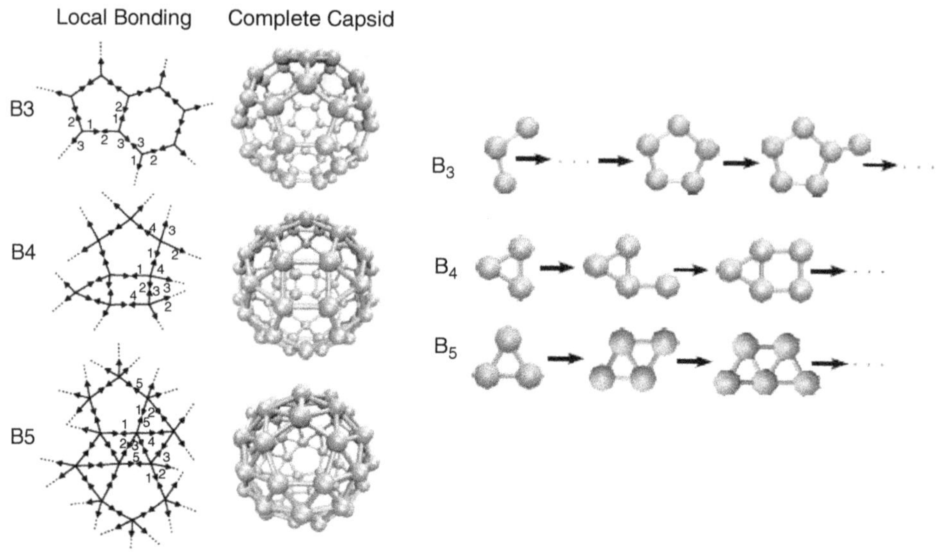

Figure 4.8.7 Left: In the left column, three models are shown, respectively, corresponding to four and five pair interactions for each subunit. The right column depicts the resulting capsid structures and the bonds resulting from complementary pairs. The growth mechanisms by formation of single (B_3 and B_4) or multiple (B_5) bonds are depicted on the right part of the figure (reproduced with permission from reference [55] ©2006 Biophysical Society).

A detailed molecular description is not possible. For modelling, each capsomer is represented by a spherical exclusion volume and three, four or five interaction sites leading to, respectively, B_3, B_4 and B_5 models (Figure 4.8.7). Subunits interact with their neighbours and the total capsid potential energy is a sum of short range pair interactions. Binding and unbinding rates of capsomers are orders of magnitude faster than the overall time required for assembling the whole capsid from its separated components [59]. Simulations show that different mechanisms can be invoked. Once dimers are formed in B_5 capsids, each monomer can form two or more bonds and the growth occurs primarily through addition of monomers. For B_3 and B_4 capsid models, growth occurs through assembling of clusters created by formation of single bonds.

4.8.3 Optical detection of very large gas-phase biomolecular systems

Macroions confined in quadrupole ion traps (see Section 3.2.1.2) can be optically detected by either elastic light scattering (ELS) or laser-induced fluorescence (LIF). Reference [6] provides a comprehensive review. The intensity of the scattered light at a forward scattering angle varies as the sixth power of the particle radius leading to a size detection limit of ~100 nm. In the case of naturally fluorescing particles with high quantum yields, there is no size limit and even atomic ions confined in Paul traps are easily detected [60]. Otherwise, dye-labelling is a convenient way to observe confined ions [61, 62].

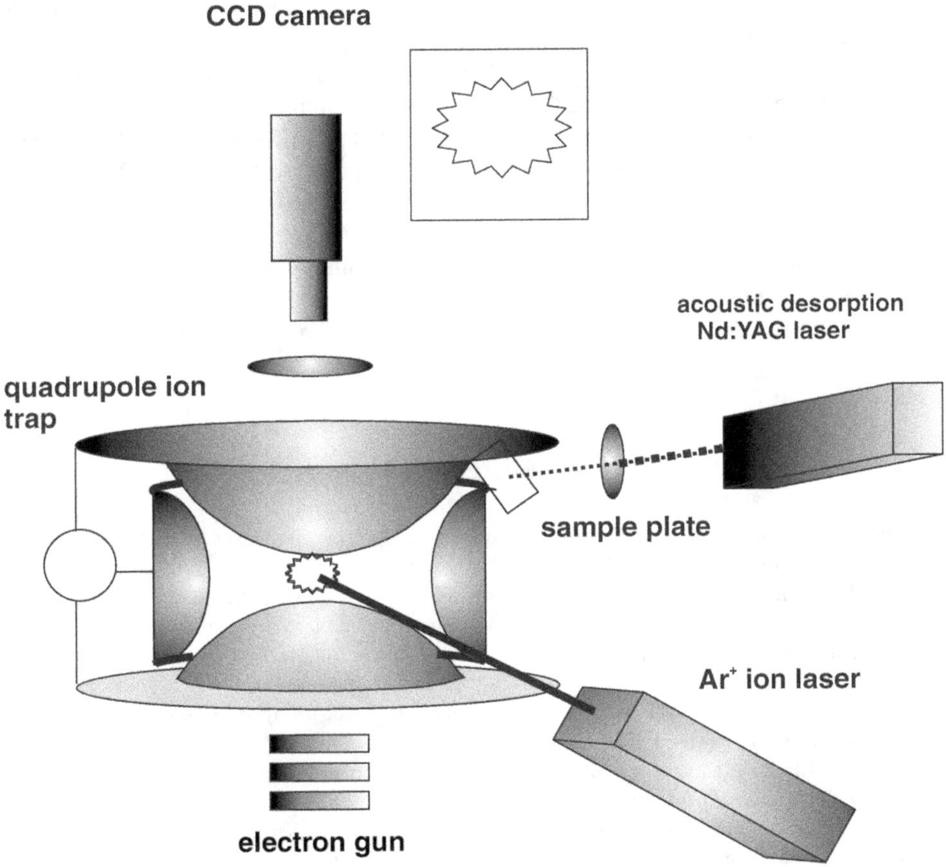

Figure 4.8.8 Optical detection of charged particles generated by laser-induced acoustic desorption and introduced in a quadrupole ion trap. Star-like oscillating trajectories are monitored through observation of the scattered light. Electron bombardment allows progressive reduction of the particle charges and determination of their mass.

Bacteria are unicellular organisms whose constituents can be analysed by mass-spectrometry. They can be introduced in the gas-phase from MALDI sources [11] and trapped in quadrupole ion traps [6]. However, their extremely large masses ($> 10^{10}$ Da) forbid the use of conventional ionization-based detection schemes. Their observation takes benefit of the very large dimensions within the same order of magnitude as visible light wavelengths. By monitoring the light scattered by a single confined bacterium, one can follow its trajectory in a trap (Figure 4.8.8) and deduce its mass-to-charge ratio from the equation of motion (see Section 3.2.1.2). A mass resolution of 1,000 can be reached for values of m/Ze larger than 10^9. For example, the mass of a dry *E. coli* is equal to 82.3 ± 0.3 fg.

Usual sources such as ESI [63] and MALDI [64] can be used for production of very large gas-phase biomolecular systems up to intact cells [63] but can lead to the partial destruction of weakly bound edifices. An interesting alternative is offered by laser-induced acoustic desorption (LIAD) [13] for setting particles into the gas-phase in absence of matrix or passage at atmospheric pressure. The studied sample is deposited on a bare silicium wafer that is then illuminated on its backside by a frequency-doubled Nd:Yag laser beam. The shock-wave provoked by the laser on one side of the wafer is transformed into an acoustic wave that propagates and desorbs ions of both signs from the biological sample situated on the other side. Ions are introduced in quadrupole ion trap and illuminated by a CW argon laser. Star-like ion trajectories in the trap are detected by observation of the forward scattered light. Particles with masses between 10^{10} and 10^{11} Da such as vaccinia viruses (ca. 0.33 μm long and 0.27 μm diameter), *E. coli* bacteria (ca. 1 μm long and 0.6 μm diameter) and red blood cells (ca. 1.7 μm thick and 7.3 μm diameter) have been studied. Interestingly, normal and anaemic (lacking haemoglobin) red cells can easily be distinguished. Mass-spectrometric techniques can today be applied for drug screening and drug discovery directly in intact biological cells [13, 63].

REFERENCES

1. Heck AJR, van den Heuvel RHH: Investigation of intact protein complexes by mass spectrometry. *Mass Spectrometry Reviews* 2004, 23:368–389.
2. Ullom JN, Franck M, Gard EE, Hom JM, Labov SE, Langry K, Magnotta F, Stanion KA, Hack CA, Benner WH: Discrimination between bacterial spore types using time-of-flight mass spectrometry and matrix-free infrared laser desorption and ionization. *Analytical Chemistry* 2001, 73: 2331–2337.
3. Fenselau C, Demirev PA: Characterization of intact microorganisms by MALDI mass spectrometry. *Mass Spectrometry Reviews* 2001, 20:157–171.
4. Benesch JLP, Robinson CV: Mass spectrometry of macromolecular assemblies. *Current Opinion in Structural Biology* 2006, 16:245–251.
5. Hogan CJ, Kittleson EM, Ramaswami B, Chen DR, Biswas P: Charge reduced electrospray size spectrometry of mega- and gigadalton complexes: whole viruses and virus fragments. *Analytical Chemistry* 2006, 78:844–852.
6. Peng WP, Cai Y, Chang HC: Optical methods for mass spectrometry of macroions. *Mass Spectrometry Reviews* 2004, 23:443–465.
7. Suurmeijer TM, Boers AL: Electron trajectories and inhomogeneity effects in a Daly detector. *Journal of Physics E-Scientific Instruments* 1971, 4:885–888.
8. Wenzel RJ, Matter U, Schultheis L, Zenobi R: Analysis of megadalton ions using cryodetection MALDI time-of-flight mass spectrometry. *Analytical Chemistry* 2005, 77:4329–4337.
9. Chen R, Cheng XL, Mitchell DW, Hofstadler SA, Wu Q, Rockwood AL, Sherman MG, Smith RD: Trapping, detection, and mass determination of coliphage T4 DNA ions of 10^8 by electrospray ionization Fourier transform ion cyclotron mass spectrometry. *Analytical Chemistry* 1995, 67: 1159–1163.
10. Fusternau SD, Benner WH, Thomas JJ, Brugidou C, Bothner B, Siudzak G: MS of an intact virus. *Angewandte Chemie-International Edition* 2001, 40:541–544.
11. Lay JO: MALDI-TOF mass spectrometry of bacteria. *Mass Spectrometry Reviews* 2001, 20: 172–194.
12. Siudzak G: Probing viruses with mass spectrometry. *Journal of Mass Spectrometry* 1998, 33: 203–211.
13. Peng WP, Yang C, Kang MW, Tzeng YK, Nie Z, Chang HC, Chang W, Chen CH: Laser-induced acoustic desorption mass spectrometry of single bioparticles. *Angewandte Chemie-International Edition* 2006, 45:1423–1426.

14. Takats Z, Wiseman JM, Gologan B, Cooks RG: Electrosonic spray ionization. A gentle technique for generating folded proteins and proteins complexes in the gas phase and for studying ion-molecule reactions at atmospheric pressures. *Analytical Chemistry* 2004, 76:4050–4058.
15. Wiseman JM, Takats Z, Gologan B, Davisson VJ, Cooks RG: Direct characterization of enzyme–substrate complexes by using electrosonic spray ionization mass spectrometry. *Angewandte Chemie-International Edition* 2005, 44:913–916.
16. Kaltashov IA, Eyles SJ: *Mass spectrometry in biophysics*: Wiley, Hoboken, NJ, 2005.
17. Boyer PD: The binding change mechanism for ATP synthase – some probabilities and possibilities. *Biochemica and Biophysica Acta* 1993, 1140:215–250.
18. Antes I, Chandler D, Wang HL, Oster G: The unbinding of ATP from F_1-ATPase. *Biophysical Journal* 2003, 85:695–706.
19. Meier T, Morgner N, Matthies D, Keis S, Pogoryelov D, Cook GM, Dimroth P, Brutschy B: Mass determination of *c* rings from ATP synthases. *Molecular Microbiology* 2007.
20. Badman ER, Hoaglund-Hyzer CS, Clemmer DE: Monitoring structural changes of proteins in an ion trap over similar to 10–200 ms: unfolding transitions in cytochrome *c* ions. *Analytical Chemistry* 2001, 73:6000–6007.
21. Breuker K, McLafferty FW: The thermal unfolding of native cytochrome *c* in the transition from solution to gas phase probed by native electron capture dissociation. *Angewandte Chemie-International Edition* 2005, 44:4911–4914.
22. Daidone I, Amadei A, Roccatano D, Di Nola A: Molecular dynamics simulation of protein folding by essential dynamics sampling: folding landscape of horse heart cytochrome *c*. *Biophysical Journal* 2003, 85:2865–2871; Lange C, Hunte C: Crystal structure of the yeast cytochrome bc_1 complex with its bound substrate cytochrome *c*. *Proceedings of the National Academy of Sciences of the United States of America* 2002, 99:2800–2805.
23. Morgner N, Barth HD, Brutschy B, Kleinschroth T, Müller W, Ludwig B: A new type of laser mass spectrometry utilized to analyse membrane proteins. *Biochemistry* 2007.
24. Loo JA, Berhane B, Kaddis CS, Wooding KM, Xie YM, Kaufman SL, Chernushevich IV: Electrospray ionization mass spectrometry and ion mobility analysis of the 20S proteasome complex. *Journal of the American Society for Mass Spectrometry* 2005, 16:998–1008.
25. de la Mora JF, Thomson BA, Gamero-Castano M: Tandem mobility mass spectrometry of electrosprayed tetraheptyl ammonium bromide clusters. *Journal of the American Society for Mass Spectrometry* 2005, 16:717–732.
26. Ruotolo B, Giles K, Campuzano I, Sandercock AM, Bateman RH, Robinson CJ: Evidence for macromolecular protein rings in the absence of bulk water. *Science* 2005, 310:1658–1661.
27. De Martino GN, Slaughter CA: The proteasome, a novel protease regulated by multiple mechanisms. *Journal of Biological Chemistry* 1999, 32:22113–22126.
28. Maddaford SP, Motamed M, Turner KB, Choi MSK, Ramnauth J, Rakhit S, Hudgins RR, Fabris D, Johnson PE: Identification of a novel non-carbohydrate molecule that binds to the ribosomal A-site RNA. *Bioorganic & Medicinal Chemistry Letters* 2004, 14:5987–5990.
29. Lilley DMJ: The ribosome functions as a ribozyme. *ChemBioChem* 2001, 2:31–35.
30. Ikeda Y, Taira K: Biologically important reactions catalyzed by RNA molecules. *The Chemical Record* 2002, 2:307–318.
31. Rostom AA, Fucini P, Benjamin DR, Juenemann R, Nierhaus KH, Hartl U, Dobson CM, Robinson CV: Detection and selective dissociation of intact ribosomes in a mass spectrometer. *Proceedings of the National Academy of Sciences of the United States of America* 2000, 97: 5185–5190.
32. Videler H, Ilag LL, McKay ARC, Hanson CL, Robinson CV: Mass spectrometry of intact ribosomes. *FEBS Letters* 2005, 579:943–947.
33. Ilag LL, Videler H, McKay AR, Fucini P, Nierhaus KH, Robinson CV: Heptameric $(L12)_6/L10$ rather than canonical pentameric complexes are found by tandem MS of intact ribosomes from

thermophilic bacteria. *Proceedings of the National Academy of Sciences of the United States of America* 2005, 102:8192–8197.
34. Sobott F, Robinson CV: Characterising electrosprayed biomolecules using tandem-MS – the non-covalent GroEL chaperonin assembly. *International Journal of Mass Spectrometry* 2004, 236:25–32.
35. Rostom AA, Robinson CJ: Detection of the intact GroEl chaperonin assembly by mass spectrometry. *Journal of the American Chemical Society* 1999, 121:4718–4719.
36. van Duijn E, Simmons DA, van den Heuvel RHH, Bakkes PJ, van Heerikhuizen H, Heeren RMA, Robinson CV, van der Vies SM, Heck AJR: Tandem mass spectrometry of intact GroEL – substrate complexes reveals substrate-specific conformational changes in the *trans* ring. *Journal of the American Chemical Society* 2006, 128:4694–4702.
37. van Duijn E, Bakkes PJ, Heeren RMA, van den Heuvel RHH, van Heerikhuizen H, van der Vies SM, Heck AJR: Monitoring macromolecular complexes involved in the chaperonin-assisted protein folding cycle by mass spectrometry. *Nature Methods* 2005, 2:371–376.
38. Aquilina JA, Benesch JLP, Bateman OA, Slingsby C, Robinson CV: Polydispersity of a mammalian chaperone: mass spectrometry reveals the population of oligomers in αB-crystallin. *Proceedings of the National Academy of Sciences of the United States of America* 2003, 100: 10611–10616.
39. Augusteyn RC: α-Crystallin: a review of its structure and function. *Clinical & Experimental Optometry* 2004, 87:356–366.
40. Peterson JJ, Young MM, Takemoto LJ: Probing α-crystallin structure using chemical cross-linkers and mass spectrometry. *Molecular Vision* 2004, 10:857–866.
41. Bothner B, Siudzak G: Electrospray ionization of a whole virus: analyzing mass, structure, and viability. *ChemBioChem* 2004, 5:258–260.
42. Forterre P: The origin of viruses and their possible roles in major evolutionary transitions. *Virus Research* 2006, 117:5–16.
43. Adamson CS, Jones IM: The molecular basis of HIV capsid assembly – five years of progress. *Reviews in Medical Virology* 2004, 14:107–121.
44. Tito MA, Tars K, Valegard K, Hajdu J, Robinson CV: Electrospray time-of-flight MS of the intact MS2 virus capsid. *Journal of the American Chemical Society* 2000, 122:3550–3551.
45. Kriwacki R, Reisdorph N, Siudzak G: Protein structure characterization with mass spectrometry. *Spectroscopy* 2004, 18:37–47.
46. Lanman J, Prevelige PE: High-sensitivity mass spectrometry for imaging subunit interactions: hydrogen/deuterium exchange. *Current Opinion in Structural Biology* 2004, 14:181–188.
47. Lanman J, Lam TT, Emmett MR, Marshall AG, Sakalian M, Prevelige PE: Key interactions in HIV-1 maturation identified by hydrogen–deuterium exchange. *Nature Structural & Molecular Biology* 2004, 11:676–677.
48. Siudzak G, Bothner B, Yeager M, Brugidou C, Fauqeu CM, Hoey K, Chang CM: Mass spectrometry and viral analysis. *Chemistry and Biology* 1996, 3:45–48.
49. Bothner B, Taylor DJ, Jun B, Lee KK, Siudzak G, Schultz CP, Johnson JE: Maturation of a tetravirus capsid alters the dynamic properties and creates a metastable complex. *Virology* 2005, 334:17–27.
50. Bacher G, Szymansli WW, Kaufman SL, Zöllner P: Charge-reduced nano-electrospray ionization combined with differential mobility analysis of peptides, proteins, glycoproteins, noncovalent protein complexes and viruses. *Journal of Mass Spectrometry* 2001, 36:1038–1052.
51. Perham M, Chen M, Ma J, Wittung-Stafshede P: Unfolding of heptameric co-chaperonin protein follows "fly casting" mechanism: observation of transient nonnative heptamer. *Journal of the American Chemical Society* 2005, 127:16402–16403.
52. Shoemaker BA, Portman JJ, Wolynes PG: Speeding molecular recognition by using the folding funnel: the fly-casting mechanism. *Proceedings of the National Academy of Sciences of the United States of America* 2000, 97:8868–8873.

53. Reyes CM, Nifosi R, Frankel AD, Kollman PA: Molecular dynamics and binding specificity analysis of the bovine immunodeficiency virus BIV tat–TAR complex. *Biophysical Journal* 2001, 80:2833–2842.
54. Lee H, Darden TA, Pedersen LG: An *ab initio* quantum mechanical model for the catalytic mechanism of HIV-1 protease. *Journal of the American Chemical Society* 1996, 118:3946–3950.
55. Carloni P, Rothlisberger U, Parrinello M: The role and perspective of *ab initio* molecular dynamics in the study of biological systems. *Accounts of Chemical Research* 2002, 35:455–464.
56. Carloni P: Density functional theory-based molecular dynamics of biological systems. *Quantitative Structure–Activity Relationships* 2002, 21:166–172.
57. Szymanski WW, Kaufman SL, Zöllner P, Blass D, Allmaier G: Charge-reduced nano-electrospray ionisation combined with differential mobility analysis of peptides, proteins, glycoproteins, non covalent protein complexes and viruses. *Journal of Mass Spectrometry* 2001, 36:1038–1052.
58. Valéry C, Paternostre M, Robert B, Gulik-Krzywicki T, Naranyan T, Dedieu JC, Keller G, Torres ML, Cherif-Cheikh R, Calvo P, *et al.*: Biomimetic organization: octapeptide self-assembly into nanotubes of viral capsid-like dimension. *Proceedings of the National Academy of Sciences of the United States of America* 2003, 100:10258–10262.
59. Hagan MF, Chandler D: Dynamics pathways for viral capsid assembly. *Biophysical Journal* 2006, 91:42–54.
60. Gudjons T, Arbes F, Benzing M, Kurth F, Werth G: Spatial fluorescence distribution and laser cooling of Ca^+ in a Paul trap. *Physica Scripta* 1995, 59:396–402.
61. Iavarone AT, Parks JH: Conformational change in unsolvated Trp-cage protein probed by fluorescence. *Journal of the American Chemical Society* 2005, 127:8606–8607.
62. Iavarone AT, Meinen J, Schulze S, Parks JH: Fluorescence probe of polypeptide conformational dynamics in gas phase and in solution. *International Journal of Mass Spectrometry* 2006, 253: 172–180.
63. Kulkarni MJ, Vinod VP, Umasankar PK, Patole MS, Rao M: Intact cell matrix-assisted laser desorption/ionization mass spectrometry as a tool to screen drugs in vivo for regulation of protein expression. *Rapid Communications in Mass Spectrometry* 2006, 20:2769–2772.
64. Siuzdak G: An introduction to mass spectrometry ionization. In *The expanding role of mass spectrometry in biotechnology*. MCC Press, San Diego, CA, 2003.

– 5 –

From Gas-Phase to Solution

GENERAL FEATURES

Gas-phase or crystal conditions are obviously not those required for sustaining life. Living organisms are mainly composed of water that enables them to maintain their osmotic, metabolic, thermal and electrochemical balance and transport nutrients within cells [1, 2]. Water is vastly present in cells (between 50 and 80%). Cells are enclosed by a lipid bilayer membrane, separating them from other cells and from the extra-cellular medium [3]. A lipid bilayer membrane is impermeable to water and ions. However, under many circumstances, such entities need to be rapidly and selectively transported across it, in response to an extra- or intracellular signal. The water-transporting task is accomplished by aquaporin water channel proteins [4]. In cells, biomolecules are in presence of water and other molecules such as DNA, RNA proteins, sugars and membranes at high concentration. The interior of living cells is thus not simply comparable to bulk water. Except in separate compartments of cells where a high water content is required (e.g. vacuoles for generating turgescence in plants), the cellular medium has a gel-like consistency. In vivo processes take place in an environment in which biomolecules occupy a considerable fraction (between 10 and 40%) of the total volume. The terms "crowded" or "volume-occupied", rather than "concentrated", are used to describe the cellular medium because no single molecular species is necessarily present at a high concentration [5–8].

It is thus most probable that studying isolated molecules in the gas-phase or molecules only surrounded by a single or double solvation hydration layer or even, to a less extent, biomolecules in bulk water is an oversimplification. One is nevertheless led to consider hydration of biomolecules [9–11] as an approximation of reality in experimental observations obtained by means of different condensed-phase methods such as radio-crystallography [10, 12, 13], NMR [14, 15], IR spectroscopy [16], fluorescence [17] or neutron scattering [9, 18, 19]. Water molecules in immediate contact with proteins (interstitial or interfacial water) can have long (>1 ns) residence times [14, 20, 21]. These "quasi-bound" water molecules exist because of the long lifetimes of the hydrogen bonds established between polar amino acid side-chains lying at the protein surface and water molecules [22].

In active sites of enzymes [23, 24], the situation is not too far from gas-phase conditions since only a very small number of water molecules are present [25, 26]. Those water molecules play a crucial role [27] and have been generally conserved through evolution [28]. A QM/MM study (Figure 5.1.1, see also Section 1.6) has been conducted on the catalytic

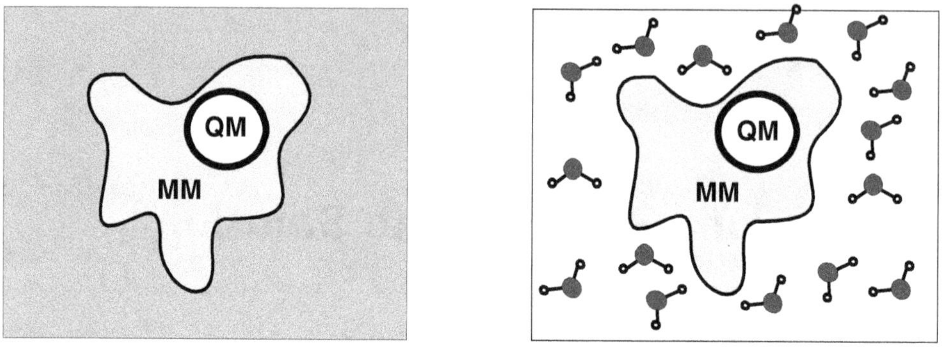

Figure 5.1.1 A region of interest (circle), such as the active site of an enzyme, is treated quantum mechanically (QM). The rest of the protein is treated by molecular mechanics (MM). Left: The solvent is described as a continuous medium (implicit water). Right: Water molecules are treated explicitly.

mechanism of the caspase-3 enzyme, a cysteine protease [27] involved in cellular death (apoptosis). This modelling study suggests a new mechanism involving a rupture of the C—S bond of cysteine concerted with a proton transfer to the side-chain of the aspartic acid substrate (Figure 5.1.2).

Even in gas-phase, a single water molecule can act as a catalyst. This has been experimentally demonstrated in the case of the reaction of OH radicals with acetaldehyde [29]. In this reaction, there is no energy barrier between reactants ($CH_3CHO + OH$) and products ($CH_3O + H_2O$). The acceleration of the reaction in presence of a single water molecule is explained by quantum calculations conducted at the CCSD(T) level [29]. A submerged or "reef" barrier between bare reactants and products slows down the reaction by introducing a bottleneck. This reef disappears when the reactant becomes acetaldehyde hydrogen-bonded to water (compare H-bond patterns in Figures 5.1.2 and 5.5.2).

Interactions of water molecules with polar and ionic groups at protein surfaces are essentially electrostatic. Water molecules are either quasi-bound or free to rotate in their confining cavities. In the last case, even the most long-ranged of these interactions, the orientationally averaged charge–dipole interaction, falls off as r^{-4}. Thus, already in the second hydration layer, water–water interactions dominate strongly over water–protein interactions and as a result, water molecules that are not in direct contact with the protein surface behave very much like bulk water.

Experimental studies have been devoted to establishment of links between gas-phase and condensed-phase properties of biomolecules using circular dichroism and Raman activity [30, 31], electrospray ionization [32, 33], fluorescence [34, 35] and X-ray spectroscopy in liquid beams [36]. Since infrared spectroscopy is an important tool in structural studies, we will present three different methodological approaches for linking gas-phase and liquid-phase IR spectra. The first approach uses quantum molecular dynamics with explicit solvent molecules (see Section 5.1.2) while the second and the third reconstruct liquid spectra, respectively, either as spectra of the considered species surrounded by their first hydration shell (see Section 5.5.4) or as a weighted sum of individual spectra of gas-phase mono-hydrated species (see Section 5.6).

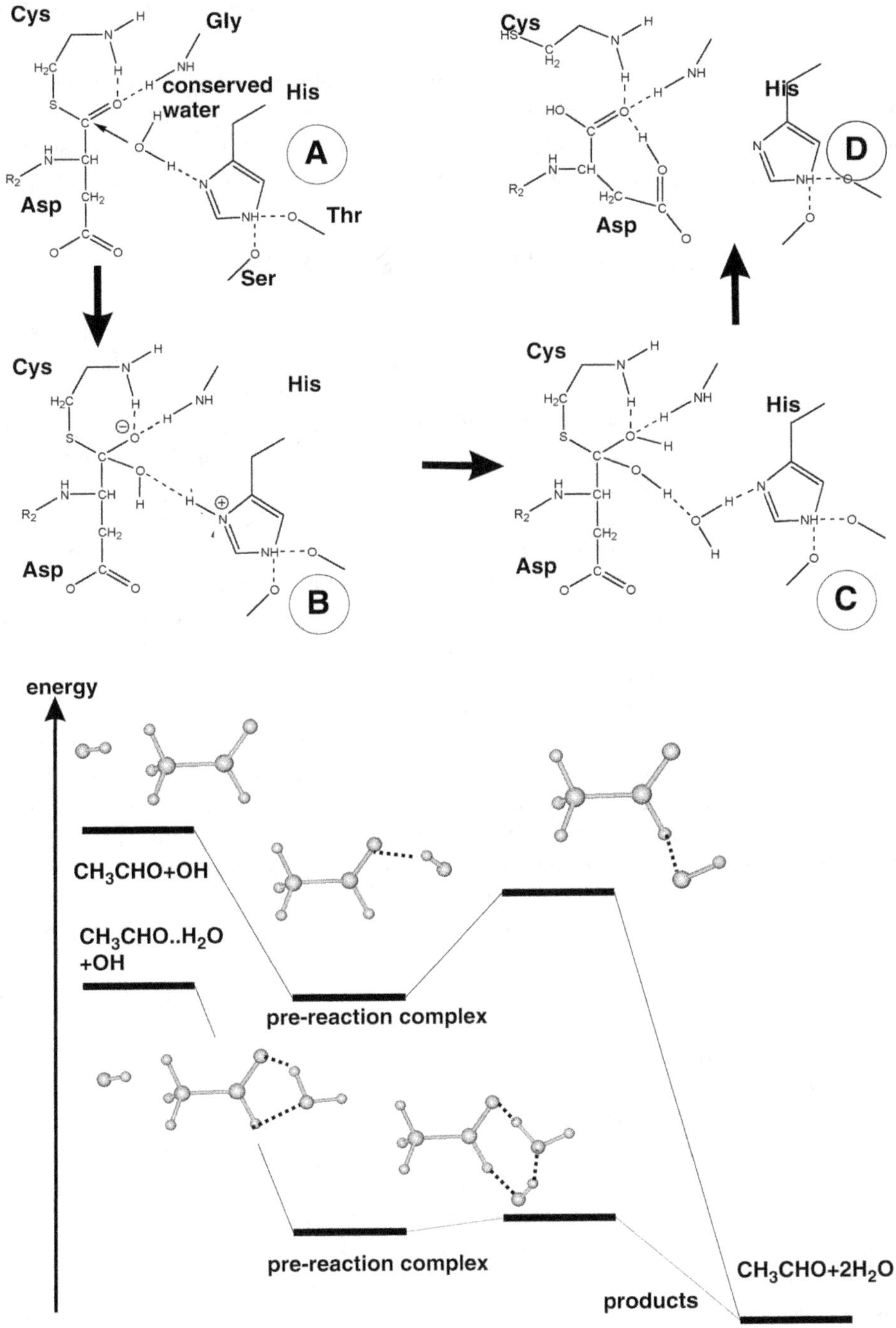

Figure 5.1.2 Top: Reaction mechanism of a cysteine protease (caspase-3) emphasizing the role of a catalytic water molecule conserved through evolution. Following the attack by this hydrolytic molecule, the C—S bond of cysteine dissociates in a concerted proton transfer to the side-chain of the aspartic acid substrate. Bottom: Catalysis of the reaction of acetaldehyde with an OH radical. Even in gas-phase, a single water molecule provides a considerable acceleration. This is attributed to the disappearance of the submerged "reef" barrier along the top reaction path.

In this chapter, we will first briefly describe different modelling approaches of solvation. We will then consider water and other solvents used in biochemistry mostly emphasizing the gas-phase cluster point of view. We will then review gas-phase studies providing information on structures of hydrated building blocks: nucleobases, peptides, sugars and neurotransmitters.

5.1 MODELLING OF SOLVATION

In the case of biomolecules, the notion of solvent includes not only water and other solutions but also the presence of large biomolecular entities of the "crowded" cellular medium. A detailed review of the theoretical approaches used in modelling solvent effects in large biomolecular systems can be found in reference [37]. A general review of modelling of proteins and large biomolecules can be found in reference [38]. The solvent has several effects on the solute (the considered biomolecule). Among them, one can distinguish the solvent influence on (a) electronic properties, for example dipole moments [39], (b) nuclear conformations [40], (c) infrared spectra [41] and (d) chemical structure, for example tautomerism [42–44]. Pure quantum calculations (QM) [45], mixed quantum and classical mechanics (QM/MM) [46] and classical molecular mechanics (MM) [38, 47] can be used for modelling the dynamic behaviour of biomolecules in a solvent. The solvent molecules can be treated individually (explicit solvent) or as a continuous medium (implicit solvent).

In a molecular dynamics simulation, a large amount of computation time can be spent in calculations only concerning solvent molecule–solvent interactions and a large number of methods have been developed in which the solvent is treated as a continuous medium. Specific interactions, solvation models and long-range order are then mostly ignored. On the contrary, it is possible to explicitly take into account a large number of solvent molecules [48]. In addition to a more or less realistic description of specific interactions such as hydrogen bonds, the great role of individual dipoles of the solvent molecules is better taken into account for calculating local dielectric screening effects than by simply using a constant or varying empirical dielectric constant [49].

In the case of liquid water, three types of water molecules must mainly be considered: (a) the strongly bound water molecules inside protein cavities or in the minor groove of DNA, (b) the more weakly bound water molecules that interact with the biomolecule surface and (c) the bulk water molecules. Often, one introduces the notion of "solvent accessible surface area" (or "SASA"). In such models, the local solvent contribution to the mean force exerted on a biomolecule can be taken as proportional to the area of the groups of atoms that are accessible to short- or medium-range interaction with the solvent. The contribution of the first solvation shell can then be simply modelled by means of an analytical function of the SASA only containing two atomic parameters. In a molecular dynamics simulation, the solvent degrees of freedom are then eliminated and the computational effort is 30% larger than in a "gas-phase" calculation but 10–50% less than in a simulation involving explicit water molecules [50].

5.1.1 Quantum calculation of microhydrated structures

Full quantum calculations concerning biomolecules in presence of solvent molecules can still only be conducted for rather small molecular sizes. When molecules in presence of only one or two water molecules are considered, quantum calculations can be performed at high

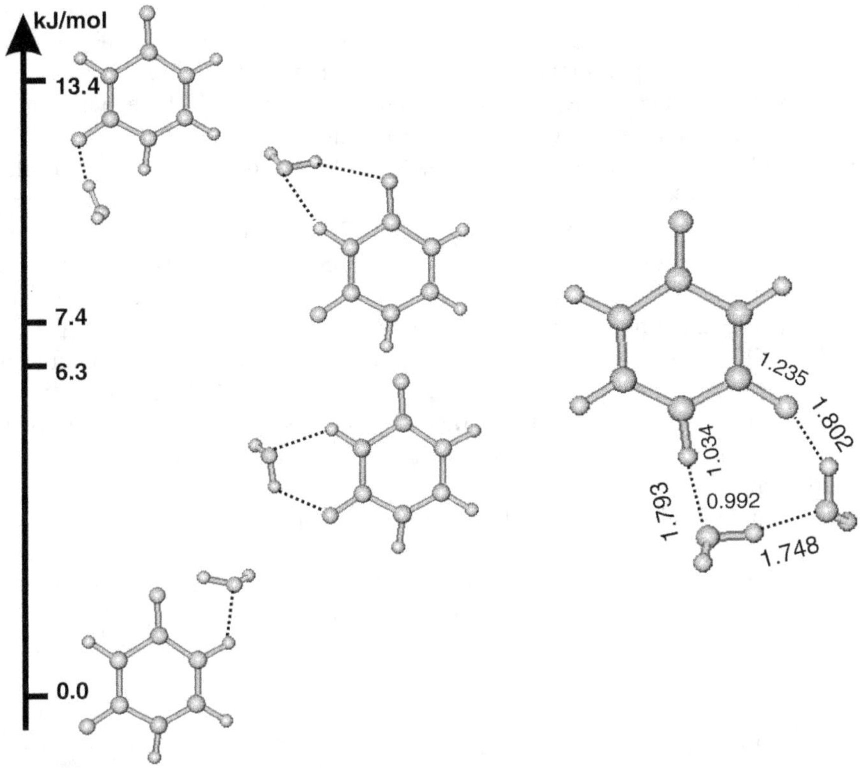

Figure 5.1.3 Left: Lowest energy configurations of the binary uracil–water complex. Right: Lowest energy configuration of the uracil–(water)$_2$ complex optimized at the B3LYP/6-31G + + G* level (distances are in Angstrom) (adapted with permission from reference [48] ©2001 American Chemical Society).

levels of theory. For example, the relative energies of the different tautomers of hydrated cytosine and adenine [42, 51] have been calculated at the CCSD(T) level using a cc-pVTZ basis set and at the MP2 level with an aug-cc-pVQZ basis set. Calculations of the preferential hydration sites and of the solution infrared spectrum of hydrated uracil have been conducted to the DFT level with a 6-31++G* basis set [52] and the MP2/6-311G++G** level [53]. The uracil solvation shell is then composed of seven water molecules with a water trimer and two water dimers (Figure 5.1.3) lying in the nucleobase plane.

For larger molecular systems such as pentapeptides, DFT calculations can be conducted at the B3LYP/6-31 +G** level [54]. In the case of very large biomolecular systems such as progesterone, new fast methods scaling linearly with size may be employed if one wishes to keep the appealing features of quantum calculations [55].

5.1.2 Quantum molecular dynamics simulation with explicit solvent

The quantum molecular dynamics Car–Parinello method (see Section 1.5.5.3) allows direct comparison between predicted infrared spectra in gas-phase and in an aqueous solution [41] as well as the study of enzymatic reactions [56] at a biologically relevant temperature. Let us consider the preceding case of hydrated uracil. The IR absorption spectrum computed at the harmonic level with the configuration deduced from the quantum chemistry calculation at 0°K does not reproduce satisfactorily the experimental IR spectrum of uracil in water obtained at room temperature. One cannot ignore the solvent environment and anharmonicities. A Car–Parinello molecular dynamics (CPMD) calculation conducted at 300°K with 49 water molecules provides a satisfying agreement with experiment. The first hydration shell obtained from CPMD at 300°K is different from that predicted from the quantum chemistry calculation at 0°K (Figure 5.1.4). The surrounding solvent favours the establishment of a tri-dimensional network of H-bonded water molecules. On the average, the first hydration shell of uracil contains six water molecules, each C=O and N—H group being, respectively, H-bonded to two and one water molecules. The H_2O molecules bonded to C=O groups are outside the uracil plane in order to minimize the water network disruption. Formation of dimers as found in the 0°K calculation does not exist any longer and there is no specific arrangement of water molecules in the first hydration shell.

Another example is given by the infrared absorption spectrum of *N*-methylacetamide presented in Figure 5.1.5 under gas-phase or liquid water conditions. *N*-methylacetamide is a widely used model of the peptide bond [57] (see Section 5.5.1). Simulations are conducted starting from configurations obtained from classical force-field calculations and further using CPMD [58] with the BLYP DFT functional and time steps of 0.12 fs.

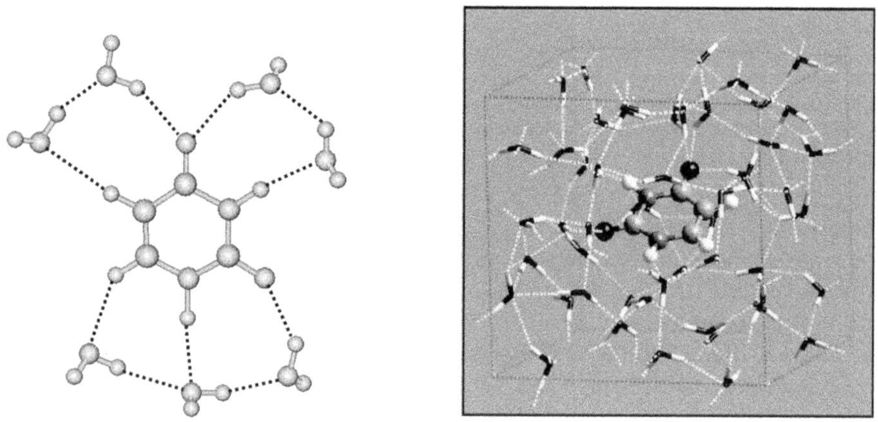

Figure 5.1.4 Left: Lowest energy configuration of the first hydration shell of uracil at 0°K optimized at the B3LYP/6-31G + + G* level (distances are in Angstrom). This configuration is planar (adapted with permission from reference [48] ©2001 American Chemical Society). Right: Instantaneous configuration of uracil surrounded by 49 water molecules sampled from a quantum molecular dynamics calculation at 300°K. Note that the two results are quite different. The 0°K planar solvation shell comprises seven water molecules while the 300°K solvation shell is three-dimensional and comprises six water molecules [48]. In the last case, on average, C=O and N—H groups are, respectively, bound to two and one water molecules. The 0°K hydration shell is probably a local minimum which is not representative of biological conditions (reproduced with permission from reference [59] ©2003 American Chemical Society).

Two hydration layers are taken into account. In contrast with hydrated cluster calculations that only provide a set of "frozen" configurations with generally a few water molecules or, at most, a full hydration layer, the hydration pattern, as usual in liquid studies, is presented as a set of radial distribution functions (rdf) (Figure 5.1.6). The rdf $g(r)$ are defined as follows [37]. N_s is the number of solvent molecules found when sampling the spherical layer located between the distances r and $r + dr$ from the solute. ρ_s is the pure solvent density.

$$g(r) = \frac{\langle N_s(r, r + dr) \rangle}{4\pi\rho_s r^2 \, dr} \tag{5.1.1}$$

Hydration patterns in the gas-phase and the liquid water are rather similar. The preferential residence sites of water are the hydrophilic sites: the hydrogen donor amide N—H and the acceptor C=O carbonyl. The C=O and N—H intramolecular distances are increased (~0.03 Å) when going from gas-phase to liquid water due to the formation of intermolecular hydrogen bonds between these groups and the bound water molecules. In aqueous solution, the N—C backbone distance is very much shortened (~0.025 Å) while, on the contrary, the *N*-methyl bond length increases from the gas-phase to the liquid-phase.

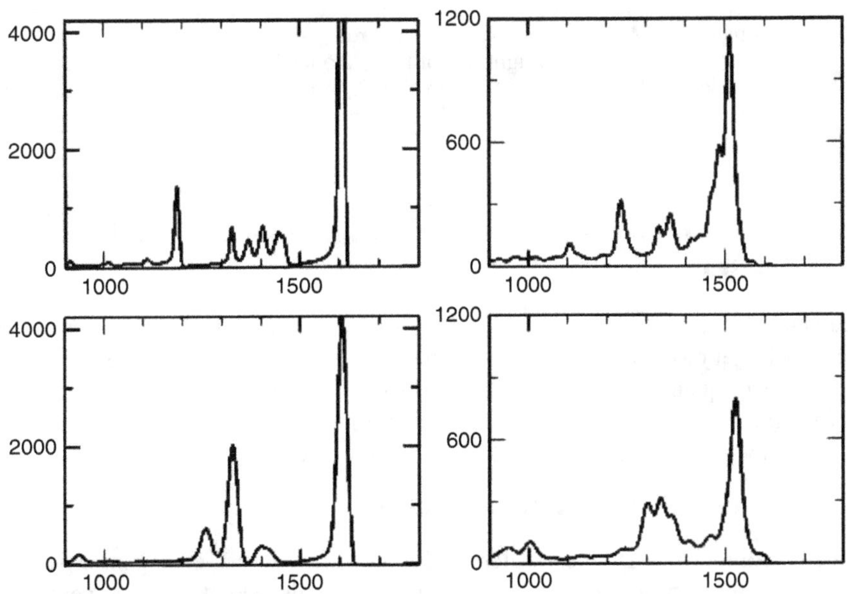

Figure 5.1.5 Car–Parrinello ab initio dynamics simulation of the infrared spectra of gas-phase *trans-N*-methylacetamide (NMA) (top left), gas-phase *cis*-NMA (bottom left), aqueous *trans*-NMA (top right) and aqueous *cis*-NMA (bottom right) in the 1,000–1,800 Cm^{-1} region (reproduced with permission from reference [41] ©2005 American Chemical Society).

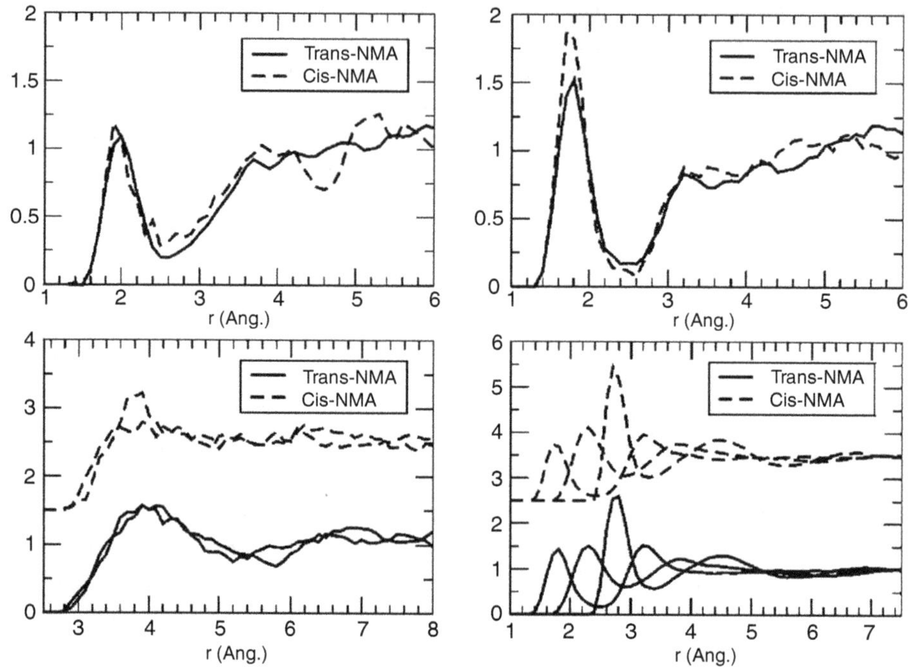

Figure 5.1.6 *N*-Methylacetamide (NMA)–water and water–water radial distribution functions (rdf) calculated from Car–Parrinello molecular dynamics simulations. The *trans* (solid lines) and *cis* (dashed lines) isomers of NMA are considered. Top left: rdf of the H atom of the amide N—H group and the oxygen atom of water O_w. Top right: rdf of the C atom of the amide C=O group and the H atom of water H_w. Bottom left: rdf of the O atom and O_w (the two C=O groups are represented). Bottom right: Solvent rdf of $O_w \cdot O_w$, O_w–H_w and H_w–H_w (reproduced with permission from reference [41] ©2005 American Chemical Society).

5.1.3 Classical molecular dynamics simulation with explicit solvent molecules

Replacement of quantum calculations by force-field simulations allows the treatment of considerably larger biomolecular systems embedded in very large numbers of solvent molecules. High-level quantum calculations of properties of water are limited to rather small sizes (typically less than 20 molecules) and simple models have been devised to incorporate large number of water molecules in force-field molecular dynamics simulations.

5.1.3.1 Simple models of water molecules

For reducing computational times, water molecules can be described by simplified models [59]. In simple interaction site models, each water molecule is kept in a rigid geometry and the interaction between molecules is described by a Lennard–Jones + Coulomb term (see Section 1.1). Those models can be improved by introducing some flexibility through harmonic stretching of O—H bonds and harmonic angle bending of the H—O—H angle and even further by explicitly including polarization and many-body effects. Familiar models are those of the transferable intermolecular potential (TIP) family. They are improvements of a three-interaction site model containing a rigid water monomer with a charge $-q$ on the oxygen atom and $+1/2q$ on hydrogens located at the atomic centres and a single Lennard–Jones (6–12) interaction site located on the oxygen atom. The TI3P, TI4P and TI5P [60] models provide better simulations of the bulk water properties. Other simple models of water have been proposed [61]. The Mercedes–Benz model is a two-dimensional model. Each water molecule is here modelled by a disk that interacts with other waters through a Lennard–Jones interaction and an orientation-dependent hydrogen bonding interaction in three radial arms arranged as in the Mercedes–Benz car company logo (see Figure 5.3.1) [62, 63]. A β-hairpin peptide with 139 atoms can be immersed in a water sphere with a radius of 20 Å containing 1,060 molecules [64] (for comparison, a β-hairpin simulation has been conducted with an implicit solvent [65]). The used force-fields are either conventional force-fields such as Charmm and Amber (see Section 1.3) or specially designed force-fields [66]. These force-fields allow, for example the full simulation of the folding process of the Trp-cage protein [67] (see Section 4.2.6).

5.1.4 Molecular dynamics simulation with implicit solvent

Typical simulations concerning very large biomolecular systems [68] can concern system sizes $\sim 10^5$–10^6 atoms and the dynamics can be followed during amount of time of tens of nanoseconds requiring 10^6–10^7 time steps. The microscopic structure of the solvent is then often ignored and the solvent is considered as a dielectric continuum [69]. When conceptually bringing a solute from the gas-phase into a solvent treated as a continuum, one must first create a cavity by breaking the cohesive forces between solvent molecules. In the second step, the solute is brought into this cavity and interacts through van der Waals forces, repulsive and dispersive. Then, the charge distribution of the solute is modified by the influence of its surrounding solvent (e.g. zwitterion formation in the case of hydrated peptides; see Section 5.5.4). In turn, the solute charge distribution "polarizes" the solvent. These two last steps are electrostatic steps that must be treated with a self-consistent procedure. Widely used models are the Onsager model and its variations, the Poisson–Boltzmann (PB) and the generalized-Born (GB) models [70–73]. Discussion of these quantum self-consistent reaction field (QM-SCRF) models [74] can be found in references [75, 76]. The basic assumption made in the Onsager model is that the biomolecule is placed in a spherical cavity inside the solvent. The solvent is simply described as a homogeneous polarizable medium of constant dielectric constant. The biomolecule dipole moment induces a dipole moment of opposite direction in its surrounding medium. Polarization of the medium in turn polarizes the charge distribution in the solvent. The Onsager reaction field model treats this mutual polarization in a self-consistent manner. The reaction field E_R is proportional to the biomolecular dipole moment μ and inversely proportional to the third power of the radius a of the solute cavity; ϵ_s is the solvent dielectric constant (78.39 for water).

$$\vec{E}_R = \frac{2(\varepsilon_s - 1)}{(2\varepsilon_s + 1)a^3}\vec{\mu} \tag{5.1.2}$$

The Onsager model is the simplest of a family of related models. In the PCM model, the cavity surrounding the biomolecule is the union of a series of interlocking atomic spheres. In the IPCM model [75], the cavity is an isodensity surface of the molecule. The SCI-PCM model takes into account the full coupling between the cavity and the biomolecular electron density. The COSMO (conductor-like screening model) is a dielectric solvent continuum model in which the solute molecule is embedded in a cavity adopting a molecular shape and surrounded by a dielectric medium with a given dielectric constant [40]. This model can be used, for example for prediction of basicities of amino acids and drugs [77].

Those implicit solvent models must be used with some caution for small systems. For example, the predicted relative energies of the neutral and zwitterionic forms of a series of neurotransmitters have been compared with an explicit solvent model and a SCI-PCM model [78]. Continuous solvent and explicit solvent molecule representations can be combined [51, 79, 80]. The "bound" or "interstitial" water molecules of the first solvation shell are then explicitly and accurately taken into account while the bulk water is simply represented as a dielectric continuum. For example, the vibrational absorption spectrum of the zwitterionic form (see Section 5.5.3.2) of L-alanine in solution has been calculated considering the amino acid surrounded by explicit water molecules only, embedded in a SCRF reaction field (Onsager model) only and surrounded by explicit water molecules embedded in a SCRF reaction field [81]. Only this last combination provides a satisfying method for treating the effects of solvent on IR spectra. It may be attributed to the fact that several water molecules from the first hydration shell move to the second hydration shell due to a non-satisfying value of the "bound" water dipole moments in the Onsager model.

In between accurate but computer-time consuming methods using explicit solvent representations and efficient but approximate implicit solvent representations, there exist improved implicit solvent models that aim to provide an accurate modelling of the continuum dielectric solvent by imposing an exact description of the macroscopic laws of electrostatics. Using this approach, the Car–Parinello quantum simulation with implicit solvent has been applied to molecules with intermediate size such as the octa-alanine peptide. In that case, it is possible to follow the unfolding of the peptide from an initially imposed α-helical configuration to its stable horse-shoe configuration [82].

5.1.5 Comparison between models

A systematic comparison between theoretical methods, ranging from purely classical force-fields, through mixed quantum mechanics/molecular mechanics (QM/MM), up to fully quantum mechanics Car–Parinello (CPMD) calculations has been conducted on a dipeptide in water [83]. The glycine–alanine peptide is neutral when isolated and becomes a zwitterion in water (see Section 5.5.3.2). The main difference between the two used force-fields, GROMOS and Amber, corresponds to their treatment of hydrogen atoms. Those atoms are described explicitly in Amber while apolar ones are not represented in GROMOS. The dipeptide is immersed, respectively, in 586, 584 and 62 water molecules in force-field QM/MM and CPMD simulations. The respective simulation durations for collecting data are 1.6 ns, 12 and 6 ps. Most of the solvation patterns are similar, except for the NH_3^+ and COO^- terminals and the amide N—H groups. Force-field simulations provide poor representations of the solvation shells around the charged terminals while QM/MM provides satisfying results as long as van der Waals interactions between the QM and MM regions are included. Using a TIP3P representation of water molecules and neglecting a quantum description of the solvent saves a sizable amount of computation time. It then allows the treatment of systems larger than those that can be dealt within a full quantum dynamics simulation.

5.2 SOLVENTS

In this chapter, we consider properties of liquid solvents with the microscopic point of view of isolated gas-phase clusters studies. The different "frozen" configurations adopted by small assemblies of molecules can be considered as "snapshots" of the transient structures observed in the real liquid situation. We will first focus the attention on the hydrogen-bond network of water and the variation of its dipole moment as a function of cluster size. We will then consider two important hydrated species, the hydrated electron and the protonated water clusters. Finally, we will provide references concerning solutions used in biochemistry.

5.2.1 Water

The water molecule is highly polarizable. When isolated, its dipole moment is equal to 1.855 D ($1\,D = 3.336 \times 10^{-30}$ C/m) while in liquid water it becomes equal to 2.6 ± 0.14D. Water has also a great ability to form hydrogen bonds as both an acceptor (A) and a donor (D). In bulk, almost all water molecules are involved in four hydrogen bonds with their neighbours. In gas-phase clusters, many water molecules lie on the cluster surface and are only involved in a single, two or three hydrogen bonds. The spectroscopic difference between O—H bond vibrational frequencies of free or bound water molecules provides a unique signature of cluster structures.

5.2.1.1 Neutral water clusters

Properties of neutral water clusters can be found in references [84–93]. The two basic units are the water monomer and dimer [94–99]. Some of their properties are given in Tables 5.2.1 and 5.2.2 and Figure 5.2.1.

The lowest energy configurations of the neutral water trimer [92, 105, 106], tetramer [84] and pentamer [107, 108] are cyclic [87, 109] (Figure 5.2.2). The first cluster with a

Table 5.2.1

Structural parameters of the isolated water molecule computed at the (a) DFT/B3LYP level with a basis set built by adding to the standard 6-31G set two sp and two d shells of diffuse functions on the oxygen centres and one s and two p shells on the hydrogen ones [100]; (b) CCSD(T)/TZ2P(f,d) level [101, 102]

Structural parameters	Predicted DFT	Predicted ab initio	Experimental
R(OH) (Å)	0.962	0.959	0.959
HOH angle (°)	105.0	104.2	105.0
Dipole moment μ (D)	1.910	–	1.855
Polarizability α (Å^3)	1.293	–	1.47
ν_1 (cm^{-1})	1635.4	1659	1648.0
ν_2 (cm^{-1})	3805.4	3833	3832.0
ν_3 (cm^{-1})	3905.5	3943	3943.0

Table 5.2.2

Structural parameters of the neutral water dimer and its anion calculated at the DFT level with optimized basis sets (see Figure 5.2.1) [100] and at the CCSD(T)/TZ2P(f,d) level [102, 104]

	R (Å)			
	OH_1	OH_2	$OH_{3/4}$	O—O
$(H_2O)_2$[a]	0.970	0.961	0.963	2.952
$(H_2O)_2$[b]	0.964	0.957	0.958	2.912
$(H_2O)_2^-$[a]	0.974	0.962	0.970	2.872
	Angles (°)			
	α_1	α_2	β_1	β_2
$(H_2O)_2$[a]	95.4	105.4	105.3	121.8
$(H_2O)_2$[b]	95.5	104.9	104.8	123.4
$(H_2O)_2^-$[a]	88.2	104.5	104.5	144.0

[a]At the DFT level with optimized basis sets (see Figure 5.2.1) [100].
[b]At the CCSD(T)/TZ2P(f,d) level [102, 104].

lowest energy configuration that is not cyclic is the hexamer [110, 111] (Figure 5.2.2). The cage configuration with eight hydrogen bonds is the most stable and is formed preferentially in supersonic expansions. The cyclic hexamer cluster, although higher in energy, can nevertheless be formed at extremely low temperature by passing helium droplets through water vapour in a pick-up cell (see Section 3.1.4) since different nucleation mechanisms can lead to different isomers [112, 113].

The dipole moment of the isolated water molecule is equal to 1.855 D while that of water molecules in condensed phase (ice or liquid) rises up to 2.6 D. The variation of this dipole moment as a function of water cluster size [39] has been studied by means of high-level quantum calculations (MP2/aug-cc-VDZ) [103]. As shown in Table 5.2.3, there is no clear variation of the cluster dipole moments as a function of size. The calculation provides access to the individual monomer dipole moments inside the clusters which increase regularly from the isolated water molecule value up to 2.7 D in the octamer and reach 2.76 D for 160 water molecules in a lattice structure similar to ice I_h.

5.2.1.2 Water cluster anions and the hydrated electron

Solvent molecules such as water or ammonia possess large dipoles and do not have low-lying valence orbitals susceptible to accept an excess electron. Nevertheless, an electron can be trapped by these solvents in the liquid-phase. The "hydrated" or "solvated" electron e_s^- can be generated by radiolysis or photolysis of water. For example, the primary pathways for decomposition of liquid water by high-energy electrons or γ-rays are [114]: $H_2O \rightarrow H + OH$ and $H_2O \rightarrow H^+ + e_s^- + OH$. A sizable fraction of electrons considered in radiation chemistry are hydrated electrons (see Chapter 4.7). The solvated electron e_s^- has a characteristic intense absorption spectrum centred ~720 nm and exhibits an extremely fast (tens of femtoseconds) dynamics [115, 116]. The gas-phase equivalents of hydrated

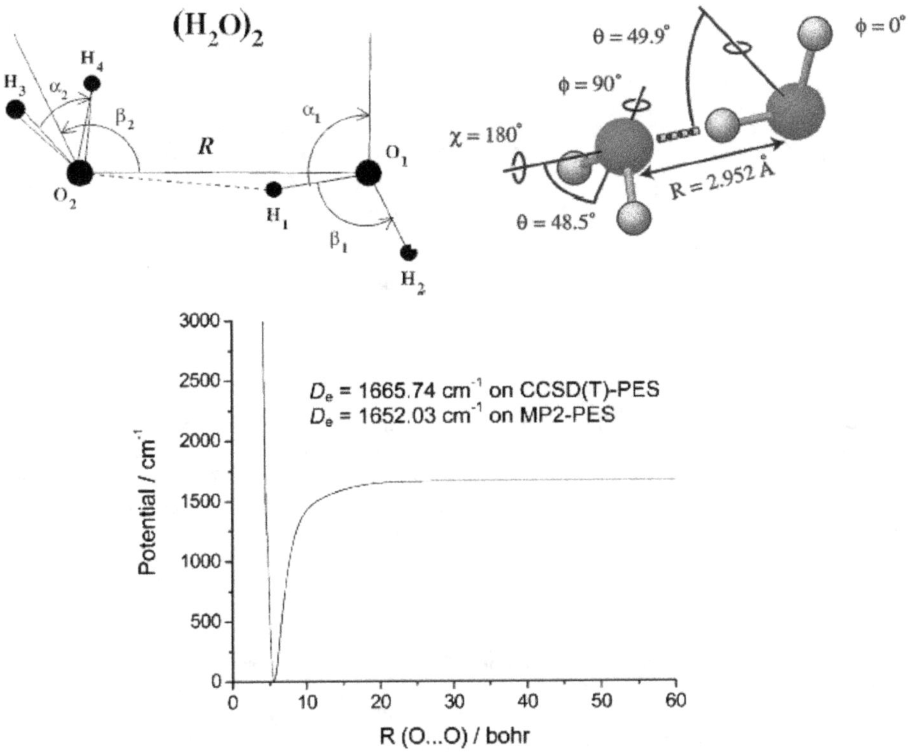

Figure 5.2.1 Top left: Geometrical parameters of the water dimer (see Table 5.2.1) (reproduced with permission from reference [100] ©1998 American Chemical Society). Top right: Experimental geometry of the water dimer determined by terahertz spectroscopy (reproduced with permission from reference [86] ©2001 National Academy of Sciences). Bottom: One-dimensional potential energy curve of the water dimer along the dissociation coordinate (reproduced with permission from reference [103] ©2006 American Chemical Society).

electrons are water anions $(H_2O)_N^-$ [117–119]. Any mass-spectrum of water anions, whichever production method is used, exhibits prominent "magic" peaks corresponding to $N = 2$, 6 and 7 and a continuous distribution above $N = 10$. All other water anions, except for the monomer anion, have been observed but with considerably weaker intensities [120, 121].

The observation of magic structures cannot be simply attributed to the existence of neutral cluster configurations with dipole moments larger than the critical value of ca. 2.5 D [122] but can be attributed to the relative stability of the neutral and negatively charged clusters. The neutral cluster stabilities result from the balance between the binding energy brought by the establishment of hydrogen bonds and geometrical strains. The neutral clusters with $N = 2$, 6, 7 and 11 are less stable than the other clusters [123].

In the case of large water cluster anions, it has been predicted that two different types of excess electron localization should exist. For values of N smaller than 32, simulations have predicted that excess electrons should be in very diffuse orbitals at the surface of the neutral cores similar to a dipole-bound anion (see Section 2.3.2.2). Above $N = 32$, internal states corresponding to electrons inside neutral cavities should be more stable [124, 125]. These

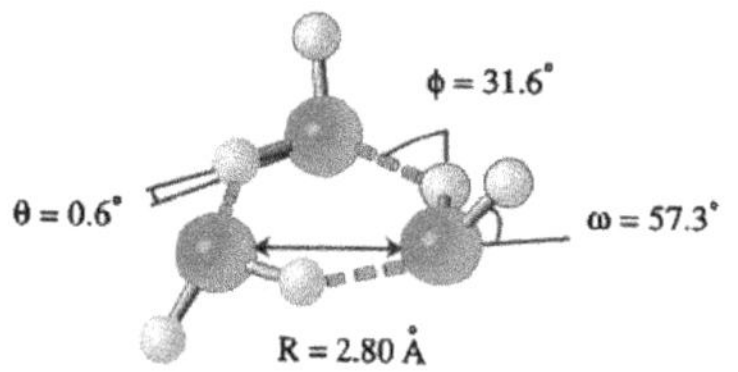

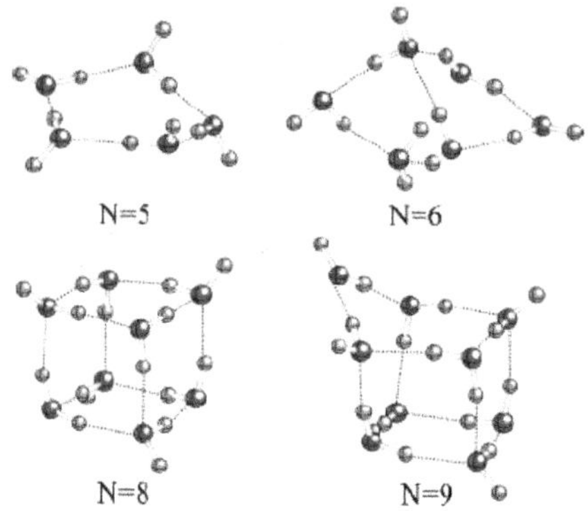

Figure 5.2.2 Left: Geometry of the water trimer determined by terahertz spectroscopy [86]. The mean O—O distance is 2.85 Å significantly shorter than in the dimer due to three-body forces (reproduced with permission from reference [86] ©2001 National Academy of Sciences). Right: Lowest energy conformations of neutral water clusters identified experimentally (reproduced with permission from reference [114] ©2005 American Chemical Society).

Table 5.2.3

Calculated dipole moments of neutral water clusters [39]

Water cluster	Dimer	Trimer cyclic	Tetramer cyclic	Pentamer cyclic	Hexamer cyclic	Hexamer cage	Hexamer prism
Cluster dipole (D)	2.683	1.071	0	0.927	0	1.904	2.701

early predictions have opened the route to CW [126] and femtosecond [127] photoelectron experiments showing that indeed several different types of isomers exist. Three types of isomers with different electron binding properties are observed.

Above $N = 11$, the most stable anions are produced in anion sources under rather warm conditions corresponding to liquid-like cluster. Those anions possess enough internal energy to rearrange following electron attachment in order to reach their lowest energy configurations. Those anions correspond to "internal" states of excess electrons. The other anions are formed from much colder neutral clusters and are less stable. Only a small local reorganization of the water molecules can take place around the excess electron initial trapping site. The anions do not have enough energy to undergo a complete reorganization. Excess electrons are then trapped in diffuse or "surface" orbitals.

Hydrated electrons are condensed-phase species (Figure 5.2.3) and surface-state water cluster anions cannot be considered as their gas-phase equivalent. In fact, only vertical detachment energies of the embryonic isomers converge to the bulk value (Figure 5.2.5). Nevertheless, high-level calculations of the water dimer and hydronium–water cluster provide insights on the spectroscopic properties of the hydrated electron. Ab initio and DFT calculations have been performed for the electronic ground state and the lowest excited singlet state of the water dimer leading to OH and H_3O radicals [129].

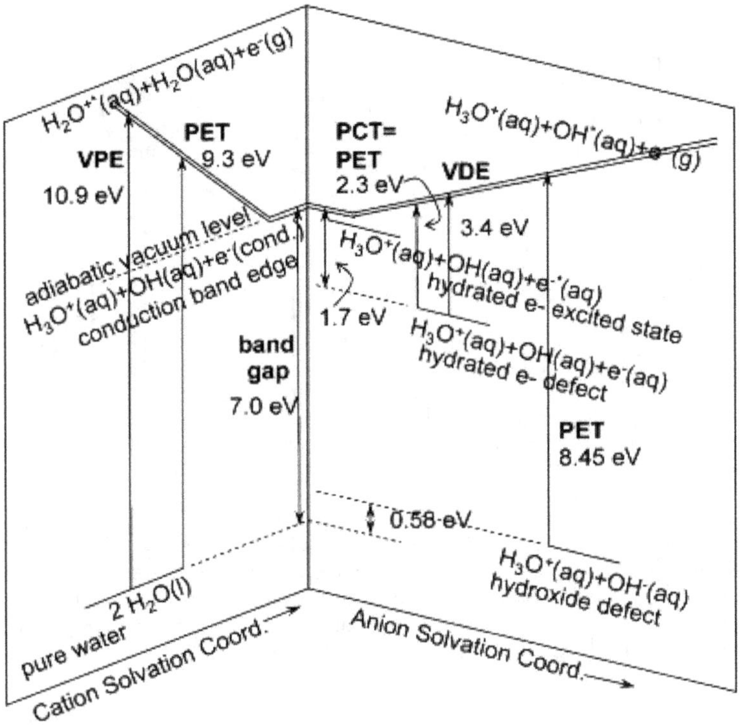

Figure 5.2.3 Energy diagram of bulk water as a function of the solvent reorganization about charge species. PCT, VPE and VDE are, respectively, the photoconductivity threshold, the vertical photoionization energy and the vertical detachment energy (reproduced with permission from reference [114] ©2005 American Chemical Society).

The hydrated electron can be viewed as an itinerant H_3O or H_5O_2 radical in water [129–132]. The hydronium H_3O is a marginally stable hypervalent radical and then corresponds to a shallow minimum on the $H + H_2O/H_2 + OH$ potential energy surface. The hydronium radical is stabilized by solvation and geometries of $H_3O{\cdot}(H_2O)_N$ clusters have been calculated up to two solvation shells. The ground state S_0 of the water dimer (Figure 5.2.4) correlates with the $OH^-{\cdot}H_3O^+$ ion pair while the first singlet excited state S_1 correlates with the $OH{\cdot}H_3O$ biradical complex. In the S_1 excited state of the water dimer, along the OH coordinate, the hydrogen transfer $H_2O{\cdots}H_2O \rightarrow H_3O{\cdots}OH$ path is a repulsive $\pi\sigma^*$ state without any barrier (Figure 5.2.6). The π orbital remains on the OH while the σ^* orbital is transferred with the proton to form a Rydberg-like H_3O with an electron occupying a diffuse orbital mostly outside the H_3O^+ core. This highest singly occupied Rydberg-like molecular orbital of the $H_3O{\cdot}(H_2O)_N$ clusters detaches from the H_3O^+ cation core when N increases. The hydronium radical thus seems to dissolve in water into a solvated hydronium cation and a solvated electron cloud without any barrier. The $H_3O{\cdot}(H_2O)_N$ cluster IR absorption spectra possess a very intense OH stretch absorption line that seems to be a fingerprint of the hydrated electron.

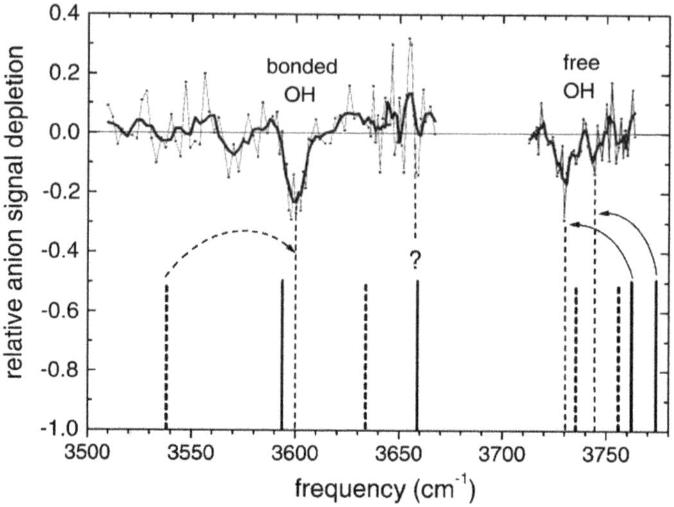

Figure 5.2.4 Experimental infrared spectrum of mass-selected neutral water dimers obtained from dipole-bound signal depletion spectroscopy. The experimental values are indicated by thin dash lines. Anharmonic (without scaling) and scaled harmonic theoretical frequencies calculated at the MP2/6-31 + + G(2d,2p) level are, respectively, indicated by full and dash bars (reproduced with permission from reference [97] ©2004 Royal Society of Chemistry).

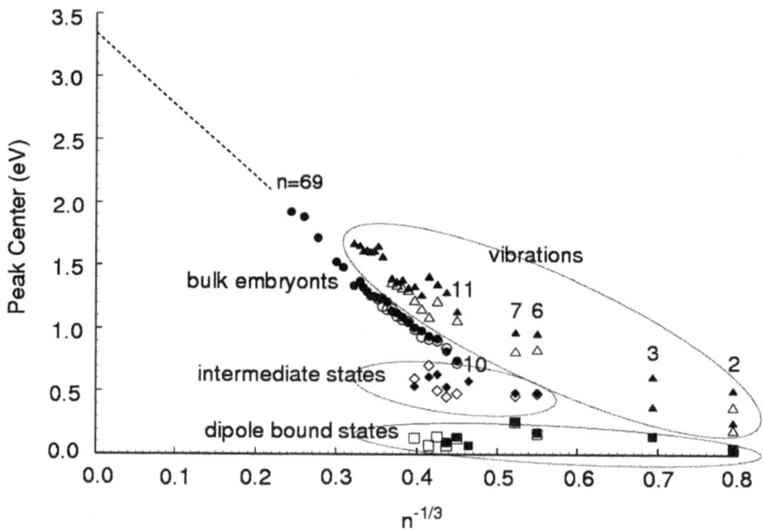

Figure 5.2.5 Plot of vertical detachment energies of water cluster anions $(H_2O)_n^-$ as a function of $n^{-1/3}$. On the right part, one can observe the sparce n values corresponding to the "magic" distribution of water anions. Water cluster anions can be grouped in three families. The very weakly bound electrons (squares) correspond to dipole-bound like anions seen from $n = 2$ to 16. From $n = 6$ to 16, intermediate states (diamonds) correspond to "surface-state" isomers. The more strongly bound electrons (circles) are observed above $n = 11$ and are embryonic bulk hydrated electrons with internal excess electrons. Triangles correspond to deuterated clusters (reproduced with permission from reference [126] ©2006 American Chemical Society).

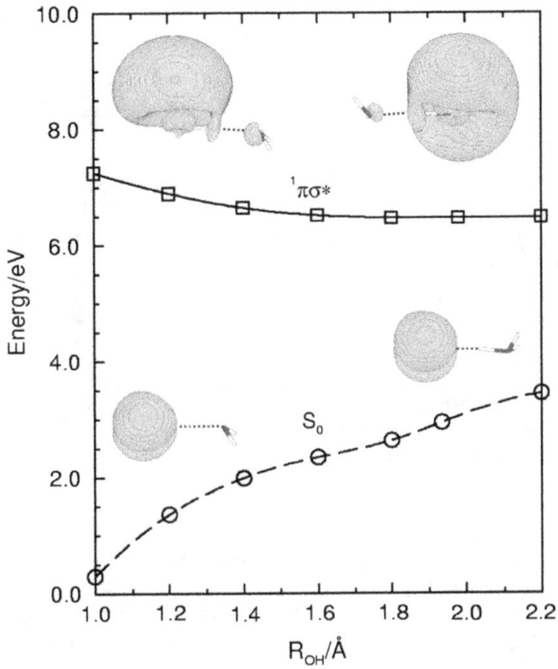

Figure 5.2.6 Potential energy curves of the S_0 and S_1 ($\pi\sigma^*$) electronic states of the water dimer, calculated along the reaction path for hydrogen transfer $H_2O...H_2O \rightarrow H_3O...OH$ (reproduced with permission from reference [129] ©2002 Royal Chemical Society).

5.2.1.3 Superoxide anion

Molecular oxygen is fundamental for a vast majority of organisms. Its hydrated anion $O_2^-{\cdot}(H_2O)_N$, also known as the superoxide anion, intervenes in the fight against external bacterial attacks. It is a short-lived species in an aqueous medium but has a long lifetime in cellular membranes and can easily be produced in the gas-phase [133]. From $O_2^-{\cdot}H_2O$, hydrogen peroxide H_2O_2 is produced and further attacks biomolecules [134]. It has thus often been held as responsible for ageing processes [135]. The $O_2^-{\cdot}H_2O$ structure has been determined by high-level calculations [136] and infrared spectroscopy [137]. The water molecules are rigidly tightened by ionic hydrogen bonds in a tetrahedral structure to the lobes of the π^* highest occupied orbital of O_2^-.

5.2.1.4 Protonated water clusters

Protonated water clusters are gas-phase models of protonated bulk water [138–145]. Mass-spectrometry alone cannot provide much information on the structure of $H^+{\cdot}(H_2O)_N$ clusters, except for the observation of very stable configurations such as that of the $H^+{\cdot}(H_2O)_{21}$ cluster. Quantum calculations [145, 146] and infrared spectroscopy of mass-selected ions, on the contrary, allow investigation of the structure of the "core" ion of the protonated water clusters [142, 147–149]. This core can be either the "Eigen" H_3O^+ protonated hydronium ion

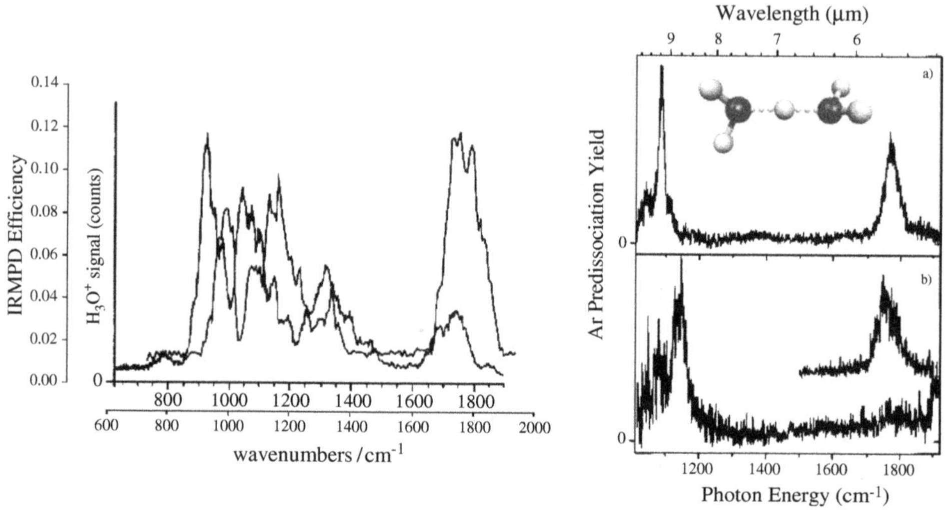

Figure 5.2.7 Left: Comparison between two experimental infrared multiphoton dissociation (IRMPD) spectra (reproduced with permission from reference [148] ©2004 American Chemical Society). Right: Infrared absorption spectrum of the protonated water dimer determined by the messenger method (IR-induced evaporation of argon atoms) (reproduced with permission from reference [152] ©2004 American Physical Society).

with the proton bound to a single water molecule or the "Zundel" $H_2O\cdots H^+\cdots OH_2$ ion with the proton shared between two water molecules [150].

The infrared spectroscopy of the Zundel $H_2O\cdots H^+\cdots OH_2$ ion (Figure 5.2.7) has been studied experimentally [147, 148, 151, 152] and theoretically [145, 146]. IR studies of the free OH stretch frequency (the IR region corresponding to the hydrogen-bonded water is too congested for easy interpretation) show that up to $N = 10$, the $H^+\cdot(H_2O)_N$ cluster structures are built from hydrogen-bonded water chains originating from the H_3O^+ or the $H_2O\cdots H^+\cdots OH_2$ ionic core [153]. For cluster sizes between $N = 10$ and 20, a two-dimensional network develops and a three-dimensional network appears above. The very stable "magic" cluster structure for $N = 21$ is highly symmetric. Calculations show that this structure is probably derived from the neutral dodecahedral cluster with the proton on the surface with an Eigen-like structure [149]. The H_3O^+ favours a specific-site hydrogen atom bonded to three acceptor–donor–donor (ADD) water molecules [145] (Figure 5.2.8).

Proton shuttling is a fundamental mechanism in enzymes. Water molecules then can act as temporary proton acceptors (Figure 5.1.2). Polarizable proton models such as ppF3C have thus been devised for investigation of biochemical reactions. In such models, water molecules alone are described with partial charges (0.41 for H atoms and -0.82 for O) and Lennard–Jones potentials. The interaction of a proton with a water molecule is described by a potential function with angular dependence: $E = a\exp[-br](1 + \alpha\cos^2\theta)$, where r is the distance of the proton from the water molecule and θ an Euler angle. Description of protonable sites of large molecular systems allows them to become polarizable and bonds can then break and form again.

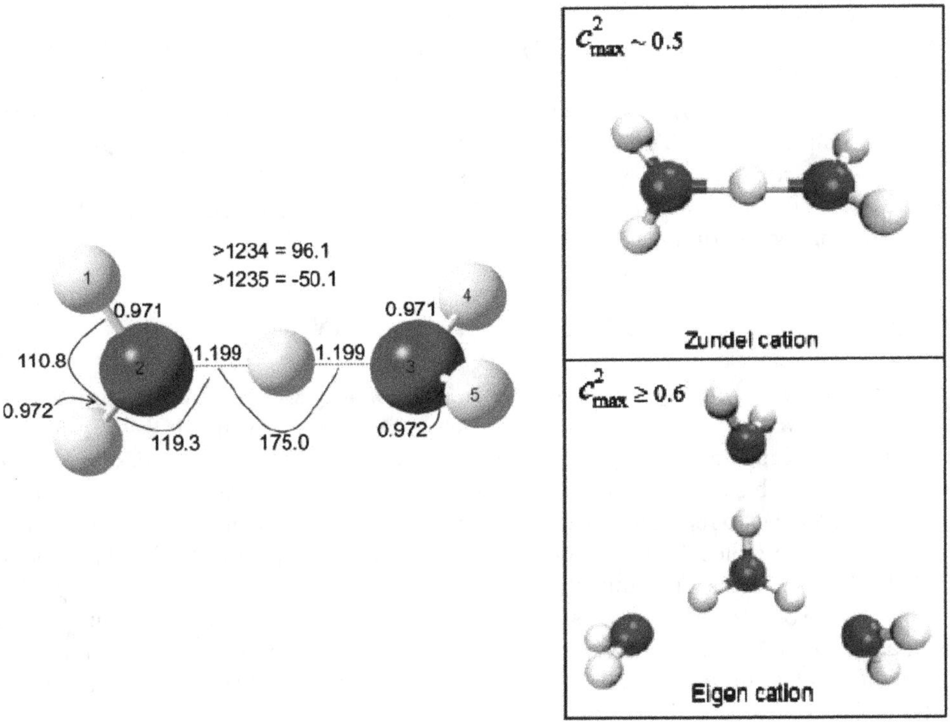

Figure 5.2.8 Left: Structure of the Zundel proton-bound water dimer calculated at the B3LYP/6-31+G* level (reproduced with permission from reference [148] ©2004 American Chemical Society). Right: Structure of the Zundel and Eigen cations (reproduced with permission from reference [154] ©2006 American Chemical Society).

5.2.2 Solvents and solvent–water mixtures

The weak bonds that ensure the secondary and tertiary structures of proteins can be disrupted by the presence of solvents. This brings proteins from their native state to denaturated forms. Several factors such as temperature, pH, large concentrations of salts and immersion in organic solvents facilitate the unfolding denaturation process. Urea and dimethylsulphoxide (DMSO) are commonly used organic solvents for denaturation. Their pure liquids, mixtures and clusters with water have been the subject of many studies. In the case of urea, it has been suggested that the denaturation process arises from an accumulation of urea in excess near the protein surface. The urea molecules then establish hydrogen bonds with the protein polar groups. Since water molecules are then displaced from the first hydration shell, hydrophobic groups can become more exposed to the solvent. Molecular dynamics [155–157] and gas-phase studies of neat urea [158–161] or urea water complexes have been conducted as well as those of DMSO [162–165] or acetone–water mixtures [166].

Another important mixture is octanol–water. The partitioning of a biomolecule or a drug between a polar aqueous phase and a non-polar phase plays an important role in many biological processes. In order to predict properties of drugs (see Lipinski rules in the conclusion), the partition coefficient of a potential drug between octanol and water is used in quantitative structure–activity analysis (QSAR) in pharmacology. The 1-octanol molecules are amphiphilic with a hydrophilic hydroxyl group and a lipophilic alkyl tail. Molecular dynamics simulations [167] show that the octanol–water solution structure is very heterogeneous. Snapshots taken during dynamic simulations show that the size distribution of homogeneous and mixed hydrogen-bonded clusters widely varies from only 2 solvent molecules to more than 60 molecules. This heterogeneity allows the octanol–water phase to mimic the complexity of membrane or body fluids.

5.3 HYDROPHOBICITY

Structures and interactions of biomolecules in solution are governed by the formation of ionic, van der Waals and hydrogen bonds as well as by hydrophobicity [168–173]. In bulk water, each molecule interacts with four neighbours and establishes a network of hydrogen bonds. "Hydrophobic forces" have been invoked as the major drive for protein folding. Those "forces" are a mean to represent the result of disruption of the H-bond network induced by the presence of a molecule that cannot form hydrogen bonds with water (apolar system) [174, 175]. In proteins, for example some amino acid side-chains such as those of phenylalanine, valine or leucine do not engage into H-bonds and are called hydrophobic in contrast, for example with serine, threonine or asparagine (polar amino acids) that do [176]. Hydrophobic molecules still exert weak van der Waals attractive forces between themselves and water molecules. If the hydrophobic solute system has a small size, the H-bond network is modified but still maintained and the solute is said to be "wet". One of the roles of the solute is then a reduction of the configuration space available for H-bonding and corresponds to an entropic effect. The hydrophobic effect results from the competition between enthalpic and entropic contributions and thus varies with temperature [177] (Figures 5.3.1 and 5.3.2).

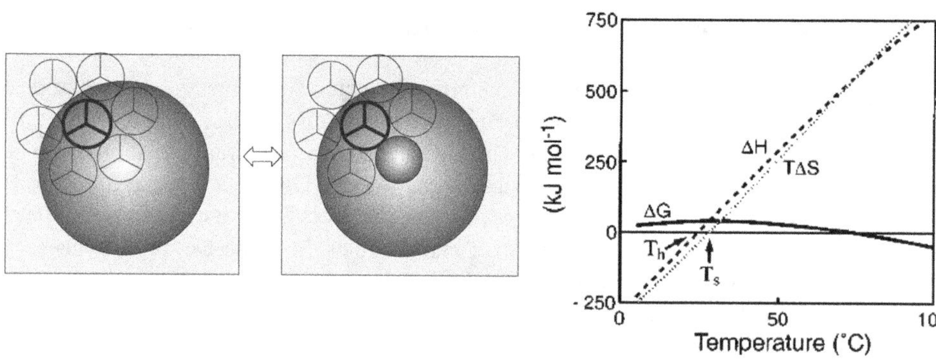

Figure 5.3.1 Left: Schematic representation of the transfer of a non-polar solute into water. Water molecules are represented by the Mercedes–Benz model (see Section 5.1.3). The environment of a first solvation shell test water molecule (in bold) is modified by the replacement of a water molecule (left) by a non-polar molecule (in grey, right) (adapted with permission from reference [172] ©2005 American Chemical Society). Right: Protein unfolding free energy as a function of temperature (reproduced with permission from reference [170] ©2002 American Chemical Society).

When the size of the hydrophobic solutes increases (>1 nm), the H-bonded water network breaks apart and the solute becomes "dry". For solutes that are initially separated in water, there is a tendency to minimize the number of strong H-bonds between water molecules that have broken. Solutes then tend to gather as if they are driven by forces called hydrophobic forces [178, 179]. For example, proteins fold in order to bring most of their hydrophobic residues in their interior. It must be noted that there is no hydrophobic potential corresponding to those forces and that those forces are not additive. Hydrophobicity is a collective phenomenon and cannot be represented by pairwise interactions. Hydrophobic forces are fundamental in the formation of biomolecular structures [180–182] but vanish when molecular systems are totally dehydrated in gas-phase.

In between the two extreme situations encountered by molecules in bulk water and totally isolated in the gas-phase, partially hydrated systems containing hydrophobic regions have been experimentally studied [183, 184]. For that purpose, electrospray sources can be modified in order to control the amount of water molecules added to ions [183, 185–187]. Completely desolvated ions produced in a conventional source are then hydrated by mixing with a bath gas of precisely controlled composition containing the solvating molecules. The obtained mixture is then expanded into the vacuum system of a mass spectrometer for mass analysis. Fenn and coworkers have studied hydration of electrosprayed singly and doubly charged ions of two hydrophobic peptide antibiotic molecules, cyclosporine A (CSA) and gramicidin (GRS) and of their non-covalent 1:1 complexes. The structures of isolated CSA [188] which strongly depends on the nature of solvent and that of GRS [189] are known in condensed-phase. The mass-spectra of the separated molecules in presence of water show that the CSA ions remain nearly dry (3–5 attached water molecules) in the gas-phase, whereas the GRS ions are nearly fully hydrated (15–17 water molecules) under the same conditions. This can be understood in terms of hydrophobicity of each amino acid (Figure 5.3.3) [190]. CSA only contains very hydrophobic amino acids (Leu, Val and Phe) while the decapeptide GRS contains four hydrophilic amino acids (Pro and Orn). Often,

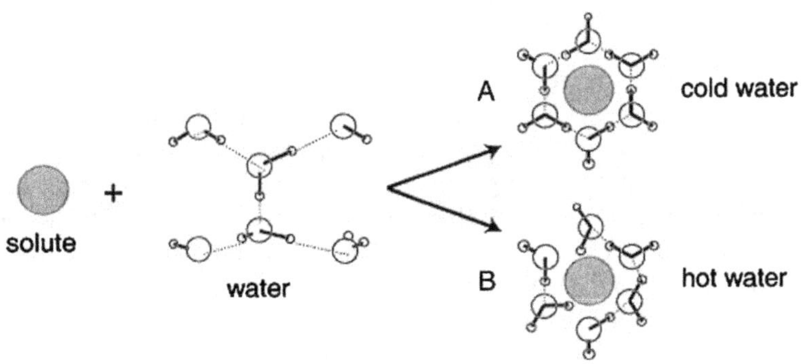

Figure 5.3.2 Schematic representation of a hydrophobic molecule (grey circle) in presence of water molecules. At room temperature (A), the presence of the small hydrophobic molecule does not disrupt the hydrogen bond network of the water molecules. The water molecules adopt only a few orientations (low entropy) to avoid wasting hydrogen bonds and most of them are fully hydrogen-bonded (low-energy). In hot water (B), more conformations become accessible for the water molecules (higher entropy) at the cost of breaking hydrogen bonds (higher energy) (reproduced with permission from reference [170] ©2002 American Chemical Society).

Phe Ile Leu Trp Cys Val Met Tyr Ala His Gly Thr Pro Ser Glu Asp Arg Asp Glu Lys

....➤

highly hydrophobic **highly hydrophilic**

Figure 5.3.3 Approximate comparison between hydrophobicities of amino acids.

peptides with both hydrophilic and hydrophobic residues (called amphipathic peptides) possess antiviral, antitumoural and antimicrobial activity due to their aptitude to disrupt membranes. In mass-spectra, CSA—GRS complex ions are surrounded by only seven water molecules, rather similarly to CSA ions alone. The interpretation given by Fenn is that the preferential hydration sites of GRS are shielded from access when GRS is complexed with CSA. CSA would bind on the side where the hydrophilic and positively charged ornithines (products of the action of the arginase enzyme on L-arginine) are situated.

This gas-phase investigation of binding of two intermediate size peptides in presence of water can be compared to the results of systematic analysis of condensed-phase data concerning protein–protein interfaces [191, 192] or DNA–protein interfaces [193]. Those studies show that there is a striking difference between interior of proteins that are strongly hydrophobic and the interfaces between proteins that are much less hydrophobic and intervene in molecular recognition [194]. Two important features of recognition are the existence of conformation changes induced by the binding ("induced-fit") and the exclusion of solvent around key residues. Changes are not directly accessible by means of mass-spectrometry studies but spectroscopic gas-phase studies (IR, FRET) may possibly offer the opportunity to investigate those changes in absence of water or in presence of a well-controlled number of water molecules as in the experimental study described above.

5.4 HYDRATION OF NUCLEOBASES, OLIGONUCLEOTIDES, DNA AND RNA

Several effects influence structures of DNA and RNA in solution as compared to gas-phase. Nucleotide chains, through variations in torsion angles of the sugar–phosphate backbone and reorientations of the nucleobases, adapt their structures in order to optimize the interactions of their hydrophilic groups with the three-dimensional network of water molecules. The repulsion of negatively charged phosphates is attenuated by the presence of counter ions and the screening due to the large dielectric constant of water. The DNA double helix conformation is thus dependent on the degree of hydration. At low humidity, the A form is favoured while the B form is found at high humidity [195]. The Z form appears with alternating purine–pyrimidine sequences and high salt concentrations. The non-specific hydration of nucleobases ("hydrophobic" effect) can be treated by using continuum representations of the solvent. On the contrary, the specific hydration pattern must be treated at a high level of theory. In this chapter, the hydration pattern of isolated nucleobases will first be considered together with the possibility of tautomerization and further followed by the influence of water on hydrogen bonding and stacking in pairs. The gas-phase stability of double helices in oligomers as a function of strand length will then be examined.

5.4.1 Hydration of nucleobases

In solution, water molecules located at more than 10 Å from DNA can still play a role in the three-dimensional structure. Gas-phase studies only focus the attention on the primary hydration shell at the nucleotide and base-pair level. The water molecules in that shell are crucial since DNA ligands such as nucleotides, amino acids or drugs are in direct contact and either replace those molecules or establish water-mediated interactions.

5.4.1.1 Hydration sites and tautomerization of isolated nucleobases

Hydration of isolated nucleobases [196] has been studied by means of different experimental methods such as electron ionization [197], UV spectroscopy [198], IR cavity ring-down spectroscopy [199], IR/UV spectroscopy [200, 201], photoelectron anion spectroscopy [202–204], helium nanodroplets [205], femtosecond spectroscopy [206] and Rydberg electron transfer [207]. Structures and energetics of hydration sites have been investigated theoretically from ab initio [80, 208–212], DFT [52] and molecular dynamics calculations [213, 214] with up to 32 water molecules [215]. A comprehensive review can be found in reference [216]. We must recall that most experimental spectroscopic investigations are conducted at low temperatures and corresponding calculations at "0°K". As shown in Figure 5.1.4, the idealized conditions of gas-phase studies probably do not represent the real conditions found in biological media. Hydration layers are usually considered with very small numbers of water molecules. Those water molecules interact with both nucleobases and other water molecules but cannot form the three-dimensional hydrogen bond networks found in bulk water.

Nucleobases are not isolated in DNA and not all possible hydration sites are available due to the linking of bases to the backbone (in N1 or N9 position) and pairing (see Figure 5.4.1). Radio-crystallographic studies of DNA show that usually pyrimidine bases are only associated with a single water molecule found in the major groove (structure A of Figure 5.4.1) or the minor groove (structures C and D). This water molecule can also form a hydrogen bond to another base through a second water molecule. Since in the biological medium, water molecules do not necessarily remain indefinitely on preferred hydration sites, it is also important to estimate the energetic barriers between those sites.

Different tautomers of isolated nucleobases can coexist in the gas-phase [217] (see Chapter 4.1). Some of them, such as enol tautomers involving deprotonation of the N1 position in cytosine, are not biologically relevant. The presence of water molecules can modify the relative different tautomer energies [42, 44, 51, 212, 218, 219]. Hydration Gibbs energies of the different nucleobases have been calculated by means of a continuous solvent model [40].

In water, amino-keto tautomers of guanine are energetically favoured due to the higher dipole moment of the keto form. In the gas-phase, IR/UV studies [210] show that three guanine–water conformers coexist in a supersonic beam expansion. One of them (g17 + W in Figure 5.4.2) is in the N7H-keto form with the water molecule simultaneously bound to the C=O and N7H groups. The most biologically relevant keto N9H is absent. The two other conformers are in the 9-enol form. One of them (g901 + W) has a water molecule bound to OH of the enol group and atom N7. The other is in either the g901a + W or g902 + W configuration displayed in Figure 5.4.2 with a water molecule bound to the N3 atom and N9H group.

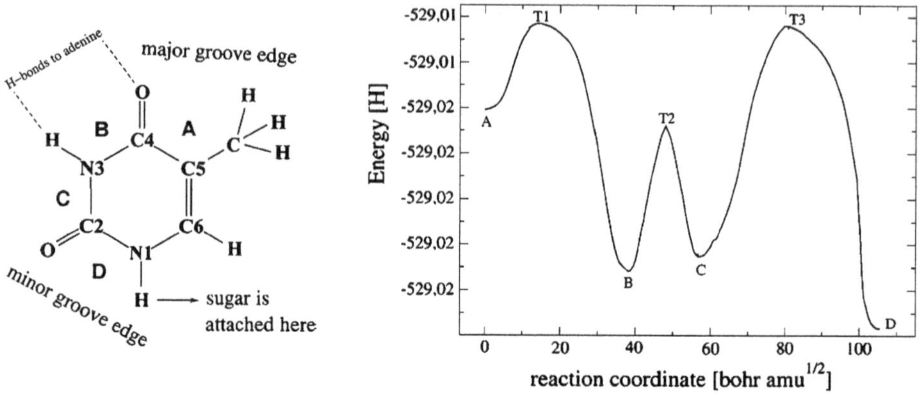

Figure 5.4.1 Left: Hydration sites of isolated thymine. Right: Relative energies and transition states of the different configurations of thymine–water complexes calculated at the RI-MP2/aug-cc-pVTZ level (reproduced with permission from reference [212] ©2006 American Chemical Society).

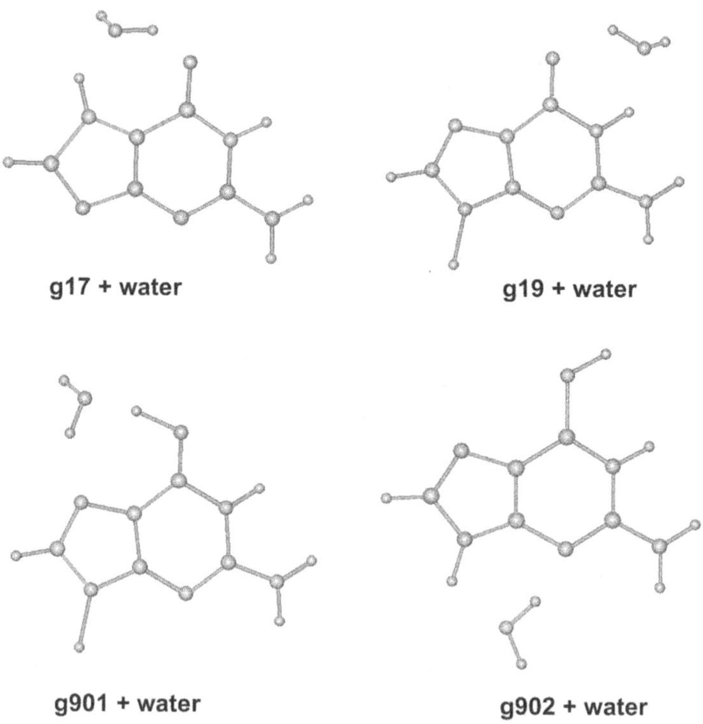

Figure 5.4.2 Four most stable tautomers of guanine and optimized structures of guanine–water complexes calculated at the RI-MP2/cc-pVTZ level. Absolute values of the respective enthalpies of formation of the monohydrated complexes are, respectively (in kJ/mol), 43.6, 38, 33 and 32.85 for g17 + W, g19 + W, g901 + W and g902 + W (adapted with permission from reference [210] ©2005 Royal Chemical Society).

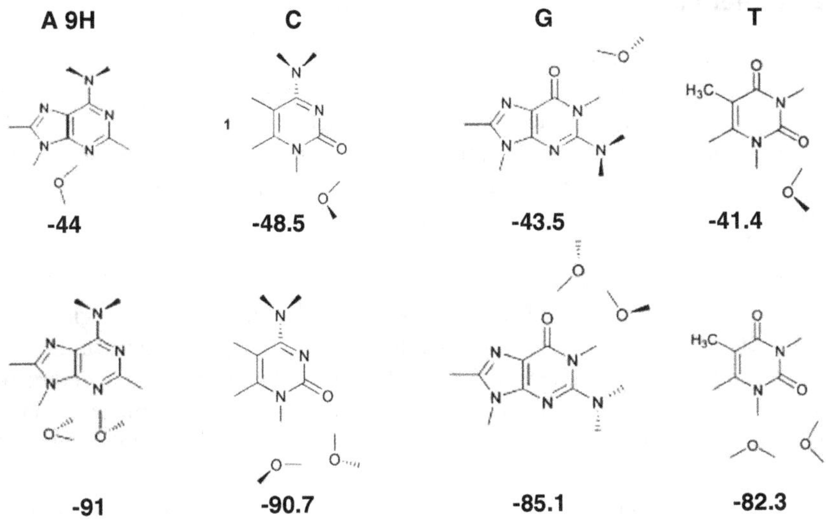

Figure 5.4.3 Most stable structures of mono- and dihydrated nucleobases optimized at the RI-MP2/TVZPP level (energies are in kJ/mol with respect to the most stable bare nucleobase tautomer) (reproduced with permission from reference [216] ©2006 Royal Chemical Society).

There is thus a large difference between gas-phase where four tautomers of guanine are observed and solution where a single tautomer is observed (Figure 5.4.3). The gas-phase study of guanine–water tautomers is complex due to the fact that complexes involving several water molecules have too short lifetimes for being observed in the nanosecond time range (see Chapter 4.1).

Hydration shells of nucleobase cations have been studied by means of R2PI [198]. At low temperatures corresponding to supersonic beam conditions, the first hydration shells of adenine and thymine cations are complete with only four water molecules while the hydration shell of the adenine dimer cation involves seven or eight water molecules.

5.4.1.2 Metal cation–nucleobase binding in presence of water

Binding of metal cations to DNA can induce the formation of base mispairings. It can also influence charge-transfer processes when DNA is considered as a molecular wire for molecular electronics. Binding of metal cations to nucleobases has been experimentally [220, 221] and theoretically [45] investigated. Nucleobases being strongly polar, the preferential binding sites of metal cations are in the negative potential regions. For example, the prominent cation binding sites of guanine are the N7 and O6 atoms. A metal cation such as Mg^+ or Zn^+ is hydrated by six water molecules and can interact with a molecule such as guanine according to three main patterns [45]. Inner shell binding corresponds to the inclusion of guanine into the first shell via one of its electron donor atoms N7 or O6. The preferential choice between those two atoms then varies according to the electronic properties of the metal. In the bidentate inner shell binding mode, both N7 and O6 atoms are bound and replace two of the six water molecules. In the outer shell binding scheme, the metal hydration pattern is not modified and the guanine molecule is then simply attached to this metal–water complex.

5.4.1.3 Dynamic behaviour of excited hydrated nucleobases

Excited state potential surfaces and dynamic behaviours of electronically excited nucleobases are modified by the presence of water molecules [206, 222–224]. The order of excited states can vary as a function of added water molecules. $\pi\pi^*$ states with rather low dipole moments (respectively, 2 and 4 D for L_a and L_b) are less stabilized than $\pi\sigma^*$ states with high dipole moments (12 D) [225]. On the contrary, $n\pi^*$ states are destabilized due to unfavourable interactions between non-bonding doublets of the nucleobases and water.

Transients observed in femtosecond pump/probe experiments of 9-methyl-adenine–$(H_2O)_n$ complexes (the protecting methyl at N9 prevents formation of non-biologically relevant hydrates) at 267/400 nm exhibit two times constants. The very short time and "long" time constants, respectively, decrease steadily from 140 and 1,240 fs for the bare molecule down to 100 and 700 fs for $n = 9$ [224]. The relative weights of the "long" time constant strongly decrease when water molecules are added (Figure 5.4.4). The interpretation is the following. The stabilization of the $\pi\pi^*$ states and the destabilization of the $n\pi^*$ states in presence of the polar water molecules lead to an increase of the barrier between these states. The preferential de-excitation pathway thus becomes preferentially relaxation of the L_b state

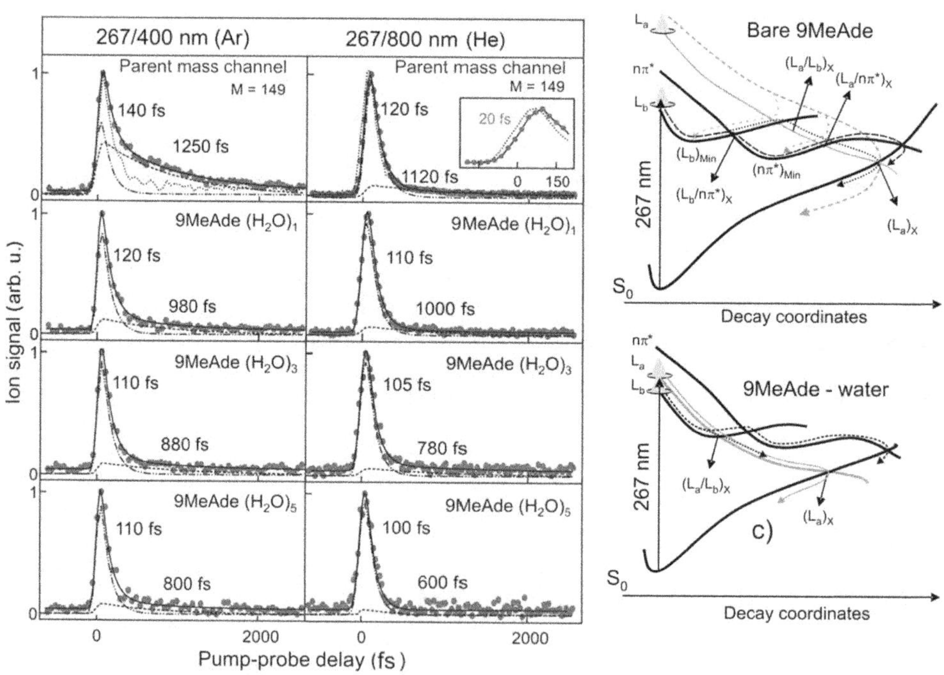

Figure 5.4.4 Left: Transients observed in femtosecond 267/400 nm pump/probe experiments of 9-methyl-adenine–$(H_2O)_n$ clusters. Note the shortening of the long time constant and the strong decrease of the corresponding weight. Right: Potential energy curves displaying the major and minor de-excitation pathways of the L_b state via the L_a/L_b and $L_a/n\pi^*$ conical intersections (reproduced with permission from reference [224] ©2006 Royal Chemical Society).

through the L_a/L_b conical intersection with a time constant of ca. 100 fs followed by the conical intersection of L_a with ground state S_0. The "long" time constant corresponds to the now minor de-excitation channel involving the $L_a/n\pi^*$ conical intersection and the $n\pi^*$ state with a time constant of ca. 1,000 fs.

5.4.2 Hydration of base pairs

In gas-phase, nucleobases can pair in either hydrogen-bonded or stacked configurations (Figure 1.1.6). Calculations predict that the presence of a very small number of water molecules can strongly modify the situation [226–228]. In presence of only two water molecules, 20% of G—C base pairs are predicted to be in stacked configuration instead of H-bonded configuration [228]. Base pairing of guanine has been experimentally studied by means of IR/UV spectroscopy [201, 210]. In absence of water, two configurations are energetically close and coexist in the gas-phase G—G pair. A single water molecule is sufficient to stabilize one specific base pair relative to the other one. Although calculations predict that some of the G—G pairs should come into stacking configurations, only hydrogen-bonded pairs remain observable (Figure 5.4.5).

5.4.3 Hydration of oligonucleotides

In DNA and RNA crystals, base pairs are each surrounded by 22 hydration sites [13, 229, 230]. Water molecules with residence times in the first hydration shell of the helices range between 0.5 and 1 ns and are essentially found in the vicinity of the phosphate groups and in the minor groove. The Dickerson dodecamer [231] possessing a palindromic complementary sequence and the 16-mer duplex containing the same central sequence d(CGCGCGAATTCGCGCG) have been studied in simulations. Starting from crystal structures, force-field simulations run in water (with 2,068 or 3,524 molecules) yield stable structures close to the experimental condensed-phase structure. On the contrary, in absence of water, the two strands remain bound as shown in gas-phase experiments using electrospray sources [232–235] but after 100 ns simulations, the obtained structures become very distorted. Although a large number of canonical hydrogen bonds are lost, 60–90% of those bonds found in water still exist

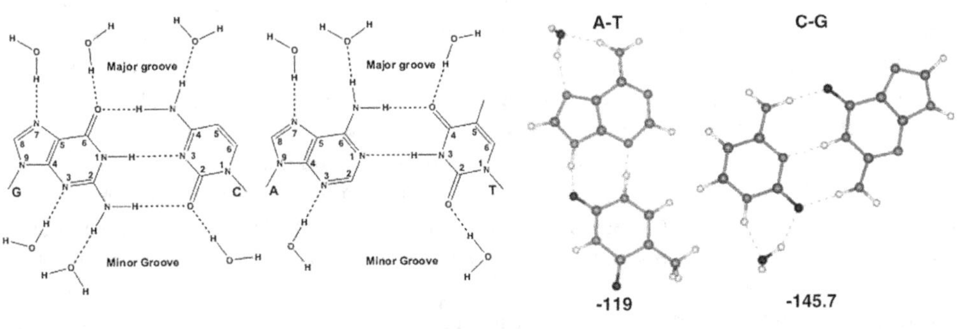

Figure 5.4.5 Left: Possible hydration sites of isolated Watson–Crick base pairs. Right: Most stable structures of monohydrated base pairs optimized at the RI-MP2/TZVPP level (energies are in kJ/mol) (adapted with permission from reference [216] ©2006 Royal Chemical Society).

at 298 K. The gas-phase DNA structure thus retains several structures from the solution structure. The relationship between gas-phase and solution-phase stabilities of 9-mer DNA duplexes has been studied by mass-spectrometry [236, 237]. Electrosprayed 5′-GGT-TXTTGG-3′–3′-CCAAYACC-5′ duplexes with X/Y = G, C, A or T with and without single internal mismatches with respect to canonical A—T or G—C pairing have been submitted to collision-induced dissociation (CID). The gas-phase stability has then been quantified through measurement of the collision energy required to dissociate half of double-stranded duplexes into single strands and compared to solution-phase stability measurements [238]. Four duplexes (GG, AC, TC and CC) exhibit enhanced gas-phase stability, two duplexes (TA and AT) are unstable in gas-phase relative to solution and the remaining (GC, CG, GT, AG, TG, TT, GA, CT, AA and CA) exhibit a linear correlation between the gas-phase and solution-phase stabilities.

QM/MM (see Section 1.6) calculations of gas-phase oligonucleotide structures show that stacking interactions depend on the number and the nature of the nucleobases in single-stranded oligonucleotides [239]. The hydrogen bonding between nucleobases belonging to the same strand or to different strands is different. This is due to the non-planearity of the amine $-NH_2$ groups of the bases that tend to be involved in π-stacking interactions instead of hydrogen bonds. Guanines tend to interact through hydrogen bonds while adenines favour stacking interactions. In double-strand oligonucleotides, the $-NH_2$ groups are more planar.

G-quadruplexes (four guanine linked through hydrogen bonds in presence of cations) are a key structure of telomers and have been studied by mass-spectrometry [234, 240, 241] and ion mobility [242]. Their large stability is justified by the stacking and hydrogen bonding between the bases and the interactions of the G carbonyl group with the cations. Very long molecular dynamics simulations extending up to 0.5–1 μs [243] show that gas-phase G-quadruplex remains stable in the gas-phase only if the suitable cations are introduced. The gas-phase structures are then, exceptionally, indistinguishable from the structures obtained from simulations run into water.

5.4.4 Hydration of DNA and RNA

The role of solvents in inducing conformations has been investigated through molecular dynamics simulations of DNA and RNA conducted by means of force-fields with explicit water molecules. Very long molecular dynamics simulations (up to 0.5 μs) of duplex DNA show that, even at high temperatures, removing of water molecules from DNA does not lead to total disruption of the double helix [244]. The two strands remain bound and the “dry” structure retains some helicity. The possibility of spontaneous transitions A → B forms induced by water has been investigated [239] showing that conformation A does not survive in water. In a denaturating solution containing, for example 50% pyridine and 50% water, C—G pairs first disappear and once some base pairs are opened, the denaturation process becomes irreversible.

The ribosome where proteins are synthesized after matching of anticodons of tRNAs to the genetic message contained in the mRNA codons is a complex of RNA and proteins (see Chapter 4.2). The rRNA–protein complex structure is maintained by a number of highly specific hydration sites with long-residing water molecules [245].

5.5 HYDRATION OF AMINO ACIDS, PEPTIDES AND PROTEINS

The secondary structure of proteins strongly depends on the formation of hydrogen bonds between amide groups. Statistically, the vast majority of water molecules that are bound to a protein establish hydrogen bonds as donors to C═O groups and to a less extent to N—H groups, except for charged COO^- groups of Asp and Glu. Those water molecules can engage into hydrogen bonding with amide groups and thus induce important structural changes. Water then acts as a "lubricant" [169]. Many water molecules, similar in amount to individual amino acids, also remain buried in the core of the proteins, so forming structurally important hydrogen-bonded linkages [246]. In this chapter, amide–water and side-chain water clusters will first be examined. Amino acid–water interactions and photodynamic processes will be considered. Finally, gas-phase studies of hydration of peptides and proteins will be presented (Figure 5.5.1).

5.5.1 Hydration of the backbone

The formamide molecule and its methylated derivatives [247] (Figure 5.5.2) are models of the peptide bond linking amino acid residues in peptide and proteins. They have been experimentally studied in the gas-phase [98, 248–254] as well as in solution [255] and simulations [256–259]. Their complexation with water has been studied experimentally by means of microwave [260] and Rydberg electron transfer [97, 261] spectroscopies and calculations [60, 262]. The most stable hydration site (A in Figure 5.5.3) corresponds to establishment of two strong hydrogen bonds involving both the N—H and C═O bonds with the same water molecules when those two groups are on the same side of the molecule, that is for F and NMF and NMA in *cis* configuration. Less stable sites (B and B′) correspond to establishment of weaker bonds with the C═O group and a C—H or C—CH_3

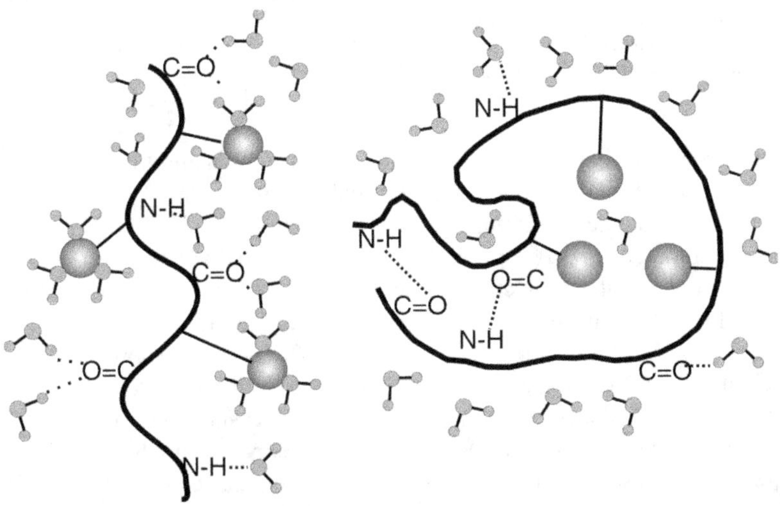

Figure 5.5.1 Sketch of hydration sites of an unfolded (left) and folded protein (right) in presence of water. Circles represent hydrophobic residues.

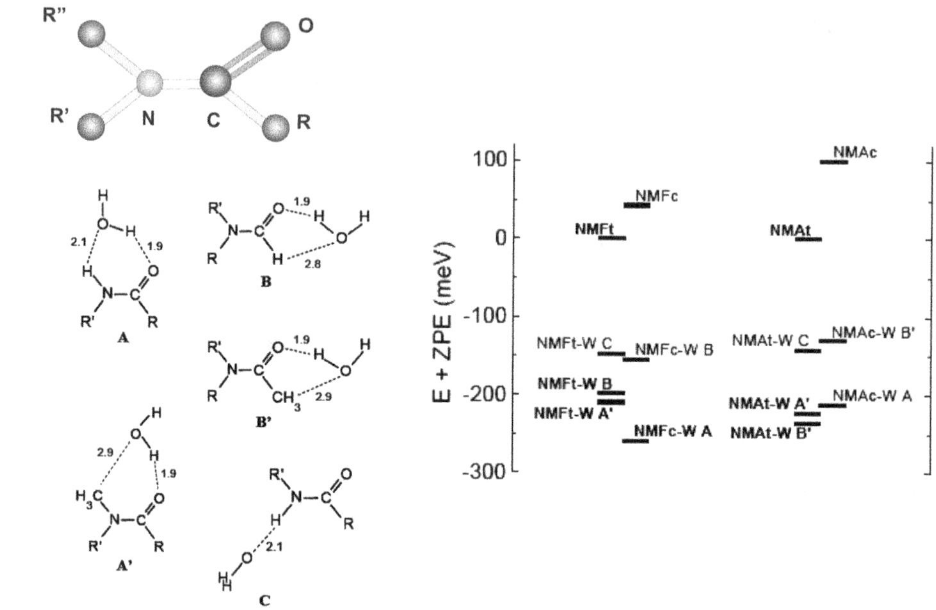

Figure 5.5.2 Left: (Top) Sketch of amide molecules mimicking the amide bond. Formamide (F) corresponds to R═R′═R″═H, *trans-N*-methylformamide (NMFt) to R═R′═H and R″═CH_3, *trans-N*-methylformamide (NMFt) to R═R′═H and R″═CH_3, *cis-N*-methylformamide (NMFc) to R═R″═H and R′═CH_3, *trans-N*-methylacetamide (NMAt) to R═R″═CH_3 and R′═H, *cis-N*-methylacetamide (NMAc) to R═R′═CH_3 and R″═H and *N,N*-dimethylformamide (DMF) to R═H and R′═R″═CH_3. (Bottom) Hydration sites of the formamide molecule and its methylated derivatives. Hydrogen bond distances are given in Angstrom. Right: Relative calculated total energies (corrected from zero-point energy) of the NMF and NMA molecules and their monohydrated complexes. Species in bold have been observed in the gas-phase. Note that for the isolated NMF, the *trans* configuration is more stable than the *cis* one and that the situation is reversed in presence of a single water (see in Chapter 4.7 a similar situation in melatonin–(water)$_2$ complexes) (reproduced with permission from reference [261] ©2002 EDP Sciences).

group. A water molecule can also act as a proton acceptor of an amide N—H bond (configuration C). This N—H···O hydrogen bond is weaker than C═O···O or double hydrogen bonds leading to less stable configurations (see Figure 5.5.3).

At very low temperature, water molecules are bound to stable hydration sites. When the temperature is raised, water molecules can overcome the energy barriers and move from one site to another. By using stimulated emission pumping with different excitation wavelengths (see Section 2.1.4.4), Zwier and coworkers [263, 264] have shown that it is possible to selectively move a water molecule bound to the C═O hydration site of formanilide (formamide substituted with a phenyl chromophore) to the N—H site or vice versa (Figure 5.5.3). This method allows the experimental determination of energy thresholds E_{th} for isomerization in both directions. These energies are estimated to be equal, respectively, to $750\,cm^{-1} \leq E_{th}(N{-}H \rightarrow C{=}O) \leq 988\,cm^{-1}$ and $796\,cm^{-1} \leq E_{th}\,(C{=}O \rightarrow N{-}H) \leq 988\,cm^{-1}$. The relative energies E_{min} of the two conformers (water bound to either C═O or N—H) can then be deduced. Their difference is estimated to be equal to $-238\,cm^{-1} \leq E_{min}\,(C{=}O) - E_{min}$

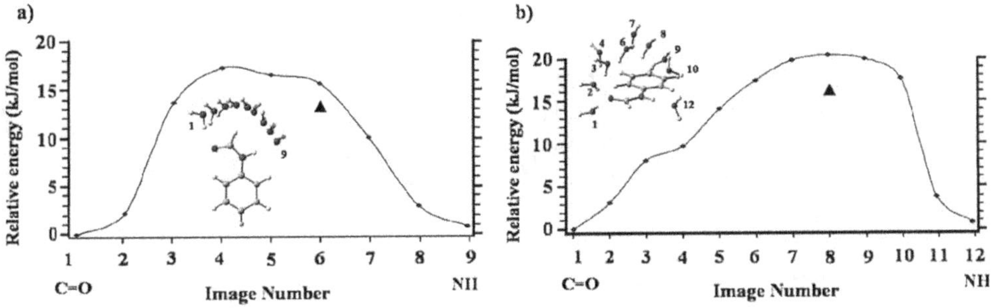

Figure 5.5.3 Left: Minimum energy in plane pathway for the transfer of a water molecule from the C=O hydrogen binding site of formanilide to the N—H binding site calculated with DFT B3LYP/ 6-31 + G*. Right: Out of plane pathway (reproduced with permission from reference [263] ©2005 American Association for the Advancement of Science).

(N—H) ≤ 192 cm^{-1}. This small value eases the water shuttling between hydrogen-bonding sites.

Molecular dynamics simulations have been conducted in the case of *N,N*-dimethylformamide (DMF)–water solutions [265]. DMF molecules are interesting because they cannot engage into strong C=O···H—N hydrogen bonding as formamide does. Most of DMF molecules only engage into hydrogen bonding with a single water molecule but some of them can engage into two weak C—H···O hydrogen bonds with two water molecules. The mixed dimer DMF···H_2O complex is more stable than the DMF$(H_2O)_2$ complex.

Proton exchange between a hydronium ion and an amide (see Section 3.3.4) raises the question whether the preferred protonation site is the C=O or the N—H group. This problem has been considered in gas-phase by comparing the IR spectra of $[HC(O)NH_2]_3H^+$ and $[HC(O)NH_2]_3NH_4^+$ cluster ions [266]. The spectral signatures of the N—H^+···O hydrogen bonds show that O-protonation occurs in preference to N-protonation in formamide. This experimental work can be related to CPMD simulations of amide hydrolysis in aqueous solution [56]. The initial step that determines the base-catalyzed hydrolysis of *N*-methylacetamide is the attack of the amide C atom by a hydroxide ion. This attack is a simple addition of the hydroxyl group and is followed by protonation of the amide N atom. The most strongly favoured pathway for dissociation of the intermediate complex is oxygen deprotonation and fragmentation into an amine and carboxyl anion.

5.5.2 Hydration of amino acid side-chains

Since water molecules tend to preferentially bind to amino groups of amino acids, hydration of side-chains of amino acids is more easily studied in complexes composed of water and molecules mimicking those side-chains. Among those complexes, the most widely investigated are those containing phenol (side-chain of tyrosine) [267–270], indole (tryptophan) [271–281] and imidazole (histidine) [282–288]. The indole–water complex has different configurations (Figure 5.5.4). In the two lowest energy configurations, the acceptor water is hydrogen-bonded to the N9H group. In secondary minima lying 3 kJ/mol above,

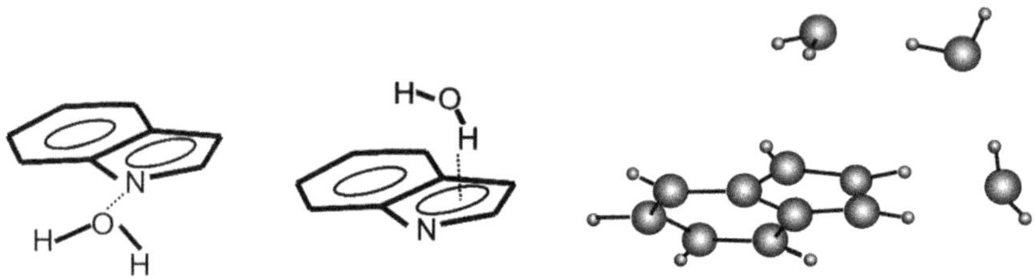

Figure 5.5.4 Left: Conventional (right) and non-conventional (π-type, left) hydrogen-bonded complexes of indole with water (adapted with permission from reference [280] ©1999 American Chemical Society). Right: Structure of a neutral indole–(water)$_3$ complex displaying both conventional and π-hydrogen bonds (adapted with permission from reference [275] ©2000 American Chemical Society).

the water molecule becomes bound in π-type hydrogen bonds. Only the lowest energy configuration is observed in the indole–(water)$_1$ complex, but both hydrogen-bonding situations are observed in the indole–(water)$_3$ complex (Figure 5.5.4). The non-conventional O—H···π hydrogen bonds are not only gas-phase features but are also observed in very high-resolution X-ray protein crystal structures [12].

5.5.3 Hydration of amino acids

5.5.3.1 Hydration of neutral amino acids

Gas-phase structures of glycine–water neutral complexes have been investigated by means of Fourier-transform microwave spectrometry [289]. In isolated glycine (see Section 1.5.3), several conformers are predicted [290–292]. Conformer I has a *cis*-COOH configuration and is stabilized by a bifurcated (N—H···O═C) hydrogen bond between the amine and the carbonyl groups. Conformer II has a N···H—O hydrogen bond and conformer III has a *trans*-COOH configuration with a N···H—O hydrogen bond. Experimentally, this third conformer is not observed because it easily relaxes to conformer I due to collisions in jet expansions. The glycine–water conformers, called Ia and IIIa, respectively, originate from conformers I and III but only the former is experimentally observed. Conformer Ia retains the N—H···O═C bond and the *cis*-COOH configuration of the non-hydrated conformer I. In this conformer, the water molecule acts both as a donor and an acceptor and is bound to the carboxylic group through two hydrogen bonds forming a close structure (Figure 5.5.5).

Tryptophan is the major chromophore of proteins [21]. Due to the great importance of solvation on its fluorescence [293, 294], several studies have been devoted to its hydrated complexes. Hydration patterns of tryptophan [21, 295–298] and phenylalanine [299, 300] have been studied in depth. Neutral hydrated tryptophan complexes can be produced by either laser desorption or expansion of a moist argon jet. The bare tryptophan (Trp) lowest energy conformations are labelled 1a, 2a and 2b. Water molecules have several binding possibilities (see Figure 5.5.6). The most favourable conformations of Trp—$(H_2O)_{N=1,2}$ correspond to water binding to the carbonyl group of tryptophan rather than to its indole side-chain. However, they are not related to attachment of a water molecule to the most stable conformer 1a of the bare molecule. Attachment to the higher lying 2a and 2b conformations is more favourable since it breaks the intramolecular OH → NH_2 bond between the amino

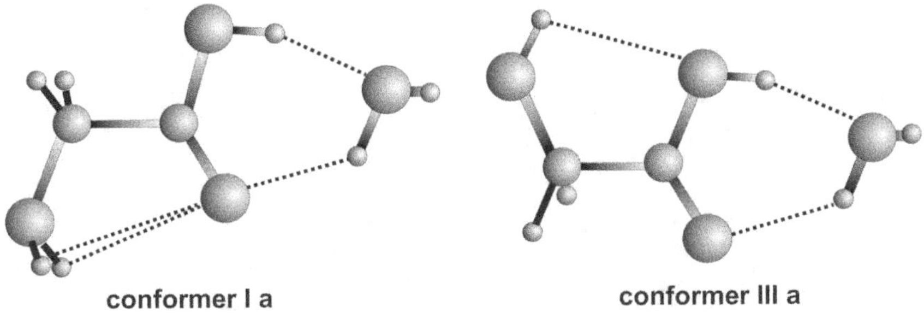

Figure 5.5.5 The two lowest energy conformers of the glycine–water complex (adapted with permission from reference [289] ©2006 Wiley).

and the carbonyl groups and enables the formation of the intramolecular $OH \rightarrow O_w$ and $OH_w \rightarrow O$ bonds. This leads to a reversal of conformer stability 1a → 2a/2b between the bare and hydrated molecules. The most stable structure of Trp—$(H_2O)_{N=3}$ corresponds to the building of a water bridge across the carbonyl group and between this group and the amino group. In all hydrated complexes, the most stable amino/carboxylic (N- and C-terminals) conformation is altered by hydration.

5.5.3.2 Hydration of protonated amino acids

Hydration of protonated valine has been investigated at very low temperature [301]. In bare valine, the isopropyl residue side-chain does not establish hydrogen bonds with the N- and C-terminal groups. In hydrated complexes of valine, it does not either interact with water molecules. Protonation occurs at the amino group and removes the lone pair of the nitrogen atom. The N-terminus is restricted to the role of hydrogen-bonding donor and thus cannot accept the proton of the C-terminus as it usually occurs in isolated unprotected neutral amino acids. In the two experimentally observed lowest energy conformers of $Val{\cdot}H^+{\cdot}H_2O$, the water molecule is bound to the charged amino group. The second water molecule of the lowest energy conformer of $ValH^+{\cdot}(H_2O)_2$ is bound to the OH group of the C-terminal while the first one remains bound to the charged N-terminal ammonium (Figure 5.5.7). This situation persists in more hydrated species. In $ValH^+{\cdot}(H_2O)_3$ and $ValH^+{\cdot}(H_2O)_4$, a water molecule remains bound to the C-terminus OH while the solvation shell of the ammonium grows.

This observed trend of solvation of the charged terminal in the case of protonated valine seems to be a general feature for amino acids with neutral side-chains in contrast with amino acids containing charged side-chains [36, 302–304]. In the case of bare protonated tryptophan [298, 305], the dissociation channel along the $\pi\sigma^*$ state leads to a H-atom loss.

Protonation and de-protonation of ionizable amino acids such as histidine or glutamic acid that can act as proton donors or acceptors are crucial for enzymatic activity [306]. Modelling of the proton transfer can be performed by means of CPMD [307, 308] and also by means of a less computer-time consuming approach. The multistate empirical valence bond (MS-EVB) method is capable of handling proton shuttling in the $HA + H_2O \leftrightarrows A^- + H_3O^+$ process [154]. In the MS-EVB approach, the system of interest is described by a matrix whose diagonal elements are valence bond states computed at the level of classical

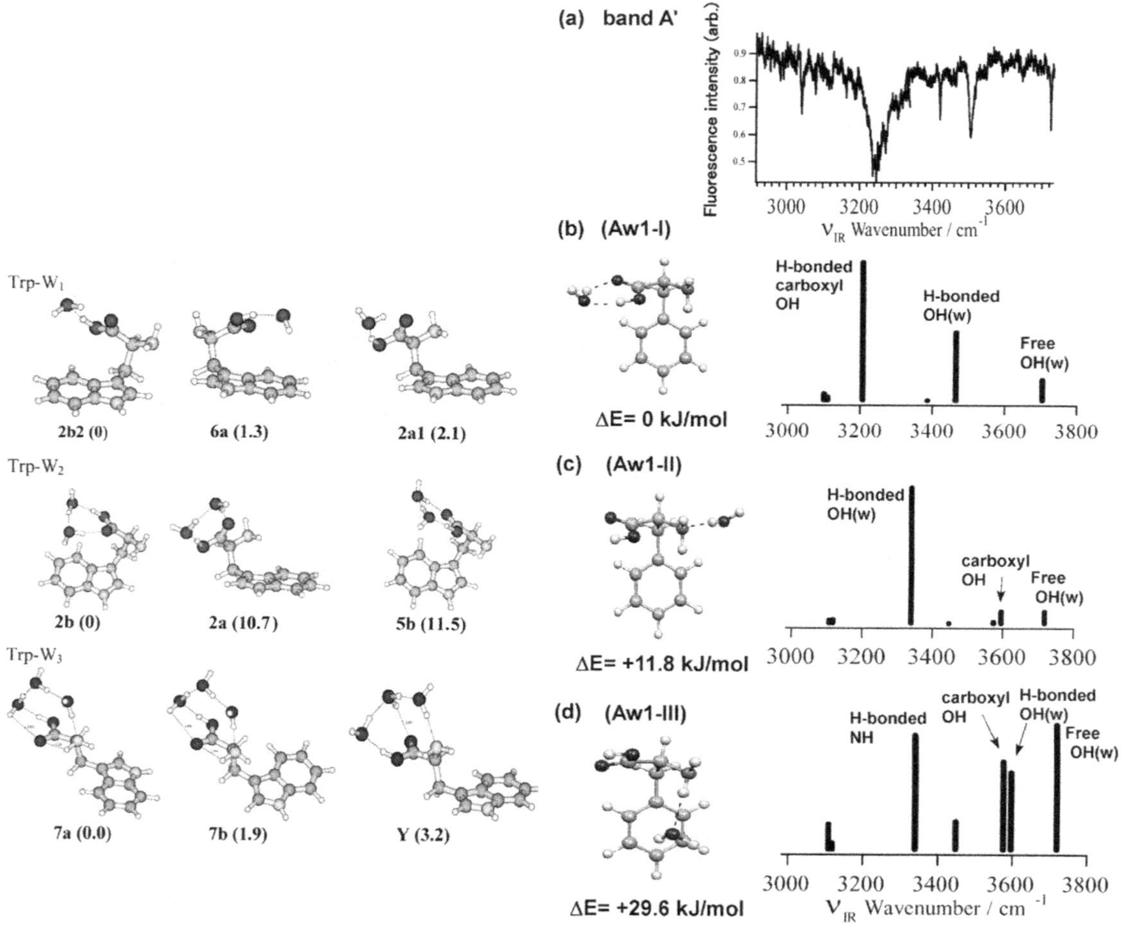

Figure 5.5.6 Left: Configurations of the lowest energy conformers of the singly, doubly and triply hydrated tryptophan. (Energies are given in kJ/mol) (reproduced with permission from reference [296] ©2002 Royal Chemical Society). Right: Lowest energy conformers of the singly hydrated phenylalanine and their predicted IR/R2PI spectra compared to the experimental spectrum (top) (reproduced with permission from reference [300] ©2006 Royal Chemical Society).

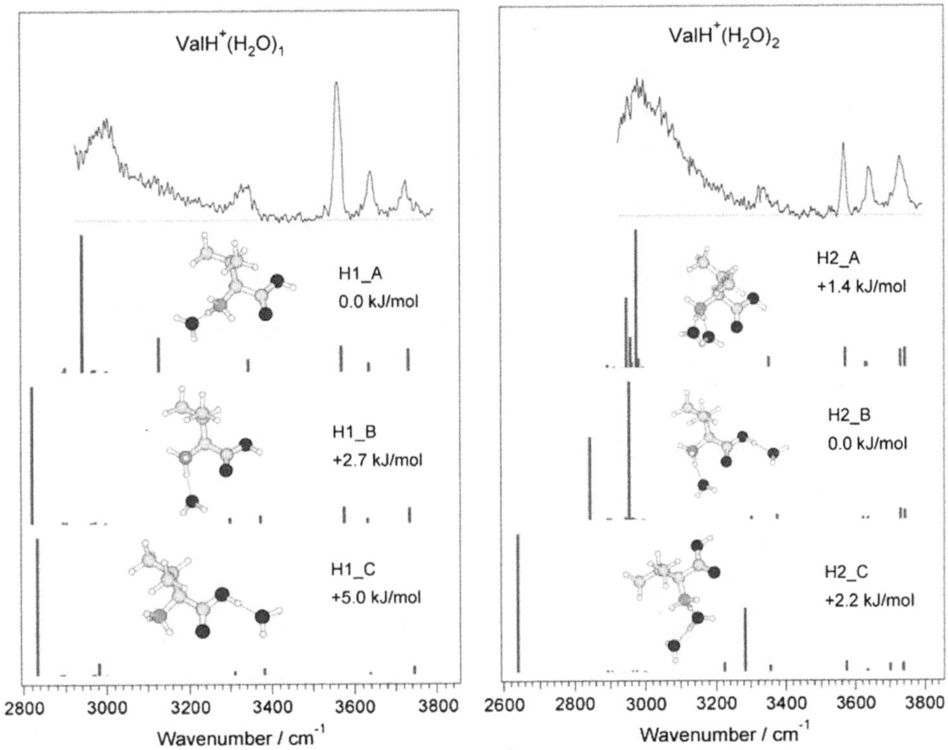

Figure 5.5.7 Experimental and calculated (B3LYP/6-31 + + G** level) infrared absorption spectra of the $ValH^+ \cdot (H_2O)$ (left) and $ValH^+ \cdot (H_2O)_2$ (right) complexes in the N—H and O—H stretch regions (reproduced with permission from reference [301] ©2006 American Chemical Society).

mechanical force-fields. The off-diagonal coupling elements describe transitions between reactants and products and are parametrized from quantum mechanical calculations concerning small fragments of the system. The protonation/de-protonation process of histidine in water has been studied by means of the MS-EVB method [154]. During the initial step, the excess proton localized on histidine forms a Zundel cation (see Section 5.2.1.4) with the closest solvating water. This is followed by formation of an Eigen cation (see Section 5.2.1.4) in the first solvation shell of the amino acid (Figure 5.5.8).

5.5.3.3 Zwitterion formation

When isolated in gas-phase, the 20 natural amino acids are in their canonical neutral form. In presence of a polar environment such as water [36, 309] or in a KBr matrix [310], the zwitterionic form, into which the NH_2 and COOH groups become NH_3^+ and COO^-, becomes energetically favoured (see Section 5.1.5). The question of the minimum number of water molecules required for a switch from the neutral to the zwitterionic form of amino acids is a widely considered topic [311–323]. For example, the neutral form of glycine is more stable than the zwitterionic form by ~90 kJ/mol in the gas-phase, whereas those two forms become nearly isoenergetic with three to five attached water molecules [324]. In the

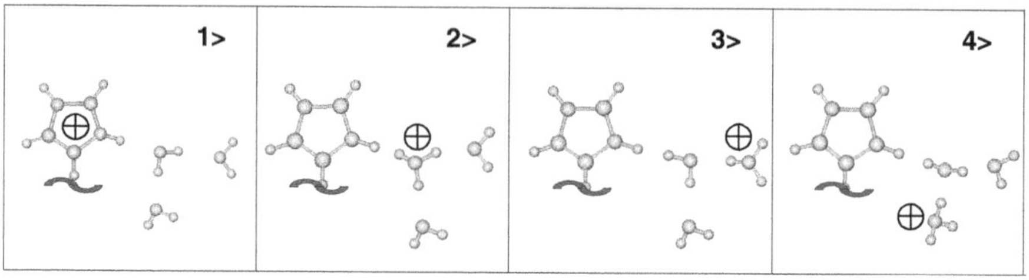

Figure 5.5.8 Schematic representation of the protonation/de-protonation of histidine in water. In state $\|1\rangle$ of the His $\cdot$ H^+—$(H_2O)_3$ model complex, the excess proton is localized on His. A temporary Zundel action is formed in state $\|2\rangle$ of His—H_3O^+—$(H_2O)_2$ and the excess proton further moves into bulk water as an Eigen cation in $\|3\rangle$ and $\|4\rangle$ (reproduced with permission from reference [156] ©2006 American Chemical Society).

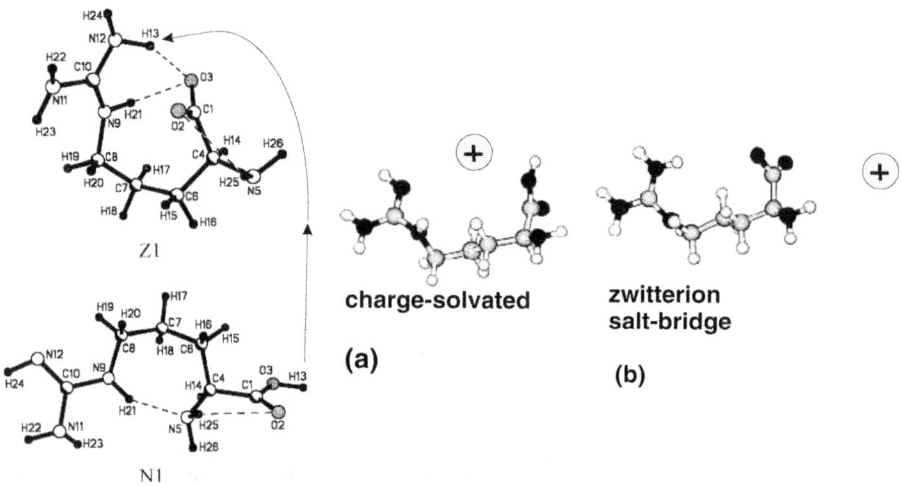

Figure 5.5.9 Left: Lowest energy structures of neutral (N1) and zwitterionic (Z1) forms of arginine calculated at the MP2/6-31 + + G(d,p) + 5(s,p) level. At the CCSD level, the zwitterionic form lies 11.7 kJ/mol above the neutral form (reproduced with permission from reference [328] ©2001 Elsevier). Right: Arginine interacting with a cation can adopt a solvated (a) or a zwitterion (b) structure (reproduced with permission from reference [332] ©2001 American Chemical Society).

proton exchange process, the proton affinity of proton acceptors plays an important role [325]. Arginine [322, 326–331] has a proton affinity ~165 kJ/mol higher than glycine and the difference between its neutral and zwitterionic forms comes down to ~15 kJ/mol (Figure 5.5.9).

The zwitterionic form of amino acids and peptides can be stabilized by electrostatic interactions brought by the presence of charges brought by metal cations [327, 333–340], protons [322, 341–343] and excess electrons [327, 344–347]. The problem of zwitterion formation in the gas-phase has also been considered in the case of de-protonated trinucleotides [348].

5.5.3.3.1 Experimental studies of zwitterion formation

Several experimental methods have been used for investigation of zwitterion formation in the gas-phase. As usual, they are based on comparison between experimentally measured parameters and predictions of those parameters in different conformations. Betaine that is an obligatory zwitterion can then constitute a reference [349]. The record of infrared spectra of sodiated amino acids by means of resonant IRMPD (see Section 2.1.3.2.5) has provided a direct vibrational signature of zwitterion formation in the Na^+–proline complex while the Na^+–glycine complex remained in its neutral (non-zwitterionic) form [336]. The hydrogen/deuterium mechanism (see Section 3.3.4) has been studied in sodiated amino acids. The H/D exchange reaction is preceded by a conversion from the charge-solvated (CS) to the zwitterionic structure in presence of a single water molecule [350].

IRMPD spectroscopy of $(Trp + K)^+$ions confined in a FT-ICR cell confirms that non-deuterated species possess two observable isoenergetic CS conformers. The zwitterionic conformer lying 15 kJ/mol above is not observed. When those ions are exposed to CH_3OD during several seconds, partial deuteration is spectroscopically monitored on both the amino and carboxyl terminals of the CS conformers through observation of the O—D bending mode, C═O stretch of CCOD and reduction of intensity of the NH_2 scissoring mode. Surprisingly, a line at 1,680 cm^{-1} also appears. Its attribution to the antisymmetric CO stretch of the de-protonated carboxylate provides the signature of zwitterion formation induced by the H/D exchange process (see Section 3.3.4).

BIRD experiments provide structural information deduced through kinetic data for loss of a single water molecule from hydrated lithiated or sodiated complexes of amino acids [325, 338, 339]. The neutral form is favoured in the lysine and glutamine complex cases. In glutamine, the metal ion is bound to the amine nitrogen and carbonyl oxygen from the amide backbone and the amide oxygen of the side-chain. In contrast, the zwitterionic form is favoured in the case of α-methyl–proline.

The stepwise hydration of valine–(Li, Na or K)–$(H_2O)_n$ cations shows that an abrupt change occurs between $n = 2$ and 3. In presence of two water molecules, the Li^+ ion is bound to the nitrogen and carbonyl oxygen of the non-zwitterionic valine. Water molecules then only interact with the metal ion. In contrast, the structure of the Li^+–valine ion changes significantly since the metal ion becomes then bound to the zwitterionic terminal COO^- of valine and two of the water molecules [351].

The existence of CS and salt-bridge structures has been examined in ion mobility experiments [333, 352]. The distinction between those structures without any direct spectroscopic or collisional dynamic signature can sometimes turn out to be difficult. A comparison between structure predictions of five different protonated trimers of trifluoroacetic acid complexed with two base molecules by means of different force-field simulations and DFT calculations on one side and BIRD or ion mobility measurements on the other [341] is presented in Table 5.5.1.

Dipole moment and susceptibility measurements (see Section 3.3.3) provide tests of the neutral or zwitterionic structure of peptides. The measured susceptibilities of WA_n–NH_2 ($n = 1$–5) protected Ac-WA_n–NH_2 ($n = 3, 5$) neutral peptides in absence of water have shown that those species are not in the zwitterionic form [353]. A check of the method was provided by measurements with protected Ac-WA_n–NH_2 ($n = 3, 5$) that cannot form zwitterions.

Table 5.5.1

Structures predicted from different force-fields or DFT calculations and experimentally observed by means of BIRD

Molecule	Amber	MMFF	Charmm	BIRD	DFT
MTBD	None	CS	SB	SB	Ion pair
TBD	SB	CS	SB	SB	Ion pair
DBU	SB	CS or SB	SB	Uncertain	Ion pair
DBN	SB	CS	CS or SB	CS	Neutral pair
TMG	SB	CS or SB	CS or SB	CS	Neutral pair

Note: CS and SB, respectively, mean charge-solvated or salt-bridge structures [341].

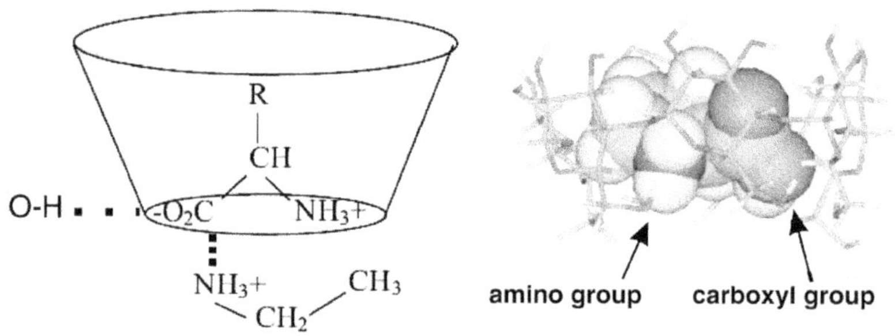

Figure 5.5.10 Left: Proposed structure of a gas-phase zwitterionic amino acid in the cavity of a cyclodextrin in presence of n-propylamine as a reactant. The positive charge of the protonated amine interacts with the carboxyl group of the amino acid. Hydrogen bonding of the amino acid with the cyclodextrin lower rim stabilizes the zwitterion. Right: Low-energy structure of a $[\text{CD–proline} + H]^+$ complex deduced from a force-field simulation. Proline is fully enclosed in the CD (reproduced with permission from reference [354] ©2006 Elsevier).

Stabilization of zwitterionic forms has been studied in host–guest structures. For that purpose, cyclodextrins (CD) have been used as hosts of amino acids in presence of the neutral base n-propylamine acting as a reactant to remove the proton from the amino acid carboxylic group. CD are cyclic sugars with a torus shape and are used as simple models for enzyme–substrate interactions or potential drug-delivery devices. They are composed of six to eight α-1,4 glucose subunits (see Chapter 4.4). They thus contain a large number of O—H groups that can stabilize the amino acid zwitterionic forms by establishing hydrogen bonds with the COO^- group (Figure 5.5.10). In a FT-ICR mass spectrometer, the reaction time of the protonated complex of CD–proline complex with n-propylamine is 30 μs at a pressure of n-propylamine of 1.7×10^{-7} Torr. Formation of a zwitterion requires a specific interaction of both the N- and C-terminals with the CD simultaneously and the size of the CD becomes an important factor since the CD cavity must be large enough. For example, in the case of b-CD with intermediate size (seven α-1,4 glucose subunits), only 6 amino acids (Gly, Ala, Cys, Asp, Val, Ser and Thr) among the 20 do not form zwitterions. Interestingly, the reaction turns out to be enantioselective with a selectivity $k_L/k_D \sim 4$ (see Chapter 4.6).

Table 5.5.2

Relative zero-point corrected electronic energies (in kJ/mol) of three neutral (N) and zwitterionic (ZW) structures of gas-phase neutral arginine calculated at different levels of theory

Method	ZW1	ZW2	ZW3	N1	N2	N3
B3LYP	15.35	17.40	5.27	10.50	9.67	0.00
MP2	14.10	10.78	4.74	13.19	10.45	0.00
CCSD	17.99	14.73	11.89	7.32	7.49	0.00

Note: B3LYP results were obtained with the 6-31 + + G(d,p) basis set supplemented with the diffuse 2(s,p) diffuse set. MP2 and CCSD results were obtained with the 6-31 + + G(d,p) + 5(s,p) basis set [328].

5.5.3.3.2 Modelling of zwitterion formation

Predictions of zwitterion formation have also been addressed by means of quantum chemistry calculations in the case of glycine and histidine [355], aspartic acid [342], serine and cysteine [328, 356]. The relative energies of the canonical neutral and zwitterionic forms strongly depend on the level of theory as shown in Table 5.5.2.

Dipole-bound excess electrons attached to amino acids or amino acid-solvent clusters stabilize the zwitterionic form of amino acids relative to their canonical forms. This has been predicted [357] and experimentally verified [344] in the case of arginine. Arginine plus an excess electron indeed exist as a zwitterion in low-temperature gas-phase environment as demonstrated by photoelectron spectroscopy. Comparison between theory and experiment is somewhat less clear in the case of glycine–water anions. Gly—$(H_2O)_{N=1-2}^-$ anions have been experimentally produced by charge exchange between water anions and glycine [358, 359]. According to the experimental spectral signatures, only canonical (non-zwitterionic similar to the isolated species)) forms were observed. In the case of Gly—$(H_2O)_{N\geq5}^-$ anions obtained by means of a discharge in a supersonic expansion [360], photoelectron spectra were interpreted as spectroscopic signatures of zwitterion structures. Calculations of neutral and anionic structures have been conducted [361, 362]. In neutral Gly—$(H_2O)_N$ complexes, proton transfer leading to zwitterion formation is assisted by the formation of a chain of water molecules that establish bridges between the carbonyl COOH and the amide NH_2 groups [362]. The "neutral zwitterion" contains the charged NH_3^+ and COO^- groups. The situation is quite different in the "anion zwitterion" that already contains the COO^- group but into which electron attachment takes place on the NH_3^+ group. That group becomes a Rydberg-like hypervalent group —NH_3 (see Chapter 4.7) with a rather low (<1 eV) vertical detachment energy (VDE). The geometries of the "neutral" and "anion" zwitterions are thus very different. Calculations show that VDEs of zwitterion anions are thus large (≈3 eV), indeed much larger than those experimentally determined in photo-detachment studies of Gly—$(H_2O)_{N\geq5}^-$ complexes [360].

Barriers for interconverting neutral and zwitterionic forms play an important role and have been computed in the intermediate case of Gly—$(H_2O)_3^-$ anions [362]. The zwitterionic Gly—$(H_2O)_3^-$ anions are predicted to be the most stable forms of the Gly—$(H_2O)_3^-$ system in the gas-phase. On the contrary, the canonical rather than the zwitterionic neutral form of Gly—$(H_2O)_3$ is predicted to be the most stable neutral form.

In the case of sodiated glycine, the barrier between canonical and zwitterionic structures is calculated as, respectively, equal to 71.6 kJ/mol at the B3LYP/6-31G* [325] (Figure 5.5.11)

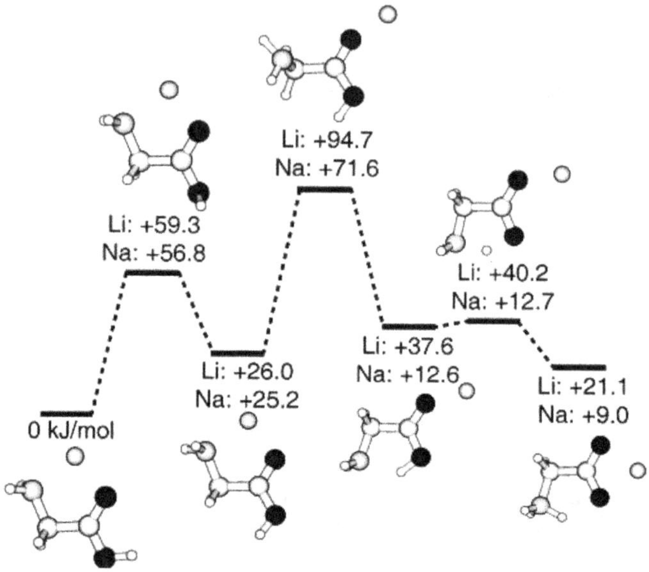

Figure 5.5.11 Potential energy pathway for conversion of Gly·Li^+ from its lowest energy non-zwitterionic form (left) to its lowest energy zwitterionic form (right) calculated at the B3LYP/6-31G* level of theory. Relative energies (kJ/mol) include zero-point energy and ΔH (298°K) corrections. Values are also included for the relative energies of the comparable Gly·Na^+ structures (reproduced with permission from reference [325] ©2006 American Chemical Society).

and 78.2 kJ/mol at the MP2/6-31G* [363] levels of theory. This barrier is comparable to the binding energy of a water molecule to Na^+–glycine. In order to allow comparison to experiments, free energies would be valuable. Even better, quantum molecular dynamics (see Chapter 4.2) would provide a clear picture of the proton exchange in the gas-phase in presence of water molecules and in solution.

5.5.3.4 Dynamic behaviour of photoexcited hydrated peptides

Tryptophan is a crucial chromophore of proteins [364] (see Section 2.2). The semi-empirical method PM3 (see Section 1.4) has been employed for the simulation of a REMPI (see Section 2.1.4) experiment involving neutral tryptophan [320]. Transient zwitterion formation appears within 10 ps in this simulation and a fast energy flow is observed between the vibrational ring modes and the amino acid backbone. With the advent of femtosecond experiments, TD-DFT and CC2 calculations [365, 366] (see Section 1.7), these predictions might be quantitatively tested.

The UV spectrum of bare protonated tryptophan is very broad (300 cm^{-1}), even at very low temperature [298] in agreement with the extremely short excited state lifetime (380 fs) [305]. Although de-excitation (quenching) of excited states should be very important in solution, fluorescence of tryptophan is still observable even at low pH. The UV spectra of protonated tryptophan ions bound to one and two water molecules have been recorded at 10°K [298, 365]. As shown in Figure 5.5.12, the UV spectrum drastically

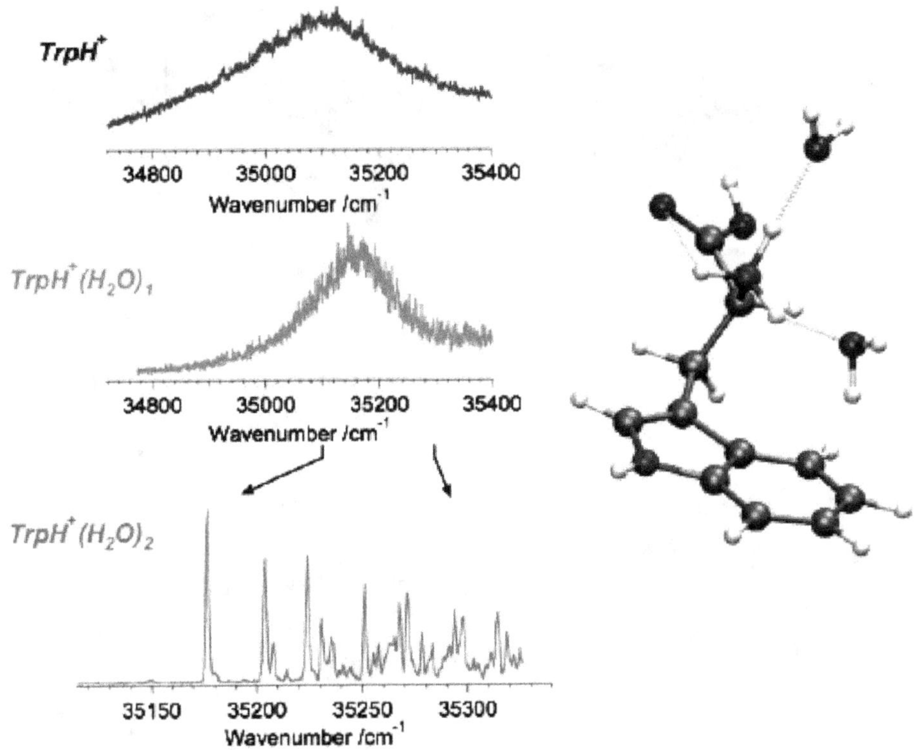

Figure 5.5.12 Left: UV spectrum of protonated tryptophan ions bound to zero, one and two water molecules recorded by monitoring the photofragmentation yield at 10°K. Right: Lowest energy structure of the protonated Trp–$(H_2O)_2 \cdot H^+$ion (reproduced with permission from reference [298] ©2006 American Chemical Society).

changes when a second water molecule is added. The extremely short lifetime of the bare molecule is due to electron transfer from the indole chromophore to the $NH_3{}^+$ ammonium terminal ("through-space" effect). The initially photoexcited $\pi\pi^*$ state is coupled to the dissociative $\pi\sigma_{NH}{}^*$ state. One of the hydrogens of the $NH_3{}^+$ is bound to the carbonyl terminal group (Figure 5.5.12). When the two other hydrogen atoms are engaged with a water molecule, the dissociation channel leading to ultrafast H-loss and responsible for ultrafast de-excitation in the bare and monohydrated ions then becomes close.

It is possible to increase the backbone separation between the ammonium and the indole chromophore by introducing additional bonds in order to test the possible influence of a "through-bond" that might occur in proteins with a protonated N-terminal. This has been performed in a study of H^+·GlyTrp and H^+·GlyGlyTrp ions. In fact, those ions exhibit UV spectra similar to that of the bare molecule, ruling out a "through-bond" effect. In fact, the presence of new bonds introduces flexibility. The ammonium terminal then still interacts directly with the indole chromophore and the "through-space" effect is maintained.

Hydration dynamics of proteins have been studied in the femtosecond time range corresponding to motions of water molecules [367]. Proteins containing a single tryptophan

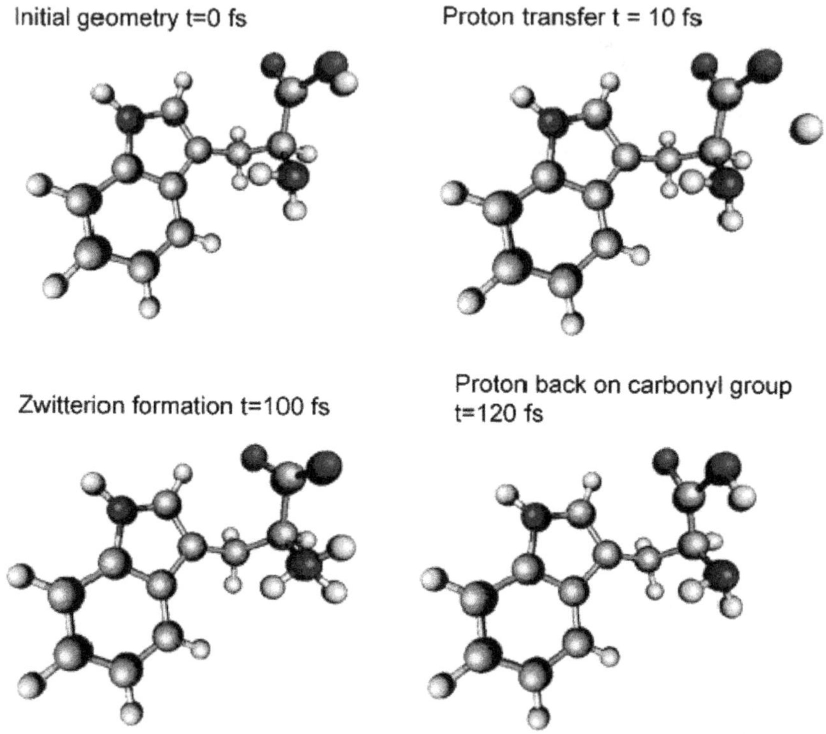

Figure 5.5.13 Snapshots of the single-photon ionization dynamics of neutral tryptophan. Proton transfer between the backbone carboxyl and amino groups takes place and a zwitterion is formed temporarily within 100 fs. At 120 fs, the proton jumps back to the initial oxygen of the hydroxyl group of the C-terminal. Compare to the proton transfer in the dialanine peptide studied by Car and Parinello molecular dynamics in Figure 4.2.15 (reproduced with permission from reference [320] ©2005 American Chemical Society).

residue (e.g. mellitin) were chosen and their fluorescence decay was fitted with three time constants. Interpretation of those time constants must be taken with caution (the problem of fitting multiexponential transients is tackled in Section 2.2) but seems to indicate that excited state fluorescence emission of tryptophan can be quenched by either nearby protein groups (amide bonds and side-chains) or prosthetic groups such as haems. According to the local environment, the fluorescence decays vary between femtoseconds and nanoseconds with a very fast component attributed to reorganization of the solvent molecules (Figure 5.5.13).

5.5.4 Hydration of peptides and proteins

A large number of molecular dynamics simulations and reviews concerning hydration of peptides and proteins have been published and we only here provide an extremely small number of references that cannot account for the whole available literature. The problems and their references are as follows: origin of β-hairpin stability in solution [368], influence

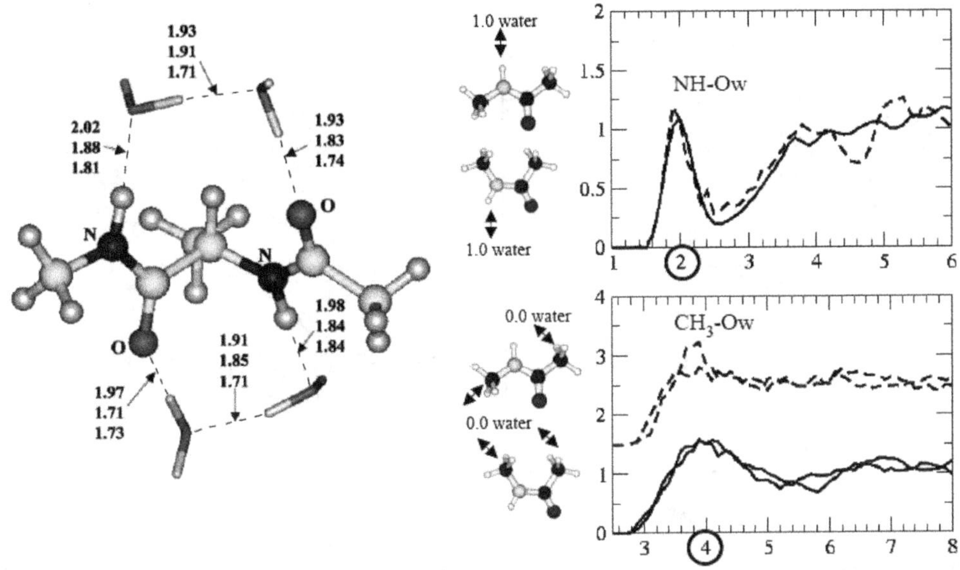

Figure 5.5.14 Left: Hydration sites of the *N*-acetyl-L-alanine-*N*′-methylamine model peptide. The three values displayed for hydrogen bonds, respectively, correspond to: (top) quantum calculations (QM) conducted at the HF level and effective fragment potential (EPF) theory [379] for water molecules; (middle) QM for waters and molecular mechanics (MM) for the dipeptide and (bottom) QM calculation at the DFT B3LYP/6-31G* for the whole system (reproduced with permission from reference [380] ©2002 Elsevier). Right: Quantum dynamic calculations (CPMD) of radial distribution functions (rdf) of aqueous *N*-methylacetamide (see also Figure 5.1.5). Distances are in Angstrom. Amide N—H and C═O groups engage in hydrogen bonds with the solvent and the first peaks of the N—H···O_{water} and C═O···H_{water} rdf correspond to short distances (2 Å). On the contrary, an exclusion sphere surrounds the hydrophobic methyl groups (reproduced with permission from reference [41] ©2005 American Chemical Society).

of solvation on the helix-forming tendency of non-polar amino acids [369], interplay between hydrophilicity and hydrophobicity [370], removal of water [371] from the neighbourhood of non-polar groups, the role of water in protein activity [372], partitioning of solvent effects and intrinsic interactions in biological recognition [373] and hydrogen bond patterns in the hydration of proteins [374]. The relevance of determining gas-phase structures of peptides is considered in reference [375]. A recent review covering a variety of experimental techniques and theoretical studies of water interactions with proteins can be found in reference [376].

In the case of small hydrated peptides, calculation of structures and simulations conducted for the interpretation of infrared spectra [41, 377, 378] can be performed with explicit water molecules. Those molecules preferentially establish hydrogen bonds with the C═O and N—H groups in competition with water dimer formation (Figure 5.5.14).

The ability of large peptides and proteins to bind water molecules can be grasped through comparison between experimental determination of conformations in BIRD [381, 382], fluorescence [383] and ion mobility [54, 184, 384–392] experiments interpreted by means of modelling [11, 20, 374, 393, 394]. Peptides and proteins can be first set into gas-phase

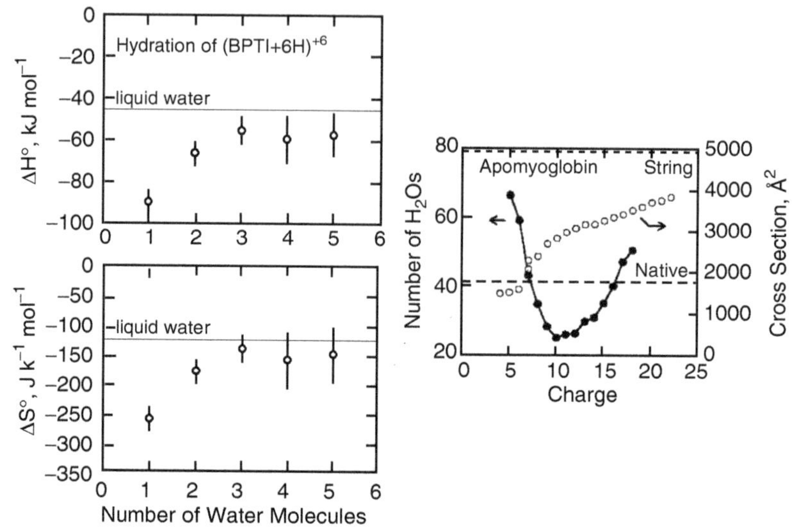

Figure 5.5.15 Left: Variation of enthalpy ($\Delta H°$) and entropy ($\Delta S°$) for the adsorption of the first water molecules by bovine pancreatic trypsin inhibitor (BPTI). Right: Average number of water molecules adsorbed on apomyoglobin $(M + mH)^{m+}$ions exposed to water vapour in an ion mobility set up (see Section 3.3.1) (reproduced with permission from reference [387] ©1999 American Chemical Society).

as de-hydrated species and then rehydrated by progressively adding water molecules one by one. Equilibrium constants measured as a function of temperature then provide enthalpies and entropies (Figure 5.5.15). In the case of bovine pancreatic trypsin inhibitor (BPTI), the $\Delta H°$ and $\Delta S°$ values of the first water molecule are more negative than for subsequent molecules. This water molecule is bound in a tetrahedral site of the protein that is not hydrated in a crystal structure. Simulations show that this tetrahedral coordination of water stiffens the low-frequency modes of the protein inducing a large entropy change for binding of the first water molecule.

Hydrogen–deuterium exchange in peptides is an important mass-spectrometry tool [395–397] (see Section 3.3.4). This H/D exchange can performed either in solution prior to vaporization or in gas-phase inside a mass-analyser. If D_2O molecules are used as exchange reagents, a relay mechanism is invoked where D_2O molecules bind to a charged site acting as a hydrogen atom donor and a basic site acting as a deuterium atom acceptor. One major difference between solution and gas-phase H/D exchange is that deuterated molecules can influence the peptide conformation in the gas-phase as has been seen above. In a combined study using H/D exchange in a quadrupole trap and ion mobility measurements of protonated peptide structures, the role of conformation on H/D exchange has been tested on the AAARAA·H^+peptide [54]. Two conformations are preferential: a solvation structure (CS) with the charge only on the R (arginine) guanidinium side-chain or a salt-bridge structure with both the R side-chain and the N-terminal protonated and a de-protonated C-terminus (a similar case is found in the RGD peptide, see Chapter 4.2). In the bare AAARAA·H^+ peptide, the lowest conformation is CS. The SB conformation, slightly more compact, is located 20 kJ/mol above. When a water molecule is added, the SB structure is more stabilized than the CS structure but still lies 7.5 kJ/mol above it. This trend continues when adding more water molecules and in bulk water salt-bridge structures become the most stable. A low-energy transition state connects those CS and SB conformations and facilitates the H/D exchange. The reactants are AAARAA$\cdot$$H^+$ and H_2O while the products are AAARAA$\cdot$$D^+$ and HOD. The relay mechanism converts a CS form to a SB form through a probably short-lived transition state lying only 16.7 kJ/mol above the lowest energy CS structure. The ion trap experiment shows that 5,000 collisions are required and the H/D exchange is not complete. This can be attributed to the fact that although the mechanism is energetically favourable, it implies a large structural change. In CS, the N- and C-terminals are widely separated (in a nearly T-shape conformation with the R side-chain) while in SB, on the contrary, the N- and C-terminals are very near and the entropy change introduces a bottleneck.

For larger systems than the preceding pentapeptides, the short-lived conformations occurring during folding and unfolding processes of proteins in solution can be followed through H/D exchange using snapshots obtained by mass-spectrometry techniques [398], in contrast with NMR H/D exchange studies which provide data representing averages of entire protein conformations [399, 400]. Hydration modifies gas-phase structures of bare peptides and proteins but it is also dependent on their initial conformations. This has been tested in a series of measurements where a lysine or a histidine has been added to alanine-containing protected (Ac) peptides. Alanine has a strongly hydrophobic side-chain while those of lysine and histidine are strongly basic and thus are the favoured protonation sites. According to the location of the basic side-chain in a peptide, the interaction between positive charge and the macrodipole formed by the amide dipoles can be favourable or unfavourable. In the $AcA_{20}K\cdot H^+$ peptide where lysine (K) is at the N-terminus, the protonated lysine interacts favourably with the macrodipole and stabilizes the formation of a α-helix. On the contrary, In the $AcK\cdot A_{20}\cdot H^+$ peptide where lysine is at the C-terminus, the protonated lysine interacts unfavourably with the macrodipole and destabilizes the formation of an α-helix and a globular peptide is instead obtained. When water is added in the drift tube (see Section 3.3.1), the globular $AcK\cdot A_{20}\cdot H^+$ peptide can bind up to two water molecules while the helical $AcA_{20}K + H^+$ peptide remains unhydrated (Figure 5.5.16).

In the globular $AcKA_{20} + H^+$ peptide, simulations show that water prefers to bind in folds of the peptide surface where sites with multiple hydrogen-bonding possibilities are available. An initially mobile water molecule explores the peptide and moves until it finds a favourable binding site. Such a site is, for example formed by the acetyl (Ac) carbonyl group and those of A14 and A9. In the helical $AcA_{20}K + H^+$ peptide case, a water molecule initially in the hydrophobic middle part of the peptide is either desorbed or migrates to one of the ends where hydrogen-bonding sites become available, the positive charge at the C-terminal being preferred.

General trends for hydration of peptides and proteins are the following [389]. The first water molecules are usually more strongly bound than the following ones. The first water molecules bind more strongly to the higher charged states. In arginine-containing peptides, water binds less strongly to a protonated guanidinium group than to a protonated amino group. The charged sites are not necessarily the most favourable hydration sites since interaction with water requires breaking of other bonds, in particular with C=O backbone groups.

a) b) c)

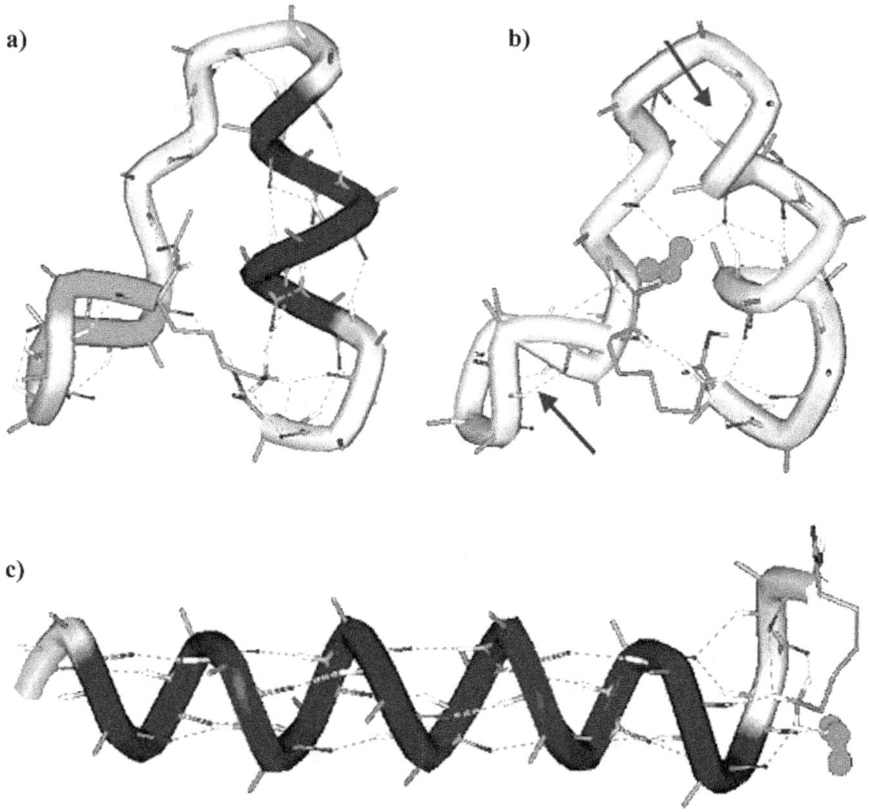

Figure 5.5.16 Conformations of $AcA_{20}K{\cdot}H^+$ and $AcKA_{20}{\cdot}H^+$ peptides: (a) the lowest energy conformation of $AcKA_{20}{\cdot}H^+$ is globular; (b) conformation of globular $AcKA_{20}{\cdot}H^+$ with the highest water-binding energy; (c) conformation of helical $AcA_{20}K{\cdot}H^+$ with the highest water-binding energy (reproduced with permission from reference [389] ©2002 American Chemical Society).

5.6 HYDRATION OF SACCHARIDES

Structures of mono- and oligosaccharides, unlike those of many proteins, are strongly dependent on the presence of water molecules due to the existence of a large number of hydroxyl groups [401–403] (see Chapter 4.3). The first hydration shell often forms integral parts of the conformation. When a sugar (see Chapter 4.3) becomes hydrated [404, 405], an extra water molecule is inserted in the cooperatively hydrogen-bonded chain. The water molecule extends the H-bonded chain by introducing two strong hydrogen bonds bridging the weakest link. For example, in α-D-xylopyranoside, the water molecule is inserted between OH4 and OH3 to form the cooperative chain OH4 $\rightarrow$ OH^{Water} $\rightarrow$ OH3 $\rightarrow$ OH2 $\rightarrow$ O1. The same water insertion pattern is observed in several other monosaccharide–water complex structures (Figure 5.6.1). This seems to be one of the general molecular recognition patterns for binding monosaccharide ligands [171, 406, 407].

The glycoaldehyde–water is the smallest hydrated sugar complex and has been studied by means of Fourier-transform microwave spectroscopy [408]. Bare glycoaldehyde possesses

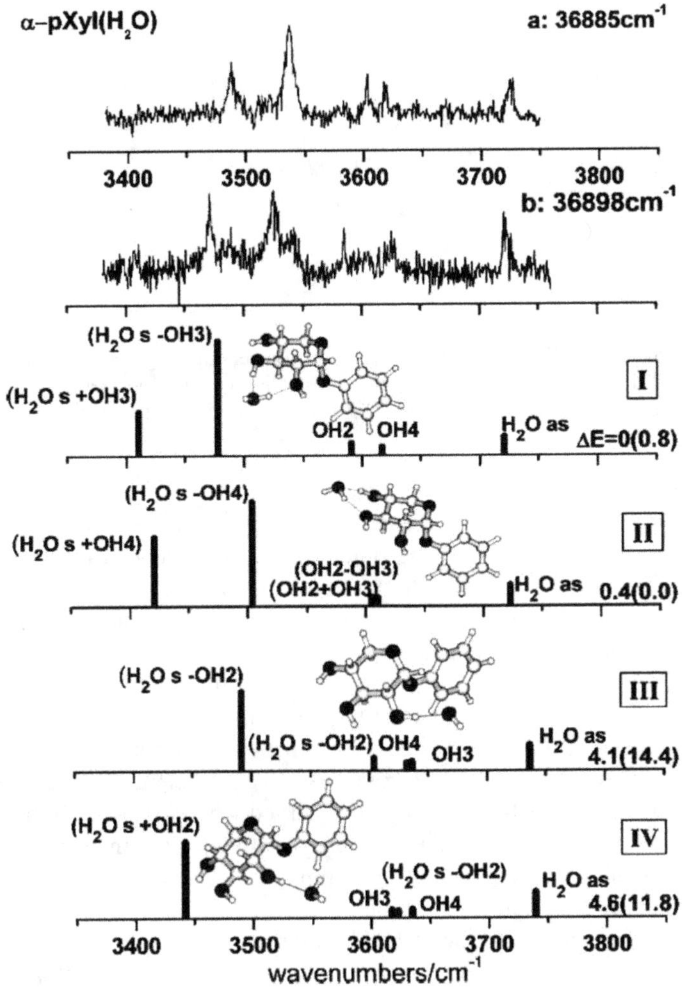

Figure 5.6.1 Experimental and computed infrared-hole burning spectra of α-D-phenylxyloside–water complexes. The spectra of the four lowest energy hydrated structures are calculated at the B3LYP/6-31 + G* level. Relative energies in kJ/mol^{-1} calculated at the MP2/6-311 + + G**//B3LYP/6-31 + G* are given in the respective figures (energies in parentheses correspond to B3LYP/6-31 + G* calculations, note the order inversion between I and II). Note also the insertion of the water molecule in the hydroxyl network in structures I and II (reproduced with permission from reference [402] ©2005 Royal Chemical Society).

four identified conformers. The most stable denoted CC (*cis* conformation around the C—C and C—O bonds) is stabilized by an intramolecular H-bond between the H atom of the unique hydroxyl group and the oxygen atom of the carbonyl group. Among the 17 identified conformers of the glycoaldehyde–water complex, the 3 lowest energy conformers, stabilized by two H-bonds, are represented in Figure 5.6.2. Only the two lowest energy structures are accessible, respectively, CC-W-1 (0 kJ/mol) and CC-W-2 (2.12 kJ/mol), and are associated to the CC conformer of the bare molecule. The only difference between these conformers is the position of the free hydrogen atom of water.

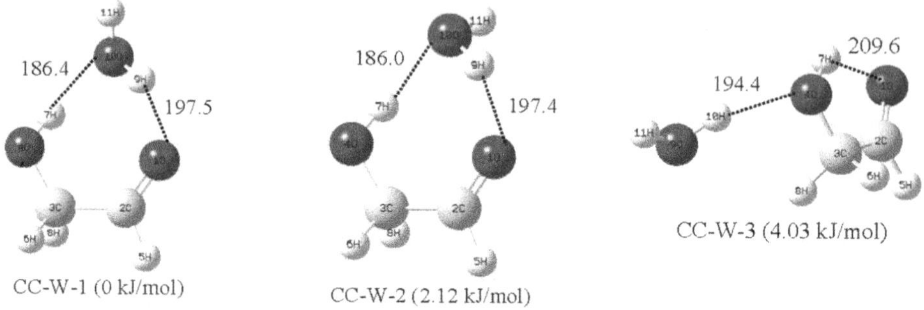

Figure 5.6.2 Structures of the three lowest energy conformers of the glycoaldehyde–water complex calculated at the B3LYP/6-311 + + G(2df,p) and G3MP2B3 levels. Bond lengths are in picometres (reproduced with permission from reference [408] ©2006 American Chemical Society).

The high-resolution spectrum (Figure 5.6.3) presents for all observed lines pairs of doublets separated by a few tens of kilohertz. Those series of lines are associated with two vibrational sublevels denoted O^+(ground state) and O^- (excited state). This splitting is due to a tunnelling effect: the water molecule shuttles from one side of the glycoaldehyde molecule to the opposite one while keeping intact the two intermolecular H-bonds. This conformational flexibility has been investigated by means of the two-dimensional potential energy surface along the hydroxyl group of glycoaldehyde and the water free OH group (Figure 5.6.3). This tunnelling can proceed through two paths: an energetically forbidden one through the TS1 transition state (17.7 kJ/mol) and a low-energy one through TS2 (4.4 kJ/mol) and TS3 (5 kJ/mol) involving the CC-W-2 conformer. This tunnelling effect probably takes place in many carbohydrates but can only be observed when the pseudorotation motion of the water molecule does not perturb the structure of minima close in energy.

In a sugar molecule, each constituting group plays a structural role [401] (see Chapter 4.3). For example, the exocyclic hydroxymethyl group C_6 plays a key role in the conformational flexibility of monosaccharides which contain it through its interaction with OH4. This exocyclic group adapts its conformation to optimize interactions with surrounding water molecules. As also observed in phenylalanine [299], a single water molecule is then already sufficient to affect the conformational landscape. When this exocyclic hydroxymethyl group C_6 is replaced in fucose by a methyl [409], a larger number ($n > 3$) of water molecules are needed to form the cooperative hydrogen bond chain OH2 → OH3 → OH4 → $OH^{Water1} \cdots OH^{Watern} \rightarrow$ O1. It has thus been suggested that one of the water molecules "mimics" the missing exocyclic hydroxymethyl group.

The problem of establishing the link between gas-phase and liquid-phase structures through measurements and calculations of infrared absorption spectra has already been considered in Sections 5.1.2 (CPMD) and 5.5.4 (first hydration shell). Another possibility is offered by the reconstruction of the IR spectrum of a hydrated species as a weighted sum of IR spectra of individual monohydrated complexes of this given species [31]. This study has been conducted in the case of the lactose disaccharide and its two monosaccharide units, glucose [410] and galactose. The structures and IR/R2PI spectra of the low-lying different conformers of the singly hydrated species are known [411]. Their relative energies at 0°K and free energies at 300°K are computed as well as the Raman and Raman optical activity

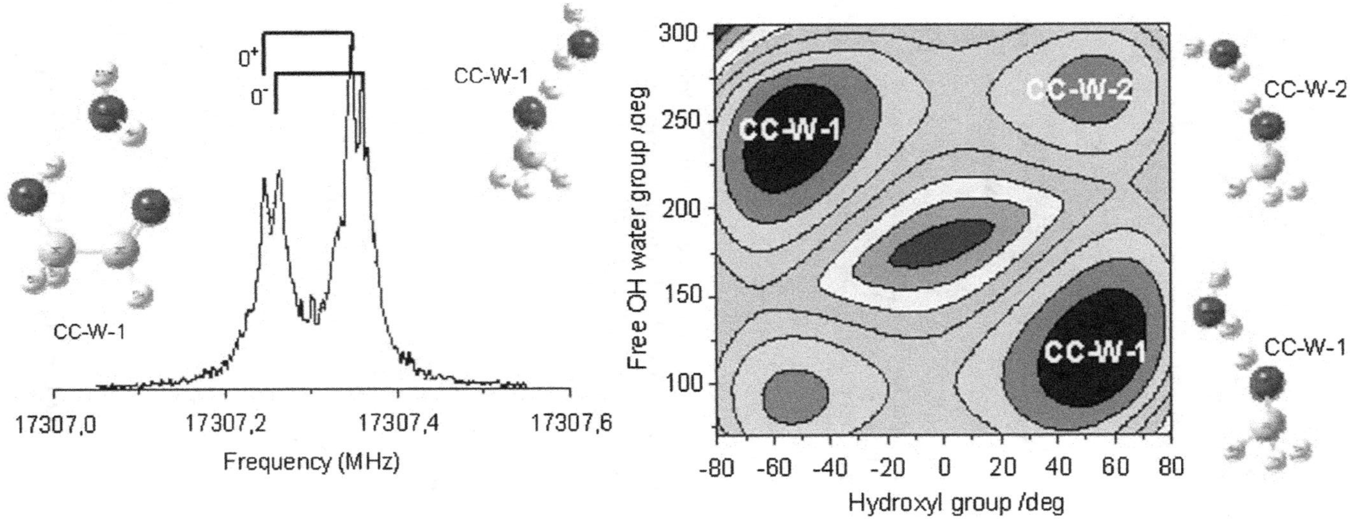

Figure 5.6.3 Left: High-resolution microwave spectrum of the glycoaldehyde–water complex. A large amplitude motion of the water molecule is responsible for the observed line splitting. Right: Potential energy surface of the glycoaldehyde–water complex calculated at the B3LYP/6-31G* level along the free water OH group and the glycoaldehyde hydroxyl group coordinate. The tunnelling motion of the water molecule between the mirror structures of CC-W-1 proceeds along the lowest energy pathway TS2, CC-W-2 and TS3 (reproduced with permission from reference [408] ©2006 American Chemical Society).

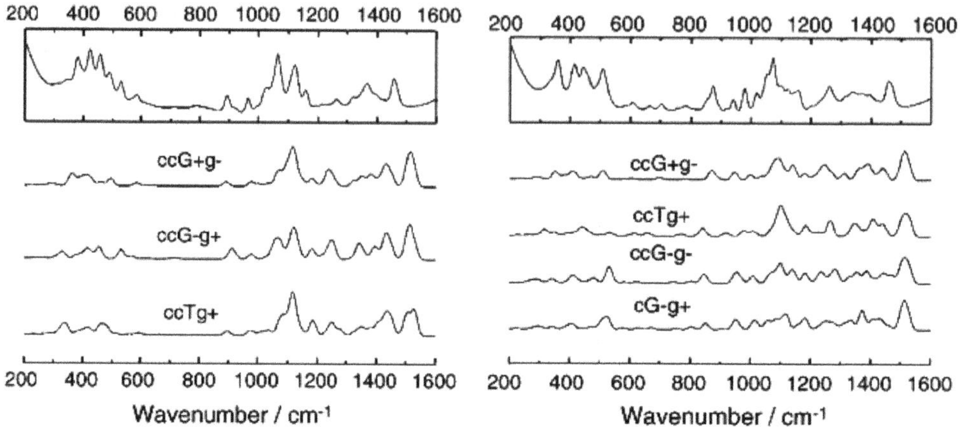

Figure 5.6.4 Mid-infrared spectra of methyl-β-glucopyranoside (left) and methyl-β-galactopyranoside (right). The upper spectra are experimental Raman spectra recorded at 298°K. The lower spectra are unscaled "gas-phase" spectra of individual low-lying conformers computed at the MP2/6-311 + + G**//B3LYP/6-31 + G* level (reproduced with permission from reference [31] ©2006 Elsevier).

(ROA) spectra of each low-lying conformer populated in the free and hydrated species. Those calculated spectra (Figure 5.6.4) then constitute the building blocks ("basis sets"). The comparison between their computed–weighted sum and the experimental Raman and ROA spectra of the different species recorded in water at 300°K provides an estimate of the different conformer populations in aqueous solution.

In the case of the two monosaccharide units, glucose and galactose, the addition of a single water molecule induces a considerable conformational change. The water molecule inserts at a position where it can replace a weak intramolecular interaction (between hydroxyl groups or between a hydroxyl group and the cycle oxygen) by two strong hydrogen bonds. This thus enhances the stability of the monosaccharide conformers to which the water molecule chooses to bind and that are not necessarily the lowest energy ones. As shown by the good match between the experimental mid-infrared spectra and the reconstructed spectra (Figure 5.6.5), these altered conformational preferences in the singly hydrated glucose and galactose observed in the gas-phase at very low temperature appear to remain in the aqueous solution at 298°K. The situation is different for the disaccharide lactose which resists conformational modification when a single water molecule is added in the gas-phase. In solution, the role of the glycosidic linkage (see Section 4.3.2) becomes important since its flexibility reinforces the role of entropy [405] instead of explicit insertion of water molecules as in monosaccharides.

5.7 HYDRATION OF NEUROMOLECULES

The presence of water can strongly modify conformations of highly flexible neurotransmitters and other neuromolecules. In this chapter are presented experimental R2PI spectroscopy

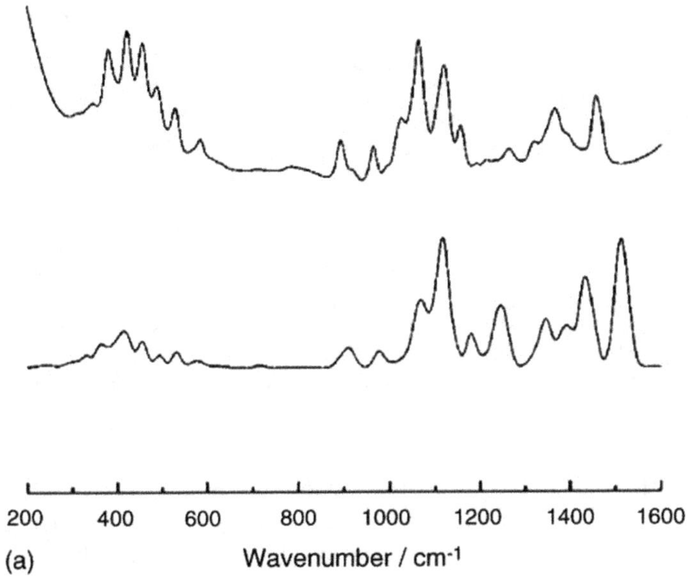

Figure 5.6.5 Comparison between the experimental mid-infrared spectrum of methyl-β-glucopyranoside (upper trace) and the simulated spectrum (lower trace) obtained by summing the computed "gas-phase" spectra of conformers ccG + g- and cG-g + with equal weight and convoluted with a 25 cm^{-1} linewidth (reproduced with permission from reference [31] ©2006 Elsevier).

studies of hydrated ephedrine and its diastereoisomer pseudoephedrine (see Chapter 4.4). Those of melatonin are then discussed followed by the rotationally resolved laser-induced fluorescence spectroscopy study of hydrated tryptamine. We then consider modelling of solvation of flexible molecules and mass-spectrometric studies of large neuropeptide conformations.

The conformational landscape of ephedrine, a chiral amine indirectly acting upon adrenergic receptors (see Chapter 4.4), is only slightly modified by hydration [412]. On the contrary, its diastereoisomer pseudoephedrine changes from an extended conformation of the bare molecule to a folded conformation in its singly hydrated complex. In hydrated clusters, water molecules can either insert into the intramolecular OH → N hydrogen bond of the monomers or add to the alcohol group. This observed difference between the hydrated diastereoisomeric complex behaviours arises from the balance between the different intra- and intermolecular bonding or steric interactions leading to the observed conformers of each species (Figure 5.7.1).

The neural hormone melatonin (*N*-acetyl-5-methoxytryptamine) principally synthesized by the pineal gland has an indole chromophore and a flexible peptide-like side-chain. It has five potential hydration sites: the carbonyl and methoxy oxygens, the amide and indole NH and the π-cloud of indole. Water is bound to the carbonyl group in the four melatonin–(water)$_1$ complexes formed in a supersonic expansion and observed by IR/R2PI [414]. The influence of hydration on the gas-phase structure of melatonin appears strikingly in the doubly hydrated complexes. Surprisingly, a *trans* → *cis* isomerization of the amide is

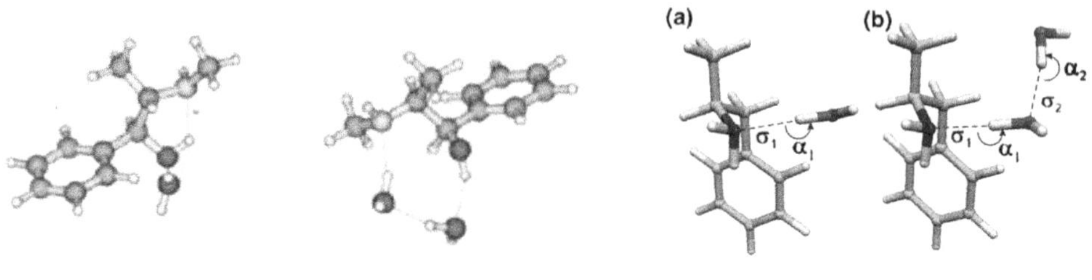

Figure 5.7.1 Left: Addition of a single water molecule to pseudoephedrine and insertion of two water molecules into ephedrine (structures of the bare molecules are shown in Figure 4.4.2) (reproduced with permission from reference [413] ©2002 Royal Chemical Society). Right: Lowest energy structures of (a) the monohydrated and (b) dehydrated complexes of amphetamine complexes calculated at the B3LYP/6-311 + + G(d,p) level (reproduced with permission from reference [413] ©2006 Elsevier).

observed in the flexible side-chain but the water dimer bridge is established between the C=O and the indole N—H groups (and not the amide group) in the melatonin–(water)$_2$ complexes. This *trans* → *cis* isomerization induced by hydration has already been observed in amide models (see Section 5.5.1). In the case of melatonin, it is supposed that the formation of the *cis*-amide favours the reorientation of the flexible side-chain into a more stabilized configuration involving the water bridge.

The establishment of two interactions involving a water molecule in a complex strongly stabilizes a given conformer with respect to the others. This is also demonstrated in a rotationally resolved laser-induced fluorescence study of tryptamine and its monohydrated complex [415]. Six conformers are identified in the bare molecule and their relative energies are comprised within only 6.8 kJ/mol. In the monohydrated complex, a single conformer is observed since the relative energy of the second lowest conformer is above 11.3 kJ/mol. The origin of this conformer selection induced by hydration is explained by the establishment of a weak hydrogen-bonding interaction between the oxygen atom of the water molecule and the H2 atom of the pyrrole ring of tryptamine (2.38 Å apart) that does not exist in the other conformers in addition to the strong hydrogen bond between a water hydrogen and the amino group (Figure 5.7.2).

Different possibilities have been used for modelling hydration in the case of neuromolecules. For example, acetylcholine and its two agonists, muscarine and nicotine [416], GABA [417] and tyramine, dopamine, norepinephrine and epinephrine [78, 418] have been studied with polarizable continuum solvent PCM models (see Section 5.1).

The torsional path integral method (see Section 1.5.3) has been applied to the hydrated complexes (one to four water molecules) of the parent 2-amino-1-phenyl-ethanol (APE), a structural analogue of flexible noradrenaline [419] (Figure 5.7.3). This method allows a gradual increase of the complex temperature and the computation of the relative corresponding parent conformer populations taking into account anharmonicity and quantum effect. Moreover, quantum effects can be studied. The bare molecule has only two populated conformers at 100°K. The addition of a single molecule causes 20% of the overall population to shift to a third parent conformer. The population of this third conformer, respectively, is raised to 40 and 60% with the addition of a second and a third water molecule. With the addition of a fourth water molecule, the population of the two lowest energy

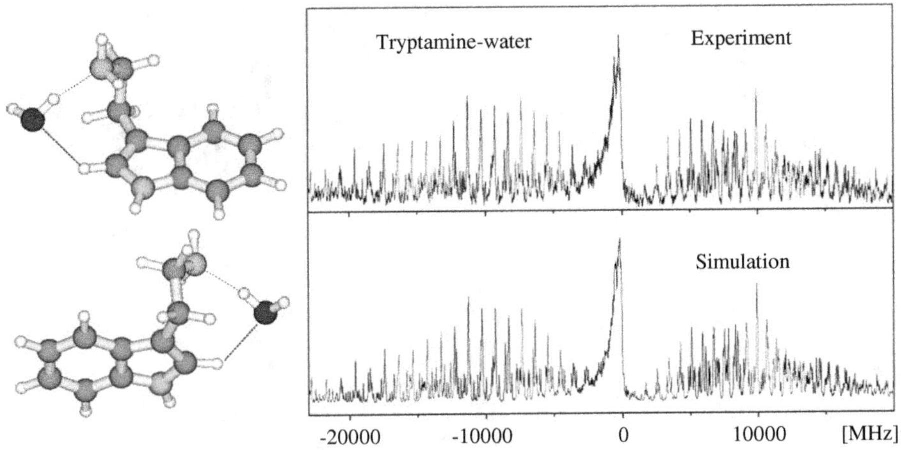

Figure 5.7.2 Left: Structure of the unique observed tryptamine–water complex. Note the weak hydrogen bond between the oxygen atom of water and the pyrrole ring. Right: Experimental and simulated rotationally resolved laser-induced fluorescence spectra of the single observed tryptamine–water complex (reproduced with permission from reference [415] ©2005 American Chemical Society).

conformers increases back to 45% of the total conformer population. The situation becomes very different at 300°K since no substantial change of parent conformer populations is observed when the number of added water molecules increases. This striking difference is naturally due to the respective dominant roles of internal energy at low temperature and entropy at high temperature. In agreement with experimental observations [420], the simulation predicts a dramatic shift in relative conformer internal energies and thus conformer populations with the addition of a fourth water molecule. A droplet made of hydrogen-bonded water molecules is formed at the tail of the APE parent molecule (Figure 5.7.3).

For large peptides, MS/MS mass-spectrometry (see Section 3.2.2) can provide information concerning the relationship between their secondary and tertiary structures in gas-phase and in solution [367, 421–424]. For example, the CID fragmentation pattern of the neuropeptide Y has been recorded in an ion trap and modelled [367]. Neuropeptide Y is one of the most abundant peptide in mammalian brains and plays a role in learning, memory and pain. The structure of the YPSKPDNPGEDAPAEDMARYYSALRHYINLITZQRY-NH_2 peptide determined by NMR in solution contains a stable α-helix at the C-terminal and a flexible N-terminal 12 residues part (Figure 5.7.4). The kinetics of CID fragmentation of a peptide can be modelled by attributing to each peptide linkage an activation energy E_a and to each amino acid a gas-phase basicity GB. A rate constant for each fragmentation pathway is then calculated with an Arrhenius expression. Peptide backbone cleavages involve a rearrangement of the peptide conformation in the vicinity of the fragmentation site. This rearrangement is affected by the secondary structure since not only peptide bonds but also non-covalent bonds, when present in the cleavage region, must be broken. An extra energy, called conformational stabilization energy and designated ΔE_b, must then be added as a parameter in the kinetic model. In gas-phase, a typical hydrogen bond has a bond strength of 20–25 kJ/mol and the amount of energy required to break

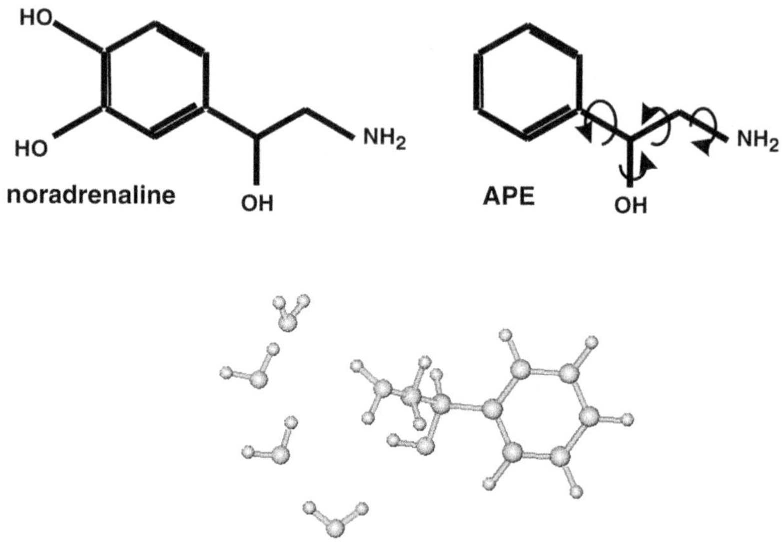

Figure 5.7.3 Top: Noradrenaline and its analogue 2-amino-1-phenyl-ethanol (APE) with its torsional angles describing the flexible chain. Bottom: Structure of the APE–$(water)_4$ complex displaying the water droplet along the flexible chain (reproduced with permission from reference [419] ©2004 American Chemical Society).

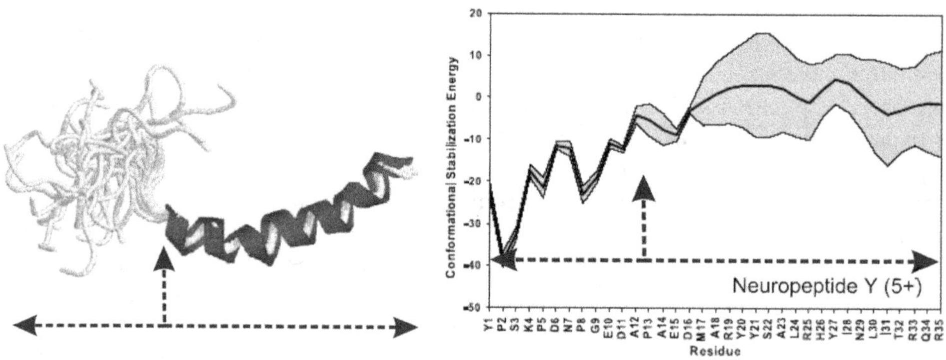

Figure 5.7.4 NMR solution structure (left) and conformational stability map (right) of the pentaprotonated neuropeptide Y. Arrows show the separation between the flexible and the α-helix regions (reproduced with permission from reference [367] ©2006 Elsevier).

a peptide bond is ca. 170 kJ/mol. During fragmentation, only a few H-bonds are broken in the vicinity of the cleavage site while the other H-bonds remain intact. The model is used to predict the abundance of each product ion and the parameters (conformational stabilization energies and gas-phase basicities) are adjusted by fitting the experimental CID mass-spectrum and its prediction. A plot of the adjusted conformational stabilization energies is presented in Figure 5.7.4 in the case of the Y peptide. The α-helical region

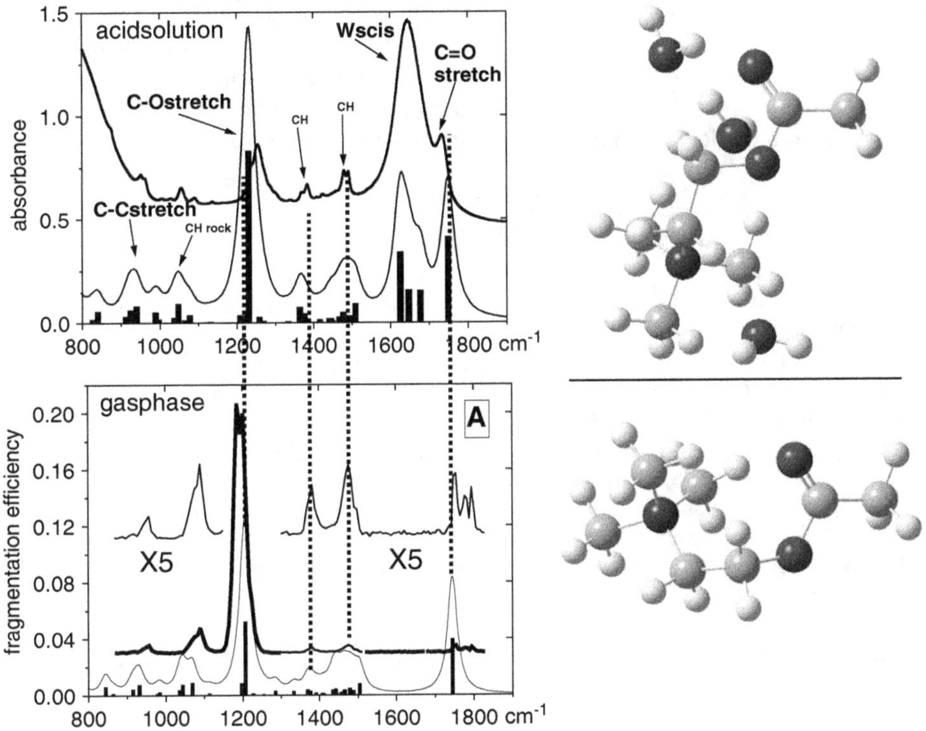

Figure 5.7.5 Left: Comparison between experimental infrared spectra of the protonated acetylcholine neurotransmitter in the gas-phase and in an acidic solution (solid lines). Right: Optimized lowest energy structures of the protonated acetylcholine neurotransmitter in the gas-phase (bottom) and complexed with three water molecules (top).

(P13–Y36) is more stabilized by the C═O···H—N bonds than the flexible N-terminal region (Y1–A12). No significant cleavages are experimentally observed by CID between residues A18 and R35.

This approach relies on the "local" network of non-covalent bonds as the dominant process modifying the peptide linkage cleavage and is thus unable to tackle the overall peptide structure as in a strongly folded peptide or protein. However, there is a good correlation between the solution structure determined by NMR and the results provided by the conformational stability maps. It is worthwhile noting that there exist mass-spectrometry methods such as the one previously presented or ECD (see Chapter 4.7) and photo-induced dissociation [423, 425] that provide information about secondary and tertiary structures through interpretation of fragmentation patterns.

The acetylcholine neurotransmitter is very flexible (see Chapter 4.3). The infrared absorption spectra of the protonated species in the gas-phase obtained by resonant IRMPD and in an acid solution are displayed in Figure 5.7.5. The presence of water molecules interacting with the C═O group of acetylcholine and the positively charged group surrounding the nitrogen atom [426] strongly modifies the acetylcholine structure. A similar study conducted on one of acetylcholine agonist, nicotine [416, 427, 428], shows a dramatic change of the most favourable protonation site when going from gas-phase to solution [429]. However, static calculations of lowest energy conformers and the corresponding predicted infrared spectra at 0°K do not provide a satisfactory match between predicted and experimental spectra recorded at 300°K in solution. Quantum molecular dynamics (CPMD, see Section 1.5.5.3) should provide a much better agreement (see Section 5.4).

REFERENCES

1. Pla SK, Zewail AH: Dynamics of water in biological recognition. *Chemical Reviews* 2004, 104:299–2123.
2. Michalarias I, Gao X, Ford RC: Recent progress on our understanding of water around biomolecules. *Journal of Molecular Liquids* 2004, 117:107–116.
3. Lague P, Roux B, Pastor RW: Molecular dynamics simulations of the Influenza hemagglutinin fusion peptide in micelles and bilayers: conformational analysis of peptide and lipids. *Journal of Molecular Biology* 2005, 354:1129–1141.
4. Agre P: Aquaporin water channels. *Angewandte Chemie-International Edition* 2004, 43: 4278–4290 (Nobel lecture).
5. Elcock AH: Atomic-level observation of macromolecular crowding effects: escape of a protein from the GroEL cage. *Proceedings of the National Academy of Sciences of the United States of America* 2003, 100:2340–2344.
6. Zorrilla S, Rivas G, Lillo MP: Fluorescence anisotropy as a probe to study tracer proteins in crowded solutions. *Journal of Molecular Recognition* 2004, 17:408–416.
7. Zorrilla S, Rivas G, Acuna AU, Lillo MP: Protein self-association in crowded protein solutions: a time-resolved fluorescence polarization study. *Protein Science* 2004, 13:2960–2969.
8. Sear RP: The cytoplasm of living cells: a functional mixture of thousands of components. *Journal of Physical Condensed Matter* 2005, 17:S3587–S3595.
9. Bizzari AR, Cannistraro S: Molecular dynamics of water at the protein–solvent interface. *Journal of Physical Chemistry B* 2002, 106:6617–6633.
10. Halle B, Davidovic M: Biomolecular hydration: from water dynamics to hydrodynamics. *Proceedings of the National Academy of Sciences of the United States of America* 2003, 100: 12135–12140.
11. Modig K, Liepinsh E, Otting G, Halle B: Dynamics of protein and peptide hydration. *Journal of the American Chemical Society* 2004, 126:102–114.
12. Steiner T: Hydrogen bonds from water molecules to aromatic acceptors in very-high resolution protein crystal structures. *Biophysical Chemistry* 2002, 95:195–201.
13. Auffinger P, Westhof E: Water and ion binding around RNA and DNA (C,G) oligomers. *Journal of Molecular Biology* 2000, 300:1113–1131.
14. Otting G, Liepinsh E, Wuttrich K: Protein hydration in aqueous solution. *Science* 1991, 254: 974–980.
15. Baldus M: Solid-state NMR spectroscopy: molecular structure an organization at the atomic level. *Angewandte Chemie-International Edition* 2006, 45:1186–1188.
16. Ge Y, Hochstrasser RM: Femtosecond two-dimensional infrared spectroscopy: IR-Cosy and THIRSTY. *Physical Chemistry Communications* 2002, 5:17–26.
17. Yeung ES: Dynamics of single biomolecules in free solution. *Annual Review of Physical Chemistry* 2004, 55:97–126.
18. Gaigeot MP, Leulliot N, Ghomi M, Jobic H, Coulombeau C, Bouloussa O: Analysis of the structural and vibrational properties of RNA building blocks by means of neutron inelastic scattering and density functional theory calculations. *Chemical Physics* 2000, 261:217–237.

19. Ruffle SV, Michalarias I, Li JC: Inelastic incoherent neutron scattering studies of water interacting with biological macromolecules. *Journal of the American Chemical Society* 2002, 124: 565–572.
20. Bagchi B: Water dynamics in the hydration layer around proteins and micelles. *Chemical Reviews* 2005, 105:3197–3219.
21. Hansia P, Vishveshwara S, Pal SK: Tryptophan–water interaction in monellin: hydration patterns from molecular dynamics simulation. *Chemical Physics Letters* 2006, 420:512–517.
22. Bandyopadhyay S, Chakraborty S, Bagchi B: Secondary structure sensitivity of hydrogen bond lifetime dynamics in the protein hydration layer. *Journal of the American Chemical Society* 2005, 127:16660–16667.
23. Gao J, Truhlar DG: Quantum mechanical methods for enzyme kinetics. *Annual Review of Physical Chemistry* 2002, 53:467–505.
24. Li ZQ, Lazaridis T: Thermodynamics of buried water clusters at a protein–ligand interface. *Journal of Physical Chemistry B* 2006, 110:1464–1475.
25. Imai T, Hiraoka R, Kovalenko A, Hirata F: Water molecules in a protein cavity detected by a statistical-mechanical theory. *Journal of the American Chemical Society* 2005, 127: 15334–15335.
26. Imai T, Harano Y, Kinoshita M, Kovalenko A, Hirata F: A theoretical analysis on hydration thermodynamics of proteins. *Journal of Chemical Physics* 2006, 125:024911–024918.
27. Sulpizi M, Folkers G, Rothlisberger U, Carloni P, Scapozza L: Applications of density functional theory-based methods in medicinal chemistry. *Quantitative Structure–Activity Relationships* 2002, 21:173–181.
28. Deerfield DW, Carter CW, Persen LG: Models for protein–zinc ion binding sites. II. The catalytic sites. *International Journal of Quantum Chemistry* 2001, 83:150–165.
29. Vöhringer-Martinez E, Hansmann B, Hernandez H, Francisco JS, Troe J, Abel B: Water catalysis of a radical molecule gas-phase reaction. *Science* 2007, 315:497–501.
30. Macleod NA, Butz P, Simons JP, Grant GH, Baker CM, Tranter GE: Structure, electronic circular dichroism and Raman optical activity in the gas phase and in solution: a computa-tional and experimental investigation. *Physical Chemistry Chemical Physics* 2005, 7: 1432–1440.
31. Macleod NA, Johannessen C, Hecht L, Barron LD, Simons JP: From the gas phase to aqueous solution: vibrational spectroscopy, Raman optical activity and conformational structure of carbohydrates. *International Journal of Mass Spectrometry* 2006, 253:193–200.
32. Loo A, Xhe JX, Cody WL: Higher order structure in the gas phase reflects solution structure. *Journal of the American Chemical Society* 1998, 120:4542–4543.
33. Prakash H, Mazumdar S: Direct correlation of the crystal structure of proteins with the maximum positive and negative charges of gaseous proteins ions produced by electrospray ionization. *Journal of the American Society for Mass Spectrometry* 2005, 16:1409–1421.
34. Iavarone AT, Parks JH: Conformational change in unsolvated Trp-cage protein probed by fluorescence. *Journal of the American Chemical Society* 2005, 127:8606–8607.
35. Iavarone AT, Meinen J, Schulze S, Parks JH: Fluorescence probe of polypeptide conformational dynamics in gas phase and in solution. *International Journal of Mass Spectrometry* 2006, 253:172–180.
36. Messer BM, Cappa CD, Smith JD, Drisdell WS, Schwartz CP, Cohen RC, Saykally RJ: Local hydration environments of amino acids and dipeptides studied by X-ray spectroscopy of liquid microjets. *Journal of Physical Chemistry B* 2005, 109:21640–21646.
37. Orozco M, Luque FJ: Theoretical methods of the solvent effect in biomolecular systems. *Chemical Reviews* 2000, 100:4187–4225.
38. Adcock SA, McCammon JA: Molecular dynamics: survey of methods for simulating the activity of proteins. *Chemical Reviews* 2006, 106:1589–1615.

39. Gregory JK, Clary DC, Liu K, Brown MG, Saykally RJ: The water dipole moment in water clusters. *Science* 1997, 275:814–817.
40. Fanfrlik J, Rejnek J, Hanus M, Hobza P: Hydration Gibbs energies of nucleic acid bases determined by Gibbs energy perturbation, continuous and hybrid approaches. *Collection of Czechoslovak Chemical Communications* 2005, 70:1756–1768.
41. Gaigeot MP, Vuilleumier R, Sprik M, Borgis D: Infrared spectroscopy of *N*-methylacetamide revisited by *ab initio* molecular dynamics simulations. *Journal of Chemical Theory and Computation* 2005, 1:772–789.
42. Trygubenko SA, Bogdan TV, Rueda M, Orozco M, Luque FJ, Sponer J, Slavicek P, Hobza P: Correlated *ab initio* study of nucleic acid bases and their tautomers in the gas phase, in a microhydrated environment and in aqueous solution – Part 1. Cytosine. *Physical Chemistry Chemical Physics* 2002, 4:4192–4203.
43. Balta B, Aviyente V: Solvent effects on glycine. II. Water-assisted tautomerization. *Journal of Computational Chemistry* 2004, 25:690–703.
44. Kryachko ES, Nguyen MT, Zeegers-Huyskens T: Theoretical study of uracil tautomers. 2. Interaction with water. *Journal of Physical Chemistry A* 2001, 105:1934–1943.
45. Sponer JE, Sychrovsky V, Hobza P, Sponer J: Interactions of hydrated divalent metal cations with nucleic acid bases. How to relate the gas phase data to solution situation and binding selectivity in nucleic acids. *Physical Chemistry Chemical Physics* 2004, 6:2772–2780.
46. Boresch S, Ringhofer S, Höchtl P, Steinhauser O: Towards a better description and understanding of biomolecular solvation. *Biophysical Chemistry* 1999, 78:43–68.
47. Mo Y, Gao J: Polarization and charge-transfer effects in aqueous solution via *ab initio* QM/MM simulations. *Journal of Physical Chemistry B Letters* 2006, 110:2976–2980.
48. Gaigeot MP, Ghomi M: Geometrical and vibrational properties of nucleic acid constituents interacting with explicit water molecules as analyzed by density functional theory calculation. 1. Uracil + $n_w H_2O$ ($n_w = 1, \ldots, 7$). *Journal of Physical Chemistry B* 2001, 105:5007–5017.
49. Giese TJ, York DM: Many-body force-field models based solely on pairwise Coulomb screening do not simultaneously reproduce correct gas-phase and condensed phase polarizability limits. *Journal of Chemical Physics* 2004, 120:9903–9906.
50. Fraternali F, van Gusteren WF: An efficient mean solvation force model for use in molecular dynamics simulations of proteins in aqueous solutions. *Journal of Molecular Biology* 1996, 256:839–848.
51. Hanus M, Ryjacek F, Kabelac M, Kubar T, Bogdan TV, Trygubenko SA, Hobza P: Correlated *ab initio* study of nucleic acid bases and their tautomers in the gas phase, in a microhydrated environment and in aqueous solution. Guanine: surprising stabilization of rare tautomers in aqueous solution. *Journal of the American Chemical Society* 2003, 125:7678–7688.
52. Gaigeot MP, Kadri C, Ghomi M: Analysis of uracil hydration by means of vibrational spectroscopy and density functional calculations. *Journal of Molecular Structure* 2001, 565:469–473.
53. Choi MY, Dong F, Miller RE: Multiple tautomers of cytosine identified and characterized by infrared laser spectroscopy in helium nanodroplets: probing structure using vibrational transition moment angles. *Philosophical Transactions of the Royal Society of London. Series A – Mathematical Physical and Engineering Sciences* 2005, 363:393–412.
54. Wyttenbach T, Paizs B, Barran P, Breci L, Liu DF, Suhai S, Wysocki VH, Bowers MT: The effect of the initial water of hydration on the energetics, structures, and H/D exchange mechanism of a family of pentapeptides: an experimental and theoretical study. *Journal of the American Chemical Society* 2003, 125:13768–13775.
55. Schütz M: A new, fast, semi-direct implementation of linear scaling local coupled cluster theory. *Physical Chemistry Chemical Physics* 2002, 4:3941–3947.
56. Zahn D: Car–Parinello molecular dynamics simulations of base-catalyzed amide hydrolysis in aqueous solution. *Chemical Physics Letters* 2004, 383:134–137.

57. Whitfield TW, Martyna GJ, Allison S, Bates SP, Vass H: Structure and hydrogen bonding in neat *N*-methylacetamide: classical molecular dynamics and Raman spectroscopy studies in a liquid of peptide fragments. *Journal of Physical Chemistry B* 2006, 110:3624–3637.
58. Tuckerman M, Marx D, Klein ML, Parrinello M: On the quantum nature of the shared proton in hydrogen bonds. *Science* 1997, 275:817–820.
59. Guimaraes CRW, Barreiro G, de Oliveira CAF, de Alencastro RB: On the application of simple explicit water models to the simulations of biomolecules. *Brazilian Journal of Physics* 2004, 34:126–136.
60. Li Y, Li H, Pan H, Han S: Structures and hydrogen bonding analysis of *N,N*-dimethylformamide and *N,N*-dimethylformamide–water mixtures by molecular dynamics simulations. *Journal of Physical Chemistry A* 2003, 107:1574–1583.
61. Udier-Blagovic M, De Tirado PM, Pearlman SA, Jorgensen WL: Accuracy of free energies of hydration using CM1 and CM3 atomic charges. *Journal of Computational Chemistry* 2004, 25:1322–1332.
62. Urbic T, Vlachy V, Dill KA: Confined water: a Mercedes–Benz model study. *Journal of Physical Chemistry B* 2006, 110:4963–4970.
63. Hribar B, Southall NT, Vlachy V, Dill KA: How ions affect the structure of water. *Journal of the American Chemical Society* 2002, 124:12302–12311.
64. Higo J, Galzitskaia OV, Ono S, Nakamura H: Energy landscape of a β-hairpin peptide in explicit water studied by multicanonical dynamics. *Chemical Physics Letters* 2001, 337:169–175.
65. Zagrovic B, Sorin EJ, Pande V: α-Hairpin folding simulations in atomistic detail using an implicit solvent model. *Journal of Molecular Biology* 2001, 313:151–169.
66. Lamoureux G, Harder E, Vorobyov IV, Roux B, MacKerell AD: A polarizable model of water for molecular dynamics simulations of biomolecules. *Chemical Physics Letters* 2006, 418:245–249.
67. Schug A, Wenzel W: All-atom folding of the trp-cage protein with an adaptive parallel tempering method. *Europhysics Letters* 2004, 67:307–313.
68. Tomasi J, Mennuci B, Cammi R: Quantum mechanical continuum solvation models. *Chemical Reviews* 2005, 105:2999–3093.
69. Ramirez FJ, Tunon I, Silla E: Aminoacid chemistry in solution: structural properties and vibrational dynamics of serine using density functional theory and a continuum solvent model. *Chemical Physics* 2004, 303:85–96.
70. Still WC, Tempcyk A, Hawley RC, Hendrickson T: Semi-analytical treatment of solvation for molecular mechanics and dynamics. *Journal of the American Chemical Society* 1990, 112:6127–6129.
71. Ishizuka T, Terada T, Nakamura S, Shimizu K: Improvement of accuracy of free-energy landscapes calculated with generalized Born model by using numerical solutions of Poisson's equation. *Chemical Physics Letters* 2004, 393:546–551.
72. Couch VA, Cheng N, Nambiar K, Fink W: Structural characterization of helices of implicitly solvated poly-alanine. *Journal of Physical Chemistry B* 2006, 110:3410–3419.
73. Tan C, Yang L, Luo R: How well does Poisson–Boltzmann implicit solvent agree with explicit solvent? A quantitative analysis. *Journal of Physical Chemistry B* 2006, 110:18680–18687.
74. Cho KH, No KT, Scheraga HA: Ion pair interactions in aqueous solution: self-consistent reaction field (SCRF) calculations with some explicit water molecules. *Journal of Physical Chemistry A* 2000, 104:6505–6509.
75. Hassan SA, Mehler EL: A general screened Coulomb potential based implicit solvent model: calculation of secondary structures of small peptides. *International Journal of Quantum Chemistry* 2001, 83:193–202.
76. Soteras I, Curutchet C, Bidon-Chanal A, Orozco M, Luque FJ: Extension of the MST model to the IEF formalism: HF and B3LYP parametrizations. *Journal of Molecular Structure THEOCHEM* 2005, 727:29–40.

77. Eckert F, Klamt A: Accurate prediction of basicity in aqueous solution with COSMO-RS. *Journal of Computational Chemistry* 2006, 27:11–19.
78. Nagy PI, Takacs-Novak K: Tautomeric and conformational equilibria of biologically important (hydroxyphenyl) alkylamines in the gas phase and in aqueous solution. *Physical Chemistry Chemical Physics* 2004, 6:2838–2848.
79. Hanus M, Kabelac M, Rejnek J, Ryjacek F, Hobza P: Correlated *ab initio* study of nucleic acid bases and their tautomers in the gas phase, in a microhydrated environment, and in aqueous solution. Part 3. Adenine. *Journal of Physical Chemistry B* 2004, 108:2087–2097.
80. Rejnek J, Hanus M, Labelac M, Ryjacek F, Hobza P: Correlated *ab initio* study of nucleic acid bases and their tautomers in the gas phase, in a microhydrated environment and in aqueous solution. Part 4. Uracil and thymine. *Physical Chemistry Chemical Physics* 2005, 7: 2006–2017.
81. Frimand K, Bohr H, Jalkanen KJ, Suhai S: Structures, vibrational absorption and vibrational circular dichroism spectra of l-alanine in aqueous solution: a density functional theory and RHF study. *Chemical Physics* 2000, 255:165–174.
82. Levy N, Borgis N, Marchi M: A dielectric continuum model of solvation for complex solutes. *Computational Physics Communications* 2005, 169:69–74.
83. Hugosson HW, Laio A, Maurer P, Rothlisberger U: A comparative theoretical study of dipeptide solvation in water. *Journal of Computational Chemistry* 2006, 27:672–684.
84. Liu K, Cruzan JD, Saykally RJ: Water clusters. *Science* 1996, 271:929–933.
85. Buck U, Huisken F: Infrared spectroscopy of size-selected water and methanol clusters. *Chemical Reviews* 2000, 100:3863–3890.
86. Keutsch FN, Saykally RJ: Water clusters: untangling the mysteries of the liquid, one molecule at a time. *Proceedings of the National Academy of Sciences of the United States of America* 2001, 98:10533–10540.
87. Anderson P, Steinbach C, Buck U: Vibrational spectroscopy of large water clusters of known size. *European Physical Journal D* 2003, 24:53–56.
88. Xantheas SS, Burnham CJ, Harrison RJ: Development of transferable models for water. II. Accurate energetics of the first few water clusters from first principles. *Journal of Chemical Physics* 2002, 116:1493–1499.
89. Xantheas SS, Apra E: The binding energies of the D-2d and S-4 water octamer isomers: high-level electronic structure and empirical potential results. *Journal of Chemical Physics* 2004, 120:823–828.
90. Fanourgakis GS, Apra E, Xantheas SS: High-level *ab initio* calculations for the four low-lying families of minima of (H_2O)(20). I. Estimates of MP2/CBS binding energies and comparison with empirical potentials. *Journal of Chemical Physics* 2004, 121:2655–2663.
91. Slipchenko MN, Kuyanov KE, Sartakov BG, Vilesov AF: Infrared intensity in small ammonia and water clusters. *Journal of Chemical Physics* 2006, 124:241101.
92. Han JX, Takahashi LK, Lin W, Lee E, Keutsch FN, Saykally RJ: Terahertz vibration–rotation–tunnelling (VRT) spectroscopy of the d_6-water trimer: complete characterization of the 2.94 THz torsional band. *Chemical Physics Letters* 2006, 423:344–351.
93. Lenz A, Ojamäe L: Theoretical IR spectra for water clusters $(H_2O)_n$ ($n = 6$–22, 28, 30) and identification of spectral contributions from different H-bond conformations in gaseous and liquid water. *Journal of the Physical Chemistry* 2006, 110:13338–13393.
94. Leforestier C, Gatti F, Fellers RS, Saykally RJ: Determination of a flexible (12D) water dimer potential via direct inversion of spectroscopic data. *Journal of Chemical Physics* 2002, 117: 8710–8722.
95. Braly LB, Liu K, Brown MG, Keutsch FN, Fellers RS, Saykally RJ: Terahertz laser spectroscopy of the water dimer intermolecular vibrations. II. $(H_2O)_2$. *Journal of Chemical Physics* 2000, 112:10314–10326.

96. van Duijneveldt-van de Rijdt JGCM, Mooij WTM, van Duijneveldt FB: Testing the quality of some water–water potentials. *Physical Chemistry Chemical Physics* 2003, 5:1169–1180.
97. Lucas B, Lecomte F, Reimann B, Barth HD, Gregoire G, Bouteiller Y, Schermann JP, Desfrancois C: A new infrared spectroscopy technique for structural studies of mass-selected neutral polar complexes without chromophore. *Physical Chemistry Chemical Physics* 2004, 6:2600–2604.
98. Csonka IP, Ruzsinczky A, Perdew JP: Proper Gaussian basis sets for density functional studies of water dimers and trimers. *Journal of Physical Chemistry B* 2005, 109:21471–21475.
99. Bukowski R, Szalewicz K, Groenenboom G, van der Avoird A: Interaction potential for water dimer from symmetry-adapted perturbation theory based on density functional description of monomers. *Journal of Chemical Physics* 2006, 125:044301.
100. Bouteiller Y, Desfrancois C, Schermann JP, Latajka Z, Silvi B: Calculation of electronic affinity and vertical detachment energy of the water dimer complex using the density functional theory. *Journal of Chemical Physics* 1998, 108:7967–7972.
101. Klopper W, van Duijnevelt JGCM, van Duijnevelt FB: Computational determination of equilibrium geometry and dissociation energy of the water dimer. *Physical Chemistry Chemical Physics* 2000, 2:2227–2235.
102. Xu X, Goddard WA: Bonding properties of the water dimer: a comparative study of density functional theories. *Journal of Physical Chemistry A* 2004, 108:2305–2313.
103. Huang XC, Braams BJ, Bowman JM: *Ab initio* potential energy and dipole moment surfaces of $(H_2O)_2$. *Journal of Physical Chemistry A* 2006, 110:445–451.
104. Tschumper GS, Leininger ML, Hoffman BC, Valeev EF, Schaefer HF, Quack M: Anchoring the water dimer potential energy surface with explicitly correlated computations and focal point analyses. *Journal of Chemical Physics* 2002, 116:690–701.
105. Keutsch FN, Saykally RJ, Wales DJ: Bifurcation tunneling dynamics in the water trimer. *Journal of Chemical Physics* 2002, 117:8823–8835.
106. Keutsch FN, Cruzan JD, Saykally RJ: The water trimer. *Chemical Reviews* 2003, 103:2533–2577.
107. Harker HA, Viant MR, Keutsch FN, Michael EA, McLaughlin RP, Saykally RJ: Water pentamer: characterization of the torsional-puckering manifold by terahertz VRT spectroscopy. *Journal of Physical Chemistry A* 2005, 109:6483–6497.
108. Miyake T, Aida M: Hydrogen bonding patterns in water clusters: trimer, tetramer and pentamer. *Internet Electronic Journal Molecular Design* 2003, 2:24–32.
109. Brown MG, Keutsch FN, Saykally RJ: The bifurcation rearrangement in cyclic water clusters: breaking and making hydrogen bonds. *Journal of Chemical Physics* 1998, 109:9645–9647.
110. Liu K, Brown MG, Carter C, Saykally RJ, Gregory JK, Clary DC: Characterization of a cage form of the water hexamer. *Nature* 1996, 381:501–503.
111. Ludwig R: Hexamers: from covalently bound organic structures to hydrogen bonded water clusters. *ChemPhysChem* 2000, 1:53–56.
112. Nauta K, Miller RE: Formation of cyclic water hexamer in liquid helium: the smallest piece of ice. *Science* 2000, 287:293–295.
113. Burnham CJ, Xantheas SS, Miller MA, Applegate BE, Miller RE: The formation of cyclic water complexes by sequential ring insertion: experiment and theory. *Journal of Chemical Physics* 2002, 117:1109–1122.
114. Garrett BC, Dixon DA, Camaioni DM, Chipman DM, Johnson MA, Jonah CD, Kimmel GA, Miller JH, Rescigno TN, Rossky PJ, *et al.*: Role of water in electron-initiated processes and radical chemistry: issues and scientific advances. *Chemical Reviews* 2005, 105:355–389.
115. Bragg AE, Verlet JRR, Kammrath A, Cheshnovsky O, Neumark DM: Electronic relaxation dynamics of water cluster anions. *Journal of the American Chemical Society* 2005, 127:15283–15295.
116. Bragg AE, Verlet JRR, Kammrath A, Cheshnovsky O, Neumark DM: Hydrated electron dynamics: from clusters to bulk. *Science* 2004, 306:669–671.

117. Lee GH, Arnold ST, Eaton JG, Sarkas HW, Bowen KH, Ludewigt C, Haberland H: Negative-ion photoelectron-spectroscopy of solvated electron cluster anions, $(H_2O)_N^-$ and $(NH_3)_N^-$. *Zeitschrift Fur Physik D – Atoms Molecules and Clusters* 1991, 20:9–12.
118. Haberland H, Langosch H: Mass spectroscopic study of some novel water clusters – $(H_2O)_N^+$ – N greater than 3. *Zeitschrift Fur Physik D – Atoms Molecules and Clusters* 1986, 2:243–247.
119. Coe JV, Lee GH, Eaton JG, Arnold ST, Sarkas HW, Bowen KH, Ludewigt C, Haberland H, Worsnop DR: Photoelectron-spectroscopy of hydrated electron cluster anions, $(H_2O)_N^-$ = 2–69. *Journal of Chemical Physics* 1990, 92:3980–3982.
120. Hammer NI, Roscioli JR, Johnson MA, Myshakin EM, Jordan KD: Infrared spectrum and structural assignment of the water trimer anion. *Journal of Physical Chemistry A* 2005, 109: 11526–11530.
121. Diken EG, Robertson WH, Johnson MA: The vibrational spectrum of the neutral $(H_2O)_6$ precursor to the "magic" $(H_2O)_6^-$ cluster anion by argon-mediated, population-modulated electron attachment spectroscopy. *Journal of Physical Chemistry A* 2004, 108:64–68.
122. Desfrancois C, Gregoire G, Schermann JP: Electrons in diffuse orbitals. *Few-Body Systems* 2004, 34:169–174.
123. Lee HM, Suh SB, Tarakeshwar P, Kim KS: Origin of the magic numbers of water clusters with an excess electron. *Journal of Chemical Physics* 2005, 122:044309.1-1-044309.6.
124. Barnett RN, Landman U, Scharf D, Jortner J: Surface and internal excess electron-states in molecular clusters. *Accounts of Chemical Research* 1989, 22:350–357.
125. Barnett RN, Landman U, Cleveland CL, Jortner J: Electron localization in water clusters. 2. Surface and internal states. *Journal of Chemical Physics* 1988, 88:4429–4447.
126. Coe JV, Arnold ST, Lee GO, Bowen KH: Photoelectron spectra of hydrated electron clusters: fitting line shapes and grouping isomers. *Journal of Chemical Physics* 2006, 125:014315.
127. Verlet JRR, Bragg AE, Kammrath A, Cheshnovsky O, Neumark DM: Observation of large water-cluster anions with surface-bound excess electrons. *Science* 2005, 307:93–96.
128. Khan A: Solvated electron in $(H_2O)_{20}^-$ dodecahedral cavity: calculated stretch frequencies and vertical dissociation energy. *Chemical Physics Letters* 2005, 401:85–88.
129. Sobolewski AL, Domcke W: Hydrated hydronium: a cluster model of the solvated electron? *Physical Chemistry Chemical Physics* 2002, 4:4–10.
130. Muguet FF, Gelabert H, Gauduel Y: Electron propagator study of the excitation spectrum of the solvated hydronium radical. *Journal de Chimie Physique et de Physico-Chimie Biologique* 1996, 93:1808–1827.
131. Muguet FF: MCSCF vibrational spectra of the symmetric and asymmetric dihydronium cations. *Journal of Molecular Structure (THEOCHEM)* 1996, 368:173–196.
132. Muguet F, Bassez MP, Robinson GW: Aquated electrons, H_2O^- anions, and OH^-/H_3O units – reply. *Journal of Physical Chemistry* 1988, 92:7262–7263.
133. Desfrancois C, Schermann JP: Solvent effects in thermal electron-attachment to clusters containing molecular-oxygen. *Zeitschrift Fur Physik D – Atoms Molecules and Clusters* 1993, 26:S344–S346.
134. Wysokinski R, Bienko DC, Michalska D, Zeegers-Huyskens T: Theoretical study of the interaction between cytosine and hydrogen peroxide. *Chemical Physics* 2005, 315:17–26.
135. Rhee SG: H_2O_2, a necessary evil for cell signaling. *Science* 2006, 312:1882–1883.
136. Bell AJ, Wright TG: Structures and binding energies of $O_2^-{\cdot}H_2O$ and $O_2{\cdot}H_2O$. *Physical Chemistry Chemical Physics* 2004, 6:4385–4390.
137. Weber JM, Kelley JA, Robertson WH, Johnson MA: Hydration of a structured excess charge distribution: infrared spectroscopy of the O_2^- $(H_2O)_n$ ($1 < n < 5$ clusters). *Journal of Chemical Physics* 2001, 114:2698–2706.
138. Yeh LI, Okumura M, Myers JD, Price JM, Lee YT: Vibrational spectroscopy of the hydrated hydronium cluster ions $H_3O^+{\cdot}(H_2O)_N$ ($N = 1, 2, 3$). *Journal of Chemical Physics* 1989, 91:7319–7330.

139. Okumura M, Yeh LI, Myers JD, Lee YT: Infrared-spectra of the solvated hydronium ion – vibrational predissociation spectroscopy of mass-selected $H_3O^+ \cdot (H_2O)_N$. *Journal of Physical Chemistry* 1990, 94:3416–3427.
140. Cheng HP: Water clusters: fascinating hydrogen-bonding network, solvation shell structures and proton motion. *Journal of Physical Chemistry A* 1998, 102:6201–6204.
141. Christie RA, Jordan KD: Theoretical investigation of the $H_3O^+(H_2O)_4$ cluster. *Journal of Physical Chemistry A* 2001, 105:7551–7558.
142. Headrick JM, Diken EG, Walters RS, Hammer NI, Christie RA, Cui J, Myshakin EM, Duncan MA, Johnson MA, Jordan KD: Spectral signatures of hydrated proton vibrations in water clusters. *Science* 2005, 308:1765–1769.
143. Hodges MP, Wales DJ: Global minima of protonated water clusters. *Chemical Physics Letters* 2000, 324:279–288.
144. James T, Wales DJ: Protonated water clusters described by an empirical valence bond potential. *Journal of Chemical Physics* 2005, 122:134306–134317.
145. Singh NJ, Park M, Min SK, Suh SB, Kim KS: Magic and antimagic protonated water clusters: exotic structures with unusual dynamic effects. *Angewandte Chemie-International Edition* 2006, 45:3795–3800.
146. Sauer J, Döbler J: Gas-phase infrared spectrum of the protonated water dimer: molecular dynamics simulations and accuracy of the potential energy surface. *ChemPhysChem* 2005, 6:1706–1710.
147. Asmis KR, Pivonka NL, Santambrogio G, Brummer M, Kaposta C, Neumark DM, Woste L: Gas-phase infrared spectrum of the protonated water dimer. *Science* 2003, 299:1375–1377.
148. Fridgen TD, McMahon TB, MacAleese L, Lemaire J, Maitre P: Infrared spectrum of the protonated water dimer in the gas phase. *Journal of Physical Chemistry A* 2004, 108:9008–9010.
149. Shin JW, Hammer NI, Diken EG, Johnson MA, Walters RS, Jaeger TD, Duncan MA, Christie RA, Jordan KD: Infrared signature of structures associated with the $H^+(H_2O)_n$ ($n = 6$ to 27) clusters. *Science* 2004, 304:1137–1140.
150. Zwier TS: The structure of protonated water clusters. *Science* 2004, 304:1119–1120.
151. Diken EG, Headrick JM, Roscioli JR, Bopp JC, Johnson MA, McCoy AB: Fundamental excitations of the shared proton in the $H_3O_2^-$ and $H_5O_2^+$complexes. *Journal of Physical Chemistry A* 2005, 109:1487–1490.
152. Headrick JM, Bopp JC, Johnson MA: Predissociation spectroscopy of the argon-solvated $H_2O_5^+$ "zundel" cation in the 1000–1900 cm^{-1} region. *Journal of Chemical Physics* 2004, 112:11523–11526.
153. Miyazaki M, Fujii A, Ebata T, Mikami N: Infrared spectroscopic evidence for protonated water clusters forming nanoscale cages. *Science* 2004, 304:1134–1137.
154. Maupin CM, Wong KF, Soudackov AV, Kim S, Voth GA: A multistate empirical valence bond description of protonable amino acids. *Journal of Physical Chemistry A* 2006, 110:631–639.
155. Zhang Z, Zhu Y, Shi Y: Molecular dynamics simulations of urea and thermal-induced denaturation of S-peptide analogue. *Biophysical Chemistry* 2001, 89:145–162.
156. Guo Y, Clark DS: Activation of enzymes for non-aqueous biocatalysis by denaturating concentrations of urea. *Biochemistry and Biophysics Acta* 2001, 154:406–411.
157. Kallies B: Coupling of solvent and solute dynamics – molecular dynamics simulations of aqueous urea solutions with different intramolecular potentials. *Physical Chemistry Chemical Physics* 2002, 4:86–95.
158. Godfrey PD, Brown RD, Hunter AN: The shape of urea. *Journal of Molecular Structure* 1997, 413:405–414.
159. Skurski P, Simons J: An excess electron bound to urea. III. The urea dimer as an electron trap. *Journal of Chemical Physics* 2002, 116:6118–6125.
160. Skurski P, Simons J: An excess electron bound to urea oligomers. II. Chains and ribbons. *Journal of Chemical Physics* 2001, 115:10731–10737.

161. Lecomte F, Lucas B, Gregoire G, Schermann JP, Desfrancois C: Urea and methylurea dipole-bound anions. *Physical Chemistry Chemical Physics* 2003, 5:3120–3125.
162. Carles S, Desfrancois C, Schermann JP, Berges J, Houee-Levin C: Rydberg electron-transfer spectroscopy and *ab initio* studies of dimethylsulfoxide–water neutral and anion dimers. *International Journal of Mass Spectrometry* 2001, 205:227–232.
163. Nonose S, Iwaoka S, Tanaka H, Okai N, Shibakusa T, Fuke K: Structures and reactions of biomolecular ions produced with electrospray ionization. *European Journal of Physics D* 2003, 24:335–338.
164. Mancera RL, Chalaris M, Refson K, Samios J: Molecular dynamics simulation of dilute aqueous DMSO solutions. A temperature-dependence study of the hydrophobic and hydrophilic behaviour around DMSO. *Physical Chemistry Chemical Physics* 2004, 6:94–102.
165. Li OS, Wu G, Yu Z: The role of methyl groups in the formation of hydrogen bond in DMSO–methanol mixtures. *Journal of the American Chemical Society* 2006, 128: 1438–1439.
166. Kopsel C, Moltgen H, Schuch H, Auweter H, Kleinermanns K, Martin HD, Bettermann H: Structure investigations on assembled astaxanthin molecules. *Journal of Molecular Structure* 2005, 750:109–115.
167. Chen B: Microscopic structure and solvation in dry and wet octanol. *Journal of Physical Chemistry B* 2006, 110:3555–3563.
168. Dill KA: Dominant forces in protein folding. *Biochemistry* 1990, 29:7133–7154.
169. Sorenson JM, Hura G, Soper AK, Pertsemlidis A, Head-Gordon T: Determining the role of hydration forces in protein folding. *Journal of Physical Chemistry B* 1999, 103:5413–5426.
170. Southall NT, Dill KA, Haymet ADJ: A view of the hydrophobic effect. *Journal of Physical Chemistry B* 2002, 106:521–533.
171. Hunter CA: Quantifying intermolecular interactions; guidelines for the molecular recognition toolbox. *Angewandte Chemie-International Edition* 2004, 43:5310–5324.
172. Xu HF, Dill KA: Water's hydrogen bonds in the hydrophobic effect: a simple model. *Journal of Physical Chemistry B* 2005, 109:23611–23617.
173. Dill KA, Truskett TM, Vlachy V, Hribar-Lee B: Modeling water, the hydrophobic effect, and ion solvation. *Annual Review of Biophysical Biomolecular Structure* 2005, 34:173–199.
174. Chandler D: Two faces of water. *Nature* 2002, 437:640–647.
175. Chandler D: Interfaces and the driving force of hydrophobic assembly. *Nature* 2005, 437:640–647.
176. Levy Y, Onuchic JN: Water and proteins: a love–hate relationship. *Proceedings of the National Academy of Sciences of the United States of America* 2004, 101:3325–3326.
177. Fisicaro E, Compari C, Braibanti A: Entropy/enthalpy compensation: hydrophobic effect, micelles and protein complexes. *Physical Chemistry Chemical Physics* 2004, 6:4156–4166.
178. Pratt LR: Molecular theory of hydrophobic effects: "She is to mean to have her name repeated". *Annual Review of Physical Chemistry* 2002, 53:409–436.
179. Pratt LR, Pohorille A: Hydrophobic effects and modeling of biophysical aqueous solution interfaces. *Chemical Reviews* 2002, 102:2671–2692.
180. Kegel WK, van der Schoot P: Competing hydrophobic and screened-Coulomb interactions in hepatitis B virus capsid assembly. *Biophysical Journal* 2004, 86:3905–3913.
181. Sandelin E: On hydrophobicity and conformational specificity of proteins. *Biophysical Journal* 2004, 86:23–30.
182. Yang C, Sharp KA: Hydrophobic tendency of polar group hydration as a major force in type I antifreeze protein. *Proteins – Structure Function and Bioinformatics* 2005, 59:266–274.
183. Zhan D, Fenn JB: Gas phase hydration of eletrospray ions from small peptides. *International Journal of Mass Spectrometry* 2002, 120:1–10.
184. Wyttenbach T, Liu DF, Bowers MT: Hydration of small peptides. *International Journal of Mass Spectrometry* 2005, 240:221–232.

185. Zhan D, Rosell J, Fenn JB: Solvation studies of electrospray ions – method and early results. *Journal of the American Society for Mass Spectrometry* 1998, 9:1241–1247.
186. Nonose S, Iwaoka S, Mori K, Shibata Y, Fuke K: Structure and reactions of hydrated biomolecular cluster ions. *European Physical Journal D* 2005, 34:315–319.
187. Wincel H: Hydration of gas-phase protonated alkylamines, amino acids and dipeptides produced by electrospray. *International Journal of Mass Spectrometry* 2006, 251:23–31.
188. Stevenson CL, Tan MM, Lechuga-Ballesteros D: Secondary structure of cyclosporine in a spray-dried liquid crystal by FTIR. *Journal of Pharmaceutical Sciences* 2003, 92:1832–1843.
189. Tishcenko GN, Andrianov VI, Vainstein BK, Woolfson MM, Dodson E: Channels in the gramicidin S-with-urea structure and their possible relation to transmembrane ion transport. *Acta Crystallographica D* 1997, 53:151–159.
190. Kyte J, Doolite R: A simple method for displaying the hydropathic character of a protein. *Journal of Molecular Biology* 1982, 157:105–132.
191. Conte L, Ghotia L, Janin J: The atomic structure of protein–protein recognition sites. *Journal of Molecular Biology* 1999, 285:2177–2198.
192. Sheinerman FB, Norel R, Honig B: Electrostatic aspects of protein–protein interactions. *Current Opinion in Structural Biology* 2000, 10:153–159.
193. Nadassy K, Wodak SJ, Janin J: Structural features of protein–nucleic acid recognition site. *Biochemistry* 1999, 38:1990–2017.
194. Deremble C, Lavery R: Macromolecular recognition. *Current Opinion in Structural Biology* 2005, 15:171–175.
195. Albiser G, Lamiri A, Premilat S: The A—B transition: temperature and base composition effects on hydration of DNA. *International Journal of Biological Macromolecules* 2001, 28:199–203.
196. Hu X, Li HY, Liang W, Han S: Explore water's protective and assistance role for uracil: a changing role of water at different temperature. *Chemical Physics Letters* 2005, 417:321–326.
197. Kim SK, Lee W, Herschbach DR: Cluster beam chemistry hydration of nucleic acid bases, ionization potentials of hydrated adenine and thymine. *Journal of Physical Chemistry* 1996, 100: 7933–7937.
198. Kim NJ, Kim YS, Jeong G, Ahn TK, Kim SK: Hydration of DNA base cations in the gas phase. *International Journal of Mass Spectrometry* 2002, 219:11–21.
199. Casaes RN, Paul JB, McLaughlin RP, Saykally RJ, van Mourik T: Infrared cavity ringdown spectroscopy of jet-cooled nucleotide base clusters and water complexes. *Journal of Physical Chemistry A* 2004, 108:10989–10996.
200. Piuzzi F, Mons M, Dimicoli I, Tardivel B, Zhao Q: Ultraviolet spectroscopy and tautomerism of the DNA base guanine and its hydrated form in a supersonic jet. *Chemical Physics* 2001, 270:205–214.
201. Abo-Riziq A, Crews B, Grace L, de Vries MS: Microhydration of guanine base pairs. *Journal of the American Chemical Society* 2005, 127:2374–2375.
202. Hendricks JH, Lyapustina SA, deClercq HL, Snodgrass JT, Bowen KH: Dipole bound, nucleic acid base anions studied via negative ion photoelectron spectroscopy. *Journal of Chemical Physics*, 104:7788–7791.
203. Hendricks JH, Lyapustina SA, de Clercq HL, Bowen KH: The dipole bound-to-covalent anion transformation in uracil. *Journal of Chemical Physics* 1998, 108:8–11.
204. Schiedt J, Weinkauf R, Neumark DM, Schlag EW: Anion spectroscopy of uracil, thymine and the amino-oxo and amino-hydroxy tautomers of cytosine and their water clusters. *Chemical Physics* 1998, 239:511–524.
205. Choi MY, Miller RE: Multiple isomers of uracil–water complexes: infrared spectroscopy in helium nanodroplets. *Physical Chemistry Chemical Physics* 2005, 7:3565–3573.
206. Kang H, Lee KT, Kim SK: Femtosecond real time dynamics of hydrogen bond dissociation in photoexcited adenine–water clusters. *Chemical Physics Letters* 2002, 359:213–219.

207. Periquet V, Moreau A, Carles S, Schermann JP, Desfrancois C: Cluster size effects upon anion solvation of *N*-heterocyclic molecules and nucleic acid bases. *Journal of Electron Spectroscopy and Related Phenomena* 2000, 106:141–151.
208. van Mourik T, Price SL, Clary DC: *Ab initio* calculations on uracil–water. *Journal of Physical Chemistry A* 1999, 103:1611–1618.
209. Jalbout AF, Adamowicz L: Dipole-bound anions of adenine–water clusters. *Ab initio* study. *Journal of Physical Chemistry A* 2001, 105:1033–1038.
210. Crews B, Abo-Riziq A, Grace L, Callahan M, Kabelac M, Hobza P, de Vries MS: IR—UV double resonance spectroscopy of guanine–H_2O clusters. *Physical Chemistry Chemical Physics* 2005, 7:3015–3020.
211. Close DM, Crespo-Hernandez CE, Gorb L, Leszczynski J: The influence of microhydration on the ionization energy thresholds of uracil and thymine. *Journal of Physical Chemistry A* 2005, 109:9279–9283.
212. Frigato T, Svozil D, Jungwirth P: Valence- and dipole-bound anions of the thymine–water complex: *ab initio* characterization of the potential energy surfaces. *Journal of Physical Chemistry A* 2006, 110:2916–2923.
213. van Mourik T, Benoit DM, Price SL, Clary DC: *Ab initio* and diffusion Monte Carlo study of uracil–water, thymine–water, cytosine–water, and cytosine–(water)(2). *Physical Chemistry Chemical Physics* 2000, 2:1281–1290.
214. Gaigeot MP, Sprik M: *Ab initio* molecular dynamics computation of the infrared spectrum of aqueous uracil. *Journal of Physical Chemistry B* 2003, 107:10344–10358.
215. Gaigeot MP, Sprik M: *Ab initio* molecular dynamics study of uracil in aqueous solution. *Journal of Physical Chemistry B* 2004, 108:7458–7467.
216. Kabelac M, Hobza P: Hydration and stability of nucleic acid bases and base pairs. *Physical Chemistry Chemical Physics* 2007, 9:903–917.
217. Fonseca Guerra C, Bickelhaupt FM, Sahai S, Wang F: Adenine tautomers: relative stabilities, ionization energies, and mismatch with cytosine. *Journal of Physical Chemistry A* 2006, 110: 4012–4020.
218. Kryachko ES, Nguyen MT, Zeegers-Huyskens T: Theoretical study of tautomeric forms of uracil. 1. Relative order of stabilities and their relation to proton affinities and deprotonation enthalpies. *Journal of Physical Chemistry A* 2001, 105:1288–1295.
219. Chandra AK, Michalska D, Wysokinsky R, Zeegers-Huyskens T: Theoretical study of the acidity and basicity of the cytosine tautomers and their 1:1 complexes with water. *Journal of Physical Chemistry A* 2004, 108:9593–9600.
220. Rodgers FM, Armentrout PB: Non covalent interactions of nucleic acid bases (uracil, thymine, and adenine) with alkali metal ions. Threshold collision-induced dissociation and theoretical studies. *Journal of the American Chemical Society* 2000, 122:8548–8558.
221. Pedersen DB, Zgierski MZ, Denomme S, Simard B: Photo-induced dehydrogenation in a gas-phase metal–DNA complex: Al–cytosine. *Journal of the American Chemical Society* 2002, 124:6686–6692.
222. Zhang RB, Zeegers-Huyskens T, Ceulemans A, Nguyen MT: Interaction of triplet uracil and thymine with water. *Chemical Physics* 2005, 316:35–44.
223. Ritze HH, Lippert H, Samoylova E, Smith VR, Hertel IV, Radloff W, Schultz T: Relevance of $\pi\sigma^*$ states in the photoinduced processes of adenine, adenine dimer, and adenine–water complexes. *Journal of Chemical Physics* 2005, 122:224320–224320-9.
224. Canuel C, Elhanine M, Mons M, Piuzzi F, Tardivel B, Dimicoli I: Time-resolved photoelectron and photoion fragmentation spectroscopy study of 9-methyladenine and its hydrates: a contribution to the understanding of the ultrafast decay radiationless decay of excited DNA bases. *Physical Chemistry Chemical Physics* 2006, 8:3978–3987.

225. Perun S, Sobolewski AL, Domcke W: Photostability of 9H-adenine: mechanisms of the radiationless deactivation of the lowest excited singlet states. *Chemical Physics* 2005, 313: 107–112.
226. Desfrancois C, Carles S, Schermann JP: Weakly bound clusters of biological interest. *Chemical Reviews* 2000, 100:3943–3962.
227. Sivanesan D, Subramanian V, Nair BU: Solvent effect on DNA base stacked dimers: an isodensity polarizable continuum approach. *Journal of Quantum Chemistry* 2001, 84:750–758.
228. Kabelac M, Hobza P: At nonzero temperatures, stacked structures of methylated nucleic acid base pairs and microhydrated nonmethylated nucleic acid base pairs are favored over planar hydrogen-bonded structures: a molecular dynamics simulations study. *Chemical European Journal* 2001, 7:2067–2074.
229. Auffinger P, Westhof E: Melting of the solvent structure around a RNA duplex: a molecular dynamics simulation study. *Biophysical Chemistry* 2002, 95:203–210.
230. Auffinger P, Westhof E: Hydration of RNA base pairs. *Journal of Biomolecular Structure & Dynamics* 1998, 16:693–707.
231. Drew HR, Wing RM, Takano T, Broka C, Tanaka S, Ikatura K, Dickerson RE: Structure of a B-DNA dodecamer: conformation and dynamics. *Proceedings of the National Academy of Sciences of the United States of America* 1981, 78:2179–2183.
232. Schnier PD, Klassen JS, Strittmatter EE, Williams ER: Activation energies for dissociation of double strand oligonucleotide anions: evidence for Watson–Crick base pairing *in vacuo*. *Journal of the American Chemical Society* 1998, 120:9605–9613.
233. Gabelica V, Rosu F, Houssier C, De Pauw E: Gas phase thermal denaturation of an oligonucleotide duplex and its complexes with minor groove binders. *Rapid Communications in Mass Spectrometry* 2000, 14:464–467.
234. Rosu F, Gabelica V, Houssier C, Colson P, De Pauw E: Triplex and quadruplex DNA structures studied by electrospray mass spectrometry. *Rapid Communications in Mass Spectrometry* 2002, 16:1729–1736.
235. Gabelica V, De Pauw E, Rosu F: Interaction between antitumor drugs and a double-stranded oligonucleotide studied by electrospray ionization mass spectrometry. *Journal of Mass Spectrometry* 1999, 34:1328–1337.
236. Pan S, Sun X, Lee JK: Stability of complementary and mismatched DNA duplexes: comparison and contrast in gas versus solution phases. *International Journal of Mass Spectrometry* 2006, 253:238–248.
237. Pan S, Sun X, Lee JK: DNA stability in the gas versus solution phases: a systematic study of sixteen duplexes with varying length, sequence, and charge level. *Journal of the American Society for Mass Spectrometry* 2006, 17:1383–1396.
238. Gaffney BL, Jones RA: Thermodynamic comparison of the base pairs formed by the carcinogenic lesion O6-methylguanine with reference both to Watson–Crick pairs and mismatched pairs. *Biochemistry* 1989, 28:5881–5889.
239. Perez A, Blas JR, Rueda M, Lopez-Bes JM, de la Cruz X, Orozco M: Exploring the essential dynamics of B-DNA. *Journal of Chemical Theory and Computation* 2005, 1:790–800.
240. Guittat L, De Cian A, Rosu F, Gabelica V, De Pauw E, Delfourne E, Mergny JL: Ascididemin and meridine stabilise G-quadruplexes and inhibit telomerase in vitro. *Biochimica et Biophysica Acta – General Subjects* 2005, 1724:375–384.
241. Rosu F, Gabelica V, Shin-ya K, De Pauw E: Telomestatin-induced stabilization of the human telomeric DNA quadruplex monitored by electrospray mass spectrometry. *Chemical Communications* 2003:2702–2703.
242. Gidden J, Baker ES, Ferzoco A, Bowers MT: Structural motifs of DNA complexes in the gas phase. *International Journal of Mass Spectrometry* 2005, 240:183–193.

243. Rueda M, Luque FJ, Orozco M: G-quadruplexes can maintain their structure in the gas phase. *Journal of the American Chemical Society* 2006, 128:3608–3619.
244. Rueda M, Kalko SG, Luque FJ, Orozco M: The structure and dynamics of DNA in the gas phase. *Journal of the American Chemical Society* 2003, 125:8007–8014.
245. Reblova K, Spackova N, Koca J, Leontis NB, Sponer J: Long-residency hydration, cation binding, and dynamics of loop E/helix IV rRNA A-L26 protin complex. *Biophysics Journal* 2004, 87:3397–3412.
246. Kurkal V, Daniel RM, Finney JL, Tehei M, Dunn RV, Smith JC: Low frequency enzyme dynamics as a function of temperature and hydration: a neutron scattering study. *Chemical Physics* 2005, 317:267–273.
247. Healey AD, Nam J: A theoretical analysis of substituted formamide conformers. *Journal of Molecular Structure* 2002, 589:423–429.
248. Sugarawa Y, Hamada Y, Hirakawa AY, Tsuboi M: Occurrence of the *cis*-peptide in gaseous *N*-methylformamide as revealed by rotational structure in its infrared spectrum. *Chemical Physics Letters* 1979, 67:186–188.
249. Suenram RD, Fraser GT, Lovas FJ, Gillies CW, Zozom J: Microwave-spectrum, structure, and electric-dipole moment of the Ar–formamide van der Waals complex. *Journal of Chemical Physics* 1988, 89:6141–6146.
250. Desfrancois C, Periquet V, Carles S, Schermann JP, Adamowicz L: Neutral and negatively-charged formamide, *N*-methylformamide and dimethylformamide clusters. *Chemical Physics* 1998, 239:475–483.
251. Maeyama T, Mikami N: Characteristic distributions of negatively charged *N*-monosubstituted amide clusters generated by electron attachment in supersonic expansions. *Physical Chemistry Chemical Physics* 2004, 6:2725–2731.
252. Maeyama T, Negishi Y, Tsukuda T, Yagi I, Mikami N: Electron localization in negatively charged formamide clusters studied by photodetachment spectroscopy. *Physical Chemistry Chemical Physics* 2006, 8:827–833.
253. Lucas B, Gregoire G, Lecomte F, Reimann B, Schermann JP, Desfrancois C: Infrared spectroscopy of mass-selected neutral molecular systems without chromophore: the formamide monomer and dimer. *Molecular Physics* 2005, 103:1497–1503.
254. Seydou M, Modelli A, Lucas B, Konate K, Desfrancois C, Schermann JP: Electron attachment to strongly polar clusters – formamide molecule and clusters. *European Physical Journal D* 2005, 35:199–205.
255. Han S, Pan H, Wang CL, Li H, Xu Z: The methyl C—H blueshift in *N,N*-dimethylormamide–water mixtures probed by two-dimensional Fourier-transform infrared spectroscopy. *Journal of Chemical Physics* 2006, 124:244502.
256. Hobza P, Sponer J: MP2 and CCSD(T) calculations on H-bonded and stacked formamide···formamide and formamidine···formamidine dimers. *Journal of Molecular Structure* 1996, 388: 115–120.
257. Skarmoutos I, Samios J: Molecular dynamics of *cis/trans N*-methylformamide liquid mixture using a new optimized all atom rigid force field. *Chemical Physics Letters* 2004, 384:108–113.
258. Cordeiro JMM, Cordeiro MAM, Bosso ARSA, Politi JRS: Theoretical studies on the stability of *N*-methylformamide in both liquid and gas phases. *Chemical Physics Letters* 2006, 423:67–70.
259. Aparicio-Fernandez S, Hall KR, Balbuena PB: Theoretical study on the properties of linear and cyclic amides in gas phase and water solution. *Journal of Physical Chemistry A* 2006, 110:9183–9193.
260. Fraser GT, Suenram RD, Lovas FJ: Electric-dipole moments of H_2O–formamide and CH_3OH–formamide. *Journal of Molecular Structure* 1988, 189:165–172.
261. Lecomte F, Lucas B, Gregoire G, Schermann JP, Desfrancois C: Structures of amide–water neutral complexes from dipole-bound anion formation. *European Physical Journal D* 2002, 20:449–457.

262. Bende A, Suhai S: BSSE-corrected geometry and harmonic and anharmonic vibrational frequencies of formamide–water and formamide–formamide dimers. *International Journal of Quantum Chemistry* 2005, 103:841–853.
263. Clarkson JR, Baquero E, Shubert VA, Myshakin EM, Jordan KD, Zwier TS: Laser-initiated shuttling of a water molecule between H-bonding sites. *Science* 2005, 307:1443–1446.
264. Zwier TS: Laser probes of conformational isomerization in flexible molecules and complexes. *Journal of Physical Chemistry A* 2006, 110:4133–4150.
265. Shin S, Kurawaki A, Hamada Y, Shinya K, Ohno K, Tohara A, Sato M: Conformational behavior of *N*-methylformamide in the gas, matrix, and solution states as revealed by IR and NMR spectroscopic measurements and by theoretical calculations. *Journal of Molecular Structure* 2006, 791:30–40.
266. Wu CC, Jiang JC, Hahndorf I, Chaudhuri C, Lee YT, Chang HC: Characterization of protonated formamide-containing clusters by infrared spectroscopy and *ab initio* calculations: I. O-protonation. *Journal of Physical Chemistry A* 2000, 104:9556–9565.
267. Benoit DM, Clary DC: Quantum simulation of phenol–water clusters. *Journal of Physical Chemistry A* 2000, 104:5590–5599.
268. Luchow A, Spangenberg D, Janzen C, Jansen A, Gerhards M, Kleinermanns K: Structure and energetics of phenol$(H_2O)_n$, $n \leq 7$: quantum Monte Carlo calculations and double resonance experiments. *Physical Chemistry Chemical Physics* 2001, 3:2771–2780.
269. Schumm S, Gerhards M, Kleinermanns K: Franck–Condon simulation of the S-1 $\rightarrow$ S-0 spectrum of phenol. *Journal of Physical Chemistry A* 2000, 104:10648–10655.
270. Tsui HHY, van Mourik T: *Ab initio* calculations on phenol–water. *Chemical Physics Letters* 2001, 350:565–572.
271. Weinkauf R, Schiedt J: Energetics of photoinduced electron transfer in the indole O_2 cluster in gas phase: possible consequences for photoexcited tryptophan in solution. *Photochemistry and Photobiology* 1997, 66:569–575.
272. Helm RM, Clara M, Grebner TL, Neusser HJ: Hydrogen bonding in the indole–water complex: a high resolution UV study of the hydrogen donor conformer. *Journal of Physical Chemistry A* 1999, 102:3268–3272.
273. van Mourik T, Price SL, Clary DC: *Ab initio* calculations on indole–water, 1-methylindole–water and indole–$(\text{water})_2$. *Chemical Physics Letters* 2000, 331:253–261.
274. Blanco S, Lopez JC, Alonso JL, Ottaviani P, Caminati W: Pure rotational spectrum and model calculations of indole–water. *Journal of Chemical Physics* 2003, 119:880–886.
275. Carles S, Desfrancois C, Schermann JP, Smith DMA, Adamowicz L: Structures and electron affinities of indole–$(\text{water})_N$ clusters. *Journal of Chemical Physics* 2000, 112:3726–3734.
276. Carney JR, Dian BC, Florio GM, Zwier TS: The role of water bridges in directing the conformational preferences of 3-indole–propionic acid and tryptamine. *Journal of the American Chemical Society* 2001, 123:5596–5597.
277. Carney JR, Zwier TS: Infrared and ultraviolet spectroscopy of water-containing clusters of indole, 1-methylindole, and 3-methylindole. *Journal of Physical Chemistry A* 1999, 103: 9943–9957.
278. Dian BC, Longarte A, Zwier TS: Hydride stretch infrared spectra in the excited electronic states of indole and its derivatives: direct evidence for the $^1\pi\sigma^*$ state. *Journal of Chemical Physics* 2003, 118:2696–2706.
279. Georgiev S, Neusser HJ: Mass analyzed threshold ionization of hydrogen bonded clusters of biological molecules: the 3-methylindole C_6H_6 complex. *Journal of Electron Spectroscopy and Related Phenomena* 2005, 142:207–213.
280. Mons M, Dimicoli I, Tardivel B, Piuzzi F, Brenner V, Millie P: Site dependence of the binding energy of water to indole: microscopic approach to the side chain hydration of tryptophan. *Journal of Physical Chemistry A* 1999, 103:9958–9965.

281. Nguyen TV, Yi JT, Pratt DW: On the energy landscapes of 3-indole acetic acid and 3-indole propionic acid. A study of side chain flexibilities in their S_0 and S_1 electronic states. *Physical Chemistry Chemical Physics* 2006, 8:1049–1056.
282. Carles S, Lecomte F, Schermann JP, Desfrancois C: Gas-phase experimental and theoretical studies of adenine, imidazole, pyrrole, and water non-covalent complexes. *Journal of Physical Chemistry A* 2000, 104:10662–10668.
283. Hockridge MR, Robertson EG, Simons JP: Ion-dip spectroscopy and structure of 5-phenyl imidazole, 4-phenyl imidazole and its 1:1 hydrate. *Chemical Physics Letters* 1999, 302:538–548.
284. Hockridge MR, Robertson EG, Simons JP, Borst DR, Korter TM, Pratt DW: The $S_1 \leftarrow S_0$ electronic transitions of 4- and 5-phenyl imidazole in the gas phase. *Chemical Physics Letters* 2001, 334:31–38.
285. Jockusch RA, Williams ER: Binding energies of proton-bound dimers of imidazole and *n*-acetylalanine methyl ester obtained by blackbody infrared radiative dissociation. *Journal of Physical Chemistry A* 1998, 102:4543–4550.
286. Rannulu NS, Amunugama R, Yang ZB, Rodgers MT: Influence of s and d orbital occupation on the binding of metal ions to imidazole. *Journal of Physical Chemistry A* 2004, 108:6385–6396.
287. Rannulu NS, Rodgers MT: Solvation of copper ions by imidazole: structures and sequential binding energies of $Cu^+(\text{imidazole})_x$, $x = 1$–4. Competition between ion solvation and hydrogen bonding. *Physical Chemistry Chemical Physics* 2005, 7:1014–1025.
288. Talbot FO, Simons JP: Infrared ion dip and ultraviolet spectroscopy imidazole, its tautomer, 5-phenyl imidazole, and its multiply hydrated clusters. *European Physical Journal D* 2002, 20:389–398.
289. Alonso JL, Cocinero EJ, Lessari A, Sanz L, Lopez JC: The glycine–water complex. *Angewandte Chemie-International Edition* 2006, 45:3471–3475.
290. Miller TF, Clary DC, Meijer A: Collision-induced conformational changes in glycine. *Journal of Chemical Physics* 2005, 122:244323.1–244323.13.
291. Miller TF, Clary DC: Quantum free energies of the conformer of glycine on an *ab initio* potential energy surface. *Physical Chemistry Chemical Physics* 2004, 6:2563–2571.
292. Shemesh D, Gerber RB: Different chemical dynamics for different conformers of biological molecules: photoionization of glycine. *Journal of Chemical Physics* 2005, 122:241104.1–241104.4.
293. Zhang L, Kao YT, Qiu W, Wang L, Zhong D: Femtosecond studies of tryptophan fluorescence dynamics in proteins: local solvation and electronic quenching. *Journal of Physical Chemistry B* 2006, 110:18097–18103.
294. Gao F, Bren N, Burghardt TP, Hansen S, Henchman RH, Taylor P, McCammon JA, Sine SM: Agonist-mediated conformational changes in acetylcholine-binding protein revealed by simulation and intrinsic tryptophan fluorescence. *Journal of Biological Chemistry* 2005, 280:8443–8451.
295. Peteanu LA, Levy DH: Spectroscopy of complexes of tryptamine and 3-indole propionic acid with various solvents. *Journal of Physical Chemistry* 1988, 92:6554–6561.
296. Snoek LC, Kroemer RT, Simons JP: A spectroscopic and computational exploration of tryptophan–water cluster structures in the gas phase. *Physical Chemistry Chemical Physics* 2002, 4:2130–2139.
297. Carcabal P, Kroemer RT, Snoek LC, Simons JP, Bakker JM, Compagnon I, Meijer G, von Helden G: Hydrated complexes of tryptophan: ion dip infrared spectroscopy in the 'molecular fingerprint' region, 100–2000 Cm^{-1}. *Physical Chemistry Chemical Physics* 2004, 6: 4546–4552.
298. Boyarkin OV, Mercier SR, Kamariotis A, Rizzo TR: Electronic spectroscopy of cold, protonated tryptophan and tyrosine. *Journal of the American Chemical Society* 2006, 128:2816–2817.
299. Lee KT, Sung J, Lee KJ, Kim SK, Park YD: Resonant two-photon ionization study of jet-cooled amino acid: L-phenylalanine and its monohydrated complex. *Journal of Chemical Physics* 2002, 116:8251–8254.

300. Ebata T, Hashimoto T, Ito T, Inokuchi Y, Altunsu F, Brutschy B, Tarakeshwar P: Hydration profiles of aromatic amino acids: conformations and vibrations of L-phenylalanine–$(H_2O)_n$ clusters. *Physical Chemistry Chemical Physics* 2006, 8:4783–4791.

301. Kamariotis A, Boyarkin OV, Mercier SR, Beck RD, Bush MF, Williams ER, Rizzo TR: Infrared spectroscopy of hydrated amino acids in the gas phase: protonated and lithiated valine. *Journal of the American Chemical Society* 2006, 128:905–916.

302. Jockusch RA, Lemoff AS, Williams ER: Effect of metal ion and water coordination on the structure of a gas-phase amino acid. *Journal of the American Chemical Society* 2001, 123:12255–12265.

303. Lemoff AS, Bush MF, Wu CC, Williams ER: Structures and hydration enthalpies of cationized glutamine and structural analogues in the gas phase. *Journal of the American Chemical Society* 2005, 127:10276–10286.

304. Lemoff AS, Wu CC, Bush MF, Williams ER: Binding energies of water to doubly hydrated cationized glutamine and structural analogues in the gas phase. *Journal of Physical Chemistry A* 2006, 110:3662–3669.

305. Kang H, Dedonder-Lardeux C, Jouvet C, Martrenchard S, Gregoire G, Desfrancois C, Schermann JP, Barat M, Fayeton JA: Photo-induced dissociation of protonated tryptophan $TrpH^+$: a direct dissociation channel in the excited states controls the hydrogen atom loss. *Physical Chemistry Chemical Physics* 2004, 6:2628–2632.

306. Garcia-Viloca M, Gao J, Karplus M, Truhlar DG: How enzymes work: analysis by modern theory and computer simulations. *Science* 2004, 303:186–195.

307. Carloni P, Rothlisberger U, Parrinello M: The role and perspective of *ab initio* molecular dynamics in the study of biological systems. *Accounts of Chemical Research* 2002, 35:455–464.

308. Carloni P: Density functional theory-based molecular dynamics of biological systems. *Quantitative Structure–Activity Relationships* 2002, 21:166–172.

309. Messer BM, Cappa CD, Smith JD, Wilson KR, Gilles MK, Cohen RC, Saykally RJ: pH dependence of the electronic structure of glycine. *Journal of Physical Chemistry B* 2005, 109:5375–5382.

310. Fischer G, Jacob R, Cao X: The infrared spectra and structures of the valyl–glycine zwitterion isolated in a KBr matrix. *Chemical Physics* 2001, 263:243–253.

311. Yu D, Armstrong DA, Rauk A: Hydrogen bonding and internal rotation barriers of glycine and its zwitterions (hypothetical) in the gas-phase. *Canadian Journal of Chemistry-Revue Canadienne de Chimie* 1992, 70:1762–1772.

312. Tajkhorshid E, Jalkanen KJ, Suhai S: Structure and vibrational spectra of the zwitterion L-alanine in the presence of explicit water molecules: a density functional analysis. *Journal of Physical Chemistry B* 1998, 102:5899–5913.

313. Yamabe S, Ono N, Tsuchida N: Molecular interactions between glycine and H_2O affording the zwitterion. *Journal of Physical Chemistry A* 2003, 107:7915–7922.

314. Wang WZ, Pu XM, Zheng WX, Wong NB, Tian AM: Some theoretical observations on the 1:1 glycine zwitterion–water complex. *Journal of Molecular Structure THEOCHEM* 2003, 626:127–132.

315. Park SW, Ahn DS, Lee S: Dynamic paths between neutral alanine–water and zwitterionic alanine–water clusters: single, double and triple proton transfer. *Chemical Physics Letters* 2003, 371:74–79.

316. Kirschner KN, Lewin AH, Bowen JP: Molecular mechanics force-field development for amino acid zwitterions. *Journal of Computational Chemistry* 2003, 24:111–128.

317. Jeon IS, Ahn DS, Park SW, Lee S, Kim B: Structures and isomerization of neutral and zwitterion serine–water clusters: computational study. *International Journal of Quantum Chemistry* 2005, 101:55–66.

318. Ahn DS, Park SW, Jeon IS, Lee MK, Kim NH, Han YH, Lee S: Effects of microsolvation on the structures and reactions of neutral and zwitterion alanine: computational study. *Journal of Physical Chemistry B* 2003, 107:14109–14118.

319. Marino T, Toscano M, Russo N, Grand A: Structural and electronic characterization of the complexes obtained by the interaction between bare and hydrated first-row transition-metal ions (Mn_2^+, Fe_2^+, Co_2^+, Ni_2^+, Cu_2^+, Zn_2^+) and glycine. *Journal of Physical Chemistry B* 2006, 110:24666–24673.
320. Shemesh D, Gerber RB: Classical trajectory simulations of photoionization dynamics of tryptophan: intramolecular energy flow, hydrogen-transfer processes and conformational transitions. *Journal of Physical Chemistry A* 2006, 110:8401–8408.
321. Rozman M, Srzic D, Klasinc L: Gas-phase interaction of protonated lysine with water. *International Journal of Mass Spectrometry* 2006, 253:201–206.
322. Bush MF, O'Brien JT, Prell JS, Saykally RJ, Williams ER: Infrared spectroscopy of cationized arginine in the gas-phase: direct evidence for the transition from nonzwitterionic to zwitterionic structure. *Journal of the American Chemical Society* 2007, 129:1612–1662.
323. Baccarelli I, Grandi A, Gianturco FA, Lucchese RR, Sanna N: Electron-attachment resonances of glycine zwitterions from quantum scattering calculations: modeling macrosolvation effects. *Journal of Physical Chemistry B* 2006, 110:26240–26247.
324. Kassab E, Langlet J, Evleth E, Akacem Y: Theoretical study of solvent effect on intramolecular proton transfer of glycine. *Journal of Molecular Structure THEOCHEM* 2000, 531:267–282.
325. Lemoff AS, Bush MF, O'Brien JT, Williams ER: Structures of lithiated and structural analogues in the gas phase: effects of water and proton affinity on zwitterionic stability. *Journal of Physical Chemistry A* 2006, 110:8433–8442.
326. Price WD, Jockusch RA, Williams ER: Is arginine a zwitterion in the gas phase? *Journal of the American Chemical Society* 1997, 119:11988–11989.
327. Zheng WJ, Xu SJ, Radisic D, Stokes S, Li X, Bowen KH: On the interaction of electrons with betaine zwitterions. *Journal of Chemical Physics* 2005, 122:101103.1–101103.3.
328. Skurski P, Gutowski M, Barrios R, Simons J: Non-ionic and zwitterionic forms of neutral arginine – an *ab initio* study. *Chemical Physics Letters* 2001, 337:143–150.
329. Julian RR, Beauchamp JL, Goddard WA: Cooperative salt bridge stabilization of gas-phase zwitterions in neutral arginine clusters. *Journal of Physical Chemistry A* 2002, 106:32–34.
330. Chapo CJ, Paul JB, Provencal RA, Roth K, Saykally RJ: Is arginine zwitterionic or neutral in the gas phase? Results from IR cavity ringdown spectroscopy. *Journal of the American Chemical Society* 1998, 120:12956–12957.
331. Ling L, Yu W, Huang Z, Lin Z, Haranczyk M, Gutowski M: Gaseous arginine conformers and their unique intramolecular interactions. *Journal of Physical Chemistry A* 2006, 110:12282–12291.
332. Julian RR, Hodyss R, Beauchamp JL: Salt bridge stabilization of charged zwitterionic arginine aggregates in the gas phase. *Journal of the American Chemical Society* 2001, 123:3577–3583.
333. Wyttenbach T, Witt M, Bowers MT: On the stability of amino acid zwitterions in the gas phase: the influence of derivatization, proton affinity, and alkali ion addition. *Journal of the American Chemical Society* 2000, 122:3458–3464.
334. Strittmatter EF, Lemoff AS, Williams ER: Structure of cationized glycine, $Gly{\cdot}M_2^+$($M = Be$, Mg, Ca, Sr, Ba), in the gas phase: intrinsic effect of cation size on zwitterion stability. *Journal of Physical Chemistry A* 2000, 104:9793–9796.
335. Rogalewicz F, Ohanessian G, Gresh N: Interaction of neutral and zwitterionic glycine with Zn_2^+in gas phase: *ab initio* and SIBFA molecular mechanics calculations. *Journal of Computational Chemistry* 2000, 21:963–973.
336. Kapota C, Lemaire J, Maitre P, Ohanessian G: Vibrational signature of charge solvation vs salt bridge isomers of sodiated amino acids in the gas phase. *Journal of the American Chemical Society* 2004, 126:1836–1842.
337. Ai HQ, Bu YX, Li P, Zhang C: The regulatory roles of metal ions ($M^{+/2+} = Li^+$, Na^+, K^+, Be^{2+}, Mg^{2+}, and Ca^{2+}) and water molecules in stabilizing the zwitterionic form of glycine derivatives. *New Journal of Chemistry* 2005, 29:1540–1548.

338. Lemoff AS, Bush MF, Williams ER: Structures of cationized proline analogues: evidence for the zwitterionic form. *Journal of Physical Chemistry A* 2005, 109:1903–1910.
339. Lemoff AS, Bush MF, Williams ER: Binding energies of water to sodiated valine and structural isomers in the gas phase: the effect of proton affinity on zwitterion stability. *Journal of the American Chemical Society* 2003, 125:13576–13584.
340. Remko M, Rode BM: Effect of metal ions (Li^+, Na^+, K^+, Ca^+, Ni^+, Cu^+, and Zn^+) and water coordination on the structure of glycine and zwitterionic glycine. *Journal of Physical Chemistry A* 2006, 110:1960–1967.
341. Wong RL, Williams ER, Counterman AE, Clemmer DE: Evaluation of ion mobility spectroscopy for determining charge-solvated versus salt-bridge structures of protonated trimers. *Journal of the American Society for Mass Spectrometry* 2005, 16:1009–1019.
342. Sang-aroon W, Ruangpomvisuti V: Conformational study of cationic, zwitterionic, anionic species of aspartic acid, water-added forms and their protonation. A DFT method. *Journal of Molecular Structure THEOCHEM* 2006, 181–187.
343. Ahn S, Cong X, Lebrilla CB: Zwitterion formation in gas-phase cyclodextrin complexes. *Journal of the American Society for Mass Spectrometry* 2005, 16:166–175.
344. Xu SJ, Zheng WJ, Radisic D, Bowen KH: The stabilization of arginine's zwitterion by dipole-binding of an excess electron. *Journal of Chemical Physics* 2005, 122:091103.1–091103.3.
345. Kass SR: Zwitterion–dianion complexes and anion–anion clusters with negative dissociation energies. *Journal of the American Chemical Society* 2005, 127:13098–13099.
346. Gutowski M, Skurski P, Simons J: Dipole-bound anions of glycine based on the zwitterion and neutral structures. *Journal of the American Chemical Society* 2000, 122:10159–10162.
347. Tian SX: Dipole-bound anions of β-alanine: canonical and zwitterionic conformers. *Journal of Physical Chemistry A* 2006, 110:4256–4259.
348. Gidden J, Bowers MT: Gas-phase conformations of deprotonated trinucleotides ($dGTT^-$, $dTGT^-$, and $dTTG^-$): the question of zwitterion formation. *Journal of the American Society for Mass Spectrometry* 2003, 14:161–170.
349. Dunbar RC: BIRD (blackbody infrared radiative dissociation): evolution, principles, and applications. *Mass Spectrometry Reviews* 2004, 23:127–158.
350. Rozman M, Bertosa B, Klasinc L, Srzic D: Gas-phase H/D exchange of sodiated aminoacids: why do we see zwitterions? *Journal of the American Society for Mass Spectrometry* 2006, 17:29–36; Polfer NC, Dunbar RC, Oomens J. Observation of zwitterion formation in the gas-phase H/D exchange with CH_3OD: solution-phase structures in the gas phase. *Journal of the American Society for Mass Spectrometry* 2007, 18:512–516.
351. Jockusch RA, Lemoff AS, Williams ER: Hydration of valine–cation complexes in the gas phase: on the number of water molecules necessary to form a zwitterion. *Journal of Physical Chemistry A* 2001, 105:10929–10942.
352. Wyttenbach T, Witt M, Bowers MT: On the question of salt bridges of cationized amino acids in the gas phase: glycine and arginine. *International Journal of Mass Spectrometry* 1999, 183:243–252.
353. Antoine R, Broyer M, Dugourd P, Breaux G, Hagemeister FC, Pippen D, Hudgins RR, Jarrold MF: Direct probing of zwitterion formation in unsolvated peptides. *Journal of the American Chemical Society* 2003, 125:8996–8997.
354. Cong X, Czerwieniec G, McJimpsey E, Ahn SH, Troy FA, Lebrilla CB: Structural relationships in small molecule interactions governing gas-phase enantioselectivity and zwitterionic formation. *Journal of the American Society for Mass Spectrometry* 2006, 17:442–452.
355. Murray JS, Peralta-Inga Z, Politzer P: Computed molecular surface electrostatic potentials of the nonionic and zwitterionic forms of glycine, histidine, and tetracycline. *International Journal of Quantum Chemistry* 2000, 80:1216–1223.

356. Pecul M: Conformational structures and optical rotation of serine and cysteine. *Chemical Physics Letters* 2006, 418:1–10.
357. Rak J, Skurski P, Simons J, Gutowski M: Low-energy tautomers and conformers of neutral and protonated arginine. *Journal of the American Chemical Society* 2001, 123:11695–11707.
358. Diken EG, Hammer NI, Johnson MA: Preparation and photoelectron spectrum of the glycine molecular anion: assignment to a dipole-bound electron species with a high-dipole moment, non-zwitterionic form of the neutral core. *Journal of Chemical Physics* 2004, 120:9899–9902.
359. Diken EG, Headrick JM, Johnson MA: Photoelectron spectroscopy of the glycine $(H_2O)_{1,2}^-$ clusters: sequential hydration shifts and observation of isomers. *Journal of Chemical Physics* 2005, 122:224317.1–224317.6.
360. Xu SJ, Nilles M, Bowen KH: Zwitterion formation in hydrated amino acid, dipole bound anions: how many water molecules are required? *Journal of Chemical Physics* 2003, 119:10696–10701.
361. Ahn DS, Kang AR, Lee S, Kim B, Kim SK, Neuhauser D: On the stability of glycine–water clusters with excess electron: implications for photoelectron spectroscopy. *Journal of Chemical Physics* 2005, 122:084310.1–084310.10.
362. Park SW, Im S, Lee S, Desfrancois C: Structure and stability of glycine–$(H_2O)_3$ cluster and anion: zwitterion vs. canonical glycine. *International Journal of Quantum Chemistry* 2007, 107:1316–1327.
363. Hoyau S, Ohanessian G: Interaction of alkali (Li^+–Ca^+) with glycine in the gas-phase. A theoretical study. *Chemical European Journal* 1998, 4:1561–1569.
364. Callis PR, Liu TQ: Quantitative prediction of fluorescence quantum yields for tryptophan in proteins. *Journal of Physical Chemistry B* 2004, 108:4248–4259.
365. Mercier S, Boyarkin OV, Kamariotis A, Guglielmi M, Tavernelli I, Cascella M, Rothlisberger U, Rizzo TR: Microsolvation effects on excited state dynamics of protonated tryptophan. *Journal of the American Chemical Society* 2006, 128:16938–16943.
366. Grégoire G, Jouvet C, Dedonder C, Sobolewski A: *Ab initio* study of the excited state deactivation pathways of protonated tryptophan and tyrosine. *Journal of the American Chemical Society* 2007, in press.
367. Zhang Z, Bordas-Nagy J: Peptide conformation in gas phase probed by collision-induced dissociation and its correlation to conformation in condensed phases. *Journal of the American Society of Mass Spectrometry* 2006, 17:786–794.
368. Maynard AJ, Sharman GJ, Searle MS: Origin of β-hairpin stability in solution: structural and thermodynamics analysis of the folding of a model peptide supports hydrophobic stabilization in water. *Journal of the American Chemical Society* 1998, 120:1996–2007.
369. Mayer B, Klein CT: Influence of solvation on the helix-forming tendency of non polar amino acids. *Journal of Molecular Structure THEOCHEM* 2000, 532:213–236.
370. Bellissent-Funel MC: Hydrophilic–hydrophobic interplay: from model systems to living systems. *Comptes Rendus Geoscience* 2005, 337:173–179.
371. Fernandez A, Scheraga HA: Insufficiently dehydrated hydrogen bonds as determinants of protein interactions. *Proceedings of the National Academy of Sciences of the United States of America* 2003, 100:113–118.
372. Degrève L, Brancaleoni GH, Fuzo L, MR, Mazzé FM, Namba AM, Vieira DS: On the role of water in the protein activity. *Brazilian Journal of Physics* 2004, 34:102–114.
373. Kitova EN, Bundle DR, Klassen JS: Partitioning of solvent effects and intrinsic interactions in biological recognition. *Angewandte Chemie-International Edition* 2004, 43:4183–4186.
374. Yosomiko T, Nakasako M, Yamasaki T, Shindo H, Higo J: Hydrogen-bond patterns in the hydration structure of a protein. *Chemical Physics Letters* 2005, 401:332–336.
375. Barran PE, Polfer NC, Campopiano DJ, Clarke DJ, Langridge-Smith PRR, Langley RJ, Govan JRW, Maxwell A, Dorin JR, Millar RP, *et al.*: Is it biologically relevant to measure the

structures of small peptides in the gas-phase? *International Journal of Mass Spectrometry* 2005, 240:273–284.

376. Raschke TM: Water structure and interactions with proteins surfaces. *Current Opinion in Structural Biology* 2006, 16:152–159.
377. Kumar S, Rai AK, Singh VB, Rai SB: Vibrational spectrum of glycine molecule. *Spectrochimica Acta Part A – Molecular and Biomolecular Spectroscopy* 2005, 61:2741–2746.
378. Kumar S, Rai AK, Rai SB, Rai DK, Singh AN, Sing VB: Infrared, Raman, and electronic spectra of analine: a comparison with *ab initio* calculations. *Journal of Molecular Structure THEOCHEM* 2006, 791:23–29.
379. Gordon MS, Freitag MA, Bandyopadhyay P, Jensen JH, Kairys V, Stevens WJ: The effective fragment potential method: a QM-based MM approach to modeling environmental effects in chemistry. *Journal of Physical Chemistry A* 2001, 105:293–307.
380. Nemukhin AV, Grigorenko BL, Bochenkova AV, Topol IA, Burt SK: A QM/MM approach with effective fragment potentials applied to the dipeptide–water structures. *Journal of Molecular Structure THEOCHEM* 2002, 581:167–175.
381. Iavarone AT, Jurchen JC, Williams ER: Effects of solvent on the maximum charge state and charge state distribution of protein ions produced by electrospray ionization. *Journal of the American Society for Mass Spectrometry* 2000, 11:976–985.
382. Rodriguez-Cruz SE, Klassen JS, Williams ER: Hydration of gas-phase ions formed by electrospray ionization. *Journal of the American Society for Mass Spectrometry* 1999, 10: 958–968.
383. Huang F, Hudgins RR, Nau WM: Primary and secondary structure dependence of peptide flexibility assessed by fluorescence-based measurement of end-to-end collision rates. *Journal of the American Chemical Society* 2004, 126:16665–16675.
384. Hudgins RR, Woenckhaus J, Jarrold MF: High resolution ion mobility measurements for gas phase proteins: correlation between solution phase and gas phase conformations. *International Journal of Mass Spectrometry* 1997, 165:497–507.
385. Woenckhaus J, Hudgins RR, Jarrold MF: Hydration of gas-phase proteins: a special hydration site on gas-phase BPTI. *Journal of the American Chemical Society* 1997, 119:9586–9587.
386. Li JW, Taraszka JA, Counterman AE, Clemmer DE: Influence of solvent composition and capillary temperature on the conformations of electrosprayed ions: unfolding of compact ubiquitin conformers from pseudonative and denatured solutions. *International Journal of Mass Spectrometry* 1999, 187:37–47.
387. Jarrold MF: Unfolding, refolding, and hydrations of proteins in the gas phase. *Accounts of Chemical Research* 1999, 32:360–367.
388. Mao Y, Ratner MA, Jarrold MF: One water molecule stiffens a protein. *Journal of the American Chemical Society* 2000, 122:2950–2951.
389. Kohtani M, Jarrold MF: The initial steps in the hydration of unsolvated peptides: water molecule adsorption on alanine-based helices and globules. *Journal of the American Chemical Society* 2002, 124:11148–11158.
390. Liu DF, Wyttenbach T, Barran PE, Bowers MT: Sequential hydration of small protonated peptides. *Journal of the American Chemical Society* 2003, 125:8458–8464.
391. Liu DF, Wyttenbach T, Bowers MT: Hydration of protonated primary amines: effects of intermolecular and intramolecular hydrogen bonds. *International Journal of Mass Spectrometry* 2004, 236:81–90.
392. Kohtani M, Jarrold MF: Water molecule adsorption on short alanine peptides: how short is the shortest gas-phase alanine-based helix? *Journal of the American Chemical Society* 2004, 126:8454–8458.
393. Russo D, Hura G, Head-Gordon T: Hydration dynamics near a model protein surface. *Biophysical Journal* 2004, 86:1852–1862.

394. Acevedo O, Jorgensen WL: Solvent effects and mechanism for a nucleophilic aromatic substitution from QM/MM simulations. *Organic Letters* 2004, 6:2881–2884.
395. Wang F, Freitas MA, Marshall AG, Sykes BD: Gas-phase memory of solution-phase protein conformation: H/D exchange and Fourier transform ion cyclotron resonance mass spectrometry of the N-terminal domain of cardiac troponin C. *International Journal of Mass Spectrometry* 1999, 192:319–325.
396. Nazabal A, Bonneu M, Saupe SJ, Schmitter JM: High-resolution H/D exchange studies on the HET-s218^{-295} prion protein. *Journal of Mass Spectrometry* 2005, 40:580–590.
397. Xiao H, Hoerner JK, Eyles SJ, Dobo A, Voigtman E, Melcuk AJ, Kaltashov IA: Mapping protein energy landscapes with amide hydrogen exchange and mass spectrometry: I. A generalized model for a two-state protein and comparison with experiment. *Protein Science* 2005, 14:543–557.
398. Kaltashov IA, Eyles SJ: *Mass spectrometry in biophysics*: Wiley, Hoboken, NJ, 2005.
399. Konnermann L, Simons DA: Protein-folding kinetics and mechanisms studied by pulse-labeling and mass spectrometry. *Mass Spectrometry Reviews* 2003, 22:1–26.
400. Jorgensen TJD, Gardsvoll H, Ploug M, Roepstorff P: Intramolecular migration of amide hydrogens in protonated peptides upon collisional activation. *Journal of the American Chemical Society* 2005, 127:2785–2793.
401. Simons JP, Jockusch RA, Carcabal P, Hung I, Kroemer RT, Macleod NA, Snoek LC: Sugars in the gas phase. Spectroscopy, conformation, hydration, co-operativity and selectivity. *International Reviews in Physical Chemistry* 2005, 24:489–531.
402. Hunig I, Painter AJ, Jockusch RA, Carcabal P, Marzluff EM, Snoek LC, Gamblin DP, Davis BG, Simons JP: Adding water to sugar: a spectroscopic and computational study of α- and β-phenylxyloside in the gas phase. *Physical Chemistry Chemical Physics* 2005, 7:2474–2480.
403. Behrends R, Kaatze U: Molecular dynamics and conformational kinetics of mono- and disaccharides in aqueous solution. *ChemPhysChem* 2005, 6:1133–1145.
404. Dashnau JL, Sharp KA, Vanderkooi JM: Carbohydrate intramolecular hydrogen bonding cooperativity and its effect on water structure. *Journal of Physical Chemistry B* 2005, 109:24152–24159.
405. Almond A: Towards understanding the interaction between oligosaccharides and water molecules. *Carbohydrate Research* 2005, 340:907–920.
406. Polacek R, Stenger J, Kaatze U: Chair–chair conformational flexibility, pseudorotation, and exocyclic group isomerization of monosaccharides in water. *Journal of Chemical Physics* 2002, 116:2976–2982.
407. Fernandez-Alonso MC, Canada FJ, Jimenez-Barbero J, Cuevas G: Molecular recognition of saccharides by proteins. Insights on the origin of the carbohydrate–aromatic interactions. *Journal of the American Chemical Society* 2005, 127:7379–7386.
408. Aviles Moreno JR, Demaison J, Huet TR: Conformational flexibility in hydrated sugars: the glycolaldehyde–water complex. *Journal of the American Chemical Society* 2006, 128:10467–10473.
409. Carcabal P, Patsias T, Hunig I, Liu B, Kaposta C, Snoek LC, Gamblin DP, Davis BG, Simons JP: Spectral signatures and structural motifs in isolated and hydrated monosaccharides: phenyl α- and β-L-fucopyranoside. *Physical Chemistry Chemical Physics* 2006, 8:129–136.
410. Araujo-Andrade C, Ruiz F, Martinez-Mendoza JR, Torrones H: Infrared and Raman spectra, conformational stability, *ab initio* calculations of structure, and vibrational assignment of a and b glucose. *Journal of Molecular Structure THEOCHEM* 2005, 714:143–146.
411. Carcabal P, Jockusch RA, Hunig I, Snoek LC, Kroemer RT, Davis BG, Gamblin DP, Compagnon I, Oomens J, Simons JP: Hydrogen bonding and cooperativity in isolated and hydrated sugars: mannose, galactose, glucose, and lactose. *Journal of the American Chemical Society* 2005, 127:11414–11425.
412. Butz P, Kroemer RT, Macleod NA, Simons JP: Hydration of neurotransmitters: a spectroscopic and computational study of ephedrine and its diastereoisomer pseudoephedrine. *Physical Chemistry Chemical Physics* 2002, 4:3566–3574.

413. Brause R, Fricke H, Gerhards M, Weinkauf R, Kleinermanns K: Double resonance spectroscopy of different conformers of the neurotransmitter amphetamine and its clusters with water. *Chemical Physics* 2006, 327:43–53.
414. Florio GM, Zwier TS: Solvation of a flexible biomolecule in the gas phase: the ultraviolet and infrared spectroscopy of melatonin–water clusters. *Journal of Physical Chemistry A* 2003, 107:974–983.
415. Schmitt M, Böhm M, Ratzer C, Vu C, Kalkman I, Meerts WL: Structural selection by microsolvation: conformational locking of tryptamine. *Journal of the American Chemical Society* 2005, 127:10356–10364.
416. Munoz-Caro C, Nino A, Mora M, Reyes S, Melendez FJ, Castro ME: Conformational population distribution of acetylcholine, nicotine and muscarine in vacuum and solution. *Journal of Molecular Structure THEOCHEM* 2005, 726:115–124.
417. Lorenzini ML, Bruno-Blanch L, Estin GL: Structural and electronic factors associated with the activity in the GABA-A system. *Journal of Molecular Structure THEOCHEM* 1998, 454:1–16.
418. Nagy PI, Volgyi G, Takacs-Novak K: Tautomeric and conformational equilibria of tyramine and dopamine in aqueous solution. *Molecular Physics* 2005, 103:1589–1601.
419. Miller TF, Clary DC: Predicting conformations of biomolecules: application to a noradrenaline analogue. *Journal of Physical Chemistry B* 2004, 108:2484–2488.
420. Macleod NA, Robertson EG, Simons JP: Hydration of neurotransmitters: a computational and spectroscopic study of a noradrenaline analogue, 2-amino-1-phenyl-ethanol. *Molecular Physics* 2003, 101:2199–2210.
421. Tsaprailis G, Nair H, Somogyi A, Wysocki VH, Zhong WQ, Futrell JH, Summerfield SG, Gaskell SJ: Influence of secondary structure on the fragmentation of protonated peptides. *Journal of the American Chemical Society* 1999, 121:5142–5154.
422. Ishikawa K, Nakamura T, Koga Y: Cross-checking of electrospray ionization mass spectrometry and computer simulation for the evaluation of the interaction strength of non-covalently bound enkephalins in solution. *Journal of Mass Spectrometry* 2001, 36:937–942.
423. Oh HB, Breuker K, Sze SK, Carpenter BK, McLafferty FW: Secondary and tertiary structures of gaseous protein ions characterized by electron capture dissociation mass spectrometry and photofragment spectroscopy. *Proceedings of the National Academy of Sciences of the United States of America* 2002, 99:15863–15868.
424. Adams CM, Kjeldsen F, Zubarev RA, Budnik BA, Haselmann KF: Electron capture dissociation distinguishes a single D-amino acid in a protein and probes the tertiary structure. *Journal of the American Society for Mass Spectrometry* 2004, 15:1087–1098.
425. Antoine R, Broyer M, Chamot-Rooke J, Dedonder C, Desfrancois C, Dugourd P, Gregoire G, Jouvet C, Onidas D, Poulain P, *et al.*: Comparison of the fragmentation pattern induced by collisions, laser excitation and electron capture. Influence of the initial excitation. *Rapid Communications in Mass Spectrometry* 2006, 20:1648–1652.
426. Woods AS: The mighty arginine, the stable quaternary amines, the powerful aromatics and the aggressive phosphate: their role in the noncovalent minuet. *Journal of Proteome Research* 2004, 3:478–484.
427. Lukas RJ, Ke L, Bencherif M, Eisenhour C: Regulation by nicotine of its on receptor. *Drug Development Research* 1996, 38:136–148.
428. Graton J, Berthelot M, Gal JF, Girard S, Laurence C, Lebreton J, Le Questel JY, Maria PC, Naus P: Site of protonation of nicotine and nornicotine in the gas phase: pyridine or pyrrolidine nitrogen? *Journal of the American Chemical Society* 2002, 124:10552–10562.
429. Seydou M, Grégoire G, Schermann JP, Desfrancois C: Infrared spectra of isolated and aqueous acetylcholine, muscarine and nicotine, to be published.

Conclusion

The focus of this book has been a survey of fundamental studies of biomolecular systems under isolated conditions with only glimpses in the direction of applications. The last few years have witnessed large changes that bring disciplines such as mass-spectrometry, modelling and spectroscopy in close contact. At the end of the 1990s, the first spectroscopic studies of mass-selected species interpreted by means of high-level quantum calculations were restricted to very elementary species such as isolated amino acids or isolated nucleobases. In only few years, this range has been enlarged up to pentapeptides [1] or even further [2] and oligosaccharides [3]. Infrared spectroscopy of isolated species has been extended up to large peptides and even proteins [4]. However, interpretation of experiments concerning large systems is still limited by the lack of accurate force-fields built for the interpretation of gas-phase experiments and by the poor experimental resolution in the case of such systems. Experimental temperatures now range from few degrees Kelvin for both neutral [5] and ionic [6] species up to the more biologically relevant temperature of 300°K. This allows the in-depth local exploration of potential energy surfaces in the vicinity of local minima [7, 8] and quantum dynamical methods provide guiding in exploration of free energy surfaces [9, 10]. Time-resolved studies are now conducted in the femtosecond regime with the support of excited-state quantum calculations. Setting biomolecules in gas-phase is routine with the workhorses of mass-spectrometry, that is electrospray (ESI) and matrix-assisted laser desorption/ionization (MALDI) developed well before most spectroscopic studies of small biomolecular systems. Nevertheless, there has been a need for new methods dealing with neutral or very cold ionic species. Some recent improvements have even led to new methods such as ejection from liquids exceeding some of the limits of ESI and MALDI in the case of membrane proteins (e.g. reference [11]).

We have seen that the range of sizes corresponding to studies of systems totally isolated or in presence of a few water molecules in precisely known sites now extend up to very large systems. Each size range corresponds to different experimental and theoretical approaches but bridges are being established (Figure C1). For example, ion mobility tubes are now coupled to mass-spectrometers for pre-selection of isomers of ionic species before high-resolution spectroscopy.

When potential energy surfaces of large biomolecular systems such as peptides, oligonucleotides and sugars hopefully, in presence of some water molecules representing, the first hydration layer will be fully explored, studies of interactions of those systems between themselves or with drugs will take advantage of methods already devised by medicinal chemists [12]. For example, the introduction of flexibility [13–16] in ligand–receptor binding is crucial for understanding biomolecular recognition.

It is rewarding to notice that the proverb "whatever can be understood is useless; whatever is useful is unintelligible" is sometimes only temporarily verified. It only took a little more than 3,000 years to understand the curing power of willow leaves. This curing power was already known by Egyptians but remained mysterious until the discoveries of salicin in 1829, aspirin in 1899. Finally, the work of John Vane (Nobel Prize) explained how aspirin and allied anti-inflammatory drugs block synthesis of the prostaglandins. Thus, there is a hope that fences between the field of biomolecules in the gas-phase and applications are vanishing. Mass-spectrometry and modelling are widely used tools in biochemistry and drug research for a very long time while spectroscopy of systems isolated in the gas-phase are on the verge to penetrate into those fields.

Spectrometric methods such as ion mobility or BIRD are applied to check the outcome of combinatorial chemistry [17–19] and quantifying drug–biomolecule interactions [20–22].

Although accurate simulations of large biomolecular system dynamical behaviour are still costly [24], it is possible to identify regions where drugs are likely to bind and then use QM/MM methods to predict affinities and structural modifications induced by ligand-binding [25, 26]. The situation described by Voltaire "Therapeutics is the pouring of a drug of which one knows nothing into a patient of whom one know even less" [12] is fortunately no longer true. Orally active drugs are small molecules with molecular weight from 160 to 500 with a number of heavy atoms in between 20 and 70 and can establish no more than five donor and 10 acceptor hydrogen bonds (Lipinski's rule of five) [27].

Structure-based drug design aims to provide therapeutic agents based on the respective structures of receptor targets and their ligands. Up to now, the search for drugs has been mostly oriented towards complementarity between structures of active sites and ligands. Three-dimensional structures of active sites are deduced from X-ray crystallography, NMR or protein homology modelling [28]. Drug structures can be studied with high accuracy and are within the reach of spectroscopic gas-phase studies in contrast with their too large receptors.

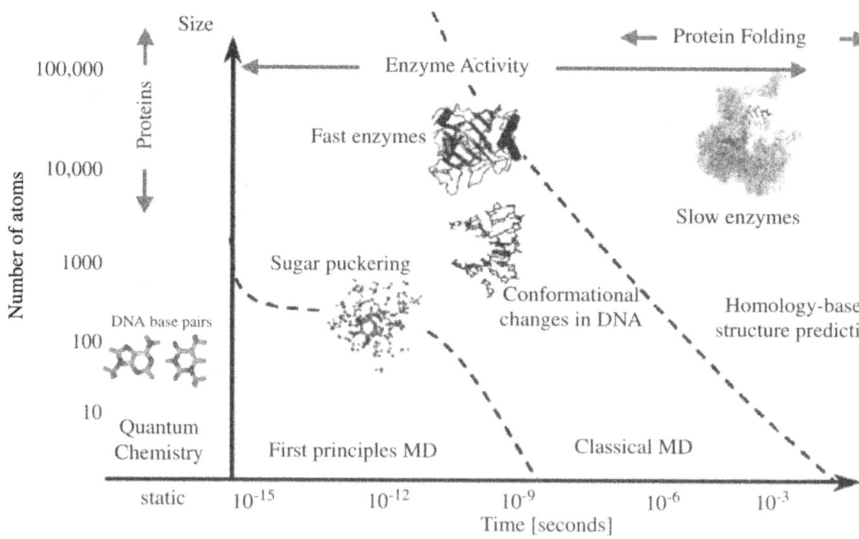

Figure C1 The size of biomolecules and timescales of biomolecular processes cover very wide ranges and require different modelling approaches. Experimental studies of enzyme activity in the gas-phase are still in their infancy while those structures covers nearly the whole mass-range (reproduced by permission from reference [23] ©Elsevier 2006).

However, most of gas-phase studies have been devoted to biomolecular building blocks and not drugs. There is in fact a striking difference between the very restricted number of biomolecular building blocks (five major nucleobases, 20 usual amino acids and a dozen of important pentoses and hexoses) leading to the construction of nearly the whole biomolecular space and the chemical space with 10^{60} different possible compounds. The "pharmaceutical" space is of course much smaller but already 4.8 million compounds and 6×10^5 structure–activity relationships for 3,000 protein targets are identified in databases [29, 30]. At least one compound satisfying Lipinski's rules has been found for 529 proteins encoded in the Human genome. Thus, gas-phase studies have been rightly and mainly devoted to building blocks of receptors providing an overview of biomolecular interactions rather than concentrating upon molecules that sometimes can be temporarily bioactive and each of them representing only an infinitesimal fraction of the drug space.

The coupling of modelling, spectroscopy and mass-spectrometry can provide new tools. For example, after the deciphering of the Human genome, there has been a great hope for the future of medicine and disciplines such as pharmacogenetics bringing together genetics, proteomics [31, 32] and clinical pharmacology. Those disciplines are supposed to transform pharmacy by a priori identifying how an individual should respond to a given drug [33]. A more accurate and predictable understanding of protein–protein [34], protein–DNA [35], protein–drug [36, 37] or DNA–drug [20, 21] interactions might then become needed.

Drugs can be issued from chemical research, designed with computer-aid [25, 38] or synthesized starting from a natural product template [39–41]. In this last case, each structure is unusual and represents a challenge for high-resolution spectroscopy since chromophores are usually absent. Nevertheless, it is still possible to couple low-resolution spectroscopy, mass-spectrometry and quantum calculations. An example is shown in Figure C2, the case

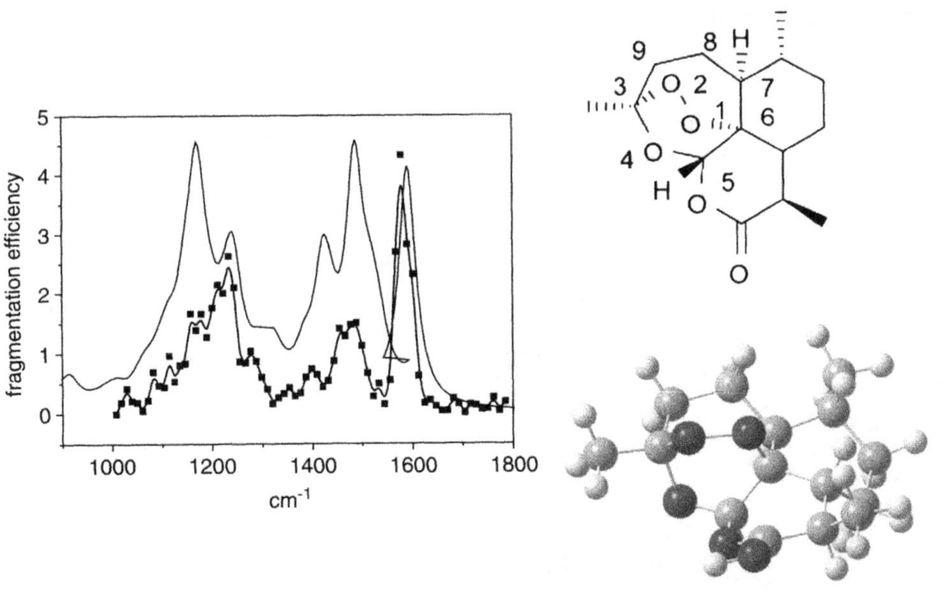

Figure C2 IRMPD spectrum and structure of protonated artemisin (by courtesy of G. Grégoire and M. Seydou).

of the powerful anti-malaria drug, artemisinin [39–41]. Artemisin is isolated from the *Artemisia annua* herb in use since 2,000 years in China as herbal tea against fever. Artemisin itself is not used but synthesized active derivatives are widely used.

The "one drug, one target concept" of drug design has evolved in the direction of "promiscuous" drugs [42]. Ligands designed to specifically bind too precisely to chosen targets can possibly induce secondary effects because of their ability to also bind to undesired targets. Instead, designing ligands capable of binding perhaps less efficiently but more safely to several targets can be more medically efficient. Such an example is provided by a series of substituted bi-indole molecules designed to bind to the HHQK sequence present in the β-amyloid peptide greatly responsible for the Alzheimer disease [37]. This amyloid peptide is widely investigated in gas-phase and modelling studies [43–45]. Screening of drugs capable of inhibiting β-amyloid protein aggregation takes advantage of mass-spectrometry [46]. Flexible docking simulations are used for interpreting mass-spectrometry, ion mobility and infrared spectroscopy studies of β-amyloid–drug interactions [36].

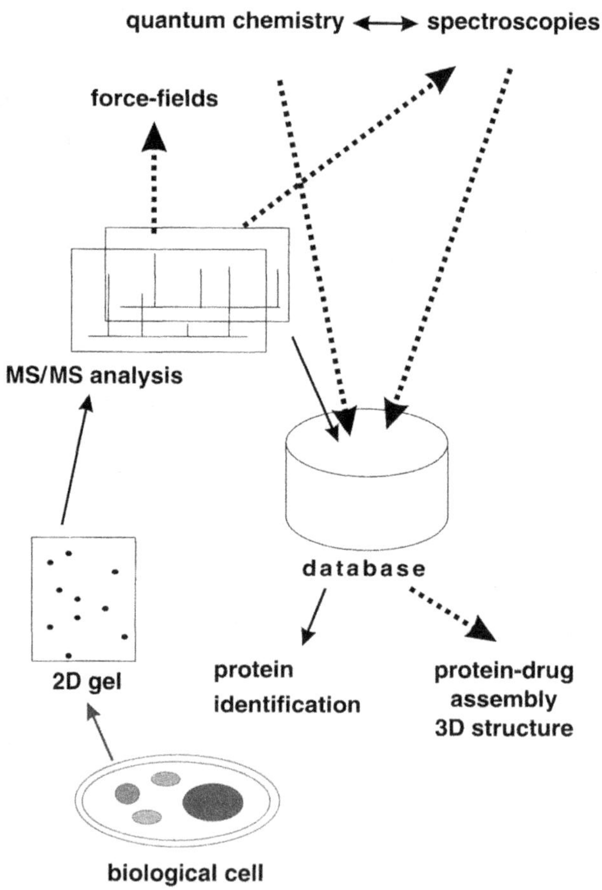

Figure C3 Coupling of a proteomic procedure (solid lines) and modelling supported by spectroscopy (dashed lines) in a database-oriented search for drugs.

New biophysical methods are introduced for the investigation of the structural biology of a whole cell [47]. Less ambitious but still promising, there is some hope that new instruments will be devised by coupling mass-spectrometers [48] and quasi-universal tools providing automated three-dimensional structure information (Figure C3). Infrared spectroscopy is one of those tools but others such as electron capture are already used for localization of ligand-binding sites [49]. Mass-spectrometry may benefit from observation of not only ionic fragments in CID MS/MS experiments but also detection of neutrals [50]. While the research field of biomolecular structures in condensed-phase is very well organized [51], spectroscopic gas-phase studies are still in the infancy with respect to the construction of a structured common knowledge basis. Attempts are made for building quantum chemistry databases [52–54]. On the contrary, the search for automated mass-spectrometric tools is very active [48, 55, 56], in particular for building tools adapted to automate protein identification [57]. It will be interesting to follow in the future if Henry Schaffer's prediction: "Chemical research migrates from experiment to computation" [58, 59] reveals itself as foresight or only euphoria. In the former case, heartening mass-spectrometry experimental works concerning tasty ethanol–water mixtures such as presented in reference [60] would unfortunately no longer take place.

REFERENCES

1. Abo-Riziq A, Bushnell JE, Crews B, Callahan M, Grace L, de Vries MS: Gas phase spectrosopy of the pentapetide FDAVS. *Chemical Physics Letters* 2006, 431:227–230.
2. Abo-Riziq A, Crews B, Callahan MP, Grace L, de Vries M: Spectroscopy of isolated gramicidin peptides. *Angewandte Chemie-International Edition* 2006, 118:5290–5293.
3. Carçabal P, Hunig I, Gamblin DP, Liu B, Jockusch RA, Kroemer RT, Snoek LC, Fairbanks AJ, Davis BG, Simons JP: Building up key segments of N-glycans in the gas phase: intrinsic structural preferences of the α(1,3) and α(1,6) dimannosides. *Journal of the American Chemical Society* 2006, 128:1976–1981.
4. Oomens J, Polfer N, Moore DT, van der Meer L, Marshall AG, Eyler JR, Meijer G, von Helden G: Charge-state resolved mid-infrared spectroscopy of a gas-phase protein. *Physical Chemistry Chemical Physics* 2005, 7:1345–1348.
5. Choi MS, Miller RE: Four tautomers of isolated guanine from infrared laser spectroscopy in helium droplets. *Journal of the American Chemical Society* 2006, 128:7230–7328.
6. Boyarkin OV, Mercier SR, Kamariotis A, Rizzo TR: Electronic spectroscopy of cold, protonated tryptophan and tyrosine. *Journal of the American Chemical Society* 2006, 128:2816–2817.
7. Robertson EG, Simons JP: Getting into shape: conformational and supramolecular landscapes in small biomolecules and their hydrated clusters. *Physical Chemistry Chemical Physics* 2001, 3:1–18.
8. Clarkson JR, Baquero E, Zwier TS: Direct measurement of the energy thresholds to conformational isomerization. II. 3-indole-propionic acid and its water-containing complex. *Journal of Chemical Physics* 2005, 122:214312.1–214312.8.
9. Marinica DC, Gregoire G, Desfrançois C, Schermann JP, Borgis D, Gaigeot MP: *Ab initio* molecular dynamics of protonated dialanine and comparison to infrared multiphoton dissociation experiments. *Journal of Physical Chemistry A* 2006, 110:8802–8810.
10. Gregoire G, Gaigeot MP, Lemaire J, Desfrançois C, Schermann JP: Resonant infrared multiphoton dissociation spectroscopy of gas-phase protonated peptides using free electron laser radiation and quantum chemistry calculations. *Physical Chemistry Chemical Physics* 2007, in press.

11. Morgner N, Barth HD, Brutschy B: A new way to detect noncovalently bonded complexes of biomolecules from liquid micro-droplets by laser mass spectrometry. *Australian Journal of Chemistry* 2006, 59:109–114.
12. Enriz RD: The legacy of the past, the reality of the present and the hopes of the future. *Journal of Molecular Structure: THEOCHEM* 2005, 731:163–172.
13. Deerfield DW, Carter CW, Pedersen LG: Models for protein-zinc ion binding sites. II. The catalytic sites. *International Journal of Quantum Chemistry* 2001, 83:150–165.
14. Casavotto CN, Kovacs JA, Abagyan RA: Representing receptor flexibility in ligand docking through relevant normal modes. *Journal of the American Chemical Society* 2005, 127:9632–9640.
15. Bonvin AMJ: Flexible protein-protein docking. *Current Opinion in Structural Biology* 2006, 6:194–200.
16. Grünberg R, Nilges M, Leckner J: Flexibility and conformational entropy in protein-protein binding. *Structure* 2006, 14:683–693.
17. Valentine SJ, Kulchania M, Barnes CAS, Clemmer DE: Multidimensional separations of complex peptide mixtures: a combined high-performance liquid chromatography/ion mobility/time-of-flight mass spectrometry approach. *International Journal of Mass Spectrometry* 2001, 212:97–109.
18. Taraszka J, Kurulugama R, Sowell RA, Valentine SJ, Koeniger S, Arnold RJ, Miller DF, Kaufman TC, Clemmer DE: Mapping the proteome of drosophila megalogaster. Analysis of embryos and adult heads by LC-IMS-MS methods. *Journal of Proteome Research* 2005, 4:25–35.
19. Barnes CAS, Clemmer DE: Assessment of purity and screening of peptide libraries by nested ion mobility TOFMS: identification of RNase S-protein binders. *Analytical Chemistry* 2001, 73:424–433.
20. Rosu F, Gabelica V, De Pauw E: Determination of affinity, stoichiometry and sequence selectivity of minor groove binder complexes with double-stranded oligodeoxynucleotides by electrospray ionization mass spectrometry. *Nucleic Acids Research* 2002, 30:e82–e91.
21. Guittat L, De Cian A, Rosu F, Gabelica V, De Pauw E, Delfourne E, Mergny JL: Ascididemin and meridine stabilise G-quadruplexes and inhibit telomerase in vitro. *Biochimica Et Biophysica Acta-General Subjects* 2005, 1724:375–384.
22. Barran PE, Roeseke RW, Pawson AJ, Sellar R, Bowers MT, Morgan K, Lu ZL, Tsuda M, Kusakabe T, Millar RP: Evolution of constrained gonadotropin-releasing hormone ligand conformation and receptor selectivity. *Journal of Biological Chemistry* 2005, 280:38569–38575.
23. Colvin ME, Cramer CJ, Dykstra CE, Jensen JH, Krimm S, Rivail JL, Thakkar AJ, Yanez M: Molecular quantum mechanics to biodynamics: essential connections. *Journal of Molecular Structure: THEOCHEM* 2006, 764:1–8.
24. Carloni P, Rothlisberger U, Parrinello M: The role and perspective of *ab initio* molecular dynamics in the study of biological systems. *Accounts of Chemical Research* 2002, 35:455–464.
25. Gohlke H, Klebe G: Approaches to the description and prediction of the binding affinity of small-molecule ligands to macromolecular receptors. *Angewandte Chemie-International Edition* 2002, 41:2644–2676.
26. Shields SJ, Oyeyemi O, Lightstone FC, Balhorn R: Mass spectrometry and non-covalent protein-ligand complexes: conformation of binding sites and changes in tertiary structure. *Journal of the American Society of Mass Spectrometry* 2003, 14:460–470.
27. Jorgensen WL: The many roles of computation in drug discovery. *Science* 2004, 303:1813–1818.
28. Homans SW: NMR spectroscopy tools for structure-aided drug design. *Angewandte Chemie-International Edition* 2004, 43:290–300.
29. Paolini GV, Shapland RHB, van Hoorn WP, Mason JS, Hopkins AL: Global mapping of pharmacological space. *Nature Biotechnology* 2006, 24:805–815.
30. Irvin JJ, Stoichet BK: ZINC – A free database of commercially available compounds for virtual screening. *Journal of Chemical Information and Modeling* 2005, 45:177–182. Kirpatrick P: Informatics: exploring pharmacological space. *Nature Reviews Drug Discovery* 2006, 5:719.

31. Burbaum J, Tobal GM: Proteomics in drug discovery. *Current Opinion in Chemical Biology* 2002, 6:427–433.
32. Hamacher M, Marcus K, Stühler K, van Hall A, Warscheid B, Meyer HE, Mannhold R, Kubiny H, Folkers G: *Proteomics in drug research (methods and principles in medicinal chemistry)*: Wiley-VCH, Weinheim, 2006.
33. Rane A: Post-genomics prospects of success in drug development and pharmacotherapy. *Pharmacogenomics* 2001, 1:6–8.
34. Armstrong JD, Pocklington AJ, Cumiskey MA, Grant SGN: Reconstructing protein complexes: from proteomics to systems biology. *Proteomics* 2006, 6:4724–4731.
35. Prabakaran P, Siebers JG, Ahmad S, Gromika MM, Singarayan M, Sarai A: Classification of protein-DNA complexes based on structural descriptors. *Structure* 2006, 14:1355–1367.
36. Grégoire G, Gillet JC, Gabelica V, Barran P, Chin WBU, Schermann JP, Desfrançois C, Weaver DF: Quantifying β amyloid peptide-drug interaction with mass spectrometry, ion mobility and IRMPD spectroscopy, in preparation.
37. Stephenson VC, Heyding RA, Weaver DF: The "promiscuous concept" with applications to Alzheimer's disease. *FEBS Letters* 2005, 579:1338–1342.
38. Aarons L, Karlsson MO, Mentré F, Rombout F, Steiner JL, van Peer A: Role of modelling and simulation in Phase I drug development. *European Journal of Pharmaceutical Sciences* 2001, 13: 115–122.
39. Pashynska V, Van den Heuvel H, Claeys M, Kosevich MV: Characterization of noncovalent complexes of antimalarial agents of the artemisin-type and Fe(III)-heme by electrospray mass spectrometry and collisional activation tandem mass spectrometry. *Journal of the American Society for Mass Spectrometry* 2004, 15:1181–1190.
40. Drew MGB, Metcalfe J, Ismail FMD: A DFT study of free radicals formed from artemisin and related compounds. *Journal of Molecular Structure: THEOCHEM* 2004, 711:95–105.
41. Cheng F, Shen J, Luo X, Zhu W, Gu JD, Ji R, Jiang H, Chen K: Molecular docking and 3-D-QSAR studies on the possible antimalarial mechanism of artemisin analogues. *Bioorganic & Medicinal Chemistry* 2002, 10:2883–2891.
42. Hopkins AL, Mason JS, Overington JP: Can we rationally design promiscuous drugs? *Current Opinion in Structural Biology* 2006, 16:1–10.
43. Urbanc B, Cruz L, Ding F, Sammond D, Khare S, Buldyrev SV, Stanley HE, Dokholian NV: Molecular dynamics simulation of amyloid β dimer formation. *Biophysical Journal* 2004, 87: 2310–2321.
44. Baumketner A, Bernstein SL, Wyttenbach T, Lazo ND, Teplow DB, Bowers MT, Shea JE: Structure of the 21–30 fragment of amyloid β-protein. *Protein Science* 2006, 15:1239–1247.
45. Bernstein SL, Wyttenbach T, Baumketnert A, Shea JE, Bitan G, Teplow DB, Bowers MT: Amyloid β protein: monomer structure and early aggregation states of A β 42 and its Pro(19) alloform. *Journal of the American Chemical Society* 2005, 127:2075–2084.
46. Cheng X, van Breemen RB: Mass spectrometry-based screening for inhibitors of β-amyloid protein aggregation. *Analytical Chemistry* 2005, 77:7012–7015.
47. Moffat K, Chiu W: Biophysical methods: bigger, finer, faster and dynamic. *Current Opinion in Structural Biology* 2005, 15:535–537.
48. Horn DM, Zubarev RA, McLafferty FW: Automated de novo sequencing of proteins by tandem high-resolution mass spectrometry. *Proceedings of the National Academy of Sciences of the United States of America* 2000, 97:10313–10317.
49. Xie YM, Yin S, Loo JA: Top-down ESI-FT-IR mass spectrometry localizes non-covalent protein-ligand binding sites. *Journal of the American Chemical Society* 2006, 128:14432–14433.
50. Lepère V, Lucas B, Barat M, Fayeton JA, Picard YJ, Jouvet C, Carçabal P, Nielsen IB, Dedonder C, Grégoire G, *et al.*: Photodissociation of protonated tryptophan: towards a full dynamics analysis from time and position resolved detection of all fragments. *PCCP* 2007.

51. Berman HM, Burley S, Chiu W, Sali A, Adzhubei A, Bourne P, Dunbrack R, Fidelis K, Franck J: Outcome of a workshop on archiving structural models of biological macromolecules. *Structure* 14:1211–1217.
52. Chass GA, Sahai MA, Law JMS, Lovas S, Farkas O, Perczel A, Rivail JL, Csizmadia IG: Toward a computed peptide structure database: the role of a universal atomic numbering system of amino acids in peptides and internal hierarchy of database. *International Journal of Quantum Chemistry* 2002, 90:933–968.
53. Chass G, Toward a computed structure database: methodology for effective molecular orbital computations *Journal of Molecular Structure: THEOCHEM* 2003, 666–667:61–67.
54. Jurecka P, Sponer J, Cerny J, Hobza P: Benchmark database of accurate (MP2 and CCSD(T) complete basis set limit) interaction energies of small model complexes, DNA base pairs, and amino acid pairs. *Physical Chemistry Chemical Physics* 2006, 8:1985–1993.
55. Horn DM, Zubarev RA, McLafferty FW: Automated reduction and interpretation of high resolution electrospray mass spectra of large molecules. *Journal of the American Society for Mass Spectrometry* 2000, 11:320–332.
56. Tabb DL, Shah MB, Hettich RL, Hurst GB: Charge states by Fourier transformation of isotope-resolved mass spectra. *Journal of the American Society for Mass Spectrometry* 2006, 17:903–915.
57. Hernandez P, Müller M, Appel RD: Automated protein identification by tandem mass spectrometry: issues and strategies. *Mass Spectrometry Reviews* 2005, 25:235–254.
58. Schaeffer HF: Computers and molecular mechanics: 1965–2001, a personal perspective. *Journal of Molecular Structure: THEOCHEM* 2001, 573:129–237.
59. Hargittai I, Hargittai M: Experiments, calculations, computations: in a shifting balance. *Journal of Molecular Structure: THEOCHEM* 2002, 592:1–6.
60. Cooper HJ, Marshall AG, Electrospray ionization Fourier transform mass spectrometric analysis of wine. *Journal of Agricultural and Food Chemistry* 2001, 49:5710–5718.

Subject Index